STRUCTURAL MECHANICS

RUSSIAN MONOGRAPHS AND TEXTS ON ADVANCED MATHEMATICS AND PHYSICS

I. A. I. KHINCHIN, A Course of Mathematical Analysis

II. S. G. MIKHLIN, Linear Integral Equations

III. P. P. KOROVKIN, Linear Operators and Approximation Theory

IV. L. E. EL'SGOL'TS, Differential Equations

V. L. A. LUSTERNIK and V. J. SOBOLEV, Elements of Functional Analysis

VI. V. L. GINZBURG, Propagation of Electromagnetic Waves in Plasma

VII. A. A. VLASOV, Many-Particle Theory and Its Application to Plasma

VIII. M. A. KRASNOSEL'SKI, Convex Functions and Orlicz Spaces

IX. Yu. V. VOROBYEV, Moments Method in Applied Mathematics

X. N. N. BOGOLIUBOV and MITROPOLSKY, Asymptotic Methods in the Theory of Nonlinear Oscillations

XI. S. D. VOLKOV, Statistical Strength Theory

XII. Z. S. AGRANOVICH and MARCHENKO, The Inverse Problem of Scattering Theory

XIII. V. F. NOZDREV, Application of Ultrasonics to Molecular Physics

XIV. S. TARG, Theoretical Mechanics: A Short Course

XV. A. S. KOMPANEYETS, Theoretical Physics

XVI. P. STEPIN, Strength of Materials

XVII. N. GLINKA, General Chemistry

XVIII. B. PAVLOV and A. TERENTYEV, Organic Chemistry

XIX. M. KORSUNSKY, The Atomic Nucleus

XX. V. V. SUSHKOV, Technical Thermodynamics

XXI. L. LEVINSON, Fundamentals of Engineering Mechanics

XXII. Y. LAKHTIN, Engineering Physical Metallurgy

XXIII. A. KASATKIN and M. PEREKALIN, Basic Electrical Engineering

A. DARKOV and V. KUZNETSOV

STRUCTURAL MECHANICS

Translated from the Russian by B. Lachinov

GORDON AND BREACH, SCIENCE PUBLISHERS
New York London Paris

GORDON AND BREACH, Science Publishers, Inc.
150 Fifth Avenue, New York, N. Y. 10011

Editorial office for Great Britain:
Gordon and Breach, Science Publishers Ltd.
8 Bloomsbury Way
London W. C. 1

Editorial office for France:
Gordon & Breach
7–9 rue Emile Dubois
Paris 14e

Distributed in Canada by:
The Ryerson Press
299 Queen Street West
Toronto 2B, Ontario

Printed in Canada

CONTENTS

INTRODUCTION

Structural mechanics is a science which studies the strength, the rigidity and the stability of engineering structures and parts thereof. The strength of materials dealing with the strength, rigidity and stability of isolated members; the theory of elasticity, which is concerned with the same problems but gives more strict solutions; the theory of plasticity which investigates the stresses and strains of plastic and elasto-plastic materials, and finally the theory of structures which studies the strength, rigidity and stability of whole structures—all form parts of this discipline.

It was Leonardo da Vinci (1452-1519), the great Italian scientist and artist, who was the first to formulate a number of valuable ideas on the strength of materials. These ideas never became widely known and remained confined to his manuscripts on mechanical research. In those days, large-scale studies of problems which form the subject of contemporary structural mechanics were utterly impossible. Only partial solutions of isolated problems related to the strength of certain structural members could be obtained.

The eminent physicist, mathematician and astronomer Galileo Galilei (1564-1642) is generally considered to be the father of scientific studies in the strength of engineering materials and structures. In those days the expansion of maritime trade called for large increases in the tonnage of cargo vessels and for improvements in their design. Dealing with these questions Galilei discovered that the ship's overall strength and sea-worthiness could not be satisfactorily ensured simply increasing the dimensions of her members in direct proportion to her size. He also proved that the dead weight to ultimate load ratio may differ for geometrically similar bodies.

Galilei's studies of beams subjected to bending led him to a number of valuable conclusions which have not lost interest up to date, but he was unable to develop a true flexural theory, as he proceeded from a false conception that the whole cross section of the beam is uniformly extended. Neither had Galilei any knowledge of the relation existing between stresses and strains. The simplest form of this relation was discovered in 1678 by Robert Hooke who expressed it as "ut tensio sic vis" ("the extension is as great as the force").

The fact that compressive stresses as well as the tensile ones exist in the cross section of a beam subjected to bending was discovered in the second half of the 18th century as the outcome of a series of tests conducted with great thoroughness. At that time the rapid development of trades and industries was constantly calling for new scientific achievements among which was the correct solution of the problem of bending put by Galilei.

Significant advances in higher mathematics and mechanics achieved in the 18th century contributed greatly to the development of studies in the strength of materials and structures. Works by J. Lagrange and L. Euler were of particular importance in this respect.

Vigorous growth of industry in the 19th century, the introduction of the steam engine, the construction of railways, bridges, dams, canals, large steamships and great buildings accelerated the studies in the strength of engineering materials and structures. The evergrowing complexity of structural forms and the pressing demand for a reduction in building costs resulted in the development of new methods of strength computation and in the formation of a new engineering science—structural mechanics (also called the theory of structures).

Trusses, arched systems, retaining walls and rigid frames form the main classes of structures dealt with by modern structural mechanics.

In their simplest form many of these structures had been already used dy the ancients, but the methods of their computation remained unknown.

At present, trusses and triangulated systems are widely used in bridge and roof construction (bridge and roof trusses) as well as in travelling cranes, tower cranes, power-line towers, aerial supports and in a large number of other structures.

Arched systems made their appearance in ancient Rome, where they were successfully used for the construction of masonry bridges and aqueducts. In the second half of the 19th century these systems became used in steel-bridge construction and in the 20th century reinforced concrete becomes the main material used for that type of structures. At present arched systems are widely used in many kinds of large-span construction work.

Retaining walls have been used to prevent the sliding down of steep slopes in various branches of engineering activities since times immemorial. Rigid frames became widespread in modern times, reinforced concrete and steel frames being currently used for the construction of single and multistoried industrial and other buildings.

The ever increasing building programs peculiar to the rapidly growing national economy of the U.S.S.R. have called to life a number of very serious problems pertaining to the design and computation of large and complicated structures subjected to extremely heavy loads.

As a result, methods of computation of complicated redundant structures have been further perfectioned and simplified to such an extent that today they are used in everyday design practice.

Important advances have also been made in the studies of thin-walled tubular sections which are frequently utilized in aircraft construction as well as in other branches of engineering. Various problems related to the stability of structures have been successfully solved. Of late the dynamics of engineering structures have been acquiring an everincreasing importance. It forms now a separate branch of the structural mechanics, same as the theory of naval and aeronautical architecture.

Among the Soviet scientists and research workers the following have made the greatest contributions to the development of structural mechanics: A. Krylov, B. Galerkin, A. Gvozdev, B. Zhemochkin, I. Rabinovich, N. Streletsky, I. Prokofyev, N. Bezukhov, N. Beljaev, V. Bolotin, K. Zavriev, A. Smirnov, S. Ponomarev, V. Vlasov, M. Filonenko-Borodich, P. Papkovich and N. Snitko.

1. KINEMATIC ANALYSIS OF STRUCTURES

1.1. *Supports*

Structural mechanics deals with unyielding systems or structures, in other words, with such structural systems no point of which can be displaced without a deformation of their elements. The immutability of such systems (their geometrical stability) with relation to the ground * is ensured dy means of supports. Reactions arising at these supports together with the applied loads form a balanced system of outer or external forces which maintain the structure in equilibrium.

Let us first examine the different types of supports which may be encountered in plane structures.

The first type as represented in Fig. 1.1 consists of two rockers (the upper and the lower one) with a pin in between permitting the rotation of the upper rocker with respect to the lower one. At the same time both rockers can move together on rollers along the *bearing plate.*

Thus, the system has two degrees of freedom, the friction developed in the bearing being usually neglected. The reaction of this type of support passes through the centre of the pin and is perpendicular to the bearing plate surface, i. e., to the surface along which the rollers may travel. Thus, only one parameter of the reaction, i. e., its magnitude, has to be known in order to determine this reaction completely.

Supports of this type are known as *free end* or *movable roller* supports. Schematically they are represented by one bar with hinged ends ** (Fig. 2.1).

The bar is conventionally considered to be of infinite length; its upper extremity may move only along a straight line, normal to its axis, a straight line being a circumference of infinite radius.

◆

* The word *ground* will hereafter refer to any rigid invariable body.

** In some cases movable supports actually consist of a vertical element with hinges at both extremities, in which case they are referred to as *pendulum supports.*

The bar is also regarded as infinitely rigid so that its strains can be completely disregarded. These two conventions fit very closely into the actual working conditions of supports of the type just described.

The second type of supports differs from the first one by the fact that the lower rocker is fixed and cannot move (Fig. 3.1). This type of bearing possesses only one degree of freedom. It is usually termed *hinged immovable* or *fixed end* support.

The reaction will still pass through the centre of the pin, but its direction may be arbitrary, and accordingly to determine it completely two parameters have to be found—its magnitude and direction (or, which is the same, the magnitude of two of its components, say, the vertical and the horizontal one).

Schematically the second type of support may be represented by two bars with hinges at their ends, the top hinge being common to both bars (Fig. 4.1). That fixes the point of application of the reaction which coincides with the top hinge, but in this case the direction of this force ramains unknown.

The directions of the bars themselves may be chosen at will as any force may be resolved into two components of any direction.

The third type of support is the *built-in end* (Fig. 5.1) whose degree of freedom is nil. The determination of the reactions developed by this support requires the knowledge of three parameters—the direction and magnitude of a force passing through any point chosen at will and the magnitude of the moment about the same point. Actually this forms a combination of the reaction of a hinged immovable support with the reactive moment.

This type of support may be represented by three bars as in Fig. 6.1. To attain perfect rigidity of the support the distance l_0 must be regarded as extremely small or the built-in end of the beam as absolutely rigid.

It is worth noting that the *number of bars in these schematic representations of supports is always equal to the number of parameters determining completely the reaction at this support.*

2.1. *Geometrical Stability of Framed Structures*

Framed or *through structures* consist of a series of separate, usually straight, members connected together by welded, riveted, bolted or other types of joints. One of the simplest two-dimensional forms of framed structures is the plane truss.

In most cases the joints of framed structures are *not hinged* and possess a certain degree of *rigidity*. The exact computation of trusses with rigid joints is extremely complicated as the system becomes many times statically indeterminate. On the other

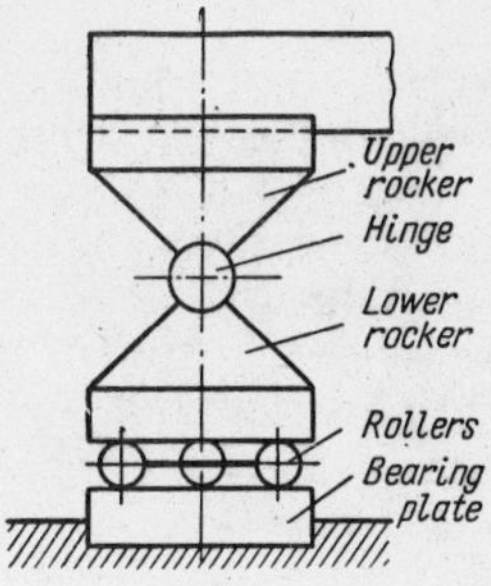

Fig. 1.1

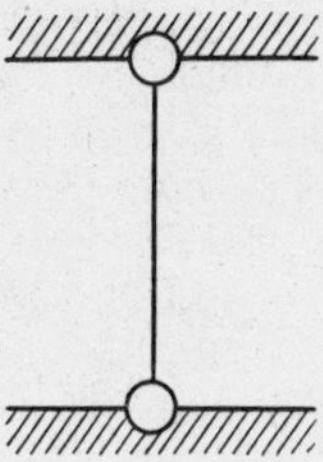

Fig. 2.1

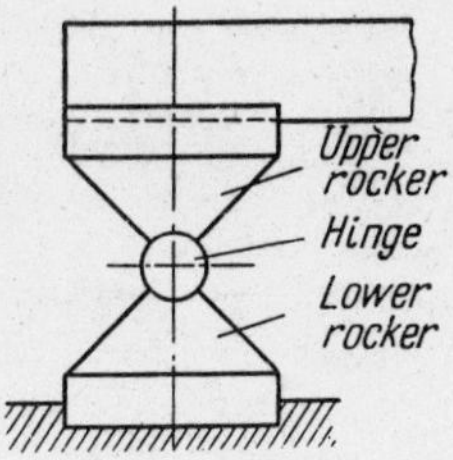

Fig. 3.1

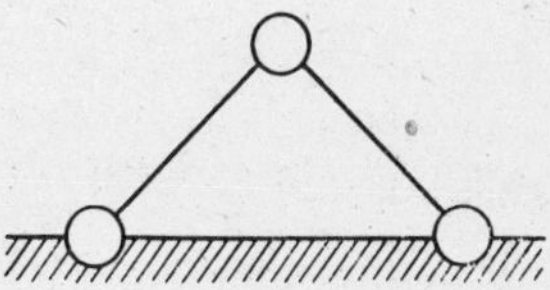

Fig. 4.1

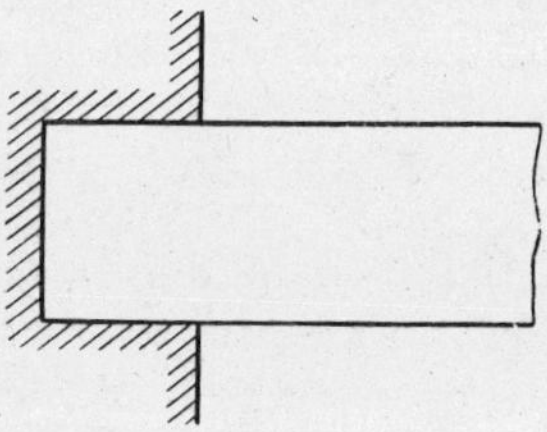

Fig. 5.1

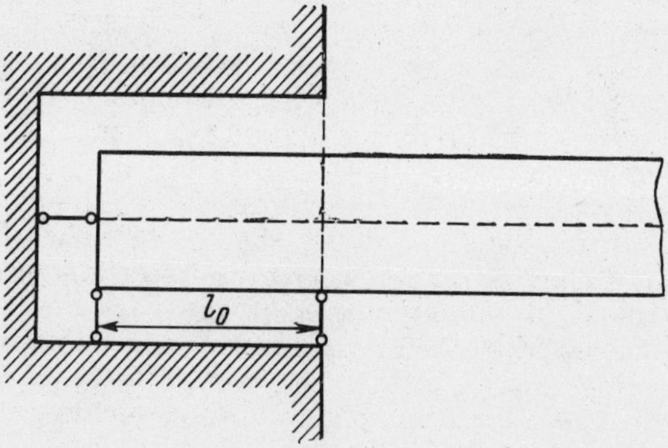

Fig. 6.1

hand, when rigid joints are conventionally replaced by hinged ones, the analysis becomes greatly simplified and under certain conditions equations provided by statics alone will suffice.

Tests carried out as well as the results of theoretical analysis indicate that in general the conventional introduction of hinges does not lead to any substantial errors in stress computations pertaining to through structures loaded with a system of forces acting at the joints. Therefore, for design purposes ordinary trusses are always regarded as being hinge-jointed.

Let us now examine a system consisting of three rigidly connected straight bars as represented in Fig. 7.1*a*. If the rigid joints are replaced by hinges, the system will continue to be *unyielding* (Fig. 7.1*b*), i. e., it will be uncapable of undergoing any distortion without the deformation of at least one of the bars.

Should, however, the quadrilateral system, shown in Fig. 8.1, undergo the same treatment, we shall immediately obtain a system whose shape can be *altered* without any deformation of its members.

The *simplest* unyielding system consisting of a number of separate pin-jointed bars is a triangle with hinges at all the three vertices (Fig. 7.1*b*).

Let us establish the rules governing the formation of geometrically stable systems comprising more than three pin-jointed bars.

In the first instance let us examine a system consisting of two bars (Fig. 9.1) placed along a straight line and connecting joint C with two fixed points A and B. If the bars AC and BC were disconnected at point C, the extremity C of bar AC would become free to move along the circular arc *m-m*, while the extremity C of bar BC—along the arc *n-n*, the two arcs having a common tangent at point C. It follows that if the extremity C of one of the two bars moves over a very short stretch along a perpendicular to AB, the other bar will offer no resistance. Thus, the system is geometrically unstable, as its shape can be altered without any change occurring in the length of its members or, in other words, without any deformation of the bars.

Hereunder we shall refer to systems consisting of two bars placed along a straight line (see Fig. 9.1) as *instantaneously unstable*, these systems becoming rigid as soon as a small shift of point C along the perpendicular to AB has been completed.

The situation would change entirely if the two bars AC and BC were not in alignment (Fig. 10.1). In this case the circumferences *m-m* and *n-n* have no common tangent, and, therefore, even the slightest displacement of joint C is impossible without a corresponding deformation of the bars.

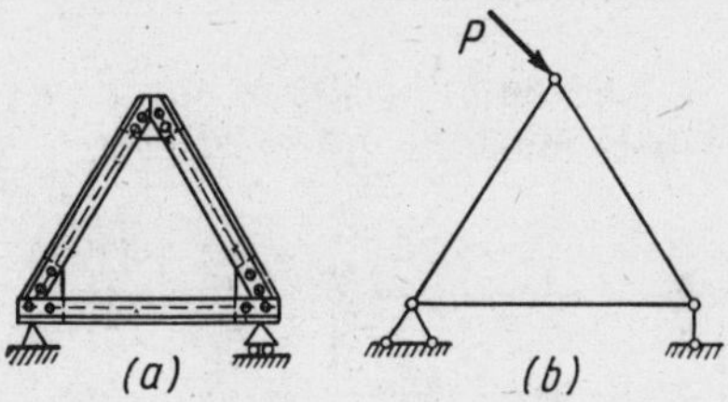

Fig. 7.1

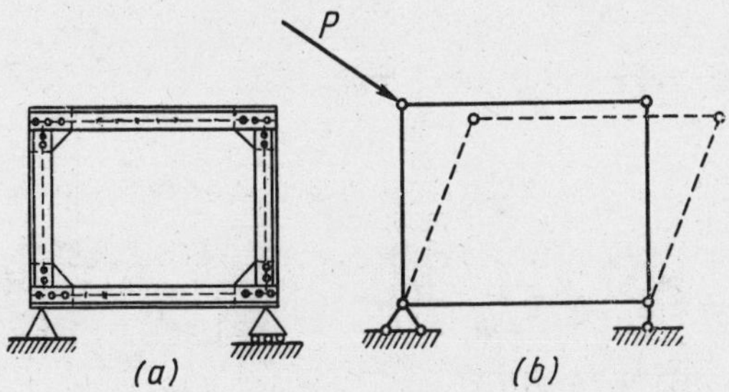

Fig. 8.1

Fig. 9.1

Fig. 10.1

It follows that each additional joint forming part of a geometrically stable system must be *attached thereto by means of two separate bars the axes of which do not lie on the same line.*

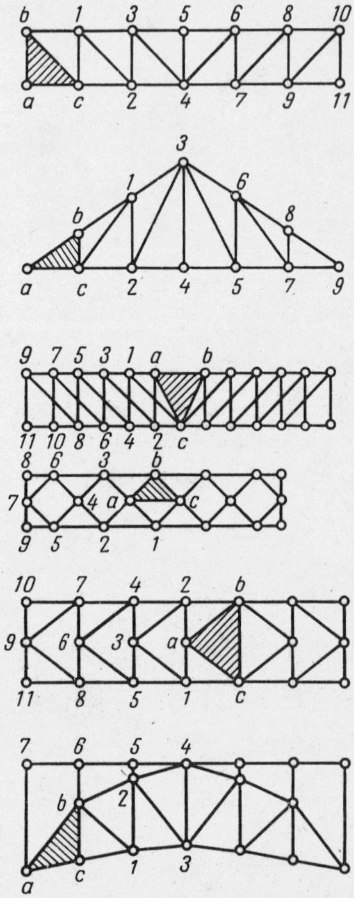

Fig. 11.1

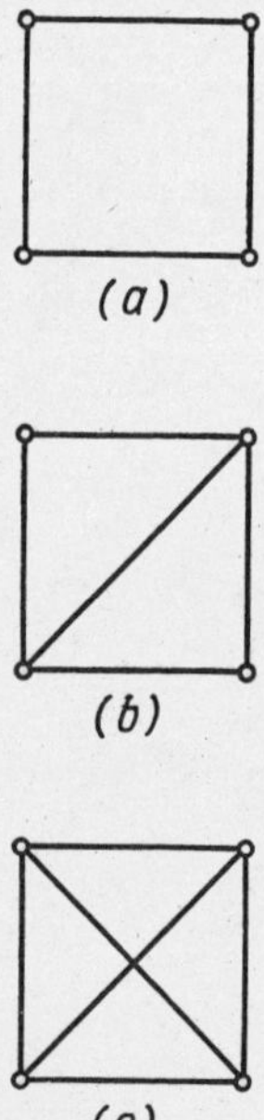

Fig. 12.1

Consequently, *any system developed from a hinged triangle by successive addition of joints, each new joint being connected to two existing ones by two bars not in alignment, will be geometrically stable* (*invariable*). Systems so formed will be called hereafter *simple* framed structures in order to distinguish them from the *complicated* ones, usually derived from the former by replacement of a number of bars or by superposition.

All the plane trusses, represented in Fig. 11.1 belong to the simple frames, each having been obtained successively adding hinged joints to a basic pin-connected triangle *abc*, in the sequence indicated. Any triangular combination of three pin-jointed bars may serve as a basis for verifying the geometrical stability of simple framed structures.

Thus, *any system consisting solely of triangles is obviously unyielding* (*geometrically stable*). This property may be checked with equal success in a reverse order, viz., by rejecting one by one all the hinged joints together with the two bars abutting to each of them. *If in the outcome a pin-jointed triangle is obtained the system is geometrically stable.*

Let us now establish the relation between the number of bars and joints forming a simple truss. As stated above, such a truss consists of one basic pin-jointed triangle, to which a number of additional joints have been successively attached, each by means of two separate bars not in alignment. Let S be the number of bars and K the number of joints. The basic triangle consists of three bars and three joints; all the other joints, numbering $(K-3)$ are attached by means of two bars each. Therefore, the total number of bars in a simple truss will be

$$S=3+2(K-3)=2K-3 \tag{1.1}$$

If the number of bars $S<2K-3$, the truss does not contain a number of bars sufficient to ensure its geometrical stability and the system will evidently be unstable. An example of such a system is furnished by a quadrangle (Fig. 12.1*a*) characterized by $S=4$ and $K=4$. Consequently

$$S=4<2K-3=2\times 4-3=5$$

This quadrangle may be converted into an unyielding system by adding a fifth diagonal bar, as shown in Fig. 12.1*b*. Should we introduce a second diagonal bar which would give a total of six bars as against four joints (Fig. 12.1*c*), this sixth bar would be redundant from the view-point of geometrical stability. This example shows that we may encounter geometrically stable systems for which $S>2K-3$.

It should be noted that the *condition* $S\geqslant 2K-3$, *though necessary, is not sufficient to ensure the geometrical stability of a hinge-connected system.* Thus, the truss represented in Fig. 13.1*a* is unstable although the number of its bars totals exactly $2K-3$. The truss shown in Fig. 13.1*b* has an even greater number of bars but still remains unstable. This is due to the fact that the right-hand panels of both these trusses consist of hinge-jointed rectangles. Furthermore, in certain cases framed structures for which the condition $S=2K-3$ is fulfilled may be instantaneously unstable.

Let us now consider the problem of connecting geometrically stable systems to the ground by means of supports.

In the majority of cases a plane structure (which may be regarded as a rigid disk or plate) will rest on two hinge supports—one movable and the other fixed (Fig. 14.1*a*). This type of connection between structure and ground is geometrically stable (unyielding). It is not essential that two of the three supporting bars should have a common hinge; indeed they may have none (Fig. 14.1*b*).

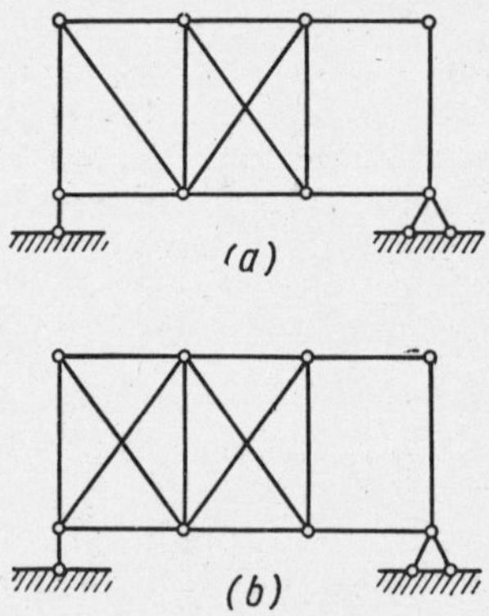

Fig. 13.1

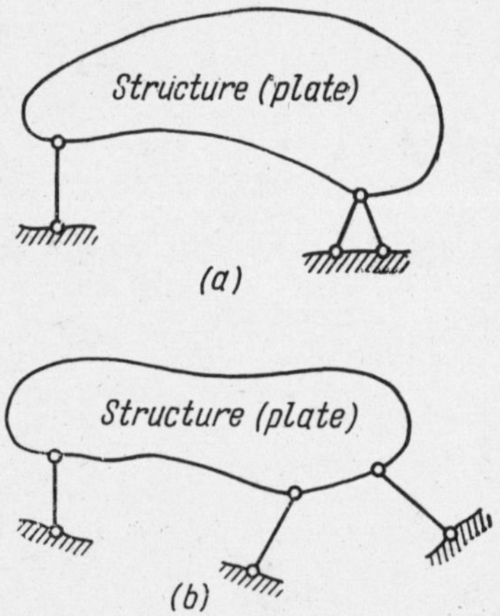

Fig. 14.1

However, should the directions of all the supporting bars intersect at one and the same point (Fig. 15.1*a*), this point will constitute an *instantaneous centre* of rotation about which the whole system will be able to accomplish an infinitely small rotary movement. (Practically such a displacement may become quite appreciable.) Once this movement accomplished, the supporting bars will no longer concur at the same point and all further displacements will become impossible without a corresponding deformation of these bars.

A system connected to the ground in the way just described will be instantaneously unstable and, therefore, such an arrangement of supports cannot be tolerated. *

On the contrary, three nonconcurring and nonparallel ** bars will always provide a geometrically stable support.

* As will be shown in Art. 6.4, even very small external loads may stress the instantaneously unstable systems very heavily.

** Parallel lines having a point of intersection in the infinity.

All the above applies equally to the connection of any two geometrically stable structures (or rigid plates) between themselves thus permitting to formulate the following rule: *two rigid plates will form a geometrically stable (unyielding) system if they are connected together by means of three bars which are not parallel and do not converge at a common point of intersection.*

If a hinge is placed at the point of intersection of any two of the three bars and is connected to the plate, the system will

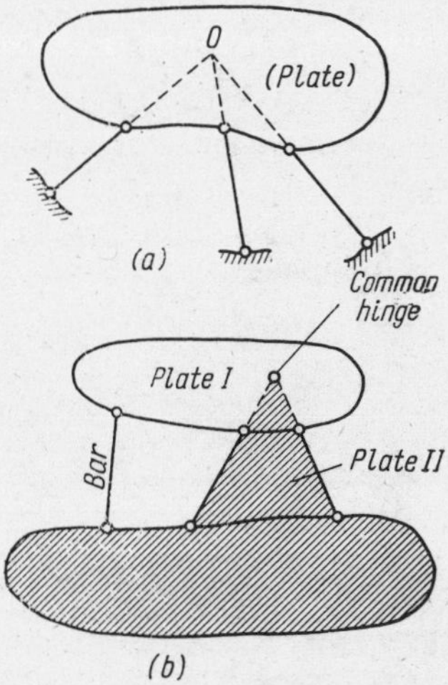

Fig. 15.1

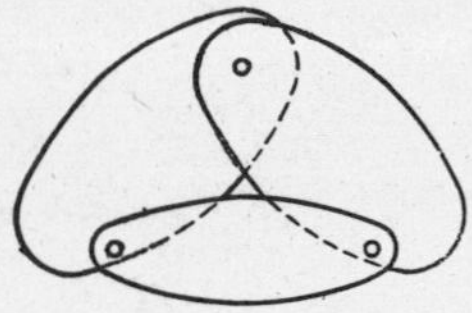

Fig. 16.1

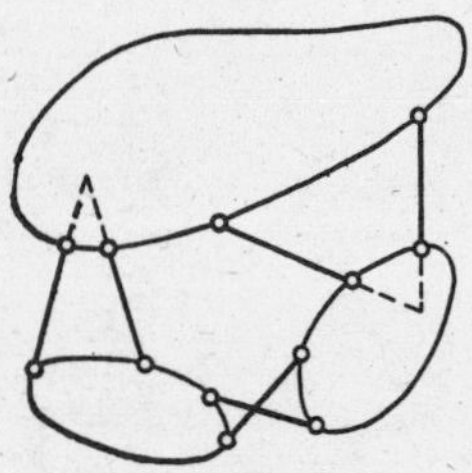

Fig. 17.1

remain unyielding and may be regarded as consisting of two separate plates connected by means of one common hinge and one bar (Fig. 15.1*b*). It follows that *two disks may be rigidly connected together using one hinge and one bar provided the direction of this bar does not pass through the centre of this hinge.*

Three plates may be connected to form one single unyielding system with the aid of three hinges placed at the vertices of a triangle, each of these hinges connecting one pair of plates (Fig. 16.1). Alternatively the same result will be obtained by placing six independent bars (Fig. 17.1), as each hinge may be replaced by two bars intersecting at the centre of this hinge.

However, the system represented in Fig. 18.1 is instantaneously unstable, the intersections of the bars connecting each pair of plates being in alignment. This system is similar to the one shown in Fig. 9.1.

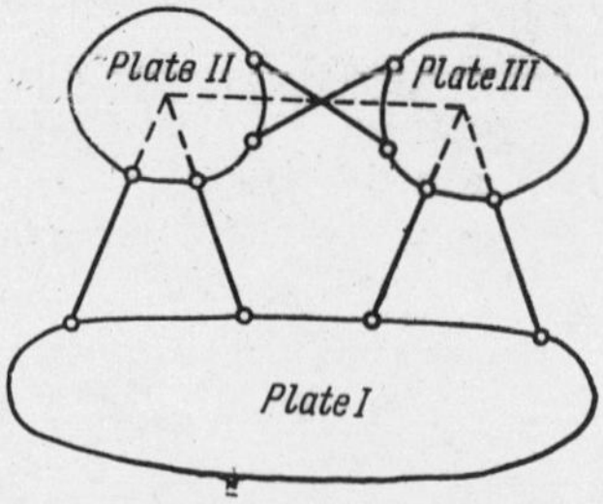

Fig. 18.1

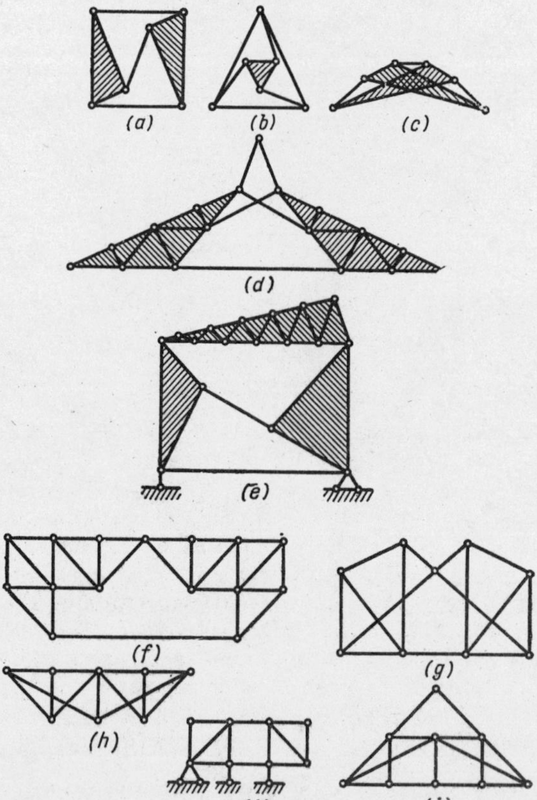

Fig. 19.1

Thus, *three rigid plates connected together with six bars, provided each pair of plates is connected by two bars and provided also the intersections of these two bars do not lie along one straight line, will always form an unyielding combination.*

Fig. 19.1 shows a number of systems constituted as just described.

A plausible arrangement of a statically determinate multispan (cantilever) beam is illustrated in Fig. 20.1 (such systems being

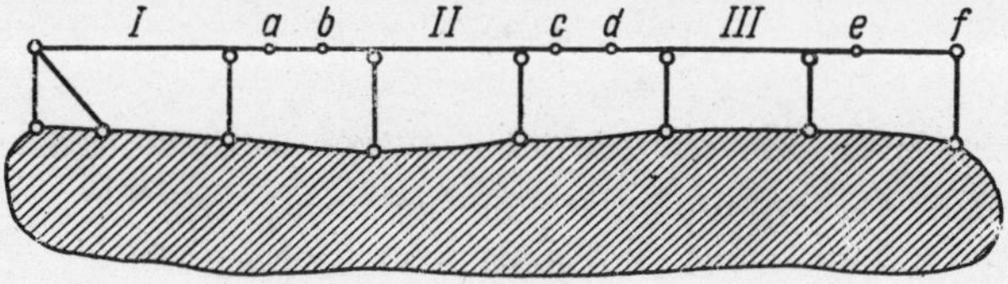

Fig. 20.1

described in greater detail in Art. 9.2). Let us check the geometrical stability of this beam. For this purpose, let us first select some unyielding portion of the structure rigidly connected to the ground and then let us see whether all the other geometrically

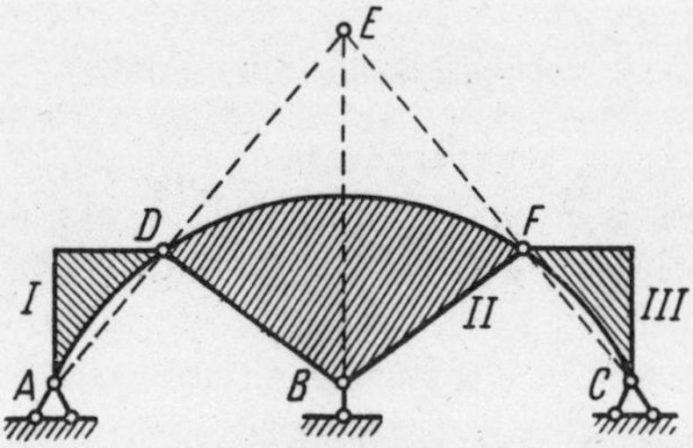

Fig. 21.1

stable parts of the structure are connected to the former by means of a sufficient number of bars. It should be kept in mind that the ground and any portion of the structure connected to it with the required minimum of three bars constitute an unyielding combination and therefore it is quite immaterial on which of the two the connecting bars will take support.

In the system under consideration bar *I* is rigidly connected to the ground with the aid of three bars which have no common point of intersection and which are not parallel. Bar *II* rests on two uprights standing directly on the ground and is attached to bar *I* by means of the insert *ab*. Bar *III* is connected to bar *II* in a similar way. Finally, the hinge *e* and an upright connect the last member *ef* to bar *III* and to the ground, respectively. Consequently the system as a whole will be geometrically stable.

Another illustration is afforded by the structure of Fig. 21.1. The lateral parts *I* and *III* may be regarded as simple stays *AD* and *CF*, and then it becomes apparent that plate *II* is connected to the ground by means of three bars (one vertical *B* and two inclined ones *AD* and *CF*) all of which intersect at one and the same point *E*. This system is, therefore, instantaneously unstable.

3.1. *Statically Determinate Framed Structures*

As has been stated, an unyielding connection of a structure with the ground may be schematically represented by three nonconcurrent bars. This type of connection is statically determinate as the number of reactive forces in these bars is equal to the number of equations furnished by statics for coplanar forces in equilibrium (for instance, $\Sigma X = 0$, $\Sigma Y = 0$, $\Sigma M = 0$).

Any plane structure will be externally statically determinate (i. e., statically determinate with reference to the supports) if the number of parameters determining the reactions at these supports is equal to three. The supports in the following examples fulfil this conditions:

(1) A combination of one fixed and one roller support for two-dimensional structures supported at two points (Fig. 22.1*a*).

(2) A combination of three roller supports for the same type of structures resting on three fulcra, provided the directions of the three reactions are neither concurrent nor parallel (Fig. 22.1*b*).

If a geometrically stable system rests on four or more supporting bars, three of which have no common point of intersection and are not parallel, the structure as a whole is statically indeterminate or redundant (Fig. 23.1). Equations provided by statics become insufficient for the analysis of such structures, additional equations based on the study of deformations or strains becoming indispensable.

Having formulated the conditions under which a structure is externally statically determinate, let us now examine those which render a framed structure internally statically determinate, i. e., such where the forces acting in all of its bars may be computed using equations of equilibrium alone.

It may be easily shown that stresses in the bars of a hinged truss subjected to concentrated loads acting at the joints will be always normal to the cross sections of these bars. Indeed,

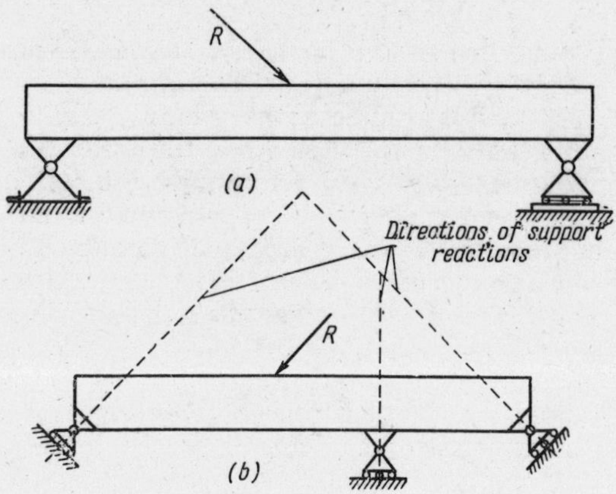

Fig. 22.1

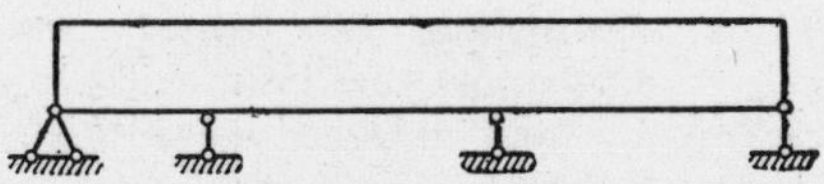

Fig. 23.1

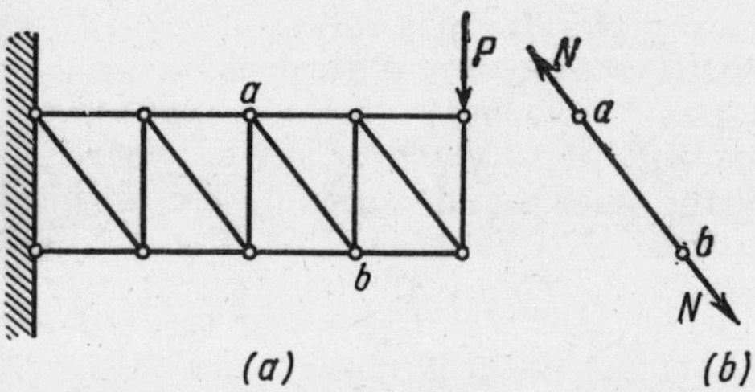

Fig. 24.1

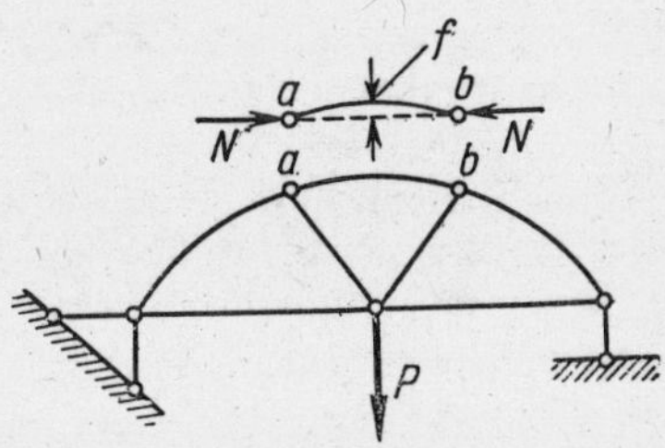

Fig. 25.1

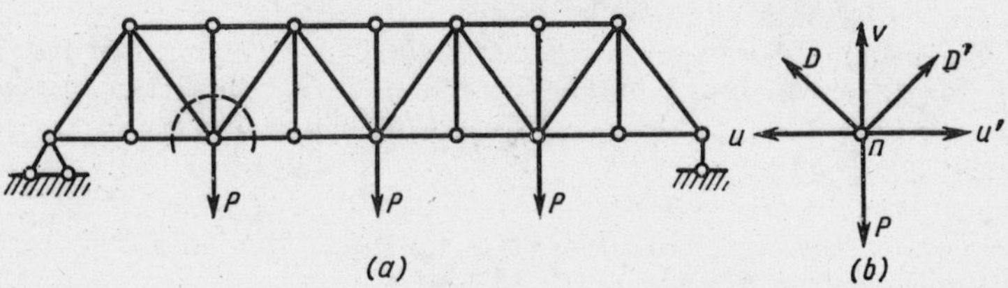

Fig. 26.1

having separated one of the bars, say, bar *ab*, let us analyze the conditions of its equilibrium (Fig. 24.1*a*, *b*).

If no external load is applied directly to this bar, its equilibrium will be ensured only in the case when the forces N acting on the bar through the hinges *a* and *b* are equal in amount but opposite in direction. These forces will always pass through the centres of the hinges since in our analysis these are assumed to be frictionless. It follows that forces N will act along a line connecting the hinge centres and, therefore, the cross sections of bar *ab* will be subjected either to direct tension or to direct compression.

Should the truss contain curved bars, these will be subjected to bending moments in addition to the normal forces just mentioned, the maximum value of these moments equalling $M_{max} = Nf$ (Fig. 25.1).

When the truss as a whole is in equilibrium under the action of external loads and reactions (Fig. 26.1*a*), each of its joints is also in equilibrium (Fig. 26.1*b*). Accordingly, tle external load applied to any joint and the internal forces in the bars converging at this joint must be balanced.

Statics will furnish each joint subjected to a system of concurrent coplanar forces with two equilibrium equations

$$\Sigma X = 0 \quad \text{and} \quad \Sigma Y = 0$$

If the truss contains K joints, we may form $2K$ equations of equilibrium which must provide for the determination of all the internal forces in the members and of the three unknown parameters of the reactions. Any other equilibrium equation which may be formed for the truss as a whole or for any part thereof can be derived from the above and consequently will contain no additional information.

Hence the truss will be statically determinate, if the number of its bars S is equal to double the number of joints K less 3

$$S = 2K - 3 \tag{2.1}$$

As will be readily observed, this is the same relation as the one giving the minimum number of bars of a geometrically stable system [(expression (1.1)].

Consequently, *any simple truss obtained by the successive addition of joints to a hinged triangle, each joint being connected by means of two bars not in alignment, is both geometrically stable and statically determinate.*

If, when counting the number of bars of a truss, those forming its supports were also taken into consideration, expression (2.1) will become

$$S_{tot} = 2K \tag{3.1}$$

This formula becomes particularly useful when the structure though being geometrically unstable (the number of its bars totalling less than $2K-3$) is connected to the ground in such a way that together they form a single unyielding and statically determinate system. A structure of that type is represented in Fig. 27.1. Here $K = 8$ while the number of bars (supporting bars are omitted) totals 12. Thus the structure is unstable for

$$S = 12 \quad \text{while} \quad 2K-3 = 13$$

However, S_{tot} (including the supporting bars) is equal to 16 which satisfies equation (3.1) and therefore the whole system

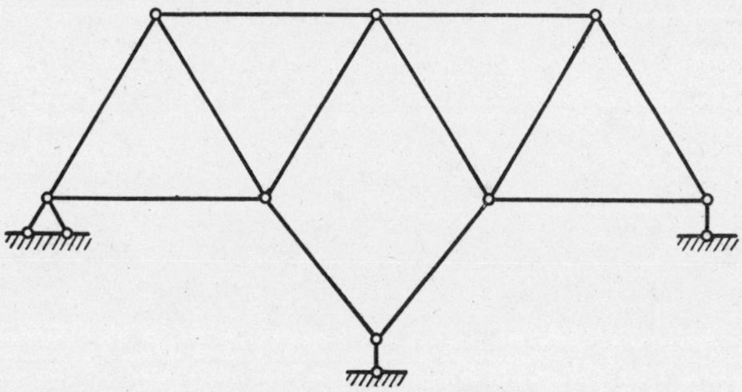

Fig. 27.1

may be (and in this case actually is) both geometrically stable and statically determinate. *

In a statically determinate system all the bars are absolutely indispensable to ensure its geometrical stability, in other words, in such a structure there is not a single superfluous (redundant) member.

When a geometrically stable system contains more bars than is strictly necessary it becomes *statically indeterminate* or *redundant.*

The theory of structures analyzes only geometrically stable systems both *statically determinate* and *statically indeterminate* or *redundant.*

The reader is invited to find out on his own to which of these two categories the structures represented in Fig. 28.1 belong.

◆

* The analysis of such systems is considered in detail later (see Art. 6.4.)

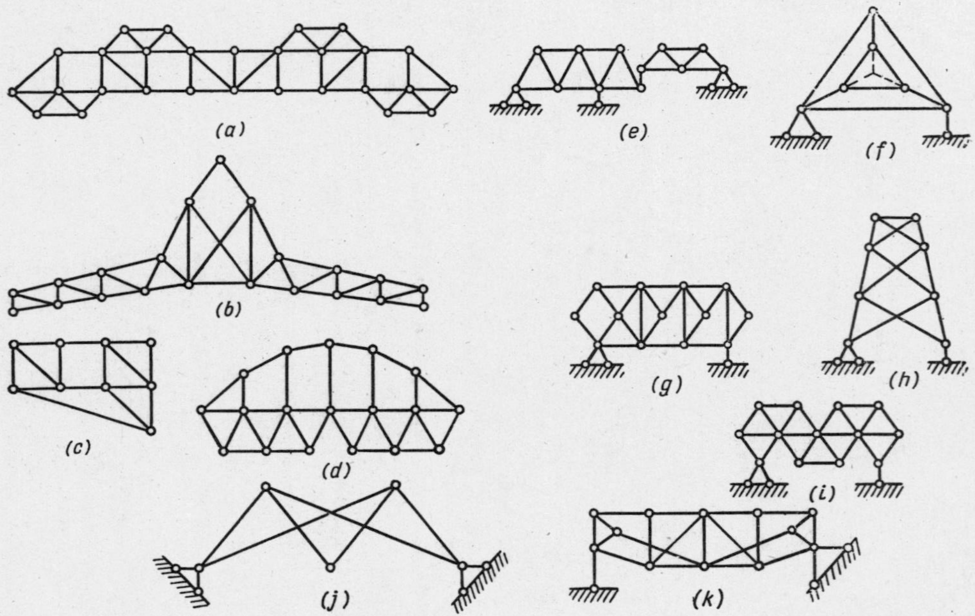

Fig. 28.1

2. BEAMS

1.2. *General*

The reader having already studied the *strength of materials* must be familiar with methods permitting the determination of stresses acting over the cross sections of statically determinate simply supported beams, as well as with the construction of diagrams showing the distribution of these stresses along a beam subjected to a system of fixed loads. The same methods are used in structural mechanics.

The following sign convention will be adopted hereunder: *The shearing force Q (or simply the shear) will be considered positive when it tends to uplift the left extremity of the right-hand portion of a beam with reference to the right extremity of the left-hand portion. The bending moment M will be reckoned positive when it tends to rotate the left extremity of the right-hand portion of a beam clockwise and the right extremity of the left-hand portion counterclockwise.*

When the loads are not at right angles with the axis of a beam, the latter will be subjected equally to forces N normal to its cross sections. *These will be regarded as positive when they cause tensile stresses and negative when these stresses are compressive.*

Positive directions of bending moments, shearing and normal forces are shown in Fig. 1.2. It will be seen that a positive bending moment causes compression of the top fibres of a beam and an extension of the lower ones, while a positive shear will tend to rotate each portion of the beam clockwise with respect to its other end.

When plotting the diagrams of shearing and normal forces their positive values should be scaled off above the x-axis and the negative ones below. It is good practice to indicate prominently on the stress diagrams the signs of the corresponding stresses. As for bending moments, their positive values shall be scaled off below the x-axis and the negative ones above it;

thus, bending moment diagrams will always appear on the side of the extended fibres of the beam. *

The sign of the shearing force can be also ascertained with the aid of the bending moment diagram, using the following rule: *The shear is positive in any cross section where the superposition of the axis of the element with the tangent to the bending*

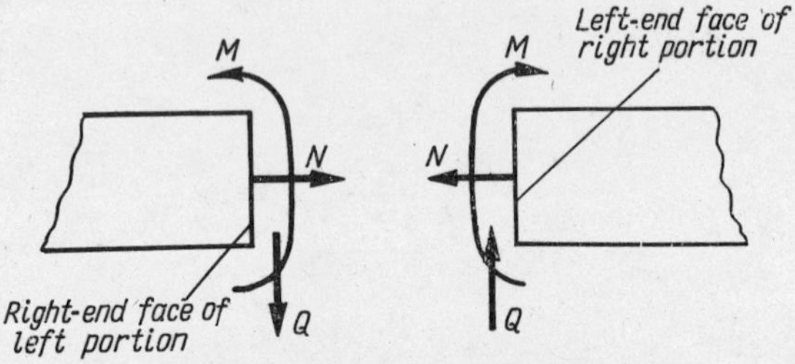

Fig. 1.2

moment diagram requires a clockwise rotation of the former, provided the angle of rotation does not exceed 90 °.

Assume, for example, that it is required to find the sign of the shearing force at cross section x of a beam, whose bending moment diagram is represented in Fig. 2.2. In this case, the axis of the beam should be rotated clockwise in order to bring it in coincidence with the tangent to the bending moment diagram (the direction of rotation is indicated by a dotted arrow), hence, the shearing force is positive. However, in cross sections close to the right-hand extremity of the beam the shear will be negative, for the superposition of the axis with the tangent would require counterclockwise rotation (see Fig. 2.2).

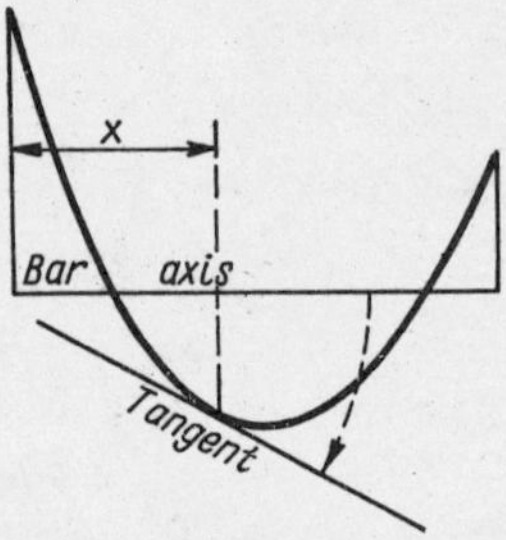

Fig. 2.2

The shear Q in any cross section is equal in amount and sign to the sum of projections of all the external forces acting to the left of this cross section on a normal to the beam axis passing through this cross section, or to the sum of projections of all the external forces to the right of the cross section on the same normal but taken with an opposite sign

$$Q = \sum_{L} Y = -\sum_{R} Y \tag{1.2}$$

◆

* In certain treatises on the strength of materials, positive bending moments are plotted on the side of compressed fibres.

the projections being reckoned positive when they are directed upwards.

The bending moment M in any cross section is equal in amount and sign to the sum of moments about the z-axis (this axis passing through the centroid of the cross section normally to the plane of the beam) of all the external forces acting to the left of the cross section or to the sum of moments of all the external forces acting to the right of this section but taken with an opposite sign

$$M = \sum_L M_z = -\sum_R M_z{}^* \tag{2.2}$$

the moments being reckoned positive when they tend to rotate the cross section clockwise.

The normal force N is equal in amount and sign to the sum of projections of all the external forces to the left of the cross section under consideration on the beam axis, or to the sum of projections (on the same axis) of all the external forces to the right of this section but taken with an opposite sign

$$N = \sum_L X = -\sum_R X \tag{3.2}$$

these projections being reckoned positive when they are directed from right to left.

There is a set of relations between the M and Q diagrams and the loading of the beam, these relations facilitating the plotting of these curves and permitting their verific tion. These relations are of great importance for they apply not only to beams but equally to bents and frames of various types.

The basic relation can be represented as follows

$$Q = \frac{dM}{dx} \tag{4.2}$$

in other words, the shear is equal to the first derivative of the bending moment in terms of dx (theorem of Zhuravsky), the sign convention as set out above remains in force for M and Q, while the positive direction of the x-axis is from left to right.

Moreover, there is equally the relation

$$q = \frac{dQ}{dx} \tag{5.2}$$

which means that the intensity of the distributed load applied normally to the beam axis is equal to the first derivative of the shear, the distributed load being reckoned positive when it is directed upwards

◆

* The index z may be omitted.

The following can be easily deducted from these two relations *:

1. Negative shears correspond to decreasing bending moment values, indicated by an increase of the bending moment diagram ordinates from left to right. Similarly decreasing bending moment diagram ordinates will signify that the corresponding shears are positive.

2. The steeper the slope of the tangent to the bending moment diagram, the greater in absolute value is the shear, for the latter is numerically equal to the natural tangent of the angle formed by the tangent to the diagram and the beam axis.

3. The bending moment will pass through a maximum or a minimum at those cross sections where the shear is nil.

4. The bending moment diagram between two concentrated loads (no distributed loads intervening) forms a straight line, generally inclined, while that of the shear reduces to a horizontal.

5. A conic parabola for bending moment diagram will correspond to a uniformly distributed load, the shear diagram becoming in that case an inclined straight line.

6. The convexity of the bending moment diagram is always turned in the direction of the distributed loads.

7. Concentrated loads cause breaks in the direction of the bending moment diagram and jumps in the shear diagram. The rises and falls in the latter case are equal in amount and direction to the magnitude of the concentrated loads as met when moving from left to right along the beam.

8. The change in the magnitude of the bending moment occurring over a certain portion of the beam length is equal to the area of the shear diagram over the same beam length provided no external moments are applied thereto.

9. The change in the magnitude of the shear occurring over a certain portion of the beam length is equal to the area of the distributed load diagram over the same beam length.

In the present chapter we shall study the methods of stress computation in cross sections of simply supported beams carrying moving loads and in those of multispan cantilever beams subjected both to fixed and moving loads. Moving loads are frequently encountered in the computation of bridges, overhead cranes and other engineering structures. An example of a moving load is furnished by a train travelling along a railway bridge, or an overhead crane moving along crane tracks, etc.

* It is deemed unnecessary to dwell in detail on the corresponding demonstrations.

Stresses and strains in the different elements of a structure depend on the position of the moving load. In order to determine the maximum design stresses, it is always necessary to know the most unfavourable position of the load or loads for the element concerned. Thus, when designing the cross section of any truss member, the moving load must be so placed as to cause the greatest possible stress in this particular member. This loading position is usually referred to as the *most unfavourable* or *dangerous*. A distinct most unfavourable load position can be always found for each truss member, every cross section of a beam, etc.

It should be noted that this remains true not only for stresses but also for reactions at the supports, for deflections and so forth.

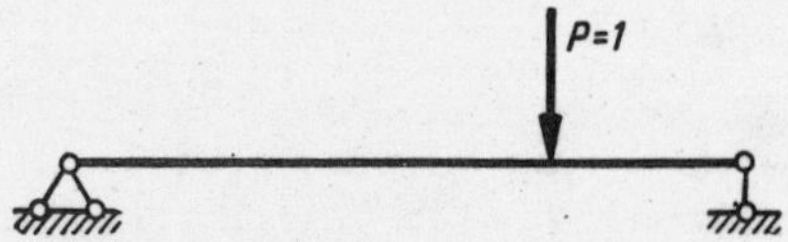

Fig. 3.2

The design of structures subjected to moving loads is greatly facilitated by the possibility of applying the *principle of superposition*. This means that the internal forces, fibre stresses and strains caused in a structure by different loads will add to one another. It follows that if some particular load increases a certain number of times, the stresses and strains set up by this load will increase in the same ratio.

It also follows that if two different groups of loads are applied to a structure, the total stress in each member will be equal to the sum of stresses caused separately by each of the two groups.*

We shall start with our analysis of the effect of moving loads with the simplest case possible—that of a single vertical unit load P moving along a simply supported beam (Fig. 3.2). Let us investigate the changes sustained by each of the parameters under consideration (reaction at the support, internal force in a truss member, bending moment in a particular cross section of a beam, the beam's deflection at a certain point, etc.) when the load $P = 1$ travels along the structure. We shall represent graphically the alterations of the parameter chosen in terms of the load position.

* The principle of superposition applies not only to the case of concentrated loads but equally to distributed load, bending moments, temperature stresses, etc. It does not apply in the case of buckling with bending, in all cases when the material does not follow Hooke's law and in some other cases.

The diagram which depicts the fluctuation of some particular parameter (say, the bending moment in a cross section of a beam) when the load $P = 1$ travels along the structure is termed the influence line for the said parameter. *

Influence lines should never be confounded with the stress diagrams. In fact, the ordinates to the latter represent the variation of the parameter under consideration (say, of the bending moment) in all cross sections of the beam for one definite position of the load, whereas those of the influence line indicate the variation of a parameter (say, of the same bending moment) in *one particular cross section* when the load unity *travels* along *the whole length* of the beam.

2.2. *Reaction Influence Lines for Simply Supported Beams with or without Overhang*

Let us assume that a unit load $P = 1$ moves along a simply supported beam AB (Fig. 4.2*a*) and let us designate by x the distance from the load to the right-hand support. This distance

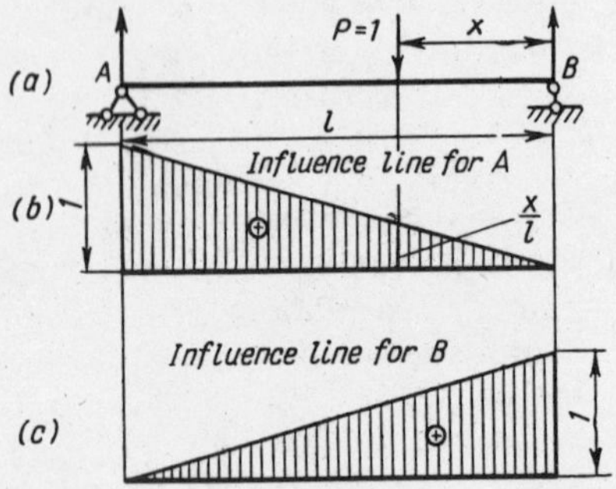

Fig. 4.2

will vary from zero, when the load is directly over this support, to l when it is over the left-hand one. In order to determine the reaction A in terms of x, we can write the equation of equilibrium of moments of all the external forces about the right-hand support.

$$\Sigma M_B = Al - Px = 0$$

◆

* Influence lines representing the variations of either stresses or strains can also be plotted for unit bending moments, external forces normal to the cross section and other types of loads moving along the structure.

whence

$$A=\frac{Px}{l}$$

However, since $P = 1$, then

$$A=\frac{1x}{l}=\frac{x}{l}$$

This equation gives us the law governing the variation of the reaction A as the load $P = 1$ shifts from one point to another.

Plotting out the relation just established we obtain the influence line for the reaction A at the left-hand support. Since this equation is of the first degree in terms of x, the influence line will be rectilinear (Fig. 4.2*b*) and

$$\text{for } x=0 \quad A=0$$

$$\text{for } x=l \quad A=\frac{l}{l}=1$$

The ordinates to the influence line for the reaction are dimensionless, for both x and l are expressed in units of length.

Some scale must be selected in order to plot the influence line. If, for instance, we adopt a scale of 1 in 1 cm, we shall lay off 1 cm over the left-hand support where $A = 1$. The ordinate to the influence line for the reaction at A measured a distance x from the right-hand support will equal $\frac{x}{l}$. This ordinate will be numerically equal to the reaction A when the distance to the load $P = 1$ as measured from the right-hand support equals x. In other words, *the ordinate to the influence line for the reaction A at a given cross section represents to scale the value of the said reaction at the instant when the unit load P is placed directly over this cross section* Accodingly, the magnitude of the reaction A corresponding to a given position of the load $P = 1$ can be obtained by simply scaling off the ordinate to the influence line at the point of load application.

When the load actually applied to the beam amounts to P_1, the reaction A will be obtained by multiplying the ordinate to the influence line at the point of loading (this ordinate, as already mentioned, representing the reaction A corresponding to a unit load) by the magnitude of force P_1. Should a number of concentrated vertical loads act on the beam simultaneously, the total reaction A will be found as the sum of separate reactions due to each of those different forces.

Let us now proceed with the construction of the influence line for reaction B. For this purpose we may equate to zero the sum of all the moments of external forces about the hinge centre at A:

$$\Sigma M_A = -Bl + P(l-x) = 0$$

leading to

$$B = \frac{P(l-x)}{l} = \frac{1(l-x)}{l} = \frac{l-x}{l}$$

This equation represents the variation of reaction B in terms of the position of load unity P. In order to trace the influence line, let us put

$$x = 0 \quad \text{then} \quad B = \frac{l}{l} = 1$$

$$\text{and} \quad x = l \quad \text{then} \quad B = \frac{l-l}{l} = 0$$

Fig. 4.2*c* represents the influence line for the reaction B. The ordinates to this line are again dimensionless and the scale should be the same as for reaction A. These ordinates represent the amount of reaction B when a unit load is applied at the cross section corresponding to the given ordinate. Hence we can determine this reaction for a load unity by simply measuring the ordinates to the influence line.

The significance of influence lines represented in Fig. 4.2*b* and *c* is greatly enhanced by the fact that they permit immediate determination of the load position causing the greatest reactions. Thus, it is readily seen that the reaction A will reach its maximum when load P_1 will stand directly over the left-hand support. The same is true for reaction B when the load P_1 is applied to the right-hand support, i. e., when it coincides with the maximum ordinate to the influence line.

It should be kept in mind, however, that *each influence line will depict solely the variations of the parameter for which it has been plotted*. Thus the influence lines for reactions A and B will convey information on these reactions respectively.

Let us consider now the influence lines for the reactions of a beam cantilevering over one of its supports as shown in Fig. 5.2*a*. The influence line for reaction A will be derived from equation

$$\Sigma M_B = Al - Px = 0$$

giving

$$A = \frac{Px}{l} = \frac{1x}{l} = \frac{x}{l}$$

This equation is identical with that obtained before for a simply supported beam, with the sole difference that in the latter equation x can vary from 0 to l while in the present case it can do so from 0 to $(l+k)$ where k is the length of the overhang. The ordinates to the influence line at pertinent points are

$$\text{for } x=0 \qquad A=0$$

$$\text{for } x=l \qquad A=\frac{l}{l}=1$$

$$\text{for } x=l+k \quad A=\frac{l+k}{l}=1+\frac{k}{l}$$

We can now proceed with the construction of the influence line for reaction A by simply laying off the ordinates obtained

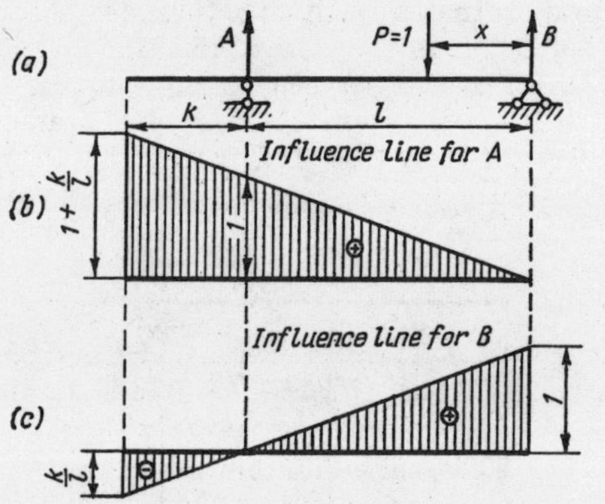

Fig. 5.2

(Fig. 5.2*b*). It should be noted that there is no real necessity to determine all the three ordinates, as the influence line is rectilinear and in this case the knowledge of only two ordinate values (say, at $x = 0$ and at $x = l$) is sufficient.

A comparison of an influence line for a beam with overhang with the influence line represented in Fig. 4.2*b* shows that the first one can be easily obtained by a simple extension of the latter until its intersection with the vertical passing through the end of the overhang.

The following equation will be used for the construction of the influence ine for reaction B

$$\Sigma M_A = -Bl + P(l-x) = 0$$

whence

$$B=\frac{P(l-x)}{l}=\frac{1\,(l-x)}{l}=\frac{l-x}{l}$$

Comparing this equation with the one relating to a simply supported beam we find that they are exactly the same, the only difference residing in the limits between which x may vary.

Let us now determine the ordinate values of this influence line

$$\text{for } x=0 \qquad B=\frac{l}{l}=1$$

$$\text{for } x=l \qquad B=\frac{l-l}{l}=0$$

$$\text{for } x=l+k \qquad B=\frac{l-(l+k)}{l}=-\frac{k}{l}$$

Plotting these ordinates as in Fig. 5.2*c* we obtain the influence line for reaction B. As in the case of reaction A, the computation of the ordinate for $x = l+k$ proves superfluous.

The influence line for the reaction B of a beam with overhang can also be derived from the one pertaining to a simple beam

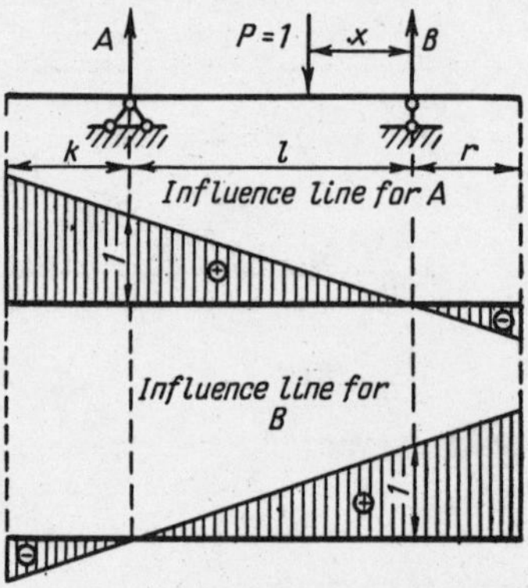

Fig. 6.2

by extending the line until its intersection with the vertical drawn through the extremity of the overhang.

The fact that some of the ordinates to the influence lines are this time negative indicates that when the load point coincides with these negative ordinates, the reaction B itself is also negative or, in other words, directed downwards.

The same procedure as described above should be followed for the construction of influence lines for the reactions of a beam cantilevering over its right-hand support.

Fig. 6.2 represents the influence lines for the reactions of a beam with two overhangs. The reader is invited to write the corresponding equations on his own.

3.2. *Bending Moment and Shear Influence Lines for Simply Supported Beams with or without Overhang*

Let us now analyze the construction of influence lines for bending moments and shearing forces induced by a moving load in a simply supported beam. We shall begin our investigation by examining the *influence line for the bending moment* in cross section I located a distance a from the left-hand support and a distance b from the right-hand one (Fig. 7.2*a*). The bending moment in this section is equal to the algebraic sum of moments of the outer forces to the left of this section about its centroid or to the sum of moments of forces to its right but taken with an opposite sign [see expression (2.2)].

As long as the load is situated to the right of section I (Fig. 7.2*a*), i. e., as long as $x \leqslant b$, the only external force to the left is the reaction A and therefore the bending moment M_I in section I is equal to Aa.

Accordingly the influence line for this bending moment may be derived from the influence line for reaction A by multiplying its ordinates by a.

Substituting for A its value found in Art. 2.2 we obtain

$$M_I = \frac{x}{l} a$$

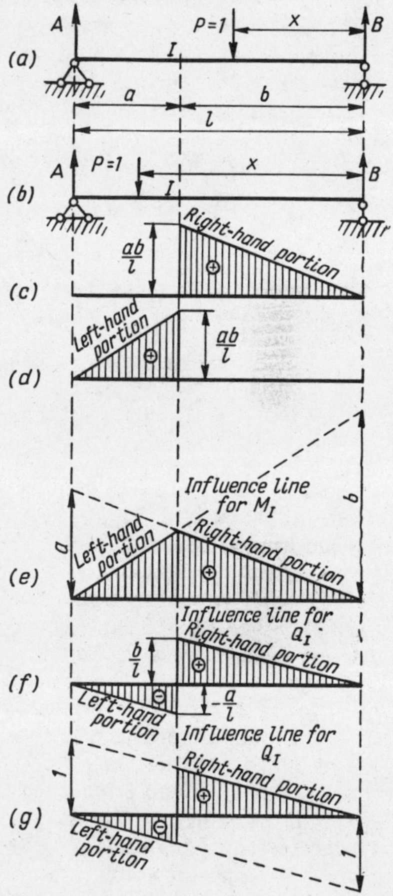

Fig. 7.2

The graphical representation of this equation requires the knowledge of two distinct values of M_I

$$\text{for } x = 0 \qquad M_I = 0$$

$$\text{for } x = b \qquad M_I = \frac{ab}{l}$$

Using these values we can now trace the *right-hand* portion of the influence line for M_I (Fig. 7.2*c*). Its ordinates will furnish the values of the bending moment in section I when the unit load is situated to the *right* of this section, i. e., when $x \leqslant b$.

When the load passes to the left of section I, i. e., when $x \geqslant b$ (Fig. 7.2*b*) it becomes more convenient to use the equations pertaining to the right-hand portion of the beam.

In that case the bending moment $M_I = +Bb$ for, although the moment of reaction B about the centroid of section I acts counterclockwise and is therefore negative, the bending moment caused by it in the beam remains positive [see expression (2.2)].

Substituting the value of B see Art. (2.2) we obtain

$$M_I = \frac{l-x}{l} b$$

For a graphical representation of this expression, we shall once again find two values of M_I

$$\text{for } x = b \qquad M_I = \frac{l-b}{l} b = \frac{ab}{l}$$

$$\text{for } x = l \qquad M_I = \frac{l-l}{l} b = 0$$

These data permit the construction of the *left-hand* portion of the influence line for M_I (Fig. 7.2*d*). Its ordinates will furnish the bending moment values for section I when load unity P is to the *left* of the section, i. e., when x varies from b to l. *

If the left- and the right-hand portions of the influence line (Fig. 7.2*c* and *d*) are now brought together (Fig. 7.2*e*) *they will intersect under cross section I*. Should these lines be extended until they meet with the verticals passing through the supports they would intercept thereon the following ordinates: over the left-hand one an ordinate equal to a, and over the right-hand one an ordinate equal to b (Fig. 7.2*e*). This can be easily proved by substituting 0 and l for x in the expressions of the right- and left-hand portions of the influence line respectively. Therefore, in practice the M_I influence line is frequently constructed in the following way: *ordinate a is plotted over the left-hand support and ordinate b over the right-hand one, a and b being respectively the distances from section I to these two supports. This being done two straight lines connecting each of these ordinates with the zero*

◆

* Ordinates of positive bending moments are directed upwards. Accordingly, when the bending moment influence line is above the beam axis the lower fibres of the beam are extended.

ordinate point at the base of the other are traced, the two lines intersecting exactly under cross section I.

The above procedure may be simplified as follows: *first draw the line corresponding to any of the two portions of the influence line (say, to the right-hand one), and then connect its point of intersection with the vertical passing through the section concerned with the zero point at the other support* (in our case at the left-hand one).

The ordinates to the bending moment influence line are expressed in units of length. This for example may be seen from the fact that the ordinate over the left support is taken equal to the length a. Hence the same scale may be adopted for both the beam length and the bending moment influence line.

Any ordinate to the influence line for M_I will furnish the value of the bending moment in section I when the unit load is situated over this particular ordinate. Accordingly the determination of the bending moment in section I for a given position of load $P = 1$ requires solely the measurement of the influence line ordinate at the load point.

It should be borne in mind that the influence line for M_I expresses the variation of the bending moment only in section I. If it were required to find the law governing the variation of the bending moment in some other section, a new influence line corresponding to that particular section should be constructed.

Let us now examine the construction of the *shear influence line* for section I. As already stated, the shear in any section is equal to the algebraic sum of vertical projections of all external forces acting to the left of the section concerned, or to the same sum taken with the opposite sign and pertaining to the external forces to the right of this section [see expression (1.2)].

Examining two unit load positions, one when the load is to the right of section I and the other when it is to the left of same, we find:

(1) In the first case, i. e., when $x \leqslant b$ (see Fig. 7.2a) the equilibrium equation relative to the left-hand portion of the beam furnishes

$$Q_I = A = \frac{x}{l}$$

Graphical representation of this relation requires the computation of two distinct values of Q_I

$$\text{for } x = 0 \qquad Q_I = 0$$

$$\text{for } x = b \qquad Q_I = \frac{b}{l}$$

Using these values we can construct the *right-hand portion of the Q_I influence line* (Fig. 7.2f), its ordinates giving the values of the shear in section I when the unit load is to the right of this section, i. e., when $x \leqslant b$.

(2) In the second case, i. e., when $x \geqslant b$ (Fig. 7.2*b*) the same considerations as above give $Q_I = -B$ [although reaction B is directed upwards and is therefore positive, it must be taken with the minus sign, in accordance with expression (1.2)].

Since $$B = \frac{l-x}{l}$$

$$Q_I \text{ becomes} -\frac{l-x}{l}$$

Computing two distinct values of Q_I

$$\text{for } x = b \quad Q_I = -\frac{l-b}{l} = -\frac{a}{l}$$

$$\text{for } x = l \quad Q_I = -\frac{l-l}{l} = 0$$

and plotting them as in Fig. 7.2*f* we obtain the left-hand portion of the shear influence line (as the ordinates are negative they are plotted downwards).

Had we extended the influence lines obtained until their intersection with the verticals passing through the supports the corresponding intercepts would equal: at the left support +1, and at the right one —1. This can be easily proved by substituting $x = 0$ and $x = l$ in the equations relative to the right-hand and to the left-hand portions of the influence line, respectively.

It follows that the shear influence line can be constructed as indicated in Fig. 7.2*g* by *plotting the ordinates +1 (upwards) and —1 (downwards) along the verticals passing through the left-hand and the right-hand supports respectively and by joining each of the two points so obtained with the base point at the other support*. It is obvious that these two lines will be parallel. This being done, a vertical is traced through the section under consideration as in Fig. 7.2*g*.

The ordinates of the shear influence lines are dimensionless, hence their scale may be the same as in the case of abutment reaction influence lines.

Ordinates to the shear influence lines represent the shear values in section I arising from unit load P acting in the section corresponding to the said ordinate. Therefore, the amount of the shearing forces in section I for a given position of the unit load P can be obtained by simply measuring the ordinate of the shear influence line at the load point.

If the ordinate at load point is negative, the shear in the section will also be negative for this position of the load. The ordinates to the Q_I influence line represent the shear variation only in section I. Should it be required to find the shear variation in some other section, a new influence line would have to be constructed.

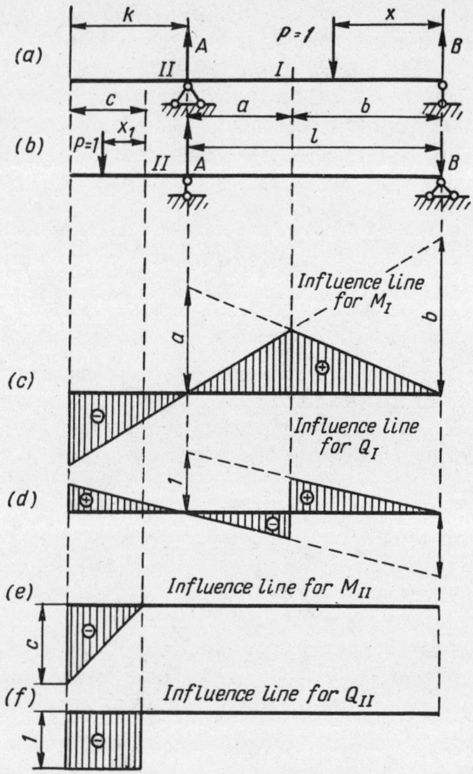

Fig. 8.2

Let us now investigate bending moment and shear influence lines for a beam cantilevering over the left support, as shown in Fig. 8.2*a*.

Construction of the influence lines for a cross section located between the supports A and B remains exactly the same as in the previous case, i. e., as in the case of a simply supported beam with no overhang. Two load points—one to the right of the section and one to its left should be considered, the bending moments and the shears being expressed in both cases through the reactions A and B.

Since the equations of the abutment reaction influence lines are the same for a simply supported beam with or without overhang (see Art. 2.2), it is obvious that the corresponding equations for the bending moment and shear will also be the same for both types of beams, with the sole difference that x, which varied in the first case from 0 to l, will now vary from 0 to $(l + k)$.

This will affect the construction of shear and bending moment influence lines in the same way as those for the reactions, in other words, these lines will simply have to be extended to the left extremity of the overhang (Fig. 8.2*c* and *d*). The reader is once again invited to check himself analytically the influence lines so obtained.

Now let us see what happens in section *II* situated a distance c from the left end of the overhang (Fig. 8.2*a*). Once again we must consider two positions of the unit load P.

(1) *The load point is to the right of section II* (Fig. 8.2*a*). In this case there are no external forces to the left of section *II* and therefore the bending moment in this section is nil. The appropriate portion of the influence line is represented in Fig. 8.2*e* by a horizontal line coinciding with the x-axis, as all the ordinates are nil along the whole stretch from section *II* to the support at B.

(2) *The load is to the left of section II* (Fig. 8.2*b*). In this case there is only one force to the left of the section, hence the bending moment in section *II* will be

$$M_{II} = -1 \cdot x_1$$

where x_1 is the distance from the load point to section *II*. This distance may vary from 0 (when the load point coincides with the section concerned) to c (when the load reaches the end of the overhang).

For these two extreme values of x_1 we have, respectively, $M_{II} = 0$ and $M_{II} = -1 \cdot c$. This portion of the influence line is represented in Fig. 8.2*e*, negative ordinates being plotted downwards. Thus, Fig. 8.2*e* represents the bending moment influence line for section *II* for any position of the load.

Let us proceed with the construction of the shear Q_{II} influence line for section *II*.

(1) *As long as the load remains to the right of the section* there are no forces whatsoever to its left and therefore $Q_{II} = 0$. The corresponding part of the influence line (from section *II* to the support at B) is represented in Fig. 8.2*f* by a horizontal stretch coinciding with the x-axis.

(2) *When the unit load is to the left of section II*, the shear $Q_{II} = -1$ which means that the shear remains constant irre-

spective of the position of the load point, provided it lies to the left of the section. This portion of the influence line is represented in Fig. 8.2*f* by a line parallel to the x-axis, negative shears being plotted downwards. Fig. 8.2*f* gives the shear influence line for section *II* in its entirety.

As will be noticed, the bending moment and shear influence lines for sections selected within the overhang differ very substantially from those relating to sections situated between the supports.

In Fig. 9.2 we have represented the bending moment influence lines for a number of sections of a beam cantilevering over both supports, sections *II* and *VI*, coinciding with the left-hand and right-hand supports, respectively. The shear influence lines for the same sections are presented in Fig. 10.2. Two sections correspond to each support, sections *IIa* and *VIa*, being immediately to the left thereof, and sections *IIb* and *VIb* immediately to the right. It will be noted that shear influence lines for sections *IIa* and *IIb* as well as for sections *VIa* and *VIb* are quite different.

Problem. It is required to construct the bending moment and shear influence lines for section m of a beam represented in Fig. 11.2*a*, the unit load P travelling from r to s.

Solution. First construct the influence line for reaction A. Beam rs being rigidly connected to beam AB by means of the stanchion pq, the reaction A will equal (see Fig. 11.2*b*)

$$A=\frac{l-x}{l}$$

The bending moment in section m for any position of the unit load on beam rs will equal

$$M_m=Aa$$

Accordingly, the bending moment influence line for section m will be geometrically similar to that of the left-hand reaction, the latter's ordinate values being multiplied by a constant factor equal to a. This influence line is represented in Fig. 11.2*c*.

The shear influence line will be obtained through the same procedure and will differ in no respect from that for reaction A.

4.2. *Influence Lines for Simple Cantilever Beams*

Let us first find the influence line for reaction A at the support (Fig. 12.2*a*). The equilibrium equation for the vertical projections of the external forces gives

$$\Sigma Y=-1+A=0$$

hence

$$A=1$$

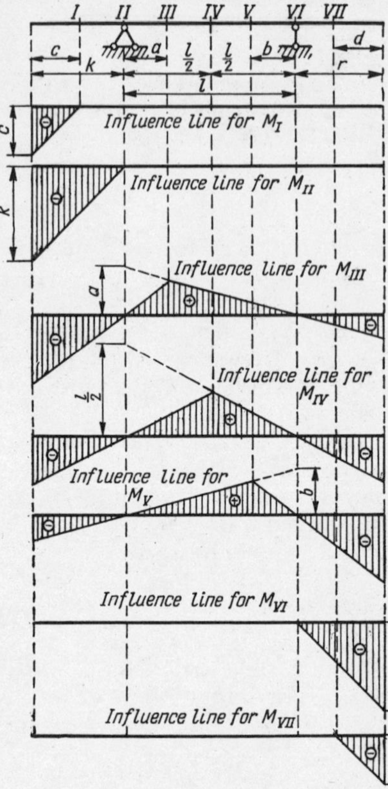

Fig. 9.2

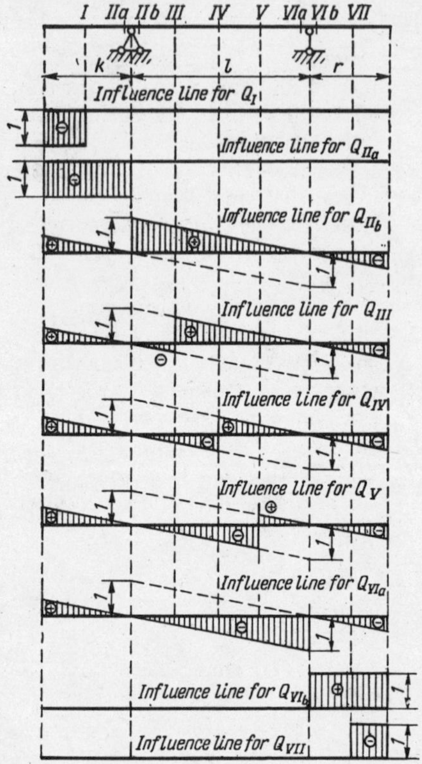

Fig. 10.2

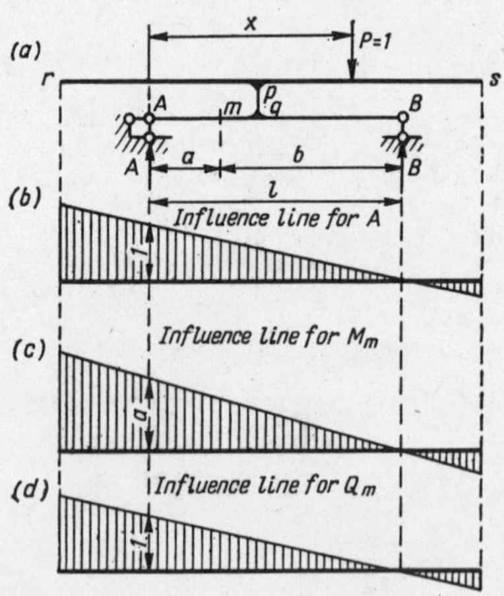

Fig. 11.2

Accordingly, for any position of the point of application of the unit load P, the reaction remains equal to 1. Adopting an appropriate scale as explained in Art. 2.2, we can trace the influence line shown in Fig. 12.2*b*. In order to find the bending moment influence line for section I located a distance c from the left-hand extremity of the beam, we shall proceed in the same way as in the case of a beam with overhang represented in Fig. 8.2, i. e.,

(1) when the load is to the right of section I (solid line in Fig. 12.2*a*)

$$M_I = 0$$

(2) when the load is to the left of section I (dash line in Fig. 12.2*a*)

$$M_I = -1 \cdot x_1 = -x_1$$

where x_1 is the distance from load point to section I;

$$\text{for } x_1 = 0 \quad M_I = 0$$

$$\text{for } x_1 = c \quad M_I = -c$$

The corresponding influence line is represented in Fig. 12.2*c*. It is quite similar to the bending moment influence line for a section within the cantilevering portion of a beam with an overhang (see Fig. 8.2*e*).

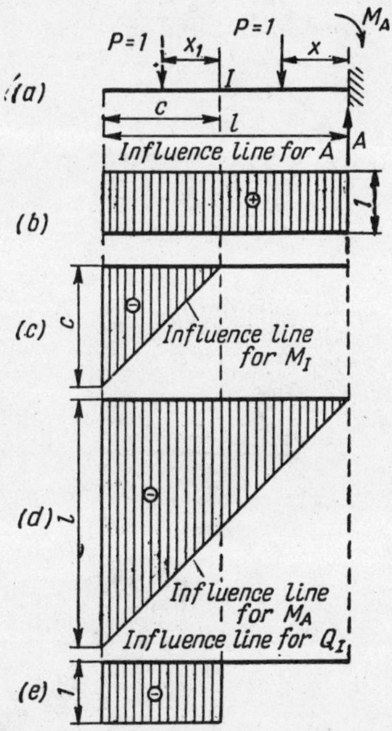

Fig. 12.2

If section I is chosen directly at the support A (c being equal to l), we shall obtain the influence line of the fixed-end moment M_A. This line is shown in Fig. 12.2*d*.

For obtaining the shear influence line we shall proceed as described in the previous article.

(1) When the unit load is to the right of section I the shear is nil, no forces existing to the left of this section.

(2) When the unit load is to the left of section I the shear $Q_I = -1$, which means that the ordinates to the influence line will remain constant and equal to -1 over the whole stretch from section I to the left extremity of the beam.

The shear influence line Q_I is represented in Fig. 12.2*e*. It has exactly the same shape as the one for section II in the cantilevering part of the beam with overhang shown in Fig. 8.2*f*.

5.2. *Influence Lines in Cases of Indirect Load Application*

Thus far we have been considering cases when the external loads were applied directly to the beams. In practice, especially in bridge construction, the loads are usually transmitted to the *main* beam or *girder* by secondary or floor beams, which in their turn support *auxiliary* beams or *stringers* (Fig. 13.2*a*). The stringers are single-span simply supported beams, each stringer span being called *a panel* and each point where a floor beam bears on a girder—*a panel point.*

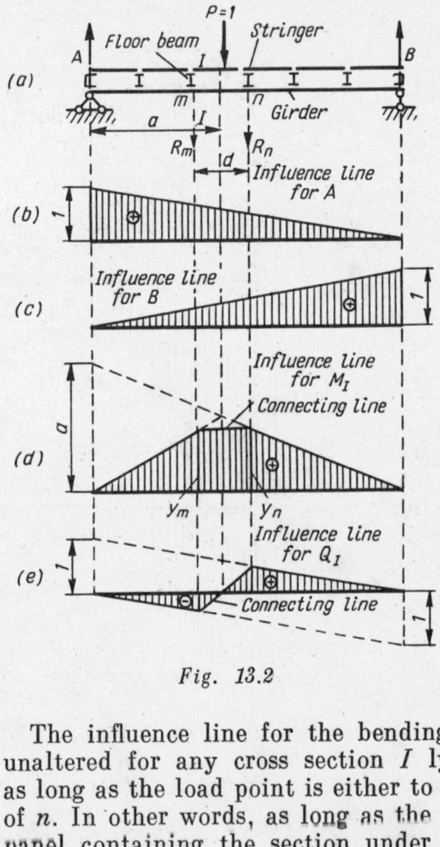

Fig. 13.2

When the load is applied to the stringer somewhere between panel points *m* and *n*, it will be transmitted to the girder only at these two points. This mode of transmission will have no effect on the girder abutment reactions as will be readily seen from the equilibrium equation of moments about any one of the supports. Hence the influence lines for the reactions will be exactly the same as if the load were applied directly to the girder (Fig. 13.2*b* and *c*).

The influence line for the bending moment will also remain unaltered for any cross section *I* lying within the panel *mn*, as long as the load point is either to the left of *m* or to the right of *n*. In other words, as long as the load is situated outside the panel containing the section under consideration, the bending moment influence line may be drawn in the same way as in the case of direct load application. This is easily confirmed by the corresponding expressions of the bending moments. Thus, in section *I* situated a distance *a* from the left support $M_I = Aa$ when the load point is between *n* and *B* and $M_I = B\ (l-a)$ when the load point is between *m* and *A*. These two expressions coincide exactly with those obtained in Art. 3.2 for ordinary beams, and therefore having constructed the bending moment influence line for section *I* as explained above, we may shade the areas bounded by portions *Am* and *Bn* of this line indicating thus that these portions are definite (Fig. 13.2*d*).

However, when the load is within the panel *mn*, its effect will be transmitted to the girder at panel points *m* and *n*, its components

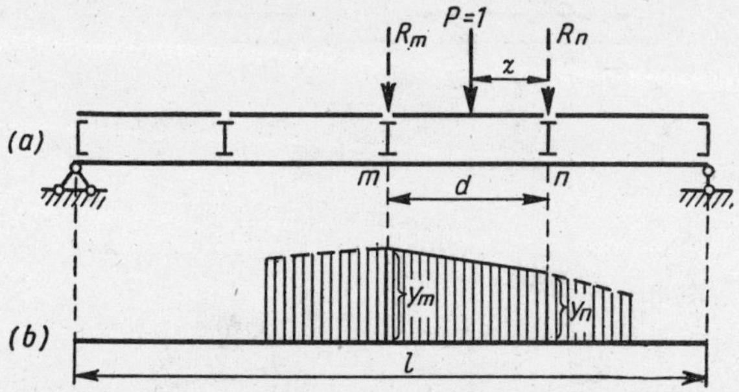

Fig. 14.2

R_m and R_n, shown in dash lines in Fig. 14.2*a*, being equal to the corresponding reactions of the stringer beam.

In order to find the shape of the influence line when the load is within the panel containing the section, let us find the value of any function S_I set up in section *I* by a unit load situated as stated above. Assuming that y_m and y_n are the ordinates to the influence line at the corresponding panel points (Fig. 14.2*b*) and using the method of superposition we can write the following equation

$$S_I = \Sigma Py = R_m y_m + R_n y_n$$

where

$$R_m = \frac{P \cdot z}{d} = \frac{1 \cdot z}{d} = \frac{z}{d}$$

and

$$R_n = \frac{P\,(d-z)}{d} = \frac{1\,(d-z)}{d} = \frac{d-z}{d}$$

where d = panel length

z = distance from the load point to the right-hand panel point.

Substituting the values of R_m and R_n in the first equation, we obtain

$$S_I = \frac{z}{d}\, y_m + \frac{d-z}{d}\, y_n$$

Accordingly, when the load is situated between the panel points m and n, the function S_I varies linearly with z from $S_I = y_n$ for $z = 0$ to $S_I = y_m$ for $z = d$.

Hence the influence line for such a function is a straight line connecting the panel point ordinates y_m and y_n. It follows that in the case of the influence line for bending moment M_I we must simply connect by a straight line the ordinates at panel points

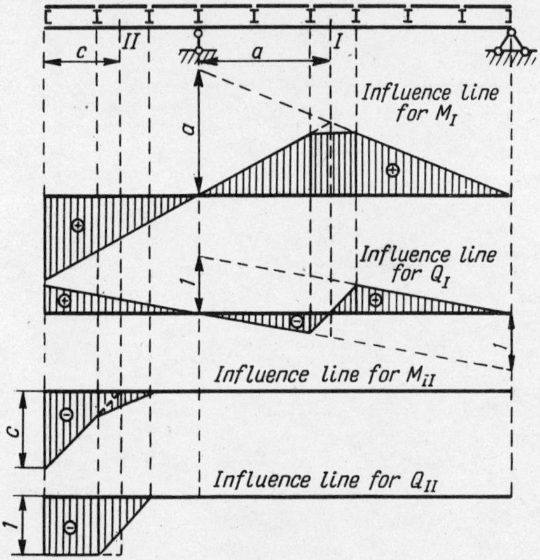

Fig. 15.2

m and n determined previously, obtaining thus the influence line represented in Fig. 13.2*d*.

The construction of the shear influence line for section *I* is quite similar. From *A* to *m* and from *n* to *B* the ordinates to this line will be exactly the same as if the load were applied directly to the girder. Within the panel *mn* which contains the cross section *I* the influence line will be represented dy a straight line connecting the ordinates at panel points (Fig. 13.2*e*).

Thus, when the load is applied through an intermediate beam the influence line may be constructed in the following sequence:

(1) first draw the line as though the load were applied directly to the main beam or girder;

(2) this being done, find the intersection of the line with the ordinates passing through the panel points pertaining to the

panel which contains the cross section under consideration and connect these intersection points by a straight line.

Fig. 15.2 represents the influence line for M and Q corresponding to sections I and II of a beam with an overhang.

Influence lines for the reactions have been omitted as they do not differ in any respect from those of a beam subjected to direct loading.

6.2. *Determination of Forces and Moments with the Aid of Influence Lines*

The *construction* of influence lines having been discussed in detail in the previous articles of this chapter, let us now examine the *determination* of forces and moments with the use of these

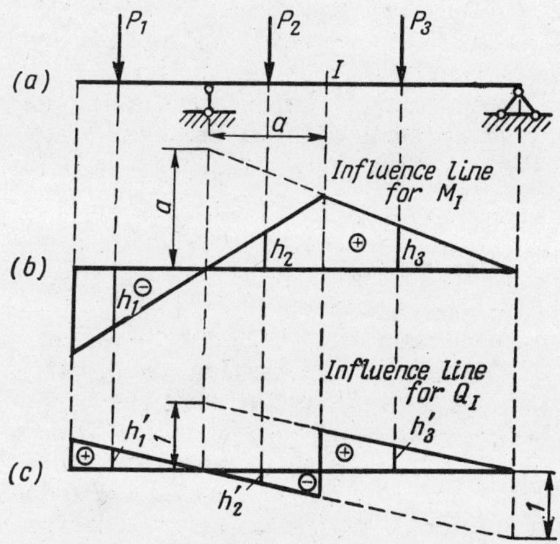

Fig. 16.2

lines (they can also be used for the determination of strains, deflections, and other deformations).

Two cases will be considered:

(a) concentrated loads and

(b) uniform loads.

Case of concentrated loads. As already explained in Art. 2.2, the determination of any function caused by a load P_1 requires the measurement of the ordinate to the influence line for this function and its multiplication by the magnitude of load. If the structure carries several loads at a time (Fig. 16.2*a*), the full value of the function in a section will be obtained by measuring the ordinate under each load, these ordinates being then multiplied by the magnitude of the corresponding loads and the products summed up.

Thus, in order to obtain the bending moment in section I (the influence line for M_I is represented in Fig. 16.2*b*) we must multiply the load P_1 by the ordinate h_1 (since this ordinate is negative, the product P_1h_1 will also be negative), the load P_2 by the ordinate h_2 and the load P_3 by the ordinate h_3. The bending moment resulting from the combined action of loads P_1, P_2 and P_3 will equal

$$M_I = \Sigma Ph = -P_1h_1 + P_2h_2 + P_3h_3$$

The ordinates to the bending moment influence line being measured in length units, say, metres, if the loads are measured in tons, the product Ph representing the bending moment will be expressed in tons multiplied by metres.

A similar procedure may be used for the determination of the shearing force Q_I in section I (the influence line for Q_I is represented in Fig. 16.2*c*)

$$Q_I = P_1h'_1 - P_2h'_2 + P_3h'_3$$

where h'_1, h'_2 and h'_3 are the ordinates to the shear influence line under the loads P_1, P_2 and P_3.

The ordinates to the shear influence line are dimensionless and therefore the product Ph' giving the shear value will be expressed in the same units as the load P.

Thus, in order to compute any function (abutment reaction, bending moment, shear, internal force in any truss member, etc.) arising under the action of several concentrated loads the ordinates to the corresponding influence line must be measured at all the load points, they must be then multiplied by the respective loads, the products so obtained being finally summed up.

Problem. Using the influence line for the bending moment M_I represented in Fig. 17.2*b* determine the value of this moment in section I of the beam shown in Fig. 17.2*a*. This beam carries three concentrated loads the amounts of which are also indicated in the same figure. Ordinate values at load points are shown on the influence line, but they can also be scaled off the drawing or calculated.

Solution. The bending moment in section I equals

$$M_I = -P_1h_1 + P_2h_2 + P_3h_3 = -4 \times 0.5 + 8 \times 1.0 + 10 \times 1.0 = 16.0 \text{ ton metres}$$

The first term of the right-hand part of the equation is preceded by a minus sign because the ordinate h_1 being negative.

Case of uniform loads. The sequence of operations is illustrated by the following example: a uniform load of intensity q is distributed along a certain length of a beam represented in Fig. 18.2a and it is required to determine the bending moment in section I

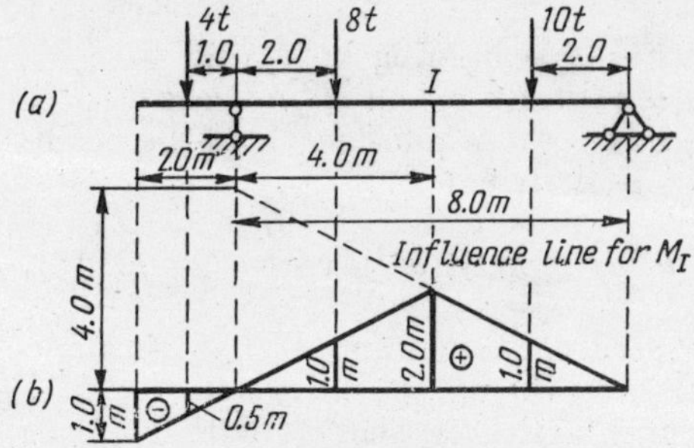

Fig. 17.2

(the influence line for M_I is shown in Fig. 18.2b). Let us replace the uniform load acting along an infinitely small length dx by a concentrated load qdx (Fig. 18.2a). The moment in section I

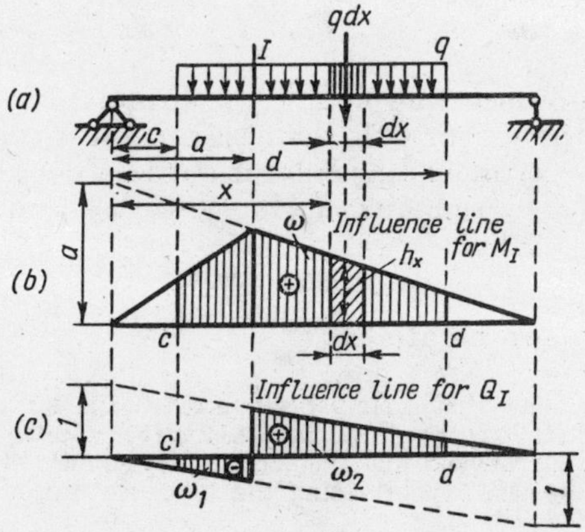

Fig. 18.2

due to this load will amount to $qdxh_x$ where h_x is the influence line ordinate under the load. Proceeding in the same way we can replace the whole load distributed along the beam by an infinitely great number of concentrated loads qdx and the bending moment in section I due to all of these loads will be then obtained by a summation of all the products $qdxh_x$ or

$$M_I=\int_c^d q\,dxh_x=q\int_c^d h_x\,dx$$

the load intensity q remaining constant.

The integration limits c and d indicate that the summation must be carried over the whole length of the beam section, along which the load is distributed.

The term $\int_c^d h_x\,dx$ represents the area bounded by the influence line, the ordinates corresponding to the limits of loading and the x-axis (this area being shaded vertically in Fig. 18.2*b*) for $h_x dx$ is an elementary area shaded with slanting lines in the same figure. If we denote the whole area by ω the bending moment in section I will be

$$M_I=q\omega$$

Thus, in order to determine the amount of any function arising in a given section as a result of the application of a uniform load, the intensity of this load must be multiplied by the area bounded by the influence line, the x-axis and the ordinates passing through the load limits.

When the influence line within the load limits changes sign the areas will be taken with their signs. Thus, the total shear in section I (the corresponding influence line is drawn in Fig. 18.2*c*) will be obtained by summing up the areas ω_1, and ω_2

$$Q_I=q\,(\omega_1+\omega_2)$$

ω_1 being reckoned negative.

Problem 1. Assume that a simply supported beam is uniformly loaded over the whole of its length with an intensity q (Fig. 19.2*a*). It is required to find the bending moment and the shear in the middle of the beam and the reaction at the left support using the influence lines represented in Fig. 19.2*b*, *c* and *d*.

Solution. As this load is spread over the whole length of the beam the areas bounded by the influence lines must be calculated for the entire span.

Determination of reaction A. The area bounded by the influence line being

$$\omega_1=\frac{1}{2}\cdot l\cdot 1=\frac{l}{2}$$

the abutment reaction equals

$$A=q\omega_1=\frac{ql}{2}$$

Determination of the bending moment M_I. The area bounded by the influence line being

$$\omega_2=\frac{1}{2}\cdot l\cdot\frac{l}{4}=\frac{l^2}{8}$$

the bending moment will equal

$$M_I=q\omega_2=\frac{ql^2}{8}$$

Determination of the shear Q_I. The influence line consists of two

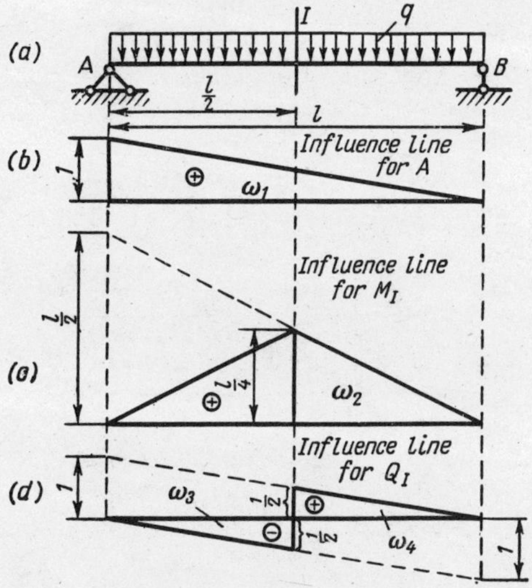

Fig. 19.2

portions bounding areas equal in size but opposite in sign

$$\omega_3=-\frac{1}{2}\cdot\frac{l}{2}\cdot\frac{1}{2}=-\frac{l}{8};\quad \omega_4=+\frac{l}{8}$$

Therefore

$$Q_I=q\,(\omega_3+\omega_4)=q\left(-\frac{l}{8}+\frac{l}{8}\right)=0$$

Problem 2. Determine with the aid of influence lines the bending moment and the shear in section I of a simply supported beam with on overhang loaded as indicated in Fig. 20.2*a*.

Solution. Start by drawing the influence lines for the bending moment and shear in section I (Fig. 20.2*b* and *c*).

Determination of the bending moment M_I. The ordinate h_1 to the bending moment influence line under the concentrated load P is equal to -1.6 m and the area ω_1 under the influence line of the uniformly loaded stretch totals

$$\omega_1=\frac{1}{2}\times 8\times 1.6=6.4\ \text{m}^2$$

Therefore the required bending moment will amount to

$$M_I = -Ph_1 + q\omega_1 = -3\times 1.6 + 2\times 6.4 = 8.0 \text{ ton metres}$$

Determination of the shear Q_I. The ordinate h_2 to the influence line for Q_I under the load P equals 0.2 whilst the area bounded by the influence line over the uniformly loaded portion of the beam is

$$\omega_2 = \frac{1}{2}\times 8\times 0.8 = 3.2 \text{ m}$$

Accordingly

$$Q_I = Ph_2 + q\omega_2 = 3\times 0.2 + 2\times 3.2 = 7.0 \text{ tons}$$

We shall now show that the *function S of any load* (whether concentrated or distributed) *acting over a straight portion of*

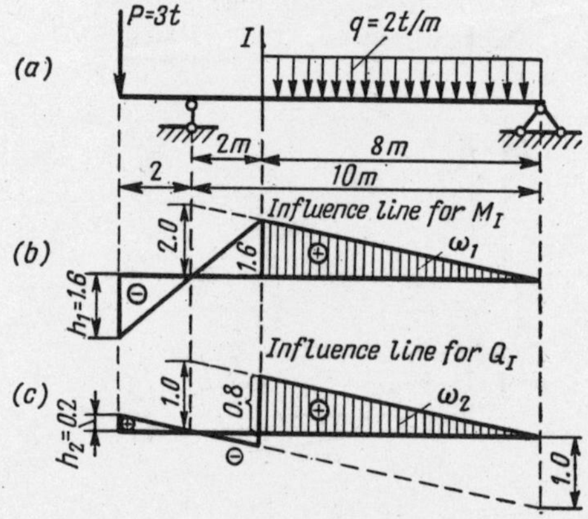

Fig. 20.2

an influence line will be equal to the resultant R of this system of loads multiplied by the ordinate h_0 corresponding to this resultant. In effect let us consider the influence line for function S presented in Fig. 21.2 and a set of concentrated loads $P_1, P_2, \ldots, P_n$ with a resultant R situated over the straight portion cd of this influence line. Then

$$S = \Sigma Ph = P_1h_1 + P_2h_2 + P_3h_3 + \ldots + P_nh_n$$

Let us express the ordinates $h_1, h_2, \ldots$, etc., in terms of their distances a_1, a_2, etc., to the point of intersection O of the line cd with the x-axis (Fig. 21.2)

$$h_1 = a_1 \tan\alpha;\ h_2 = a_2 \tan\alpha;\ h_3 = a_3\tan\alpha, \ldots,$$
$$h_n = a_n \tan\alpha$$

Substituting these values in the formula giving the value of function S we obtain

$$S = (P_1a_1 + P_2a_2 + P_3a_3 + \dots + P_na_n) \tan \alpha$$

As will be readily seen, the expression in parentheses represents the moment of the loads P_1, P_2, P_3, . . ., P_n about point O,

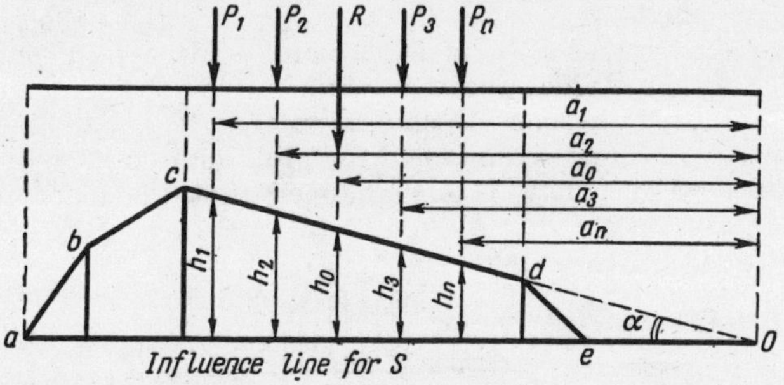

Fig. 21.2

the moment being equal to the moment of their resultant about the same point, i. e., to Ra_0 (Fig. 21.2). Consequently, $S = Ra_0 \tan \alpha = Rh_0$.

7.2. *Determination of the Most Unfavourable Position of a Load*

We have just seen how the influence lines for various functions (abutment reactions, bending moment, shears, etc.) may be used for the determination of the value of the appropriate function for any given position of a load.

We shall now endevour to find the position of the load corresponding to the maximum value of the function considered. Such a position is usually termed the most *unfavourable* or *dangerous position.*

Hereafter the maximum positive values of the function will be denoted by A_{max}, M_{max}, Q_{max}, etc., whilst the maximum negative values by A_{min}, M_{min}, Q_{min}, etc.

1. *Case of a single concentrated moving load.* In this case the position of the load producing the maximum value of function S is found very easily. It coincides with the position of the maximum ordinate to the influence line. By multiplying the amount of the load by this ordinate we shall obtain the maximum value of the function under consideration. Positive maximums of the

function are furnished by the largest positive ordinates whilst the negative maximums or minimums—by the largest negative ordinates.

2. *Case of a set of concentrated moving loads.* In this case we must find such a position of the given set of loads which would provide the maximum value of ΣPh, where h stands for the ordinate to the influence line corresponding to the respective load P. When the number of loads is not very great the problem is solved dy trial, the set of loads being shifted from one position to another. When the maximum value of the function is sought the loads are made to coincide alternately with the maximum positive ordinates and when the minimum one is required—with the negative ones. It may happen that the loads will be

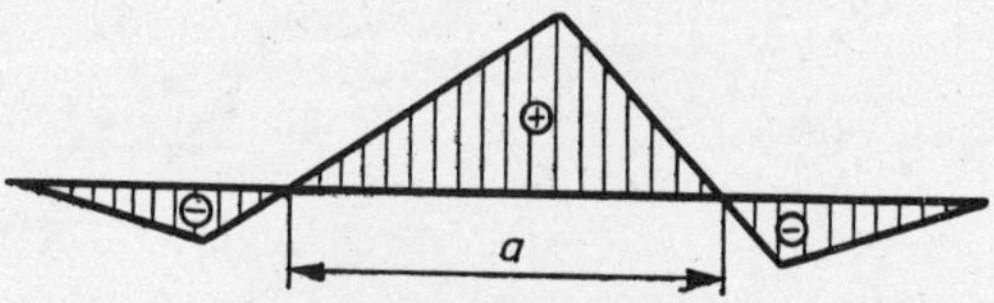

Fig. 22.2

simultaneously situated over the positive and negative portions of the influence line. Such a case would arise, for instance, if it were desired to find S_{max} for an influence line represented in Fig. 22.2 due to a set of loads (say, a locomotive) whose total length would exceed the length a corresponding to the positive part of the line.

Fig. 23.2 shows the most unfavourable position of a twin-axle bogie with equal wheel loads for various influence lines. Fig. 23.2*a* represents the loading corresponding to M_{Imax}. In this case the greatest value of ΣPh is obtained when the left wheel coincides with the maximum positive ordinate. Should we bring the right wheel over this ordinate, the left one would shift to ordinate mn, and as the latter is smaller than m_1n_1 the sum ΣPh would also be smaller than in the first case. Any other position of the set of loads considered would equally lead to a smaller value of the bending moment.

Fig. 23.2*b* shows the position of the same set of loads providing for M_{Imin}.

Fig. 23.2*c* and *d* indicates the load positions corresponding to the maximum and minimum values of the shearing force. In the first case it is assumed that the left-hand load stands an infinitesimal distance to the right of section I and therefore its amount must be multiplied by the ordinate ab (Fig. 23.2*c*). In the

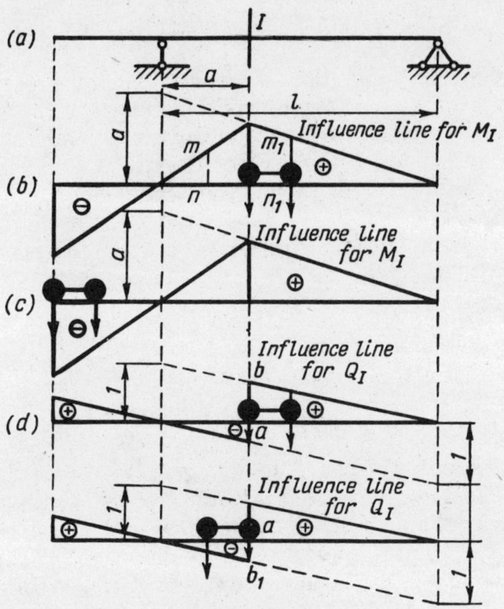

Fig. 23.2

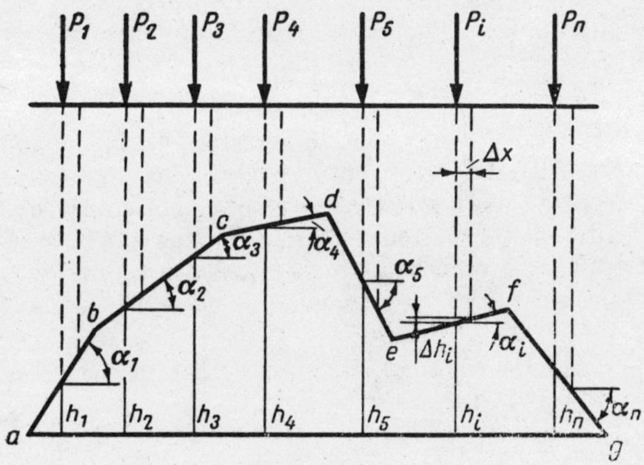

Fig. 24.2

second case it is assumed that it is the right-hand load which is infinitely close to section *I* from its left and therefore the amount of this load must be multiplied by the ordinate ab_1 (Fig. 23.2*d*).

Let us consider now the influence line for a function S consisting of a number of straight portions intersecting at points a, b, c, d, e, f, and g and a set of concentrated loads as indicated in Fig. 24.2, the loads being in position *I*. As will be seen, none of these loads stand over the vertices mentioned above. Assuming that the whole set of loads is shifted over a distance x to the right (position *II*), the ordinate h_i, corresponding to a load P_i, will be increased by

$$\Delta h_i = \Delta x \tan \alpha_i \,*$$

while the increment of function S will equal

$$\Delta S = \sum_{i=1}^{i=n} P_i \cdot \Delta h_i = \sum_{i=1}^{i=n} P_i \cdot \Delta x \tan \alpha_i = \Delta x \cdot \sum_{i=1}^{i=n} P_i \cdot \tan \alpha_i \quad (6.2)$$

Should we shift the set of loads again by Δx to the right (position *III*) the new increment of function S would still be given by the expression (6.2). Assume now that position *II* corresponds to the maximum value of the function S (in other words, that this position is the most unfavourable or the most dangerous one). In that case the increment ΔS will be positive when the set of loads is shifted from position *I* to position *II* and negative when the loads move from position *II* to position *III*. Thus, when the set of loads passes through its most unfavourable position, the increment of the function S (and accordingly the sum $\sum_{i=1}^{i=n} P_i \tan \alpha_i$) must change sign. **

As will be easily seen from expression (6.2), a change in sign of the increment ΔS may occur only when one or more loads which were previously situated over one rectilinear portion of the influence line have shifted to an adjacent portion. ***

It follows that a *dangerous position of the set of loads will occur when one or more load points coincide with the ordinates passing*

* In Fig. 24.2 the angles α_1, α_2, α_3, α_4, and α_i are positive whilst the angles α_5 and α_n are negative.

** The same remains true for minimum values of the function S.

*** It may happen that having reached its maximum, the function remains constant during the passage of certain loads from one of the apices to the next one. In that case a maximum will exist even though ΔS is nil and no load is at an apex, but the rule just mentioned still holds good, for initially this maximum occurred when the critical load (or loads) stood over an apex (or apices) of the influence line.

through the apices of the influence line. This important remark greatly facilitates the search for the most unfavourable position of the loads, as it reduces the number of trials to the cases when one or several load points stand over the said apices.

Hereafter both the load and the apex in the influence line over which this load must stand to induce a maximum of the function under consideration will be termed *critical.*

Let us assume now that position *II* is the most unfavourable one and that it occurs when the critical load P_3 stands over the critical apex c of the influence line (Fig. 24.2). In that case the increment ΔS must be positive when the system of loads shifts towards the right from position *I* to position *II* and it must be negative as soon as the load P_3 passes to the right of point c. For the same reason, the sum $\sum_{i=1}^{i=n} P_i \cdot \tan \alpha_i$ must be positive when the loads stand to the left of the dangerous position and becomes negative as soon as they have shifted to the right of the latter [see exp. (6.2)]. We must also have $P_3 \cdot \tan \alpha_2 > P_3 \cdot \tan \alpha_3$ which leads to $\alpha_2 > \alpha_3$. Thus, the slope of that portion of the influence line which is to the left of the critical apex must be greater than the slope of the portion situated immediately to the right of this apex. This condition is satisfied in Fig. 25.2*a* only. It follows that a *critical point in the influence line will always coincide with one of its convex apices or peaks*, the same remaining true in the case when the minimum value of a function is sought. This again reduces the number of trials necessary to find a dangerous position for a given set of loads.

It should be noted that the intersection points of an influence line which form peaks when the maximum value of a function is sought cease being such when its minimum is required, and vice versa. Thus, in Fig. 25.2*c* points c, e, ang g of the influence line form peaks when S_{max} is sought, while points a, b, d, and f would become such were the minimum of S required. In order to ascertain the nature of the extreme points a and g of the influence line the x-axis should be extended in both directions (as shown by dash lines in Fig. 25.2*c*), these portions being considered as part of this influence line with zero ordinates.

We have already stated that when seeking S_{max} the sum $\sum_{i=1}^{i=n} P_i \cdot \tan \alpha_i$ *is positive when the set of loads is situated to the left of its most unfavourable position and negative when this set has shifted to the right of the latter.* It is clear that when S_{min} is required, the sum $\sum_{i=1}^{i=n} P_i \cdot \tan \alpha_i$ *will be negative when the loads are to the left of their dangerous position and positive when they are to its right.* This also simplifies the determination of the most unfavourable loading.

As an example, let us find the most unfavourable position of a system of loads shown in Fig. 26.2 providing for S_{max} when the influence line for S consists of three rectilinear portions as shown in Fig. 27.2. The tangents of the angles formed by these three portions with the x-axis are

$$\tan\alpha_1 = +\frac{1}{8};\ \tan\alpha_2 = -\frac{0.25}{4} = -\frac{1}{16};\ \tan\alpha_3 = -\frac{1}{8}$$

We know that the most unfavourable position of the loading cannot occur without at least one of the loads coinciding with the peak b or c of the influence line *.

The loads being shifted from right to left the sum $\sum_{i=1}^{i=n} P_i \tan\alpha_i$ will remain negative as long as all the loads are situated over portions bc and cd of the influence line forming negative angles with the x-axis. As stated before when the loads pass through a dangerous position, this sum must change sign and become positive. Accordingly, we must continue to move the loads in the same direction, i. e., from right to left until this sum becomes positive. Let us consider the loading represented in Fig. 28.2. So long as the loads remain to the right of this position

$$\sum_{i=1}^{i=n} P_i\cdot\tan\alpha_i = (3.5+3.5)\frac{1}{8} - (3.5+3.5+3.5)\frac{1}{16} - \\ -(3+3)\frac{1}{8} = -\frac{17}{32} < 0$$

But as soon as they shift to the left this sum becomes

$$\sum_{i=1}^{i=n} P_i\cdot\tan\alpha_i = (3.5+3.5+3.5)\frac{1}{8} - (3.5+3.5)\frac{1}{16} - \\ -(3+3)\frac{1}{8} = \frac{1}{8} > 0$$

This means that the passage of the set of loads from a position slightly to the right from the one indicated in Fig. 28.2 to a position slightly to its left causes a change in the sign of the increment ΔS from negative to positive. Therefore, the posi-

* Apices a and d do not form peaks and therefore the passage of a load over one of these two points is of no danger.

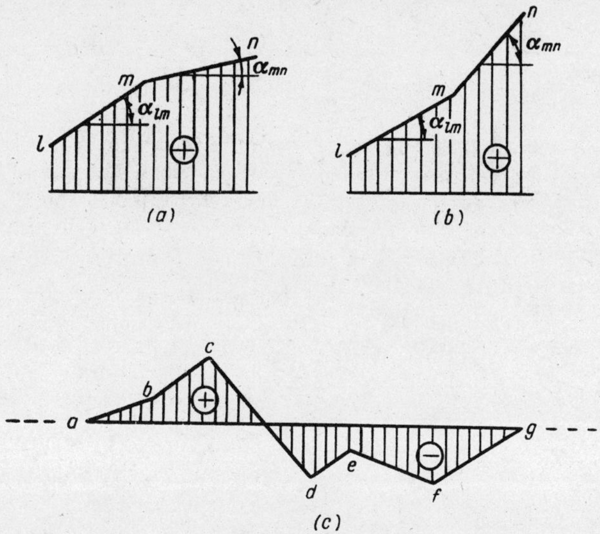

Fig. 25.2

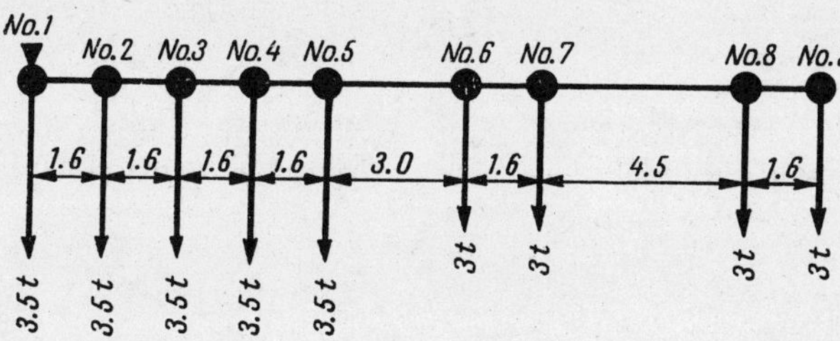

Fig. 26.2

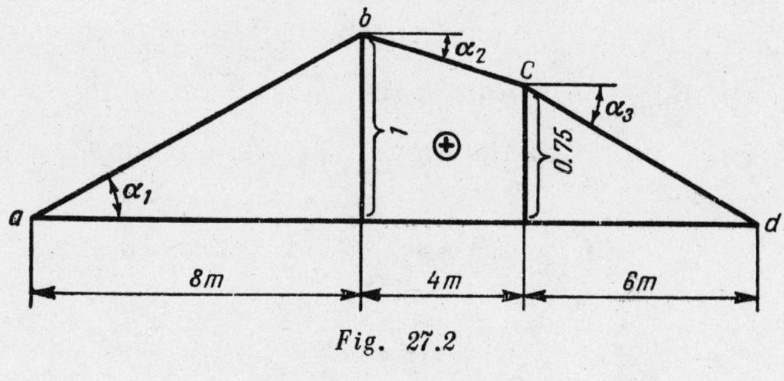

Fig. 27.2

tion represented in Fig. 28.2 is a dangerous one and load *3* is a critical load.

Suppose now that the loads *8* and *9* (see Fig. 26.2) which are still beyond the limits of our structure when the first dangerous loading occurs are considerably greater than all the other loads and total 15 tons each. In that case if the train of loads is shifted further to the left so that loads *8* and *9* would reach portions *bc* and *cd* of the influence line the sum $\sum_{i=1}^{i=n} P_i \cdot \tan\alpha_i$ would again become negative, and at the moment one of these loads passes

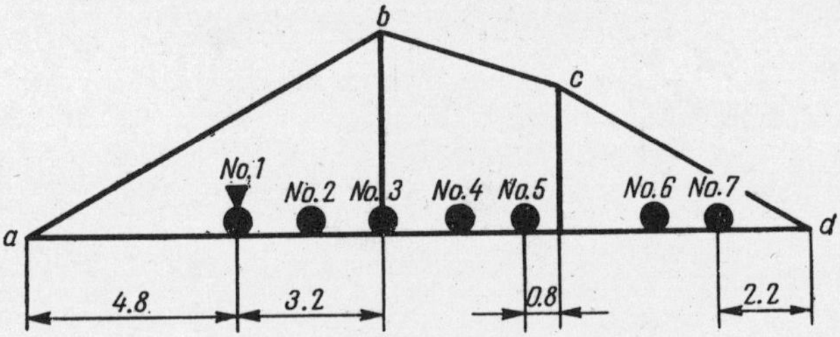

Fig. 28.2

the peak *b* it will change sign again. Accordingly there would be a second dangerous position of the set of loads considered, for which the value of S_{max} should be again calculated. The larger of the two maximums should be adopted for design purposes.

Let us now consider the case when the influence line forms a triangle as represented in Fig. 29.2. Let P_{cr} denote the critical load, $\sum_L P$—the sum of the loads situated over the left-hand portion of the influence line, and $\sum_R P$—the sum of these loads over the right-hand one.

We have shown previously that when the set of loads is to the left of its dangerous position, the sum $\sum_{i=1}^{i=n} P_i \cdot \tan\alpha_i$ is positive and when it shifts to the right the sum becomes negative. In other words, in the present case

$$\tan\alpha_L (\sum_L P + P_{cr}) + \tan\alpha_R \sum_R P > 0$$

and

$$\tan\alpha_L \sum_L P + \tan\alpha_R (\sum_R P + P_{cr}) < 0$$

Substituting in the above expressions $\frac{h_{max}}{a}$ for $\tan\alpha_L$ and $-\frac{h_{max}}{b}$ for $\tan\alpha_R$ (see Fig. 29.2) and carrying out some elementary transformations, we obtain

$$\frac{\sum\limits_L P + P_{cr}}{a} > \frac{\sum\limits_R P}{b}$$

and

$$\frac{\sum\limits_L P}{a} < \frac{P_{cr} + \sum\limits_R P}{b}$$

Adding to both sides of the first expression $\frac{\sum\limits_L P + P_{cr}}{b}$ and to both sides of the second one $\frac{\sum\limits_L P}{b}$ we get

$$\left(\sum_L P + P_{cr}\right)\frac{a+b}{ab} > \frac{\sum\limits_L P + P_{cr} + \sum\limits_R P}{b}$$

and

$$\sum_L P\,\frac{a+b}{ab} < \frac{\sum\limits_L P + P_{cr} + \sum\limits_R P}{b}$$

Substituting l for $(a + b)$ and denoting the sum of all the loads by ΣP, these expressions will easily reduce to

$$\sum_L P + P_{cr} > \Sigma P \cdot \frac{a}{l} \tag{7.2}$$

and

$$\sum_L P < \Sigma P \cdot \frac{a}{l} \tag{8.2}$$

These two inequalities show that *the critical load is the one which renders the sum* $\sum\limits_L P + P_{cr}$ *greater than* $\sum P \frac{a}{l}$, *provided that* $\sum\limits_L P$ *is smaller than the latter.*

In most cases the moving load (say, a locomotive) may enter the structure (say, a bridge) from both sides, and to each of these cases there will correspond its own maximum value of the function S. In order to obtain the larger of these two values the front wheels of the locomotive, which are usually the heaviest, should be placed over the left-hand portion of the influence line when $a < b$ (see Fig. 29.2) and over its right-hand portion when $a > b$.

Problem. It is required to find the most dangerous position of a train of loads shown in Fig. 31.2 with respect to the influence line for function S represented in Fig. 30.2 and characterized by $l = 8$ m, $a = 2$ m and $h_{max} = 1$.

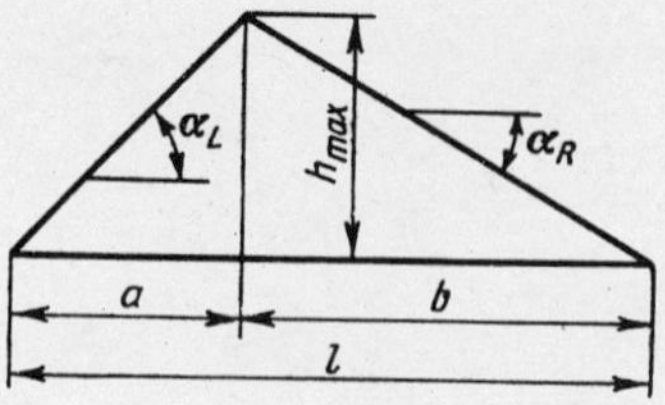

Fig. 29.2

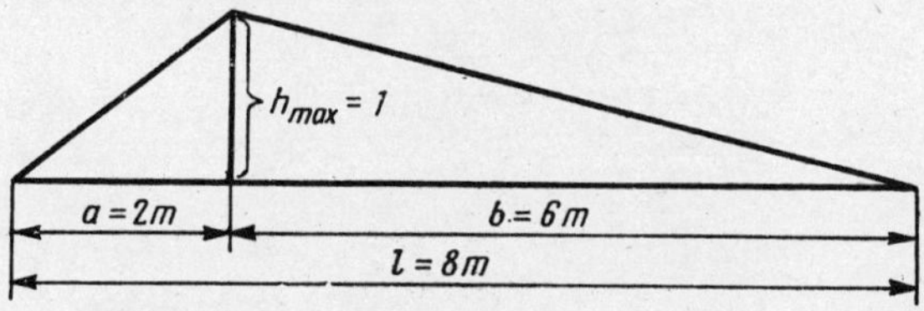

Fig. 30.2

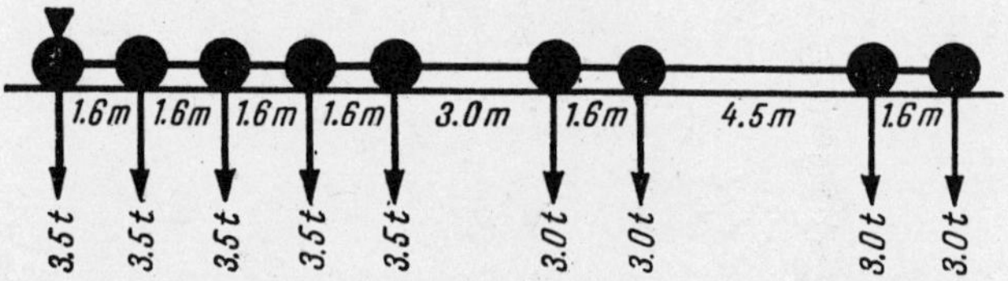

Fig. 31.2

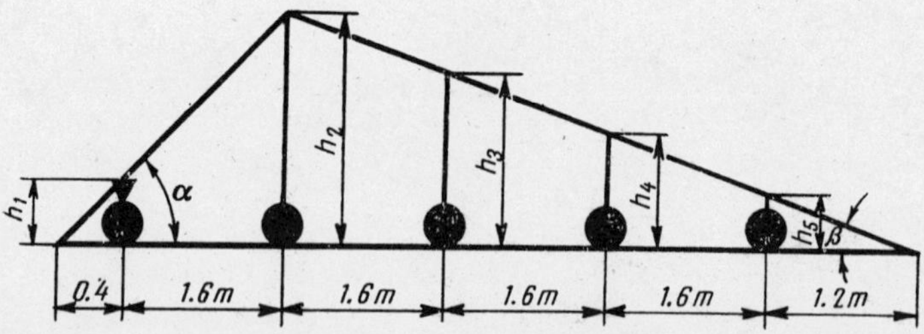

Fig. 32.2

Solution. The sum of loads which can find place on a span 8 metres long equals $R = 5 \times 3.5 = 17.5$ tons.

Shifting the train of loads from right to left and making use of the inequalities (7.2) and (8.2), we shall find

$$\sum_L P + P_{cr} > 17.5 \times \frac{2}{8} = 4.375 \text{ tons}$$

and

$$\sum_L P < 4.375 \text{ tons}$$

This shows that the second load is the critical one for only in that case both of the inequalities become satisfied. Effectively, let $\sum_L P = 3.5$ tons and $P_{cr} = 3.5$ tons, then

$$\sum_L P + P_{cr} = 3.5 + 3.5 = 7.0 > 4.375 \text{ tons}$$

and

$$\sum_L P = 3.5 < 4.375 \text{ tons}$$

The most unfavourable position of the train of loads thus found is indicated in Fig. 32.2. In order to find the value of S_{max} corresponding to this loading let us find the ordinates h_1, h_2, h_3, h_4, and h_5

$$h_1 = 0.4 \tan \alpha = 0.4 \frac{1}{2} = 0.2$$

$$h_2 = h_{max} = 1$$

$$h_3 = (1.6 \times 2 + 1.2) \tan \beta = (1.6 \times 2 + 1.2) \frac{1}{6} = 0.733$$

$$h_4 = (1.6 + 1.2) \tan \beta = (1.6 + 1.2) \frac{1}{6} = 0.467$$

$$h_5 = 1.2 \tan \beta = 1.2 \times \frac{1}{6} = 0.2$$

wherefrom

$$S_{max} = \Sigma Ph = P \Sigma h = 3.5\,(0.2 + 1 + 0.733 + 0.467 + 0.2) = 9.1$$

If the ordinates of the influence line were measured in metres, the function $S_{max} = 9.1$ would be expressed in ton metres. On the other hand, if these ordinates were dimensionless the result obtained would be expressed in tons.

3. *Case of a moving uniformly distributed load.* In Art. 6.2 we have seen that the value of a function S induced by a uniformly distributed load is equal to the product of the intensity of that load q by the area ω bounded by the influence line and the ordinates passing through the limits of the load, i. e., $S = q\omega$. The intensity of the load q being constant, the maximum value of the function will correspond to the maximum of ω which in its turn will occur when the load q will occupy the whole of that portion of the structure, over which the influence line does not change sign. *

◆

* It is assumed that the loads may be distributed over a stretch of any length.

In Fig. 33.2 we have represented the most unfavourable cases of loading for a beam with overhang carrying a uniform load. Case *a* indicates the load position for M_{max}; case *b* for

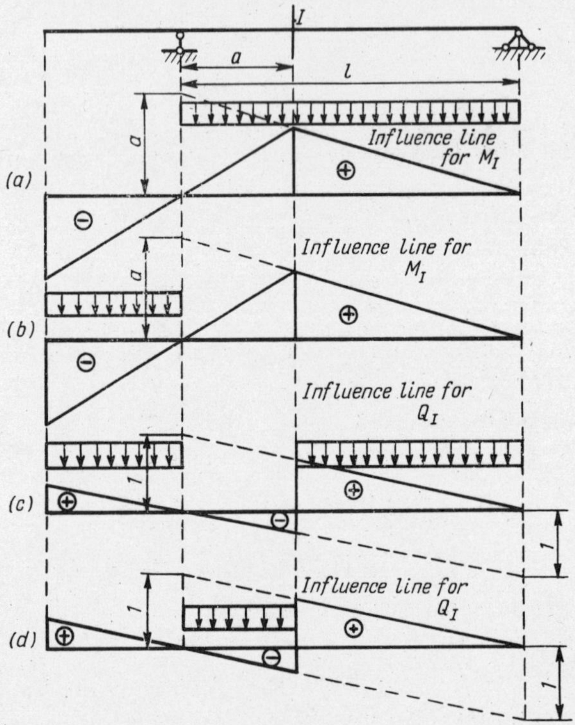

Fig. 33.2

M_{min}; case *c* for Q_{max} and case *d* for Q_{min}. It will be noted that in all the four cases these portions of the beam which correspond to the positive or negative parts of the influence line are fully loaded.

8.2. *Determination of Maximum Moments and Forces Using Equivalent Uniform Loads*

We have seen that the determination of the maximum value of a function by the direct application of a train of concentrated loads to the influence line involves a considerable amount of calculations due to the necessity of finding the most unfavourable position of the loads. In the case of triangular influence lines, all the operations may be considerably simplified through the use of so-called *equivalent loads*, whose values can be taken from appropriate tables and graphs.

The *equivalent load* may be defined as a *uniformly distributed load which will induce in a given member (or section) of the structure under consideration the same force or moment as the corresponding system of concentrated loads in their most unfavourable position.*

Denoting by q_{eq} the intensity of the equivalent load and by Ω the area bounded by the influence line, we may write the following equation

$$\Sigma P_i h_i = q_{eq}\Omega$$

from which it may be seen that there will be always only one definite value of the equivalent load for each particular loading. Indeed, solving the above equation for q_{eq} we obtain

$$q_{eq} = \frac{\Sigma P_i h_i}{\Omega}$$

In our example of Art. 7.2 we have found that for the train of loads considered the maximum value of a certain function S totalled 9.1. In this case

$$q_{eq} = \frac{9.1}{\frac{1\times 8}{2}} = \frac{9.1}{4} = 2.275 \text{ tons per metre}$$

It might seem that this leads us exactly nowhere, for in order to find an equivalent load we must first determine the maximum value of the function by trial. In reality this is not so. For a triangular influence line the intensity of an equivalent load for a given set of concentrated loads is independent of the actual value of the ordinates to the influence line and will alter only with a change in the length of that portion of the structure which carries the loads and with a variation in the position of the influence line apex with respect to its extremities This permits computation and tabulation (or representation in the form of graphs) of equivalent load intensities pertaining to typical loading schemes and to the more common shapes of triangular influence lines.

Let us call similar two influence lines when the ordinates of one of them may be obtained by multiplying those of the other by a constant factor, and let us show that the intensity of the equivalent loads for two similar lines remains the same.

Fig. 34.2*a* and *b* represents two such lines; the base lengths of these lines are the same while their ordinates differ by a constant factor equal to n.

The equivalent load for line b is

$$q'_{eq} = \frac{\Sigma P'_i h'_i}{\Omega'} \quad \text{with} \quad h'_i = nh_i$$

and

$$\Omega' = 0.5nh\,(a+b) = n\Omega$$

where h_i and Ω are the ordinate and the area of the influence line represented in Fig. 34.2a, respectively.

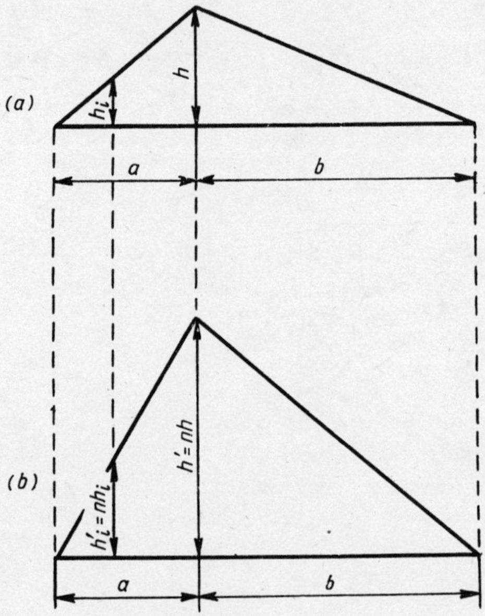

Fig. 34.2

Substituting Ω' and h' by their values expressed in terms of Ω and h we find

$$q'_{eq} = \frac{n\Sigma P_i h_i}{\Omega \cdot n} = \frac{\Sigma P_i h_i}{\Omega} = q_{eq}$$

for line a.

We have thus proved that the intensity of the equivalent load depends on three factors only: (1) the distribution and magnitude of the loads; (2) the length of the loaded portion; (3) the position of the apex of the influence line over the span (or over the loaded portion of the structure).

Intensities of equivalent loads computed for a standard train H_1 (Fig. 35.2) * used in the U.S.S.R for designing railway bridges are tabulated hereunder.

As will be observed, the table contains the values of equivalent loads for various lengths of the loaded portion (up to 200 m) and for three different positions of the influence line apex, namely

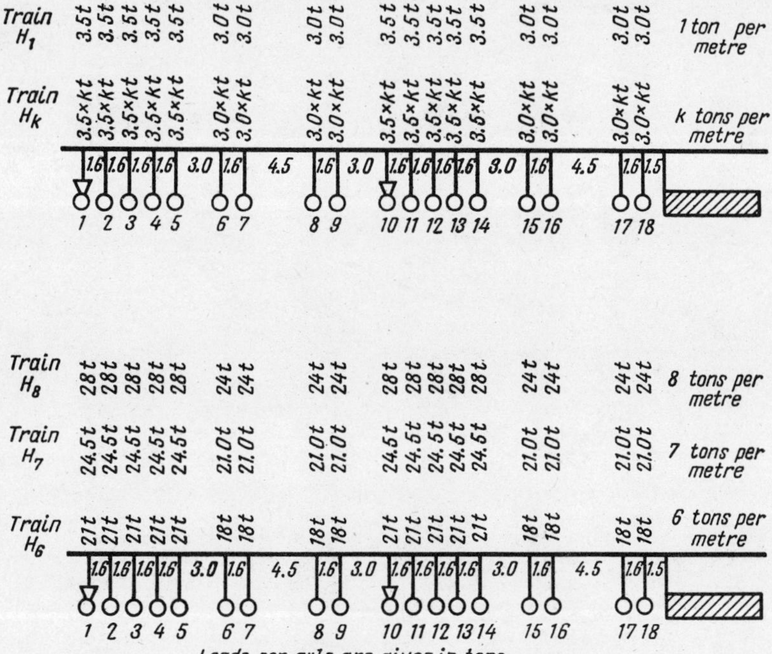

Fig. 35.2

when the latter is over the edge, at quarter span and at mid-span. When the apex falls at some intermediate point, the value of the equivalent load may be obtained by interpolation.

By multiplying all the axle loads of the standard H_1 train by a factor k which characterizes the class of loading, we shall obtain loading schemes for different classes of trains. Thus. the design of trunk lines is carried out for trains of class 7 or 8 (H_7 or H_8) whilst lines of local importance are designed for H_6 and H_7 trains.

* The distances between loads in Fig. 35.2 are given in metres.

Table 1.2

Equivalent Loads per Running Metre of Track for Standard H_1 Train in Tons

Length of the loaded portion in m	Type and sketch of influence line		
	Apex at the extremity	Apex at quarter span	Apex in the middle
	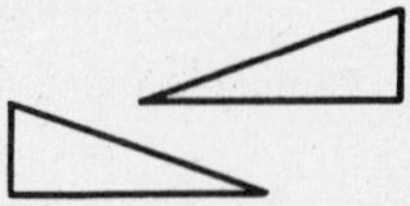	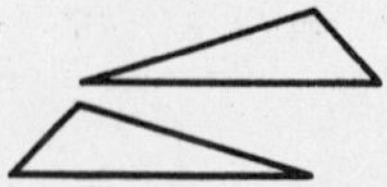	
1	2	3	4
1	7.00	7.00	7.00
2	4.20	3.50	3.50
3	3.42	3.01	2.51
4	3.15	2.57	2.45
5	2.91	2.41	2.41
6	2.80	2.26	2.26
7	2.71	2.26	2.26
8	2.63	2.28	2.28
9	2.51	2.23	2.23
10	2.42	2.16	2.16
12	2.29	2.05	1.98
14	2.16	1.97	1.88
16	2.03	1.88	1.82
18	1.95	1.77	1.79
20	1.88	1.69	1.74
40	1.77	1.61	1.59
45	1.73	1.56	1.52
50	1.70	1.55	1.46
60	1.65	1.52	1.44
70	1.61	1.46	1.44
50	1.58	1.43	1.43
60	1.51	1.37	1.37
70	1.46	1.33	1.32
80	1.41	1.29	1.27
90	1.37	1.26	1.22
100	1.34	1.24	1.18
110	1.32	1.22	1.15
120	1.29	1.20	1.13
130	1.27	1.18	1.11
140	1.26	1.16	1.10
150	1.24	1.15	1.08
160	1.23	1.14	1.07
170	1.21	1.12	1.06
180	1.20	1.11	1.06
190	1.19	1.10	1.05
200	1.18	1.09	1.05

Problem 1. Using the method of equivalent loads determine the stress produced by a standard H_7 train in member *4-5* of a single track bridge truss represented in Fig. 36.2*a*, the corresponding influence line being shown in Fig. 36.2*b*.

Solution. In order to obtain the maximum value of the tension induced the whole portion of the span corresponding to the positive ordinates of the influence line should be loaded; the length of this portion totals 8.0 m. In

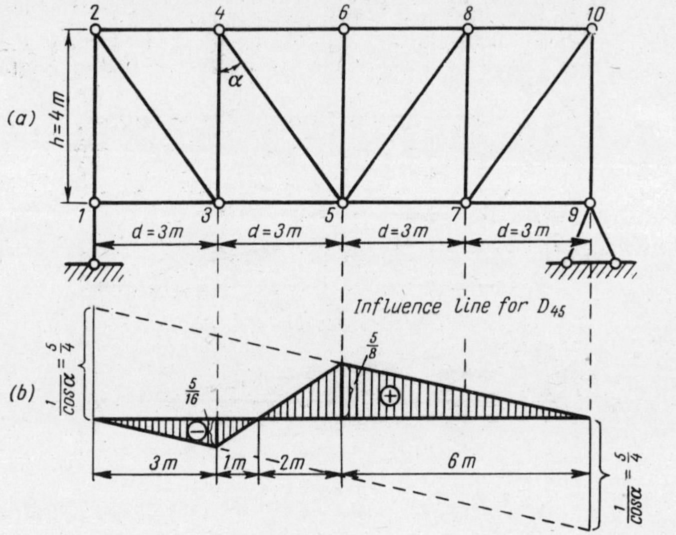

Fig. 36.2

column 3 of Table 1.2 we find that the equivalent load for train H_1 would equal in that case 2.28 tons per metre. The maximum tension produced by an H_7 train will then be derived from the following expression

$$2D_{45}=kq_{eq}\Omega=7\times 2.28\times\frac{5}{8}\times\frac{8}{2}=39.90 \text{ tons*}$$

whence

$$D_{45}=0.5\times 39.90=19.95 \text{ tons}$$

The maximum compression will be obtained by loading the whole stretch over the negative portion of the influence line. From Table 1.2 we find that for a length $l=4$ m the equivalent load is 2.57 tons per metre (see column 3). The maximum compression will then be given by

$$2D_{45}=kq_{eq}\Omega=7\times 2.57\times\frac{5}{16}\times\frac{4}{2}=11.24 \text{ tons}$$

◆

* The factor 2 in front of the left-hand term of this expression is due to the fact that the equivalent loads in Table 1.2 are given for one track, i. e., for both trusses of a single-track bridge.

whence

$$D_{45}=0.5\times 11.24=5.62 \text{ tons}$$

Problem 2. Using the method of equivalent loads find $M_{I\,max}$, $Q_{I\,max}$, and $Q_{I\,min}$ arising in an end supported plate girder bridge (Fig. 37.2*a*) during the passage of an H_7 train. The influence lines for bending moment and shear in section I of one girder are represented in Fig. 37.2*b* and *c*. The bridge is again a single track one the rails being fixed to stringers and cross beams supported by two parallel girders. Accordingly the equivalent load for one girder will be half of that given in Table 1.2.

Solution. Determination of $M_{I\,max}$. The length of the loading should be taken equal to the whole span of the girder, i. e., to 10.0 m. As the apex

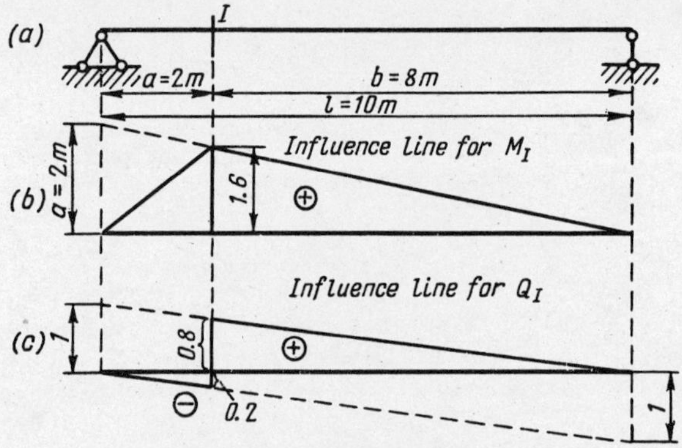

Fig. 37.2

of the influence line falls between the quarter span point and the end of the girder the equivalent load must be found by interpolation.

For an H_1 train

$$q_a=q_0+\left(q_{l/4}-q_0\right)\frac{a}{1/4l}=2.42+(2.16-2.42)\frac{2.0}{2.5}=$$

$$=2.42-0.26\times 0.8=2.21 \text{ tons per metre}$$

where $q_0=$ equivalent load for a loaded length of 10.0 m with the influence line apex over the left extremity of the girder (found in Table 1.2)

$q_{l/4}=$ same equivalent load but for an influence line with the apex at quarter-span (also found in Table 1.2)

$q_a=$ equivalent load for the case when the influence line apex is at $\frac{1}{5}$ of the girder span (as in Fig. 37.2*b*)

$a=$ distance from the influence line apex to the nearest end of the girder

$l=$ length of loading equal in this particular case to the whole span of the girder.

For an H_7 train the equivalent load will amount to

$$q_a\times 7=2.21\times 7=15.47 \text{ tons per metre}$$

For one girder it will reduce to one half, i. e., $q_1=7.735$ tons per metre.

The area under the bending moment influence line for section I equals

$$\omega_1=\frac{1}{2}\times 10\times 1.6=8\ \text{m}^2$$

Consequently

$$M_{I\,max}=q_1\omega_1=7.735\times 8.0=61.88\ \text{ton metres}$$

Determination of $Q_{I\,max}$. In order to find the maximum positive shear in section I the load should cover the entire positive portion of the Q_I influence line (Fig. 37.2*c*). The length of this portion is equal to 8 metres and the influence line apex is over its left extremity. Table 1.2 yields the following value for the equivalent load corresponding to the standard H_1 train

$$q_0=2.63\ \text{tons per metre}$$

For an H_7 train this load will increase sevenfold and will total $7\times 2.63 = 18.41$ tons per metre, while for one girder this should be halved, i. e., $q_2 = 9.2$ tons per metre.

The area bounded by the positive portion of the shear influence line equals

$$\omega_2=\frac{1}{2}\times 8\times 0.8=3.2\ \text{m}$$

and therefore

$$Q_{I\,max}=q_2\omega_2=9.2\times 3.2=29.44\ \text{tons}$$

Determination of $Q_{I\,min}$. The greatest negative shear in section I will occur when that portion of the girder where the ordinates to the shear influence line are negative (Fig. 37.2*c*) is loaded in its entirety. This portion is 2 metres long and the influence line apex is at its right-hand extremity. For this case we find in Table 1.2 an equivalent load corresponding to a standard H_1 train equal to 4.2 tons per metre. For an H_7 train this must be increased by 7 or to $7\times 4.20 = 29.4$ tons per metre, and for one girder it reduces to $q_3 = 0.5\times 29.4 = 14.7$ tons per metre.

The area under the negative portion of the influence line equals

$$\omega_3=-\frac{1}{2}\times 2\times 0.2=-0.2\ \text{m}$$

Hence,

$$Q_{I\,min}=q_3\omega_3=-0.2\times 14.7=-2.94\ \text{tons}$$

9.2. *Multispan Statically Determinate Beams*

By *multispan statically determinate cantilever beam* we understand a geometrically stable structure consisting of a series of simply supported beams with or without overhangs connected together by means of hinged joints. Such beams might be also called *multispan hinged beams*. The multispan cantilever beams also belong to this class of beams constituting a particular case thereof.

Single beams constituting these structures might be either of plate girder or trussed construction or both. The theory of the multispan statically determinate beams has been developed in 1871 by the eminent Russian engineer G. Semikolenov.

Beams of this type are usually more economical than a series of disconnected simply supported beams spanning the same opening. This may be illustrated by the following example: assume that two equal and adjacent spans AB and BC 10 metres long each have to be bridged over, the design load being evenly distributed and equal to 2 tons per metre. In the first instance let

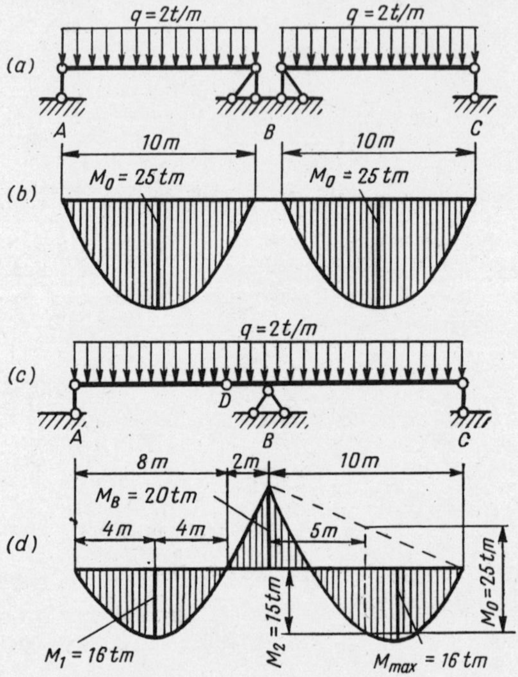

Fig. 38.2

us try separate simply supported beams (Fig. 38.2*a*). The bending moments at midspan of each beam will amount to

$$M_0 = \frac{ql^2}{8} = \frac{2 \times 10^2}{8} = 25 \text{ ton metres}$$

The diagrams of these bending moments are represented in Fig. 38.2*b*.

Now let us envisage a double-span hinged beam, and let us use a beam with a two-metre overhang BD across the span BC (Fig. 38.2*c*), this overhang being hinge-connected to the end of an 8 metre beam AD. The maximum bending moments (positive and negative) in the most dangerous cross sections of these two beams will be:

At midspan of beam AD

$$M_1 = \frac{2\times 8^2}{8} = 16 \text{ ton metres}$$

Over the support B of beam CD (beam AD transmitting a concentrated load $P = 8$ tons through hinge D) the bending moment

$$M_B = -\left(8\times 2 + \frac{2\times 2^2}{2}\right) = -20 \text{ ton metres}$$

In the middle of the span BC the bending moment will amount to

$$M_2 = \frac{2\times 10^2}{8} - \frac{20}{2} = 15 \text{ ton metres}$$

but this is no longer a dangerous section, for the maximum moment must coincide with zero shear and the latter will occur at a distance x from the right-hand support, this distance being derived from the following equation

$$Q_x = -C + qx = 0$$

where C is the right-hand abutment reaction equal to

$$C = \frac{-8\times 2 - 2\times 2\times 1 + 2\times 10\times 5}{10} = 8 \text{ tons}$$

and therefore

$$-8 + 2x = 0 \quad x = 4 \text{ metres}$$

The bending moment in this section will be

$$M_{max} = 8\times 4 - 2\times 4\times 2 = 16 \text{ ton metres}$$

The bending moment diagram for the double-span hinged beam is represented in Fig. 38.2*d*. It will be observed that in absolute value the bending moments in this beam are smaller than in each of the separate beams considered in the first place and therefore the double-span beam is obviously more economical.

The use of continuous beams also leads to a substantial reduction of bending moments as compared with single beams, but the multispan statically determinate beams present certain additional advantages: (a) their relatively short members are well suited for prefabrication, transportation and installation, using standard hoisting equipment; (b) all the forces induced therein are statically determinate and will not be influenced by any settlement of the supports. The above considerations have led to fairly wide use of multispan cantilever beams in engineering structures.

Statically determinate multispan beams may always be obtained introducing a number of hinges into a similar continuous beam. As will be shown later the number of hinges must be equal to the degree of redundancy of the continuous beam.

Fig. 39.2a represents a five-span continuous beam whose constraints at the supports may be schematically replaced by seven hinged bars. In order to determine the forces acting in these bars we have only three independent equilibrium equations and therefore the stress computation for this beam cannot be

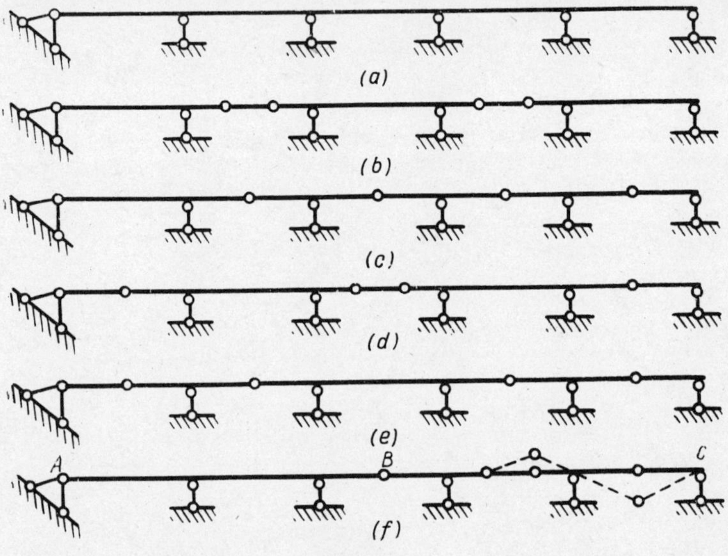

Fig. 39.2

carried out with the aid of statics alone. This beam has a degree of a redundancy] equal to four.

If we denote by C the number of constraints at the supports, then the degree of redundancy n of the beam will be equal to $n = C - 3$. Applying this formula to the beam in Fig. 39.2a we shall obtain $n = 7 - 3 = 4$.

Each hinge introduced either in the span or over a support of a continuous beam provides for one additional equation of statics, this equation expressing that the sum of moments of all the external forces applied to the beam either to the right or to the left of the hinge about its centre equals zero.

Hence, when the number of hinges introduced into the beam equals its degree of redundancy, the beam becomes statically determinate for all the unknowns may be obtained in that case with the aid of the equations of statics alone.

The hinges must be distributed along the beam in such a way that each part of the structure should become statically determinate and remain geometrically stable. *

Several ways of transforming the continuous beam represented in Fig. 39.2*a* into a statically determinate one are illustrated in Fig. 39.2*b*, *c*, *d* and *e*. Fig. 39.2*f* gives an example of an unsatisfactory hinge distribution for although their number in this case also equals four and therefore corresponds to the degree of redundancy of the initial beam, portion AB of the transformed beam still remains statically indeterminate while portion BC has become unstable (The possible displacements of that portion of the beam are shown in dash lines.)

A continuous beam with one built-in end is represented in Fig. 40.2*a*. It should be remembered that a built-in end is equivalent to three support constraints as represented schematically in Fig. 41.2. Accordingly the total number of constraints of the beam is $C = 7$ and its degree of redundancy is $n = C - 3 = 7 - 3 = 4$.

Thus, in order to transform this beam into a statically determinate one, four hinges should be introduced as illustrated in Fig. 40.2*b*. A continuous beam with two built-in ends is represented in Fig. 42.2*a*: the right-hand end of this beam still retains one degree of freedom as it can move horizontally. Therefore at this end the number of restraints is equal to two as indicated in Fig. 43.2. Thus, the total number of constraints of this beam is $C = 8$ and its degree of redundancy equals $n = 8 - 3 = 5$. In order to make this beam statically determinate it would be necessary to introduce five hinges. One way of distributing these hinges is shown in Fig. 42.2*b*.

The best way to find out whether a multispan beam of that type is stable or not and also to get a clear picture of its work under load is to represent schematically the interaction of its separate parts. Assume for instance that it is required to investigate the stability of the beam represented in Fig. 44.2*a*. The interaction of its elements is represented schematically in Fig. 44.2*b*, where all the intermediate hinges are replaced by fixed hinged supports connecting the appropriate beam members. This schematic drawing shows clearly that the whole beam is geometrically stable, for each of its constituent members is a simple beam with or without overhang connected to the ground

◆

* A method of investigating the geometrical stability of a multispan hinged beam was presented in Art. 2.1.

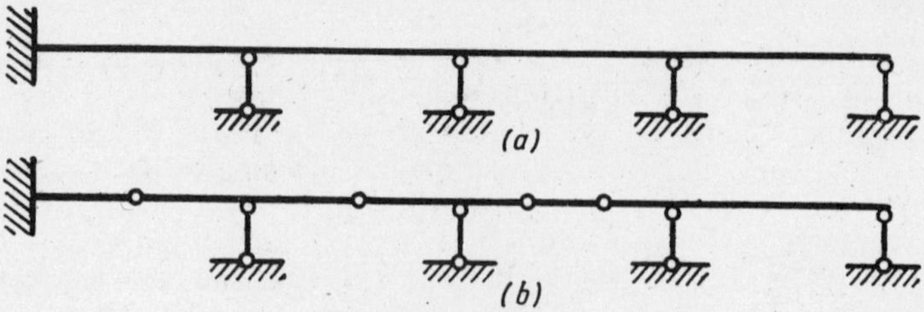

Fig. 40.2

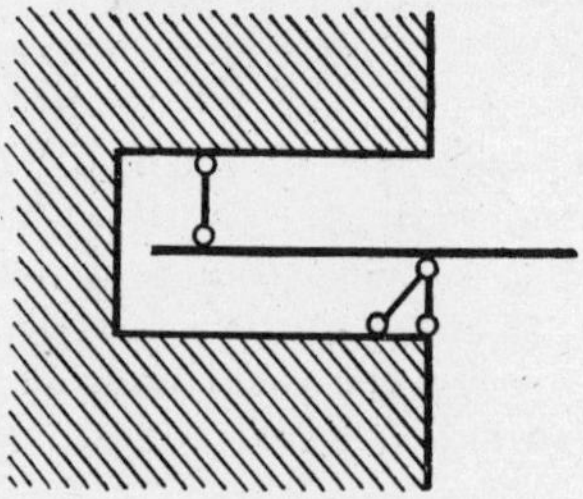

Fig. 41.2

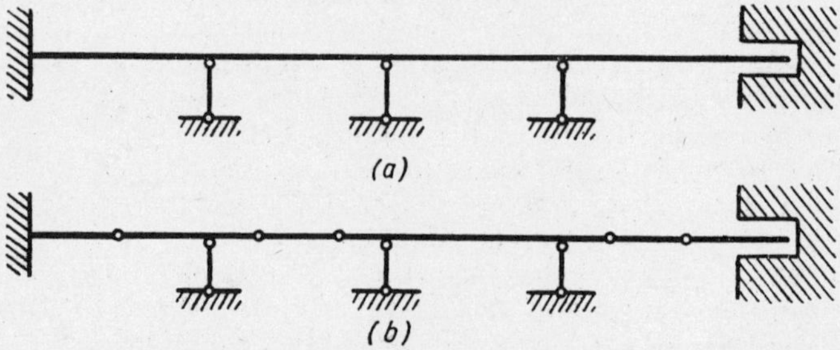

Fig. 42.2

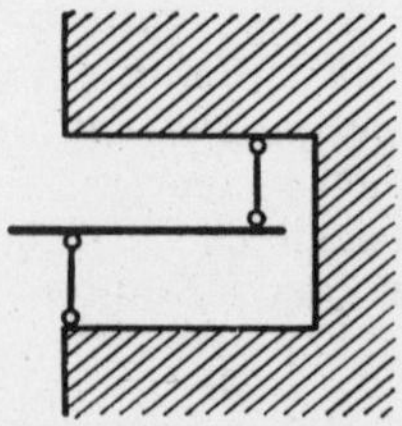

Fig. 43.2

or to another part of the structure whose stability is ensured by means of three nonconcurrent bars.

Effectively, beam ABE is connected to the ground by means of three supporting bars and is therefore geometrically stable. The beam ECF is supported by two bars at its end E and rests on a vertical bar standing directly on the ground at C. The supports of beam FD are exactly similar, thus ensuring its stability.

The examples presented above lead to the establishment of the following rules relative to the distribution of hinges in beams which have no built-in ends:

(1) *there may be no more than 2 hinges in each span*;

(2) *there must be no hinges in the spans adjacent to the one provided with 2 hinges*;

(3) *spans containing one hinge only may follow each other with the exception of one of the extreme spans where there should be no hinge at all.*

Thus far we have considered cases where all the supports but one were free to move in a horizontal direction. Let us now examine cases where two or more supports are fixed and will allow no horizontal displacement. In this case the introduction of ordinary hinges becomes insufficient for the transformation of the continuous beam into a statically determinate one. This would require the installation of mobile hinges which offer no resistance to horizontal displacements. One of these hinges is represented schematically in Fig. 45.2. An example of a statically determinate beam with three fixed supports and two movable hinges is given in Fig. 46.2*a*, the interaction of its elements being schematically indicated in Fig. 46.2*b*.

The reader is invited to establish on his own the relation between the number of fixed supports and of mobile hinges in a statically determinate multispan beam.

The most commonly used multispan hinged beams are represented in Figs. 47.2*a* and 48.2*a*.

The first one is characterized by alternating double-hinged spans and spans devoid of any hinges. * It consists thus of a series of beams with two overhangs supporting 'suspended' simple beams. The second beam is characterized by the presence of a hinge in each of its spans with the exception of the last one; the interaction of its elements is represented schematically in Fig. 48.2*b*.

◆

* The usual three-span cantilever bridge belongs to this type of structures. Each of its outer spans is anchored down at the shore and overhangs into the central span about one third of its length. The suspended span, resting on the "cantilever arms", occupies the remaining third of the central span.

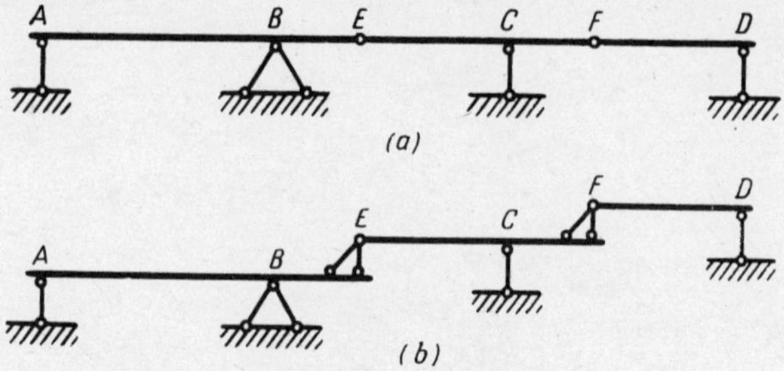

Fig. 44.2

Fig. 45.2

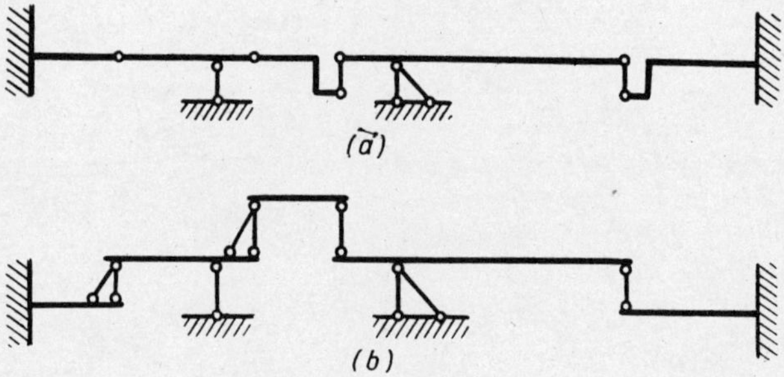

Fig. 46.2

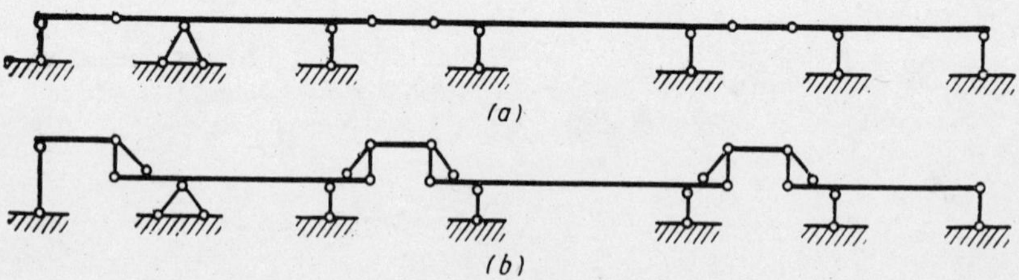

Fig. 47.2

It should be noted that the favourable effect of the overhangs may be taken advantage of not only in ordinary solid web beams

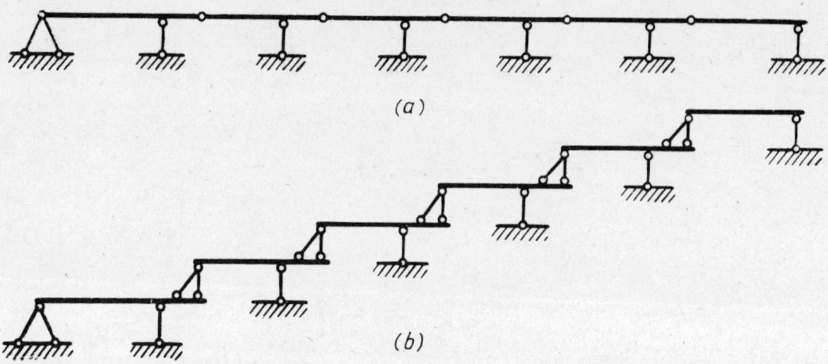

Fig. 48.2

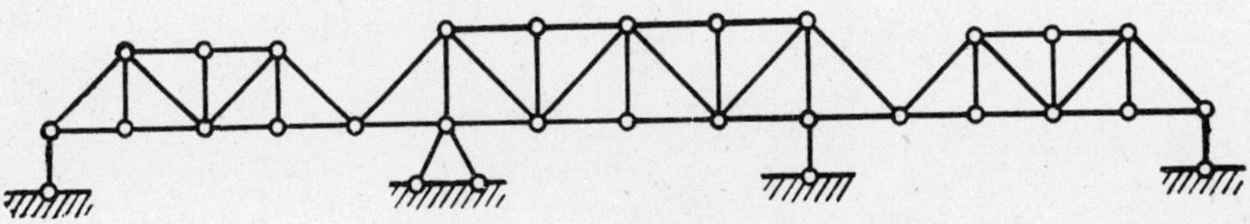

Fig. 49.2

but also in trussed systems such as indicated in Fig. 49.2. The reactions of such a system will be found in exactly the same way as for an ordinary statically determinate multispan beam.

10.2. *Determination of Moments and Forces Induced by a System of Fixed Loads in Multispan Statically Determinate Beams*

The design of statically determinate multispan beams will be now illustrated using as an example the hinged beam represented in Fig. 50.2*a*. Fig. 50.2*b* contains the schematic drawing of the interaction of its separate members.

The reactions R_A, R_B, R_C, R_D will be reckoned positive when directed upwards, while the forces R_1, R_2, and R_3 arising in the hinges from the interaction of the different elements of the beam will be considered such (see Fig. 50.2*b*) when the upper element exerts a downward pressure on the lower one. Fig. 50.2*c*, *d*, *e*, *f* shows all the separate elements of the beam as well as all the forces acting on these elements. All the reactions and forces indicated in these drawings are positive.

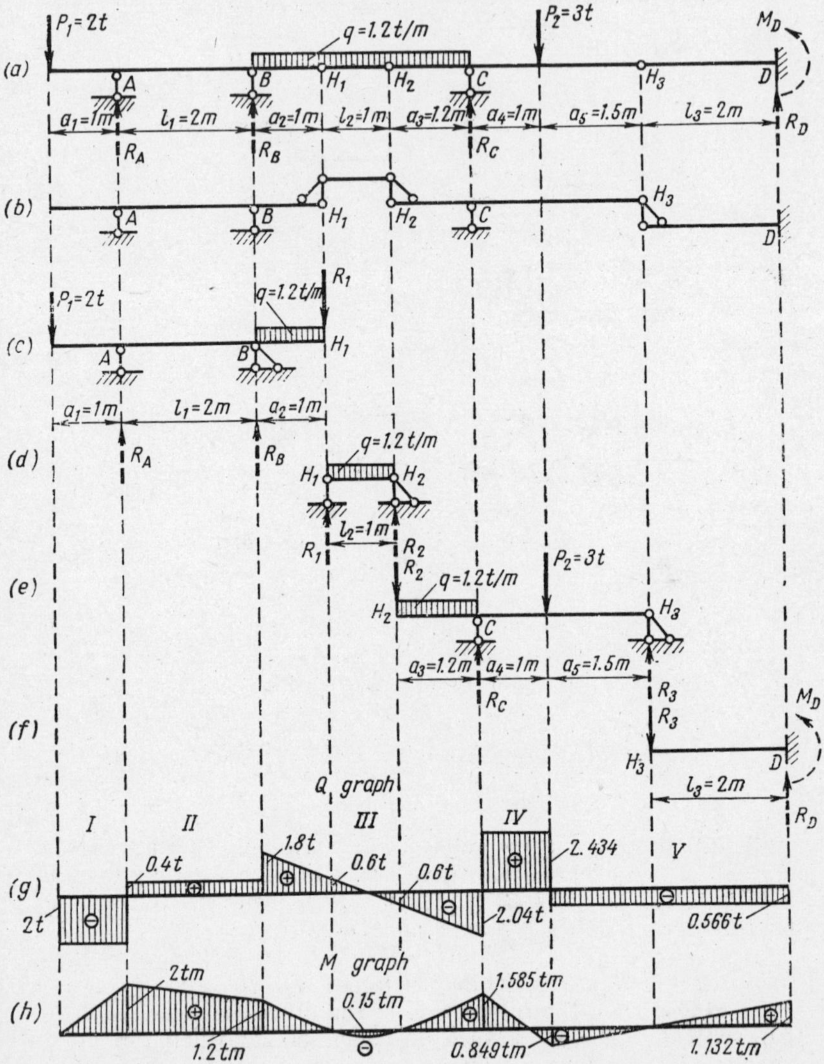

(a)
$P_1=2t$
$q=1.2t/m$
$P_2=3t$
M_D
A
B
H_1
H_2
C
H_3
D
$a_1=1m$
$l_1=2m$
$a_2=1m$
$l_2=1m$
$a_3=1.2m$
$a_4=1m$
$a_5=1.5m$
$l_3=2m$
R_A
R_B
R_C
R_D
(b)
(c)
R_1
(d)
R_2
(e)
R_3
(f)
Q graph
I
II
III
IV
V
(g)
2t
0.4t
1.8t
0.6t
0.6t
2.04t
2.434
0.566t
M graph
(h)
2tm
1.2tm
0.15tm
1.585tm
0.849tm
1.132tm

Fig. 50.2

We shall start with determining the reactions R_1 and R_2 of the upper most simply supported element H_1H_2 spanning a length of 1 m. This element is subjected to a uniform load whose intensity q equals 1.2 tons per metre and also to the reactions at the hinges (see Fig. 50.2d) totalling

$$R_1 = R_2 = \frac{ql_2}{2} = \frac{1.2 \times 1}{2} = 0.6 \text{ ton}$$

Next we shall determine the reactions of the element ABH_1 situated just below the element H_1H_2 and constituting a simply supported beam with two overhangs. This beam is subjected to the action of loads and reactions indicated in Fig. 50.2c. From the equilibrium of moments we obtain

$$\Sigma M_A = -P_1a_1 + qa_2\left(l_1 + \frac{a_2}{2}\right) + R_1(l_1 + a_2) - R_Bl_1 = 0$$

wherefrom

$$R_B = \frac{-P_1a_1 + qa_2\left(l_1 + \frac{a_2}{2}\right) + R_1(l_1 + a_2)}{l_1} =$$

$$= \frac{1}{2}\left[-2 \times 1 + 1.2 \times 1\left(2 + \frac{1}{2}\right) + 0.6(2 + 1)\right] = 1.4 \text{ tons}$$

The equilibrium of the moments about point B yields

$$\Sigma M_B = -P_1(a_1 + l_1) + qa_2 \cdot \frac{a_2}{2} + R_1a_2 + R_A \cdot l_1 = 0$$

and thus

$$R_A = \frac{P_1(a_1 + l_1) - \frac{qa_2^2}{2} - R_1a_2}{l_1} =$$

$$= \frac{2(1+2) - 1.2 \times \frac{1^2}{2} - 0.6 \times 1}{2} = 2.4 \text{ tons}$$

The values of the two reactions just found can be checked using the equilibrium equation of the vertical components

$$\Sigma Y = -P_1 - qa_2 - R_1 + R_A + R_B =$$

$$= -2 - 1.2 \times 1 - 0.6 + 2.4 + 1.4 = -3.8 + 3.8 = 0$$

Thus, the values of reactions R_A and R_B are correct.

Consider now reactions R_C and R_3 of the simply supported beam with overhang H_2CH_3; the forces and the reactions acting on this beam are shown in Fig. 50.2e. The equilibrium equation furnishes again

$$\Sigma M_C = -R_2a_3 - qa_3\frac{a_3}{2} + P_2a_4 - R_3(a_4 + a_5) = 0$$

wherefrom

$$R_3 = \frac{-R_2a_3 - 0.5qa_3^2 + P_2a_4}{a_4 + a_5} =$$

$$= \frac{-0.6\times1.2 - 1.2\times0.5\times1.2^2 + 3\times1}{1+1.5} = 0.566 \text{ ton}$$

The other equilibrium equation gives

$$\Sigma M_3 = -R_2(a_3 + a_4 + a_5) - qa_3\left(\frac{a_3}{2} + a_4 + a_5\right) - $$
$$- P_2a_5 + R_C(a_4 + a_5) = 0$$

wherefrom

$$R_C = \frac{R_2(a_3 + a_4 + a_5) + qa_3\left(\frac{a_3}{2} + a_4 + a_5\right) + P_2a_5}{a_4 + a_5} =$$

$$= \frac{0.6(1.2+1+1.5) + 1.2\times1.2\left(\frac{1.2}{2} + 1 + 1.5\right) + 3\times1.5}{1+1.5} = 4.474 \text{ tons}$$

These two reactions will be checked as above

$$\Sigma Y = -R_2 - qa_3 - P_2 + R_C + R_3 = -0.6 - 1.2\times1.2 - 3 +$$
$$+ 0.566 + 4.474 = -5.04 + 5.04 = 0$$

which shows that all the computations were carried out correctly.

Next comes the turn of the cantilever beam H_3D loaded at its free end by the vertical pressure R_3 (see Fig. 50.2*f*). From the equilibrium of the moments we obtain

$$\Sigma M_D = -R_3l_3 - M_D = 0$$

leading to

$$M_D = -R_3l_3 = -0.566\times2 = -1.132 \text{ ton metres}$$

The negative value of the moment obtained indicates that this moment acts in a direction opposite to the one indicated in Fig. 50.2*f*. From the equilibrium equation

$$\Sigma Y = -R_3 + R_D = 0$$

we get

$$R_D = R_3 = 0.566 \text{ ton}$$

Having determined all the reactions at the supports and all the pressures exerted by the separate elements of the beam on each other, we may now proceed with the determination of shears Q and bending moments M acting in the various cross sections of the beam and with the construction of the corresponding diagrams. There are two ways of carrying out these computations.

(1) The shearing forces Q and the bending moments M for the multispan statically determinate beam under consideration (Fig. 50.2*a*) may be determined in the same way as for an ordinary statically determinate beam taking into consideration only the loads applied and the reactions at the supports but disregarding the interaction pressures at the hinges. *

If carried out correctly, these computations must show that the bending moments at all the hinges are nil. The values or expressions of the shearing forces Q and the moments M may be then used for the construction of the corresponding graphs.

(2) The shearing forces and the bending moments may be determined separately for each of the elements constituting the multispan beam allowing the construction of the Q and M graphs for each of these elements (Fig. 50.2*c*, *d*, *e* and *f*). Putting together these graphs will give the corresponding diagrams pertaining to the full length of the beam

The first of the methods just described may by recommended for beams with a reduced number of spans whilst the second one is better suited for the beam consisting of a large number of elements.

In our case the first of the two methods will be used for the construction of the shear diagram. Disregarding the intermediate hinges, the beam under consideration may be divided into five portions characterized by different expressions for the shearing forces. These portions are denoted by corresponding ciphers in Fig. 50.2*g*. Let x represent the distance from the cross section considered to the left extremity of the beam, the following equations for each of the portions mentioned will then be obtained.

Portion I

$$(0 \leqslant x \leqslant 1.0\text{m}): Q^{I} = \sum_{L} Y = -P_1 = -2 \text{ tons}$$

Portion II

$$(1\text{m} \leqslant x \leqslant 3\text{m}): Q^{II} = \sum_{L} Y = -P_1 + R_A = -2 + 2.4 = 0.4 \text{ ton}$$

Portion III

$$(3\text{m} \leqslant x \leqslant 6.2\text{m}): Q^{III} = \sum_{L} Y = -P_1 + R_A + R_B - q(x-3) =$$

$$= -2 + 2.4 + 1.4 - 1.2(x-3) = 5.4 - 1.2x$$

◆

* These interactions have been already taken care of in the determination of the reactions at the supports.

Portion IV

$$(6.2\text{m} \leqslant x \leqslant 7.2\text{m}): Q^{IV} = -\sum_R Y = -R_D + P_2 =$$

$$= -0.566 + 3 = 2.434 \text{ tons}$$

Portion V

$$(7.2 \leqslant x \leqslant 10.7\text{m}): Q^{V} = -\sum_R Y = -R_D = -0.566 \text{ ton}$$

The values of the shears thus obtained for all the five portions of the beam will furnish the shear diagram represented in Fig. 50.2*g*.

The bending moment diagram will be obtained by the second of the two methods described. The corresponding graph for the element ABH_1 will be derived from the moments due to the actions of the force $P_1 = 2$ tons, to the reactions at the supports $R_A = 2.4$ tons and $R_B = 1.4$ tons, to the uniformly distributed load $q = 1.2$ tons per metre, and to the interaction force $R_1 = 0.6$ ton (see Fig. 50.2*c*). This graph will be rectilinear along the left-hand overhang $a_1 = 1$ metre and over the span $l_1 = 2$ metres, no distributed load acting along these parts. At the left extremity of the beam the bending moment will be nil, at the support A it will total $-P_1a_1 = -2$ ton metres and over the support B it equals $-P_1(a_1 + l_1) + R_Al_1 = -2(1 + 2) + 2.4 \times 2 = -1.2$ ton metres.

Within the portion BH_1 (right-hand overhang) the bending moment diagram will be concave, for this portion of the beam is subjected to a distributed load acting in a downwards direction. At the right-hand extremity of the element ABH_1 the bending moment will again equal zero. The data so obtained yield the diagram represented in Fig. 50.2*h*.

Using the same procedure we shall obtain the bending moment diagram H_2CH_3 (Fig. 50.2*e*). At both extremities of the beam (hinges H_2 and H_3) the bending moments will equal zero. Under the load P_2 the moment will equal $R_3a_5 = 0.566 \times 1.5 = 0.849$ ton metres and over the support C it will amount to

$$R_3(a_5 + a_4) - P_2a_4 = 0.566(1.5 + 1.0) - 3 \times 1 = -1.585 \text{ ton metres}$$

Over the left-hand overhang the graph will be curvilinear while between the supports it will be represented by a straight line. These data will be again used for the construction of the bending moment graph pertaining to the element H_2CH_3 (Fig. 50.2*h*).

The bending moment diagram for the element H_1H_2 will be bounded by a conic parabola exactly similar to the one obtained for a uniformly loaded similarly supported beam (Fig. 50.2*e*). Its maximum ordinate will equal $\frac{ql_2^2}{8}=\frac{1.2\times1^2}{8}=0.15$ ton metre. The diagram for the element H_3D will be bounded by a

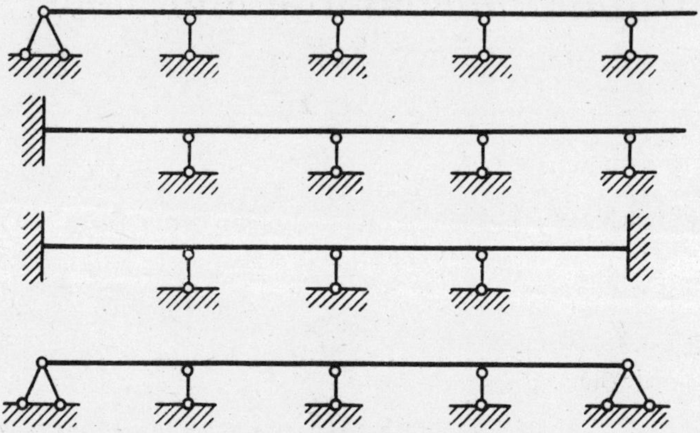

Fig. 51.2

straight line passing through zero at point H_3 and through the top of the ordinate $M_D = -1.132$ m at the wall as shown in Fig. 50.2*h* for the corresponding element.

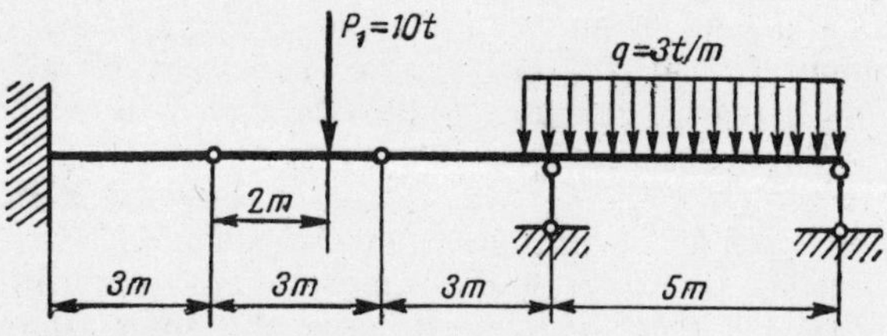

Fig. 52.2

All these separate graphs when placed together will furnish the bending moment diagram for the full length of the beam appearing in Fig. 50.2*h*.

The reader is invited to check the Q and the M diagrams using the expressions mentioned in Art. 1.2.

Four different continuous beams are shown in Fig. 51.2. It is suggested that the reader should find several alternative schemes of rendering each beam statically determinate by introducing intermediate hinges. He is also invited to carry out all the computations leading to the construction of the bending moment and shear diagrams for the beam of Fig. 52.2 and to find the length

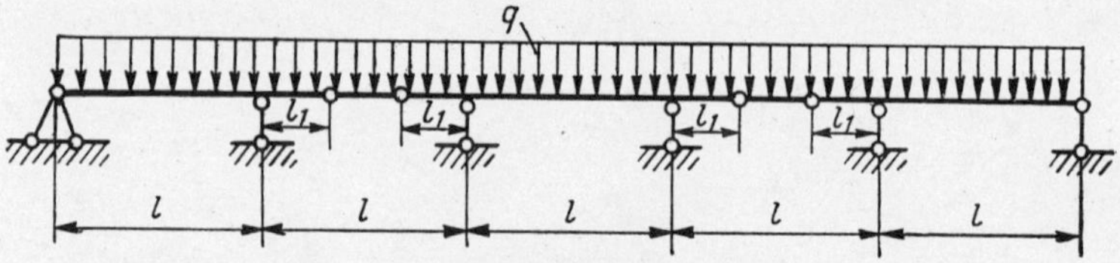

Fig. 53.2

of the overhang l_1 which would equalize the bending moments at mid-length of the three central spans of the beam represented in Fig. 53.2.

11.2. *Influence Lines for Multispan Statically Determinate Beams*

In Art. 5.2 of the present chapter we have shown that when the load is transmitted through secondary beams (stringers) the influence line for the main end-supported beam remains rectilinear. We shall show now that in this respect influence lines for multispan statically determinate beams are quite similar to those just mentioned.

Assume that it is required to draw the influence lines for reactions A, B and C of beam AC represented in Fig. 54.2*a*.

The element CD of this beam is freely supported at one end, its other end being hinge-connected to the end D of the cantilever beam AD. When the unit load is applied to the element CD the reactions at points D and C will be exactly the same as in the case of a simply supported beam, but when the load shifts to beam AD the reactions at points D and C become nil. Accordingly, the influence line for reaction C will have the shape indicated in Fig. 54.2*c*.

As regards the reaction at support A its value will be the same as for an ordinary beam with overhang as long as the load unity is applied between points A and D. When this load is applied at point D the reaction at A will be directed downwards and will reach its maximum negative value. When the load unity moves along the element DC the pressure exerted at hinge D will equal $\frac{1x}{l_2}$, in other words, it will have the same value as though it were transmitted to the same point through a stringer and cross beam. Accordingly, the influence line for reaction A of the element DC will be rectilinear with a zero ordinate at point C. This influence line is represented in Fig. 54.2*d* while that for the reaction at point *B* is shown in Fig. 54.2*e*.

Let us consider now the construction of the influence line for the shears in sections *I* and *II* of the structure schematically represented in Fig. 55.2*a*.

Section *I* will be subjected to the action of the shearing force only when the unit load *P* is applied between abutment *1* and joint *3*. When this load is applied at joint *2*, it is fully transmitted to the overhang of the main beam with the shear in section *I* then becoming equal to —1. When the load unity shifts to the left or to the right of point *2* the pressure at this joint will decrease becoming nil when the load reaches point *1* or point *3*, the value of the said pressure diminishing proportionally to the distance of the load from one of these two points. Accordingly, the influence line will be triangular in shape with an ordinate at section *I* = = —1 (Fig. 55.2*b*).

In section *II* the shearing force will be exactly the same as in the case of a direct application of the load as long as the latter is situated between points *2* and *5* or *6* and *10*. The corresponding portions of the influence line will therefore be represented by the lines ac_1 and c_2b which cut the verticals passing through points *A* and *B* at +1 and —1, respectively. Between points *5* and *6* the influence line must remain straight, its ordinates h_5 and h_6 having already been found and therefore we only have to join points c_1 and c_2. When the load unity is applied to the terminal beams *1-2* or *10-11* the value of the shearing force in section *II* will vary from h_2 (or h_{10}) to zero, the latter value corresponding to the case when the load reaches the abutment. The variation of any function being linear when the load shifts along a secondary beam, we may simply connect the ordinates at points *2* and *10* with the points of zero ordinate *1* and *11* (Fig. 55.2*c*).

Let us now consider the construction of influence lines for statically determinate beams of more than two spans. In such cases it is always recommended to begin with tracing the interaction scheme.

Fig. 56.2*a* represents such a beam, the interaction scheme of its four elements being shown in Fig. 56.2*b*. Let us first construct the influence line for the reaction at support *A* (Fig. 56. 2*c*). For that part of the beam from its left extremity to the hinge H_1 the construction will be carried out exactly in the same way as for a simply supported beam with two overhangs (see Art. 2.2). When the load unity is applied somewhere between points H_1 and H_2 the reaction *A* will be equal to the ordinate *ab* multiplied by the pressure P_1 exerted by the element H_1H_2 on the beam ABH_1. This pressure varies linearly from $P_1 = 1$ when the unit

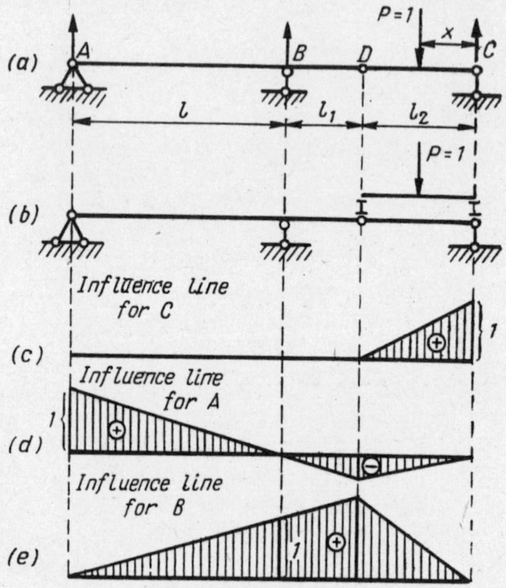

Fig. 54.2

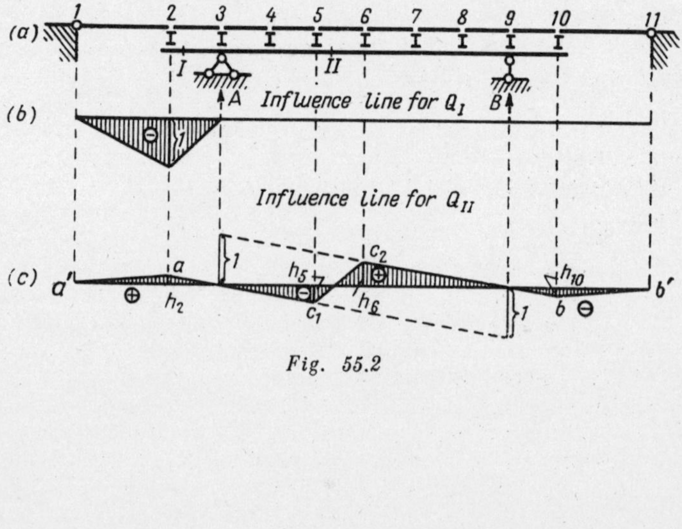

Fig. 55.2

load is applied at hinge H_1 to zero when it reaches hinge H_2 and therefore the influence line over the portion H_1H_2 of the beam may be obtained by simply connecting the ordinate b over hinge H_1 with a point of zero ordinate at the hinge H_2. Once the load has shifted to the right of the hinge H_2, the reaction at point A will equal zero * and therefore the ordinates of the influence line

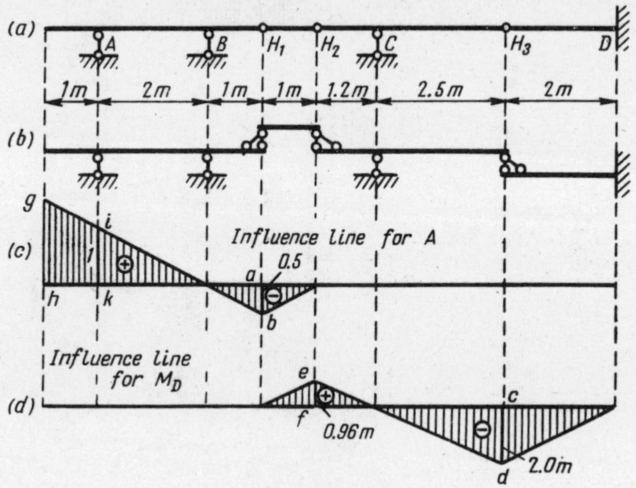

Fig. 56.2

from H_2 to D will also equal zero. The similitude of triangles will permit us to find the ordinates to the pertinent points of our influence line

$$\frac{\overline{gh}}{\overline{ik}}=\frac{1+2}{2} \text{ whence } \overline{gh}=\overline{ik}\cdot\frac{3}{2}=1\times\frac{3}{2}=1.5$$

$$\frac{\overline{ab}}{\overline{ik}}=\frac{1}{2} \text{ whence } \overline{ab}=\overline{ik}\cdot\frac{1}{2}=1\times\frac{1}{2}=0.5$$

Let us now construct the influence line for the bending moment acting over section D of our beam (Fig. 56.2*d*). When the load travels along portion H_3D the construction of the influence line will be exactly the same as for a cantilever beam with a built-in end (see Art. 4.2). Passing to portion H_3C we notice that the pressure R_3 varies proportionally to the distance of the unit load from point C reaching zero when the load is over this point; therefore the

◆

* This follows from the equilibrium of element H_1H_2.

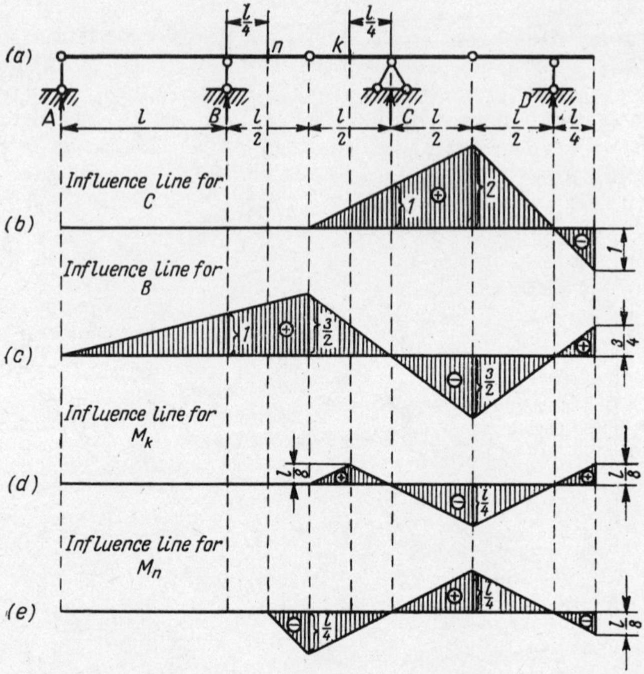

Fig. 57.2

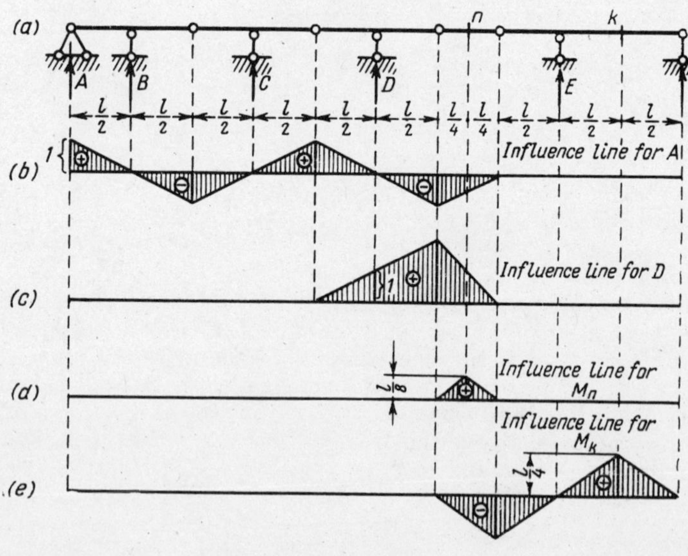

Fig. 58.2

influence line over this portion will be represented by a line connecting point d with a point of zero ordinate at C. Point e under hinge H_2 will be obtained by extending this line until its intersection with the vertical passing through this hinge, and the last portion of the line between hinges H_2 and H_1 will be obtained by connecting point e with a point of zero ordinate at the hinge H_1. The similitude of triangles permits the computation of the ordinate ef as follows

$$\frac{ef}{cd}=\frac{1.2}{2.5} \quad \text{whence} \quad \overline{ef}=\overline{cd}\cdot\frac{1.2}{2.5}=2\times 0.48=0.96 \text{ m}$$

It is apparent that the influence line for any function in any section of a multispan statically determinate beam may be constructed following the procedure outlined hereunder:

(1) The influence line corresponding to that portion of the beam which contains the section under consideration is constructed exactly in the same way as for a simply supported beam (with or without overhangs).

(2) The ordinate obtained at the point where the beam member containing the section meets with the adjacent one is then connected with a point of zero ordinate under the second support of this latter element. The same procedure may be followed in order to obtain the influence line over the more distant elements of the beam.

(3) The ordinates to the pertinent points of the influence line may be derived from the similitude of triangles which constitute it.

The reader is invited to check the influence lines represented in Figs. 57.2 and 58.2.

12.2. *Bending Moments and Shearing Forces Induced by Fixed Loads in Statically Determinate Bents, Knee Frames and Beams of Polygonal Design*

The determination of reactions arising at the supports of statically determinate bents and beams of polygonal design, the computation of internal forces acting over their cross sections and the tracing of Q, N and M diagrams are carried out in the same way as for ordinary rectilinear beams. All the formulas, sign conventions and equilibrium equations mentioned in Art. 1.2 remain valid. When dealing with knee frames or other structures comprising vertical elements it is good practice to decide beforehand which extremity of such an element will be considered as the left-hand one and to mark this extremity by some conventional sign (for instance, an asterisk). The following examples will illustrate the construction of Q, N and M diagrams for structures in question.

Problem 1. Required the Q, N and M diagrams for a beam represented in Fig. 59.2*a*.

Solution. Having decided to consider the lower extremity of the element AB as the left-hand one, mark it with an asterisk. The beam consisting of two elements, use expressions (1.2) through (3.2) for the determination of the shearing and normal forces and of the bending moments in each of these elements.

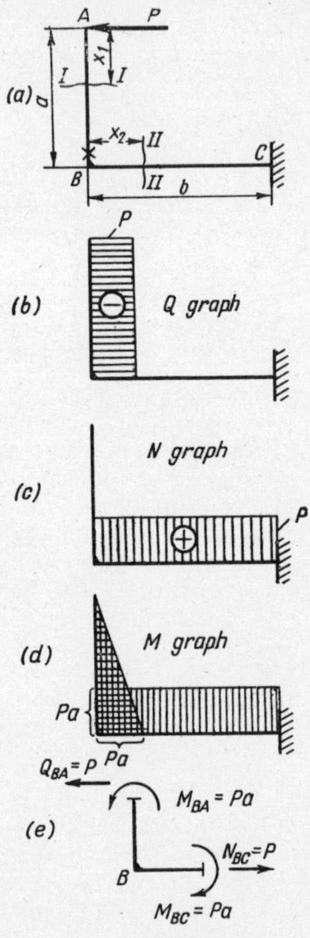

Fig. 59.2

Element I. The internal forces acting over a cross section a distance x_1 from the upper end of the element AB will be

$$Q^I=-\sum_R Y=-P \qquad N^I=-\sum_R X=0$$

$$M^I=\sum_R M=-(-Px_1)=Px_1$$

Element II. The internal forces acting over any section a distance x_2 from the left end of the element BC will equal

$$Q^{II}=\sum_L Y=0 \qquad N^{II}=\sum_L X=P$$

$$M^{II}=\sum_L M=-Pa$$

Graphs obtained with the aid of the above expressions are reproduced in Fig. 59.2*b*, *c* and *d*. It should be noted that the expressions obtained for M^I and Q^I do not satisfy expression (4.2) of Art. 1.2 derived from the theorem of Zhuravsky. Indeed,

$$\frac{dM^I}{dx_1}=\frac{d(Px_1)}{dx_1}=P=-Q^I$$

$$\text{instead of } \frac{dM^I}{dx_1}=Q^I$$

This is due to the fact that in section I of the beam positive values of the abscissas were measured downwards, in other words, from right to left, while the relation $Q=\dfrac{dM}{dx}$ remains true only when positive abscissas are measured from left to right.

Let us now check the equilibrium of joint B. Separating it from the other parts of the structure and applying at the cuts the internal forces computed above we obtain the following equilibrium equations (Fig. 59.2*e*)

$$\Sigma M_B=-M_{BA}+M_{BC}=-Pa+Pa=0$$
$$\Sigma Y=0$$
$$\Sigma X=-Q_{BA}+N_{BC}=-P+P=0$$

which shows that all the internal forces were computed correctly.

It should be remembered that equilibrium equations must be satisfied whatever the number of bars meeting at one joint, provided all the external loads applied directly to this joint are duly taken care of.

Problem 2. Required to trace the Q, N and M diagrams for a knee frame represented in Fig. 60.2*a*.

Solution. Choosing once again the lower ends of the vertical elements as their left-hand extremities, mark them with asterisks. Subdivide the knee frame itself into four separate portions and write for each expressions (1.2) through (3.2) giving the shearing and normal forces and the bending moments.

Portion I

$$Q^{I}=\sum_{L} Y=qx_1 \qquad N^{I}=\sum_{L} X=0 \qquad M^{I}=\sum_{L} M=qx_1\frac{x_1}{2}=\frac{qx_1^2}{2}$$

Portion II

$$Q^{II}=-\sum_{R} Y=0 \qquad N^{II}=-\sum_{R} X=-qa \qquad M^{II}=-\sum_{R} M=-\frac{qa^2}{2}$$

Portion III

$$Q^{III}=-\sum_{R} Y=P=qa; \qquad N^{III}=-\sum_{R} X=-qa$$

$$M^{III}=-\sum_{R} M=-\frac{qa^2}{2}-P(x_3-a)=qa\left(\frac{a}{2}-x_3\right)$$

Portion IV

$$Q^{IV}=-\sum_{R} Y=-qa; \qquad N^{IV}=-\sum_{R} X=-P=-qa$$

$$M^{IV}=-\sum_{R} M=qa\left(x_4-\frac{a}{2}\right)-Pa+M=qa\left(x_4-\frac{a}{2}\right)-qa^2+qa^2=$$

$$=qa\left(x_4-\frac{a}{2}\right)$$

The diagrams obtained using the above expressions are represented in Fig. 60.2*b*, *c* and *d*. Fig. 60.2*e* represents joint B subjected to the internal forces and moments acting at the cuts. It will be easily observed that all the equilibrium equations for this joint are satisfied:

$$\Sigma M_B=-qa^2-\frac{qa^2}{2}+\frac{3}{2}qa^2=0$$

$$\Sigma X=+qa-qa=0 \qquad \Sigma Y=qa-qa=0$$

Problem 3. Required to construct the Q, N and M graphs for the statically determinate frame represented in Fig. 61.2*a*.

Solution. Determine reactions R_A, R_B and H_B shown in Fig. 61.2*a* utilizing the well-known equilibrium expressions

$$\Sigma M_B=R_A 2a+Pa-q2aa=0$$

wherefrom remembering that $P=qa$ we obtain

$$R_A=\frac{2qa^2-qa^2}{2a}=q\frac{a}{2}$$

$$\Sigma M_A=-R_B 2a+q2aa+Pa=0$$

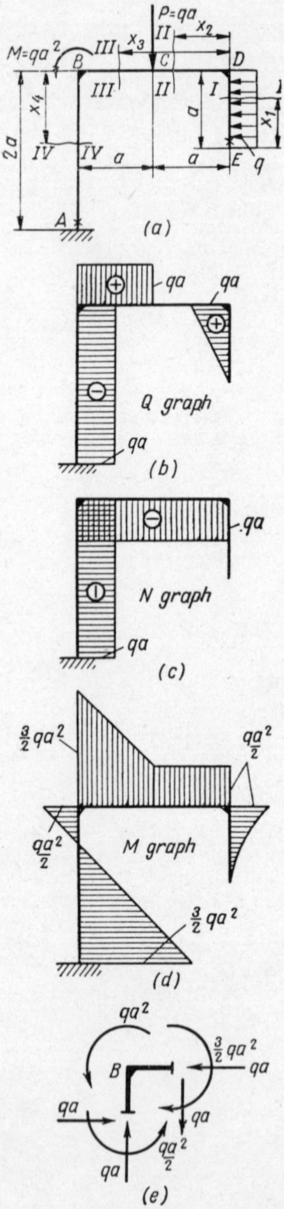

Fig. 60.2

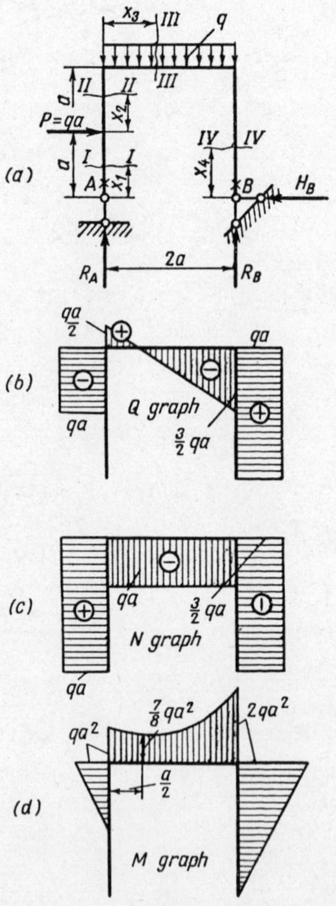

Fig. 61.2

and thus

$$R_B = \frac{2qa^2 + qa^2}{2a} = \frac{3}{2} qa$$

$$\Sigma X = P - H_B = 0$$

giving

$$H_B = P = qa$$

Mark again the lower ends of the vertical elements by an asterisk as in Fig. 61.2*a* considering them to form the left-hand extremities and subdivide the beam into four portions for each of which the following expressions are readily obtained.

Portion I

$$Q^I = 0; \quad N^I = -R_A = -q\frac{a}{2} \qquad M^I = 0$$

Portion II

$$Q^{II} = -P = -qa; \quad N^{II} = -R_A = -q\frac{a}{2}; \quad M^{II} = -Px_2 = -qax_2$$

giving for $x_2 = 0$; $M^{II} = 0$ and for $x_2 = a$ $M^{II} = -qa^2$

Portion III

$$Q^{III} = R_A - qx_3 = q\left(\frac{a}{2} - x_3\right); \; N^{III} = -P = -qa;$$

$$M^{III} = R_A x_3 - Pa - \frac{qx_3^2}{2} = q\left(\frac{a}{2}x_3 - a^2 - \frac{x_3^2}{2}\right)$$

when $x_3 = 0$ $\quad Q^{III} = q\frac{a}{2} \quad M^{III} = -qa^2$

when $x_3 = a$ $\quad Q^{III} = -q\frac{a}{2} \quad M^{III} = -qa^2$

when $x_3 = 2a$ $\quad Q^{III} = -\frac{3}{2}qa \quad M^{III} = -2qa^2$

The shearing force $Q^{III} = q\left(\frac{a}{2} - x_3\right)$ becomes nil when $x_3 = \frac{a}{2}$ and accordingly the bending moment will pass in this section through a maximum or a minimum

$$M^{III} = q\left(\frac{a^2}{4} - a^2 - \frac{a^2}{8}\right) = -\frac{7}{8}qa^2$$

Portion IV

$$Q^{IV} = H_B = qa; \quad N^{IV} = -R_B = -\frac{3}{2}qa; \quad M^{IV} = H_B x_4 = qax_4$$

when $x_4 = 0$ $\quad M^{IV} = 0$

when $x_4 = 2a$ $\quad M^{IV} = 2qa^2$

The corresponding diagrams for Q, N and M are represented in Fig. 61.2*b*, *c* and *d*.

Problem 4. Required to construct the Q, N and M diagrams for a beam represented in Fig. 62.2*a*,

Solution. Replace the inclined load P by its vertical and horizontal components P_y and P_x

$$P_x = P_y = P \cdot \cos 45° = 0.707P$$

and determine reaction R_A which will suffice in the case under consideration

$$\Sigma M_B = R_A 2l + P_y 1.707l + P_x 0.707l = 0$$

wherefrom

$$R_A = \frac{-(1.707P_y + 0.707P_x)\, l}{2l} = \frac{-1.707 - 0.707}{2} \times 0.707P = -0.853P$$

The negative value of this reaction indicates that it is directed downwards. For each of the three portions of the beam the following equations giving

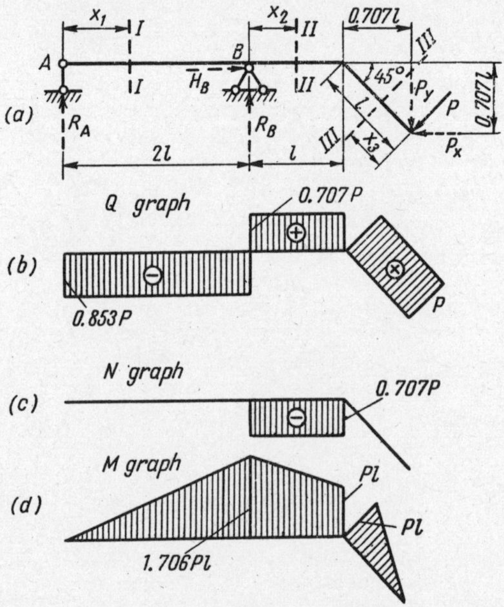

Fig. 62.2

the values of the shearing and normal forces and of the bending moments may be now written as

Portion I

$$Q^I = R_A = -0.853P; \; N^I = 0; \; M^I = R_A x_1 = -0.853Px_1$$

$$\text{when } x_1 = 0 \quad M^I = 0 \quad \text{when} \quad x_1 = 2l \quad M^I = -1.706Pl$$

Portion II

$$Q^{II} = P_y = 0.707P; \; N^{II} = -P_x = -0.707P$$

$$M^{II} = -P_x 0.707l - P_y (1.707l - x_2) = -0.707P (0.707l + 1.707l - x_2) =$$
$$= -P (1.707l - 0.707x_2)$$

$$\text{when } x_2 = 0 \quad M^{II} = -1.707Pl \quad \text{when } x_2 = l \quad M^{II} = -Pl$$

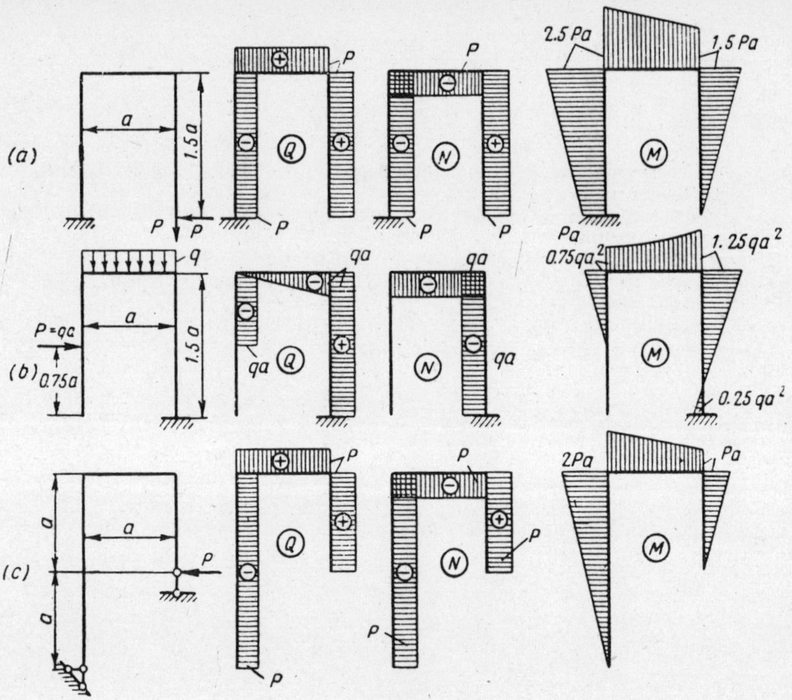

Fig. 63.2

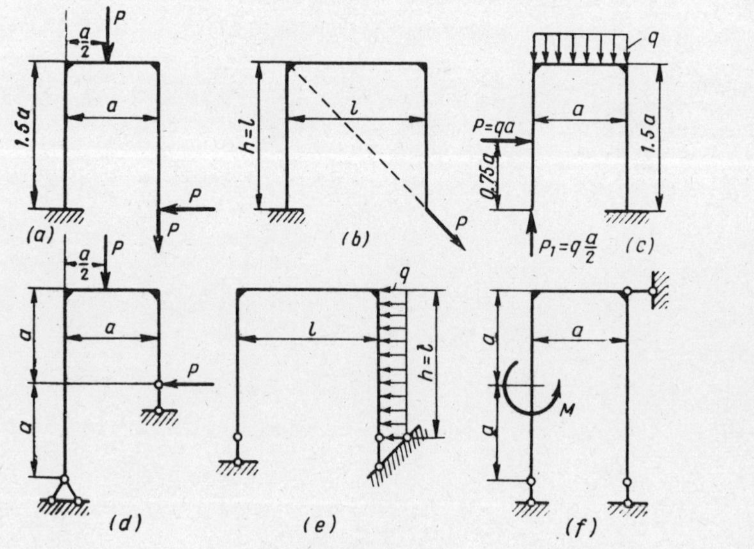

Fig. 64.2

Portion III

$$Q^{III}=P;\ N^{III}=0;\ M^{III}=-Px_3$$

The corresponding diagrams are represented in Fig. 62.2*b*, *c* and *d*. The reader is invited to

(1) check the sign of the shear diagram represented in Figs. 59.2 through 62.2 using the rule mentioned in Art. 1.2 which stipulates that the shear is positive when the axis of the beam must be turned clockwise in order to superimpose it with the tangent to the bending moment diagram,

(2) check the Q, N and M graphs represented in Fig. 63.2,

(3) trace the Q, N and M diagrams for the frames represented in Fig. 64.2.

3. THREE-HINGED ARCHES AND FRAMES

1.3. *Three-Hinged Systems*

A three-hinged system consists of two plates (*I* and *II*), connected together by means of a hinge (hinge *C* in Fig. 1.3), with two-hinged supports *A* and *B* resting on the ground. As the latter can itself be regarded as another rigid plate, it may be said that a three-hinged system consists of three plates connected together by means of three hinges, these hinges not lying on one straight line. We have seen previously (see Art. 2.1 and Fig. 16.1) that connections of this type are characteristic of geometrically stable structures.

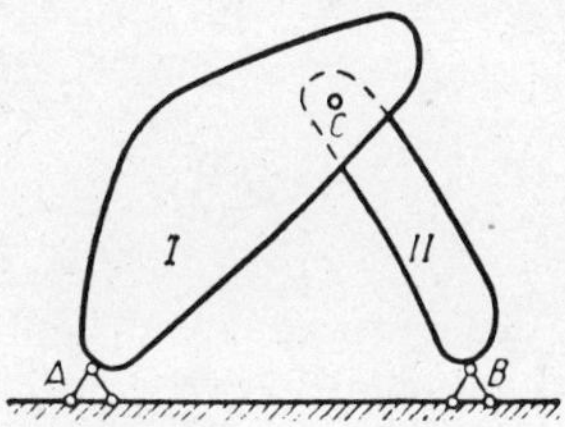

Fig. 1.3.

When the plates *I* and *II* consist of curved bars the system is called a *three-hinged arch* (Fig. 2.3*a*); in the event these bars are straight or *L*-shaped, the system will be called a *three-hinged bent* or *frame* (Fig. 2.3*b* and *c*); finally, when these plates are through structures, the system becomes a *three-hinged truss* or *spandrel arch* (Fig. 2.3*d*).

The distance l between the centres of the hinges at the supports is called the *span* of the arch while the distance f from the centre of the crown hinge to the straight line passing through the former two is called its *rise* (Fig. 2.3*a*).

A three-hinged system may or may not have a vertical axis of symmetry. In the first case (Fig. 4.3) the central hinge *C* will lie on this axis of symmetry and the hinges at the supports *A* and *B* will be at one and the same level. Nonsymmetrical systems may have their supports at different levels (Fig. 3.3).

In three-hinged systems the reactions at the supports *A* and *B* will be characterized by two parameters each—its magnitude and direction or by any two components of these reactions, say, the vertical *V* and the horizontal *H*. (These two components are frequently referred to as the vertical and horizontal reactions.)

Accordingly, the reactions of a three-hinged arch will be fully determined by four parameters, for instance, the amounts of the reactions H_A, H_B, V_A and V_B (Fig. 4.3).

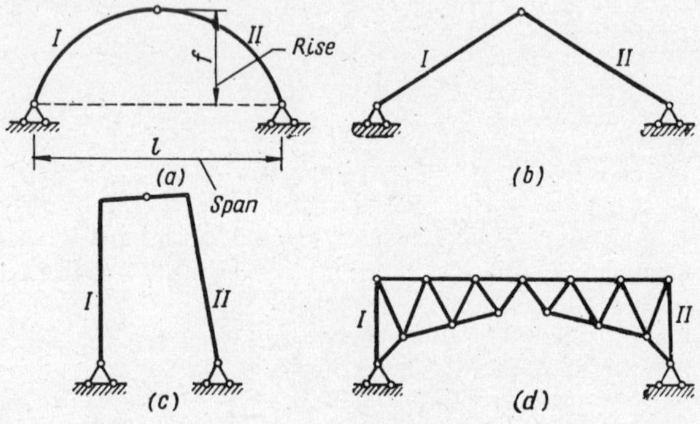

Fig. 2.3

These may be obtained from the three equilibrium equations of external forces (including the reactions) acting upon the system and from a fourth equation, expressing that the moment of all

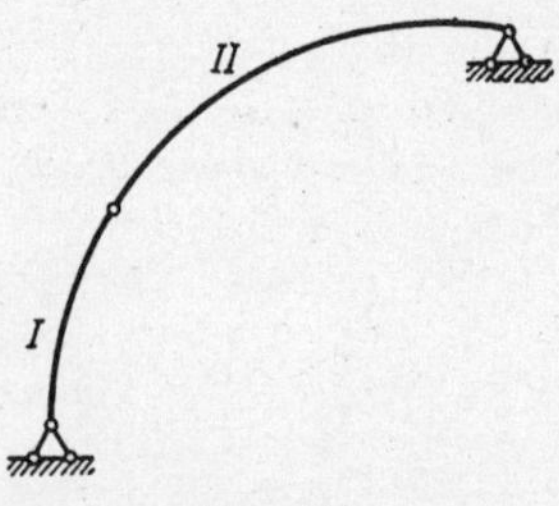

Fig. 3.3

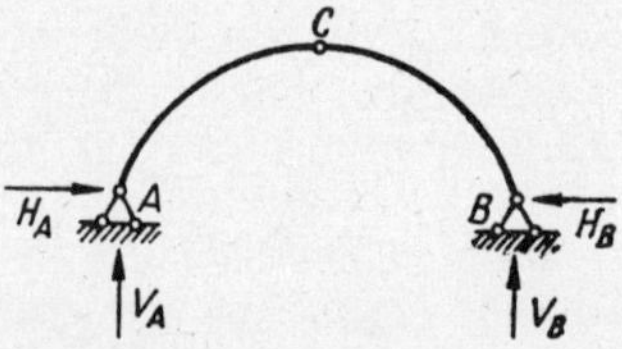

Fig. 4.3

the external forces acting to the left or to the right of the crown hinge about its centre must be nil. *

Thus, a three-hinged system is always statically determinate.

When a system of vertical loads acts on a three-hinged system the horizontal components H_A and H_B of the reactions at the

* This is due to the fact that in any hinged system in equilibrium the moment about any hinge must be equal to zero.

supports will not reduce to zero. Accordingly, the three-hinged systems usually develop a *thrust* which must be absorbed either by the supports or by some other arrangement.

It will be shown later that the bending moments and shears acting over cross sections of three-hinged arches are considerably smaller than the corresponding stresses in a simple beam covering

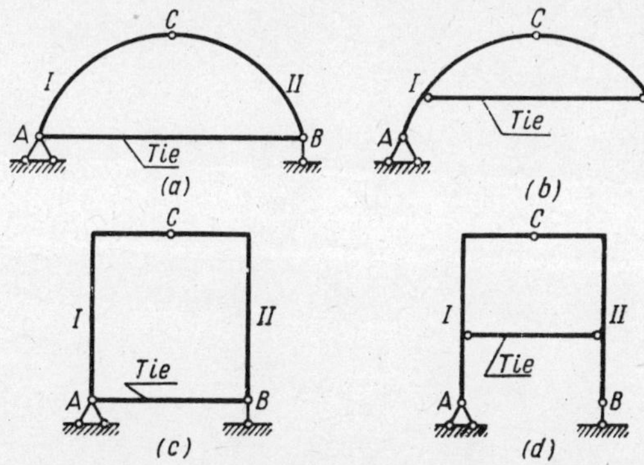

Fig. 5.3

the same span and carrying the same load. Therefore, three-hinged arches are more economical than ordinary beams, particularly for large-span structures. *

However, when the spans are small, three-hinged arches become less desirable than ordinary beams, as their construction is more complicated and the provision of hinges both at the supports and at the crown requires the use of more intricate arrangements.

In the three-hinged systems considered thus far both supports were capable of absorbing a horizontal thrust. In practice it is not unusual to encounter similar systems in which one of the hinges is movable. In this case the geometrical stability of the system

◆

* The first arched system for a large span was proposed in 1776 (i. e., some hundred years before the creation of the science of structural mechanics) by the eminent Russian engineer I. Kulibin. On the basis of general principles of theoretical mechanics, he designed an arched wooden bridge 300 m long spanning the whole of the river Neva at St. Petersburg. He was the first to determine the interaction of external forces and stresses in a three-hinged arched system and to use a funicular polygon for the determination of the shape of his arch many years before this method became widely known. A huge 30 m model of Kulibin's bridge was tested by a load of approximately 56 tons and approved by the Russian Academy of Science. The great mathematician and member of the Academy L. Euler checked all the computations and drawings of Kulibin's bridge and found them perfectly correct.

is ensured by ties established either at the level of the supports or somewhat higher (Fig. 5.3*a* represents a three-hinged tied or bowstring arch; Fig. 5.3*b*—a three-hinged arch with an elevated tie; Fig. 5.3*c*—a three-hinged tied bent, and Fig. 5.3*d*—a similar bent with an elevated tie).

2.3. *Support Reactions of a Three-Hinged Arch*

1. ANALYTICAL METHOD

As has already been stated, when a system of vertical loads is applied to a three-hinged arch (Fig. 6.3*a*) a vertical and a horizontal reaction will arise at each of the two supports making four

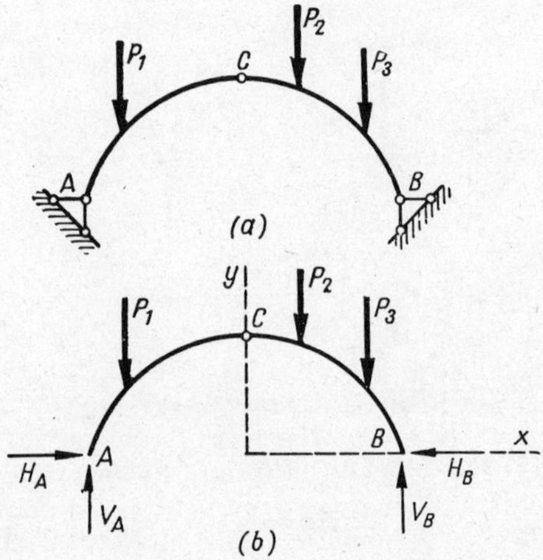

Fig. 6.3

reactions to be determined in all. Let us designate the vertical reactions by V_A and V_B and the horizontal ones by H_A and H_B, respectively (Fig. 6.3*b*).

In addition to the three equilibrium equations supplied by the statics for coplanar systems, a fourth equation can be used in the case of a three-hinged arch, this equation expressing that the bending moment at the hinge C equals zero, or in other words, that the sum of the moments of all the external forces acting to the right or to the left of this hinge about its centre is nil

$$\sum_{L} M_C = 0 \quad \text{or} \quad \sum_{R} M_C = 0$$

These four equations of statics will determine completely the four reactions at the supports.

It is recommended to avoid as much as possible simultanious solutions of several equations with several unknowns. For instance, in the case of an ordinary arch represented in Fig. 6.3*a* we may first write the equilibrium equation for the moments of all forces about hinge B which will contain only one unknown vertical reaction V_A. When this is known we may solve the equation $\sum_L M_C = 0$ expressing that the sum of moments of all forces acting

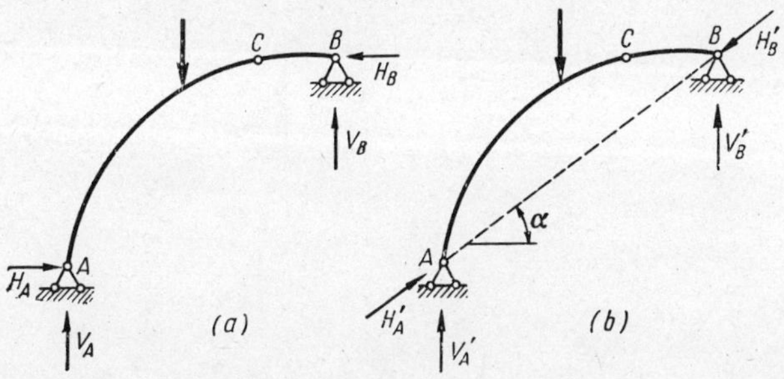

Fig. 7.3

on the left part of the arch about hinge C is nil, this equation containing the reaction V_A which has just been determined and the unknown reaction H_A. We may than proceed with the solution of an equation expressing that the moment of all external forces about hinge A is zero which will give us the value of reaction V_B and then obtain the magnitude of H_B equating to zero the projection of all the external forces on the horizontal.

The computations just described may be checked using the equations

$$\Sigma Y = 0 \quad \text{and} \quad \Sigma_R M_C = 0$$

If the two supports were at different levels as in Fig. 7.3*a*, the equation $\Sigma M_B = 0$ would contain two unknowns V_A and H_A, thus requiring the solution of a system of two equations with two unknowns. This can be easily avoided if both reactions were resolved into components one of which would follow the line connecting the two supports A and B (Fig. 7.3*b*). When these components V'_A, V'_B, H'_A and H'_B are determined, the vertical and horizontal components will be easily found using the expressions

$$V_A = V'_A + H'_A \sin\alpha; \quad V_B = V'_B - H'_B \sin\alpha$$
$$H_A = H'_A \cos\alpha; \quad H_B = H'_B \cos\alpha$$

2. GRAPHICAL METHOD

The graphical determination of the reactions requires that the resultants R_1 and R_2 of all the forces applied to the left and to the right of the central hinge should be found in the first place. The reactions induced by each of these resultants R_1 and R_2 will then be determined, their summation giving the final value of the reaction required. We may start with determining the reactions at the support caused by the application of the force R_1. In this case the reaction at the right-hand support B_1 must pass through the hinge at this support and the hinge at the crown

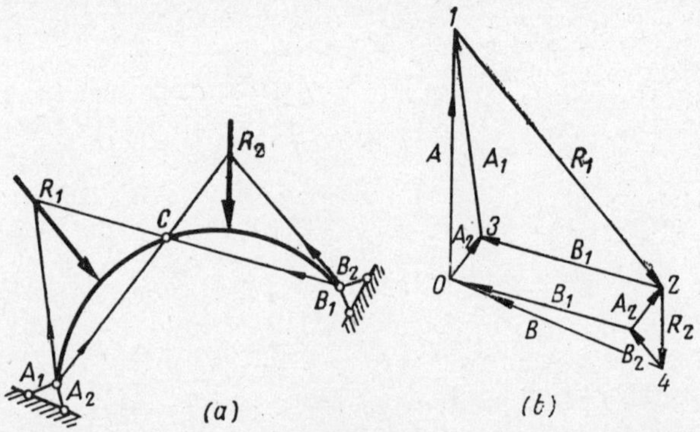

Fig. 8.3

(Fig. 8.3*a*) as otherwise the right-hand portion of the arch which is subjected solely to the reaction at B_1 and the interaction of hinge C could not remain in equilibrium. With reaction A_1 arising at the left-hand support, the arch as a whole will be in equilibrium under the action of three forces A_1, B_1, R_1.

Theoretical mechanics states that three coplanar forces acting on a body in equilibrium must necessarily concur at one and the same point. The use of this theorem enables us to find immediately the direction of reaction A_1 whereafter the force polygon (Fig. 8.3*b*) will give us the magnitude of both support reactions A_1 and B_1. The support reactions A_2 and B_2 due to the application of the right-hand resultant R_2 will be found in exactly the same way (see Fig. 8.3*a*).

The method of superposition will enable us to obtain the resultant reactions A and B at both supports. For this purpose a line parallel to the line of action of the reaction A_2 will be traced through point *3* of a force polygon (Fig. 8.3*b*) and the magnitude of reaction A_2 will be laid off along this line. The point *0* so obtained will then be connected with point *1*, thus giving the magnitude of the full reaction A at the left-hand support, the full reaction B at the right-hand support being obtained by the same method. The vertical and horizontal components V_A, H_A, V_B and H_B can be obtained thereafter in the usual way.

The graphical method of determining the reactions at the supports of a three-hinged arch carrying a number of vertical loads is illustrated in Fig. 9.3. At the outset resultants R_1 and R_2 are found using the method of force and funicular polygons whereafter the procedure followed does not differ from the one just described.

Problem 1. Using both methods described above determine the support reactions of a three-hinged arch supporting two vertical loads as indicated in Fig. 10.3*a*.

Solution. 1. *Analytical method.* Replace the support reactions by their components V_A, H_A and V_B, H_B (Fig. 10.3*b*). In order to determine the magnitude of V_A equate to zero the sum of all the forces acting on the arch about point B

$$\Sigma M_B = V_A l - P_1 (l - a_1) - P_2 (l - a_2) = 0$$

whence

$$V_A = \frac{P_1 (l - a_1) + P_2 (l - a_2)}{l} = \frac{M_B}{l} \qquad (1.3)$$

Here M_B is the moment of all the external loads about the hinge at the right-hand support.

The magnitude of H_A will be obtained from the equilibrium of the moments of all external forces acting on the left half of the arch about the crown hinge C

$$\Sigma_L M_C = V_A l_1 - H_A f - P_1 (l_1 - a_1) = 0$$

whence

$$H_A = \frac{V_A l_1 - P_1 (l_1 - a_1)}{f} = \frac{M_C^0}{f} \qquad (2.3)$$

Here M_C^0 is the moment of all the loads (except of H_A) acting on the left-hand portion of the arch about point C.

The vertical reaction V_B will be obtained by summing up and equating to zero the moments of all the external forces about hinge A

$$\Sigma M_A = -V_B l + P_2 a_2 + P_1 a_1 = 0$$

whence

$$V_B = \frac{P_1 a_1 + P_2 a_2}{l} = \frac{M_A}{l} \qquad (3.3)$$

Here M_A is the moment of all the loads about the left-hand support.

The last unknown reaction H_B will be found by projecting all the forces on the x-axis

$$\Sigma X = H_A - H_B = 0$$

whence

$$H_A = H_B = H$$

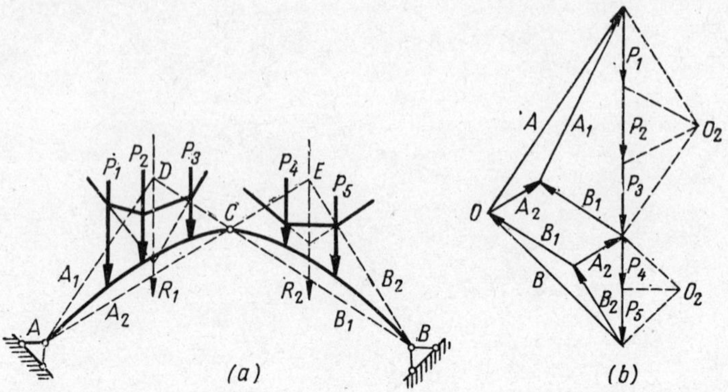

Fig. 9.3

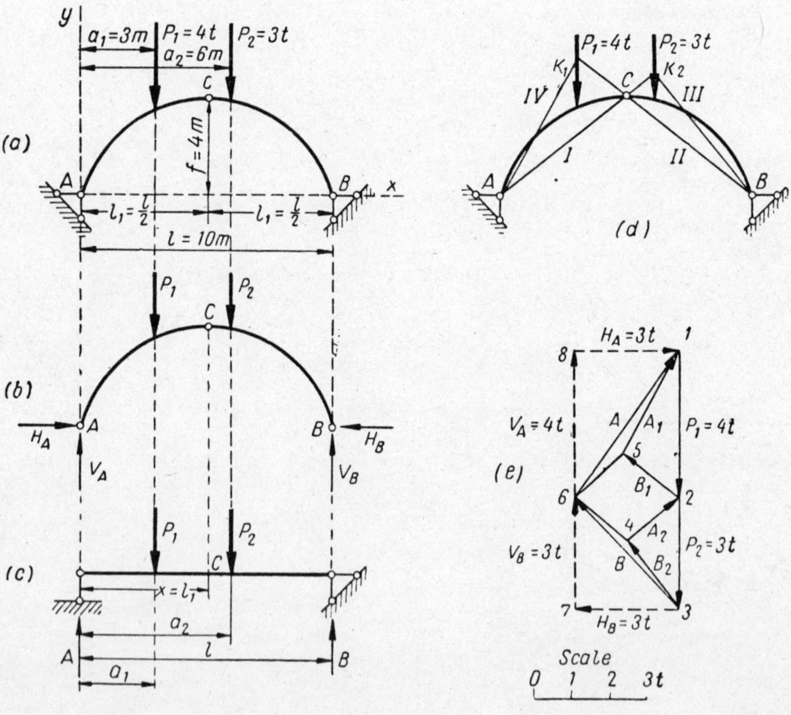

Fig. 10.3

The last formula shows that *the thrusts arising at both supports of three-hinged symmetrical arches subjected to vertical loads are equal in size and opposite in direction.*

Substituting in equations (1.3) through (4.3) the numerical values of all the parameters we obtain

$$V_A=\frac{4\,(10-3)+3\,(10-6)}{10}=\frac{28+12}{10}=4 \text{ tons}$$

$$V_B=\frac{4\times 3+3\times 6}{10}=3 \text{ tons}$$

$$H_A=H_B=H=\frac{4\times 5-4\,(5-3)}{4}=3 \text{ tons}$$

From the expressions (1.3) and (3.3) it will be observed that the vertical support reactions of three-hinged arches carrying vertical loads alone have the same values as the reactions of simply supported beams of the same span and loaded in the same way (Fig. 10.3c). The bending moment at midspan of this beam being equal to M_C^0, *the thrust at the arch supports may be obtained by dividing this bending moment by the rise of the arch* [see equations (2.3) and (4.3)].

2. *Graphical method.* Using the schematic drawing of the arch (Fig. 10.3*d*) let us connect hinges A and B with the crown hinge C, extending these lines (lines *I* and *II*) to their intersection with the direction of forces P_2 and P_1, respectively, at points K_2 and K_1. These points are then connected directly to points A (line *IV*) and B (line *III*).

Let us now lay off to scale forces P_1 and P_2 (vectors *1-2* and *2-3*) along a vertical as in Fig. 10.3*e*. Force P_1 is then resolved into two components A_1, B_1 parallel to the lines *IV* and *II* (see Fig. 10.3*d*) for which purpose rays *2-5* and *1-5* are traced through its ends. Force P_2 is resolved in the same way thus obtaining a ray *2-4* equal in amount to A_2 and parallel to the line *I* and a ray *3-4* equal in amount to B_2 and parallel to line *III*. Thereafter rays *4-6* and *5-6* are traced parallel to lines *2-5* and *2-4*, respectively. Ray *6-1* will be equal to the reaction at A and ray *3-6* to the reaction at B. The vertical and horizontal components of these reactions V_A, V_B and H_A, H_B are easily found.

Problem 2. Determine analytically the thrust of an arch represented in Fig. 11.3 uniformly loaded over the entire span with an intensity q.

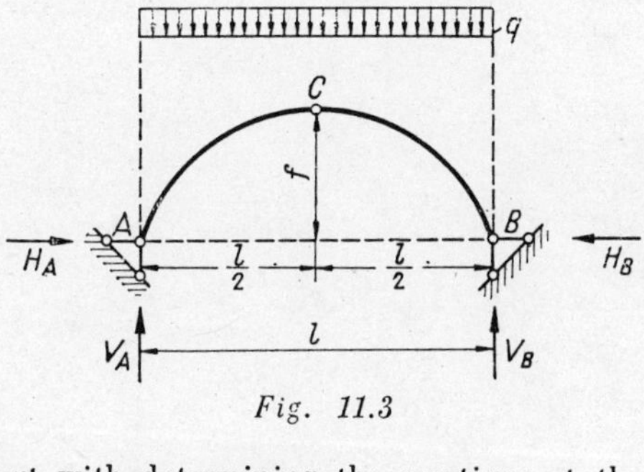

Fig. 11.3

Solution. Start with determining the reactions at the supports using the following equilibrium equations

$$\Sigma M_B=0 \quad \text{and} \quad \Sigma M_A=0$$

In the case under consideration these equations become

$$\Sigma M_B = V_A l - ql\frac{l}{2} = 0$$

$$\Sigma M_A = -V_B l + ql\frac{l}{2} = 0$$

whence

$$V_A = V_B = \frac{1}{2}\,ql$$

In the case of vertical loads alone the thrust $H_A = H_B = H$ may be determined by equating to zero the moments of all external forces acting on the left half of the arch about the crown hinge C

$$\sum_L M_C = V_A\frac{l}{2} - Hf - q\frac{l}{2}\cdot\frac{l}{4} = 0$$

whence

$$H = \frac{ql^2}{8f}$$

Problem 3. Required to determine both graphically and analytically the reactions induced at the supports of the three-hinged arch represented in Fig. 12.3*a* by an inclined force $P = 5$ tons for $\cos\alpha = 0.6$ and $\sin\alpha = 0.8$.

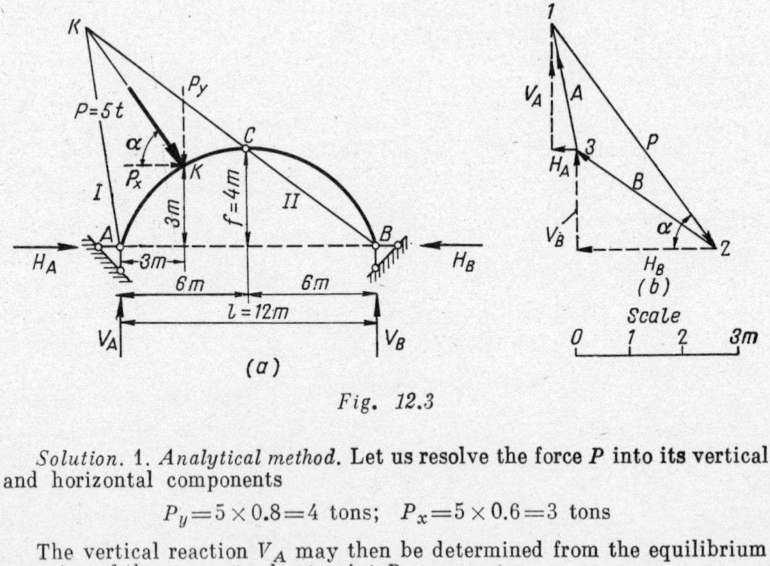

Fig. 12.3

Solution. 1. *Analytical method.* Let us resolve the force P into its vertical and horizontal components

$$P_y = 5\times 0.8 = 4 \text{ tons};\quad P_x = 5\times 0.6 = 3 \text{ tons}$$

The vertical reaction V_A may then be determined from the equilibrium equation of the moments about point B

$$\Sigma M_B = 12V_A - 9P_y + 3P_x = 0$$

whence

$$V_A = \frac{9P_y - 3P_x}{12} = \frac{36-9}{12} = \frac{27}{12} = 2.25 \text{ tons}$$

The reaction V_B will be obtained from the equilibrium of moments about point A

$$\Sigma M_A = -12V_B + 3P_y + 3P_x = 0$$

whence

$$V_B = \frac{4\times 3 + 3\times 3}{12} = 1.75 \text{ tons}$$

We may now determine the horizontal reaction H_A equating to zero the moments of all forces acting on the left half of the arch about the crown hinge C

$$\underset{L}{\Sigma M_C} = 6V_A - 4H_A - 1P_x - 3P_y = 0$$

whence

$$H_A = \frac{2.25\times 6 - 3\times 1 - 4\times 3}{4} = -0.375 \text{ ton}$$

The negative sign obtained indicates that the reaction H_A is directed towards the left. To determine the reaction H_B let us equate to zero the sum of horizontal projections of all the forces

$$\Sigma X = H_A + P_x - H_B = 0$$

whence

$$H_B = -0.375 + 3 = 2.625 \text{ tons}$$

2. *Graphical method.* Trace line *II* through hinges B and C until its intersection with the direction of force P at point K (Fig. 12.3*a*). Point K will then be connected by line *I* with the hinge A. Then lay to scale force P parallel to its direction as shown in Fig. 12.3*b* and through its ends trace rays *1-3* and *2-3* parallel to lines *I* and *II* of Fig. 12.3*a*, respectively. These two rays will represent to scale the reactions at the supports A and B; their horizontal and vertical components are H_A and H_B, V_A and V_B.

3.3. *Determination of Stresses in Three-Hinged Arches*

1. ANALYTICAL METHOD

The internal forces or stresses acting over the cross sections of a three-hinged arch consist of bending moments M, shears Q and normal forces N. They may be computed on the basis of loads and reactions acting to the left or to the right of the section considered.

We shall use the same sign conventions for the three-hinged arches as adopted in Art. 1.2 [expressions (1.2) through (3.2)] for ordinary beams, with the exception of the sign of the normal force which in this case will be reckoned positive when producing a compression.

In the computation of stresses auxiliary coordinate axis will be used for each cross section considered, the axis of abscissas u coinciding with the tangent and the axis of ordinates v with the normal to the centre line of the arch at this section. The projections of forces on these axes will be designated by U and V.

With these conventions expressions (1.2) through (3.2) become

$$Q=\sum_L V=-\sum_R V$$
$$M=\sum_L M=\sum_R M$$
$$N=\sum_L U=-\sum_R U \quad (1.3)$$

In these expressions the moments will be reckoned positive when they tend to turn the section clockwise, the components V when they are directed upwards and the components U when they

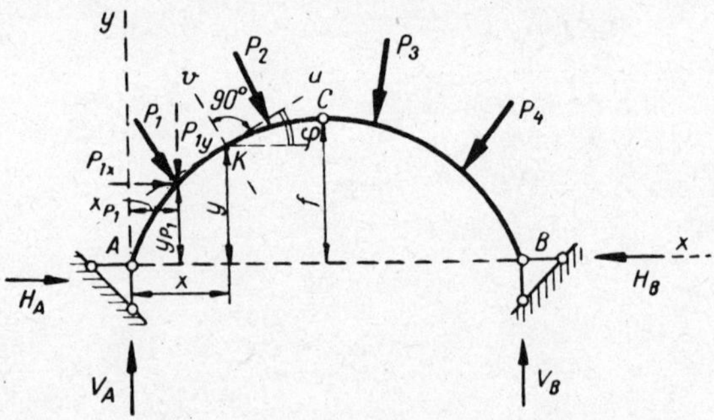

Fig. 13.3

are directed from left to right. Using expressions 1.3 let us determine the internal forces acting over a cross section K of an arch represented in Fig. 13.3

$$Q=\sum_L V=V_A\cos\varphi-H_A\sin\varphi-\sum_L P_y\cos\varphi-\sum_L P_x\sin\varphi$$
$$M=\sum_L M=V_Ax-H_Ay-\sum_L P_y(x-x_p)-\sum_L P_x(y-y_p)$$
$$N=\sum_L U=V_A\sin\varphi+H_A\cos\varphi-\sum_L P_y\sin\varphi+\sum_L P_x\cos\varphi \quad (2.3)$$

where x and y = coordinates of point K on the centre line of the arch

φ = angle between the tangent to the centre line of the arch at point K and a horizontal

P_y and P_x = vertical and horizontal components of force P, respectively

x_p and y_p = coordinates of the point of application of force P.

In the expressions for Q, M and N the summation must comprise the components P_y and P_x of all the external loads and forces applied to the arch to the left of section K. In the case of the arch represented in Fig. 13.3 only one component of force P_1 (P_{1y} or P_{1x}) will enter into each of these equations. It should be noted that the stresses Q, M and N could be expressed with equal success using the forces to the right of section K.

If vertical loads alone are applied to the arch (Fig. 14.3*a*) all the horizontal components P_x are equal to zero, the vertical

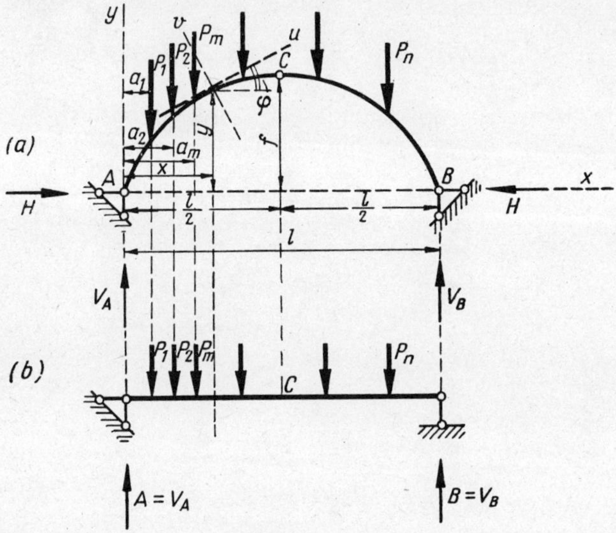

Fig. 14.3

components P_y equal P and the thrust $H_A = H_B = H$. In this case expressions (2.3) become

$$Q=(V_A-\sum_L P)\cos\varphi-H\sin\varphi$$

$$M=V_A x-\sum_L P(x-x_p)-Hy$$

$$N=(V_A-\sum_L P)\sin\varphi+H\cos\varphi$$

The expression $(V_A-\sum_L P)$ represents the shear Q^0 in the corresponding section of an end-supported "reference" beam subjected to the same loads as shown in Fig. 14.3*b* and the expression $[V_A x-\sum P(x-x_p)]$—the bending moment M^0 in the same section of the same beam. *

* Q^0 might be called the beam shearing force, and M^0 the beam bending moment.

With these designations the above expressions become

$$Q = Q^0 \cos\varphi - H\sin\varphi$$
$$M = M^0 - Hy$$
$$N = Q^0 \sin\varphi + H\cos\varphi \qquad (3.3)$$

Once the magnitudes of Q, M and N have been determined for a sufficient number of cross sections, the graphs of these functions will be easily constructed. When vertical forces alone act on the arch, any of the three sets of equations (1.3), (2.3) or (3.3) may be used, in other cases use should be made of expressions (1.3) or (2.3). It will be noted that in the event of a vertical loading each graph may be obtained by the summation of two other graphs. For instance, the bending moment diagram may be obtained by summing up the bending moment diagram M^0 for reference beam with the graph of the arch ordinates y multiplied by $(-H)$, this illustrating very clearly the extent to which the bending moments are reduced in the arches.

Problem 1. Required to determine the reactions at the supports as well as the bending moment, shear and normal force acting over section K a

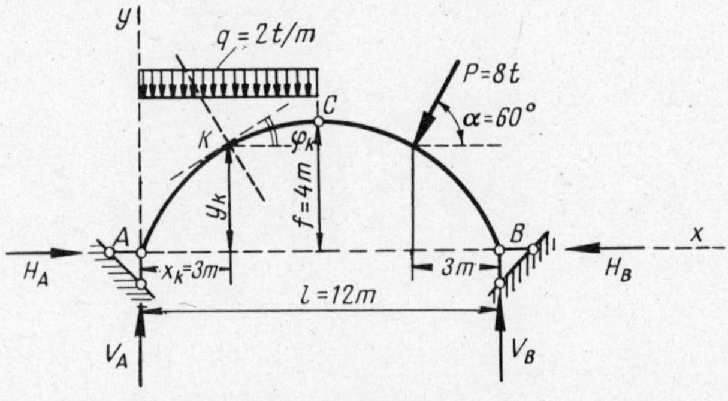

Fig. 15.3

three-hinged arch represented in Fig. 15.3. The centre line of the arch follows a conic parabola given by the equation

$$y = \frac{4f}{l^2}(l-x)\,x = \frac{4\times 4\,(12-x)\,x}{12\times 12} = \frac{(12-x)\,x}{9}$$

The abscissa x_k of point K is 3 metres.

Solution. First determine the ordinate of point K

$$y_k = \frac{(12-3)\,3}{9} = 3 \text{ metres}$$

The tangent of the angle formed by the tangent to the centre line of the arch and the axis of abscissas will be given by the first derivative of the parabola

$$\tan\varphi_x = y' = \frac{12-2x}{9}$$

for point K ($x = 3$ metres) this tangent will be given by

$$\tan\varphi_k = \frac{12-2\times3}{9} = \frac{2}{3}$$

The corresponding sine and cosine will be derived from the formulas *

$$\sin\varphi_k = \frac{\tan\varphi_k}{\sqrt{1+\tan^2\varphi_k}} = \frac{2}{3\sqrt{1+\frac{4}{9}}} = 0.555$$

$$\cos\varphi_k = \frac{1}{\sqrt{1+\tan^2\varphi_k}} = \frac{1}{\sqrt{1+\frac{4}{9}}} = 0.832$$

The reactions at the supports will be determined using the followirg equations

$$\Sigma M_B = V_A 12 - q6\times9 - P\cos\alpha 3 - P\sin\alpha 3 = 0$$

whence

$$V_A = \frac{2\times6\times9+8\,(0.5+0.866)\,3}{12} = 11.73 \text{ tons}$$

$$\Sigma Y = V_A - q6 - P\sin\alpha + V_B = 0$$

and accordingly

$$V_B = 2\times6+8\times0.866-11.73 = 7.20 \text{ tons}$$

$$\sum_L M_C = V_A 6 - q6\times3 - H_A 4 = 0$$

then

$$H_A = \frac{11.73\times6-2\times18}{4} = \frac{34.38}{4} = 8.60 \text{ tons}$$

$$\Sigma X = H_A - H_B - P\cos\alpha = 0$$

leading to

$$H_B = 8.60-8\times0.5 = 4.60 \text{ tons}$$

The bending moment in section K will amount to

$$M_K = V_A 3 - H_A 3 - q3\times\frac{3}{2} = 0.39 \text{ ton metre}$$

while the shear in the same section will total

$$Q_K = V_A\cos\varphi_k - H_A\sin\varphi_k - q3\cos\varphi_k = 0$$

and finally the normal force N_K will be

$$N_k = V_A\sin\varphi_k + H_A\cos\varphi_k - q3\sin\varphi_k = 10.34 \text{ tons}$$

Problem 2. Required to construct the diagrams of bending moments M, shears Q and normal forces N for an arch represented in Fig. 16.3*a* and fol-

◆

* Their values could also be found directly using appropriate tables.

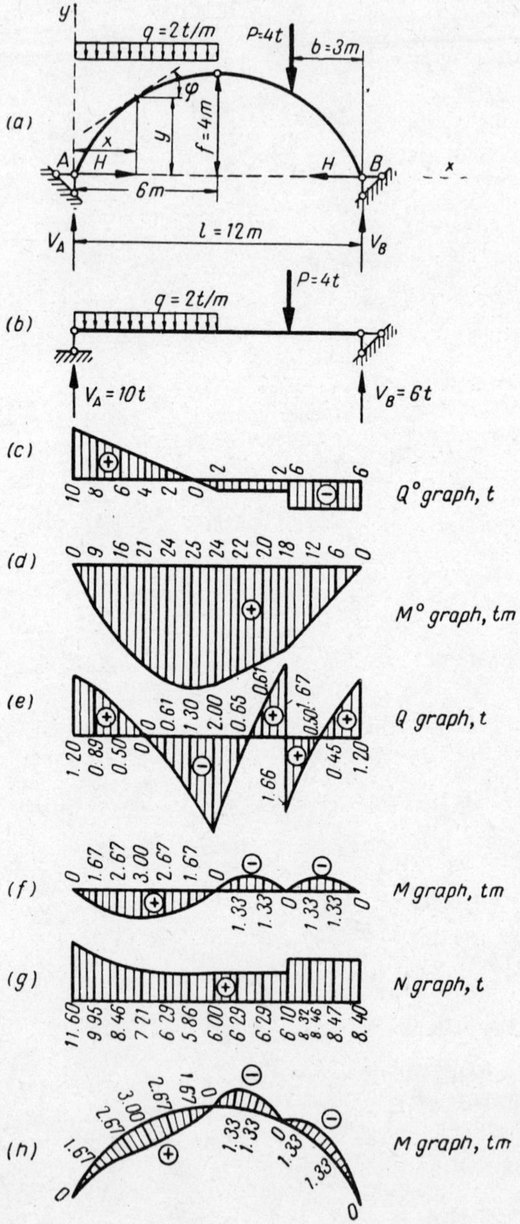
(a)
q = 2t/m
P=4t
b=3m
f=4m
6m
l = 12m
V_A
V_B
A
B
H
(b)
q = 2t/m
P=4t
V_A = 10t
V_B = 6t
(c)
Q° graph, t
(d)
M° graph, tm
(e)
Q graph, t
(f)
M graph, tm
(g)
N graph, t
(h)
M graph, tm

Fig. 16.3

lowing a conic parabola whose equation is

$$y=\frac{4f}{l^2}(l-x)\,x$$

Solution. Let us determine first the reactions at the supports V_A and V_B

$$\Sigma M_B=V_A l-\frac{ql}{2}\times\frac{3}{4}\,l-Pb=0$$

$$\Sigma Y=V_A+V_B-q\frac{l}{2}-P=0$$

whence

$$V_A=\frac{3}{8}\,ql+\frac{b}{l}\,P=10 \text{ tons}$$

$$V_B=q\frac{l}{2}+P-V_A=6 \text{ tons}$$

The thrust H will be derived from the equation

$$\underset{L}{\Sigma M_C}=V_A\frac{l}{2}-\frac{ql}{2}\times\frac{l}{4}-Hf=0$$

whence

$$H=\frac{V_A\dfrac{l}{2}-\dfrac{ql^2}{8}}{f}=6 \text{ tons}$$

The arch carrying vertical loads alone, expressions (3.3) may be used for the construction of the Q, M and N graphs required.

Fig. 16.3*b* represents a simply supported reference beam loaded in the same way as the arch and Fig. 16.3*c* and *d* represents the diagrams of the shears Q^0 and bending moments M^0. All further computations are entered in Table 1.3, column 1 containing the abscissas x of the points along the arch centre line taken at one metre increments, and column 2 containing the corresponding ordinates, calculated using the expression

$$y=\frac{4f}{l^2}(l-x)\,x=\frac{12-x}{9}\,x$$

Column 3 contains the values of tan φ computed from

$$\tan\varphi=y'=\frac{4f}{l^2}(l-2x)=2\,\frac{6-x}{9}$$

while the following three columns contain the values of φ, sin φ and cos φ. The values of Q^0 and M^0 tabulated in columns 7 and 13 are taken directly from the corresponding diagrams reproduced in Fig. 15.3*c* and *d*. Columns 8 through 12 contain the products of the shear Q^0 and the thrust H by sin φ, cos φ and the ordinates of the centre line of the arch.

The last three columns of Table 1.3 (columns 14, 15 and 16) contain the values of Q, M and N acting over the corresponding cross sections of the arch. They have been computed using formulas (3.3), which means that the magnitude of Q was obtained by summing up ciphers appearing in columns 9 and 10, the value of M—by summing up ciphers of columns 12 and 13, and the value of N—those of columns 8 and 11.

The shear, bending moment and normal force diagrams appearing in Fig. 16.3*e*, *f* and *g* have been constructed using the data contained in the last three columns of Table 1.3. * In these three diagrams the ordinates have been laid off from a horizontal axis; in addition the bending moment diagram represented in Fig. 16.3*h* has been constructed by laying off these ordinates from the curved centre line of the arch.

2. GRAPHICAL METHOD

The graphical determination of internal forces Q, M and N acting over the cross sections of three-hinged arches is carried out by constructing the so-called *funicular polygon* or *line of pressure*.

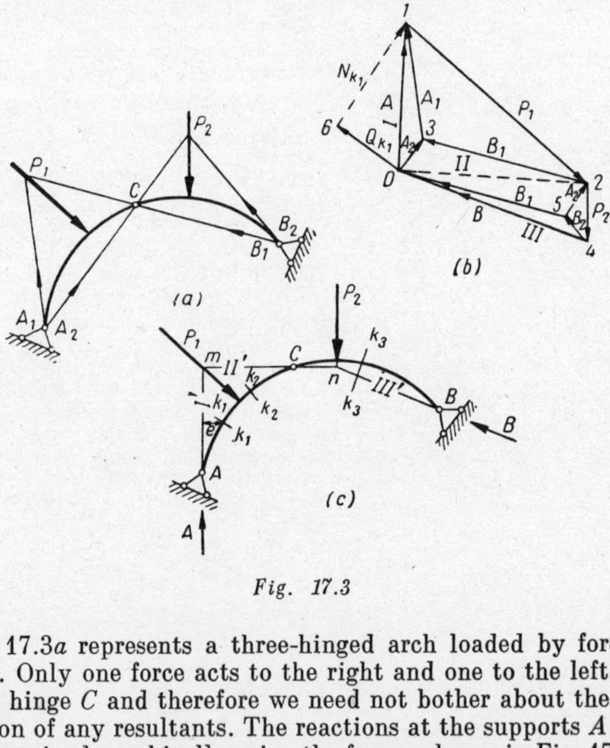

Fig. 17.3

Fig. 17.3*a* represents a three-hinged arch loaded by forces P_1 and P_2. Only one force acts to the right and one to the left of the central hinge C and therefore we need not bother about the determination of any resultants. The reactions at the supports A and B are determined graphically using the force polygon in Fig. 17.3*b*.**

Let us proceed now with the construction of the *funicular polygon* (Fig. 17.3*c*) corresponding to the force polygon already

* For convenience, different scales have been adopted for different diagrams.

** See Fig. 8.3 of Art. 2.3 for explanation.

Table 1.3

x m	y m	tan φ	φ	sin φ	cos φ	Q^0 tons	Q^0 sin φ tons	Q^0 cos φ tons	$-H$ sin φ tons	H cos φ tons	$-Hy$ ton metres	M^0 ton metres	Q tons	M ton metres	N tons
1	2	3	4	5	6	7	8	9	10	11	12	13	14	15	16
0	0	1.333	53°08′	0.800	0.600	10	8.00	6.00	—4.80	3.60	0	0	1.20	0	11.60
1	1.222	1.111	48°01′	0.743	0.669	8	5.94	5.35	—4.46	4.01	—7.33	9	0.89	1.67	9.95
2	2.222	0.889	41°38′	0.664	0.747	6	3.98	4.48	—3.98	4.48	—13.33	16	0.50	2.67	8.46
3	3.000	0.667	33°42′	0.555	0.832	4	2.22	3.33	—3.33	4.99	—18.00	21	0	3.00	7.21
4	3.556	0.444	23°56′	0.406	0.914	2	0.81	1.83	—2.44	5.48	—21.33	24	—0.61	2.67	6.29
5	3.889	0.222	12°31′	0.217	0.976	0	0	0	—1.30	5.86	—23.33	25	—1.30	1.67	5.86
6	4.000	0	0	0	1.000	—2	0	—2.00	0	6.00	—24.00	24	—2.00	0	6.00
7	3.889	—0.222	—12°31′	—0.217	0.976	—2	0.43	—1.95	1.30	5.86	—23.33	22	—0.65	—1.33	6.29
8	3.556	—0.444	—23°56′	—0.406	0.914	—2	0.81	—1.83	2.44	5.48	—21.33	20	0.61	—1.33	6.29
9	3.000	—0.667	—33°42′	—0.555	0.832	—2 —6	1.11 3.33	—1.66 —4.99	3.33	4.99	—18.00	18	1.67 —1.66	0	6.10 8.32
10	2.222	—0.889	—41°38′	—0.664	0.747	—6	3.98	—4.48	3.98	4.48	—13.33	12	—0.50	—1.33	8.46
11	1.222	—1.111	—48°01′	—0.743	0.669	—6	4.46	—4.01	4.46	4.01	—7.33	6	0.45	—1.33	8.47
12	0	—1.333	—53°08′	—0.800	0.600	—6	4.80	—3.60	4.80	3.60	0	0	1.20	0	8.40

mentioned. For this purpose we shall extend the direction of reaction A (Fig. 17.3*c*) until its intersection at m with the direction of force P_1. Through the point of intersection we shall trace the string II' parallel to ray II of the force polygon, this ray representing the resultant of the reaction A and the load P_1. Let point n be the intersection of the string II' with the line of action of the load P_2. Through this point we shall trace the string III' parallel to ray III, the latter being the resultant of reaction A and loads P_1 and P_2.

If all the operations were carried out correctly, the string II' representing the resultant of forces A and P_1 will pass through the centre of hinge C whilst the string III' whose direction must coincide with the direction of the reaction at B will pass through the pin of this support.

The funicular polygon I'-II'-III' (Fig. 17.3*c*) is frequently termed *polygon* or *line of pressure* as each string coincides with the direction of the pressure exerted by one portion of the arch on the other.

Hence these strings will coincide in direction with the resultant of all the forces acting on the arch to the left or to the right of the section considered.

This may be illustrated by Fig. 17.3*c* where to the left of section k_1-k_1 there is only the reaction at the support A and thus string I' of the funicular polygon will coincide with the direction of force A which is the resultant of all the forces to the left of the cross section considered. Passing to section k_2-k_2 we note that there are already two forces A and P_1 to its left. At the same time string II' represents the resultant of these two forces. In the case of section k_3-k_3 there will be already three forces A, P_1 and P_2 to its left, their resultant passing through the point of intersection of string II' with force P_2 as string II' is itself the resultant of forces A and P_1. Therefore, this resultant coincides with string III'.

Accordingly, *any line in the pressure polygon AmnB represents the direction of the resultant of all forces applied to the left* (*or to the right*) *of the section under consideration.* The magnitude of this resultant may be determined with the aid of the polygon of forces. Thus, in Fig. 17.3*b* the resultant of forces A and P_1 will be given by the length of ray II measured to scale.

Thus, the polygon of forces and the pressure polygon permit the determination of all the stresses in any cross section of the arch. For instance, the bending moment may be obtained by multiplying the resultant by its distance to the centroid of the section under consideration.

In section k_1-k_1 (Fig. 17.3c) the bending moment M will thus equal Ae where e is the distance to the line of action of force A measured along a perpendicular dropped on this line from the centre of gravity of the cross section.

In order to determine the shear and the normal force acting over section k_1-k_1 the resultant of all the forces to the left of this section (ray *I* or reaction A) must be resolved into two components, one parallel to the tangent to the arch centre line at this section (ray *6-1*) and the other (ray *0-6*) normal to the same line. It is clear that ray *0-6* will represent the shear Q and ray *6-1* the normal force N in our section.

The line of pressure provides a very clear picture of the work of an arch. Thus, Fig. 17.3c shows that the forces acting on the arch tend to increase the curvature of its right-hand portion where the resultant is below the centre line, while the curvature of the left-hand portion will decrease.

When a system of vertical loads P_1, P_2, P_3, etc., acts on a three-hinged arch, the construction of the pressure line will be carried out in the following sequence:

(1) first find the resultant R_1 of all the external loads applied to the left of the crown hinge;

(2) next find the resultant R_2 of all the external loads applied to the right of the same hinge;

(3) then determine the reactions A and B induced by the resultants R_1 and R_2 just determined;

(4) finally construct the force polygon and the line of pressure taking into consideration all the separate vertical loads P.

There is always only one polygon or line of pressure corresponding to any set of loads applied to a three-hinged arch. *When these loads are distributed the line of pressure becomes a smooth curve.*

If the centre line of the arch were to coincide with the pressure line pertaining to any particular set of loads, these loads will induce neither bending moments nor shearing forces in the cross section of the arch which will then be subjected to normal forces alone. This provides substantial advantages especially for masonry or concrete arches. Hereunder we shall designate by the term *rational* such a configuration of the centre line of an arch which will coincide with the line of pressure corresponding to the dead load.

It should be noted that the line of pressure can also be obtained analytically. For this purpose it would be necessary to find the magnitude of the bending moment M and the normal force N in a number of cross sections and then determine the eccentricity e using the formula $e=\frac{M}{N}$.

Having laid off these eccentricities along the normals to the centre line of the arch, the line of pressure will be obtained by simply connecting together the points obtained. The construction of a line of pressure for an arch whose reactions were determined in Fig. 9.3 is illustrated in Fig. 18.3. Fig. 19.3 represents the determination of internal forces acting over section k of this arch.

When vertical loads alone are applied to the arch, the horizontal component of any resultant of forces to the right or to the left of a section will always equal the thrust H. *

Therefore, if the resultant of all the forces acting to the left of any cross section were resolved into its vertical and horizontal components S and H (Fig. 20.3) the bending moment in this section would be equal to the thrust H multiplied by the distance measured vertically from the centroid of this section to the line of pressure (the vertical component inducing no moment in this section). *Accordingly, when vertical loads alone are applied, the vertical distances from the centre line to the line of pressure represent at a certain scale the bending moments acting over the corresponding sections of the arch.* In other words, these distances constitute a diagram of the bending moments with the sole difference that in this case the diagram will be situated on the side of the compressed fibres. Fig. 21.3 represents such a diagram pertaining to the arch shown in Fig. 19.3*a*.

Problem 3. It is required to construct graphically the pressure line of the arch analyzed in Problem 2 (see Fig. 16.3) and to determine the stresses in section k indicated in Fig. 22.3*a*.

Solution. Let us replace the uniformly distributed load applied to the left half of the arch by 6 concentrated forces amounting to 2 tons each and acting at the centres of 6 equal portions each 1 metre long. After that let us construct the force polygon using the values of the reactions computed in Problem 2 and the loads actually applied and let us trace the rays *I* through *VIII* as in Fig. 22.3*b*.

Drawing (as in Fig. 22.3*a*) a series of strings parallel to these rays we shall obtain a polygon of pressure. The area between the centre line of the arch and the line of pressure shaded vertically in the figure just mentioned represents the diagram of bending moments M. In many respects it is analogous to the diagram obtained analytically in Problem 2 and represented in Fig. 16.3*h*, but differs by the fact that in the latter case the distances had to be measured normally to the centre line of the arch and not vertically (this was reflected by a hatching normal to the centre line). Moreover, the diagram is located next to the compressed fibres instead of the extended ones as was the case in Fig. 16.3*h*. Together with the scale of lengths and forces an additional scale to which the bending moment ordinates should be measured in the graph is indicated in Fig. 22.3. This latter scale is obtained by multiplying the scale of length by the magnitude of the thrust equal in this case to 6 tons.

◆

* Each ray of the polygon of forces (Fig. 19.3*b*) has the same horizontal component equal to this thrust.

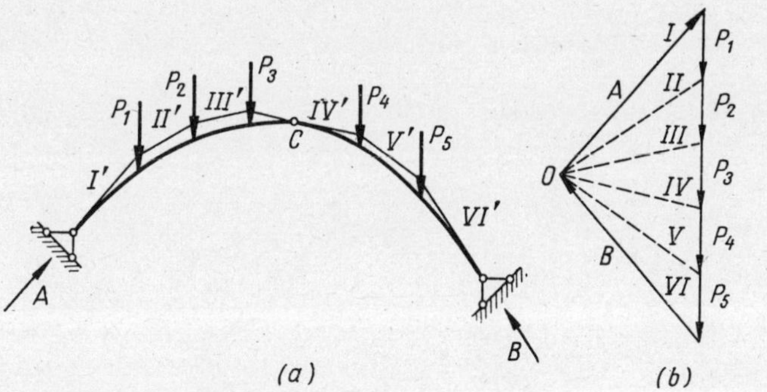

Fig. 18.3

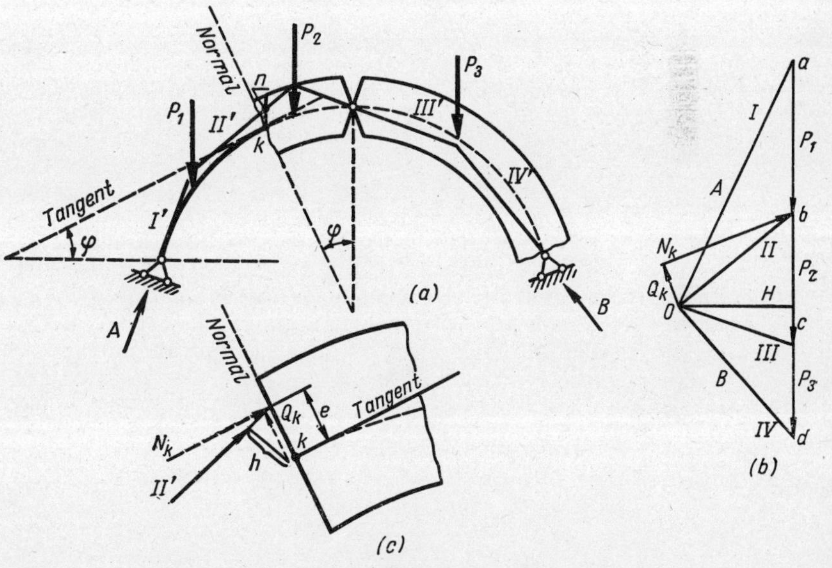

Fig. 19.3

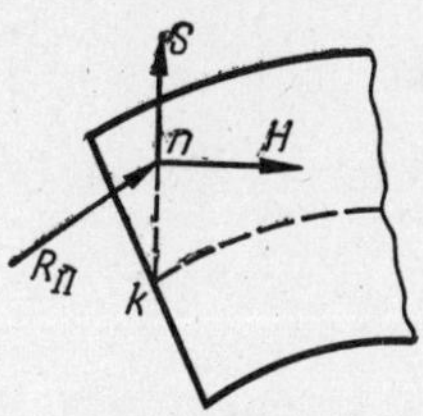

Fig. 20.3

At point k the bending moment will be obtained by measuring the corresponding ordinate in the graph which furnishes a value of three ton metres.

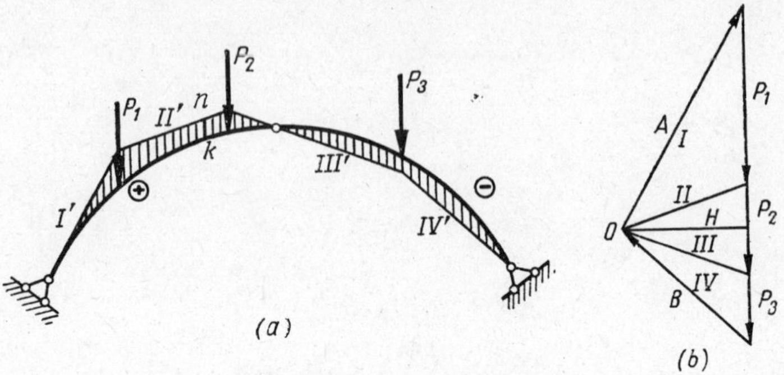

Fig. 21.3

The shear in this section will be nil as the tangent to the centre line of the arch is parallel to the polygon of pressure, while the normal force N_k is equal to the ray *IV* (Fig. 22.3*b*), i. e., to 7.2 tons.

4.3 *Maximum Economy Arches*

As stated above, we shall consider that a three-hinged arch provides for maximum economy if its centre line coincides with the line of pressure of all the dead loads acting on this arch. In that case these loads will produce no bending in the structure. Let y and η be the ordinates of the arch centre line and of the line of pressure, respectively. These ordinates are a certain function of x

$$y = f(x) \quad \text{and} \quad \eta = \varphi(x)$$

If the centre line of the arch were to provide for maximum economy as defined above, we should have

$$y = \eta$$

Let us examine the case of an arch subjected to vertical loads only (Fig. 23.3). The equilibrium equation of the moments of all forces lying to the left of any point k on the line of pressure will be

$$M_k = V_A x - H\eta - \Sigma P(x - a) = 0$$

whence

$$\eta = \frac{V_A x - \Sigma P(x - a)}{H}$$

It will be noted that the numerator in the last expression is equal to the bending moment in the corresponding section of the refe-

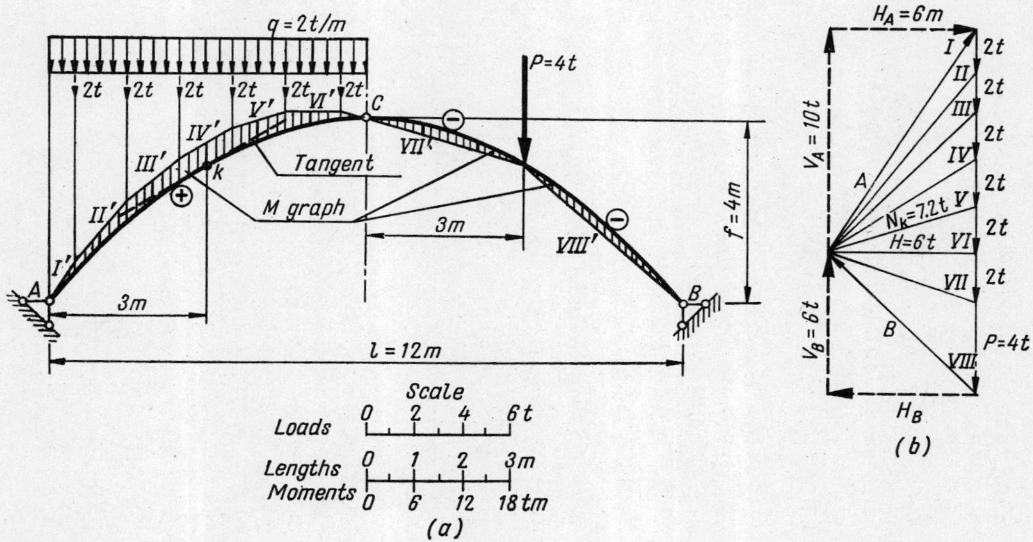

q = 2t/m
2t
2t
2t
2t
2t
2t
P=4t
I'
II'
III'
IV'
V'
VI'
VII'
VIII'
A
B
C
k
Tangent
M graph
3m
3m
f = 4m
l = 12m
Scale
Loads
Lengths
Moments
0 2 4 6t
0 1 2 3m
0 6 12 18tm
(a)
$H_A = 6m$
$V_A = 10t$
$V_B = 6t$
H_B
$N_k = 7.2t$
H=6t
P=4t
(b)

Fig. 22.3

rence beam, i. e., to M_x^0, and therefore

$$\eta=\frac{M_x^0}{H}$$

Substituting this expression in the equation $y = \eta$ we obtain the following expression for the centre line of an arch of maximum economy

$$y=\frac{M_x^0}{H}$$

Consequently, in the case of vertical loads maximum economy will be achieved if the arch centre line follows the bending moment diagram of a simply supported beam.

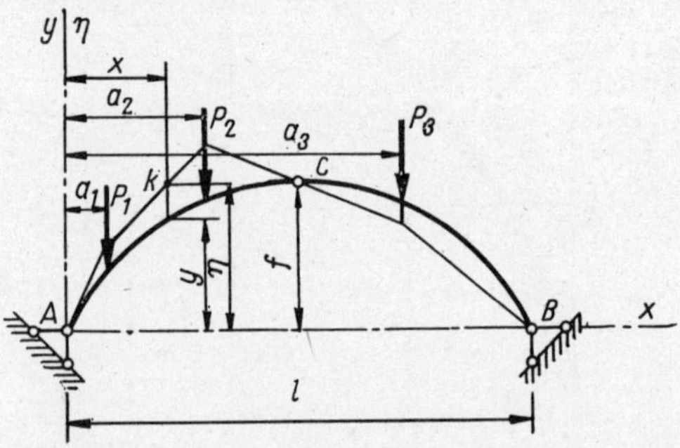

Fig. 23.3

Problem. Assume that a three-hinged arch carries a vertical load of intensity q uniformly distributed over the whole of its length, the span of the arch being l, its rise f and the central hinge being situated at the crown. It is required to determine the configuration for the centre line of such an arch, which would provide for maximum economy.

Solution. To solve this problem we shall use the expression

$$y=\frac{M_x^0}{H}$$

In the present case

$$M_x^0=\frac{ql}{2}x-qx\frac{x}{2}=\frac{qx}{2}(l-x)$$

$$H=\frac{M_C^0}{f}=\frac{q\frac{l}{2}}{2}\left(l-\frac{l}{2}\right)\frac{1}{f}=\frac{ql^2}{8f}$$

whence

$$y=\frac{\frac{qx}{2}(l-x)\,8f}{ql^2}=\frac{4f}{l^2}(l-x)\,x$$

i. e., the centre line of the arch must follow a conic parabola.

5.3. *Design of Three-Hinged Arches Subjected to Moving Loads*

1. INFLUENCE LINES FOR ABUTMENT REACTIONS

Let us assume that a three-hinged arch carries a unit load P applied a distance x from the left-hand abutment (Fig. 24.3a),

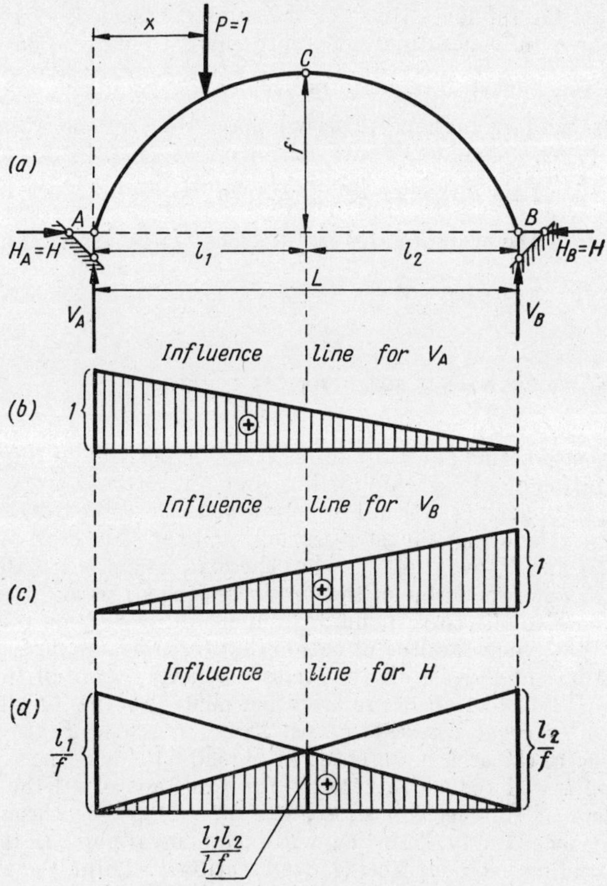

Fig. 24.3

and let us write the equilibrium equation of the moments of all the forces about the support pins

$$\Sigma M_B = V_A l - 1(l-x) = 0; \quad \Sigma M_A = -V_B l + 1x = 0$$

Solving these equations for V_A and V_B we obtain

$$V_A = \frac{l-x}{l}; \quad V_B = \frac{x}{l}$$

It will be observed that the expressions for V_A and V_B are the same as those for the reactions of a simple beam obtained in Art. 2.2. This means that the influence lines for V_A and V_B do not differ from the influence lines for the support reactions of a simple beam; these influence lines are represented in Fig. 24.3*b* and *c*.

Since the thrust H is determined by the equation $H = \frac{M_C^0}{f}$, the corresponding influence line will have the same shape as that for the beam moment M_C^0 differing from it only by a constant factor $\frac{1}{f}$. This influence line is shown in Fig. 24.3*d*. In case $l_1 = l_2 = \frac{l}{2}$ the ordinate of this influence line at a section passing through the crown equals $\frac{1}{4f}$.

2. INFLUENCE LINES FOR INTERNAL FORCES

As a preliminary step, we shall examine the methods of determining the so-called neutral points, i. e., the position of the points of application of a load which will render the internal force (bending moment, shear or normal force) nil at the section k under consideration. Denoting the stresses acting over this cross section by M_k, Q_k and N_k we shall say that the load is applied at the neutral point when the value of the corresponding stress and therefore the ordinate to the corresponding influence line become nil. It is obvious that when the line of action of a force passes through one of the abutment hinges, all the stresses at any section of the arch will be nil. In addition, there are other neutral points on the arch which are of great interest for us. Thus, if a load P is applied at point F_m of the arch represented in Fig. 25.3 the bending moment in section k will reduce to zero for the resultant of all the forces to the left of this section (i. e., reaction A) passes through its centroid. Accordingly, point F_m will be a neutral point in relation to the bending moment acting over section k. Point F_m will lie on the vertical passing through the intersection point F of lines Ak and BC.

If we consider the arch shown in Fig. 26.3, the bending moment in section k would reduce to zero only if the load P were applied at point F_m to a special bracket fixed to the arch between section k and the crown hinge, for in this case the direction of reaction A

would again pass through section k. However, if no such bracket existed, there would be no real neutral point in relation to the bending moment acting over section k. In effect if the point of application of the load were transferred upwards, so that the load would act directly on the right-hand portion of the arch. the

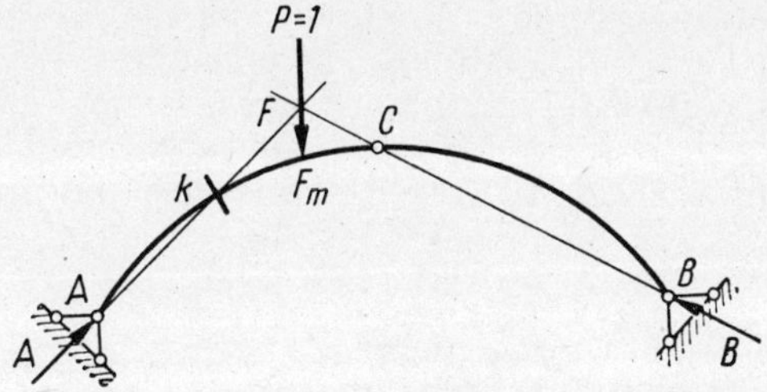

Fig. 25.3

direction of reaction A would alter, this reaction passing through the hinges A and C, and therefore the bending moment in section k would no longer equal zero.

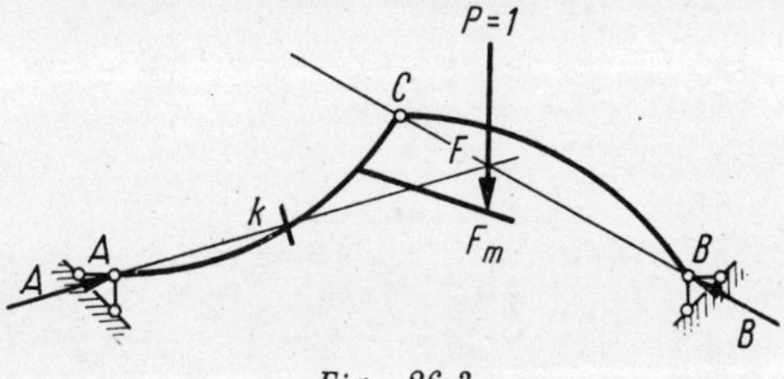

Fig. 26.3

Denoting by u_m the abscissa of the neutral point pertaining to the bending moment in section A (see Fig. 27.3) and using the similitude of the triangles AFF_1 and BFF_1 we obtain

$$\overline{FF}_1 = u_m \tan\alpha = u_m \frac{y_k}{x_k}$$

$$\overline{FF}_1 = (l - u_m)\tan\beta = \frac{(l-u_m)f}{l_2}$$

Therefore

$$\frac{u_m y_k}{x_k} = \frac{(l-u_m)f}{l_2}$$

whence

$$u_m = \frac{lfx_k}{y_k l_2 + x_k f} \tag{4.3}$$

This expression permits the analytical determination of the abscissa u_m of the neutral point or, in other words, of the point where the influence line for M_k will pass through zero.

In order to determine the neutral point corresponding to Q_k (Fig. 28.3) a line AF must be drawn through the centre of the hinge at the support A (this line being parallel to the tangent S-S at point k to the centre line of the arch) and the intersection of this line at point F with line BC must be found. If a load P were now applied at a point F_q lying on the same vertical as point F, the shearing force Q_k in section k would reduce to zero, for there would be only one force acting to the left of section k and this force would

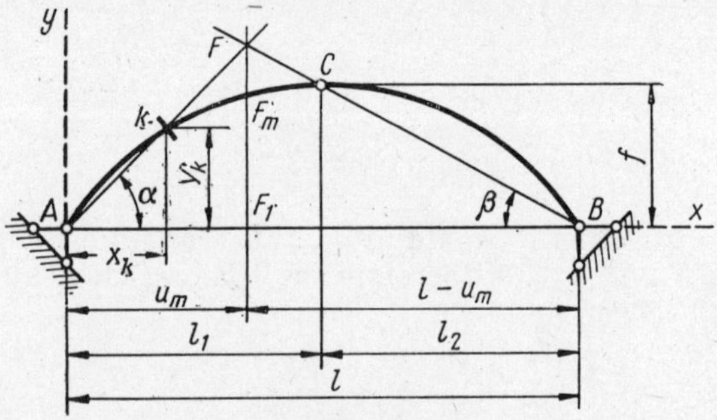

Fig. 27.3

be parallel to the tangent through this section. At cross section k_1 of the same arch (Fig. 29.3) the shear would reduce to zero only if the load P were applied to a bracket fixed to the arch between this section and the crown hinge, for the point of application of this load falls on the vertical passing through the intersection of the lines AF and BC. From Fig. 30.3 it is clearly seen that

$$\overline{FF}_1 = u_q \tan \varphi_k$$

$$\overline{FF}_1 = (l - u_q) \tan \beta$$

Therefore

$$u_q \tan \varphi_k = (l - u_q) \tan \beta$$

whence

$$u_q = \frac{l \tan \beta}{\tan \beta + \tan \varphi_k} \tag{5.3}$$

This expression permits the computation of the position of the neutral point for the shearing force in section k.

The normal force N in section k will become nil when load P is applied at point F_n (Fig. 31.3) lying on the same vertical as point F, this point being determined by the intersection of line

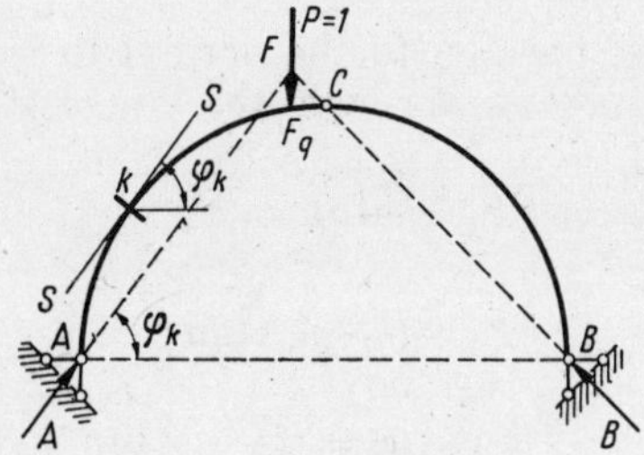

Fig. 28.3

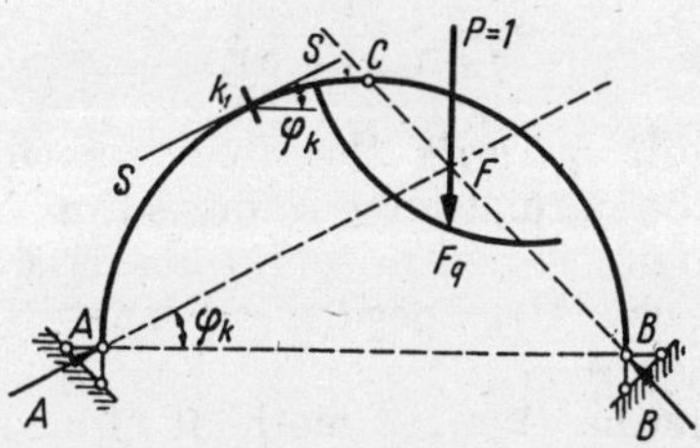

Fig. 29.3

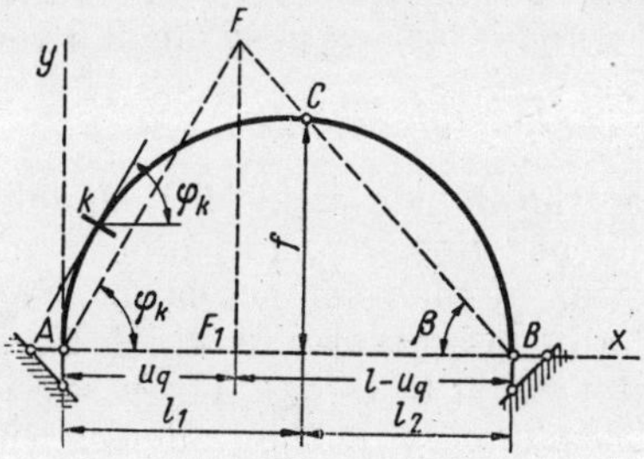

Fig. 30.3

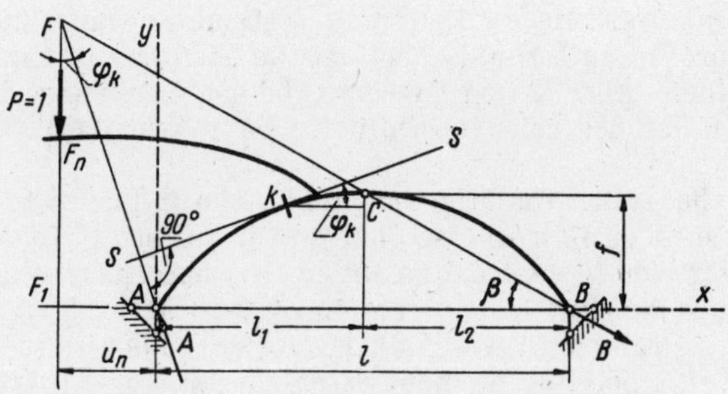

Fig. 31.3

BC with a line AF parallel to the normal to the arch centre line at section k and passing through the hinge A. From Fig. 31.3 we note that

$$\overline{FF_1} = -u_n \cot \varphi_k$$

and

$$\overline{FF_1} = (l - u_n) \tan \beta$$

whence

$$-u_n \cot \varphi_k = (l - u_n) \tan \beta$$

and accordingly

$$u_n = \frac{l \tan \beta}{\tan \beta - \cot \varphi_k} \tag{6.3}$$

This last formula permits the determination of the neutral point related to the normal force N in section k.

When the value of the neutral point abscissa obtained from formulas (4.3), (5.3) or (6.3) is negative it means that this point lies to the left of hinge A.

Let us now examine different methods of constructing influence lines for M_k, Q_k and N_k. As will be seen from expression (3.3), the bending moment acting over section k of the arch represented in Fig. 32.3*a* will amount for any position of a vertical unit load to

$$M_k = M_k^0 - Hy_k$$

This means that the influence line for M_k may be obtained by summing the influence line for the bending moment M_k^0 at the corresponding section of the reference beam (Fig. 32.3*b*) and that for the thrust H, all the ordinates of which have been multiplied by a constant factor equal to $(-y_k)$. These two influence lines are shown in Fig. 32.3*c* and *d* while the influence line for the bending moment in the arch obtained by their summation is represented in Fig. 32.3*e*. It is clear that the point of intersection d of lines a_1b and ab_1 must lie on the same vertical as the neutral point F_m, this providing a rapid check on the accuracy of the influence line obtained. Fig. 32.3*f* represents the same influence line, with the only difference that its ordinates have been laid off directly from the x-axis.

It may be shown that the area under the influence line for M_k will reduce to zero for any section k of a uniformly loaded three-hinged arch whose centre line follows a conic parabola. Indeed, the bending moment in any section of such an arch will amount to zero (see Problem in Art. 4.3). If we were to determine the magnitude of this bending moment using the influence line we would use the equality $M_k = q\Omega$, but as M_k is always zero, the area under the influence line Ω must also reduce to zero.

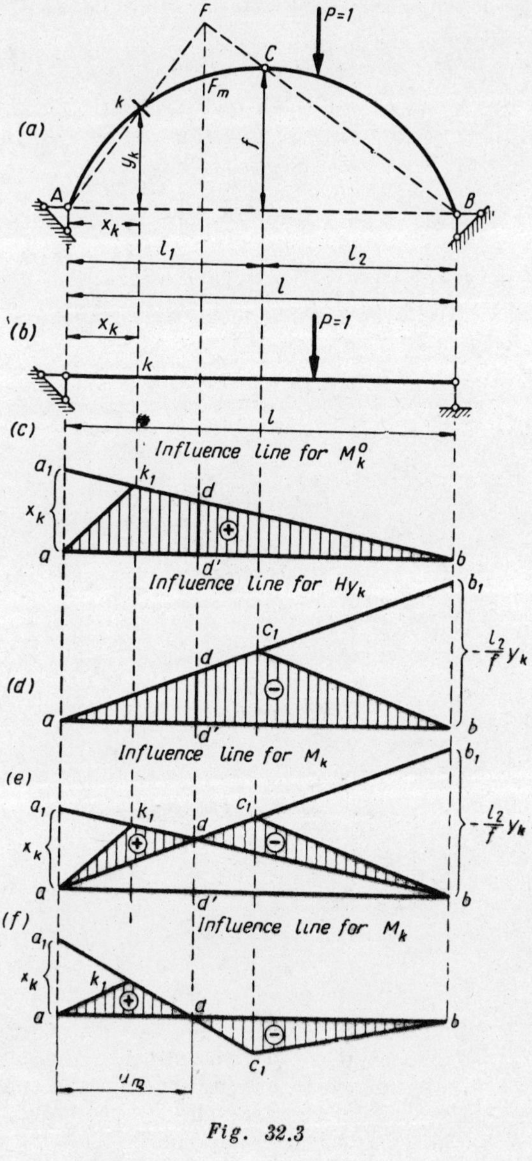

F
P=1
C
F_m
k
(a)
y_k
f
A
B
x_k
l_1
l_2
l
(b)
x_k
P=1
k
(c)
l
Influence line for M_k^o
a_1
k_1
d
x_k
a
b
d'
Influence line for Hy_k
b_1
c_1
d
$\frac{l_2}{f} y_k$
(d)
a
b
d'
Influence line for M_k
b_1
(e)
a_1
k_1
c_1
d
x_k
$\frac{l_2}{f} y_k$
a
b
d'
(f)
Influence line for M_k
a_1
x_k
k_1
a
d
b
c_1
u_m

Fig. 32.3

For the construction of the influence line for the shear Q_k (Fig. 33.3*a*) we may use the first formula of the set of expressions (3.3), viz.,

$$Q_k = Q_k^0 \cos\varphi_k - H \sin\varphi_k$$

where Q_k^0 is the shear in the corresponding section of an end-supported beam of the same span l (Fig. 33.3*b*).

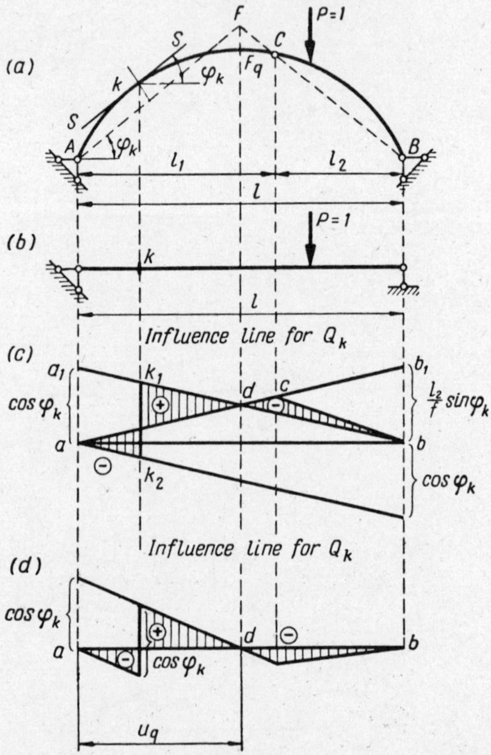

Fig. 33.3

This expression shows that the influence line Q_k may also be obtained by the summation of two influence lines, the first for Q_k^0 all the ordinates of which are multiplied by a constant factor $\cos\varphi_k$ and the second for the thrust H the ordinates of which are multiplied by $(-\sin\varphi_k)$. The influence line for Q_k obtained in this way is represented in Fig. 33.3*c* where abk_1k_2a is the influence line for $Q_k^0 \cos\varphi_k$ and the triangle acb is the influence line for

$H \sin \varphi_k$. Point d in Fig. 33.3c must fall on the same vertical as the neutral point F_q.

The same influence line is shown in Fig. 33.3d with the only difference that its ordinates have been laid off directly from the x-axis.

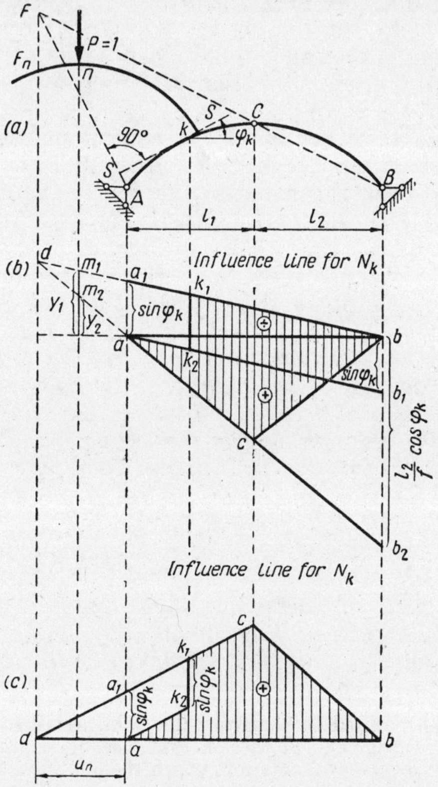

Fig. 34.3

In order to construct the influence line for the normal force N_k for cross section k of the arch we shall use the last formula of expressions (3.3)

$$N_k = Q_k^0 \sin \varphi_k + H \cos \varphi_k$$

Summing up graphically the two components ($Q_k^0 \sin \varphi_k$ and $H \cos \varphi_k$) we obtain the influence line for N_k represented in Fig. 34.3b. Here abk_1k_2a is the influence line for $Q_k^0 \sin \varphi_k$ and the triangle abc represents the influence line for $H \cos \varphi_k$. Lines a_1b and ab_2 must intersect at d in the vertical passing through the neutral point F_n.

The positive ordinates to line da_1 represent the values of $Q_k^0 \sin \varphi_k$ when the unit load P is applied to a bracket shown in Fig. 34.3*a* while the negative ordinates to line da (plotted also above ab) represent the values of $H \cos \varphi_k$ for the same position of the load. In this case the normal force N acting over section k will be obtained by subtracting the ordinate y_2 from y_1 and will be equal to m_1m_2. A corresponding influence line with its ordinates laid off directly from the axis of abscissas is presented in Fig. 34.3*c*.

Preliminary determination of the neutral points would allow direct construction of the influence lines M_k, Q_k and N_k. This method has received the name of the *neutral point method*. The procedure to be followed may be easily derived from the examination of the influence lines shown in Figs. 32.3*f*, 33.3*d* and 34.3*c*. For instance, if it were required to use this method for the construction of the influence line for M_k proceed as follows (see Fig. 32.3*f*).

(1) *Lay off along the vertical passing through the left-hand support (provided the section under consideration is in the left half of the arch) the abscissa of section k, i. e., the distance x_k.*

(2) *Mark the neutral point F_m on the x-axis.*

(3) *Connect the ordinate x_k over the left-hand support (point a_1) with the projection of neutral point (point d) on the x-axis (line a_1d).*

(4) *Find the point of intersection of this line with the vertical passing through section k (point k_1).*

(5) *Connect k_1 with the point of zero ordinate at the left-hand support (point a).*

(6) *Find the point of intersection of the line a_1d with the vertical passing through the crown hinge (point c_1).*

(7) *Connect point c_1 with the point of zero ordinate over the right-hand support (line c_1b).*

Problem. It is required to construct the influence lines for M_k, Q_k and N_k, acting over section k of a three-hinged parabolic arch dealt with in Problem 2 of Art. 3.3, and to determine with the aid of these influence lines the stresses induced in this section by the system of loads indicated in Fig. 35.3*a*. The parameters of point k are

$$x_k = y_k = 3 \text{ metres}; \ \tan \varphi_k = \frac{2}{3}; \ \sin \varphi_k = 0.555; \ \cos \varphi_k = 0.832$$

Solution. Determine graphically the position of neutral points F_m, F_q and F_n as well as their abscissas u_m, u_q and u_n and check the values of these abscissas using formulas (4.3), (5.3) and (6.3).

$$u_m = \frac{12 \times 4 \times 3}{3 \times 6 + 3 \times 4} = 4.8 \text{ metres}$$

$$u_q = \frac{12 \times 4/6}{4/6 + 2/3} = 6.0 \text{ metres}$$

$$u_n = \frac{12 \times 4/6}{4/6 - 3/2} = -9.6 \text{ metres}$$

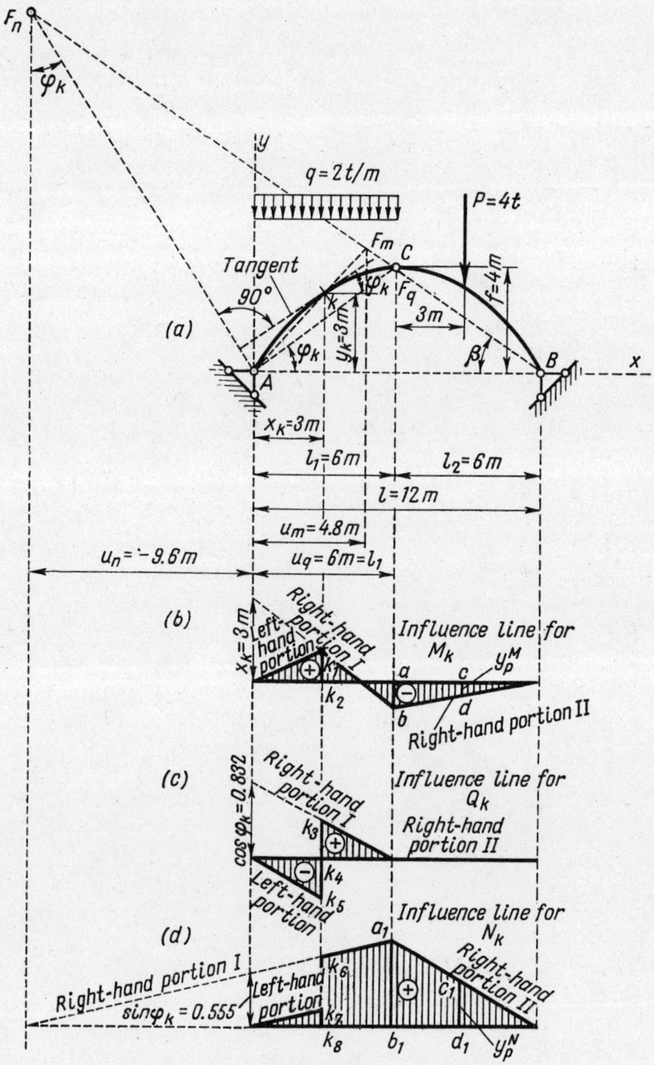

F_n
φ_k
y
q=2t/m
P=4t
F_m
C
Tangent
90°
k
φ_k
F_q
y_k=3m
f=4m
3m
(a)
φ_k
β
B
x
A
x_k=3m
l_1=6m
l_2=6m
l=12m
u_m=4.8m
u_n=-9.6m
u_q=6m=l_1
(b)
x_k=3m
Left-hand portion
Right-hand portion I
Influence line for M_k
k_1
k_2
a
c
y_p^M
b
d
Right-hand portion II
(c)
cos φ_k=0.832
Right-hand portion I
Influence line for Q_k
Right-hand portion II
k_3
k_4
k_5
Left-hand portion
(d)
Influence line for N_k
a_1
Right-hand portion I
k_6
Right-hand portion II
sin φ_k = 0.555
Left-hand portion
k_7
c_1
k_8
b_1
d_1
y_p^N

Fig. 35.3

The construction of the influence lines required may now be carried out as follows: scale off on the vertical passing through the left-hand support the lengths x_k, $\cos\varphi_k$ and $\sin\varphi_k$ as indicated in Fig. 35.3*b*, *c* and *d* and connect the ordinates obtained with the projection of the neutral point on the x-axis. After that find the intersection of the vertical passing through the crown hinge C with the lines just obtained. Connect this point of intersection with the point of zero ordinate at the right-hand support. Find also the point of intersection of the above line with the vertical passing through section k. On the bending moment influence line this point is connected directly with the zero point at the left-hand support, while in those for the shear Q_k and for the normal force N_k parallels to the line determined in the first place should be traced through the left-hand support until their intersection with the vertical mentioned above. Applying the laws of similitude to the triangles involved, we may now determine the ordinates to the pertinent points of the influence lines, the areas under these lines and the internal forces induced in section k of the arch by the given system of loads.

(a) *Influence line for* M_k

$$\frac{\overline{k_1k_2}}{x_k}=\frac{u_m-x_k}{u_m}$$

whence

$$\overline{k_1k_2}=\frac{x_k(u_m-x_k)}{u_m}=\frac{3(4.8-3)}{4.8}=1.125 \text{ metres}$$

$$\frac{\overline{ab}}{l_1-u_m}=\frac{\overline{k_1k_2}}{u_m-x_k}$$

whence

$$\overline{ab}=\overline{k_1k_2}\,\frac{l_1-u_m}{u_m-x_k}=1.125\times\frac{6-4.8}{4.8-3}=-0.75 \text{ metre}$$

$$\overline{cd}=y_p^M=\frac{\overline{ab}}{2}=-0.375 \text{ metre}$$

The area under the influence line corresponding to the distributed load equals

$$\omega_q^M=\frac{1.125\times 4.8}{2}-\frac{0.75(6-4.8)}{2}=2.25 \text{ square metres}$$

Accordingly, the bending moment M_k will amount to

$$M_k=q\omega_q^M+Py_p^M=2\times 2.25-4\times 0.375=3.0 \text{ ton metres}$$

(b) *Influence line for* Q_k

$$\frac{\overline{k_3k_4}}{l_1-x_k}=\frac{\cos\varphi_k}{l_1} \text{ whence } \overline{k_3k_4}=\frac{l_1-x_k}{l_1}\cos\varphi_k=$$

$$=\frac{6-3}{6}\times 0.832=0.416$$

$$\overline{k_4k_5}=\cos\varphi_k-\overline{k_3k_4}=0.832-0.416=0.416$$

$$y_p^Q=0;\quad \omega_q^Q=-\frac{0.416\times 3}{2}+\frac{0.416\times 3}{2}=0$$

$$Q_k=q\omega_q^Q+Py_p^Q=2\times 0+4\times 0=0$$

(c) *Influence line for N_k*

$$\frac{\overline{k_6k_8}}{x_k-u_n}=\frac{\sin\varphi_k}{-u_n} \quad \text{whence} \quad \overline{k_6k_8}=\frac{x_k-u_n}{-u_n}\sin\varphi_k=$$

$$=\frac{3+9.6}{9.6}\times 0.555=0.728$$

$$\overline{k_7k_8}=\overline{k_6k_8}-\sin\varphi_k=0.728-0.555=0.173$$

$$\frac{\overline{a_1b_1}}{l_1-u_n}=\frac{\sin\varphi_k}{-u_n} \quad \text{whence} \quad \overline{a_1b_1}=\frac{l_1-u_n}{-u_n}\sin\varphi_k=$$

$$=\frac{6+9.6}{9.6}\times 0.555=0.902$$

$$\overline{c_1d_1}=y_p^N=\frac{\overline{a_1b_1}}{2}=\frac{0.902}{2}=0.451$$

$$\omega_q^N=\frac{0.173\times 3}{2}+\frac{0.728+0.902}{2}\times 3=2.705$$

$$N_k=q\omega_q^N+Py_p^N=2\times 2.705+4\times 0.451=7.21 \text{ tons}$$

The magnitudes of M_k, Q_k and N_k just found coincide with those computed in Problem 2 of Art. 3.3 (Table 1.3 and Fig. 16.3).

6.3. *Core Moments and Normal Stresses in Three-Hinged Arches*

In any eccentrically loaded bar the normal unit stresses reach their maximum and their minimum in the outer fibres of the cross sections and, provided the material follows Hooke's law, their magnitudes may be obtained from the equation

$$\sigma=\frac{N}{F}\pm\frac{M}{W}$$

where F=area of the cross section
W=its resisting moment
N and M=normal force and bending moment acting over the section, respectively.

It is assumed that both N and M act in a plane passing through one of the principal axes of inertia of the section and normal to it.

When a moving load is applied to the arch, the use of the above formula would require that both the influence lines for N and M should be used simultaneously, these influence lines having an entirely different configuration and one of them possessing both positive and negative portions. It is therefore expedient to transform the above-mentioned formula so that it should consist of one term only.

This may be obtained by the following procedure. Let us first find the components N and Q of the resultant R of all forces acting to the left of the section involved and passing through a point s thereof (Fig. 36.3). Let us then apply at the extreme upper point of the core of this section (say, point k_1) two normal forces N equal in size and opposite in direction which will balance each other. As a result we shall have three forces N acting over this section which may be replaced by a moment equal to $N\ (e + c_1)$ and by a normal force N acting at the upper edge of the core. In this case the unit stress at the bottom fibre of the section will be given by the formula

$$\sigma_m = \frac{N\,(e + c_1)}{W_m}$$

Normal forces applied at the upper limit of the core produce no stresses in the lower fibres of the section. The product $N\ (e + c_1)$ represents the moment of the normal force applied at point s of the section about the upper point of the core k_1 and will be hereafter called the *core moment*. The core moment differs from the ordinary bending moment by the fact that its computation requires that the distance of the forces (to the left or to the right of the section) should be measured not to the centroid of the section but to the upper or the lower point of its core

The normal stress at point n may be determined in a similar way, only in this case the moment of external forces should be taken about the lower core point n and the appropriate resisting moment W_n should be used in lieu of W_m

$$\sigma_n = \frac{N\,(e - c_2)}{W_n}$$

Thus

$$\sigma_m = \frac{M^c_{k_1}}{W_m} \quad \text{and} \quad \sigma_n = \frac{M^c_{k_2}}{W_n}$$

where $M^c_{k_1}$ = moment of external forces (to the right or to the left of the section) about the upper core point k_1

$M^c_{k_2}$ = moment of the same forces about the lower core point k_2.

The above two formulas are monomial and therefore they lead to a quicker and simpler determination of the maximum normal unit stresses in the cross sections of an arch carrying a moving load. As for the influence lines of core moments they are constructed in exactly the same way as those for the bending moments.

The influence lines for the core moments and for the bending moment in section k of a three-hinged arch are represented in

Fig. 37.3, these influence lines having been constructed using the neutral point method. The small triangles shaded black on the influence lines for the core moments just next to the centroid are due to a vertical rise in the influence line for the normal force at this point (see Fig. 34.3). In practice these areas are usually ignored due to their insignificance.

Using the core moment influence lines, let us now solve the following problem:

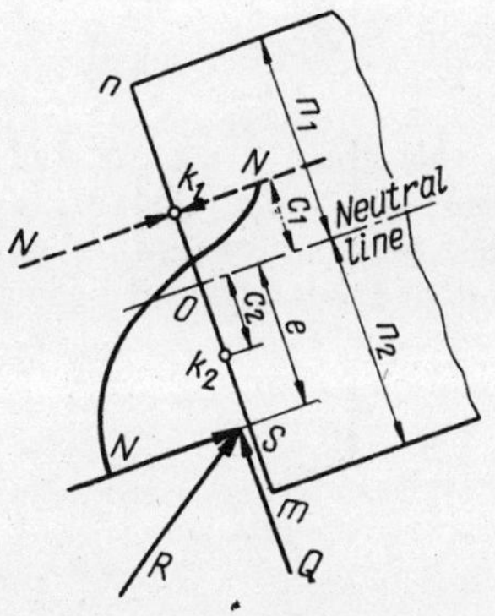

Fig. 36.3

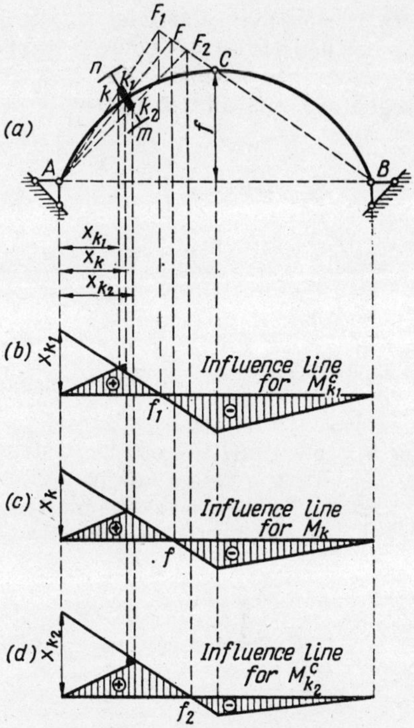

Fig. 37.3

which part of the three-hinged arch represented in Fig. 37.3*a* should be loaded (uniformly or by a train of concentrated loads) in order to obtain the maximum tensile stresses at the extrados of section k. It is obvious that the extrados will be extended only when the resultant of all external forces (the right-hand or the left-hand ones) passes below the core (Fig. 38.3). In that case the moment of the resultant about the core point k_2 will be negative. Consequently, the load or loads should be placed over the negative portion of the influence line for the core moment $M^c_{k_2}$. The loading of the positive portion of this line would cause the compression of the extrados of the arch at section k.

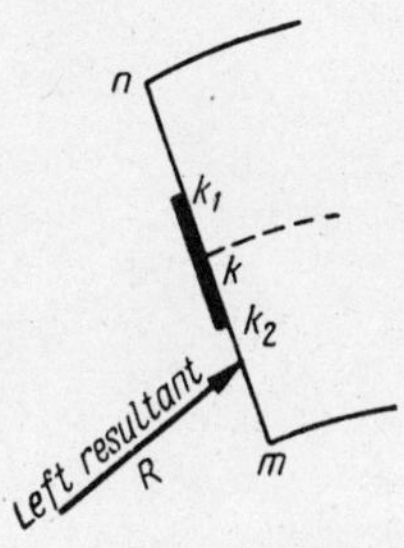

Fig. 38.3

7.3. *Analysis of Three-Hinged Tied Arches and Bents*

In the preceding articles (2.3 to 6.3) we have passed in review the methods of stress computation applicable to ordinary three-hinged arches without ties. Let us now envisage the tied three-hinged systems and in particular three-hinged arches and bents.

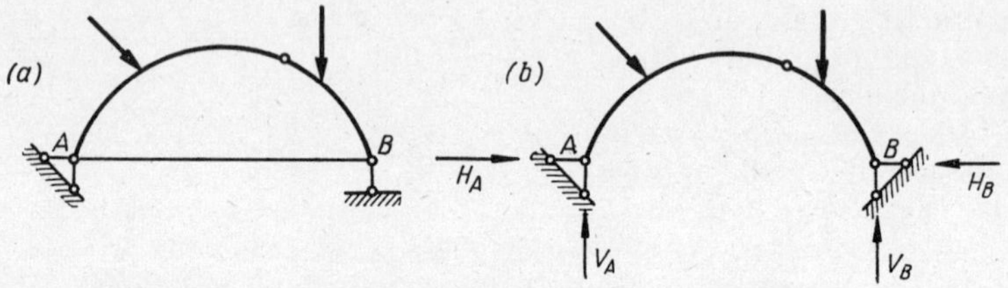

Fig. 39.3

Certain peculiarities of these structures introduce a number of changes in the stress computation methods described above.

Fig. 39.3*a* represents a bowstring arch freely supported at B, the tie precluding the horizontal movement of the abutment hinge

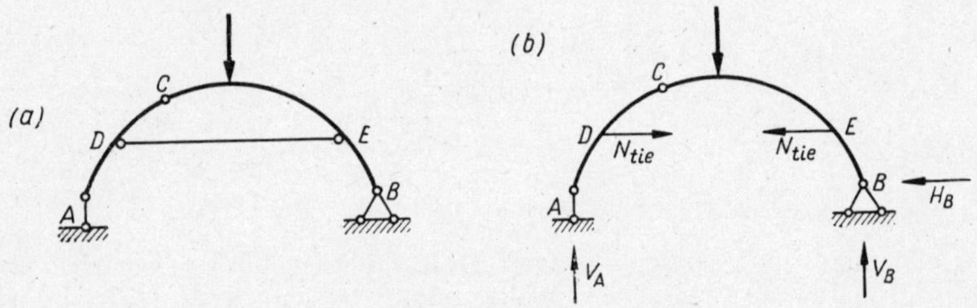

Fig. 40.3

and therefore replacing the horizontal constraint at this point. Accordingly, computation methods pertaining to ordinary three-hinged arches (Fig. 39.3*b*) will permit the determination of all the stresses in this particular case. Stresses in the cross sections of both arches will be exactly the same and the internal force in the tie will be equal to the horizontal thrust H_B. The vertical reactions V_A and V_B will also remain exactly the same. Thus, the influence lines for the abutment reactions and the stresses acting over the corresponding cross sections of a bowstring arch will not differ in any respect from those of an ordinary three-hinged arch.

Let us now consider a three-hinged arch with an elevated tie DE as shown in Fig. 40.3*a*. We may replace the tie by two horizontal forces N_t applied at points D and E and equal to the tension in the tie (Fig. 40.3*b*). The three abutment reactions V_A, V_B and H_B may then be determined as usual with the aid of three equilibrium equations of all external forces applied to the arch. These equations do not contain the above mentioned forces N_t which balance each other; their magnitude may be obtained by equating to zero the sum of all the moments of the external forces applied to the left (or to the right) half of the arch about the crown hinge C. The stresses in all the cross sections of the arch as well as the methods of constructing the corresponding influence lines may be derived from the expressions (1.3).

Problem 1. Given the arch with surelevated tie (Fig. 41.3*a*) following a conic parabola defined by the equation

$$y=\frac{4f}{l^2}(l-x)\,x$$

Required to determine the reactions V_A, V_B and H_B, the tension in the tie N_t, the internal forces M_k, Q_k and N_k and to construct the influence lines for all these forces and stresses.

Solution. The reactions are determined from the equilibrium equations of all the external forces acting on the arch

$$\Sigma M_B=V_Al-q\frac{l}{2}\,\frac{3}{4}\,l-Pb=0$$

$$\Sigma M_A=\frac{ql}{2}\cdot\frac{l}{4}+P\,(l-b)-V_Bl=0$$

$$\Sigma X=-H_B=0$$

These equations yield

$$V_A=\frac{3}{8}\,ql+\frac{Pb}{l}=\frac{3}{8}\times 2\times 12+\frac{4\times 3}{12}=10 \text{ tons}$$

$$V_B=\frac{ql}{8}+P\left(1-\frac{b}{l}\right)=\frac{2\times 12}{8}+4\left(1-\frac{3}{12}\right)=6 \text{ tons}$$

$$H_B=0$$

The tension in the tie is determined from

$$\sum_L M_C=V_A\cdot\frac{l}{2}-N_td-\frac{ql}{2}\cdot\frac{l}{4}=0$$

whence

$$N_t=\left(V_A\frac{l}{2}-\frac{ql^2}{8}\right)\frac{1}{d}=\frac{M_C^0}{d}=\frac{10\times\frac{12}{2}-\frac{2\times 12^2}{8}}{2}=12 \text{ tons}$$

where M_C^0 is the bending moment actin over section C of a simple beam shown in Fig. 41.3*b*.

The angle φ_k formed by the tangent to the centre line of the arch at point k and the x-axis and the ordinate y_k of point k are determined as follows

$$\tan\varphi=\frac{dy}{dx}=\frac{d}{dx}\left[\frac{4f}{l^2}(l-x)\,x\right]=\frac{4f}{l^2}(l-2x)$$

$$\text{for } x=x_k=3\text{ m}\quad \tan\varphi=\tan\varphi_k=\frac{2}{3}$$

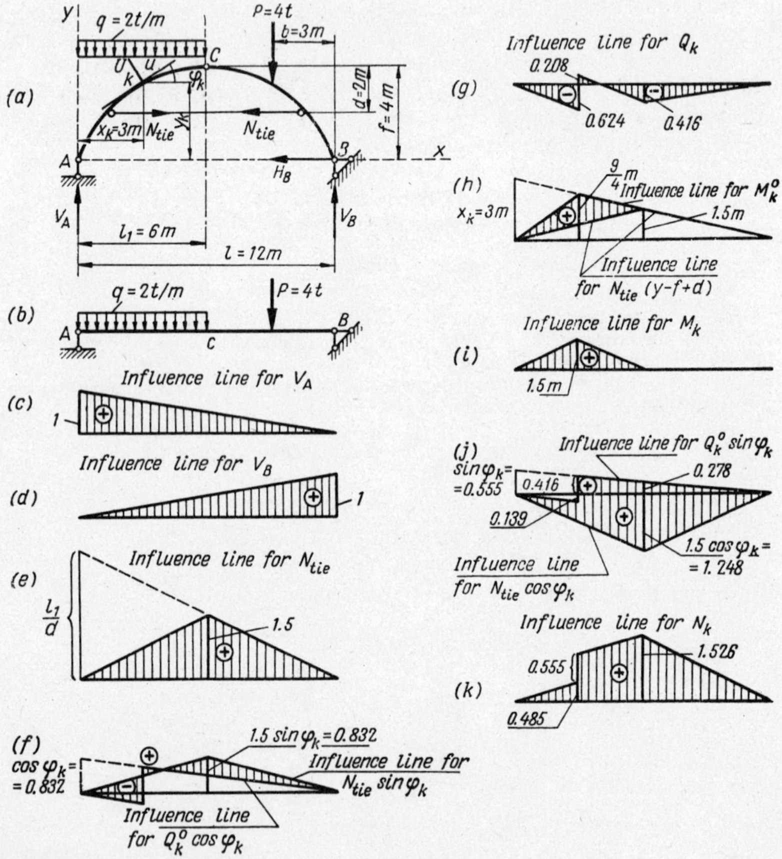

Fig. 41.3

wherefrom

$$\varphi_k=33°42';\ \sin\varphi_k=0.555\quad \cos\varphi_k=0.832$$

and

$$y_k=\frac{4f}{l^2}(l-x_k)\,x_k=\frac{4\times 4}{12^2}(12-3)\,3=3\text{ metres}$$

Substituting these values in expressions (1.3) we obtain the stresses acting over section k of the arch

$$Q_k=\sum_L V=V_A \cos \varphi_k-N_t \sin \varphi_k-qx_k \cos \varphi_k=-3.33 \text{ tons}$$

$$M_k=\sum_L M=V_A x_k-N_t (y_k-f+d)-\frac{qx^2_k}{2}=9.0 \text{ ton metres}$$

$$N_k=\sum_L U=V_A \sin \varphi_k+N_t \cos \varphi_k-qx_k \sin \varphi_k=12.20 \text{ tons}$$

The influence lines for the abutment reactions V_A and V_B shown in Fig. 41.3c and d will be the same as for an ordinary three-hinged arch. The influence line for the tie tension will be derived from the equation $N_t = \frac{M_C^0}{d}$ (see Fig. 41.3e).

The construction of influence lines for Q_k, M_k and N_k will be based on the following relations similar in every respect to those of (3.3)

$$Q_k=Q_k^0 \cos \varphi_k-N_t \sin \varphi_k$$

$$M_k=M_k^0-N_t (y_k-f+d)$$

$$N_k=Q_k^0 \sin \varphi_k+N_t \cos \varphi_k$$

These influence lines together with the intermediate graphical operations are represented in Fig. 41.3f, g, h, i, j and k.

As a check let us find the stresses in section k using the influence lines just obtained

$$Q_k=2\left(-\frac{3}{2}\times 0.624+\frac{3}{2}\times 0.208-\frac{3}{2}\times 0.416\right)-4\times\frac{1}{2}\times 0.416=$$

$$=-3.33 \text{ tons}$$

$$M_k=2\times\frac{1}{2}\times 1.5\times 6=9.0 \text{ ton metres}$$

$$N_k=2\left[\frac{3}{2}\times 0.485+\frac{3}{2}(0.485+0.555+1.526)\right]+4\times\frac{1}{2}\times$$

$$\times 1.526=12.20 \text{ tons}$$

These values coincide exactly with those found previously. All the computations may be safely regarded as correct.

Let us consider now the three-hinged bents. Their abutment reactions will be determined in exactly the same way as for the three-hinged arches, the same applying to the determination of the internal forces and to the construction of influence lines (whether by graphical or analytical methods). Exception must be made for vertical members (provided these are present) for the neutral point method cannot be applied to the construction of the bending moment and shear influence lines for the latter. However, this method remains quite suitable for horizontal and inclined members of the bent.

Problem 2. Determine the abutment reactions and the stresses in cross sections m and n passed through the uprights of the bent in Fig. 42.3*a* and draw the corresponding influence lines for section m.

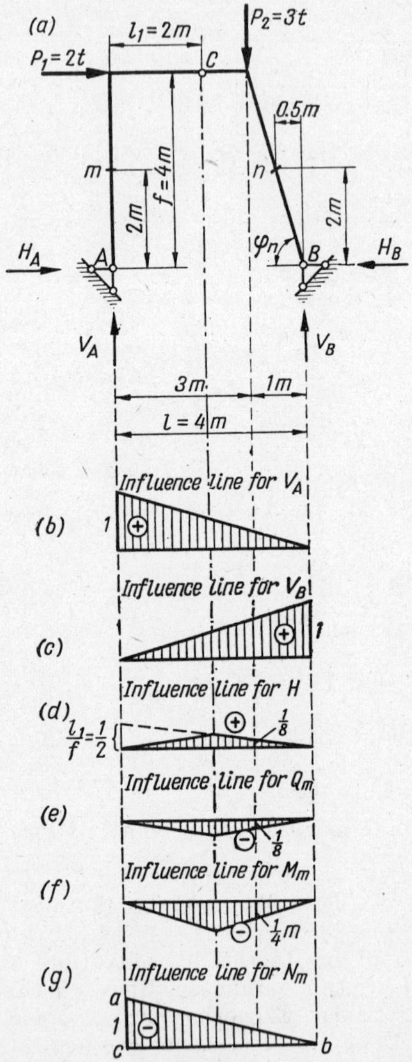

Fig. 42.3

Solution. The abutment reactions will be obtained using the equilibrium equations

$$\Sigma M_A = 4P_1 + 3P_2 - 4V_B = 0$$
$$\Sigma M_B = 4P_1 - P_2 + 4V_A = 0$$
$$\Sigma M_{C_L} = 2V_A - 4H_A = 0$$
$$\Sigma X = P_1 + H_A - H_B = 0$$

from which

$$V_B = \frac{1}{4}(4P_1 + 3P_2) = \frac{1}{4}(2\times 4 + 3\times 3) = 4.25 \text{ tons}$$
$$V_A = \frac{1}{4}(-4P_1 + P_2) = \frac{1}{4}(-2\times 4 + 3) = -1.25 \text{ tons}$$
$$H_A = \frac{2}{4}V_A = -\frac{2}{4}\times 1.25 = -0.625 \text{ ton}$$
$$H_B = P_1 + H_A = 2.0 - 0.625 = 1.375 \text{ tons}$$

Regarding the lower extremity of the uprights as their left-hand one, we may now find the stresses acting over sections m and n:

$$Q_m = -H_A = 0.625 \text{ ton}$$
$$M_m = -2H_A = 2\times 0.625 = 1.25 \text{ ton metres}$$
$$N_m = V_A = -1.25 \text{ tons}$$

and

$$Q_n = -V_B \cos\varphi_n + H_B \sin\varphi_n = -4.25\times 0.242 + 1.375\times 0.968 = 0.304 \text{ ton}$$
$$M_n = 0.5V_B - 2H_B = 0.5\times 4.25 - 2.0\times 1.375 = -0.625 \text{ ton metre}$$
$$N_n = V_B \sin\varphi_n + H_B \cos\varphi_n = 4.25\cdot 0.968 + 1.375\times 0.242 = 4.45 \text{ tons}$$

Fig. 42.3*b*, *c* and *d* represents the influence lines for the abutment reactions V_A, V_B and $H_A = H_B = H$ which do not differ in any respect from those of an ordinary arch (see Fig. 24.3).

Fig. 42.3*e*, *f* and *g* contains the influence lines for Q_m, M_m and N_m derived from the expressions $Q_m = -H_A$, $M_m = -2H_A$ and $N_m = V_A$.

The influence line for N_m might also be obtained using the neutral point method (see Fig. 34.3*c*). In this case the neutral point will coincide with the centre of hinge *B*, and therefore having laid off a length corresponding to $\sin \varphi_m = 1$ over point *A* we should connect the ordinate *a* so obtained with the neutral point, i. e., with the point of zero ordinate over hinge *B* (Fig. 42.3*g*).

4. THE TRUSSES

1.4. *Definitions and Classification of Trusses*

The *truss* is a framed structure which will continue to form an unyielding combination even when all its rigid joints are conventionally replaced by perfect hinges. As a rule, trusses are used for the same purposes as beams and girders, except that the spans they cover are usually much larger. In these cases solid web beams

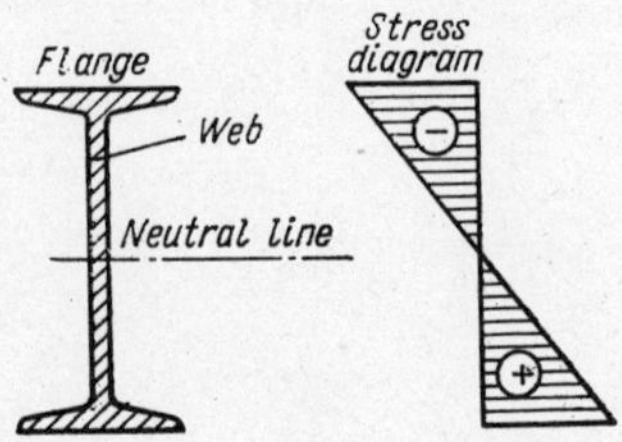

Fig. 1.4

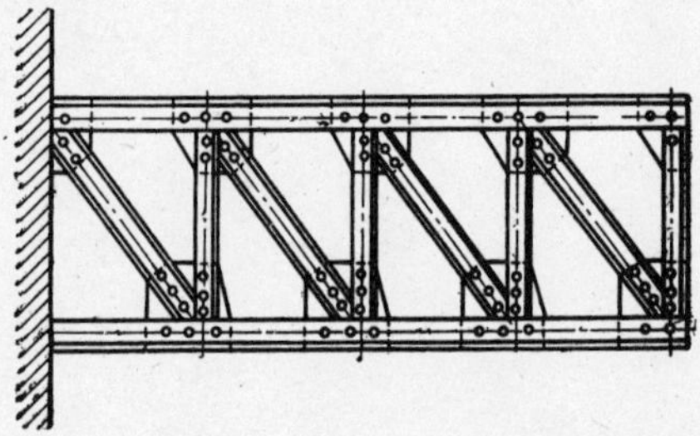

Fig. 2.4

become uneconomical due to the fact that the strength of the web can never be utilized to the full extent (unit stresses in the web being lower than in the flanges as will be seen from Fig. 1.4) and also due to the danger of web buckling which becomes more and more acute with the increase in the height of the beams.

In framed structures such as trusses (provided the loads act at the joints) all the members are subjected either to direct extension or compression which ensures a far better utilization of the materials, the stress diagram for each of these members being practically rectangular. Therefore the trusses are always much lighter than solid web beams of the same span and the same height. A typical example of a truss is shown in Fig. 2.4.

Apart from two-dimensional trusses in which all the bars are situated in one and the same plane, there exist three-dimensional or space framed structures in which the elements are situated in several planes (Fig. 3.4). However, in a great number of cases the design of three-dimensional framed structures may be reduced to the case of several plane systems.

The *span* of a truss (Fig. 4.4*a*) is the distance between its supports. The lower and upper longitudinal members form the upper and lower *chords* of the truss, while the members which connect

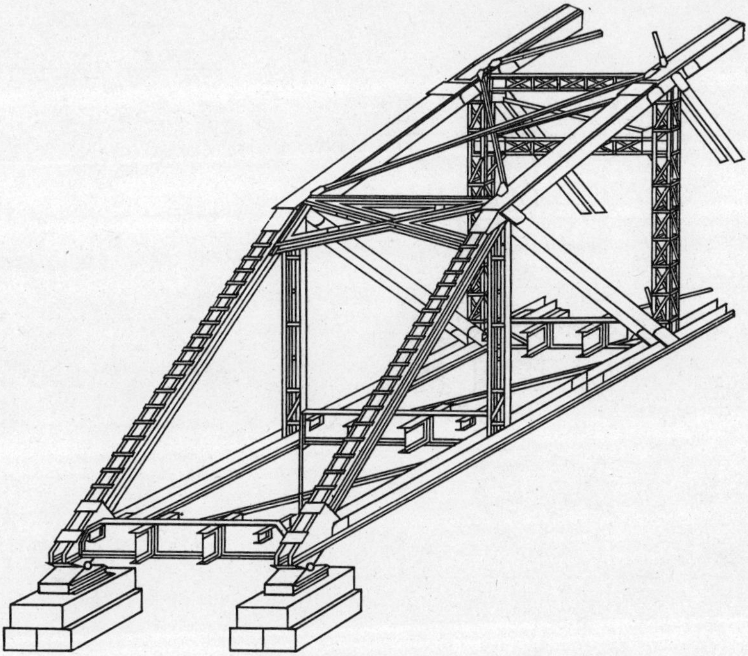

Fig. 3.4

the two chords are called the *web members*. The latter may be subdivided into *verticals* and *diagonals* or into *struts* and *ties*, the struts being always compressed and the ties extended. A *counter-brace* is a member designed to resist both tensile and compressive stresses. The end posts also called batter braces connect the upper chord to the lower one and may be regarded as belonging both to the upper chord and to the web members. The distance between two adjacent joints measured along the horizontal is usually called a *panel*, the joints themselves being frequently referred to as *panel points*.

The following five criterions may serve as a basis for the classification of trusses:

(a) the shape of the upper and lower chords;
(b) the type of the web;
(c) the conditions at the supports;
(d) the destination of the structure;
(e) the level of the floor.

In accordance with the first criterion, the trusses may be subdivided into trusses with parallel chords (Fig. 4.4*a*) and into

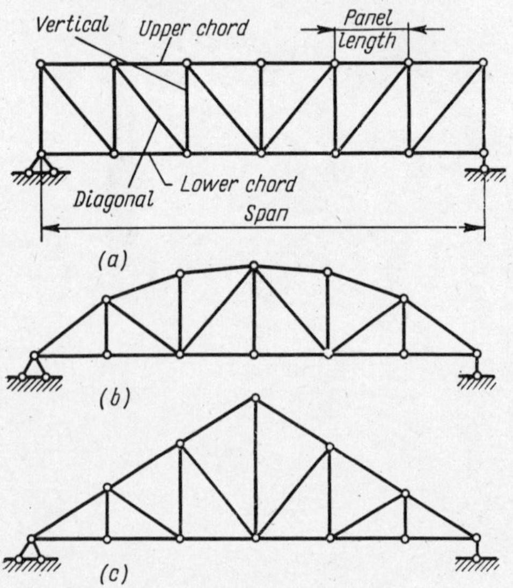

Fig. 4.4

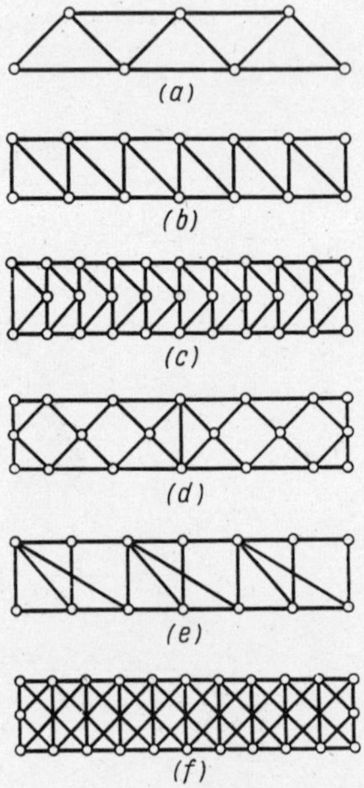

Fig. 5.4

polygonal and triangular trusses (Fig. 4.4*b* and *c*). Trusses with a parabolic upper chord (Fig. 4.4*b*) belong to the first of the last two kinds.

The second criterion permits to subdivide the trusses into those with a triangular pattern of the web * (Fig. 5.4*a*), those with a quadrangular pattern formed by verticals and diagonals ** (Fig. 5.4*b*), those in which the web members form a letter *K* (the

◆

* In the English speaking countries, where the great majority of truss types are called after the names of engineers who first introduced them on a large scale, this truss is known as the *Warren truss* (*Tr. note*).

** The more widely used of these are the *Pratt* and the *Howe* trusses, the first being characterized by extended diagonals and compressed verticals, and the second—by extended verticals and compressed diagonals (*Tr. note*).

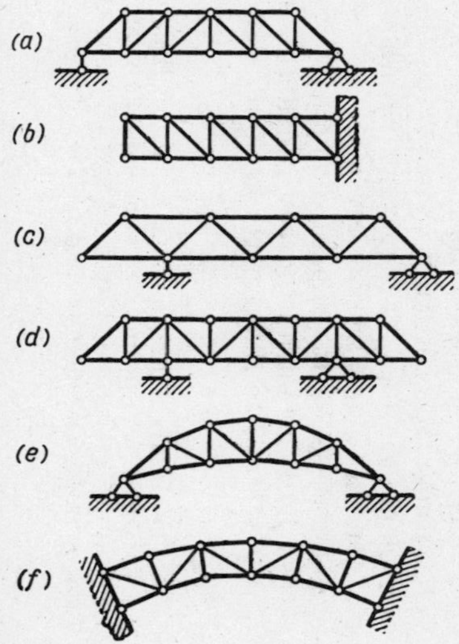

Fig. 6.4

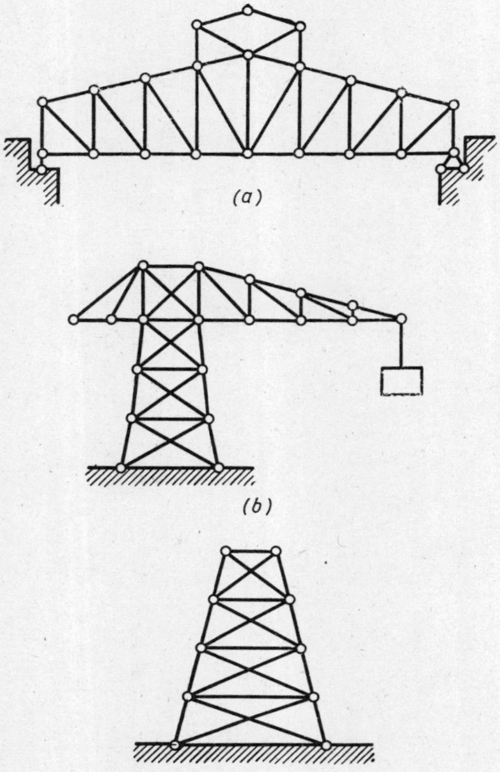

Fig. 7.4

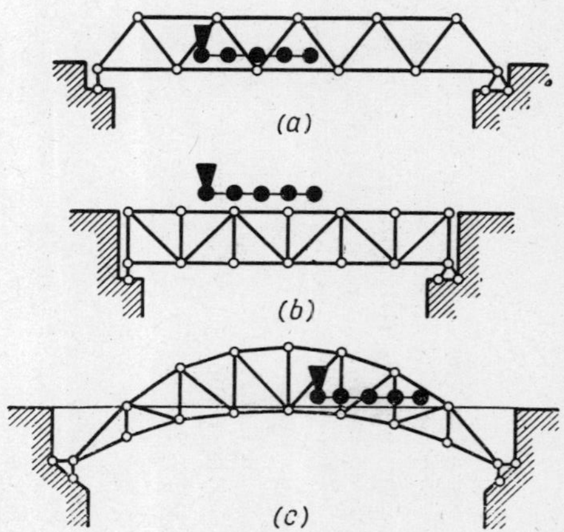

Fig. 8.4

so-called *K-truss* shown in Fig. 5.4*c*), and finally trusses, the webs of which are formed by the superposition of two or more simple grids illustrated in Fig. 5.4*d*, *e* and *f*, usually referred to as the double, or multiple trusses *.

The third criterion permits to distinguish between the ordinary end-supported trusses (Fig. 6.4*a*), the cantilever trusses with a built-in end (Fig. 6.4*b*), the trusses cantilevering over one or both supports (Fig. 6.4*c* and *d*, respectively), and finally the crescent and arched trusses in Fig. 6.4*e* and *f*.

As regards their destination the trusses may be subdivided into roof trusses (Figs. 7.4*a* and 4.4*c*), bridge trusses (Figs. 4.4*a* and 8.4) and miscellaneous trusses used in crane construction (Fig. 7.4*b*) and in the construction of various towers, bents, etc. (Fig. 7.4*c*).

In bridge construction the trusses are frequently subdivided into through-bridge trusses, in which the railway (or road) is carried directly by the bottom chord joints (Fig. 8.4*a*), the deck-bridge trusses where the upper chords or their joints carry the roadway (Fig. 8.4*b*), and finally the trusses where the deck is carried at some intermediate level (Fig. 8.4*c*).

2.4. *Direct Methods of Stress Determination in Members of Simple Trusses*

We have already seen (Art. 2.1 an 3.1) that framed structures formed by adding consecutively any number of joints to a hinge-connected triangle (each joint being connected by means of two concurrent bars) are statically determinate and form an unyielding combination. Two-dimensional framed structures formed in this way are usually called *simple trusses*.

In Article 3.1 it has been shown that $2K$ equations of statics can be written for any statically determinate truss (K being the number of its joints), with the aid of which both the abutment reactions and stresses (internal forces) in all the members can be determined. It is usual to start with the determination of the abutment reactions for which purpose three equilibrium equations are written for the truss as a whole.

The stresses in the separate members of the truss can be determined by considering the equilibrium of separate parts or joints of the structure, these parts or joints being acted upon both by

◆

* The truss in Fig. 5.4*d* is usually called the *double Warren truss* for its web may be obtained by the superposition of two simple triangular webs while the truss in Fig. 5.4*e* may be regarded as a modification of the *Post* or of the *Whipple* truss (*Tr. note*).

the external forces and the stresses in the sectioned bars. The total number of independent equilibrium equations amounts to $2K-3$.

It is very important to find such imaginary sections which will allow direct determination of stresses in the separate bars, without necessitating the simultaneous solution of several equations with several unknowns. This simplifies very considerably all the computations and at the same time enhances their accuracy.

The following two methods will usually permit the determination of the stresses in all the members of a simple truss by solving in each case one equation with a single unknown.

THE METHOD OF MOMENTS

This method is used mainly when a section can be passed through the truss in such a way as to cut three nonconcurrent members, as for example section *I-I* in Fig. 9.4*a*.*

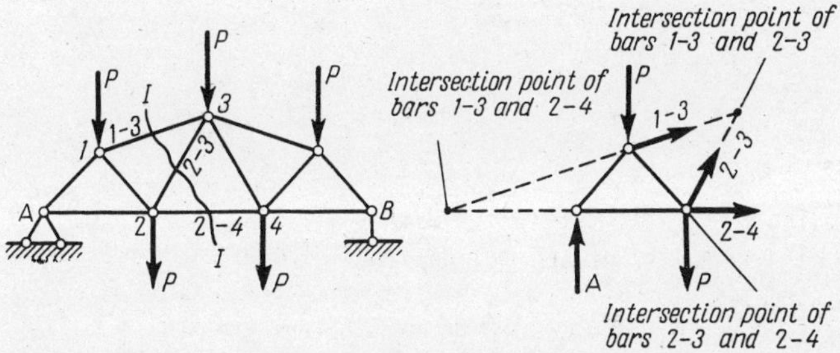

Fig. 9.4.

The axes of such members will intersect by pairs at three different points not lying on one and the same straight line (Fig. 9.4*b*).

The equilibrium equations of the moments of all forces, both internal and external, acting on the cut-off portion of the truss taken about each of these intersection points will reduce to one equation with one unknown, this unknown being the internal force acting in the bar not passing through the moment point.

Thus, in order to determine the stress acting in any member of the truss, a section should be taken across this truss cutting three nonconcurrent bars, one of these bars constituting the member in which it is desired to find the stress. In such a case the equation of all the moments about the point of intersection of the two other

* It will be shown later that this method can be applied also in certain more complicated cases.

bars will yield immediately the stress in the member under consideration.

The point of intersection of two members about which the moments are taken is usually called the origin of moments.

When writing the equilibrium equations all the stresses in the bars are conventionally reckoned positive which, with the convention of the signs adopted, means that the bars are in tension and the stresses are directed away from the joints. Therefore, when a negative stress is obtained this indicates that the member is compressed, the stress acting towards the joint.

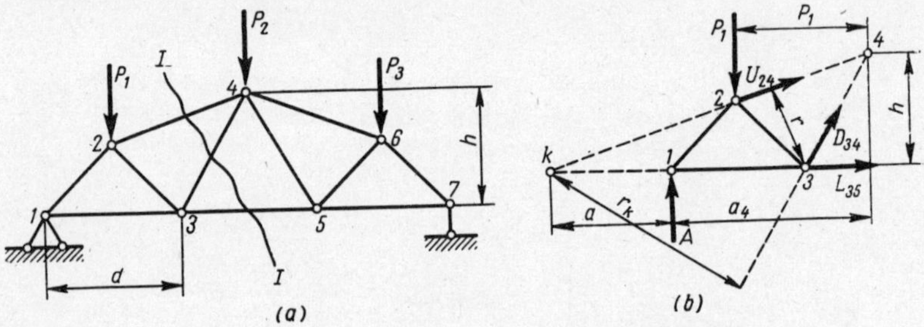

Fig. 10.4

We shall now illustrate the method of moments just described by several examples.

In these examples we shall denote by the letter U the stresses in the upper chord, by the letter L the stresses in the lower one, and by the letters D and V the stresses in the diagonals and verticals, respectively. These letters will be accompanied by ciphers indicating in each case the numbers of the joints to which the bar in question is connected.

Let us now determine the stress in the member *3-5* of the truss in Fig. 10.4*a*. For this purpose we shall pass section *I-I* cutting the member under consideration and two other members, one belonging to the upper and the other to the lower chords. It is always more convenient to consider that part of the truss acted upon by a smaller number of forces, and therefore we shall discuss here the left-hand portion of our truss which must be in equilibrium under the action of the external forces A and P_1 and of the internal stresses U_{24}, D_{34} and L_{35}, these stresses replacing the right-hand portion of the truss (Fig. 10.4*b*).

In order to determine the unknown stress L_{35} using a single equation we shall place the origin of moments at point *4* where members *2-4* and *3-4* concur.

The sum of moments of all the forces acting on the left-hand portion of the truss about point *4* is

$$\Sigma M_4 = Aa_4 - P_1 p_1 - L_{35} h = 0$$

wherefrom

$$L_{35} = \frac{Aa_4 - P_1 p_1}{h} = \frac{M_4^0}{h}$$

Here h is the lever arm of stress L_{35} about the origin of moments (in this particular case it is equal to the height of the truss), and

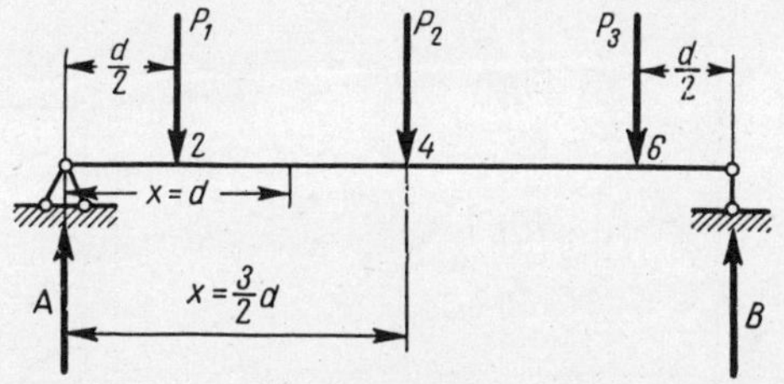

Fig. 11.4

M_4^0 is the moment of all the external forces (including the reaction) applied to the left-hand portion of the truss about joint *4*, this moment being equal to the bending moment acting over a section of a simple beam situated at the same distance from the support as the origin of moments in the truss.

If indeed the truss were replaced by a simple beam having the same span and subjected to the same loads (Fig. 11.4), the bending moment acting over a section of this beam situated at the same distance from the left-hand support as the origin of moments would be exactly equivalent to the moment of all forces applied to the left-hand portion of the truss about this origin of moments.

Thus, the stress in any member of the lower chord of a truss may be found as a quotient of the beam bending moment by the lever arm of the stress about the origin of moments.

The bending moment in a simple beam remaining always positive under any system of vertical loads, the stress L_{35} will also remain always positive, which means that the elements of the lower chord will be extended as long as the loads act downwards.

Let us now determine the stress in member *2-4* of the upper chord. In this case the origin of moments should be taken at joint *3* and the moments of all forces acting on the left-hand portion of the truss about this point should be equated to zero (see

Fig. 10.4*b*)

$$\Sigma M_3 = Ad - P_1 \frac{d}{2} + U_{24} r = 0$$

from which

$$U_{24} = -\frac{Ad - P_1 d/2}{r} = -\frac{M_3^0}{r}$$

The numerator of the fraction which we have denoted by M_3^0 is again equal to the beam bending moment acting over a section the abscissa of which is equal to *d*. As the beam moment M_3^0 is

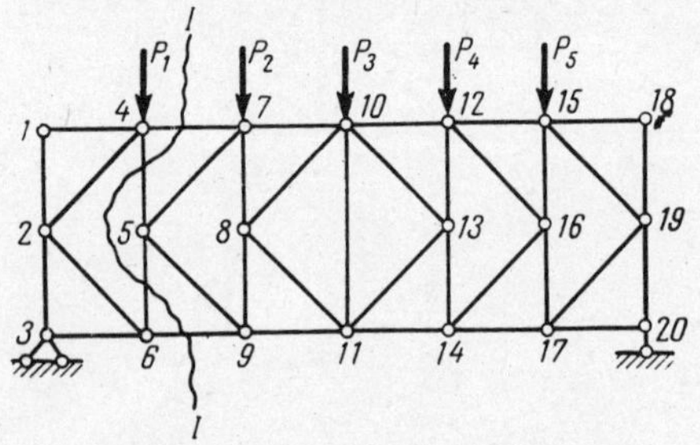

Fig. 12.4

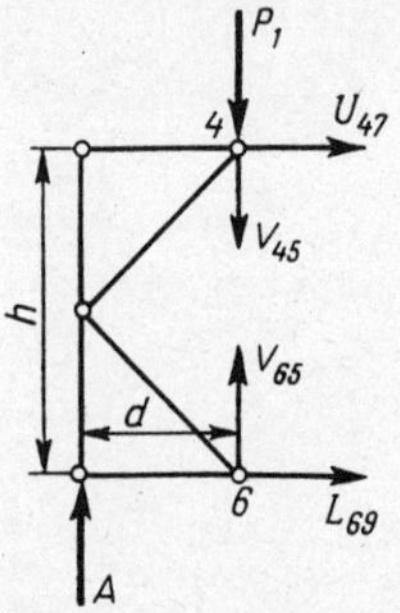

Fig. 13.4

always positive under the given system of loads and as the fraction $\frac{M_3^0}{r}$ is preceded by a negative sign, the stress U_{24} is negative, which means that member *2-4* is compressed.

It may be easily shown, using the same reasoning, that all the members of the upper chord as well as the end posts of a truss will always remain compressed under any system of vertical loads.

In order to determine the stress D_{34} induced in the diagonal *3-4*, let us equate to zero the sum of moments of all the forces acting on the left-hand part of the truss about point *k* at which the direction of bars *2-4* and *3-5* intersect well beyond the perimeter of the truss (see Fig. 10.4*b*).

$$\Sigma M_k = -Aa + P_1\left(a + \frac{d}{2}\right) - D_{34} r_k = 0$$

wherefrom

$$D_{34} = \frac{P_1\left(a + \frac{d}{2}\right) - Aa}{r_k} = \frac{M_k^0}{r_k}$$

It will be thus observed that *in the method of moments the magnitude of the stress is always expressed by the quotient of the moment of external forces acting on the left-hand portion of the truss M by the lever arm of the stress r about the same point*

$$N=\frac{M}{r} \tag{1.4}$$

Simple trusses defined above may have a more intricate pattern as represented in Figs. 12.4 and 14.4. Nevertheless, the method of moments remains applicable for the determination of stresses in their members. Indeed, if the truss in Fig. 12.4 is sectioned

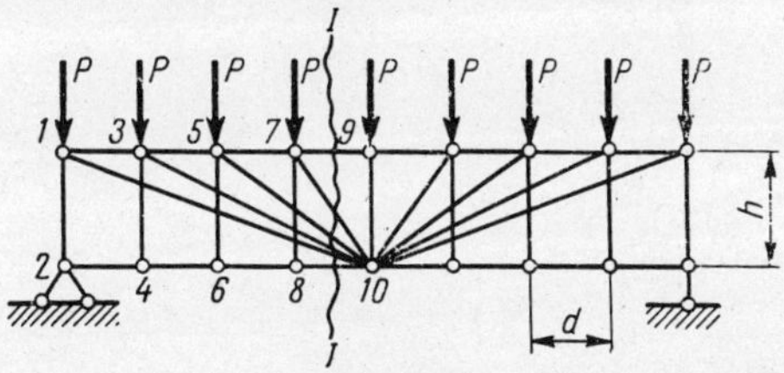

Fig. 14.4

along line *I-I*, the origin of moments may be taken at point *6* where three of the four sectioned members converge (Fig. 13.4), and therefore we shall again obtain one equation with one unknown which will yield the stress in the upper chord member *4-7*

$$\Sigma M_6=Ad+U_{47}h=0$$

wherefrom

$$U_{47}=-\frac{Ad}{h}=-\frac{M_6^0}{h}$$

If it is desired to find the internal force acting in member *6-9* of the lower chord the origin of moments should be shifted to point *4*, then

$$\Sigma M_4=Ad-L_{69}h=0$$

whence

$$L_{69}=\frac{Ad}{h}=\frac{M_4^0}{h}$$

Stresses in the upper and the lower chords of the truss shown in Fig. 14.4 can also be determined by the method of moments. Thus, in order to find the stress in bar *7-9*, section *I-I* should be passed, cutting in addition to the member considered five more bars, all converging at point *10*. If this point is taken as the origin of moments (Fig. 15.4), the equilibrium equation becomes

$$\Sigma M_{10}=A4d-4P\times 2.5+U_{79}h=0$$

from which

$$U_{79} = -\frac{4Ad - 10Pd}{h}$$

Let us now consider an even more complicated truss proposed by the eminent Russian engineer V. Shukhov for one of the large-span buildings in Moscow (Fig. 16.4). * This truss constitutes an unyielding system being composed of two basic triangles *1-4-5* and *2-3-6* connected by three nonconcurrent bars *1-2*, *3-4* and *5-6*.

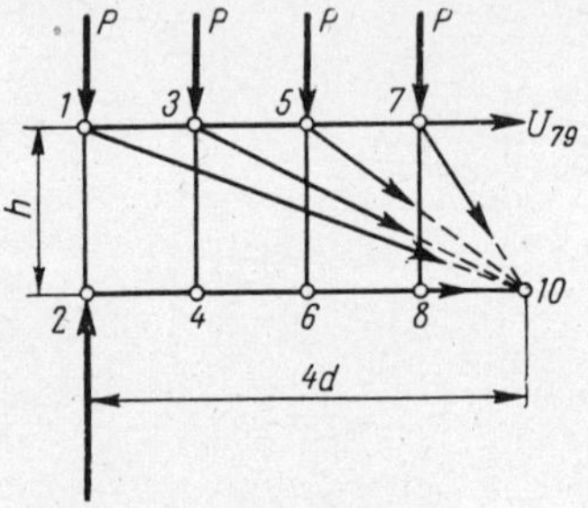

Fig. 15.4

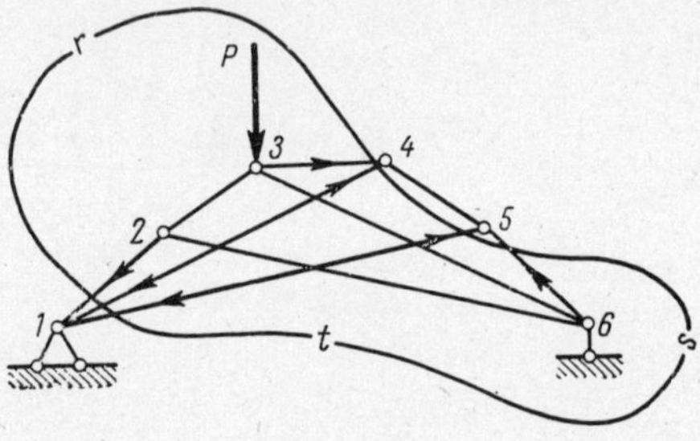

Fig. 16.4

The truss is statically determinate as the number of bars S satisfies the condition $S = 2K - 3 = 2 \times 6 - 3 = 9$. It is not possible to find a section through the Shukhov truss cutting any number of bars converging at a single point with the exception of one.

However, the section *r-s-t* which cuts bars *1-2*, *3-4* and *5-6* once and bars *1-4* and *1-5* twice permits the determination of stresses acting in bars *1-2*, *3-4* and *5-6*.

As will be seen from Fig. 17.4, the stresses in bars *1-4* and *1-5* will balance, these stresses entering the equilibrium equation twice with an opposite sign. Therefore in this section only three unknown stresses will remain U_{21}, U_{34} and U_{65} which may be easily determined by the method of moments.

Thus, in order to find the stress in bar *1-2* we shall place the origin of moments at the point of intersection of bars *3-4* and *5-6* (point k_1 in Fig. 17.4). Then

$$\Sigma M_{k1} = -U_{21} r_k - Pp - Bb_k = 0$$

wherefrom

$$U_{21} = -\frac{Bb_k + Pp}{r_k}$$

* This truss cannot be considered as belonging to the simple ones but nevertheless all the stresses in its members may be determined by the method of moments.

Similarly, point k_2 where the bars *1-2* and *5-6* intersect will be taken as the origin of moments for the determination of the stress U_{34} and point k_3 will form the origin of moments for stress U_{65} (Fig. 18.4). Thus, stresses U_{21}, U_{34} and U_{65} are determined independently using three equations, each containing only one unknown.

The stresses in all the other members will now be easily obtained by passing straight sections across any number of bars, provided that the stresses remain unknown in not more than three of them.

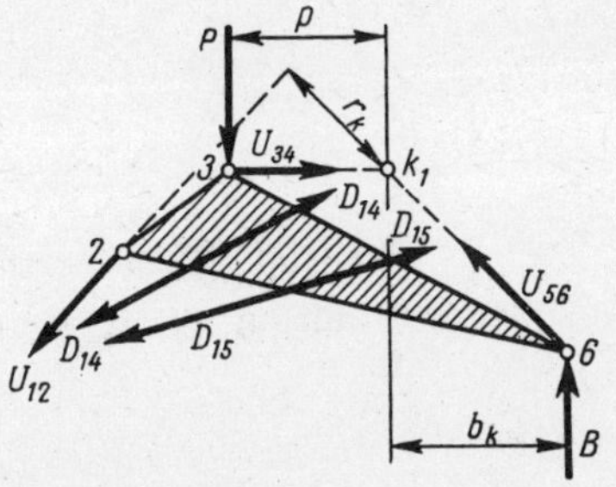

Fig. 17.4

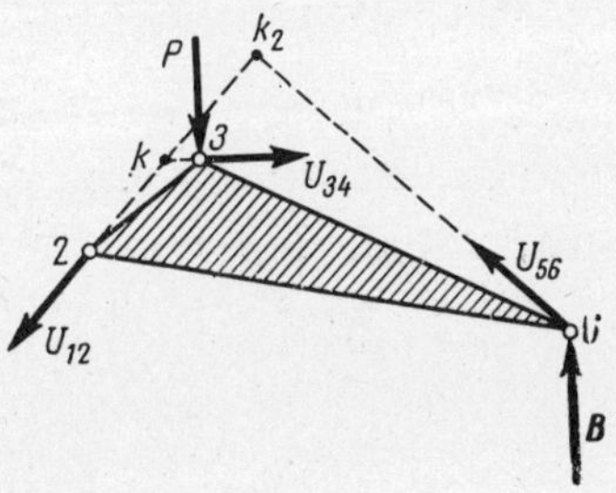

Fig. 18.4

The examples just considered lead to the following conclusions:

The method of moments is very expedient when a section may be taken cutting any number of bars converging at a single point, provided this point does not fall on the direction of the member investigated.

This method can also be used in cases when a section cuts more than three nonconcurrent bars, provided the stresses in all the bars except three are already known.

The same method may be utilized when the section crosses any number of bars, provided each bar with the exception of three is sectioned twice.

The method of moments is frequently considered as forming a particular case of the more general method of sections. Indeed when two of the sectioned members are parallel it becomes impossible to take the origin of moments at the point of their intersection and therefore the method of moments can no longer be applied. But passing a section through the truss will still permit the determination of the stresses required as we may in that case use the equilibrium equation of the vertical components of the internal and external forces (it is assumed that the chords are horizontal).

As an example, let us consider the truss represented in Fig. 19.4. Sections *I-I* and *II-II* will permit the computation of stresses

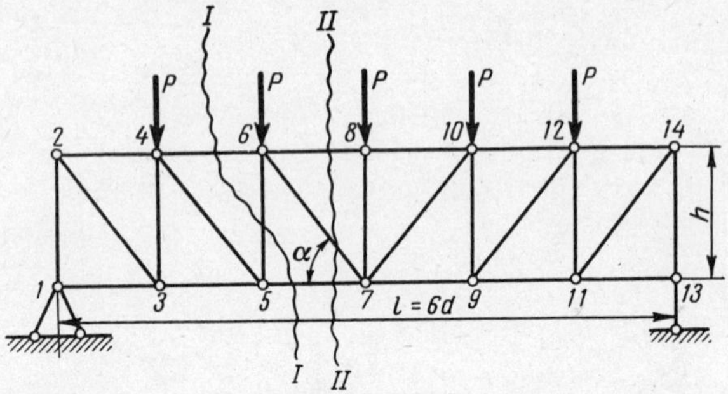

Fig. 19.4

in bars *5-6* and *6-7*, respectively. In effect projecting on the vertical all the forces (both external and internal) acting on the left-hand portion of the truss (Fig. 20.4) we obtain

$$\Sigma Y = A - P + V_{56} = 0$$

wherefrom

$$V_{56} = -(A - P) = -Q$$

where Q is the shear in the corresponding section of a simple beam of the same span.

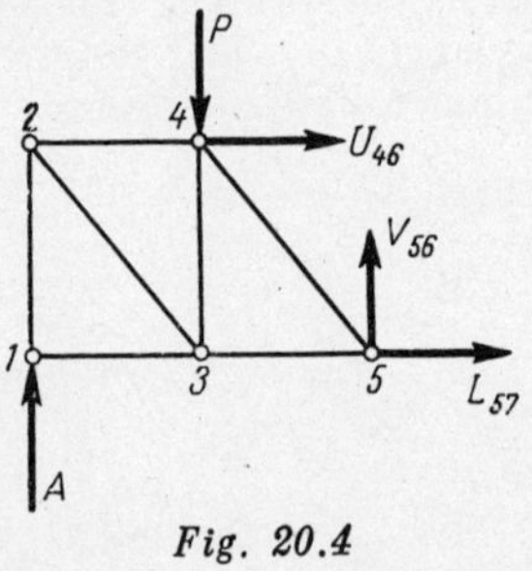

Fig. 20.4

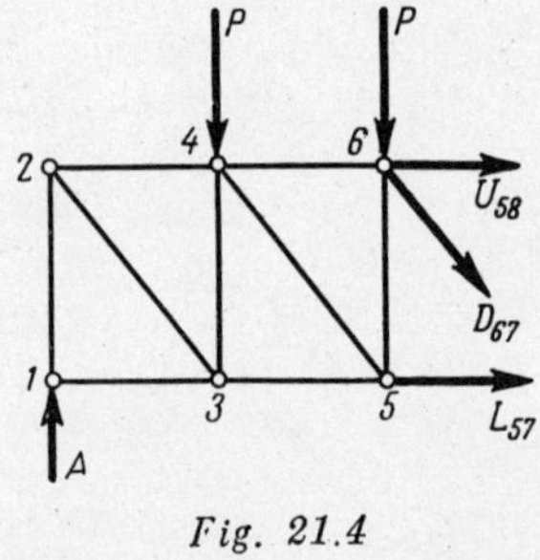

Fig. 21.4

The equilibrium of that portion of the truss to the left of section *II-II* (Fig. 21.4) will again furnish

$$\Sigma Y = A - P - P - D_{67} \sin \alpha = 0$$

from which

$$D_{67} = \frac{A - 2P}{\sin \alpha} = \frac{Q}{\sin \alpha}$$

where Q is again the shear in the corresponding section of a simple beam, this shear being equal to $(A - 2P)$.

THE METHOD OF JOINTS

In this method the equilibrium of each joint is considered separately, the joint being separated from the rest of the truss which is replaced by the stresses acting in the sectioned bars. In the case

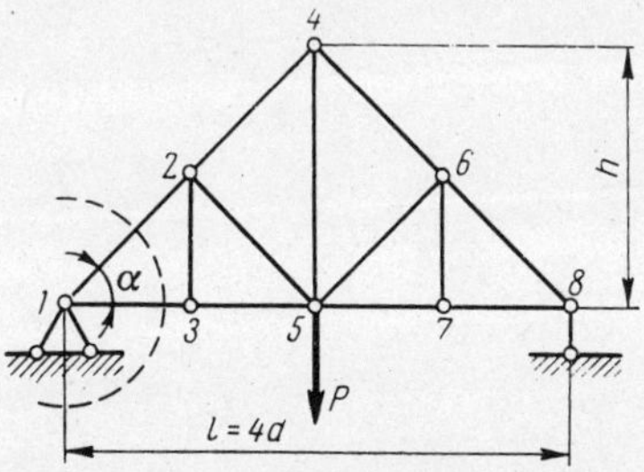

Fig. 22.4

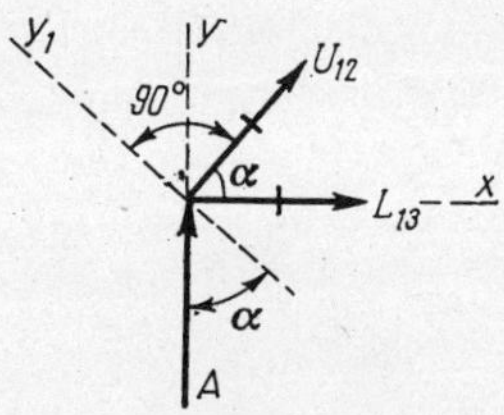

Fig. 23.4

of simple trusses the method of joints permits the successive determination of stresses acting in all the members starting with a joint formed by the meeting of two bars only.

As an illustration of the above, let us determine the stresses in the bars *1-2*, *1-3*, *2-3* and *3-5* of the truss represented in Fig. 22.4. We shall begin with considering the equilibrium of joint *1* at the left-hand support (Fig. 23.4). The projection of all the forces applied to this joint on the normal to bar *1-3* (in this case a vertical) gives

$$\Sigma Y = A + U_{12} \sin \alpha = 0$$

from which

$$U_{12} = -\frac{A}{\sin \alpha}$$

In the present case A is equal to $\frac{P}{2}$ and therefore

$$U_{12} = -\frac{P}{2 \sin \alpha}$$

The magnitude of the stress in the bar *1-3* will be obtained by projecting all the forces on a direction perpendicular to bar *1-2*, i. e., on the axis y_1

$$\Sigma Y_1 = A \cos \alpha - L_{13} \sin \alpha = 0$$

wherefrom

$$L_{13}=\frac{A\cos\alpha}{\sin\alpha}=\frac{P}{2}\cot\alpha$$

The same result could also be obtained by projecting all the forces on the x-axis leading to

$$\Sigma X=L_{13}+U_{12}\cos\alpha=0$$

wherefrom

$$L_{13}=-U_{12}\cos\alpha$$

Substituting in this expression U_{12} by its value found earlier we obtain once again

$$L_{13}=\frac{P}{2\sin\alpha}\cos\alpha=\frac{P}{2}\cot\alpha$$

In order to determine the stresses in members *3-2* and *3-5* we shall separate the joint *3* (Fig. 24.4). Equating to zero the sum of all the horizontal components we find

$$\Sigma X=-L_{31}+L_{35}=0$$

Remembering that L_{13} and L_{31} denote the same stress in bar *1-3* we obtain

$$L_{35}=L_{13}=\frac{P}{2}\cot\alpha$$

The vertical projection of all forces acting on joint *3* gives

$$\Sigma Y=V_{32}=0$$

The stress in bar *3-2* would remain nil if this bar were not at right angles with the lower chord.

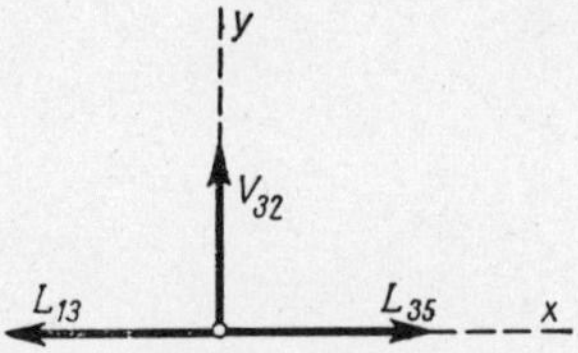

Fig. 24.4

Hence, when two out of three bars meeting at a joint lie on a straight line, the stresses in these two bars will be equal in amount and in sign and the third bar will remain idle as long as no external force is applied to this joint.

The equilibrium of joint *2* will now permit the determination of the stresses in bars *2-4* and *2-5* which will be expressed in terms of the stresses U_{21} and V_{23} already known.

It should be noted however that in the method of joints the stresses in all the members are determined consecutively, those found at a later stage being expressed in terms of those found at a previous one. Therefore, any accidental error committed in determining any particular stress will be carried along and will render inaccurate a number of subsequent computations. Another

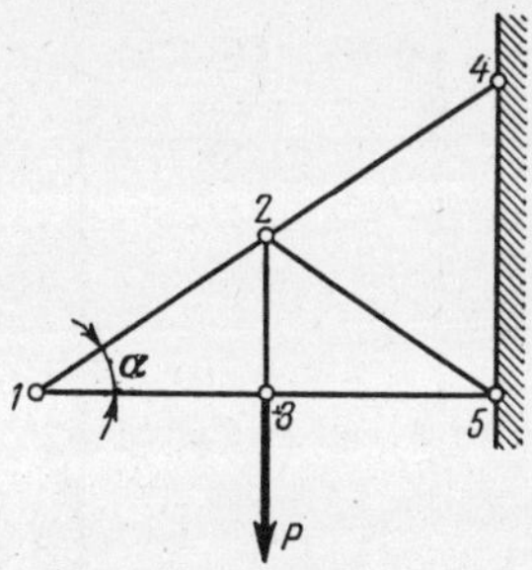

Fig. 25.4

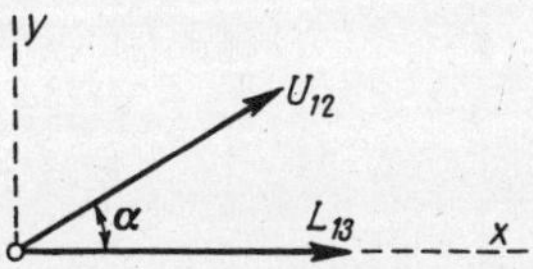

Fig. 26.4

setback of the method of joints resides in the fact that trigonometrical functions enter the equilibrium equations, thus complicating the calculus.

In certain cases the computations will be simplified if we remember that *if only two bars meet at a joint where no external force is applied the stresses in both these bars will be nil*. This will be readily confirmed by considering the equilibrium of joint *1* of the truss represented in Fig. 25.4.

Indeed the projection of all the forces acting on this joint on a vertical and on a horizontal (see Fig. 26.4) gives

$$\Sigma Y = U_{12} \sin \alpha = 0$$
$$\Sigma X = U_{12} \cos \alpha + L_{13} = 0$$

whence

$$U_{12} = L_{13} = 0$$

In the actual design of trusses all the three methods described above are frequently used together, preference being given in each particular case to the one leading more directly to the result required.

Problem 1. Compute the stresses in members *4-6*, *3-6* and *7-8* of the truss shown in Fig. 27.4.

Solution. Pass two sections as indicated in Fig. 27.4 and consider the equilibrium of the left-hand portions of the truss. All the calculations are given in Table 1.4; column 3 contains sketches of the portion of the truss under consideration, the other columns containing the corresponding equilibrium equations and their solution.

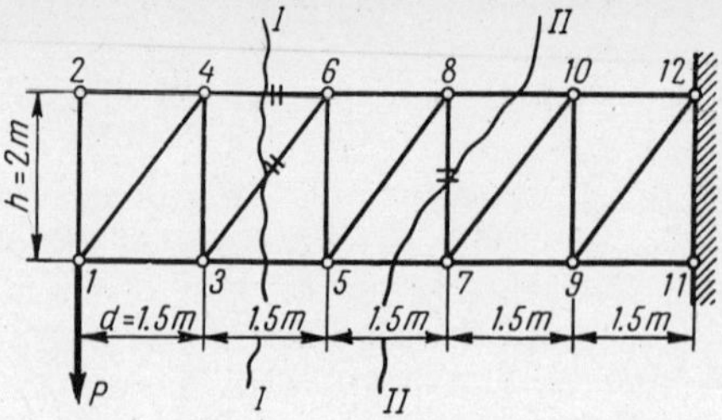

Fig. 27.4

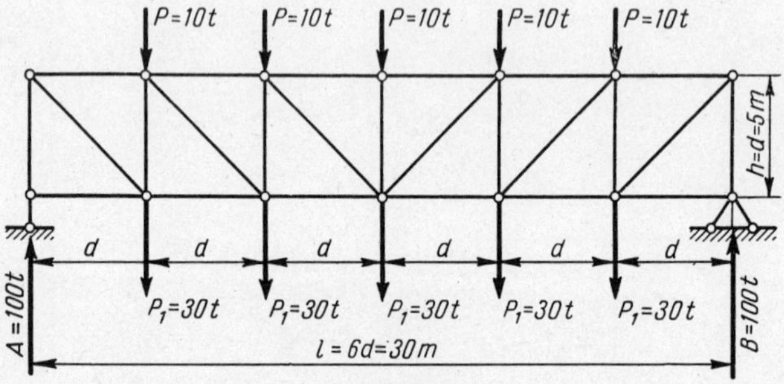

Fig. 28.4

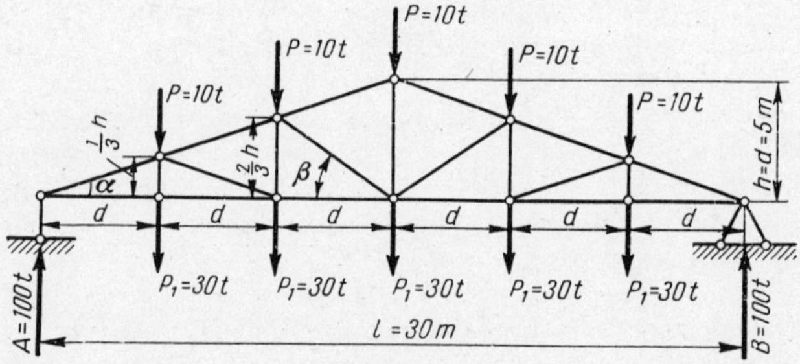

Fig. 29.4

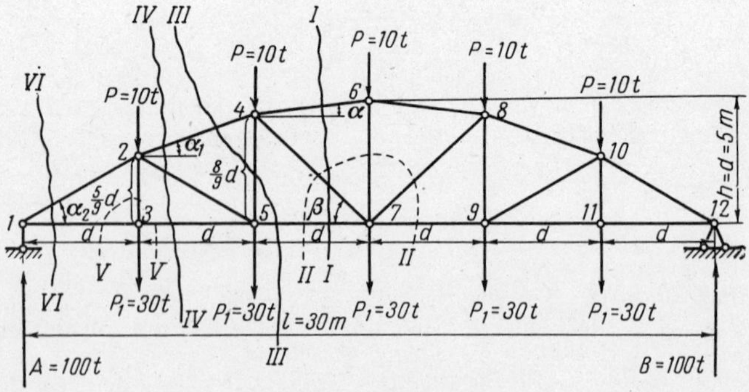

Fig. 30.4

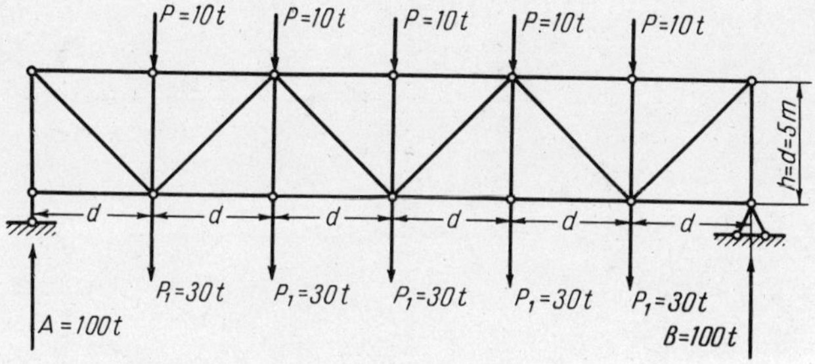

Fig. 31.4

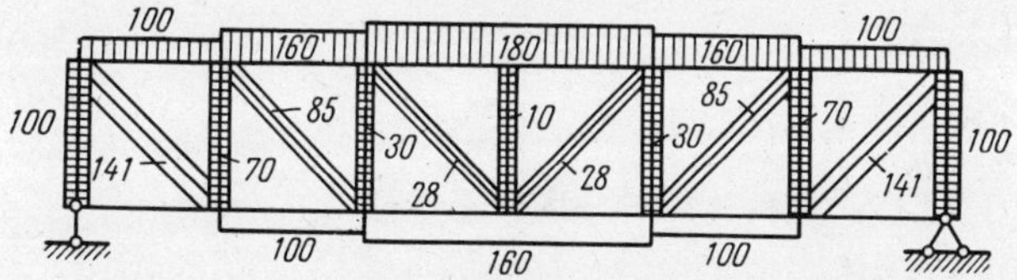

Fig. 32.4

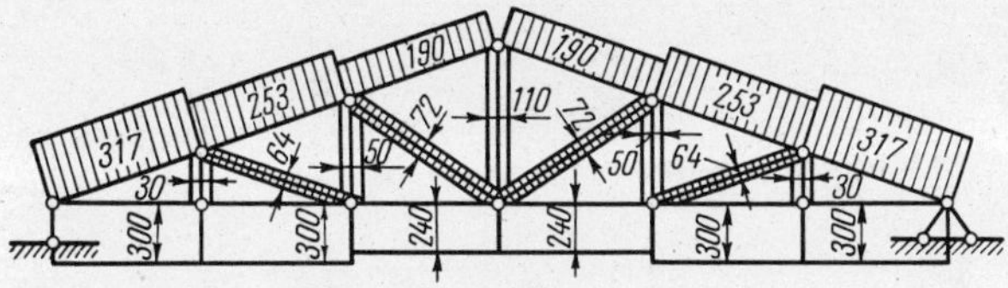

Fig. 33.4

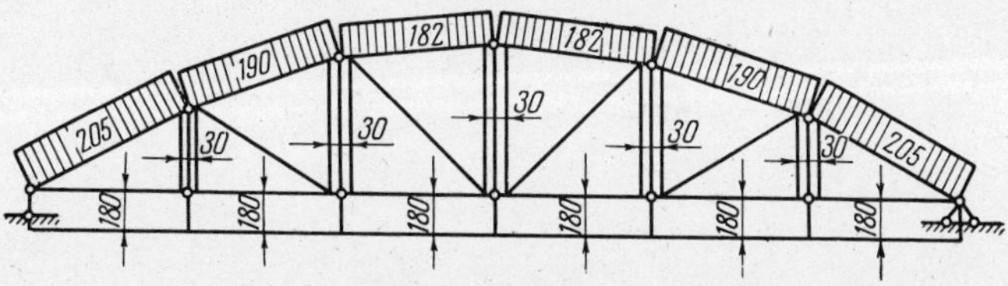

Fig. 34.4

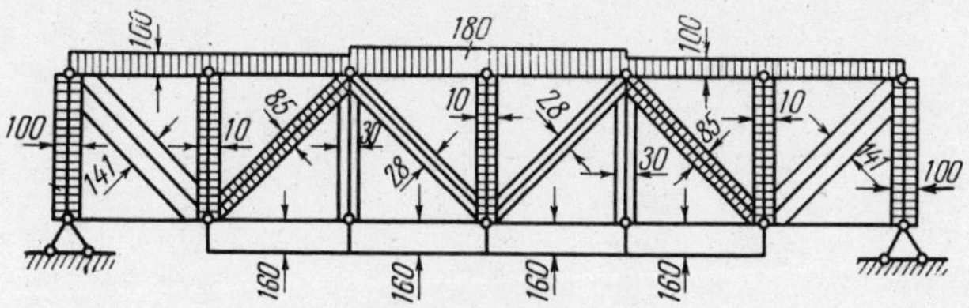

Fig. 35.4

Table 1.4

Bar. No.	Section No.	Sketch of portion considered	Equilibrium equations	Solution	Magnitude of stresses: Tension (+)	Magnitude of stresses: Compression (−)	Notes
4-6	I-I	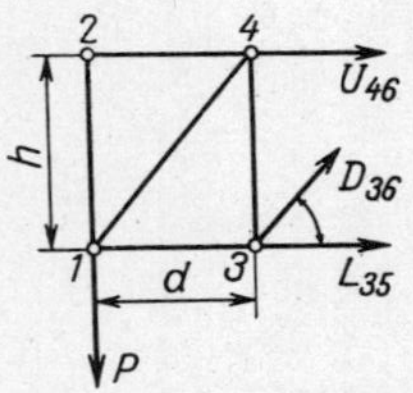	$\Sigma M_3 = -Pd + U_{46}h = 0$	$U_{46} = \frac{Pd}{h} = \frac{1.5P}{2}$	$\frac{3}{4}P$		See Fig. 27.4 $\left(h = 2\text{ m},\ d = 1.5\text{ m},\ \tan\alpha = \frac{h}{d} = \frac{2}{1.5} = \frac{4}{3};\ \sin\alpha = \frac{4}{5}\right)$
3-6	I-I		$\Sigma Y = -P + D_{36}\sin\alpha = 0$	$D_{36} = \frac{P}{\sin\alpha} = \frac{5P}{4}$	$\frac{5}{4}P$		
7-8	II-II	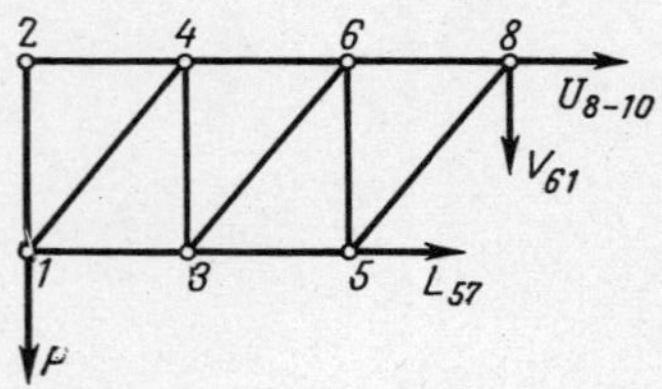	$\Sigma Y = -P - V_{78} = 0$	$V_{78} = -P$		P	See Fig. 27.4

Problem 2. Compute the stresses in all the members of the trusses represented in Figs. 28.4, 29.4, 30.4 and 31.4 and draw the corresponding diagrams. All the four trusses carry the same loads and have identical spans.

Solution. The results of all the calculations are represented in the form of graphs in Figs. 32.4, 33.4, 34.4 and 35.4, the width of the band along each truss member being in direct proportion to the magnitude of the stress. Compressions (reckoned negative) are hatched while tensions (reckoned positive) are left unshaded. (The values of all the stresses are given in tons.)

Computations pertaining to the truss in Fig. 30.4 are entered into Table 2.4 from which the sequence of all the operations is quite clear.

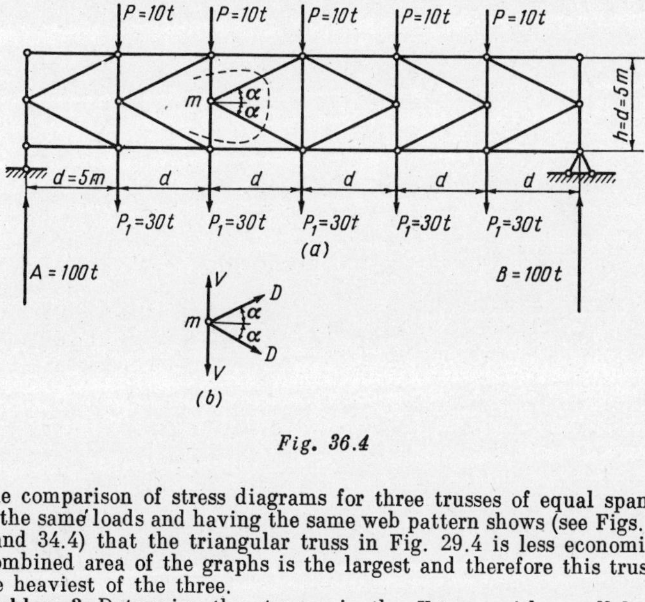

Fig. 36.4

The comparison of stress diagrams for three trusses of equal span, carrying the same loads and having the same web pattern shows (see Figs. 32.4, 33.4 and 34.4) that the triangular truss in Fig. 29.4 is less economical as the combined area of the graphs is the largest and therefore this truss will be the heaviest of the three.

Problem 3. Determine the stresses in the *K*-truss with parallel chords represented in Fig. 36.4*a*.

Solution. Examining any one of the joints at midheight of the truss where two inclined bars meet with the vertical, we find from the projection of all forces on the horizontal that

$$\Sigma X = D\cos\alpha + D'\cos\alpha' = 0$$

wherefrom

$$D\cos\alpha = -D'\cos\alpha'$$

and if $\alpha = \alpha'$ then

$$D = -D'$$

which means that the stresses in the inclined bars of one and the same panel are equal in magnitude but opposite in sign.

The determination of the stresses in all the members of the truss is omitted. The graphs of these stresses are shown in Fig. 37.4.

Table 2.4

Bar No.	Section or joint No.	Sketch of portion considered	Equilibrium equations	Solution	Magnitude of stresses: Tension (+)	Magnitude of stresses: Compression (−)	Notes
4-6			$\Sigma M_7 = A3d - 2P \times 1.5d - 2P_1 \times 1.5d + U_{46} r_1 = 0$	$U_{46} = -\frac{A3d}{r_1} + \frac{+3Pd+3P_1 d}{r_1}$		182	$r_1 = (a+3d)\sin\alpha$; $\tan\alpha = \frac{1}{9}$ (see Fig. 30.4); at the same time $\tan\alpha = \frac{8}{9} \times \frac{d}{(a+2d)}$ wherefrom $a = 6d$; $\sin\alpha = \frac{\tan\alpha}{\sqrt{1+\tan^2\alpha}}$
4-7	I-I	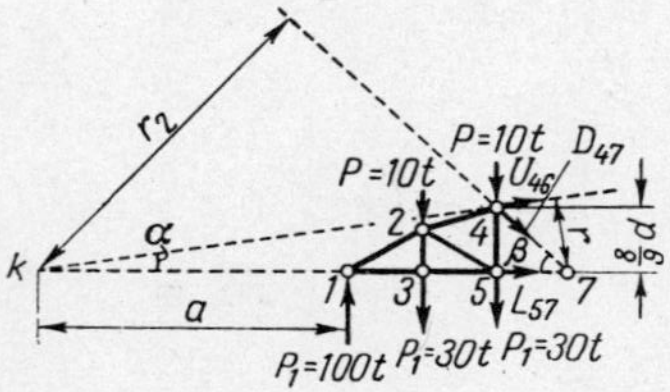	$\Sigma M_k = -A_a + 2P(a+1.5d) + 2P_1(a+1.5d) + D_{47} r_2 = 0$	$D_{47} = \frac{A_a}{r_2} + \frac{-2P(a+1.5d)}{r_2} + \frac{-2P_1(a+1.5d)}{r_2}$	0	0	$r_2 = (a+3d) \times \sin\beta$; $\tan\beta = \frac{8}{9}$ (see Fig. 30.4)

5-7			$\Sigma M_4 = A2d - Pd - P_1 d - L_{57}\frac{8}{9}d = 0$	$L_{57} = (A2d - Pd - P_1 d)\frac{9}{8}d$	180		See Fig. 30.4
7-6	II-II	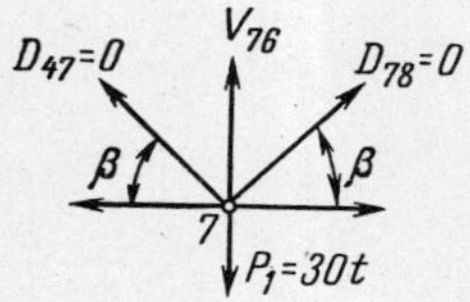	$\Sigma Y = V_{76} - P_1 = 0$	$V_{76} = P_1$	30		$D_{47} = D_{78} = 0$ (see Fig. 30.4)
2-4	III-III	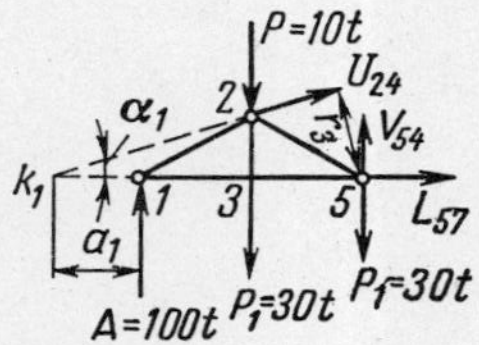	$\Sigma M_5 = A2d - Pd - P_1 d + U_{24} r_3 = 0$	$U_{24} = -\frac{A2d}{r_3} - \frac{Pd + P_1 d}{r_3}$		190	$r_3 = (a_1 + 2d) \times \sin\alpha$ $\tan\alpha_1 = \frac{3}{9}$ To find a_1 use equation $\frac{5}{9}\frac{d}{(a_1+d)} = \tan\alpha_1 = \frac{3}{9}$
5-4			$\Sigma M_{k1} = -Aa_1 + P(a_1 + d) + 2P_1(1.5d + a_1) - V_{54}(2d + a_1) = 0$	$V_{54} = \frac{P(a_1+d)}{2d+a_1} + \frac{2P_1(1.5d+a_1)}{2d+a_1} - \frac{Aa_1}{2d+a_1}$	30		

Table 2.4 (continued)

Bar No.	Section or joint No.	Sketch of portion considered	Equilibrium equations	Solution	Magnitude of stresses: Tension (+)	Magnitude of stresses: Compression (−)	Notes
2-5	IV-IV	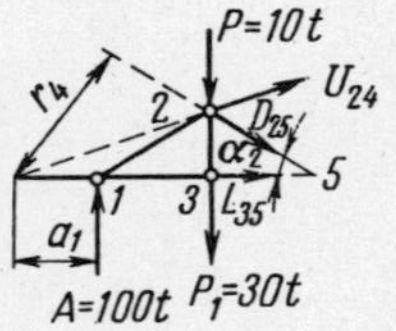	$\Sigma M_{k1} = -Aa_1 + P(d+a_1) + P_1(d+a_1) + D_{25}r_4 = 0$	$D_{25} = \frac{Aa_1 - P(d+a_1)}{r_4} - \frac{P_1(d+a_1)}{r_4}$	0	0	$r_4 = (a_1+2d) \times \sin\alpha_2$; $\tan\alpha_2 = \frac{5}{9}$ (see Fig. 30.4)
3-5			$\Sigma M_2 = Ad - L_{35}\frac{5}{9}d = 0$	$L_{35} = \frac{Ad}{5d}9$	180		
1-3	V-V	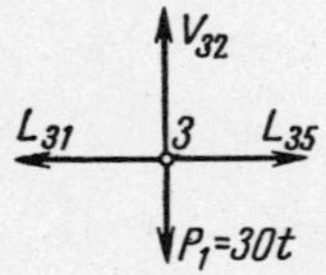	$\Sigma X = -L_{31} + L_{35} = 0$	$L_{31} = L_{35}$	180		See Fig. 30.4
3-2			$\Sigma Y = V_{32} - P_1 = 0$	$V_{32} = P_1$	30		
1-2	VI-VI	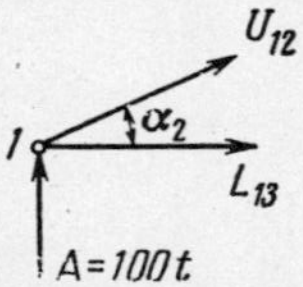	$\Sigma Y = A + U_{12} \times \sin\alpha_2 = 0$	$U_{12} = -\frac{A}{\sin\alpha_2}$		205	See Fig. 30.4

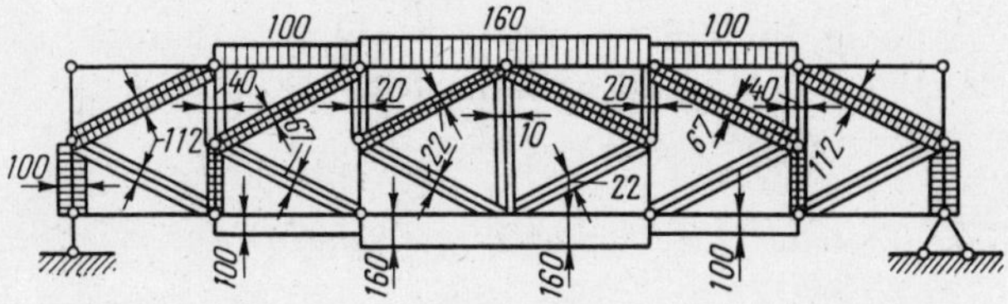

Fig. 37.4

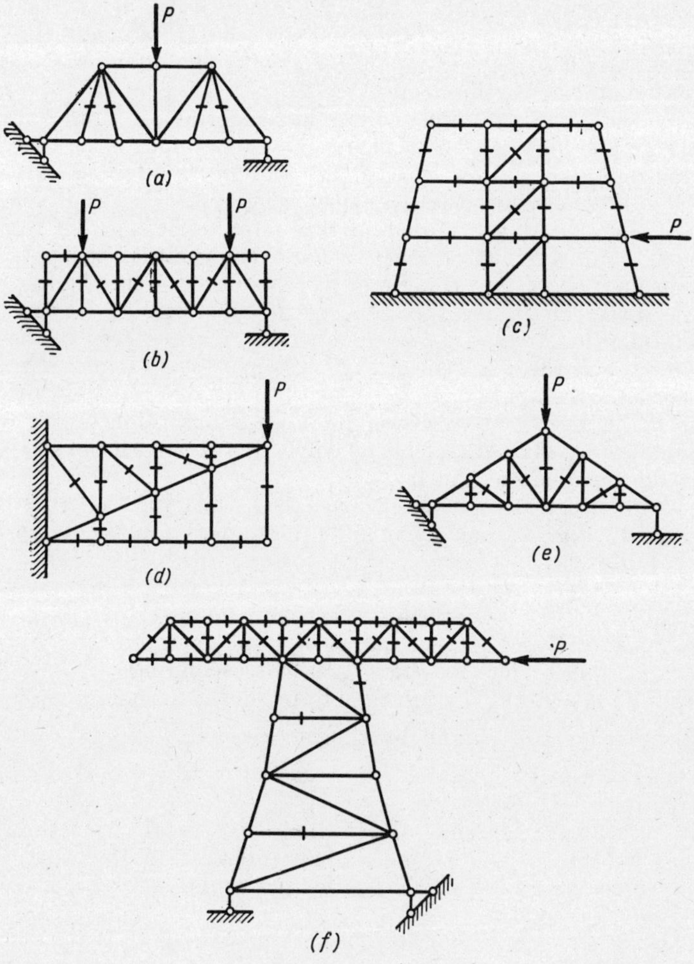

Fig. 38.4

The reader is invited to prove on his own that the stresses in all truss members marked with a dash in Fig. 38.4 are nil.

3.4. *Graphical Method of Stress Analysis in Simple Trusses*

It has been shown in the preceding article that all stresses in a simple truss may be determined analytically by the method of joints. Every truss of this type will contain at least one joint where only two bars meet and this joint should be the starting point of the operation.

/We shall now examine the graphical method of stress analysis, this method being based on the resolution of forces along two given directions. The following sequence should be adopted.

A joint where only two bars meet must be selected and the external force (reaction) applied to this joint must then be resolved along the directions of the two bars, using either the parallelogram method or the triangle of forces method. The triangle must necessarily close, for the joint is in equilibrium. Having thus obtained the stresses in two members we may proceed to the next joint, where the resultant of the stress previously found and of the external force applied to the joint (if any) must be again resolved along the directions of the next couple of bars. Continuing in the same way we shall complete the determination of all the stresses in all the members of the truss.

As an illustration of the above, let us analyze the truss represented in Fig. 39.4 acted upon at the upper chord joints by three equal forces.

Fig. 40.4*a* shows the same truss, its two supports having been replaced dy the appropriate reactions, found either graphically or analytically. Commencing with joint *1*, where two bars *1-2* and *1-3* meet, let us draw the corresponding force polygon (Fig. 40.4*b* and *c*). The joint being in equilibrium and acted upon by reaction $A = \frac{3P}{2}$ and by the stresses U_{12} and L_{13}, we must

(1) lay off at some scale the reaction A whose magnitude and direction are both known;

(2) through both ends of this line trace parallels to the directions of bars *1-2* and *1-3* until their intersection, thus completing the force polygon (which in this particular case reduces to a triangle).

The sides *bc* and *ca* of this triangle measured to the same scale as reaction A will give the magnitude of the stresses in bars *1-2* and *1-3*, respectively. The direction (sign) of these stresses will

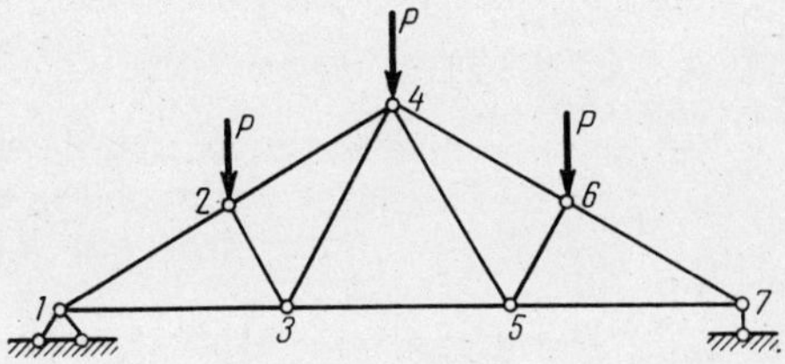

Fig. 39.4

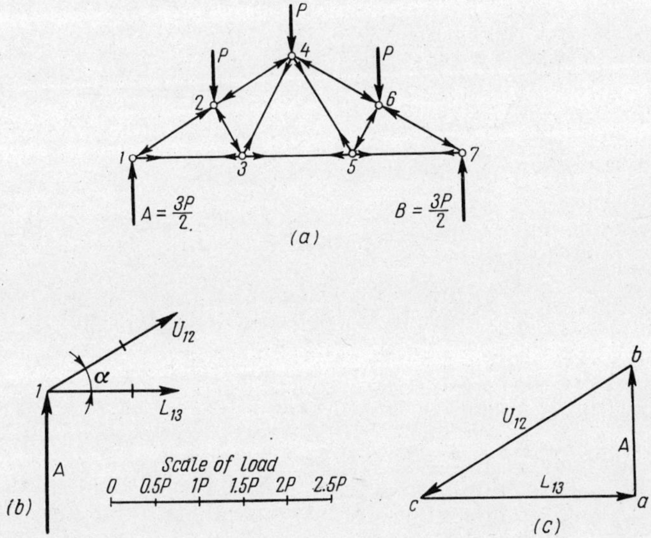

Fig. 40.4

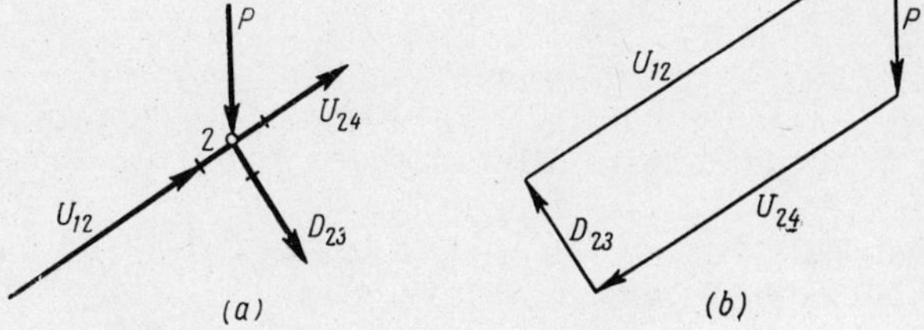

Fig. 41.4

be determined remembering that in a closed force polygon all the forces follow one another.

Thus, the reaction A being directed upwards, we shall find that the stress U_{12} acts towards the joint which means that bar *1-2* is compressed while stress L_{13} acts away from the joint and therefore bar *1-3* is extended.

We may now mark the direction of the stresses found on the schematic drawing of the truss (Fig. 40.4*a*) where arrows pointing towards the joint will indicate compression and those pointing away from the joint—tension. Arrows should be also shown at the

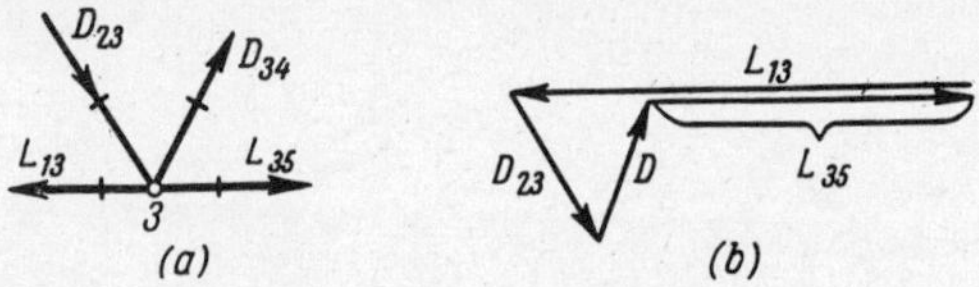

Fig. 42.4

other extremities of bars *1-2* and *1-3*, as the stresses in these bars will have to be accounted for in considering the equilibrium of joints *2* and *3*.

At joint *2* acted upon by the load P three bars, namely bars *2-1*, *2-4* and *2-3*, meet together. The stresses U_{12} and L_{13} being known as well as the load P (Fig. 41.4*a*), the construction of the force polygon (Fig. 41.4*b*) will be carried out as follows: lay off stress U_{12} acting towards the joint and the load P and then through the free ends of these lines draw two parallels to the directions of bars *2-4* and *2-3* until their intersection, thus forming a closed polygon. Fig. 41.4*b* shows that both stresses U_{24} and D_{23} are compressive.

Stresses in bars *3-4* and *3-5* may now be obtained by considering the equilibrium of joint *3* (see Fig. 40.4*a*). Four bars meet at this joint, namely bars *3-1*, *3-2*, *3-4* and *3-5*, but the stresses are unknown only in the last two. Hence they may be found by constructing a force polygon as shown in Fig. 42.4. It will be readily observed that both these stresses are tensile.

The determination of the stresses in the right-hand half of the truss is not required due to the symmetry of the structure.

The procedure just described permits the determination of the stresses in all the members of a truss by constructing successively force polygons related to each joint. Each stress will appear twice in these polygons, at first as an unknown to be found and later as a given force applied to the adjacent joint where two other unknown stresses are sought. All of these polygons may be merged together to form a single diagram called the Maxwell-Cremona or stress diagram in which each stress will be met with only once.

Such a merger is represented in Fig. 43.4. This operation is made possible by the fact that in the force polygons the forces appear in the sequence they are met with when each joint is passed around in the clockwise direction. (The opposite direction could be adopted as well, but following the tradition we shall always use the one mentioned first.) Thus, in joint *1* in Fig. 40.4*a* we meet first the reaction A followed by stress U_{12} and then by L_{13}. In

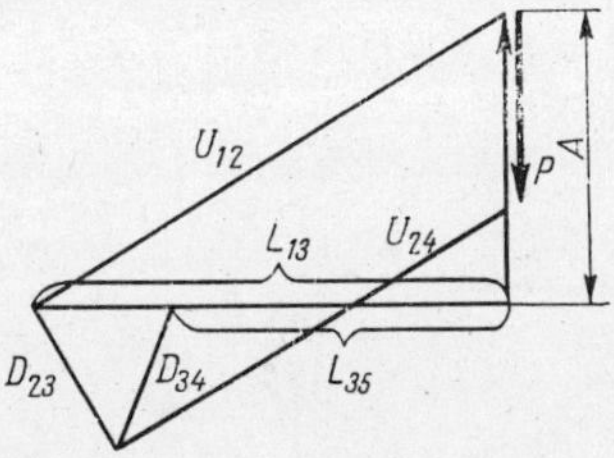

Fig. 43.4

joint *2* first comes stress U_{12}, then the load P and stresses U_{24} and D_{23}. This sequence is maintained in the force polygons in Figs. 40.4*c* and 41.4*b*.

In practice the stress diagram for the whole truss is generally obtained directly, omitting the force polygons of individual joints. This method will be explained using as an example the truss shown in Fig. 44.4. The notations to be used are as follows: letters and ciphers will denote areas bounded by the truss members (areas *a*, *b*, and *c*) or by the lines of action of the loads and reactions (areas *I*, *II*, and *III*). Each stress, load or reaction will be designated by two indices corresponding to the two areas it separates. Hence, the left-hand abutment reaction forming the boundary between areas *I* and *III* will be indicated by *III-I* (but not *I-III*, the clockwise direction being followed), the load P by *I-II* and the right-hand reaction by *II-III*. Similarly, the stress in bar *1-2* will be denoted by *I-a* when joint *1* is considered and by *a-I* for joint *2*. With these notations the numbering of joints may be completely omitted.

The construction of the stress diagram will start at joint *1* over the left-hand abutment where only two bars meet. Having laid off the magnitude of reaction *III-I* along the vertical to the scale selected we shall obtain the stresses in bars *1-2* and *1-3* by tracing through points *I* and *III* parallels to the directions of these two bars whose intersection at point *a* will permit scaling off of stresses in the upper and lower chords *I-a* and *a-III*.

In order to determine the directions (signs) of these stresses it will suffice to remember that in a closed polygon all the forces follow each other in one and the same direction. Thus, knowing the direction of the reaction *III-I* in the triangle *III-I-a-III* of the diagram we shall readily determine the directions of the stresses *I-a* and *a-III*, the first being directed from *I* towards *a*, and the second from *a* towards *III*. Marking these directions on the sketch of the truss (see Fig. 44.4) we find immediately

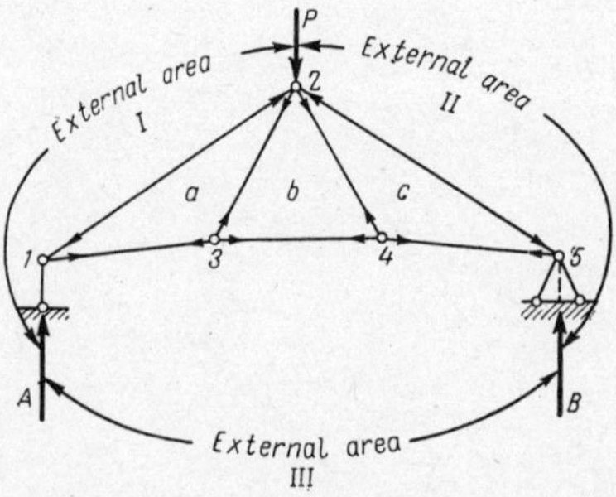

Fig. 44.4

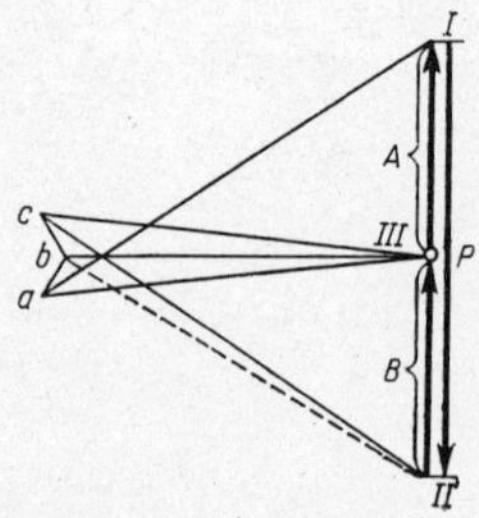

Fig. 45.4

that the stress *I-a* is directed towards joint *1* and is therefore compressive, while the stress *a-III* acts away from this joint and accordingly bar *1-3* is extended.

We may now pass to joint *3* of the lower chord acted upon by the stress *III-a* just found and by two unknown stresses *a-b* and *b-III*. These stresses will be obtained by tracing through points *III* and *a* two lines parallel to the bars *b-III* and *a-b*, the point of intersection of which shall be marked *b*. The sign of these stresses will be derived as heretofore from the direction of the stress *III-a* previously found (see Fig. 45.4). Marking these directions on the sketch of the truss (see Fig. 44.4) we note that all the bars meeting at joint *3* are extended.

The next joint to be considered is joint *2* acted upon by the load *I-II*, two stresses already found *b-a* and *a-I*, and two unknown stresses *II-c* and *c-b*. Returning to Fig. 45.4 we find that the diagram contains the two stresses *b-a* and *a-I*. Adding to these the load *I-II* (line *I-II*) we may readily find the resultant *b-II* of these three forces marked in dash line in Fig. 45.4. Resolving this force along two directions parallel to the bars *2-5* and *2-4* we shall find the stresses in these bars given by the length of segments *II-c* and *c-b*. The force polygon *b-a-I-II-c-b* indicates that bar *III-c* is compressed while bar *c-b* is extended.

Passing to the last joint *4* we find that out of the three bars meeting at this joint the stress in one bar only remains unknown. If the diagram has been constructed accurately, line *c-III* giving this stress must be parallel to bar *4-5* and must pass through point *III*, in other words, the diagram should be closed.

The force polygon for point *5* will be represented by *II-III-c-II*.

In the stress diagram the external forces were laid off in the same order as they were encountered when passing around the

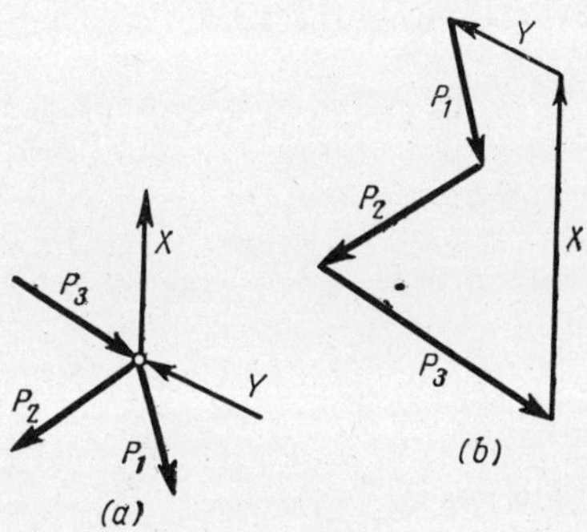

Fig. 46.4

perimeter of the truss in a clockwise direction. The force polygon of external loads and reactions must also be closed, the whole truss being in equilibrium. The closure of the external force polygon and of the stress diagram constitutes a ready check on the accuracy of all the operations.

The construction of the stress diagram is usually commenced by tracing the closed polygon of loads and reactions which must be laid off in the same order as they are met when passing around the truss in a clockwise direction. This being done, the stresses in the bars intersecting at each joint are determined graphically commencing with the joint where only two bars meet. These stresses will also be laid off in the sequence they are encountered when passing around each joint in a clockwise direction.

The construction of the force polygon for each joint should be carried out in such a way that the two unknown stresses should come last. Thus in the example given in Fig. 46.4*a* force P_1

should come first in order that the unknown stresses Y and X should come last (Fig. 46.4*b*).

In the stress diagram each line denoting an internal force belongs to two force polygons corresponding to two adjacent joints, and therefore it is not recommended to show the directions of the stresses in the diagram, these directions being different in the two cases just mentioned. Moreover, it is easy to determine the direction (sign) of each stress without going around the whole of the force polygon corresponding to the joint under consideration. Indeed, each stress in the diagram is denoted by two indices following each other in the sequence they were met when passing around the joint in a clockwise direction. This sequence will therefore be different for two adjacent joints, for instance, the stress in bar *2-3* (see Fig. 44.4) should be denoted by *a-b* if referred to the lower chord joint *3* and by *b-a* when referred to the upper chord joint *2*.

As point *a* of the force polygon represented in Fig. 45.4 lies below point *b* the stress *a-b* will be directed away from the lower joint and therefore the corresponding bar will be extended.

Problem. Required to construct a stress diagram for the truss represented in Fig. 47.4*a* and to determine the stresses in all the members of this truss.

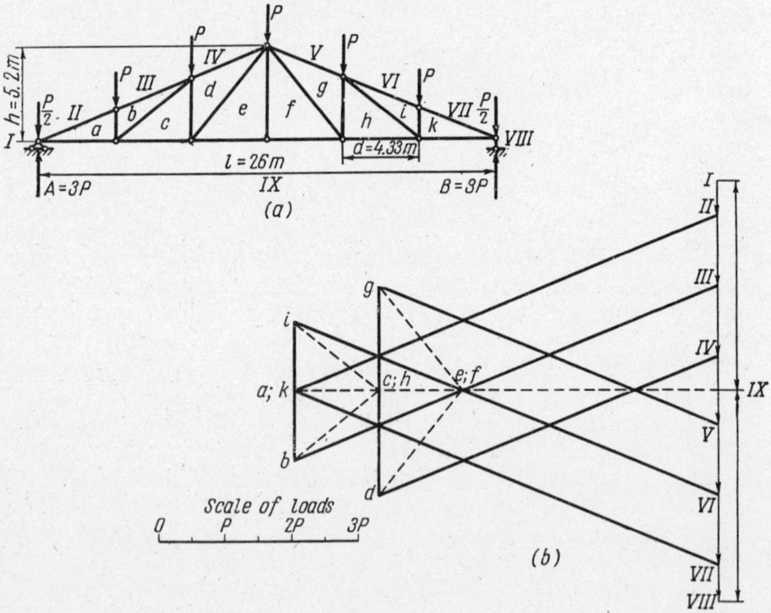

Fig. 47.4

Solution. Using the notations described above indicate by letters *a*, *b*, *c*, etc., the areas bounded by the members of the truss and by *I*, *II*, . . . , *IX* the areas separated by the directions of the loads and reactions as the truss is passed around its perimeter in a clockwise direction. A force polygon *I*-*II*- . . .-*IX*-*I* (Fig. 47.4*b*) may be then constructed commencing with the load acting over the left-hand abutment, reaction *A* (force *IX*-*I*) being laid off last. The force polygon must close as the system of external forces (loads and reactions) is balanced. *

The construction of the diagram starts at the joint directly above the left-hand abutment, by tracing a closed force polygon of all the forces acting at this joint. Forces *IX*-*I* and *I*-*II* are already set out, their resultant *IX*-*II* being directed upwards. This force must now be resolved along the directions of the lower and the upper chords, for which purpose a line parallel to bar *II*-*a* is traced through point *II* and a horizontal parallel to bar *a*-*IX* through point *IX*, their intersection being marked by the letter *a*. The sign of the stresses will be determined by the application of the rule that in a closed force polygon all the forces follow one another in the same direction. Accordingly the stress *II*-*a* must be directed from right to left and downwards and stress *a*-*IX* from left to right. In other words, the stress in bar *II*-*a* will be directed towards the joint which means that this bar is compressed and the stress in bar *a*-*IX* away from the joint indicating that this member is extended.

Table 3.4

Upper chord		Lower chord		Verticals		Diagonals	
Bar	Stress	Bar	Stress	Bar	Stress	Bar	Stress
a-*II*	—6.85*P*	*a*-*IX*	6.30*P*	*a*-*b*	—1.00*P*	*b*-*c*	1.60*P*
b-*III*	—6.85*P*	*c*-*IX*	5.05*P*	*c*-*d*	—1.50*P*	*d*-*c*	1.95*P*
d-*IV*	—5.40*P*	*e*-*IX*	3.80*P*	*e*-*f*	0	*f*-*g*	1.95*P*
g-*V*	—5.40*P*	*f*-*IX*	3.80*P*	*g*-*h*	—1.50*P*	*h*-*i*	1.60*P*
i-*VI*	—6.85*P*	*h*-*IX*	5.05*P*	*i*-*k*	—1.00*P*		
k-*VII*	—6.85*P*	*k*-*IX*	6.30*P*				

Points *b*, *c*, *d*, etc., of the stress diagram will be found as follows: point *b* by tracing through point *a* a parallel to the vertical *a*-*b* and through point *III* a parallel to the bar *III*-*b*; point *c* by drawing through point *b* a parallel to the diagonal *b*-*c* and a horizontal through point *IX*; point *d* will be formed by the intersection of a vertical passing through point *c* and of a line parallel to the upper chord member *IV*-*d* passing through point *IV*. Points *e*, *f*, *g*, *h*, *i* and *k* will be found in the same way. It will be

◆

* In the force polygon forces *VIII*-*IX* (reaction *B*) and *IX*-*I* (reaction *A*) are slightly offset towards the right in order to avoid confusion with the external loads.

noted that points e and f coincide, indicating that the stress in the vertical e-f is nil. The construction of the last diagram pertaining to the right-hand half of the truss could be omitted as the stresses in the two halves of a symmetrical truss subjected to a symmetrical system of loads will be exactly the same.

In the stress diagram of Fig. 47.4 dash lines indicate tensions and solid lines compressions. It will be seen that all the members of the lower chord and the diagonals are extended, while the upper chord members and the verticals are compressed.

The magnitudes of the stresses scaled off the diagram are tabulated above.

4.4. *Direct Method of Stress Determination in Complicated Statically Determinate Framed Structures*

Occasionally the designer will have to deal with framed structures of a considerably more intricate pattern than those formed by the successive addition of supplementary joints to a basic triangle, each of these joints being attached by means of two concurrent bars. These systems remain statically determinate and in a number of cases they may be derived from the simple

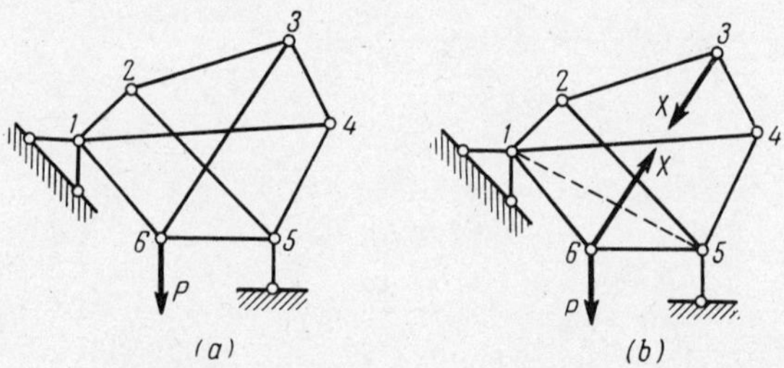

Fig. 48.4

systems by replacing one or more bars by the same number of other bars without disturbing the geometrical stability of the system as a whole.

As a rule, the analysis of such systems will require simultaneous solution of several equations with several unknowns. However, in many cases the complicated systems may be reduced to the simple ones, or to such systems which can be analyzed without solving equations with numerous unknowns, by a fictitious replacement (substitution) of bars. The additional equations permitting to solve the problem will express that the stresses in all the substitute members remain nil.

The following example will illustrate this method. Assume that it is required to find the stresses in all the members of the structure represented in Fig. 48.4*a* acted upon by some external force, say, load P applied at joint *6*. It will be immediately seen that in this structure three bars meet at each joint, hence the method of joints becomes inapplicable. At the same time the method of moments will lead to a number of equations each containing several unknowns, which is extremely undesirable. In these circumstances let us transform the structure into a simple system by replacing bar *6-3* by bar *1-5* as indicated in Fig. 48.4*b*. The structure so obtained forms an unyielding combination as it may be formed by the successive addition of joints *2*, *4* and *3* to a basic hinged triangle *1-5-6*, each of these joints being connected by means of two concurrent bars.

The analysis of such a *transformed* system is greatly simplified for the stresses in all the bars may be found, say, by the method of joints without the solution of equations with multiple unknowns.

Let us denote by X the stress induced by the load P in bar *6-3* and let us apply the same force to joints *3* and *6* of the transformed system in the direction of bar *3-6* (it is assumed that this bar is extended).

It is obvious that the stresses in all the members of both the original and the transformed systems will become exactly the same when the stress in the substitute bar *1-5* reduces to zero under the combined action of forces P and X. Indeed, the two systems will be identical for any bar may always be replaced by a force acting in the same direction and having the same magnitude as the stress in this bar, and when the stress is zero this means that the bar may be omitted without disturbing the system.

The principle of superposition enables us to express the stress in any member i of the transformed system (and accordingly of the original one, too) by

$$N_i = N_{ip} + \overline{N}_{ix} X \tag{2.4}$$

where N_{ip} = stress in the transformed system induced by the load P

$\overline{N}_{ix}$ = same stress induced by a unit load $X = 1$

The same formula applying to the substitute bar, we may write that the stress N_s in this bar equals

$$N_s = N_{sp} + \overline{N}_{sx} X = 0$$

wherefrom

$$X = -\frac{N_{sp}}{\overline{N}_{sx}} \tag{3.4}$$

Substituting the value of X thus obtained in the expression (2.4) we shall find the stresses in all the members of the system.

In more complicated cases it becomes sometimes necessary to replace two or more bars. In such cases the method just described will not dispense completely with the solution of several equations with several unknowns. The total stresses in the substitute bars will still reduce to zero, and their expressions will take the form

$$\begin{aligned} N_1 &= N_{1p} + \overline{N}_{11}X_1 + \overline{N}_{12}X_2 + \overline{N}_{13}X_3 + \dots = 0 \\ N_2 &= N_{2p} + \overline{N}_{21}X_1 + \overline{N}_{22}X_2 + \overline{N}_{23}X_3 + \dots = 0 \\ N_3 &= N_{3p} + \overline{N}_{31}X_1 + \overline{N}_{32}X_2 + \overline{N}_{33}X_3 + \dots = 0 \end{aligned} \tag{4.4}$$

where $N_1, N_2, N_3, \ldots =$ total stresses in the substitute bars *1, 2, 3*, etc.

$X_1, X_2, X_3, \ldots =$ unknown stresses in the bars which have been replaced

$\overline{N}_{11}, \overline{N}_{12}, \overline{N}_{13}, \ldots =$ stresses induced in substitute bar *1* by unit loads $X_1 = 1, X_2 = 1, X_3 = 1, \ldots$, respectively

$\overline{N}_{21}, \overline{N}_{22}, \overline{N}_{23}, \ldots =$ same stresses in substitute bar *2*, etc.

The values of the unknown stresses X_1, X_2, X_3, etc., will be in this case obtained by solving the system of equations (4.4).

In complicated structures the correct position of the substitute bar is not always clear. However, it may be found in the following way: having eliminated one bar reject one by one all the joints connected to the remaining structure by two distinct bars until a joint is found whose connections are insufficient. The additional bar needed to fix this joint with respect to the remainder of the structure will constitute the required substitute bar. If the structure so obtained still does not belong to the category of simple framed structures, another of its bars should be eliminated and further joints should be rejected until one more joint inadequately connected to the rest of the structure is found, indicating the position of the second substitute bar.

This procedure may be repeated as many times as necessary to transform the structure into a simple system.

Problem. Using the replacement method determine the stresses in all the members of a framed structure in Fig. 49.4*a* for $\sin\alpha = 0.6$ and $\sin\beta = 0.8$.

Solution. Replacing bar *3-6* by bar *1-5* as shown in Fig. 49.4*b* we obtain a simple system permitting the determination of the stress X in the replaced bar by equating to zero the stress in the substitute bar *1-5*

$$N_{15}=N_{15p}+\overline{N}_{15x}X=0$$

whence

$$X=-\frac{N_{15p}}{\overline{N}_{15x}}$$

N_{15p} and $\overline{N}_{15x}$ being the stresses induced in the substitute bar *1-5* by the load P and the unit force $X = 1$, respectively.

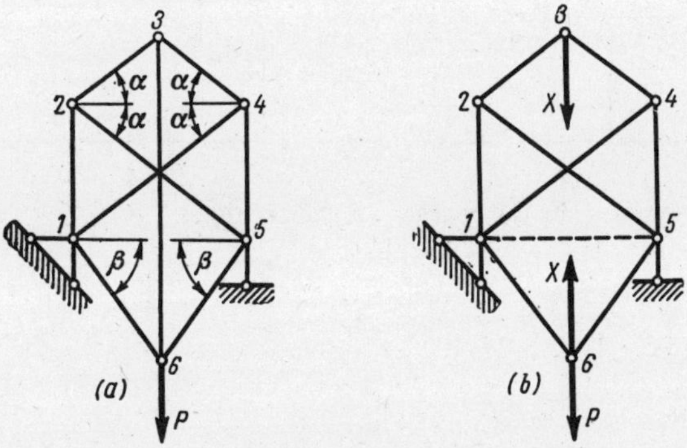

Fig. 49.4

The stresses in all the other members of truss will be found using the formula

$$N_i=N_{ip}+\overline{N}_{ix}X$$

where N_{ip} and $\overline{N}_{ix}$ are the stresses in the corresponding member of the transformed system induced by the load P and the unit force $X = 1$, respectively.

In this example the method of joints should be retained as its consecutive application to joints *3*, *2*, *4*, *6* and *5* will show immediately that only bars *1-6*, *5-6* and *1-5* of the transformed system are stressed by the load P, all the other members remaining idle.

All the computations are listed in Table 4.4. Entries into the 4th and the 5th columns have been made only after finding the stress X in the replaced bar *3-6*. Values appearing in column 4 have been obtained by multiplying those of column 2 by the magnitude of the stress X (see below), while the entries of column 5 result from the summation of figures contained in columns 3 and 4.

Table 4.4

Bar No.	Stress induced by unit force $X=1$	Stress induced by load P	Stress induced by force X	Total stress in members of the original system
2-3 or *4-3*	$-\frac{5}{6}$	0	$+\frac{15}{14}P$	$+\frac{15}{14}P$
1-2 or *5-4*	-1	0	$-\frac{9}{7}P$	$-\frac{9}{7}P$
2-5 of *4-1*	$+\frac{5}{6}$	0	$-\frac{15}{14}P$	$-\frac{15}{14}P$
1-6 of *5-6*	$-\frac{5}{8}$	$+\frac{5}{8}P$	$+\frac{45}{56}P$	$+\frac{10}{7}P$
1-5	$-\frac{7}{24}$	$-\frac{3}{8}P$	$+\frac{3}{8}P$	0

Note: The stress X in the replaced bar *3-6* equals

$$X=-\frac{N_{15p}}{\overline{N}_{15x}}=-\frac{3P\cdot 24}{8\cdot 7}=-\frac{9}{7}P$$

5.4. *Stress Distribution in Different Types of Trusses*

Stresses computed for trusses of the same span, the same height and the same number of panels and acted upon by the same system of loads, but differing in the outline of their upper chords have been illustrated in Figs. 32.4, 33.4, 34.4, 35.4 and 37.4 of Art. 2.4. Examining these figures it will be noted that in certain trusses the chord stresses increase from the abutments towards the centre line (Figs. 32.4, 35.4 and 37.4), while in other trusses they decrease (Figs. 33.4 and 34.4). In trusses of different shape but of the same web pattern the verticals and the diagonals may sustain stresses of different sign; thus, in the truss in Fig. 32.4 the diagonals are extended while in the truss in Fig. 33.4 they are compressed.

For a number of trusses the mode of stress variations in chord members, the sign of the stress in the elements of the web as well as certain other peculiarities of their performance may be predicted without detailed calculations.

As an example, let us take the three trusses represented in Fig. 50.4*a*, *b* and *c* which differ one from another only by the position of their diagonals.

In order to facilitate judgement regarding the sign of the stress induced in the different elements of these trusses by a uniformly distributed load we shall make use of an auxiliary uniformly loaded beam appearing in Fig. 50.4*d*. The M and Q diagrams for this beam are represented in Fig. 54.4*e* and *f*. The bending moment diagram shows that in the beam the lower fibres are extended and the upper ones are compressed, indicating that in a truss the upper chord will be compressed and the lower one extended. In the same beam the bending moment increases from the ends towards the middle and accordingly (the height of a truss with parallel chords remaining constant), the stresses in the chord members will also increase from the abutments towards the centre line.

Sections taken through the auxiliary beam and the trusses (section *I-I* in Fig. 50.4*a*, *b*, *c*, *d*) will help to find the signs of the stresses in the web members. The shear in section *I-I* of the beam being positive tends to lift the left-hand portion of the beam with respect to the right-hand one. Hence the sectioned diagonal of the truss shown in Fig. 50.4*a* will be extended as shown in Fig. 51.4*a* and the sectioned diagonals of the trusses of Fig. 50.4*b* and *c* will be compressed (Fig. 51.4*b* and *c*). The same reasoning will show that all the diagonals of the Pratt truss represented in Fig. 50.4*a* are extended, those of the Howe truss in Fig. 50.4*b* are compressed while in the Warren truss appearing in Fig. 50.4*c* extended diagonals will alternate with compressed ones.

Fig. 50.4*f* shows also that the shearing forces decrease towards the middle of the beam; similarly the stresses in the diagonals of the trusses will also drop towards midspan.

When the loads are applied to the upper chord of the truss in Fig. 50.4*c* its verticals *1*, *3* and *5* are compressed and verticals *2* and *4* are idle. Vice versa, if the load is applied to the lower chord verticals *1*, *3* and *5* will become idle and verticals *2* and *4* will become extended. This may be easily proved by considering the equilibrium of the appropriate joints of the truss.

The direction of the stresses in the verticals of the Pratt and the Howe trusses in Fig. 50.4*a* and *b* will be readily found considering the equilibrium of that portion of these trusses which lies to the right of section *II-II* (Fig. 52.4*a* and *b*). The shear acting in the left-hand portion being positive is directed upwards, compressing the vertical in the Pratt truss (Fig. 52.4*a*) and extending

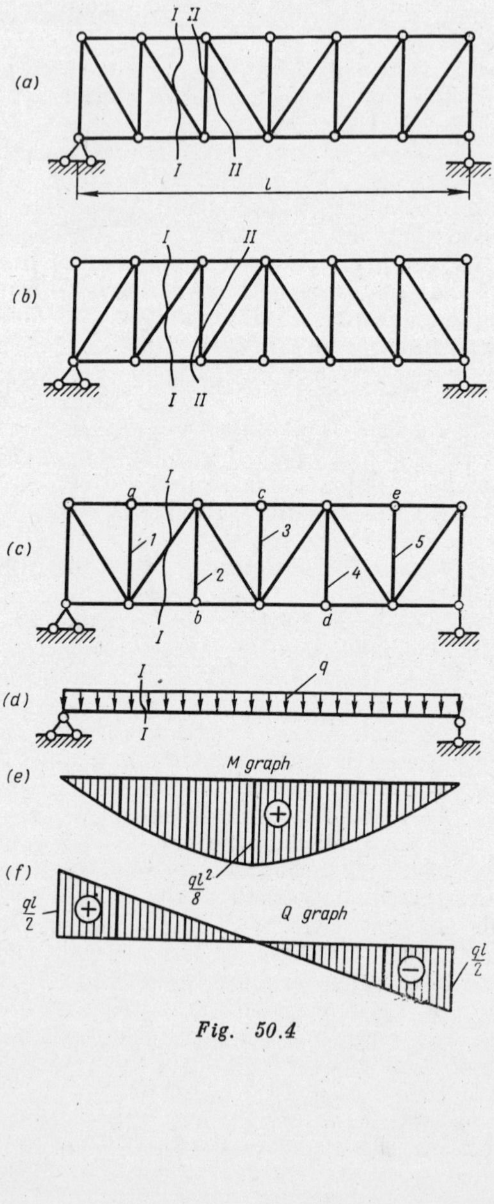
I
II
(a)
I
II
l
I
II
(b)
I
II
I
a
c
e
(c)
1
3
5
2
4
b
d
I
I
q
(d)
I
M graph
(e)
$\frac{ql^2}{8}$
(f)
$\frac{ql}{2}$
Q graph
$\frac{ql}{2}$

Fig. 50.4

it in the Howe truss (Fig. 52.4*b*). These stresses will also decrease towards the centre line of the trusses like the shearing forces in the simple beam (Fig. 50.4*f*).

All the above is readily confirmed by the stress diagrams in Figs. 32.4 and 35.4.

If the trusses were loaded differently, the stress distribution might alter considerably. For instance, if a single load were applied at midspan of a beam its bending moment and shear diagrams would be such as shown in Fig. 53.4*a*, *b* and *c*. In this case the shear remains constant along each half span. The same will apply to the stresses in the web members of the trusses.

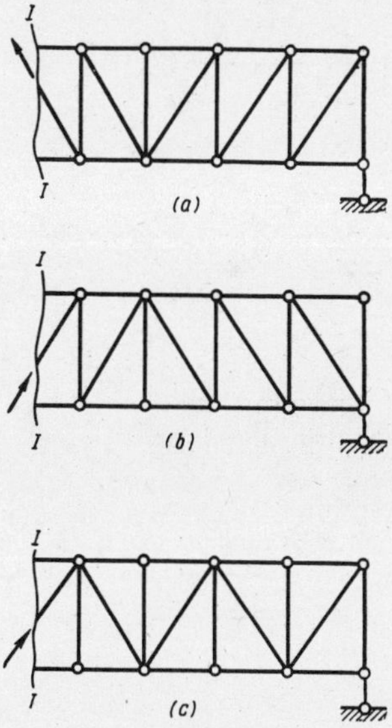

Fig. 51.4

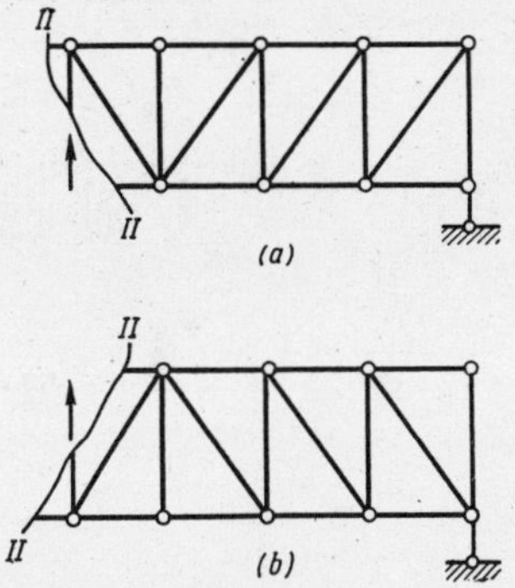

Fig. 52.4

When two symmetrical concentrated loads are applied at the hip joints of a truss (as in Fig. 54.4*a*) the stresses in all the chord members except the end ones will remain constant as may easily be deducted from the bending moment diagram of the auxiliary beam represented in Fig. 54.4*b*. At the same time the stresses in the web members will be nil (see shear diagram of a simple beam in Fig. 54.4*c*).

The analysis of stress distribution becomes considerably more complicated for trusses with nonparallel chords such as shown in Fig. 55.4*a*, *b* and *c*. When the upper chord follows exactly the bending moment diagram, the stress (extension) in the lower chord members will remain constant and the compression in the upper chord will be directly proportional to $\frac{1}{\cos\alpha}$ where α is the angle formed by the corresponding member of the chord with

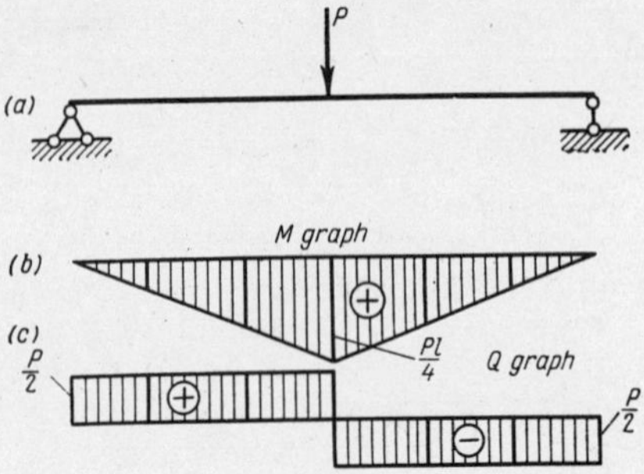

Fig. 53.4

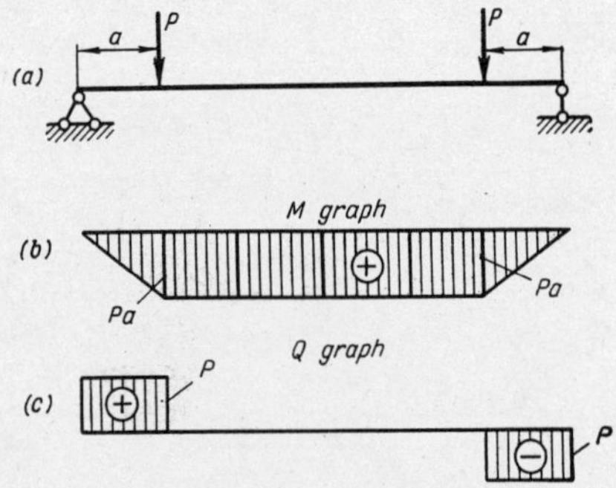

Fig. 54.4

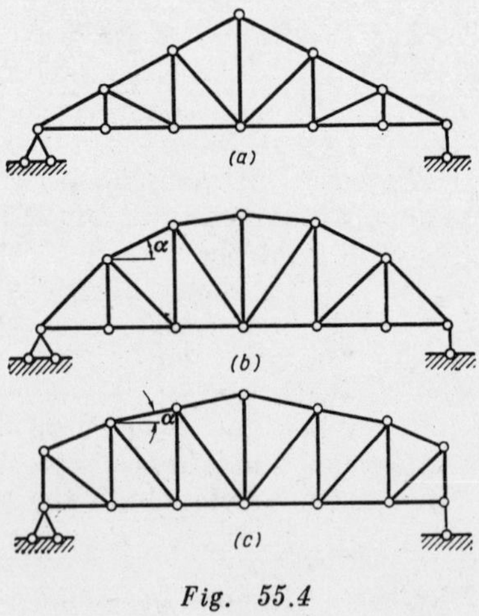

Fig. 55.4

a horizontal. Such will be the case of a uniformly loaded parabolic truss (compare the bending moment diagram in Fig. 50.4*e* with truss in Fig. 54.4*b*) or of a triangular truss carrying one concentrated load applied at its apex (compare the bending moment diagram in Fig. 53.4*b* with the truss in Fig. 55.4*a*). In these cases all the diagonals remain idle and the stresses in the verticals are either equal to the load applied at the corresponding joint (if the loads are carried by the lower chord) or become nil, when the load is applied to the top chord. The accuracy of these statements is well illustrated by the stresses computed for the parabolic truss represented in Fig. 34.4.

When the outline of the truss chords does not coincide with the bending moment diagram, only the signs of the stresses in the top and bottom chords and the mode of their variation may be still predicted fairly easily, the lower chord being always extended and the upper one compressed as long as unit stresses of the same sign continue to exist in the upper and lower fibres of the auxiliary simple beam.

Let us take for example a triangular truss acted upon by a uniform load and let us superpose the corresponding bending moment diagram in Fig. 50.4*e* on the schematic drawing of the truss as represented in Fig. 56.4. The scales should be so adjusted that maximum ordinates of both drawings coincide.

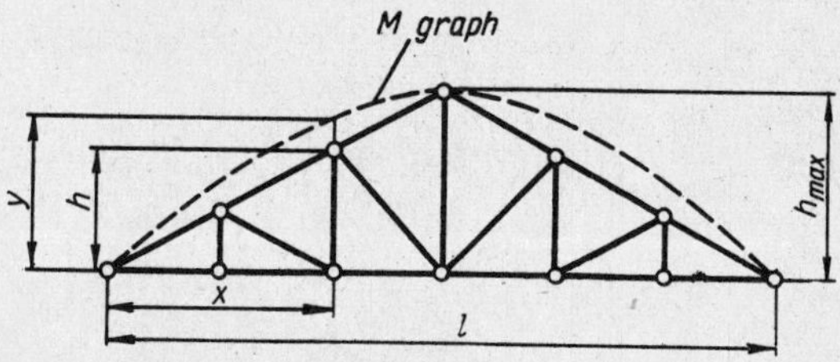

Fig. 56.4

The ordinate of the bending moment diagram a distance x from the left-hand abutment will be

$$y = h_{max} - \left(\frac{0.5l - x}{0.5l}\right)^2 h_{max} = 4x\frac{(l-x)}{l^2} h_{max}$$

At the same place the height of the truss equals

$$h = \frac{h_{max}}{0.5l} x = h_{max}\frac{2x}{l}$$

Accordingly the ratio $\frac{y}{h} = 2\left(1 - \frac{x}{l}\right)$ will decrease from abutment to centre line and so will the stresses in both the top and the bottom chords. Confirmation of this statement will be found in the diagram of stresses induced in a triangular truss by a uniform load represented in Fig. 33.4.

6.4. *Analysis of Geometrical Stability of Framed Structures*

1. SIMPLE STRUCTURES

It has been shown in Art. 2.1 that a framed structure may be instantaneously unstable even if the number of bars in each of its parts is sufficient to ensure its rigidity. Therefore, the number of bars forming a given structure cannot constitute alone a criterion of its geomertical stability.

In some cases instantaneously unstable structures can be detected fairly easily. Indeed it can be proved that in separate members of such structures finite loads will induce infinite or indefinite stresses.

Conversely it may also be shown that if any given load will produce a well defined set of finite stresses in all the members of a framed structure and that when the load is nil, all the stresses in all the members of this structure will also reduce to zero, this structure constitutes an unyielding combination. The method of investigating the rigidity of framed structures based on this property may therefore be termed the *zero load method.*

It should be noted however that *before applying this method care should be taken to ascertain that the number of bars in each part of the structure is sufficient to ensure its stability.* Otherwise erroneous conclusions may be arrived at as will be seen from the example of a hinged quadrangle represented in Fig. 57.4. Indeed the method of joints shows immediately that when no load is

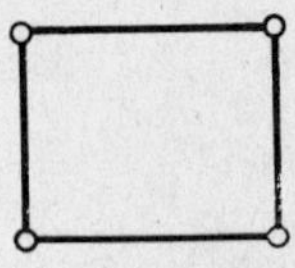

Fig. 57.4

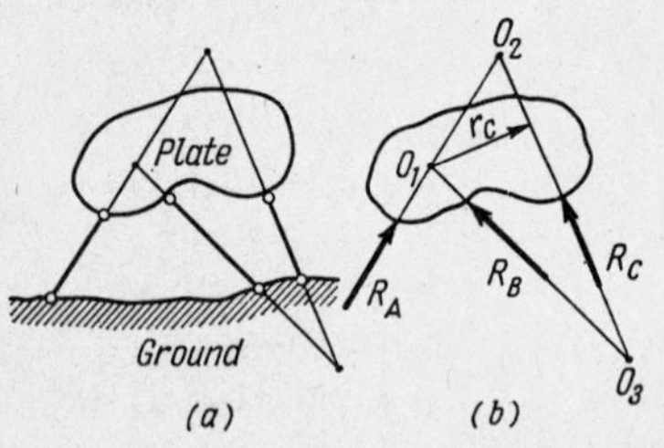

Fig. 58.4

applied the stresses in all the members are nil, but nevertheless the system is unstable.

In order to demonstrate the accuracy of the statements made above let us consider the following examples.

A plate connected to the ground by means of three nonconcurrent bars forms as we know an unyielding combination (Fig. 58.4). It is easily proved that under zero load the stresses in all the connecting bars will be necessarily nil. Indeed let us replace these bars by the corresponding reactions R_A, R_B, and R_C (Fig. 58.4*b*) and let us consider the equilibrium of the moments of all the

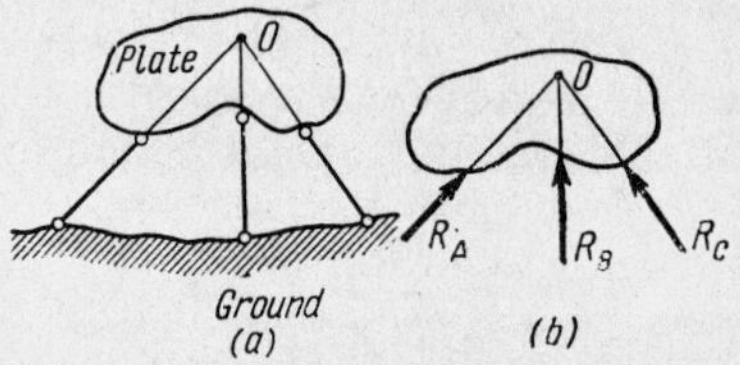

Fig. 59.4

forces acting on the plate about the point of intersection of reactions R_A and R_B (point O_1). We obtain

$$R_C r_C = 0$$

and as the lever arm $r_C \neq 0$, the reaction R_C is necessarily nil. The same reasoning will show that R_A and R_B are also nil. This serves to confirm the statement made above that *all the members of a geometrical stable system always remain idle when the structure carries no load.*

Now let us investigate the case when the plate is supported by three concurrent bars intersecting at point O (Fig. 59.4*a*). Replacing once again the bars by the corresponding reactions and equating to zero the sum of all the moments of external forces about point O we obtain the identity

$$\Sigma M_0 = R_A r_A + R_B r_B + R_C r_C = 0$$

for $r_A = r_B = r_C = 0$. Accordingly, the values of the reactions remain undetermined. The other two equilibrium equations (for instance, the equations of the force projections on the x and y-axes) will not help in finding a definite solution for they will contain three unknowns. Thus, *the stresses in an instantaneously unstable system may have no well defined value even when no load is applied.*

The same conclusion will be reached if some arbitrary value were attributed to any one of the reactions. It could then be resolved along the directions of the other two bars. The whole system being thus in equilibrium. That means that we can find any number of reaction values satisfying the equilibrium conditions, which indicates that the system is instantaneously unstable.

If the same system is subjected to some finite load P not passing through point O, the sum of moments of all external forces about this point becomes

$$\Sigma M_0 = R_A 0 + R_B 0 + R_C 0 + Pr \neq 0$$

as neither P nor r are zero. That means that the system is not in equilibrium and the plate will rotate about point O. However as soon as an infinitesimal rotation will have occurred, the three supporting bars will no longer remain concurrent and the reactions induced therein by the load P will be able to balance this load. At this particular moment the equilibrium equation about the same point O becomes

$$\Sigma M_0 = R_A r_A + R_B r_B + R_C r_C + Pr = 0$$

indicating that the reaction in at least one of the bars must be infinite, for the lever arms r_A, r_B and r_C are infinitely small. Hence, *the internal forces developed in an instantaneously unstable system acted upon by a finite load may surpass any given value* and therefore such systems cannot be used.

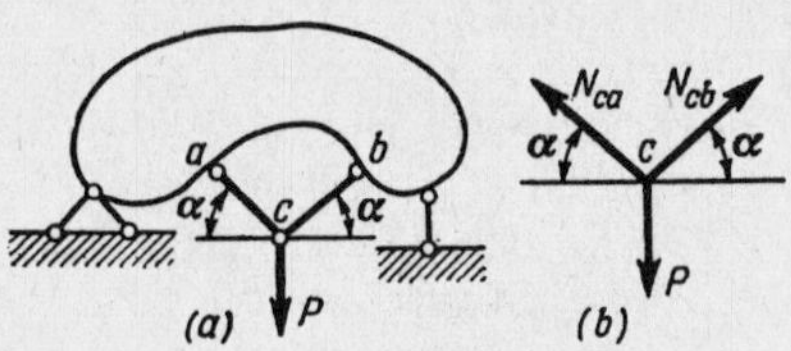

Fig. 60.4

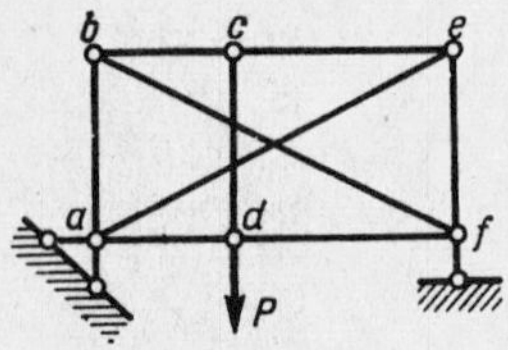

Fig. 61.4

Another example may be furnished by the geometrically stable structure in Fig. 60.4*a* consisting of a plate adequately connected to the ground and supporting joint c attached to it by two concurrent bars ac and bc. If a load P were applied to this joint (Fig. 60.4*b*), the stresses N_{ca} and N_{cb} in these bars will be given by

$$\Sigma X = -N_{ca} \cos \alpha + N_{cb} \cos \alpha = 0$$

$$\Sigma Y = N_{ca} \sin \alpha + N_{cb} \sin \alpha - P = 0$$

wherefrom

$$N_{ca} = N_{cb} = \frac{P}{2 \sin \alpha}$$

It follows that when the angles α formed by the two bars with the horizontal approach zero, the stresses in these bars will increase indefinitely proving that the system has become instantaneously unstable. Indeed, in that case joint c will be connected to the rest of the structure by two bars lying on one and the same horizontal and we know that such systems are unstable.

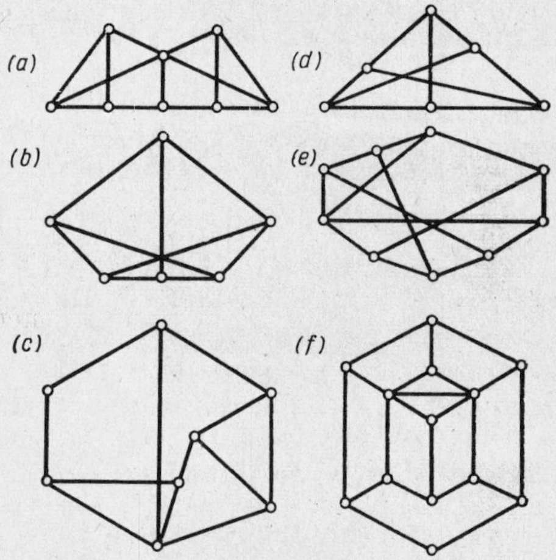

Fig. 62.4

One more example of instantaneously unstable structures is presented in Fig. 61.4. Although the number of bars in this system equals $2K-3$, the examination of equilibrium conditions at joints c and d leads to contradictory conclusions. Indeed the equilibrium of joint c requires that the stress in bar cd should be nil, while the equilibrium of joint d requires that it should equal $+P$. This controversity indicates clearly that the system is instantaneously unstable.

Thus, *if a system is provided with a number of bars sufficient to ensure its rigidity it will be instantaneously unstable, if*

(1) *finite forces induce in one or more members infinite stresses* or

(2) *the stresses cannot be determined or controversial stress values result from the consideration of different parts or joints of the structure.*

Fig. 62.4 represents a number of framed structures the stability of which the reader is invited to investigate using the zero load method. He should keep in mind that this method becomes inapplicable if the number of bars is inferior to $(2K-3)$.

2. COMPLICATED STRUCTURES

First, let us examine the case when the transformation of the complicated system into a simple one requires the replacement of but one bar.

The transformed system will consist of an elementary triangle to which a certain number of joints has been added, each connected by two concurrent bars and accordingly this system will form an unyielding combination; hence, the stress N_{ep} induced in the substitute bar by a load P will have a well defined and finite value. If a unit stress $X = 1$ directed along the bar that has been replaced induces in the substitute bar a stress $\overline{N}_{ex}$ also distinct from zero, then, in accordance with formula (3.4), the inner force X in the replaced member of the original system will equal

$$X = -\frac{N_{ep}}{\overline{N}_{ex}}$$

Since this stress is finite and well defined, the same will apply to all the other stresses induced by a load P in the original system which, as we know, proves that the system is geometrically stable.

On the other hand, if $\overline{N}_{ex} = 0$ then

$$X = -\frac{N_{ep}}{\overline{N}_{ex}} = \pm\infty \qquad \text{or } X = \frac{0}{0}$$

In other words, the stress in the replaced bar becomes either infinite or indeterminate indicating that the whole system is instantaneously unstable.

Accordingly, the expression $X = -\frac{N_{ep}}{\overline{N}_{ex}}$ constitutes a means of investigating the stability of complicated systems. When $\overline{N}_{ex} \neq 0$ the system forms an unyielding combination, and when $\overline{N}_{ex} = 0$ it is instantaneously unstable.

The above can be formulated as follows: *when the stress induced in the substitute bar of the transformed system by a unit force $X = 1$ acting along the replaced bar of the original system differs from zero,* *the system is geometrically stable, but when this stress becomes nil, the system is instantaneously unstable and unfit for practical use.*

Figs. 63.4 and 64.4 represent a certain number of original and transformed systems for which the reader is invited to check the

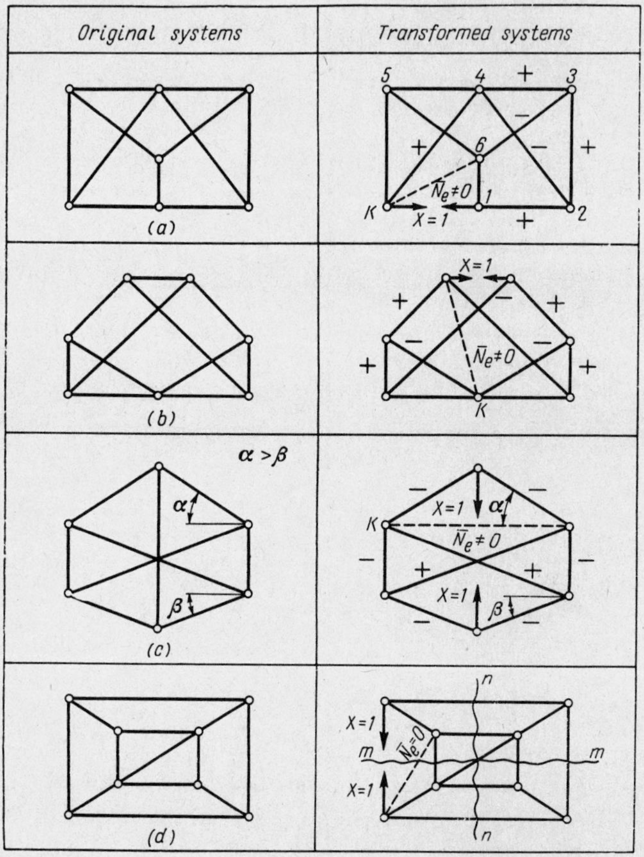

Fig. 63.4

accuracy of the value of $\bar{N}_{ex}$ indicated, and to decide accordingly whether the system is stable or not. The substitute bars are shown in dash lines.

The plus and minus signs placed against certain bars indicate the direction (sign) of the stress induced in the transformed system by a unit force $X = 1$ acting along the replaced bar of the original one. Knowing the direction of these stresses (the reader is invited to verify them) and considering the equilibrium of

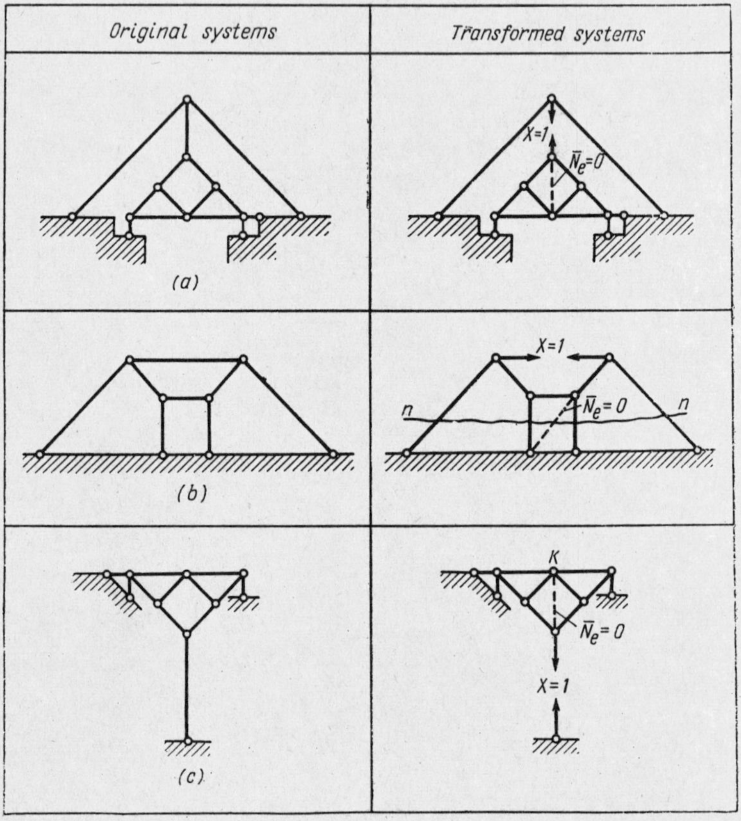

Fig. 64.4

joint K or using the method of shears or that of the moments, the reader will find in each case whether $\overline{N}_{ex}$ is nil or possesses some definite value.

Let us investigate, for instance, the system in Fig. 63.4 *a*. The equilibrium of joint *1* of the transformed system shows immediately that bar *1-2* is extended and that the stress in bar *1-6* is nil. Passing to joint *2* we see that bar *2-3* is extended and bar *2-4* is compressed. Moreover, the projection on the vertical of all the stresses acting on joint *4* will show that bar *4-K* must be extended in order to balance the push exerted by bar *2-4*. Hence the substitute bar *K-6* will be compressed, for otherwise the projections of all the forces applied to joint *K* on the horizontal will not balance and therefore the system is stable. The same result could have been arrived at by passing from joint *2* to joint *3* and then to joint *6*.

It is suggested that the reader should prove that the structure represented in Fig. 63.4*c* will become unstable when $\alpha = \beta$.

For the structure in Fig. 63.4*d* he will find that $\overline{N}_e$ is zero by taking in succession sections *n-n* and *m-m*. For the system in Fig. 64.4*b* it is easier to project on the horizontal all the forces acting above section *n-n*.

Using the same methods the reader should then investigate the stability of the structure in Fig. 65.4.

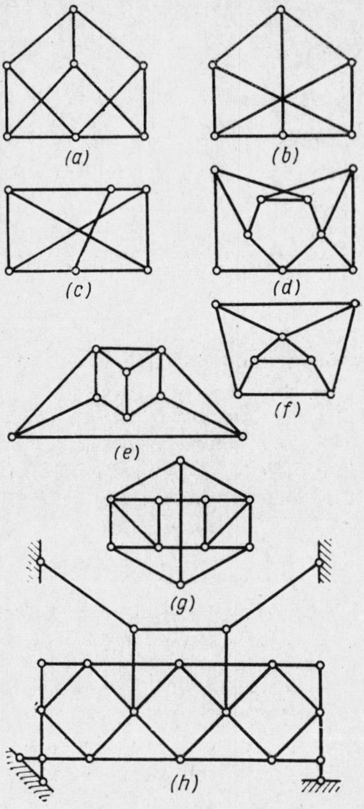

Fig. 65.4

When the transformation of a complicated system into a simple one requires the replacement of more than one bar, the equations denying the existence of a difference between the original and the transformed systems are, as we have already seen (Eq. 4.4)

$$N_1 = N_{1p} + \overline{N}_{11}X_1 + \overline{N}_{12}X_2 + \overline{N}_{13}X_3 + \ldots = 0$$

$$N_2 = N_{2p} + \overline{N}_{21}X_1 + \overline{N}_{22}X_2 + \overline{N}_{23}X_3 + \ldots = 0$$

$$N_3 = N_{3p} + \overline{N}_{31}X_1 + \overline{N}_{32}X_2 + \overline{N}_{33}X_3 + \ldots = 0$$

etc.

The stresses X_1, X_2, etc., arising in this case in the substitute bars will possess concrete values meaning that the whole system is stable only when the determinant D is different from zero, e. g., when

$$D = \begin{vmatrix} \overline{N}_{11}\overline{N}_{12}\overline{N}_{13} \\ \overline{N}_{21}\overline{N}_{22}\overline{N}_{23} \\ \overline{N}_{31}\overline{N}_{32}\overline{N}_{33} \end{vmatrix} \neq 0$$

On the contrary, when $D = 0$ the values of stresses X_1, X_2, etc., become uncertain, which indicates that the system is instantaneously unstable.

7.4. *Influence Lines for Stresses in Simple Framed Structures*

As the loads are generally applied to a truss at panel points everything that has been said in Art. 5.2 about the construction of influence lines for girders with floor beams and stringers remains true for those pertaining to trusses.

All the methods used for computing stresses induced by fixed loads (see Art 2.4) viz., the method of moments, the method of shears and the method of joints may be employed for the construction of influence lines.

The method of moments. In order to construct the influence line for the stress in bar *7-9* of the deck bridge truss in Fig. 66.4*a* we shall pass section *I-I* across three bars of the corresponding panel. When the unit load P is to the right of joint *8*, it is more convenient to consider the equilibrium of the left-hand part of the truss as the latter is acted upon solely by the abutment reaction (Fig. 66.4*b*).

Placing the origin of moments at point *6* and equating to zero ΣM of all the forces acting to the left of section *I-I* we obtain

$$\Sigma M_6 = A \cdot 3d - L_{79}h = 0$$

wherefrom

$$L_{79} = \frac{3Ad}{h}$$

Thus, when the load is applied to the right of joint *8*, the stress in bar *7-9* equals the left-hand reaction A miltiplied by a constant factor $\frac{3d}{h}$. It should be noted also that $3Ad$ is numerically equal to the bending moment M_6^0 acting over the cross section of a simple beam situated at the same distance from the supports as the origin of moments (point *6*) in the truss.

It is clear from the above that as long as the load remains to the right of point *8* the influence line for the stress L_{79} will be the same as for reaction A multiplied by $\frac{3d}{h}$. Hence the right-hand part of the influence line may be obtained by laying off $\frac{3d}{h}$ along the vertical passing through the left-hand abutment and by connecting it with a point of zero ordinate at the right-hand one (line a_1b in Fig. 66.4*d*).

When the load is to the left of joint *6* the stress L_{79} can be derived from the equilibrium equation relative to the right-hand part of the truss (Fig. 66.4*c*).

$$\Sigma M_6 = -B \cdot 5d + L_{79}h = 0$$

giving

$$L_{79} = \frac{5Bd}{h}$$

In other words, the stress in bar *7-9* equals in this case the right-hand reaction B multiplied by $\frac{5d}{h}$. Note that once again $5Bd$ is the equivalent of the simple beam bending moment M_6^0 acting over a section corresponding to point *6*. Hence for the load located to the left from joint *6* the influence line for stress L_{79} may be drawn by joining a point over the right-hand abutment having for ordinate $\frac{5d}{h}$ with a point of zero ordinate over the left-hand one (line b_1a in Fig. 66.4*d*). If all the operations have been carried out correctly lines a_1b and b_1a will intersect under joint *6*. We may now shade the area bounded by these lines between joints *6-16* and *1-6* respectively, i. e., the area *acba* in Fig. 66.4*d*.

Another way of obtaining the same influence line is based on the relation existing between the stress L_{79} and the simple beam bending moment M_6^0

$$L_{79} = \frac{M_6^0}{h}$$

This relation indicates that the influence line for stress in bar *7-9* can be obtained by dividing all the ordinates of a simple beam bending moment influence line by the height of the truss h. Incidentally, this proves once more that *lines a_1b and b_1a must intersect at a point lying in the vertical passing through joint 6* (point *c*).

The above example leads to the conclusion that the stress influence lines for end-supported trusses can be obtained using the following procedure:

1. *For the right-hand portion of the influence line lay off along the vertical passing through the left abutment (upwards or downwards depending on the sign of the stress) an ordinate $\frac{a}{h}$ where a is*

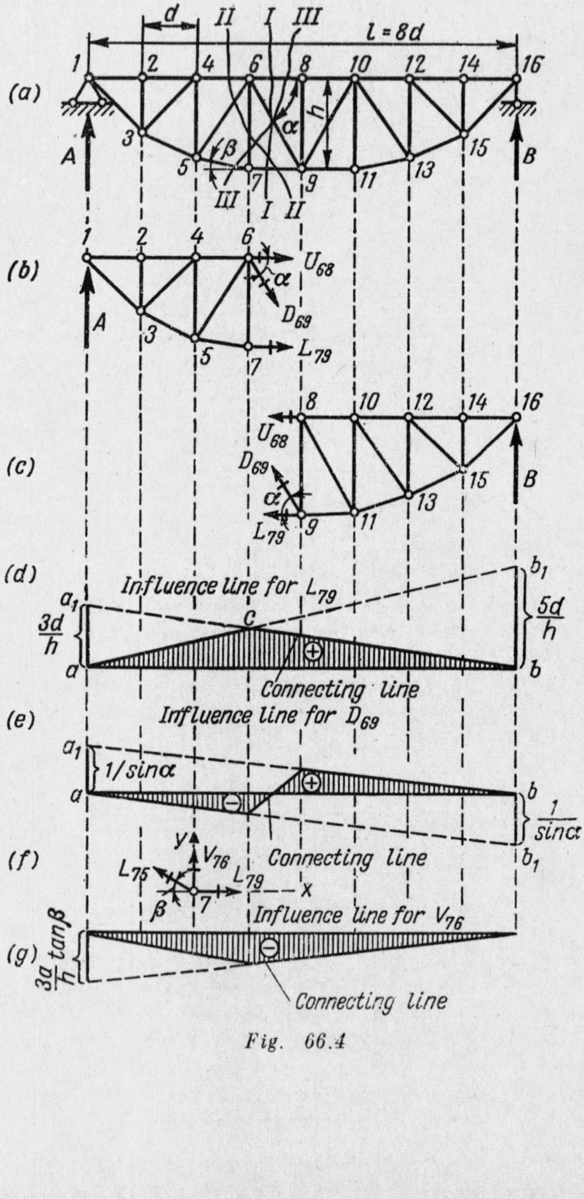

(a)
(b)
(c)
(d)
(e)
(f)
(g)
d
l = 8d
I
II
III
h
α
β
A
B
U68
D69
L79
Influence line for L79
3d/h
5d/h
a1
a
b1
b
c
Connecting line
Influence line for D69
1/sinα
Connecting line
y
V76
L75
L79
x
Influence line for V76
3a/h tanβ
Connecting line

Fig. 66.4

the distance of the origin of moments to the left-hand abutment, and h is the lever arm of the stress about the same point.

2. *Connect this ordinate with a point of zero ordinate at the right-hand abutment.*

3. *On the line so obtained mark the intersection point of the right- and of the left-hand parts of the influence line, this point lying in the vertical passing through the origin of moments.*

4. *Connect this intersection point with the point of zero ordinate over the left-hand abutment.*

5. *Connect by a straight line the two points of intersection of the above lines with the verticals bounding the panel which contains the bar under consideration.*

The sequence of all the operations would remain exactly similar if instead of laying off $\frac{a}{h}$ along the vertical passing through the abutment A we started by laying off $\frac{b}{h}$ along the one passing through the abutment B where b is the distance between this abutment and the origin of moments. Then the ordinate $\frac{b}{h}$ should be connected by a straight line with the point of zero ordinate over the left-hand abutment, the apex of the influence line should be found by projecting on this line the origin of moments, and finally the right-hand part of the influence line should be obtained by connecting this point with the point of zero ordinate at the right-hand abutment.

The method of shears. As an illustration of this method, let us draw the influence line for the stress in the diagonal *6-9* of the same truss (see Fig. 66.4*a*). The equilibrium of all the vertical projections of forces acting on the left-hand portion of the truss (Fig. 66.4*b*) when the unit load $P = 1$ travels between joints *8* and *16* requires that

$$\Sigma Y = A - D_{69} \sin \alpha = 0$$

leading to

$$D_{69} = \frac{A}{\sin \alpha}$$

When the load is situated between joints *1* and *6*, the same considerations relative to the right-hand portion of the truss (see Fig. 66.4*c*) entail

$$\Sigma Y = B + D_{69} \sin \alpha = 0$$

whence

$$D_{69} = -\frac{B}{\sin \alpha}$$

These two expressions giving the stress D_{69} in terms of the reactions show that when the load is to the right of joint *8* the influence line may be obtained by multiplying the reaction A by a constant factor $\frac{1}{\sin\alpha}$, and when it has shifted to the left of joint *6* we must apply a factor $\left(-\frac{1}{\sin\alpha}\right)$ to the influence line ordinates of reaction B.

Accordingly the construction of the corresponding parts of the influence line will consist in setting off the ordinates $+\frac{1}{\sin\alpha}$ over the left-hand abutment and $-\frac{1}{\sin\alpha}$ below the right-hand one and in connecting them with the points of zero ordinate at the other end of the truss. This will give us the lines a_1b and ab_1 in Fig. 66.4*e* respectively. Marking on the first line the position of joint *8* and on the second that of joint *6*, we obtain the right-hand (positive) and the left-hand (negative) parts of the influence line; these two points should be connected by a straight line.

It may be observed that in this case too the intersection point of the two portions of the influence line falls on the vertical passing through the origin of moments, both points being infinitely distant.

The change in the sign of the ordinates to the influence line obtained indicates that bar *6-9* will be consecutively compressed and then extended as the unit load travels along the deck from joint *1* to joint *16* as mentioned above; members designed to resist stresses of opposite sign are called counterbraces.

The method of joints may be conveniently used for the construction of the influence line for the stress in the vertical *6-7* (Fig. 66.4*a*). Both the method of moments and the method of shears would be of no avail in this case as any section through the truss would cross at least four bars (see sections *II-II* and *III-III* of the same figure).

Equating to zero the sum of vertical projections of all the forces acting at joint *7* (Fig. 66.4*f*) we obtain

$$\Sigma Y = V_{76} + L_{75}\sin\beta = 0$$

wherefrom

$$V_{76} = -L_{75}\sin\beta$$

and this is valid for any position of the load along the truss as it can never be applied directly to joint *7*, the bridge being of the deck type. Hence the influence line for the stress V_{76} could be derived from that of the stress L_{75} by multiplying its ordinates by $(-\sin\beta)$.

As for stress L_{75} it can be obtained by equating to zero the sum of horizontal projections of all the forces acting on the joint under consideration

$$\Sigma X = -L_{75} \cos \beta + L_{79} = 0$$

leading to

$$L_{75} = \frac{L_{79}}{\cos \beta}$$

Therefore

$$V_{76} = -L_{75} \sin \beta = -L_{79} \tan \beta$$

The same result can be achieved directly projecting all the forces applied to joint *7* on a normal to bar *7-5*.

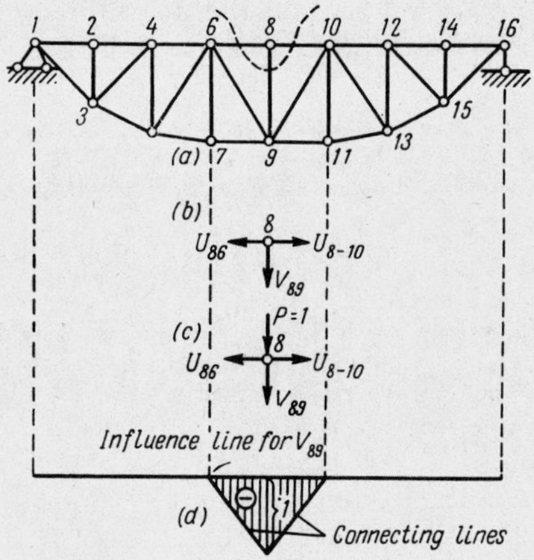

Fig. 67.4

The influence line for stress V_{76} obtained by multiplying the ordinates of the influence line for L_{79} by ($-\tan \beta$) is represented in Fig. 66.4*g*.

The influence line for the vertical *8-9* (Fig. 67.4*a*) should also be constructed using the method of joints for again any section through the truss cutting this bar will cross at least three more bars.

Considering the equilibrium of joint *8* we find immediately that (1) when the load is applied to any joint except joint *8* (Fig. 67.4*b*)

$$\Sigma Y = -V_{89} = 0$$

(2) when the load is applied to joint *8* (Fig. 67.4*c*)

$$\Sigma Y = -V_{89} - P = 0$$

and therefore

$$V_{89} = -P = -1$$

Consequently when the unit load is applied to any of the joints *1*, *2*, *4*, *6* or *10*, *12*, *14* and *16*, the vertical *8-9* remains idle, but when this load shifts to joint *8* the stress V_{89} becomes equal to 1.

Knowing the ordinates to the influence line at the relevant panel points and connecting these by straight lines we obtain the influence line required. This line represented in Fig. 67.4*d* has the shape of a triangle with a maximum ordinate equal to —1 over joint *8*. The sign of the ordinate indicates that the vertical can be only compressed and therefore constitutes a strut.

Problem 1. Draw the influence lines for the stresses in bars *7-8* and *7-9* of the Pratt truss shown in Fig. 68.4*a*.

Solution. The influence line for L_{79} will be obtained by the method of moments, adopting joint *8* as the origin of moments. The equilibrium of that portion of the truss to the left of section *k-k* (Fig. 68.4*b*) when the load is to the right of this section requires

$$\Sigma M_8 = A3d - L_{79}h = 0$$

and therefore

$$L_{79} = \frac{A3d}{h} = A\,\frac{3\times 3}{4} = 2.25A$$

Thus, the required influence line will be obtained by laying off an ordinate equal to 2.25 over the left-hand abutment, by connecting this ordinate with the zero ordinate point at the opposite end of the truss, by marking the position of the origin of moments (joint *8*) on this line and finally by drawing a line through the zero point at the left-hand abutment and the point just mentioned. The completed influence line will be of triangular shape with the apex directly under joint *8* (Fig. 68.4*c*).

The method of shears is well adapted for the construction of the influence line for the stress in bar *7-8*. Using section *n-n* (Fig. 68.4*d*) and equating to zero the projection of all the forces acting on the left-hand portion of the truss we obtain, when the load $P = 1$ is to the right of the section *n-n*,

$$\Sigma Y = A + V_{78} = 0$$

or

$$V_{78} = -A$$

Similarly when the load unity is to the left of section *n-n*, the equilibrium of the right-hand portion of the truss requires

$$\Sigma Y = B - V_{78} = 0$$

wherefrom

$$V_{78} = B$$

It should be noted that when the loads are transmitted through the upper chord (as in deck bridges) the first joint to the right of section *n-n* relative to the stress V_{78} is joint *8*, but when the loads are applied to the lower chord (through bridges) it will be joint *9*. The same will apply to joints *6* and *7*, the first being immediately to the left of section *n-n* in the case of deck bridges and the second in the case of through bridges. As the equations of equilibrium of the left- and/or the right-hand portions of the truss are independent of the level at which the loads are transmitted, the influence lines for both cases will be strictly parallel,

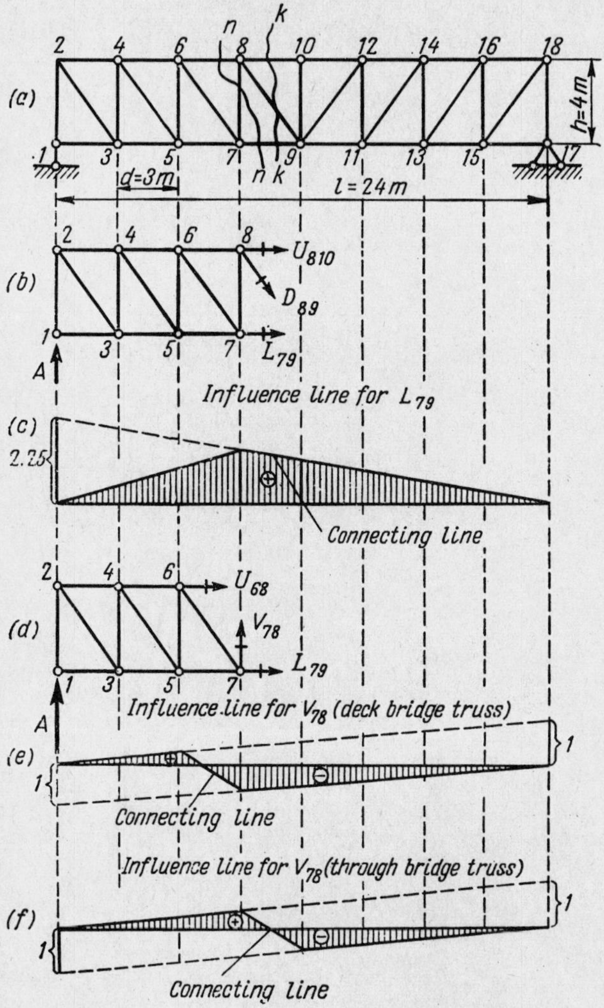

Fig. 68.4

but the position of the panel through which section n-n passes will vary leading to a displacement of the panel points corresponding to the apices of the line. The influence lines in Fig. 68.4e and f correspond to the two positions of the floor beams, the first pertaining to deck bridges and the second one to through bridges.

Problem 2. Required the influence line for stress D_{56} of the truss represented in Fig. 69.4*a*.

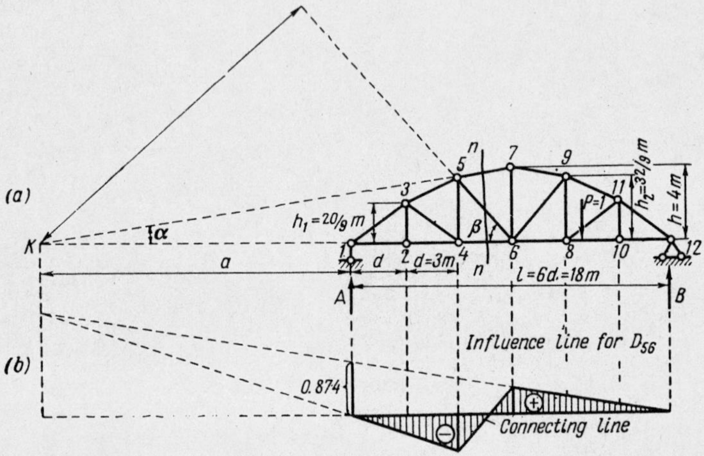

Fig. 69.4

Solution. Taking section *n-n* and using the method of moments (point K being taken as the origin) we find that when the load unity is to the right of our section

$$\Sigma M_k = -Aa + D_{56}r = 0$$

wherefrom

$$D_{56} = \frac{Aa}{r}$$

Here r is the lever arm of the stress D_{56} relative to point K, and a that of the reaction A about the same point. The distance a may be found from the triangle *K-5-4*

$$a + 2d = \frac{h_2}{\tan \alpha}$$

where h_2 is the height of the vertical *5-4* equal to $\frac{32}{9}$ metres and $\tan \alpha =$

$$= \frac{4 - \frac{32}{9}}{3} = 0.1481.$$

Hence

$$a + 2d = \frac{32}{9 \times 0.1481} = 24.0 \text{ metres} \qquad a = 18 \text{ m}$$

The lever arm r equals

$$r=(a+3d)\sin\beta$$

The angle β will be determined from

$$\tan\beta=\frac{32/9}{3}=\frac{32}{27}=1.185$$

Using tables of natural trigonometric functions we find *

$$\beta=49°50' \text{ and } \sin\beta=0.764$$

Using these values we find

$$r=(18+9)\,0.764=20.6 \text{ metres}$$

Substituting the above in the formula giving D_{56} in terms of a and r we obtain

$$D_{56}=\frac{18A}{20.6}=0.874A$$

The construction of the influence line for D_{56} will begin with its right-hand portion which will be formed by the line connecting the 0.874 ordinate over the left-hand abutment with the zero ordinate point at the other end of the truss. The left-hand portion will be obtained remembering that the directions of the two parts always intersect under the origin of moments. Within the panel containing section *n-n* a third line will connect the vertices lying under the panel points on both sides of section *n-n*. The completed influence line is shown in Fig. 69.4*b*.

Problem 3. Required the influence lines for stresses U_{75}, D_{56} and V_{76} arising in a triangular roof truss in Fig. 70.4*a* when the loads are applied to the lower chord.

Solution. Influence line for stress U_{75}. Passing section *n-n* and considering the equilibrium of the left-hand part of the truss when load unity P is to the right of the section we obtain

$$\Sigma M_6=A\cdot 3d+U_{75}r=0$$

wherefrom

$$U_{75}=-\frac{3Ad}{r}$$

The influence line represented in Fig. 70.4*b* will thus have a triangular shape with its apex directly under the origin of moments.

Influence line for stress D_{56}. Using the same section and equating to zero the sum of all the moments of forces acting on the left-hand part of the truss about point *1* we obtain, when the unit load $P=1$ is to the right of section *n-n*,

$$\Sigma M_1=D_{56}r_1=0$$

wherefrom

$$D_{56}=0$$

Hence the ordinates of the influence line will reduce to zero as long as the unit load is to the right of the panel containing bar *5-6*.

◆

* The same figures could be obtained using the formula

$$\sin\beta=\frac{\tan\beta}{\sqrt{1+\tan^2\beta}}$$

The left-hand portion of the influence line may be constructed using the equation of the equilibrium of moments pertaining to the right-hand part of the truss, the load unity P being to the left of section n-n,

$$\Sigma M_1 = -Bl - D_{56} r_1 = 0$$

wherefrom

$$D_{56} = -\frac{Bl}{r_1}$$

i. e., the stress D_{56} is equal to the right-hand abutment reaction B multiplied by $\left(-\frac{l}{r_1}\right)$.

The corresponding influence line appears in Fig. 70.4*c*.

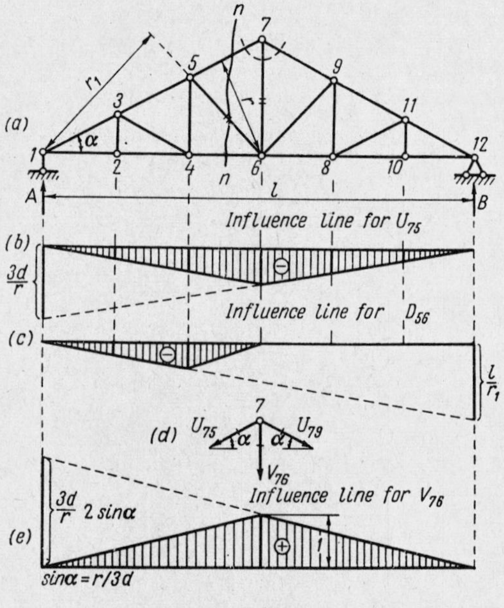

Fig. 70.4

Influence line for stress V_{76}. Using the method of joints and projecting all the forces acting on joint 7 on a horizontal we obtain

$$\Sigma X = -U_{75} \cos \alpha + U_{79} \cos \alpha = 0$$

indicating that

$$U_{75} = U_{79}$$

The projection of the same forces on a vertical gives

$$\Sigma Y = -V_{76} - 2U_{75} \sin \alpha = 0$$

wherefrom

$$V_{76} = -2U_{75} \sin \alpha$$

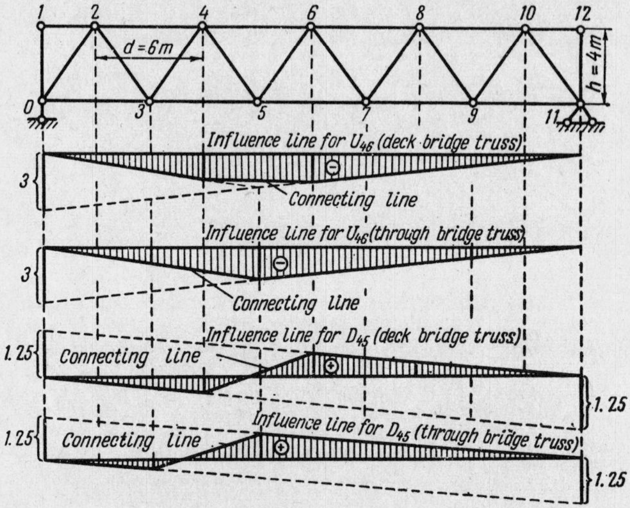

Fig. 71.4

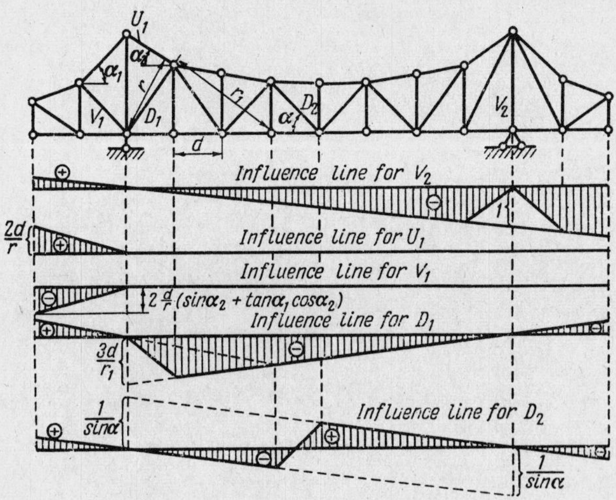

Fig. 72.4

Hence, the influence line for V_{76} may be obtained by multiplying all the ordinates of the influence line for U_{75} by a constant factor ($-2\sin\alpha$). The maximum ordinate of this influence line shown in Fig. 70.4*e* will be equal to 1.

The reader is invited to solve the following two problems on his own.

Problem 1. Prove the accuracy of the influence lines in Fig. 71.4.

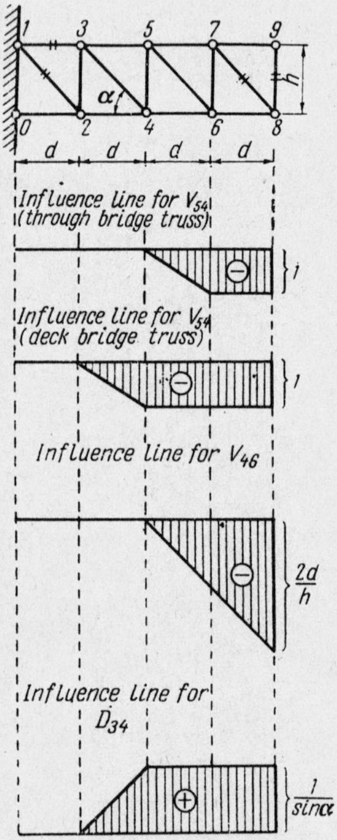

Fig. 73.4

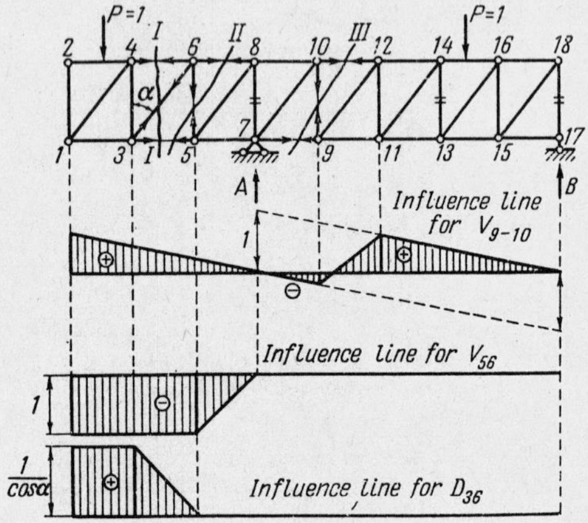

Fig. 74.4

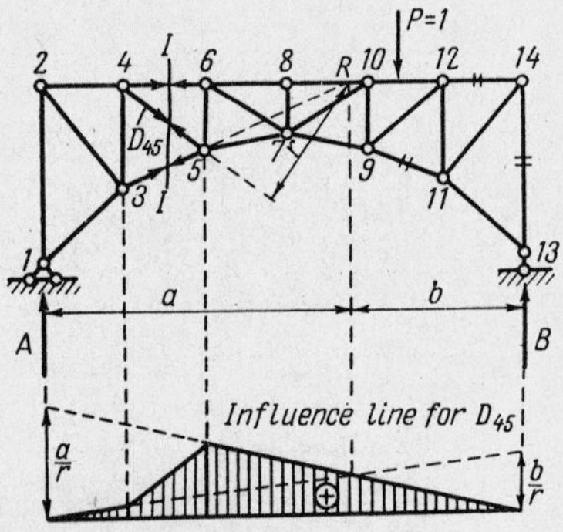

Fig. 75.4

Problem 2. (a) Prove the accuracy of the influence lines in Fig. 72.4 through 76.4 and

(b) draw the influence lines for the stresses in additional bars marked by a double dash.

Hints. (a) It is recommended to use the method of joints for the influence line for stress V_2 of the truss represented in Fig. 72.4. When the load unity P is applied to any joint with the exception of the joint over the right-hand abutment, $V_2 = -B$. When the load is over the right-hand abutment, $V_2 = 0$.

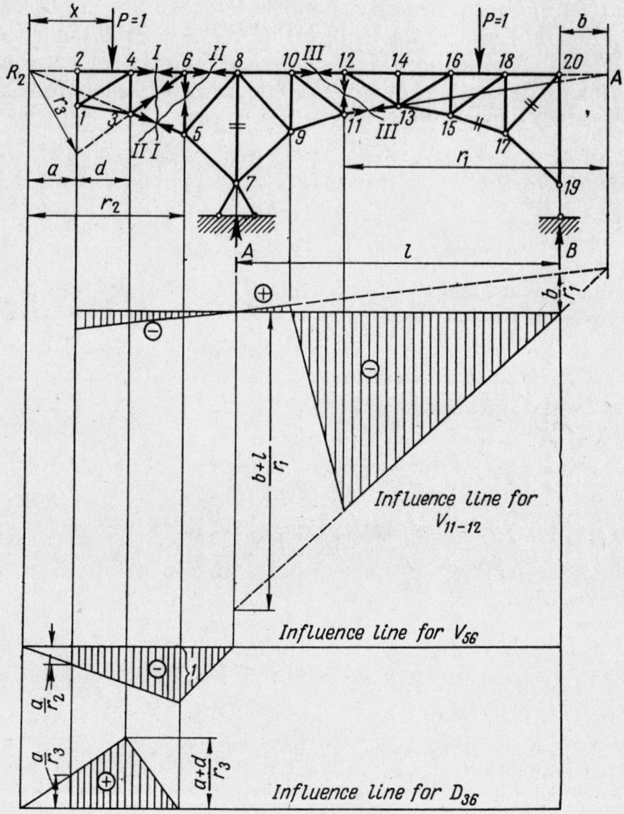

Fig. 76.4

(b) As regards the truss in Fig. 73.4 it is recommended to consider the equilibrium of that portion of the truss to the right of the section, when the load unity is to the left thereof. It is obvious that in this case the bar under consideration will remain idle.

8.4. *Influence Lines for Stresses in Complicated Framed Structures*

The design of complicated framed structures and in particular of multispan statically determinate ones may be carried out using the replacement method described in Art. 4.4, whereby the complicated truss is converted into a simple one.

As an example let us consider the truss represented in Fig. 77.4*a*. In order to obtain the influence line for the reaction *C* at the intermediate support when the load travels along the upper chord, let us replace the support C by a vertical member *6'-6* (Fig. 77.4*b*). At the joint *6'* we must then apply an external force X_c which will be equal to the reaction C when the stress in the substitute bar *6'-6* becomes nil

$$N_{6'6}=N'_{6'6}+X_c\overline{N}_{6'6}=0$$

wherefrom

$$X_c=-\frac{N'_{6'6}}{\overline{N}_{6'6}}$$

Here $N'_{6'6}$ is the stress in bar *6'-6* of the truss shown in Fig. 77.4*b* when the load unity travels along the upper chord and $\overline{N}_{6'6}$ is the stress in the same bar induced by the force $X_c = 1$.

Hence the influence line for the abutment reaction X_c may be obtained by dividing the ordinates to the influence line for stress $N'_{6'6}$ by $(-\overline{N}_{6'6})$.

The influence line for the stress $N'_{6'6}$ may be constructed using the equilibrium equations regarding joint *6'* (Fig. 78.4*a* and *b*).

$$\Sigma X=-N'_{6'5'}\cos\alpha+N'_{6'7'}\cos\alpha=0$$

$$\Sigma Y=N'_{6'5'}\sin\alpha+N'_{6'7'}\sin\alpha+N'_{6'6}=0$$

wherefrom

$$N'_{6'6}=-2N'_{6'5'}\sin\alpha$$

Substituting $N'_{6'6}$ in the expression for reaction X_c we obtain

$$X_c=\frac{2N'_{6'5'}\sin\alpha}{\overline{N}_{6'6}}$$

In order to obtain the influence line for $N'_{6'5'}$ let us pass section *II-II* (see Fig. 78.4*a*) and let us consider the equilibrium of the left-hand part of the truss assuming that unit load $P = 1$ is to the right of the section

$$\Sigma M_6=A\cdot 5d-N'_{6'5'}\frac{d}{\sin\alpha}=0$$

wherefrom

$$N'_{6'5'}=5A\sin\alpha$$

Consequently, the right-hand portion of the influence line for stress $N'_{6'5'}$ may be obtained by laying off the ordinates $5\sin\alpha=\frac{5\sqrt{2}}{2}$ above the left-hand abutment and by connecting it with the point of zero ordinate over the right-hand one.

The left-hand part of the influence line will be obtained remembering that the lines always intersect under the origin of moments (point *6*).

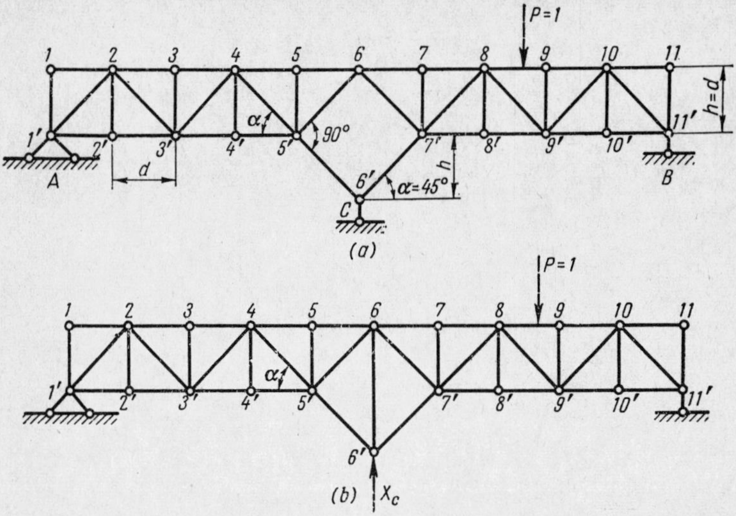

Fig. 77.4

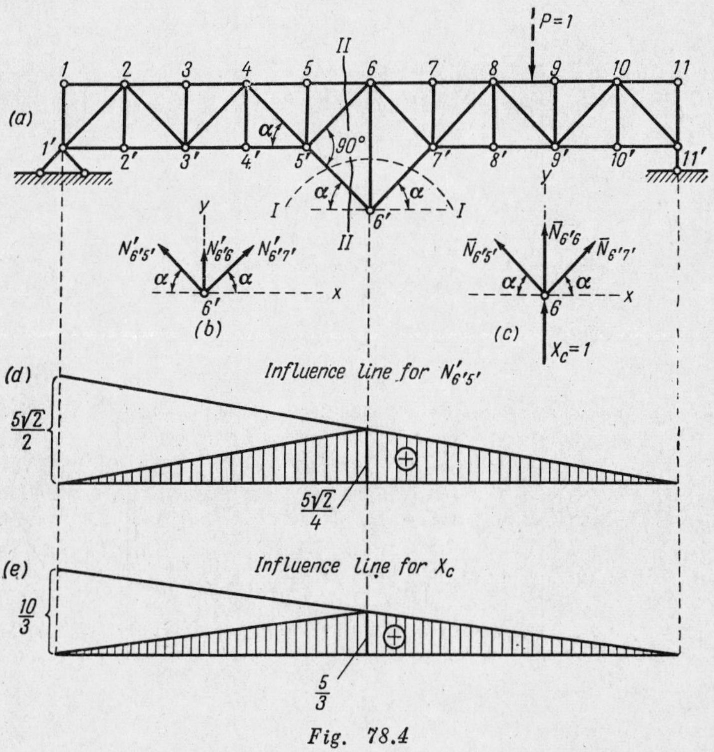

Fig. 78.4

The corresponding influence line for the simple truss (Fig. 78.4*a*) is represented in Fig. 78.4*d*.

Let us determine now the stress induced in bar *6′-6* by a force $X_c = 1$ using for that purpose the equilibrium of joint *6′* (Fig. 78.4*c*)

$$\Sigma Y = \bar{N}_{6'6} + 1 + 2\bar{N}_{6'5'} \sin\alpha = 0$$

where

$$\bar{N}_{6'5'} = -\frac{5\sqrt{2}}{4} \text{ (see Fig. 78.4}d\text{)}$$

Consequently

$$\bar{N}_{6'6} = -1 + 2\frac{5\sqrt{2}}{4} \times \frac{\sqrt{2}}{2} = \frac{3}{2}$$

therefore

$$X_c = 2N'_{6'5'}\frac{\sin\alpha}{\bar{N}_{6'6}} = \frac{2N'_{6'5'}\sqrt{2} \times 2}{3 \times 2} = \frac{2\sqrt{2}}{3}N'_{6'5'}$$

Hence, the influence line for reaction X_c will be obtained by multiplying all the ordinates to the influence line for $N'_{6'5'}$ by a constant factor equal to $\frac{2\sqrt{2}}{3}$.

The corresponding influence line is represented in Fig. 78.4*e*; with its aid the influence lines for stresses in all the other bars of the truss can be easily obtained.

9.4. *Trusses with Subdivided Panels*

When the method of moments is used the stress in any member of a truss can be expressed by the formula

$$N = \pm\frac{M}{r}$$

where M = moment of the forces to the right or to the left of the section about the origin of moments

r = lever arm of the stress N about the same point.

The above formula shows that other conditions remaining unchanged, the stress N decreases proportionally to the increase in the lever arm r. Accordingly, the increase in the height of the truss which always leads to the increase of the lever arm r will entail a reduction in the stresses induced in its elements.

Structurally it is more convenient when the diagonals form an angle close to 45° with the horizontal and therefore an increase in the height of the truss will lead to a lengthening of the panels. Thus in a truss with parallel chords the length of a panel will usually be very close to the truss height (Fig. 79.4*a*). However, panels of increased length require the use of heavier floor beams and stringers which may outweigh the economy obtained through the reduction of stresses in the truss members.

A rational solution of the problem resides in the subdivision of the panels with the introduction of secondary members, forming auxiliary king-posted beams, which will transmit the loads applied within the panel to the joints of the main truss.

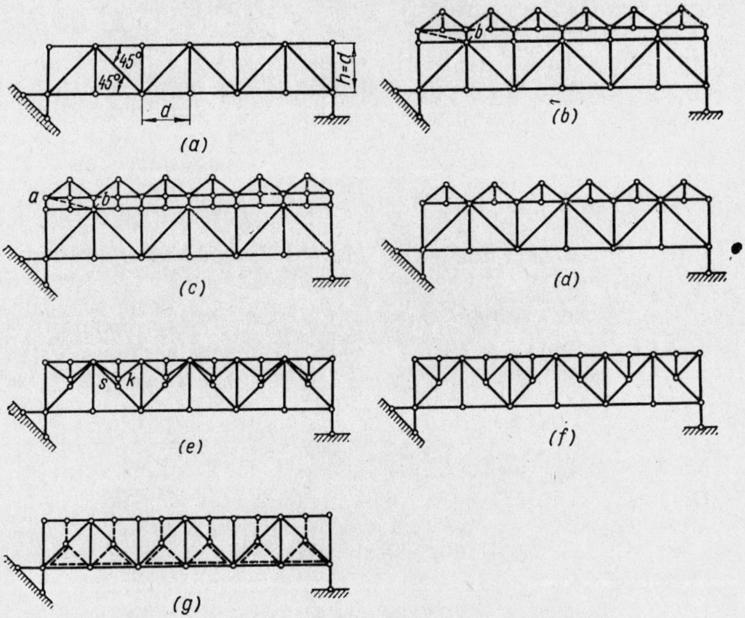

Fig. 79.4

These auxiliary systems will permit the installation of cross beams at intermediate points which provides for a considerable reduction in the weight of the floor elements. These systems will remain idle as long as the load is outside the panel which they reinforce, and will become stressed only while the load is within the limits of that panel. In Fig. 79.4*b* we have represented a deck-bridge truss, the upper chord of which is reinforced as described above. The bar *ab* is always idle, its only purpose being to ensure the stability of the combined system.

If the king-posts were extended downwards and connected to the upper chord members we would obtain the truss shown in Fig. 79.4*c*. It will be readily observed that the stresses in all the members of the latter truss are identical to those of the truss in Fig. 79.4*b*. A gradual shortening of all the vertical members connecting the auxiliary king-posted beams with the upper chord leads to the system represented in Fig. 79.4*d* in which the beams coincide with the upper chord members of the main truss. If we now turn the king-posted beams upside down we will obtain the truss shown in Fig. 79.4*e*, and if in the latter the length of *ks* becomes nil, we will finally obtain a deck-bridge truss with subdivided panels represented in Fig. 79.4*f* which in the English speaking countries is usually called a *subdivided Warren truss.**

The secondary elements represented in Fig. 79.4*f* transmit the loads applied to the upper chord to the main joints of the same chord. In other cases these elements may transmit the loads applied to the lower chord to the joints of the upper one or vice versa as for instance in Fig. 79.4*g*.

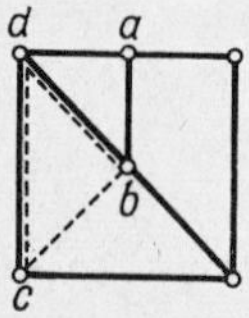

Fig. 80.4

It should be noted that auxiliary systems similar to that shown in Fig. 80.4 cannot be regarded as constituting a genuine trussed beam reinforcement, for in addition to vertical loads it will transmit equally horizontal forces to the joints of the main system.

In structures, where the secondary elements (subverticals and subdiagonals, as they are frequently called) transmit the loads to the main joints of the same chord, all the members may be regarded as belonging to three groups:

1. Members belonging to the main truss, the stresses in which are not influenced by the presence of auxiliary systems.

2. Members belonging entirely to the auxiliary systems, the stresses in which may be obtained in the same manner as for an isolated end-supported trussed beam.

3. Members belonging simultaneously to the main and the auxiliary systems. Stresses in such members will be obtained by the summation of those pertaining to the main and the auxiliary systems considered separately.

When the secondary members transmit the load from the upper chord to the lower one or vice versa, the truss members will form four distinct groups. Three of these have been just enumerated

* In Russia, trusses of that type were first used by the eminent Russian engineer and scientist, Professor L. Proskuryakov of the Moscow Institute of Railway Transportation. A bridge of this type was designed by him in 1895 and built across the river Yenisei, all the stresses in this truss having been determined with the aid of influence lines. The rigidity and the reduced weight of this bridge have placed it among the top-ranking engineering achievements of that time.

while *the fourth is constituted by such members for which the influence lines change depending on whether the load travels along one or the other chord as the performance of such members is altered by the presence of the secondary ones.*

The influence lines for the stresses in members of the fourth group will be obtained as follows: first draw the influence line for the appropriate member of the main truss both for the case of a load travelling along the upper chord and along the lower one, disregarding the presence of the secondary elements. This being done, examine the effect of the secondary members, for which purpose shift the load from joint to joint of the auxiliary system, noting with care to which member of the main truss this load is transmitted.

Problem 1. Draw the influence lines for the stresses in members *2-3*, *5-4'* and *4'-7* of the through bridge truss with subdivided panels represented in Fig. 81.4*a*.

Solution. Start with the construction of the influence line for stress V_{23}. The member *2-3* belongs to the first group and therefore the corresponding influence lines may be obtained disregarding completely the subverticals and subdiagonals (see Fig. 81.4*b*). Using the method of joints and considering the equilibrium of joint *3* we shall find a triangular influence line represented in Fig. 81.4*c*. The subdiagonal *4'-7* belongs to the second group, the stress $D_{4'7}$ may be obtained as for an isolated king-posted beam shown in Fig. 81.4*d*. In this case it is easily found that when the load unity acts at the joint *5'* the stress in bar *4'-7* will be given by the equation

$$\Sigma Y = \frac{1}{2} + D_{4'7} \sin \alpha = 0$$

or

$$D_{4'7} = -\frac{1}{2 \sin \alpha}$$

and when the load shifts to the supports, the stress $D_{4'7}$ becomes nil. The corresponding influence line is represented in Fig. 81.4*e*.

As for the stress in bar *5-4'* which belongs to the third group we shall pass a section *I-I* and assuming that the load unity is to the right of this section, we shall obtain

$$\Sigma Y = A + D_{54'} \sin \alpha = 0$$

wherefrom

$$D_{54'} = -\frac{A}{\sin \alpha}$$

This equation indicates that as long as the load is to the right of section *I-I* the influence line for $D_{54'}$, may be obtained by multiplying the ordinates to the influence line for the abutment reaction A by $\left(-\frac{1}{\sin \alpha}\right)$. Having thus obtained the right-hand portion of the influence line required, we may draw its left-hand portion using the rule that they must intersect in the vertical passing through the origin of moments and that when the load reaches the left-hand abutment the ordinate to the influence line reduces to zero. In the case considered the origin of moments is infinitely distant, the truss chords being parallel. The influence line will be completed by connecting points a and c corresponding to joints 5 and 5′ (Fig. 81.4f). It is interesting to note that for the truss of Fig. 81.4b we should connect points a and b corresponding to joints 5 and 7 eliminating thereby the triangle abc which represents the influence line for the member 5-4′ of the auxiliary system (similar to the one shown in Fig. 81.4e for the member 4′-7).

Problem 2. Required the influence line for the stress V_{34} of the through bridge truss shown in Fig. 82.4a.

Solution. The vertical under consideration belonging to the fourth group of members, we must begin with the construction of the influence lines relative to this member for loads travelling along the upper and lower chords of the main system, represented in Fig. 82.4b.

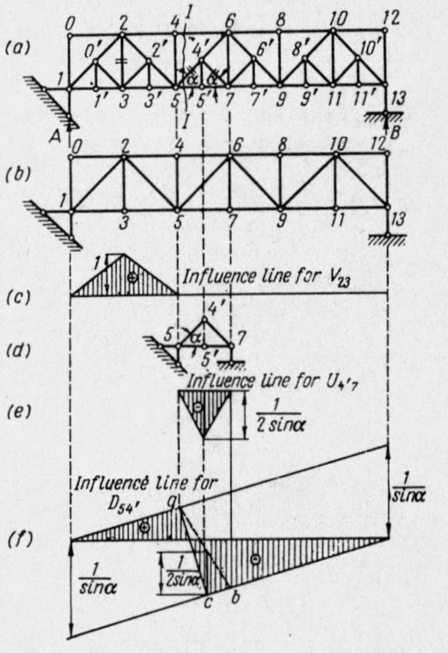

Fig. 81.4

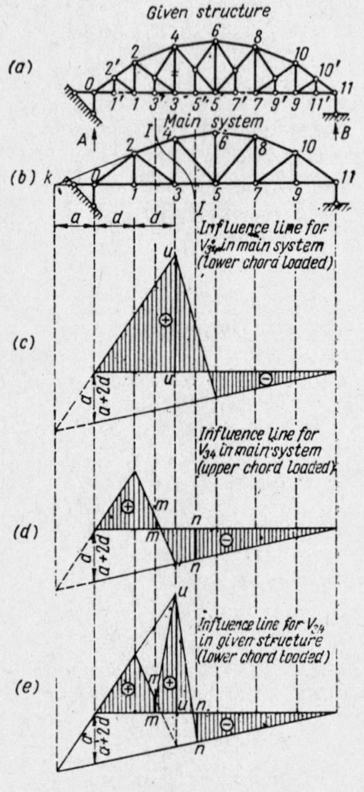

Fig. 82.4

For this purpose let us pass section I-I and write that ΣM about point k for the left-hand part of the truss equals zero when the load unity is to the right of this section

$$\Sigma M_k = -Aa - V_{34}(a+2d) = 0$$

whence

$$V_{34} = -\frac{Aa}{a+2d}$$

Connecting the ordinate $\frac{a}{a+2d}$ at the left-hand abutment with the zero ordinate at the right-hand one we shall obtain the right-hand portion of the influence line required. Its left-hand portion will be derived from the rule that the two lines always intersect under the origin of moments (point k).

In case the load travelling along the lower chord the completed influence line will be obtained tracing the connecting line through the points corresponding to joints *3* and *5* (Fig. 82.4*c*) and, in the event the loads were applied to the upper chord, through the two points corresponding to joints *2* and *4* (Fig. 82.4*d*).

These two influence lines show that when the load is to the left of joints *1* and *2* or to the right of joints *5* and *6*, the stress in the vertical *3-4* is independent of the level of load application. But when the load stands over joints *3'* or *5'* of the lower chord the secondary members will transmit it entirely to the joints of the upper one, which in effect is equivalent to the transfer of the load itself. Accordingly ordinates *m-m* and *n-n* will prevail at these moments. Nevertheless when the load moves to joint *3* all the secondary members become idle and it will be the ordinate *u-u* in Fig. 82.4*c* that will give the value of the stress V_{34}.

These ordinates will suffice for the construction of the influence line for the truss with subdivided panels. The required influence line bounds the shaded area in Fig. 82.4*e*.

The following problem should be solved by the reader on his own.

Problem. (a) Check the influence lines pertaining to the through bridge trusses in Figs. 83.4 and 84.4,

(b) draw the influence lines for stresses in the members of the same trusses marked by a double dash.

Hints. Prior to the construction of the influence line for stress V_{89} of the truss in Fig. 84.4, eliminate all the secondary members, thus finding the main system represented in Fig. 85.4*a*. Then using the method of joints find the stress V^0_{89} relative to this system

$$\Sigma Y = -V^0_{89} - 2U^0_{79} \sin \alpha = 0$$

wherefrom

$$V^0_{89} = -2U^0_{79} \sin \alpha$$

The stress in bar *8-9* of the main system is thus equal to that in bar *7-9* of the same system multiplied by a constant factor ($-2 \sin \alpha$). The influence line for the stress U^0_{79} is given in Fig. 85.4*b*. It has the shape of an isosceles triangle and its ordinate at the apex equals

$$\frac{l}{4 \times 3.5d \cos \alpha} = -\frac{18}{4 \times 3.5} \cdot \frac{\sqrt{3^2 + 0.5^2}}{3} = -\frac{3\sqrt{37}}{14}$$

The influence line for V^0_{89} will have the same shape and, provided the load travels along the lower chord, its maximum ordinate will equal (Fig. 85.4*c*)

$$\frac{3\sqrt{37}}{14} \cdot \frac{2 \times 1}{\sqrt{37}} = \frac{3}{7}$$

On the other hand, when the unit load travelling along the upper chord reaches joint *9* the equilibrium of this joint requires that

$$\Sigma Y = -V^0_{89} - 2U^0_{79} \sin \alpha - 1 = 0$$

and

$$V^0_{89} = -2U^0_{79} \sin \alpha - 1 = -\frac{4}{7}$$

This influence line is shown in Fig. 85.4*d*.

The comparison of the influence lines of Fig. 85.4*c* and *d* indicates that when the load is either to the left of joint *6* or to the right of joint *10* the stress is independent of the level of load application.

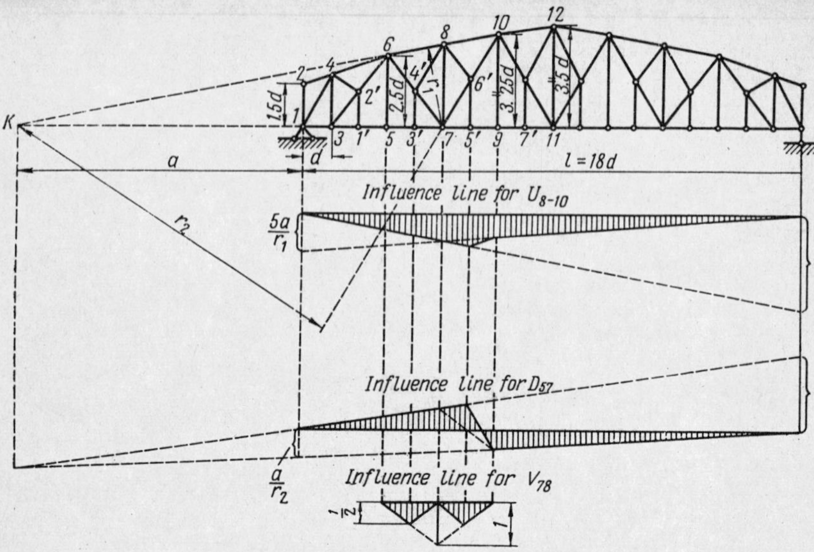

Fig. 83.4

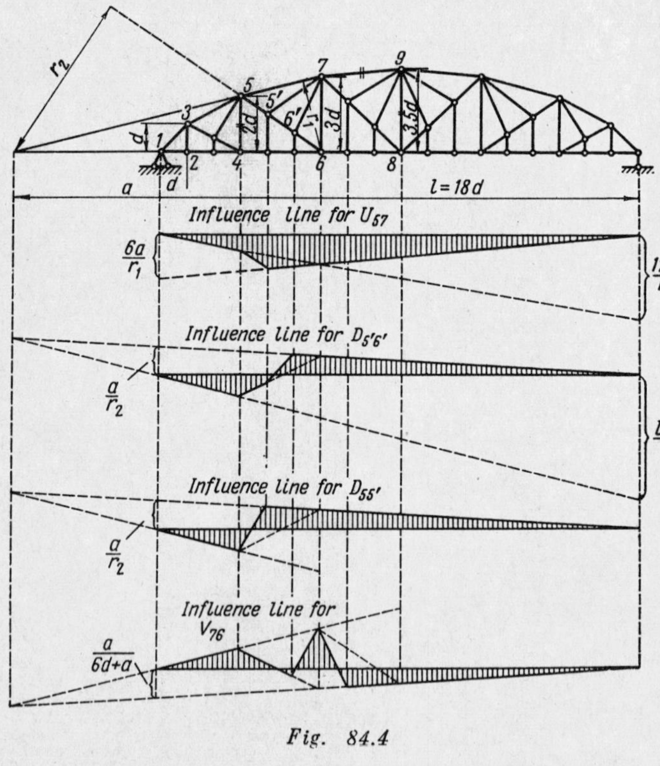

Fig. 84.4

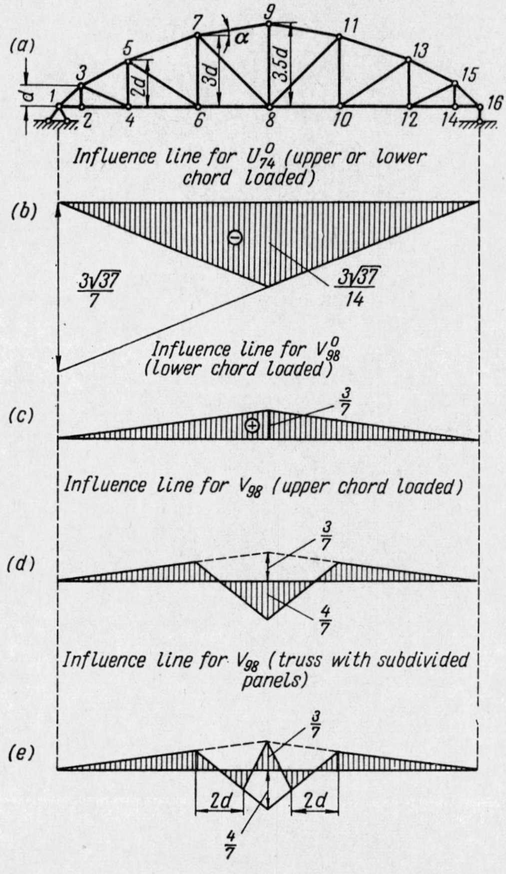

Fig. 85.4

At the same time any load applied to the secondary joints of panels *6-8* or *8-10* is transmitted to the upper chord and may be regarded as acting directly at the joints *7*, *9* or *11*.

The corresponding influence line for the truss with subdivided panels is shown in Fig. 85.4*e*.

10.4. *Thrust Developing Framed Structures*

1. TRUSSES WITH INCLINED SUPPORTS

If the vertical supporting bar representing the roller support of an ordinary truss is replaced by an inclined one, the system becomes a thrust developing truss as in addition to vertical reactions it will be characterized also by horizontal reactions at the abutments.

Let us examine the arched truss in Fig. 86.4*a*. Denoting by V_A, H_A and V_B, H_B the vertical and horizontal components of

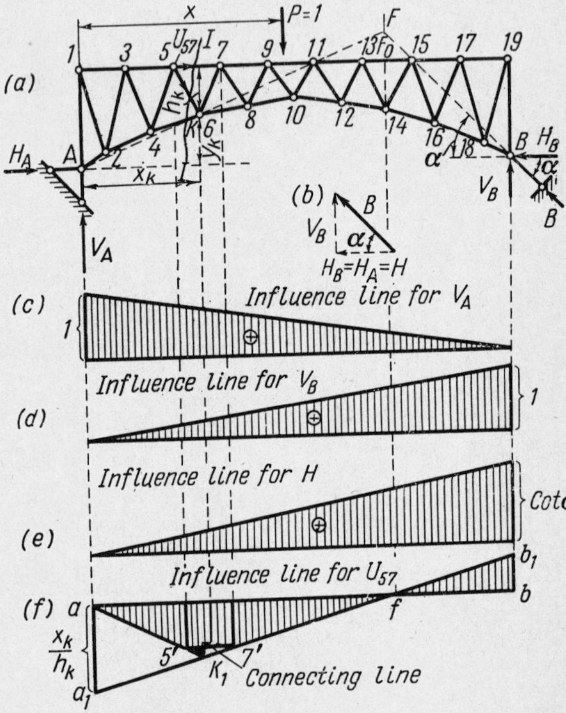

Fig. 86.4

the abutment reactions A and B respectively and by x the distance from the load unity to the left-hand abutment we shall obtain

$$H_A = H_B = H$$

Equating to zero the moments of outer forces about the hinges A and B we get, on the other hand

$$V_A = \frac{l-x}{l} \text{ and } V_B = \frac{x}{l}$$

The two latter equations are exactly the same as for an ordinary simply supported truss or beam and the corresponding influence lines are represented in Fig. 86.4*c* and *d*.

As regards the influence line for the thrust H it may be derived using the relation existing between H and V_B (Fig. 86.4*b*)

$$H = V_B \cot \alpha$$

The influence line for H obtained by multiplying all the ordinates to the influence line for V_B by $\cot \alpha$ is represented in Fig. 86.4*e*.

Let us now draw the influence line for the stress in some truss member, say in bar *5-7*. For this purpose let us pass a section *I-I* and placing the load unity to the right of this section, let us equate to zero the moments (about point k coinciding with joint *6*) of all the external forces acting on the left-hand portion of the truss

$$\Sigma M_k = V_A x_k - H y_k + U_{57} h_k = 0$$

wherefrom

$$U_{57} = -\frac{1}{h_k}(V_A x_k - H y_k) = -\frac{1}{h_k}(M_k^0 - H y_k)$$

When the unit load is applied at point F_0 lying in the same vertical with the point of intersection of lines AK and BF (point F), the stress in bar *5-7* becomes nil, for the resultant of all the forces applied to the left of section *I-I* passes through point k and the moment equation becomes

$$\Sigma M_k = U_{57} h_k = 0$$

Accordingly point F_0 is a neutral point for the stress U_{57}. At the same time the term $(M_k^0 - H y_k)$ entering the expression for U_{57} is equal to the bending moment in section k of a three-hinged arch. Hence the construction of the influence line for stress U_{57} may be carried out in the same way as that for the bending moment acting over section k of the said arch, provided all the ordinates of this latter are multiplied by $\left(-\frac{1}{h_k}\right)$. Consequently, having laid off the ordinate $\left(-\frac{x_k}{h_k}\right)$ over the left-hand abutment we must connect this ordinate with the neutral point f and then extend this line until its intersection with the vertical passing through joint B. The left part of the influence line will be obtained bearing in mind that it must pass through the zero ordinate at the left-hand abutment and must intersect with the right-hand part in a vertical passing through the origin of moments. The two lines being drawn, the positions of joint *5* should be marked on the left one and that of joint *7* on the right one, these two points being finally connected to form the completed line represented in Fig. 86.4*f*.

Let us now consider a truss with supports at different levels (Fig. 87.4*a*). We shall commence by constructing the influence lines for the reactions. For this purpose we may resolve the right-hand reaction B into its vertical and horizontal components V_B^* and H_B at a point b' situated at the same level as point A. Denoting as usual the horizontal and vertical components of reaction A by V_A and H_A and placing the unit load a distance x from the left-hand support, we may then write the equilibrium equations of the moments first about point b' and then about the centre of the hinge A

$$\Sigma M_{b'} = V_A (l_1 + l_2) - 1 (l_1 + l_2 - x) = 0$$
$$\Sigma M_A = -V_B (l_1 + l_2) + 1 \cdot x = 0$$

whence

$$V_A = \frac{l_1 + l_2 - x}{l_1 + l_2} \text{ and } V_B = \frac{x}{l_1 + l_2}$$

These two expressions are represented graphically in Fig. 87.4*c* and *d* which show that the vertical reactions of the truss vary exactly in the same way as those of a simply supported beam with a span of $l = l_1 + l_2$ (Fig. 87.4*b*).

The horizontal projection of all the forces acting on the truss shows that

$$H_A = H_B = H$$

The relation between H and V_B may be found by equating to zero the sum of their moments about hinge C (Fig. 87.4*a*)

$$\Sigma M_C = -V_B l_2 + Hf = 0$$

wherefrom

$$H = \frac{V_B l_2}{f}$$

In this expression $V_B l_2 = M_C^0$ is the bending moment acting over section C of a simply supported beam, spanning $(l_1 + l_2)$, when the load P is to the left of this section. Consequently, the thrust H equals $\frac{M_C^0}{f}$, which is exactly the same as in the case of a three-hinged arch with a span of $(l_1 + l_2)$ and a rise equal to f (Fig. 87.4*e*). It is apparent that the influence line for H obtained by multiplying all the ordinates of the line for V_B by $\frac{l_2}{f}$ will coincide with that for the thrust of an arch shown in Fig. 87.4*f*.

* The influence line for V_B will permit the determination of reaction B for any position of a vertical load using the formula $B = \frac{V_B}{\sin \alpha}$. The same result may be achieved with the aid of the influence line for H_B since $B = \frac{H_B}{\cos \alpha}$.

The influence line for the stress in bar *2-4* of the same truss may be obtained passing a section *I-I* (Fig. 88.4*a*) and writing that ΣM about point *3* equals zero when load $P = 1$ is to the right of this section

$$\Sigma M_3 = V_A a_3 - H y_3 - L_{24} r = 0$$

wherefrom

$$L_{24} = \frac{1}{r}(V_A a_3 - H y_3) = \frac{1}{r}(M_3^0 - H y_3)$$

When the direction of load P passes through point F (Fig. 88.4*a*) the stress in bar *2-4* will reduce to zero, for in this case point *3* will fall on the line of action of the resultant of V_A and H_A. Knowing the position of the neutral point and using the above expression for L_{24} the influence line is readily drawn, especially if we take heed of the analogy existing between the expressions for L_{24} and for the bending moment acting over a corresponding

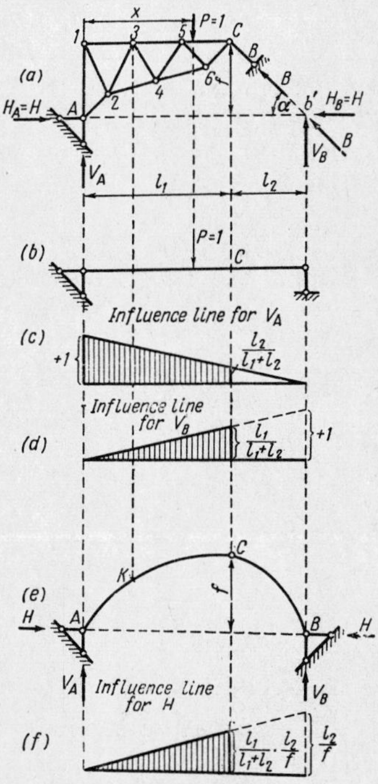

Fig. 87.4

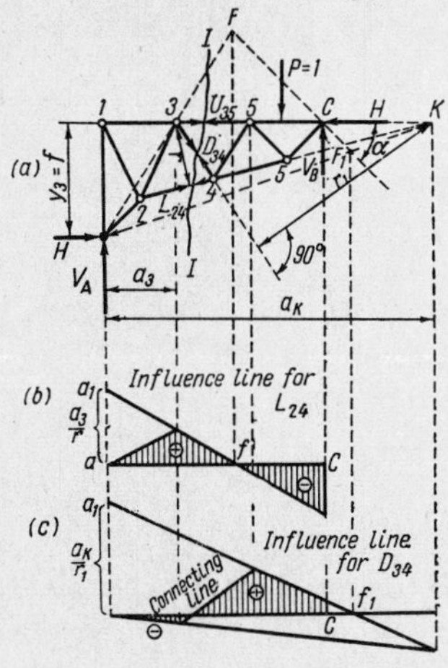

Fig. 88.4

section of a tree-hinged arch (see Fig. 32.3*f*). As the first differs from the latter only by a constant factor $\frac{1}{r}$ the influence line relative to the right part of the truss may be obtained by laying off an ordinate $\frac{a_3}{r}$ over the left support and by connecting it with the projection of neutral point on the horizontal, which gives us line a_1f (Fig. 88.4*b*). The line corresponding to the left part of the truss will be drawn using the well-known rule that the two must

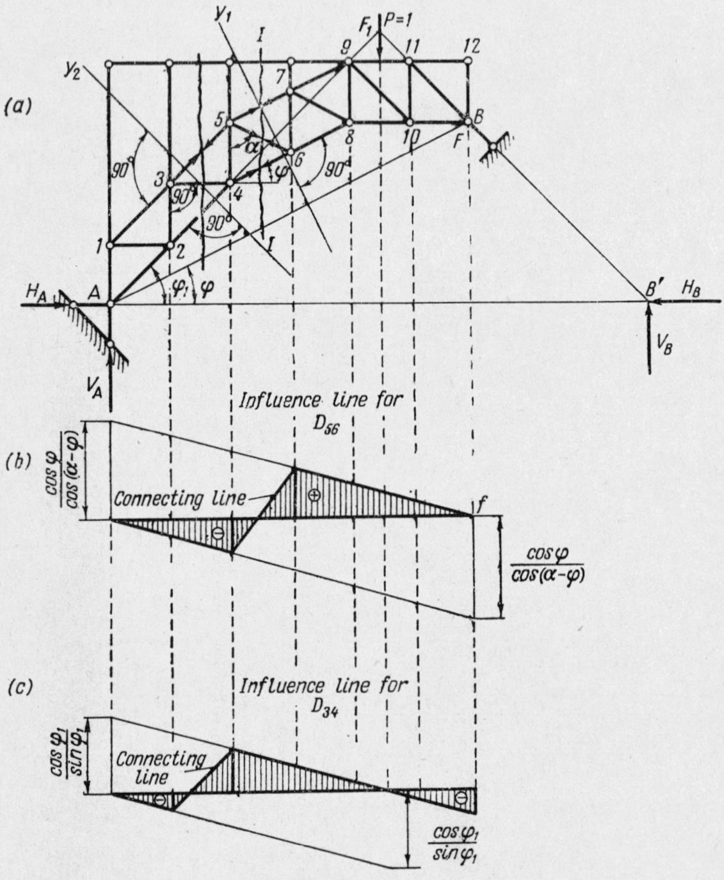

Fig. 89.4

intersect in a vertical passing through the origin of moments (point *3*). The influence line for D_{34} obtained in a similar way is shown in Fig. 88.4*c*.

Influence lines for web members of arch trusses with parallel chords can be obtained by projecting the stresses acting in all the members cut by a section on a normal to the chords, the method of moments being inoperative in this case as the chords intersect at a point infinitely distant.

For instance, the ordinates to the influence line for stress D_{56} of the truss represented in Fig. 89.4*a* can be found by equating to zero the projections of all the forces acting to the left of section *I-I* on a normal to the direction of the chord members *5-7* and *4-6**

$$\Sigma Y_1 = V_A \cos\varphi - H\sin\varphi - D_{56}\cos(\alpha-\varphi) = 0$$

where φ = angle between the chord members and the horizontal
α = angle formed by the diagonal *5-6* with the vertical.

From this equation we obtain

$$D_{56} = \frac{1}{\cos(\alpha-\varphi)}(V_A\cos\varphi - H\sin\varphi)$$

or

$$D_{56} = \frac{1}{\cos(\alpha-\varphi)}(Q_k^0\cos\varphi - H\sin\varphi)$$

where Q_k^0 is the shear in a simply supported beam of the same span.

The term $(Q_k^0 \cos\varphi - H\sin\varphi)$ being identical with the expression of the shear acting over a cross section of a three-hinged arch, the influence line relative to the right part of the truss can be constructed in the same manner as that for the shear in an arch (see Fig. 33.3*d*).

This means that an ordinate equal to $\frac{\cos\varphi}{\cos(\alpha-\varphi)}$ over the left abutment must be connected with the neutral point f determined by projecting on the x-axis the point of intersection of lines $b'B$ and AF (in this particular case line AF is parallel to the chord members *5-7* and *4-6* and passes through point B). The line relative to the left part of the truss will be parallel to the one pertaining to its right part due to the parallelism of the chords.

The completed influence line for stress D_{56} is represented in Fig. 89.4*b* while another influence line namely that for stress D_{31} obtained in a similar way is represented in Fig. 89.4*c*.

2. THREE-HINGED TRUSSED ARCHES

Were the right-hand supporting bar of the truss shown in Fig. 87.4*a* replaced by some framed system such as system *CB* (Fig. 90.4) we would obtain a three-hinged trussed arch, consisting essentially of two pin-connected trusses with immovable hinge supports.

◆

* It should be remembered that $H_A = H_B = H$ and that stresses in the horizontal deck members remain nil as long as the loads remain vertical.

Let us examine a three-hinged arch with supports at the same level represented in Fig. 91.4 *a*. Both vertical reactions and thrust for such systems are determined in exactly the same way as in the case of solid arches. Thus, for a load unity situated a distance x from the left abutment, reactions V_A and V_B will amount to

$$V_A = \frac{l-x}{l} \text{ and } V_B = \frac{x}{l}$$

while the thrust H will equal $\frac{M_C^0}{f}$ where M_C^0 is the bending moment acting over the corresponding cross section of a simply supported beam of the same span, carrying the same load.

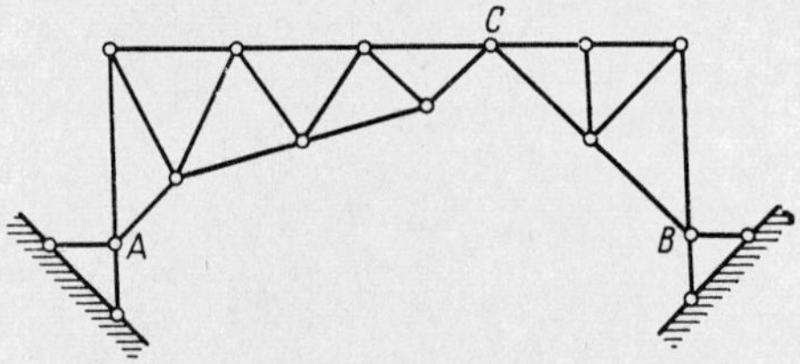

Fig. 90.4

The influence lines for the vertical reactions and the thrust are shown in Fig. 91.4*b*, *c* and *d*.

As regards to stresses induced in the separate members of the semi-arches it is clear that as long as the load is applied directly to the semi-arch under consideration the other one may be fictitiously replaced by an inclined supporting rod. In other words, the system may be reduced to the case of a simple truss with supports at different levels, which has been just examined.

The construction of influence lines for the stresses in members of such trusses is already familiar to the reader who will easily follow that of the influence line for L_{24} shown in Fig. 91.4*e*. It remains to find out what happens when the load shifts to the other semi-arch. In that case the load may be resolved into two components, the first of which is applied at the crown hinge and the other to a joint directly over the abutment at the opposite end of the arch. The ordinates to the influence line when unit load P acts over the crown hinge and the abutment are well known and equal y_c (Fig. 91.4*e*) and zero, respectively. Accordingly, in the case of bar *2-4* when the load travels from hinge C towards the right, the stress will vary linearly from y_c to 0 in accordance with the expression

$$L_{24} = \frac{l_2 - z}{l_2} y_c + \frac{z}{l_2} 0$$

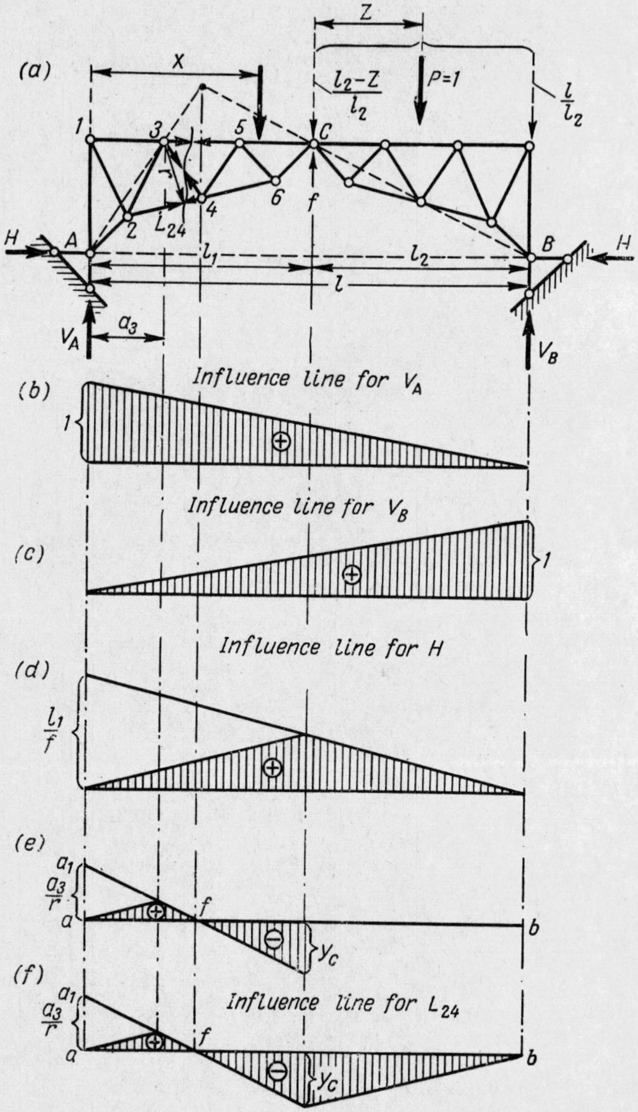

(a)
X
Z
P=1
$\frac{l_2-Z}{l_2}$
$\frac{l}{l_2}$
1
3
5
C
2
4
6
f
L_{24}
H
A
B
H
l_1
l_2
l
a_3
V_A
V_B
(b)
Influence line for V_A
1
(c)
Influence line for V_B
1
(d)
Influence line for H
$\frac{l_1}{f}$
(e)
a_1
$\frac{a_3}{r}$
a
f
b
y_c
(f)
Influence line for L_{24}
a_1
$\frac{a_3}{r}$
a
f
b
y_c

Fig. 91.4

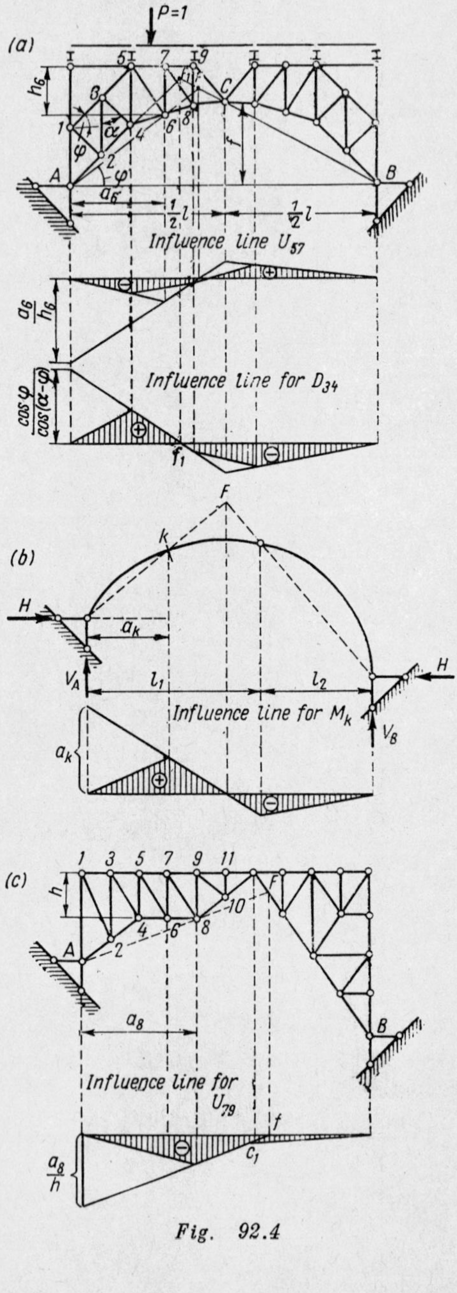
(a)
P=1
Influence line U_{57}
Influence line for D_{34}
(b)
Influence line for M_k
(c)
Influence line for U_{79}

Fig. 92.4

The corresponding influence line will be represented by the straight line connecting y_c with point b as shown in Fig. 91.4*f*.

Several influence lines for stresses in members of different trussed arches are represented in Fig. 92.4. Two of these systems have their end supports at different levels. In systems of the latter type it is more convenient to resolve the abutment reactions along a vertical and a line which connects the abutment hinges

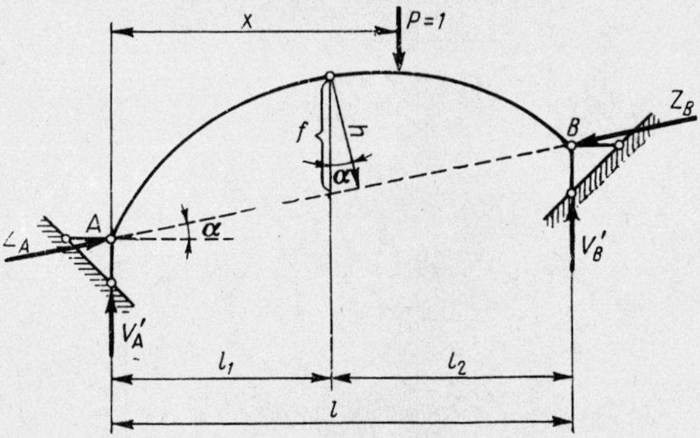

Fig. 93.4

(Fig. 93.4). The vertical reactions V'_A and V'_B will be determined from the well-known equations

$$V'_A = \frac{l-x}{l} \text{ and } V'_B = \frac{x}{l}$$

The components Z_A and Z_B of the abutment reactions when only vertical loads are involved will be given by

$$Z_A = Z_B = Z = \frac{M_C^0}{h}$$

where h is the lever arm of component Z about the crown hinge C. The thrust H is easily found from

$$H = Z \cos \alpha$$

where α is the inclination to the horizontal of the line passing through the abutment hinges A and B.

This enables us to rewrite the expression for the thrust as follows

$$H = Z \cos \alpha = \frac{M_C^0}{h} \cos \alpha = \frac{\frac{M_C^0}{h}}{\cos \alpha}$$

As the term $\frac{h}{\cos\alpha}$ represents the length of the vertical insert between the line connecting the abutment hinges and the crown

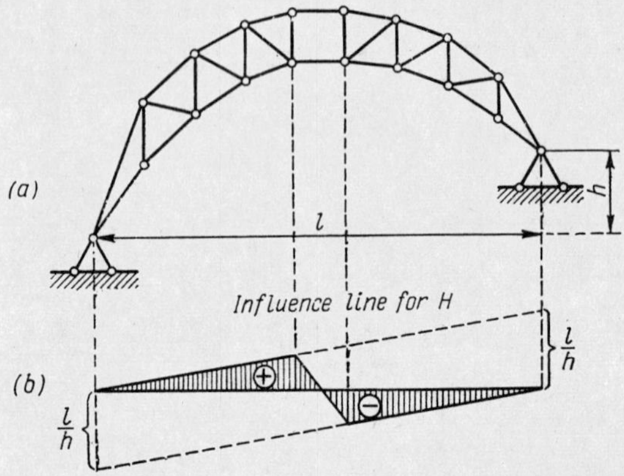

Fig. 94.4

hinge we may denote it by f whereafter the expression for H will become

$$H = \frac{M_C^0}{f}$$

The reader is invited to check the accuracy of the influence line for the thrust H of the trussed arch represented in Fig. 94.4.

11.4. *Variants of Trussed Arches*

Let us consider the trussed arch with elevated tie provided with one fixed and one roller support as shown in Fig. 95.4. This statically determinate arch may be obtained by replacing the inclined bar of the right-hand abutment support by the tie *ab* absorbing the thrust.

The method of stress analysis for similar tied arches is illustrated hereunder using as an example the structure represented schematically in Fig. 96.4*a*. The influence lines for reactions V_A and V_B are of the usual triangular shape with ordinates equal to unity under the supports as shown in Fig. 96.4*b* and *c*. The influence line for the force H in the tie (equivalent to the thrust H) will be readily determined by equating to zero the moments (about crown hinge C) of all the forces to the left of section *I-I*

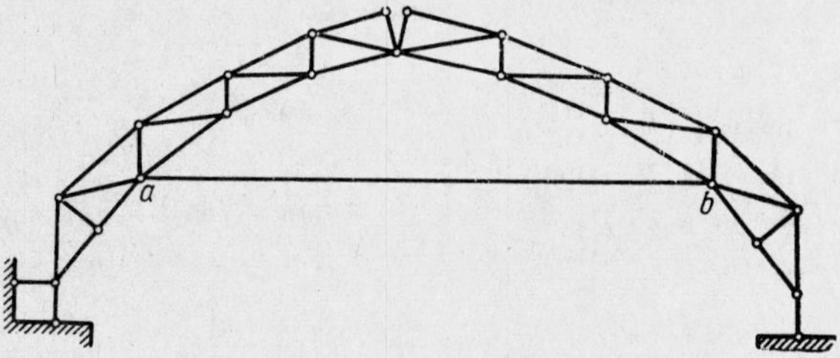

Fig. 95.4

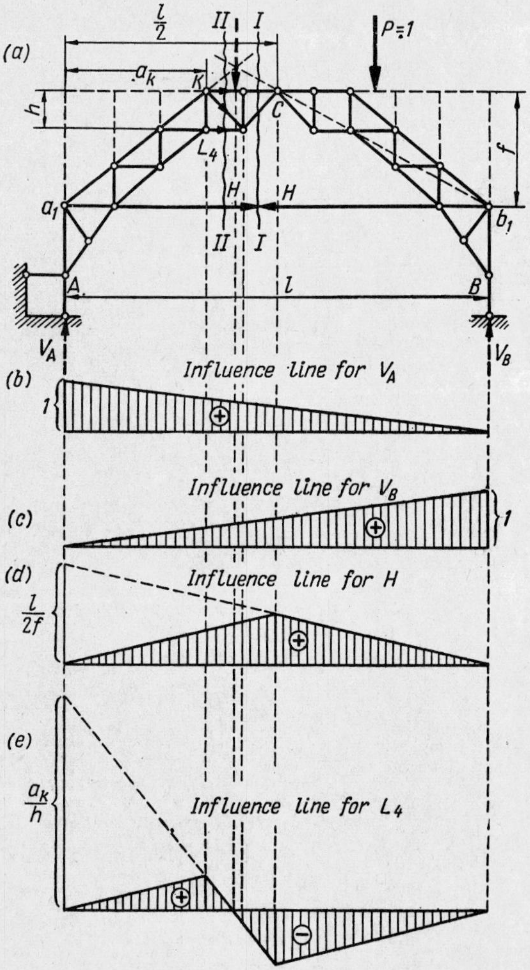

Fig. 96.4

when load unity P is to the right thereof

$$\sum_L M_C = V_A \frac{l}{2} - Hf = 0$$

wherefrom

$$H = \frac{V_A l}{2f} = \frac{M_C^0}{f}$$

When the load is applied to the left of the section the equation becomes

$$\sum_R M_C = -V_B \frac{l}{2} + Hf = 0$$

and

$$H = \frac{V_B l}{2f} = \frac{M_C^0}{f}$$

Accordingly, the influence line for the force H may be obtained by multiplying the ordinates of the bending moment M_C^0 acting over section C of a simply supported beam by a constant factor $\frac{1}{f}$. This influence line is represented in Fig. 96.4*d*.

The influence line for stress L_4 (Fig. 96.4*a*) may be obtained passing section *II-II* and writing that ΣM about point k of all forces to the left of this section equals zero

$$\Sigma M_k = V_A a_k - Hf - L_4 h = 0$$

wherefrom

$$L_4 = \frac{1}{h}(V_A a_k - Hf)$$

The neutral point corresponding to L_4 will fall on the vertical passing through the intersection of lines a_1k and b_1C. The completed influence line for L_4 is given in Fig. 96.4*e*.

The reader is invited to draw the influence lines for stresses in other members of the structure.

Let us now consider a system consisting of two pin-connected trusses and a multihinged arch, two variants of which are shown in Fig. 97.4*a* and *b*. It is easily proved that such systems constitute unyielding combinations and remain statically determinate. Methods of stress analysis for these systems will be shown using as an example the structure in Fig. 98.4*a*.

As usual the influence lines for vertical reactions V_A and V_B will be triangular in shape with ordinates equalling unity at the supports (Fig. 98.4*b* and *c*). It is easy to prove that the horizontal components of stresses acting in all the members of the multihinged arch ASB remain constant throughout the system for any given set of vertical loads applied to the trusses AC and CB. For

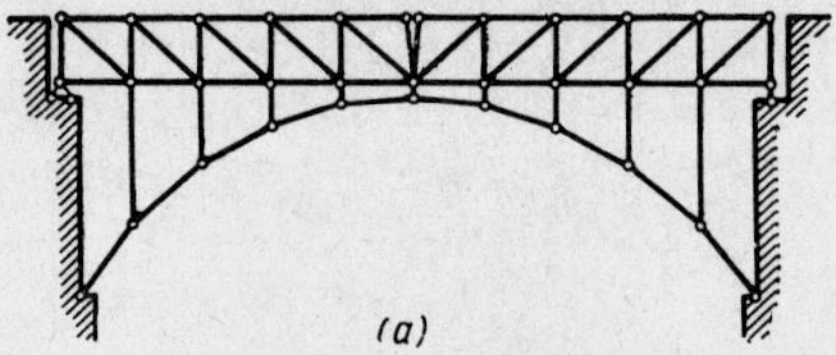

(a)

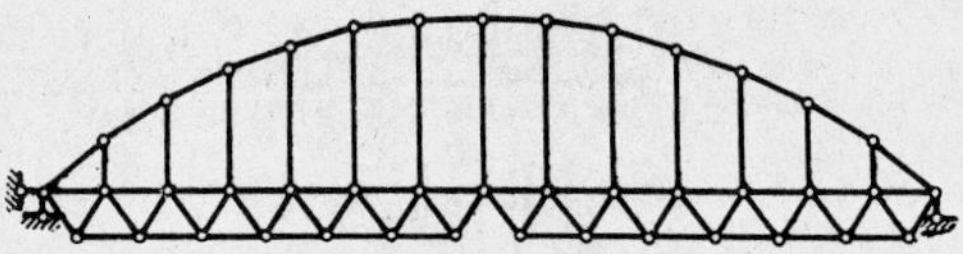

(b)

Fig. 97.4

this purpose consider the equilibrium of any joint (say, joint $n-1$ in Fig. 98.4*d*). Projecting all the forces on the horizontal we get

$$\Sigma X = -N_{n-1}\cos\alpha_{n-1} + N_n\cos\alpha_n = 0$$

wherefrom

$$N_{n-1}\cos\alpha_{n-1} = N_n\cos\alpha_n$$

where

$$N_{n-1}\cos\alpha_{n-1} = H_{n-1}$$

and

$$N_n\cos\alpha_n = H_n$$

Therefore

$$H_{n-1} = H_n = H$$

$$N_{n-1} = \frac{H}{\cos\alpha_{n-1}}; \quad N_n = \frac{H}{\cos\alpha_n}; \text{ etc.}$$

Projecting the same forces on the vertical we obtain

$$\Sigma Y = H\tan\alpha_n - H\tan\alpha_{n-1} - V_{n-1} = 0$$

wherefrom

$$V_{n-1} = H(\tan\alpha_n - \tan\alpha_{n-1})$$

The above expressions indicate that the influence lines for stresses in all the separate links of the arch as well as in all the verticals or suspensions will have the same shape as the influence line for the thrust H.

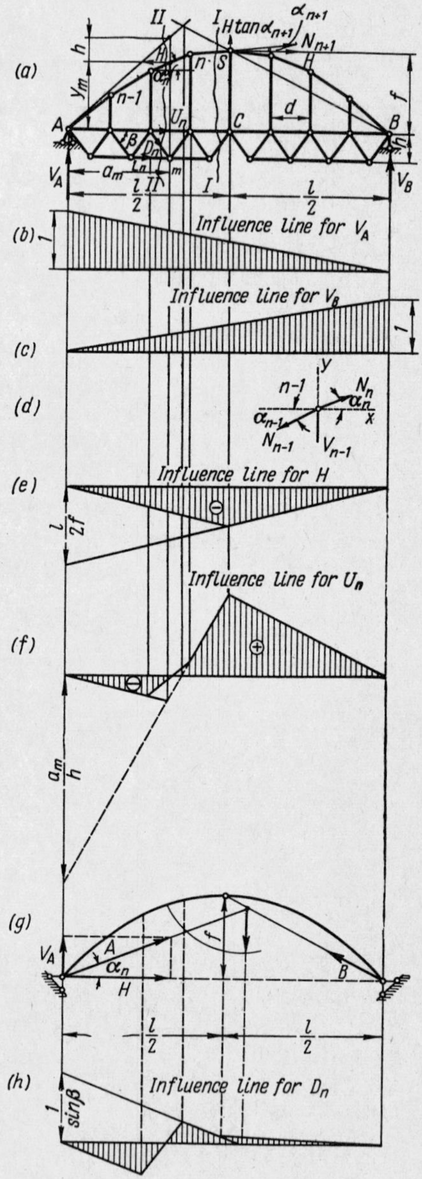
(a)
(b)
Influence line for V_A
Influence line for V_B
(c)
(d)
(e)
Influence line for H
Influence line for U_n
(f)
(g)
(h)
Influence line for D_n

Fig. 98.4

As for the latter, it may be obtained by passing section *I-I* (see Fig. 98.4*a*) and by equating to zero the sum of moments about the crown hinge *C* of all the forces applied to the left of the section, the stress N_{n+1} acting on hinge *S* having been previously resolved into two components *H* and $H \tan \alpha_{n+1}$

$$\Sigma M_C = V_A \frac{l}{2} + Hf = 0$$

wherefrom

$$H = -\frac{V_A l}{2f} = -\frac{M_C^0}{f}$$

The negative value of *H* indicates that all the links of the arch are compressed. The influence line for *H* is a triangle with its apex turned downwards and situated directly under the crown hinge *C* (Fig. 98.4*e*)

Let us now construct the influence line for stress U_n. For this purpose we shall pass section *II-II* equating to zero the moments of all the forces about point *m* when the load unity acts to the right of this section

$$\Sigma M_m = V_A a_m + U_n h - H (y_m + h) = 0$$

wherefrom

$$U_n = -\frac{1}{h} [V_A a_m - H (y_m + h)] = -\frac{1}{h} [M_m^0 - H (y_m + h)]$$

It will be observed that the term in brackets represents the bending moment acting over section *K* of a fictitious three-hinged arch of the same span whose crown hinge coordinates are equal to $\frac{l}{2}$ and *f* while those of the centroid of section *K* equal a_m and $(y_m + h)$. This will enable us to find the position of the neutral point pertaining to the influence line for U_n. For this purpose we shall first locate the centroid of section *K* along the vertical passing through the origin of moments *m* (Fig. 98.4*a*) after which the lines *AK* and *BS* may be drawn, their intersection determining the abscissa of the neutral point required. The completed influence line for U_n is shown in Fig. 98.4*f*.

In order to construct the influence line for stress D_n in one of the diagonals let us equate to zero the sum of vertical projections of all the forces to the left of section *II-II* (see Fig. 98.4*a*) when the unit load is to the right of this section

$$\Sigma Y = V_A - H \tan \alpha_n - D_n \sin \beta = 0$$

whence

$$D_n = \frac{1}{\sin \beta} (V_A - H \tan \alpha_n)$$

The latter equation shows that the required influence line can be obtained through the summation of the ordinates to the influence line for $\frac{V_A}{\sin\beta}$ with the ordinates to the influence line for the thrust H multiplied by $\left(-\frac{\tan\alpha_n}{\sin\beta}\right)$.

The neutral point method can be used for the construction of D_n influence line too. For this purpose we must first find the position of the unit load for which the expression $(V_A - H\tan\alpha_n)$ reduces to zero. In this expression V_A and H can be regarded as the vertical reaction and thrust of a fictitious arch of the same span as the actual structure and having for coordinates of the crown hinge $\frac{l}{2}$ and f (Fig. 98.4*g*).

The position of the neutral point will be derived from

$$D_n = \frac{1}{\sin\beta}(V_A - H\tan\alpha_n) = 0$$

showing that

$$\frac{V_A}{H} = \tan\alpha_n$$

The latter condition may be fulfilled only if the left-hand abutment reaction A of the three-hinged arch forms with the horizontal an angle α_n. The neutral point will be situated at the intersection of this reaction with reaction B of the fictitious arch, the latter acting necessarily along a line passing through the crown and the right-hand abutment hinges. The completed influence line for stress D_n is represented in Fig. 98.4*h*.

Let us now examine a structure in which the two trusses surmount the multihinged arch as shown in Fig. 99.4*a*. This system is geometrically stable and statically determinate, its main peculiarity residing in the fact that it takes support at four distinct points A', B', A'' and B''.

The following procedure may be recommended for the determination of the abutment reactions: the directions of the extreme links of the multihinged arch should be extended until their intersection with the verticals drawn through the centres of the abutment hinges A' and B' of the trusses. Here the reactions arising in the extreme links may be resolved along a vertical and a horizontal direction into two components V''_A, H_A, and V''_B, H_B, respectively (Fig. 99.4*a*). As already shown in the beginning of this article $H_A = H_B = H$.

Having denoted by V'_A and V'_B the reactions at the supports A' and B' and by V_A and V_B the total vertical reactions of the whole system we have

$$V_A = V'_A + V''_A; \qquad V_B = V'_B + V''_B$$

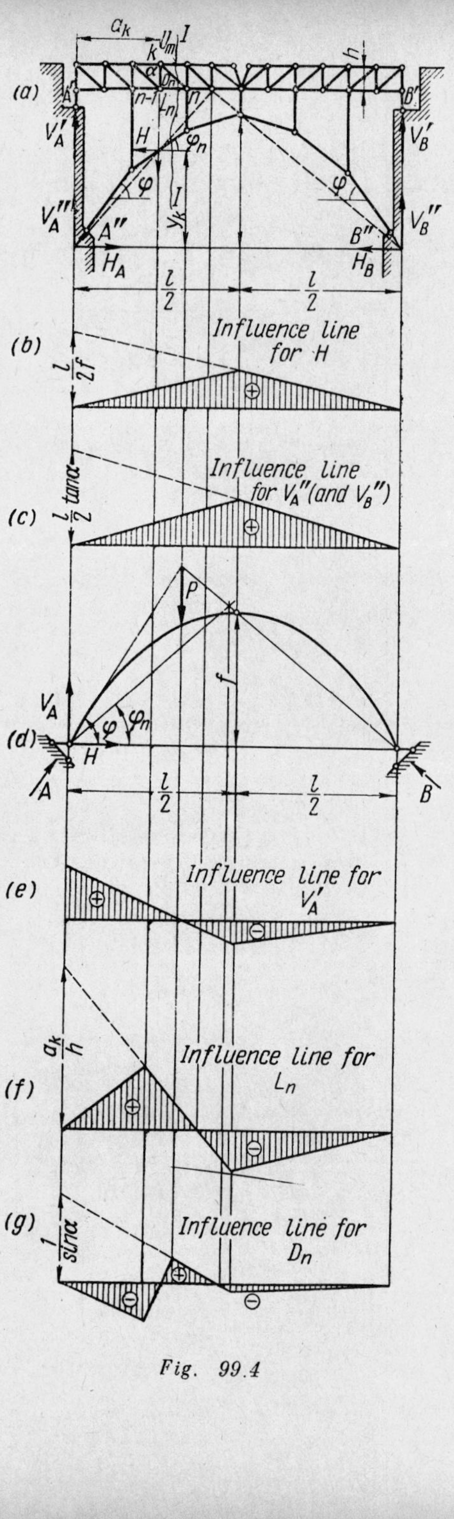
(a)
(b)
Influence line for H
(c)
Influence line for V_A'' (and V_B'')
(d)
(e)
Influence line for V_A'
(f)
Influence line for L_n
(g)
Influence line for D_n

Fig. 99.4

The equation expressing the equilibrium of the moments about the point of intersection of V_B and H_B gives

$$\Sigma M = V_A l - M_{ext} = 0$$

wherefrom

$$V_A = V'_A + V''_A = \frac{M_{ext}}{l}$$

where M_{ext} is the moment of all the external loads acting on the structure about the same point.

It follows that *the sum of the vertical components of reactions V'_A and V''_A is equal to the reaction of a simply supported beam.*

The thrust H will be conveniently determined by equating to zero the sum of moments about the central hinge S of all the forces acting on the left (or right) half of the structure when the load unity is to the right thereof

$$\Sigma M_S = V_A \frac{l}{2} - Hf - M_S^L = 0$$

wherefrom

$$H = \frac{M_S^0}{f}$$

M_S^0 being the bending moment acting at midspan of a simply supported beam carrying the same load.

Consequently, the influence line for the thrust H will have the shape of an isosceles triangle represented in Fig. 99.4*b*. The vertical reactions V''_A and V''_B can be also expressed in terms of the thrust H

$$V''_A = H \tan \varphi \quad \text{and} \quad V''_B = H \tan \varphi$$

It follows that the influence lines for these two reactions will take the shape of the triangle shown in Fig. 99.4*c*. The reaction V'_A will be deducted from

$$V'_A = V_A - V''_A = V_A - H \tan \varphi$$

This expression shows that the required influence line may be obtained by the summation of the ordinates of two other influence lines, namely, that for the abutment reaction of an end-supported beam and that for the thrust H, the latter being multiplied by a constant factor ($-\tan \varphi$). As is well known, the first of these two influence lines is a right triangle with an ordinate equal to unity over the left-hand support.

The influence line for V'_A could be also obtained by the neutral point method. The position of the neutral point is conditioned by

$$V'_A = V_A - H \tan \varphi = 0$$

showing that the ratio $\frac{V_A}{H}$ must be equal to $\tan\varphi$. This latter condition will be fulfilled when the resultant of V_A and H (e. g., the left-hand reaction of the fictitious three-hinged arch represented in Fig. 99.4*d*) will be inclined an angle φ to the horizontal. Thus, the position of the neutral point relative to the reaction V'_A will be determined by the intersection of the abutment reactions A and B of the said fictitious arch. It follows that in order to draw the influence line for V'_A by the neutral point method, an ordinate equal to unity should be laid off along the vertical passing through the left abutment; this ordinate should be then connected with the neutral point and extended until the intersection with the vertical passing through the crown hinge, the ordinate so obtained being finally connected to a point of zero ordinate at the right-hand support (Fig. 99.4*e*).

If it were required to construct the influence line for the stress acting in some chord member of the truss, say, member $(n-1)\,n$ we should proceed as follows. Having passed section *I-I* and equating to zero the moment about hinge k of all the forces acting on the left part of the truss we obtain

$$\Sigma M_k = (V'_A + V''_A)\,a_k - Hy_k - L_n h = 0$$

wherefrom

$$L_n = \frac{1}{h}\,[(V'_A + V''_A)\,a_k - Hy_k] = \frac{M_k^0}{h}$$

In this expression M_k^0 is the bending moment acting over section k of a three-hinged arch whose span l equals that of the structure involved, while the centroid coordinates equal a_k and y_k. The completed influence line obtained by this method is represented in Fig. 99.4*f*.

Let us consider now the construction of the influence line for the stress arising in one of the web members, say, in the diagonal kn of Fig. 99.4*a*. As long as the load unity remains to the right of section *I-I*, the stress D_n will be determined by the equation

$$\Sigma Y = V_A - D_n \sin\alpha - H\tan\varphi_n = 0$$

wherefrom

$$D_n = \frac{1}{\sin\alpha}\,(V_A - H\tan\varphi_n)$$

indicating that the neutral point will be located in the line of action of a load rendering $\frac{V_A}{H} = \tan\varphi_n$.

As has already been mentioned, this becomes possible when the resultant A of V_A and H, in other words, the left-hand reaction of a fictitious three-hinged arch represented in Fig. 99.4*d*, is inclined an angle φ_n to the horizontal. Hence the neutral point will be determined by the intersection of a line passing through the left-hand abutment at an angle φ_n with the horizontal and a line connecting the right-hand abutment with the crown hinge. The influence line for stress D_n will be obtained by laying off the ordinate $\frac{1}{\sin\alpha}$ over the left abutment and by connecting this ordinate with the projection of the neutral point on the x-axis.

To find that part of the influence line relative to the left portion of the semistructure, a line parallel to the first should be drawn through the zero point at the left abutment whereafter the position of the joints k and n shall be marked on these two lines and connected together. That portion of the influence line corresponding to the right half of the structure will be obtained by connecting the ordinate at the crown hinge with the zero point over the right-hand abutment. The completed line is represented in Fig. 99.4*g*.

Influence lines for any other web member or vertical connecting the multihinged arch with the truss can be obtained in a similar way.

5. SPACE FRAMEWORK

1.5. *General*

In a most general way of speaking the term *space framework* indicates three-dimensional through structures capable of resisting loads acting in different planes.

Certain of such structures may be reduced, for a given arrangement of loads, to a combination of plane structures (trusses), which simplifies greatly their design.

Thus, the bridge truss shown in Fig. 1.5*a* can be reduced to two vertical plane trusses *ABCD* and *MNFE* when the loads *P* are symmetrical about the longitudinal axis of the structure. However, if the same truss were loaded unilaterally, it should be considered as a space structure, the horizontal trusses *AMNB* and *DEFC* transmitting part of the load from one vertical truss to the other.

The three-dimensional structure of Fig. 1.5*b* supporting a water tank is Shukhov's hyperboloid which cannot be reduced to any number of plane structures and must be designed as a single unit. The same applies to the Schwedler dome illustrated in Fig. 1.5*c*.

The different members of space frameworks are usually connected together by riveted or welded joints, providing a certain degree of rigidity. However, computations taking into consideration this rigidity become exceedingly cumbersome, and therefore in actual design work such structures are always regarded as articulation-connected (differing thereby from three-dimensional framed bents in which all the joints are made and regarded rigid).

The articulations of space framed structures must allow rotation around three mutually perpendicular axes thus providing three degrees of freedom as compared to the single one of the pin joints of plane trusses. Accordingly, all the members of a space structure meeting at one joint can rotate about any line passing through the point of intersection of their axes, whilst those of a plane truss may do so only about an axis perpendicular to the plane of the truss.

On the other hand, the arrangement of the individual members of a space framework must be such that they should form an unyielding combination just as in the case of a plane one.

Consequently, a space framework is a geometrically stable structure, consisting of a number of bars situated in different

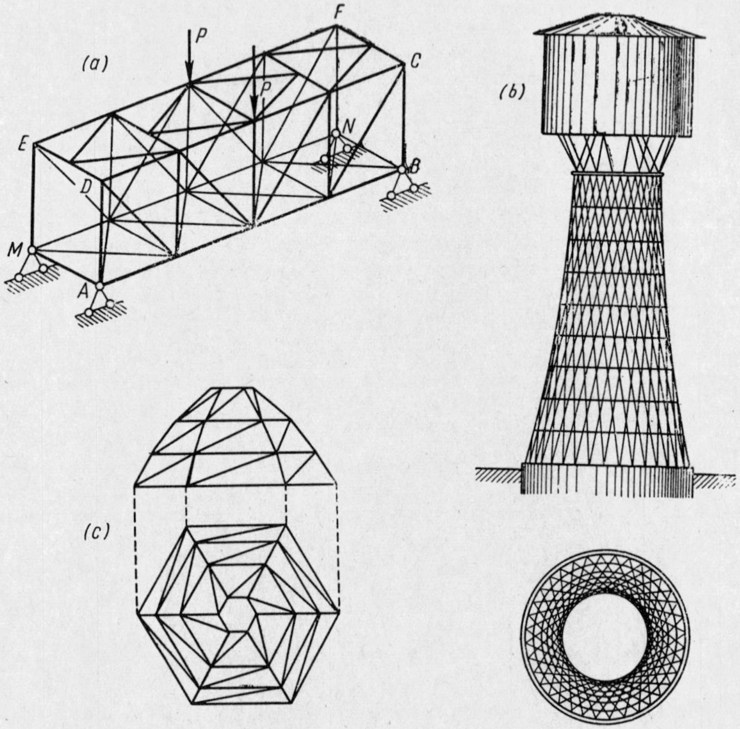

Fig. 1.5

planes and connected together by so-called universal or bell-and-socket joints. When such a structure is subjected to a system of loads acting at the joints, no flexural stresses are induced in any of its members which become directly extended or compressed.

Any system of noncoplanar forces in equilibrium must comply with six statical equilibrium equations which may be grouped together into

three equations of projections

$$\Sigma X = 0; \quad \Sigma Y = 0; \quad \Sigma Z = 0$$

and three equations of moments

$$\Sigma M_x=0;\quad \Sigma M_y=0;\quad \Sigma M_z=0$$

In statically determinate systems these equations are always sufficient for the computation of all the reactions at the supports and of all the stresses in the individual members.

It must be borne in mind that the solution of these equations becomes the easier, the smaller the number of unknowns in each of them. Therefore, it is advisable to seek such systems of equations in which each contains no more than one unknown (two at the utmost).

2.5. *Space Framework Supports*

Space frameworks are connected to their foundation or any other unyielding system using three different types of supports:

(1) the spherical movable support (Fig. 2.5),
(2) the spherical roller support (Fig. 3.5),
(3) the spherical fixed support (Fig. 4.5).

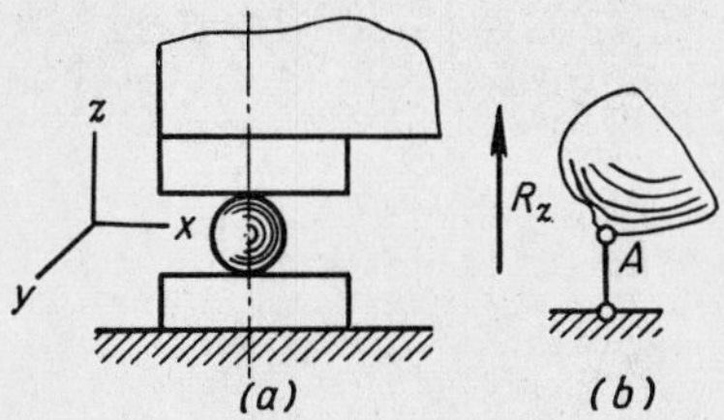

Fig. 2.5

The first consists of two flat parallel slabs with a ball in between. This type of support allows rotation about all the three axes x, y and z, as well as the displacement along any direction lying in the xy plane. Only the displacements along the z-axis (both upward and downward) are prevented. (The arrangement precluding upward displacement is not shown in Fig. 2.5.) Thus only one constraint is imposed by a support of that type, a vertical reaction R_z being developed along the direction of this constraint. The conventional schematic representation of a support of the first type is shown in Fig. 2.5*b*.

The second type of support consists in principle of two rockers, the upper and the lower, with a ball inserted in their sockets; the lower rocker bearing on rollers which lie on a slab provided with lateral ribs. Similar ribs existing on the lower surface of the rocker make any lateral displacement of the two impossible.

This type of support permits free rotation about any axis passing through the centre of the ball and a longitudinal displacement in a direction perpendicular to the roller axes. It prevents displacement along two directions, one being perpendicular to the plane of the roller axes and the other parallel to their axes, thus imposing two constraints on the body it carries. Two reactions, R_x and R_z or R_y and R_z (depending on the position of the rollers) will develop at a support of that type. Its conventional representation is shown in Fig. 3.5*b*.

The fixed spherical support (Fig. 4.5*a*), occasionally referred to simply as spherical support, consists of a pair of similar rockers with a ball, but no rollers, so that the upper rocker can only rotate about any axis passing through the centre of the ball, but cannot move in any direction. A support of this type will impose three constraints, hence, three reactions R_x, R_y and R_z may develop. Schematically this support is represented in Fig. 4.5*b*.

The minimum number of constraints necessary to maintain a body in a fixed position is always equal to the number of equilibrium equations. Therefore, in the case under consideration this number will equal six and the simplest combination of such constraints is shown in Fig. 5.5.

The body *I* is provided with a fixed sphericai support at point *C* and a roller support at point *B* which leaves the body free to rotate only about the axis *BC*. This last degree of freedom will be eliminated if a movable spherical support is added at a third point *A*, provided point *A* does not fall on the line *BC*.

If the structure proper does not constitute an unyielding combination the number of constraints at the supports should be increased accordingly. As an example, let us consider the hinged quadrangle $B_1B_2B_3B_4$ in Fig. 6.5*a*. If it were attached to the ground at three corners using as heretofore three supports of the different types described, it would conserve two degrees of freedom, for its shape could be altered in its own plane, and furthermore it could fold around one of the diagonals. The system could be made immovable by adding two constraints and effectively, four supports of the roller type as shown in Fig. 6.5*a* will provide the required stability. Indeed, point B_1 can move neither along the vertical, nor along the direction B_1B_2 due to the constraints developed by the support at this same point; at the same time the displacement along B_1B_4 is made impossible due to the presence of a horizontal constraint at point B_4. Accordingly, joint B_1 is fully immobilized. The joints B_2, B_3 and B_4 are connected to the first one and to the ground using a sufficient number of bars (constraints) to make the whole system completely stable.

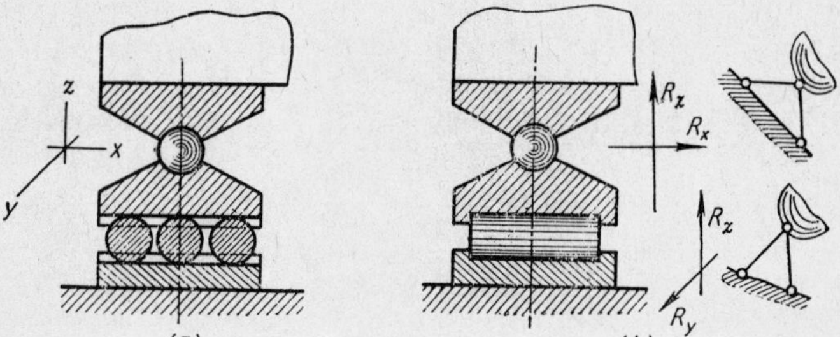

(a) (b)

Fig. 3.5

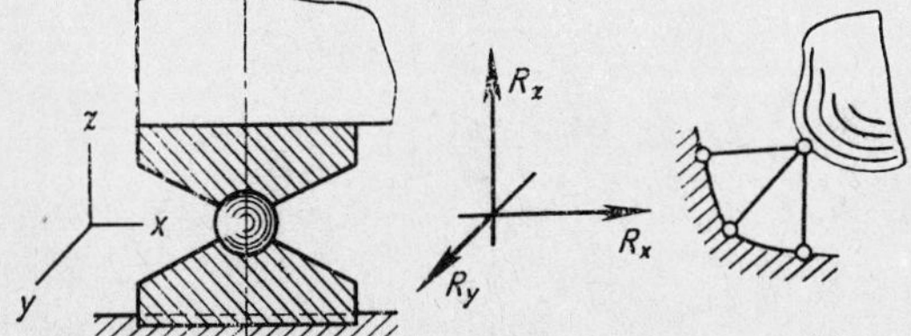

(a) (b)

Fig. 4.5

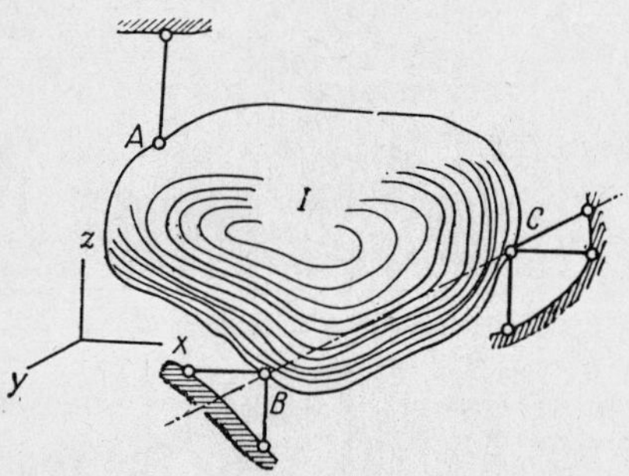

Fig. 5.5

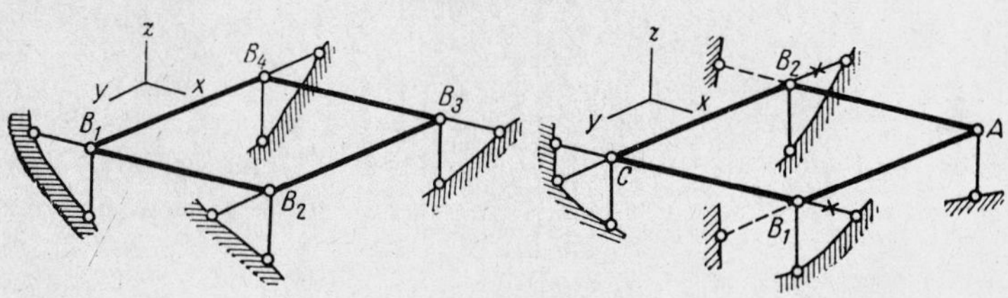

(a) (b)

Fig. 6.5

The position of the supports must be judiciously chosen, for otherwise it may happen that one part of the structure will have redundant constraints and will become statically indeterminate, while the other part will retain one or more degrees of freedom. An example of a faulty distribution of support constraints is given in Fig. 6.5*b*. The direction of the horizontal constraints at joints B_1 and B_2 coincides with that of point C, these two constraints become redundant, whereas both joints B_1 and B_2 are free to move towards A. This could be corrected by shifting the constraints, marked with a cross to new positions indicated in dash lines.

3.5. *The Formation of Statically Determinate Space Framework*

The simplest unyielding plane system is constituted by a triangle ACB shown in Fig. 7.5*a*. Let us add joint D using two bars AD and CD as indicated in Fig. 7.5*b*. The system obtained will be

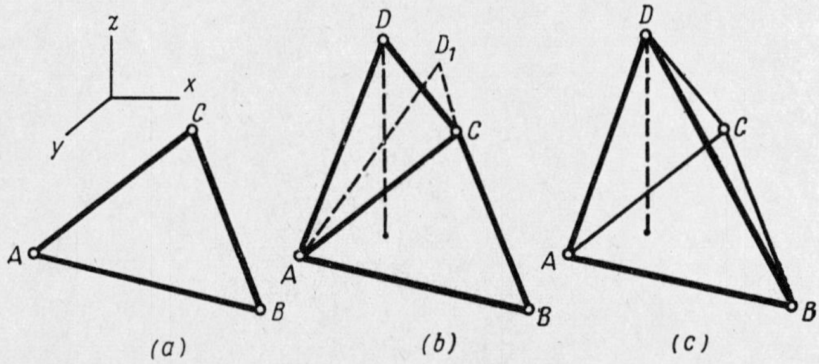

Fig. 7.5

unstable, for triangle ADC can rotate about AC. In order to obtain an unyielding combination, a third bar not lying in the plane of ADC should be introduced, say bar BD (Fig. 7.5*c*).

The pyramid so obtained is the simplest three-dimensional framed structure; additional joints, each connected to the already existing system by three separate noncoplanar bars, may be introduced to form new structures, which will remain statically determinate and unyielding.

Let us now examine the relation existing in a space framework as described above between the number of joints, the number of bars and the number of constraints at the supports. Let S be the number of bars, S_0 the number of constraints and K the number of articulated joints. The total number of the unknown stresses and reactions will then equal $(S+S_0)$ and the total number of equilibrium equations which may be used to find these unknowns is $3K$, for at each joint we may equate to zero the x, y and z projections of all forces (internal and external) applied to this joint.

Hence the number of redundant members and/or support constraints i will be given by

$$i = S + S_0 - 3K \tag{1.5}$$

When $i > 0$, the system is statically indeterminate, when $i < 0$, the system is unstable, and only when $i = 0$ the system may remain statically determinate and form an unyielding cobmination.

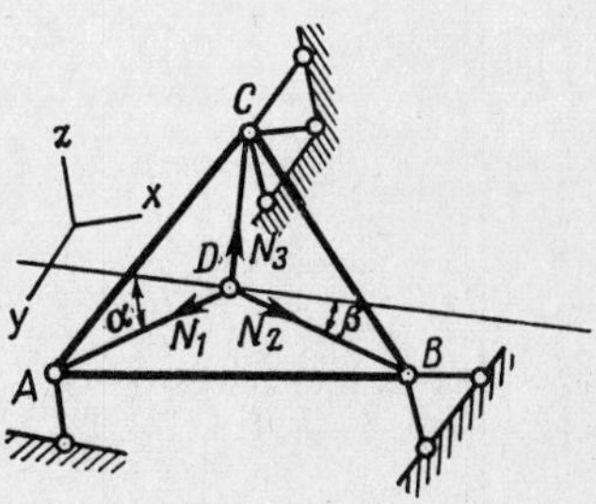

Fig. 8.5

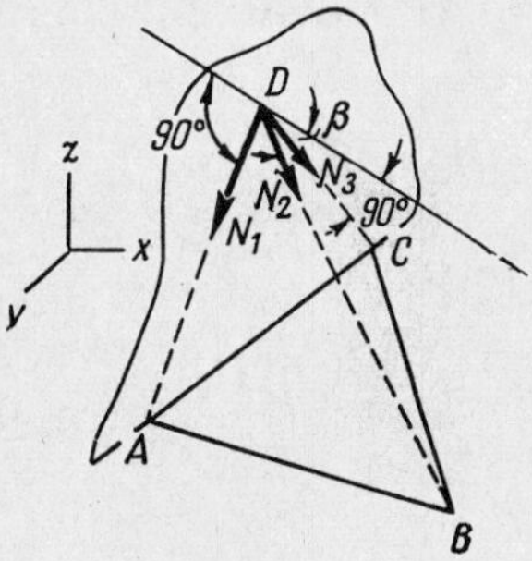

Fig. 9.5

However, this condition though necessary is not sufficient, for the equation

$$S + S_0 - 3K = 0$$

permits the determination only of the number of bars and support constraints required, but furnishes no information on their mutual position. The latter must be known in order to determine whether the system is statically determinate or not.

In the case of the simplest structure shown in Fig. 7.5*c* we have: $S = 6$; $S_0 = 6$; $K = 4$, the expression (1.5) showing that in this case $i = 6 + 6 - 3 \times 4 = 0$ and therefore the requirement stipulated above is satisfied assuming that the constraints at the support (not shown in Fig. 7.5) are the same as in Fig. 5.5. However, the same results could be obtained for the structure given in Fig. 8.5 which differs from the one just mentioned by the fact that bars AD, BD and CD lie in one and the same plane, thus making the whole structure instantaneously unstable, for joint D can move along a normal to the plane ABC.

Accordingly, having made sure that $i = 0$, the stability of the system must be examined by the method of zero load described for plane structures in Art. 6.4. As will be remembered, this method consists in the computation of stresses in all the members of the system at zero load; when these stresses are nil, the system is geometrically stable, but when they are indeterminate and may differ from zero, the system is instantaneously unstable.

In the case of the structure shown in Fig. 7.5*c* it is easy to prove that at zero load all its members remain idle. Indeed, separating joint D and projecting the stresses N_1, N_2 and N_3 acting in members AD, BD and CD respectively on a normal to plane ADC (Fig. 9.5) we obtain $N_2 \cos \beta = 0$, wherefrom $N_2 = 0$. The same reasoning shows that the stresses in all the other members of the system are also nil, which means that the structure forms an unyielding combination.

But if we apply this reasoning to joint D of the system shown in Fig. 8.5 we obtain

$$N_1 \times 0 + N_2 \times 0 + N_3 \times 0 = 0$$

which is an identity satisfied for any values of N_1, N_2 and N_3. The remaining two equilibrium equations which may be written for this joint will contain three unknowns whose values therefore remain indeterminate. This indicates clearly that the system is instantaneously unstable.

Let us now examine the structure represented in Fig. 10.5. We have in this case

$$S_0 = 8;\ S = 16;\ K = 8;\ i = 8 + 16 - 3 \times 8 = 0$$

and thus the system may be statically determinate. Applying again the zero load method we shall start by separating joint A and by projecting stresses N_1, N_2, N_3 and N_4 on a normal to plane ACB_2B_1. This leads immediately to $N_4 = 0$. Proceeding now to joint E we may easily prove that all the bars meeting at this joint remain idle. Passing consecutively to points D, C, B_1, B_2 and B_4 and considering their equilibrium, we shall find that all the other bars of our structure remain unstressed, which proves that this structure is both statically determinate and geometrically stable.

The system which we have just examined does not belong to the category of simple structures for it is impossible to dismantle it by the successive elimination of joints, each connected to the remainder of the system with the aid of three bars only.

Such systems are termed *complicated* and may be obtained by replacing one or more bars of a simple system by a corresponding number of differently situated members. For instance, if in our system we replace the diagonal AB_2 by a diagonal B_1C (shown by a dotted line) we shall be able to take the structure down by eliminating successively joints A, E, D and C, each with three connecting bars. Thus the complicated system shown in Fig. 10.5 in solid lines can be obtained by altering the position of only one bar in a simple system.

In conclusion let us examine the plane truss represented in Fig. 11.5, all the joints of which are of the universal type. If we

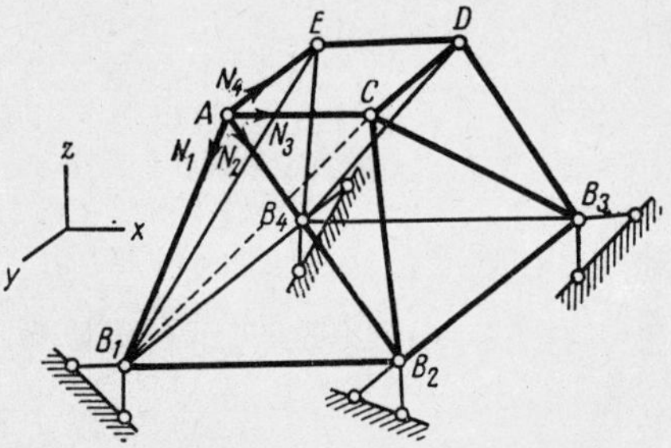

Fig. 10.5

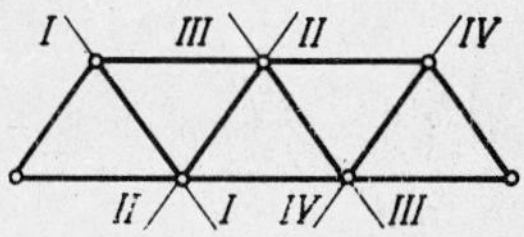

Fig. 11.5

assume that the central triangle of this truss is rigidly connected to the ground by means of 6 support constraints we have

$$S=11;\quad S_0=6;\quad K=7;\quad i=11+6-3\times 7=-4$$

Thus the system is unstable and has four degrees of freedom; indeed, it may fold along lines *I-I*, *II-II*, *III-III* and *IV-IV*.

4.5. *Stress Analysis in Space Framework*

The following three methods are in use for stress determination in statically determinate space frameworks:

(a) the method of sections,

(b) the method of bar replacement,

(c) the method of reducing the space structure to a series of plane ones.

We shall examine each of these methods in turn.

(a) *The method of sections.* This method is used for the computation of stresses in the members of simple framed structures and consists essentially in passing a section through a certain number of bars in which the stresses are sought. The portion of the structure removed is replaced by the internal forces acting along the sectioned bars, these forces being then determined with the aid of equilibrium equations. In general six equations of statics may be written for each section and therefore the number of unknown stresses determined for a single section may not exceed six.

Depending on the equilibrium equations used and on the position of the section itself this method may be subdivided into:

(1) the method of moments,

(2) the method of shears,

(3) the method of joints.

In the first of these three methods the equilibrium equations are obtained by expressing that the sum of moments of all external forces acting on a body in equilibrium about some preselected axis is always nil. As its name implies, this method is very similar to the method of moments described in Art. 2.4 for plane structures.

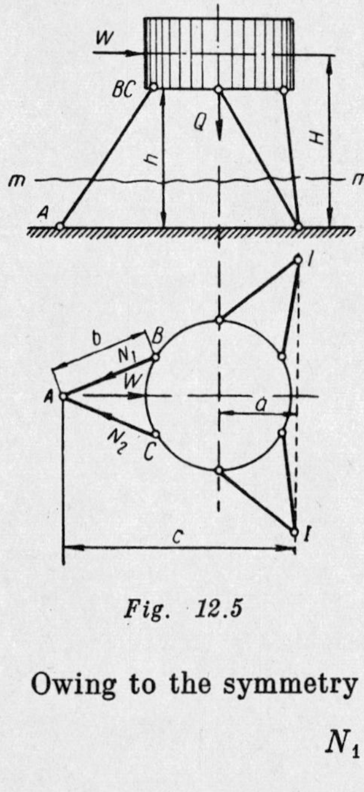

Fig. 12.5

As an illustration of this method, let us determine stresses N_1 and N_2 acting in the legs of an elevated tank appearing in Fig. 12.5. Having passed the section *m-m* we may equate to zero ΣM of all the forces acting on the upper portion of the structure about the axis *I-I*. The stresses N_1 and N_2 are regarded as applied at point *A*, where their resultant is resolved into a vertical and a horizontal component. This leads to the following equation

$$\Sigma M_1 = WH - Qa - (N_1 + N_2)c \sin \alpha = 0$$

where the angle α is given by $\tan \alpha = \frac{h}{b}$.

Owing to the symmetry of the loading, $N_1 = N_2$ and therefore

$$N_1 = N_2 = \frac{WH - Qa}{2c \sin \alpha}$$

The second method is analogous to the method of shears used in the analysis of plane structures. In this case the equilibrium equations express that the sum of projections of all external forces on some conveniently chosen axis is nil. This method will be made quite clear if we consider the cantilever truss represented in Fig. 13.5. Joints *A*, *B* and *C* of this truss are rigidly fixed by means of six support constraints (not shown in the drawing). Using expression (1.5) we find that

$$i = S + S_0 - 3K = 15 + 6 - 3 \times 7 = 0$$

and since under zero load all the bars will remain idle, which becomes immediately apparent if joints *1*, *2*, *3* and *4* are isolated in succession, the system is statically determinate and forms an unyielding combination.

In order to determine the stresses N_1 and N_2 acting in the diagonals let us pass section R and assume that the projection on the z-axis of all forces applied to the right-hand portion of the truss is nil

$$\Sigma Z = -P + (N_1 + N_2) \sin \alpha = 0$$

Taking the moments of N_1 and N_2 about the x-axis we obtain

$$\Sigma M_x = N_1 a \sin \alpha - N_2 a \sin \alpha = 0$$

and therefore

$$N_1 = N_2$$

In this case the solution of $\Sigma Z = 0$ yields

$$N_1 = N_2 = \frac{P}{2 \sin \alpha}$$

When the section passed separates only one joint we obtain the method of joints. The equilibrium equations used in this case do not differ in principle from those used in the previous one.

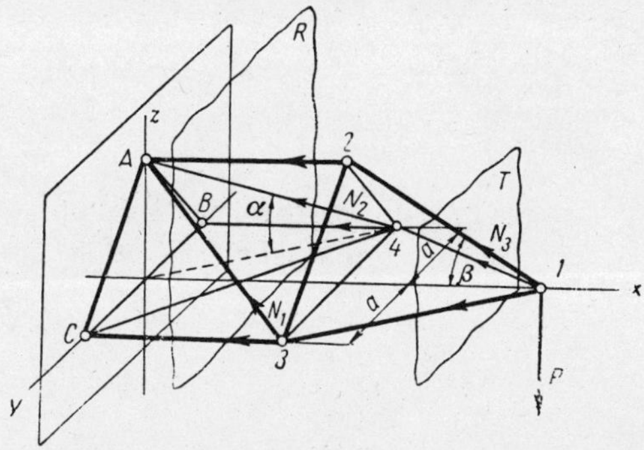

Fig. 13.5

We shall use this method to determine the stress N_3 acting in bar *1-2* of the same truss (see Fig. 13.5).

Separating joint *1* and equating to zero the sum of forces projected on the z-axis we obtain

$$\Sigma Z = -P + N_3 \sin \beta = 0$$

wherefrom

$$N_3 = \frac{P}{\sin \beta}$$

The method of joints is particularly well suited in the following cases:

1. When three bars meet at an unloaded joint. In this case as previously mentioned (see Fig. 9.5) all three bars are idle.

2. When all the bars meeting at a joint, with the exception of one, lie in the same plane. If no load is applied to such a joint or if this load acts in the said plane, the stress in the member which is outside this plane will be nil.

(b) *The method of bar replacement*. The method can be advantageously used for complicated space systems when it is impossible to pass a section cutting six bars only, thus making the method of sections practically inapplicable. The basic principle of this method derives from the fact that any complicated statically determinate system can be reduced to a simple one by replacing one or more bars.

Let us take up the case of a complicated system which can be converted into a simple one by the replacement of one bar only. Let X be the stress in the bar to be replaced. Having introduced the substitute bar, let us consider the simple structure so obtained under the action of the given set of loads and of the load X applied along the direction of the bar replaced. Denoting by N_p, N_x and $\overline{N}_x$ the stresses induced in the substitute bar by the loads actually applied, the force X and by a load unity acting in the direction of X, respectively, we may write

$$N_x = X\overline{N}_x$$

The combined stress in the substitute bar may be then expressed by $N_p + X\overline{N}_x$. As in the actual structure this bar is absent, we must equate this stress to zero

$$N_p + X\overline{N}_x = 0$$

which leads immediately to

$$X = -\frac{N_p}{\overline{N}_x} \tag{2.5}$$

Once the value of X is known, the stress in any member of the structure will be easily found using the formula

$$N_k = N_{kp} + \overline{N}_{kx}X \tag{3.5}$$

where N_{kp} = stress induced in member k of the simple structure by the actual set of loads P

$\overline{N}_{kx}$ = stress in the same member induced by the load unity $X = 1$.

The same procedure can be followed when the conversion of the given system to a simple one requires the replacement of several bars. In the latter case the determination of stresses in the bars which are being replaced will require the solution of several equations with several unknowns equal in number to that of the bars just mentioned. It will be readily seen that in this respect there is no difference between space frameworks and plane structures (see Art. 4.4).

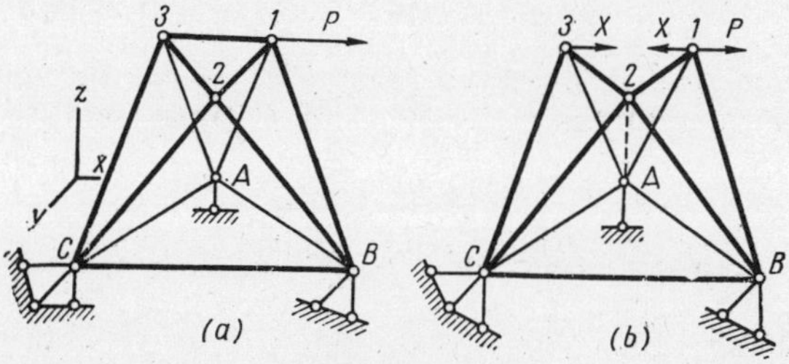

Fig. 14.5

In order to illustrate the use of the above method, let us compute the stresses in the structure shown in Fig. 14.5*a* (incidentally these stresses could be obtained directly by the method of sections).

As usual let us check first whether the system is statically determinate and stable. In the case under consideration $S = 12$, $S_0 = 6$ and $K = 6$, wherefrom

$$i = 12 + 6 - 3 \times 6 = 0$$

which shows that at least one of the basic requirements is satisfied. The demonstration that under zero load all the bars remain idle will be given later.

Replacing bar *1-3* by bar *A-2* (Fig. 14.5*b*) we obtain a simple structure for which the stress in the substitute bar may be found by the method of joints.

Starting with joint *1* we find the stress produced in bar *1-2* by the load P; passing to joint *3* we see that under the action of this load bar *2-3* remains idle; separating then joint *2* we shall easily obtain the stress in bar *A-2*.

This being done let us examine the stress arising in the same bars from the application of the unit load X.

This stress may be represented by

$$\overline{N}_x = \overline{N}'_x + \overline{N}''_x$$

where $\overline{N}'_x$ = stress in bar *A-2* induced by the load unity applied at joint *1*

$\overline{N}''_x$ = stress in the same bar induced by the same load applied at joint *3*.

Owing to the symmetry of the structure

$$\overline{N}'_x = \overline{N}''_x$$

and accordingly

$$\overline{N}_x = 2\overline{N}'_x$$

As will be readily observed from Fig. 14.5*b*, the stress $\overline{N}'_x$ is opposite in sign to the stress N_p and P times smaller than the latter.
Hence

$$\overline{N}'_x = -\frac{N_p}{P} \quad \text{and} \quad \overline{N}_x = 2\overline{N}'_x = -\frac{2N_p}{P}$$

Substituting this value in expression (2.5) we obtain

$$X = -\frac{N_p}{\overline{N}_x} = \frac{N_p P}{2N_p} = \frac{1}{2} P$$

Once X is known, the stresses in all the members of the structure are found with no difficulty.

Returning to the demonstration that under zero load all bars of the structure remain idle we can now state that for $P = 0$ the force $X = 0$ and accordingly bar *1-3* is idle. Separating joints *1*, *3* and *2* in succession we shall find immediately that the same applies to all the other bars.

The method of bar replacement can be of considerable help when investigating the geometrical stability of the structure.

Determining the stress $\overline{N}_x$ induced in the substitute bar by a load unity we may meet with two cases:

1. The stress $\overline{N}_x = 0$. Then X becomes indeterminate being expressed by $\frac{0}{0}$ which indicates that the system is instantaneously unstable.

2. The stress $\overline{N}_x \neq 0$. Then in the absence of external loads $X = 0$, and both terms of the expression

$$N_k = N_{kp} + X\overline{N}_{kx}$$

reduce to zero indicating that the system forms an unyielding combination.

(c) *The method of reducing the space structure to a series of plane ones.* This method becomes applicable when the structure is composed of distinct groups of coplanar members. In such cases all the external loads should be resolved along planes coinciding with those of the groups of bars just mentioned, whereafter each of these coplanar groups may be analyzed separately.

Let us consider, for instance, the system represented in Fig. 15.5. The number of support constraints $S_0 = 5 \times 2 = 10$, the number of bars $S = 35$ and the number of joints $K = 15$, giving

$$i = 10 + 35 - 3 \times 15 = 0$$

Separating consecutively joints *6*, *7*, *8*, *9*, *10*, *1*, *2*, *3*, *4* and *5* it is easy to prove that when $P = 0$ all the bars remain idle.

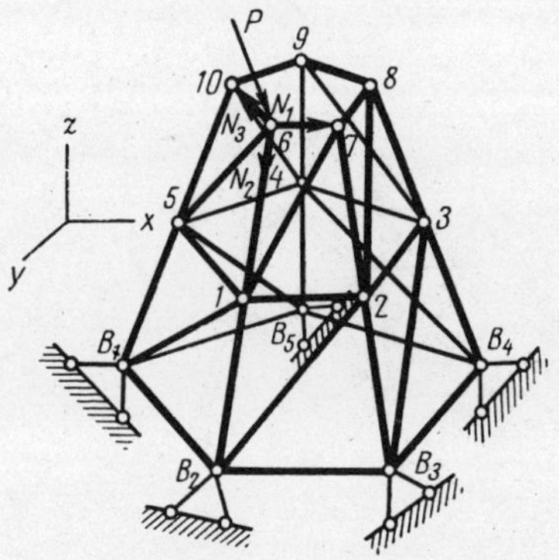

Fig. 15.5

Accordingly, the structure is both statically determinate and geometrically stable. In order to find the stresses induced by load P let us resolve this load into three components N_1, N_2 and N_3 as indicated in Fig. 15.5.

Isolating thereafter joints *7*, *8*, *2* and *3* we find that bars *7-8*, *2-3* and *3-*B_3 remain idle. Similarly, isolating joints *9*, *10*, *6* and *5* we shall prove that the same applies to bars *9-10*, *10-4*, *4-5* and *5-*B_5. Alone two plane trusses B_2*-6-7-*B_3 and B_1*-10-6-*B_2 will take up the entire load. These trusses may be designed in the usual way, component N_1 being applied to the first one, component N_3 to the second, and component N_2 being divided between the two in any arbitrary proportion.

5.5. *Examples of Stress Analysis in Space Framework*

Let us determine the stresses in the members of the central panel of the cantilever structure represented in Fig. 16.5 both by the method of section and by reducing the structure to a series of plane trusses.

(a) *Method of sections.* Start by proving that all the web members of the plane truss *5-8-12-9* remain idle. For this purpose isolate joint *9*; four bars meet at this joint but three (*1-9*, *2-9* and *10-9*) lie in the same plane, and as no external load is applied to the joint the stress in bar *5-9* must be nil. For the same reason bars *5-10*, *10-6*, *6-11*, *11-7* and *7-12* will remain idle.

This being known, pass section R cutting all the members of the panel under consideration.

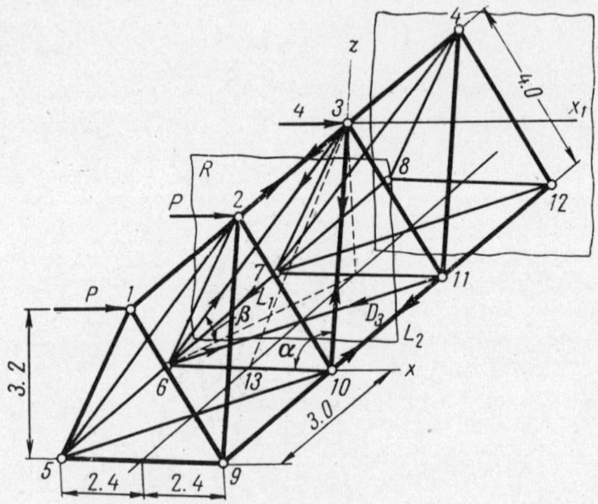

Fig. 16.5

Determine stress U_{23} by equating to zero the sum of moments of all forces acting on the left part of the truss about the x-axis coinciding with the direction of bar *6-10*

$$\Sigma M_x = U_{23}h = U_{23}\,3.2 = 0$$

wherefrom

$$U_{23} = 0$$

In order to determine stresses D_1 and D_2 in diagonals *3-6* and *3-10* project the forces acting in the left part of truss on the x- and z-axes

$$\Sigma X = 2P + (D_1 - D_2)\cos\alpha = 0$$

$$\Sigma Z = D_1 \sin\beta + D_2 \sin\beta = 0$$

wherefrom

$$D_1 = -D_2$$

and by substituting in ΣX we find

$$D_2 = \frac{P}{\cos\alpha}$$

From triangle *10-13-3*

$$\cos\alpha = \frac{2.4}{5.0} = 0.48$$

Therefore

$$D_1 = -\frac{25}{12} P$$

$$D_2 = \frac{P}{0.48} = +\frac{25}{12} P$$

In order to determine stresses L_1 and L_2 of the lower chord elements *6-7* and *10-11* write $\Sigma M_{x1} = 0$ about an axis x_1 parallel to the x-axis but passing through the centre of joint *3*

$$\Sigma M_{x_1} = -(L_1 + L_2)\, 3.2 = 0$$

giving

$$L_1 = -L_2$$

This being known, write $\Sigma M_z = 0$ about the z-axis passing through the same point

$$\Sigma M_z = -6P - 3P + 2.4L_1 - 2.4L_2 = 0$$

wherefrom

$$L_1 = \frac{9P}{4.8} = \frac{15}{8} P; \quad L_2 = -\frac{15}{8} P$$

(*b*) *Method of reduction to plane trusses.*

Start by resolving the loads along the planes of the two inclined lateral trusses. The corresponding components (Fig. 17.5) will equal

$$P_1 = P_2 = \frac{P}{2\cos\gamma} = \frac{5}{6} P$$

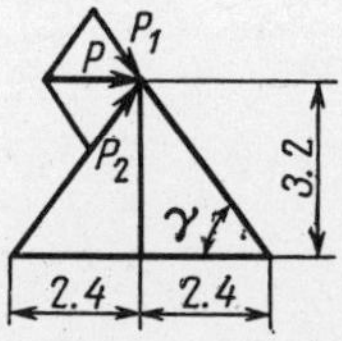

Fig. 17.5

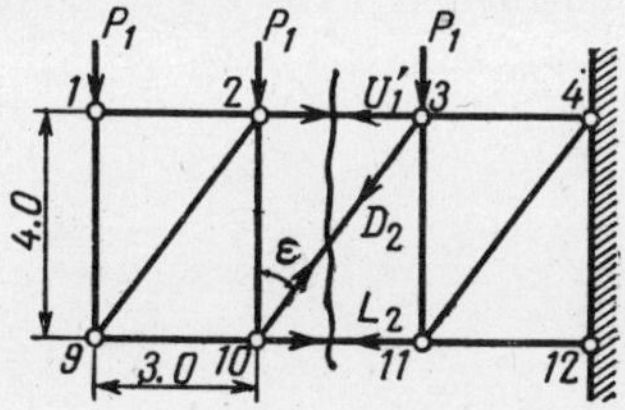

Fig. 18.5

Now consider the truss *1-4-12-9* shown in Fig. 18.5 and determine stress U_1' in the upper chord member *2-3*. As this same member belongs also to truss *5-1-4-8* (see Fig. 16.5), another stress U_1'' will be induced in it at the same time. Owing to the fact that the vertical projections of loads P_1 and P_2 are of opposite sign

$$U_1' = -U_1''$$

and therefore the resulting stress will be nil.

The stress D_2 in the diagonal will be obtained by projecting all the forces acting on the left part of the truss on a normal to the chords

$$\Sigma Y = -2P_1 + D_2 \cos\varepsilon = 0$$

As $P_1 = \frac{5}{6} P$ and $\cos\varepsilon = 0.8$

$$D_2 = \frac{2 \times 5 \times 5}{6 \times 4} P = \frac{25P}{12}$$

Obviously the stress D_1 in the corresponding diagonal of the other truss will equal

$$D_1 = -D_2 = -\frac{25P}{12}$$

Finally equating to zero ΣM about point *3* gives

$$\Sigma M_3 = -6P_1 - 3P_1 - 4L_2 = 0$$

wherefrom

$$L_2 = -\frac{9P_1}{4} = -\frac{9 \times 5}{6 \times 4} P = -\frac{15}{8} P$$

and the stress

$$L_1 = -L_2 = \frac{15}{8} P$$

Thus the stresses determined by both methods are in complete agreement.

6. KINEMATIC METHOD OF INFLUENCE LINE CONSTRUCTION

1.6. *General*

The kinematic method of influence line construction for any given function (shear, bending moment, normal force, abutment reaction, stress in a member of a truss) is based on one of the most general principles of theoretical mechanics—the *principle of virtual displacements.*

In accordance with this principle, the *total work performed by any given system of forces along virtual displacements of a body in*

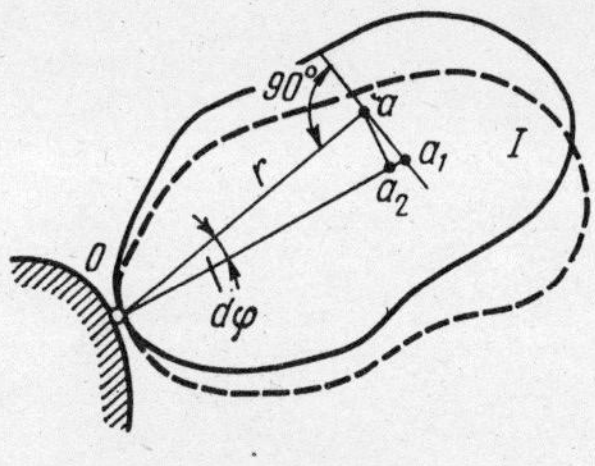

Fig. 1.6

equilibrium must be nil. Such displacements are reckoned infinitely small and therefore they may be accomplished without disturbing any of the existing internal or external constraints. At the same time the insignificance of these displacements permits the introduction of the following simplifications: when plate *I* shown in Fig. 1.6 rotates an infinitesimal angle $d\varphi$ about point O, any other point a located a distance r from point O will shift to a_2 along a circular arc; however, since the angle of rotation is very small, we may consider that point a moves along the tangent to the arc and not the arc itself, neglecting completely the distance a_1a_2. For the same reason we may also neglect the difference between $d\varphi$ and $\tan (d\varphi)$ which simplifies very considerably the construction of the virtual displacement graphs.

The kinematic theory (also called the strain energy or elastic energy theory) enables the construction of influence lines for all types of structures: beams, arches or trusses, statically determinate or redundant. The methods based on this theory are particularly well fit for certain complicated cases where they lead to quicker and more reliable results; they are also very useful for checking influence lines constructed by other methods.

Hereunder we shall describe only one of the methods derived from the above theory. This method might be termed the instantaneous centre of rotation method and is extremely simple and easy to grasp.

2.6. *Basic Principles of the Kinematic Method*

Let us consider a plate rigidly connected to the ground by means of three constraints represented by three supporting bars (Fig. 2.6*a*). Assume that it is required to construct the influence

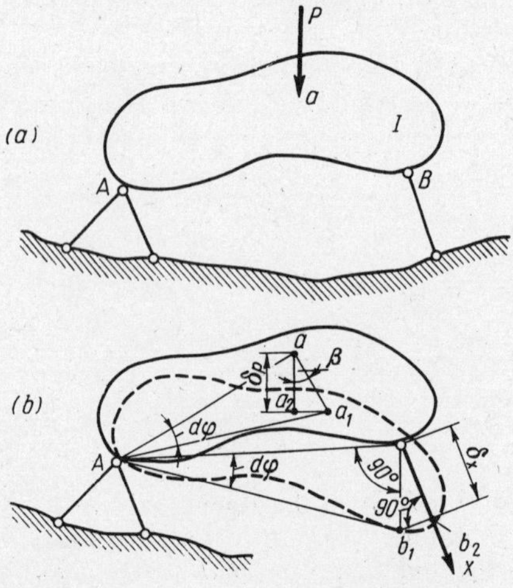

Fig. 2.6

line for the stress in one of these bars, say, bar B, when a unit load P, remaining always parallel to itself (Fig. 2.6*b*) travels along the plate.

Unlike the statical methods, which require that a section be passed across several bars, usually separating the whole structure or part thereof from the ground, the method described involves the elimination of one bar (or constraint) only and its replacement by a force X equal to the stress acting therein.

In the case under consideration the system so obtained can rotate freely about point A and is maintained in equilibrium by the load P, the force X and the reactions acting along the remaining supporting bars.

Let us apply to this system the principle of virtual displacements. As point A is rigidly connected to the ground it will constitute the centre of rotation of the plate, the only possible motion of which will consist in a rotation about this point. Suppose that the plate has turned clockwise an infinitesimal angle

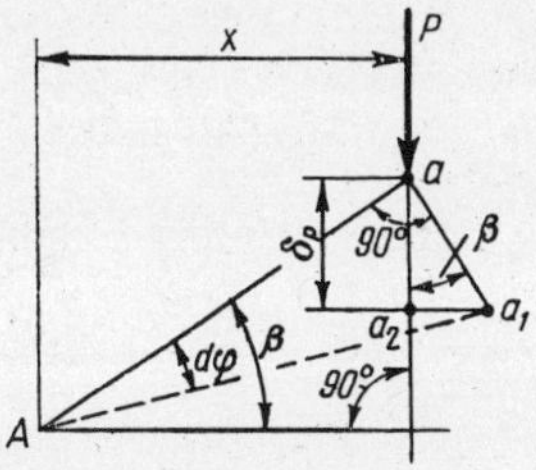

Fig. 3.6

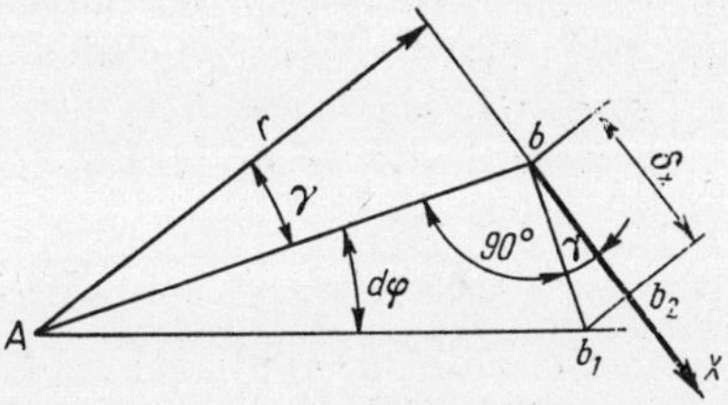

Fig. 4.6

$d\varphi$ (Fig. 2.6b) causing the load point to shift from a to a_1 and the point of application of the force X from b to b_1.

Denoting by δ_p the component of the displacement a-a_1 directed along the force P and by δ_x the component of the displacement b-b_1 directed along the force X we may write

$$P\delta_p + X\delta_x = 0$$

which expresses that the work performed by the external forces acting on a body in equilibrium remains nil. The load P being equal to unity, we draw immediately from the above

$$X = -\frac{\delta_p}{\delta_x} \qquad (1.6)$$

which represents the equation for the required influence line in its most general form.

Analyzing this expression we note that from triangle a_2aa_1 (Fig. 3.6)

$$\delta_p = aa_1 \cos\beta$$

where $aa_1 = Aa\,d\varphi$.
It follows that

$$\delta_p = Aa\cos\beta\, d\varphi$$

$Aa \cos\beta$ being equal to the lever arm of the load P about point A, let us denote it by x which gives

$$\delta_p = x d\varphi$$

Hereafter any displacement accomplished by some point of the plate in the direction of the load will be reckoned positive and any displacement accomplished in the opposite direction—negative.

It is clear that the displacements of different points will depend on the position of load P, this displacement being proportional to the lever arm x, or in other words, to the distance of the load point to the centre of rotation A.

Having determined the displacements of all the possible points of application of the load P we may represent these displacements graphically obtaining the so-called *diagram of virtual displacements* or *displacement graph of the system.*

Let us examine the term δ_x. Using the same reasoning as above we obtain from Fig. 4.6

$$\delta_x = bb_1 \cos\gamma = Ab \cos\gamma \, d\varphi = r d\varphi$$

As both the point of application of the force X and its direction remain constant, the displacement δ_x is a constant independent from the position of the load P, and may therefore be regarded as representing the *scale* to which the virtual displacements have been drawn.

Indeed, the shape of the influence line will depend solely on the numerator δ_p of the expression (1.6) but the determination of numerical values of the ordinates to this line requires that the value of the denominator δ_x be exactly known.

Thus, the *ordinates to the influence line for any function are equal to those of the graph of virtual displacements of the system, made possible by the elimination of the corresponding constraint, divided by the scale factor* δ_x.

The sequence in which the construction of the influence lines by the method of the instantaneous centre of rotation should be carried out as follows:

(1) eliminate the constraint corresponding to the function under consideration and replace it by the force X,

(2) draw the graph of virtual displacements for the mechanism obtained upon elimination of the said constraint,

(3) determine the scale factor pertaining to this graph,

(4) determine the signs of the ordinates to the influence line.

As an illustration of the above, let us consider the example of a cantilever beam appearing in Fig. 5.6*a* for which it is required to construct the influence line for reaction B.

The elimination of the right-hand support leaves the beam free to pivot about the remaining one (point A), which will therefore constitute the centre of rotation of the system. If the beam is turned counterclockwise through an angle $d\varphi$ about this centre the displacements of all the points will be represented by a straight line intersecting the beam axis at A (where the displacement is nil). The ordinates to this line will be reckoned positive to the left of A (all the points being displaced downwards, e. g., along the direction of force P) and negative to the right of it.

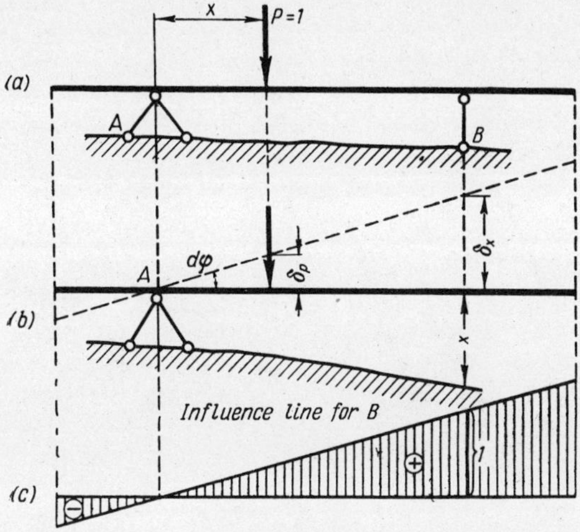

Fig. 5.6

The displacement δ_x is positive and equal to the ordinate corresponding to point B. If in our drawing we put $\delta_x = 1$, expression (1.6) will give

$$X = -\frac{\delta_p}{\delta_x} = -\frac{\delta_p}{1} = -\delta_p$$

In order to obtain the influence line for reaction B all that remains to be done is to change the sign of all the ordinates to the displacement graph as shown in Fig. 5.6*c*.

In the following articles we shall consider more complicated cases.

3.6. *Replacement of Constraints by Corresponding Forces*

As already stated, the construction of an influence line for any function starts with the elimination of the corresponding constraint which must be replaced by a force.

Let us consider in detail some of the more typical cases of constraint elimination.

(a) *Elimination of the constraint corresponding to the vertical reaction.* In this case the fixed support should be represented by two concurrent bars one of which is horizontal and the other vertical as shown in Fig. 6.6*a*. The vertical supporting bar is then eliminated and replaced by forces $X = V_A$ directed towards the hinges as shown in Fig. 6.6*b*, which corresponds to a positive reaction inducing a compressive stress in the eliminated bar.

(b) *Elimination of the constraint corresponding to a thrust.* In this case the horizontal bar is removed (Fig. 7.6) and replaced by forces $X = H_A$ again directed towards the hinges, this direction coinciding with the direction of the thrust reckoned positive.

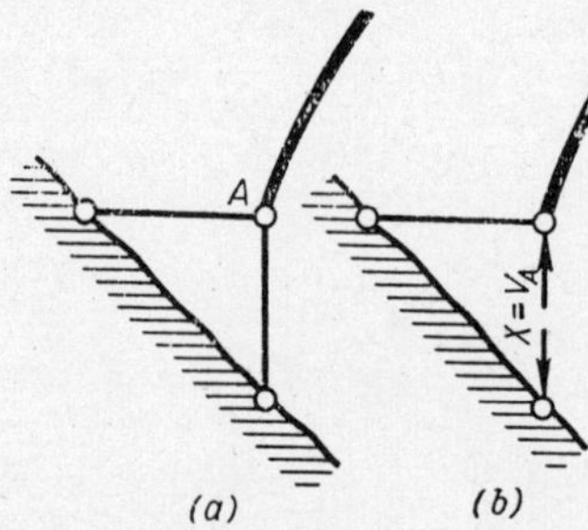

Fig. 6.6

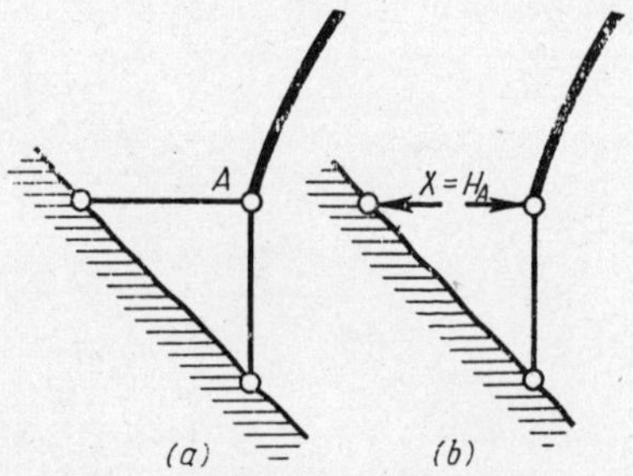

Fig. 7.6

(c) *Elimination of constraints corresponding to stresses in truss members.* It is the member for which the influence line is required that should be removed. The forces X should be directed away from the joints thus indicating that tensile stresses are reckoned positive (Fig. 8.6).

(d) *Elimination of constraints corresponding to shearing forces.* Any cross section of a beam, an arch, or a bent capable of resisting the action of a bending moment, a shearing or a normal force may be schematically replaced by a connection consisting of three bars as indicated in Fig. 9.6*b*.

The mutual position of these bars may be varied at will but they must always ensure the rigidity of the connection which implies that these three bars may never have a common point of intersection.

In the arrangement appearing in Fig. 9.6*b* the force acting in the vertical bar is equal to the shear, which follows from the equilibrium of vertical components of all forces acting to the left (or to the right) of section *I-I*

$$\Sigma Y = Q - X = 0$$

whence

$$X = Q$$

Consequently, the construction of the influence line for the shear acting over section CC reduces to the construction of that for the stress X in the vertical bar.

Upon removal of the vertical bar the two parts of the beam will have a mobile connection represented schematically in Fig. 9.6 *c*.

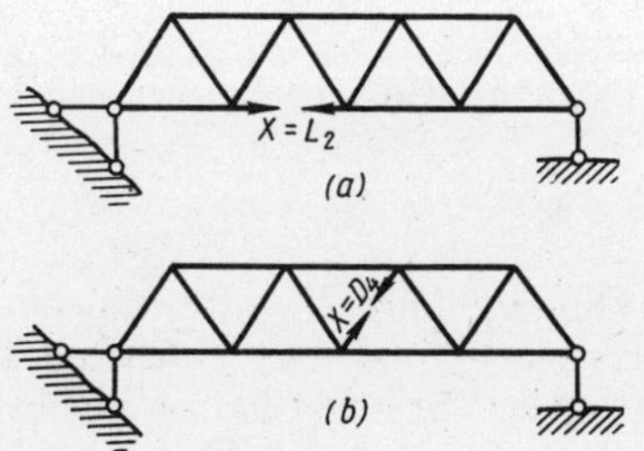

Fig. 8.6

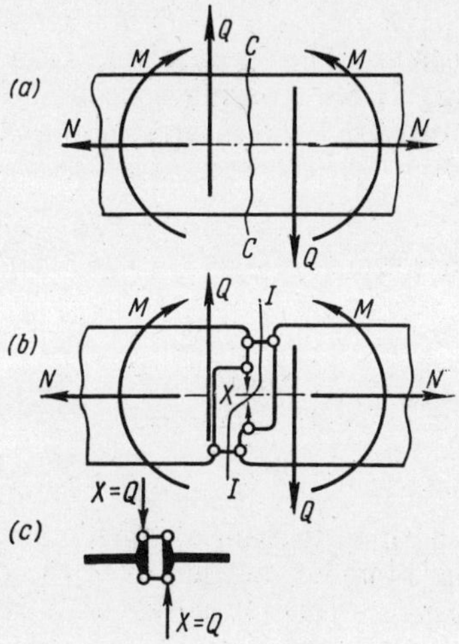

Fig. 9.6

(e) *Elimination of the constraint corresponding to a normal force.* Adopting for the connection bars a pattern represented in

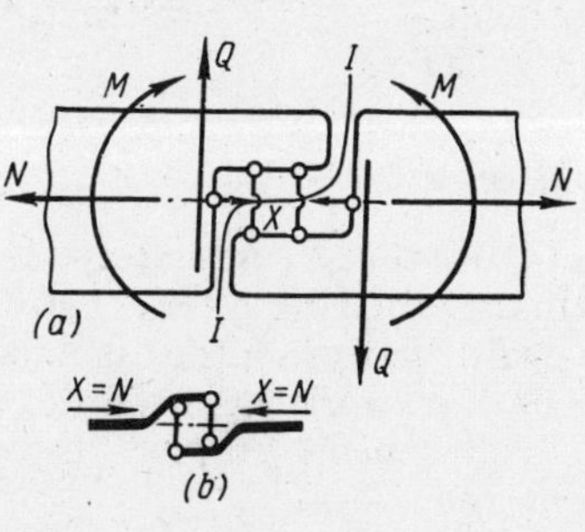

Fig. 10.6

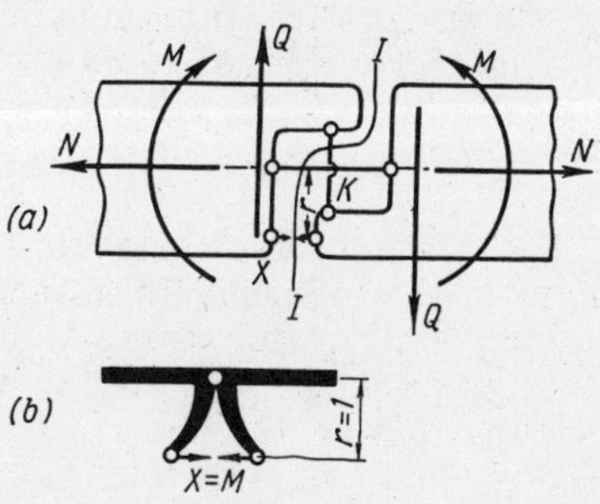

Fig. 11.6

Fig. 10.6*a* and projecting all the forces acting to the left (or to the right) of section *I-I* on a horizontal we obtain

$$\Sigma X = N - X = 0$$

wherefrom

$$X = N$$

In other words, the stress in the horizontal bar is equal to the normal force N acting in the member under consideration.

Upon removal of the horizontal bar the two parts of the member will have a mobile connection represented schematically in Fig. 10.6*b*.

(f) *Elimination of the constraint corresponding to a bending moment.* The connecting bars may be placed as indicated in Fig. 11.6*a*. Passing section *I-I* and equating to zero ΣM about point K at the intersection of the vertical rod with the one coinciding with the neutral axis of the member we obtain

$$\Sigma M_k = M - Xr = 0$$

wherefrom

$$X = \frac{M}{r}$$

If $r = 1$, the force X in the connecting rod will be numerically equal to the moment

$$X = \frac{M}{1} = M$$

and thus, instead of constructing the influence line for the bending moment acting over the cross section we may construct the influence line for the stress X induced in the lower bar of Fig 11.6. Upon elimination of this bar the connection between the two parts of the member will consist of two bars intersecting in its neutral axis which is equivalent to a hinge. Schematically this connection is represented in Fig. 11.6*b*.

All the above shows that the construction of influence lines for the usual stress functions may be reduced to the construction of those for a normal force acting in a bar.

4.6. *Construction of the Displacement Graphs*

The virtual displacement graph determines completely the shape of the influence lines, their ordinates differing by a constant factor only. For this reason it may be said that when the displacement graph is completed the main bulk of work has been done. Let us examine the construction of displacement graphs in the case of one, two and four hinge-connected plates.

(a) *Displacement graph for a single plate.* Assume that plate I with one fixed point O is acted upon by one moving load $P = 1$ and a fixed force X, the direction of the former may be arbitrary. Let plate I turn an angle $d\varphi$ about point O in a clockwise direction (on the choice of direction see Art. 6.6).

The x-axis of the graph may be chosen at will excepting parallels to the direction of force P. Thus, in Fig. 12.6*b* the x-axis is normal to the line of action of force P while in Fig. 12.6c it has been chosen horizontal. The axis of ordinates must be always taken parallel to force P. The pertinent points of the system will be denoted on the graphs by the same letters with a prime index.

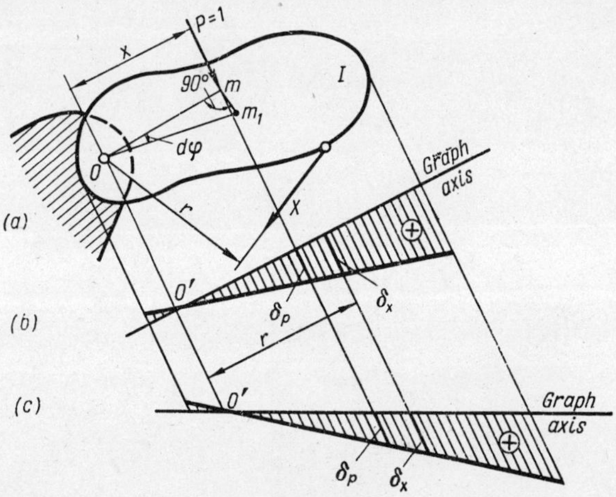

Fig. 12.6

The displacement of point m along the direction of P will equal (see Art. 2.6)

$$\delta_p = x d\varphi$$

where x is the distance of the line of action of the force to the centre of rotation (always measured along a normal to this line, regardless of the direction adopted for the x-axis of the graph).

The above expression shows clearly that in the case under consideration the displacement graph will form a straight line intersecting the x-axis at point O' where x reduces to zero.

To the right of point O' the ordinates of the graph are positive, as the direction of the displacement δ_p coincides with the direction of the force P, while to the left of this same point the ordinates will be negative, for this portion of the plate will move in an opposite direction.

The scale factor will be obtained remembering that

$$\delta_x = r d\varphi$$

It follows that for $x = r$

$$\delta_x = \delta_p$$

In other words, the scale factor is equal to the displacement graph ordinate measured a distance r from point O' (Fig. 12.6*b* and *c*).

(b) *Displacement graph for two pin-connected plates.* Let us construct the displacement graph for two plates I and II fixed at points O_1 and O_2 and connected to one another by means of hinge *1* located in line with points O_1 and O_2 (Fig. 13.6). As we know, such a system will be instantaneously unstable, hinge *1* being able to sustain infinitesimal displacements along a normal to line O_1-O_2,

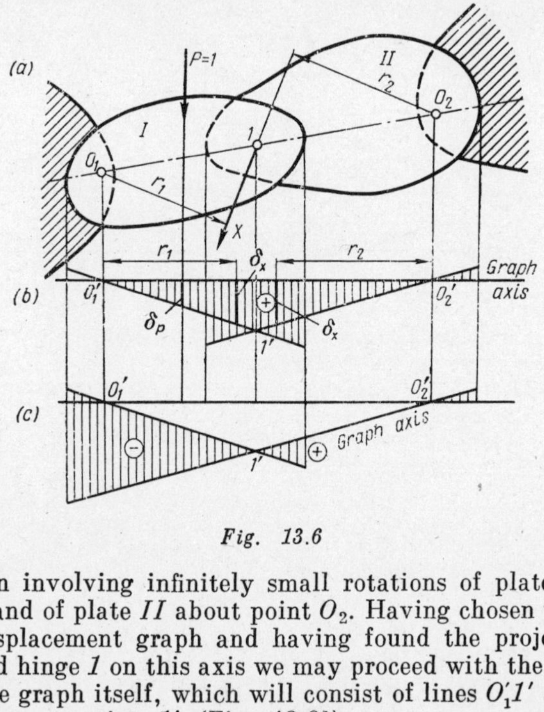

Fig. 13.6

its motion involving infinitely small rotations of plate I about point O_1 and of plate II about point O_2. Having chosen the x-axis of the displacement graph and having found the projections of O_1, O_2 and hinge *1* on this axis we may proceed with the construction of the graph itself, which will consist of lines $O_1'1'$ and O'_21' intersecting at point $1'$ (Fig. 13.6*b*).

The scale factor can be found assuming either that the displacement of point *1*, common to both plates, is caused by the rotation of plate I about point O_1, or by the rotation of plate II about point O_2. In the first case the scale factor will be given by the length of the insert between the line bounding the graph and the x-axis, measured along a parallel to the direction of force P a distance r from point O_1', and in the second by the length of a similar insert but measured a distance r_2 from point O_2'. If measured correctly, both scale factors will be exactly the same.

Thus far we have admitted that plates *I* and *II* are fixed to the ground at points O_1 and O_2 which remain immobile, thereby implicating the presence of a third unmoved plate constituted by the ground itself.

In this respect it is quite important to note that from the viewpoint of theoretical mechanics all of these three plates are perfectly equivalent. Therefore, it is absolutely immaterial which of

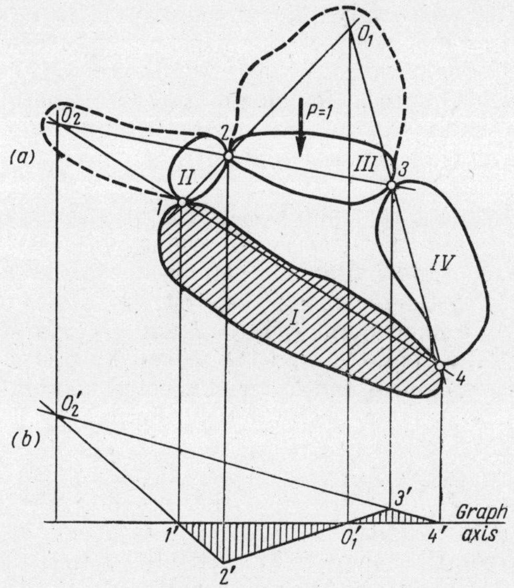

Fig. 14.6

the three will be reckoned immovable and no change whatsoever will occur in the outline of the displacement graph when the label "immobile" is shifted from one plate to another.

Indeed, if it were assumed that plate *II* is the immovable one, line $1'$-O_2' should be adopted as the axis of the displacement graph and nothing except the hatching of the graph area would alter as shown in Fig. 13.6*c*.

The importance of the above remark resides in the fact that in a number of cases the construction of the virtual displacement graph may be considerably simplified by an appropriate choice of that part of the structure which will be reckoned immovable.

(c) *Displacement graph for a system of four plates.* Let us consider a system of plates *I*, *II*, *III* and *IV* connected by means of hinges *1*, *2*, *3* and *4* (Fig. 14.6*a*), such systems being frequently encountered in practice. If we assume that plate *I* is the immovable, points *1* and *4* will lie on the axis of the graph (points *1'* and *4'* in Fig. 14.6*b*). Imparting to plate *II* an infinitesimal rotation about point *1* in a clockwise direction we shall obtain a displacement graph represented by the line *1'-2'*.

In order to complete the displacement graph for plate *III* the displacement of only one extra point is required as the displacement of point *2* is already known (point *2'*). It is very convenient to adopt as such the instantaneous centre of rotation (otherwise called the instantaneous centre of zero velocity) of this plate with reference to plate *I*, for on the graph this point will necessarily lie on the axis of zero displacements. In order to find this centre let us extend the line *1-2* until its intersection with the line *3-4* at point O_1.

It is easy to prove that point O_1 constitutes the required centre of rotation. Indeed, fictitiously enlarging plate *III* until inclusion of point O_1, and fixing this point we obtain two instantaneously unstable systems formed: the first, by plates *II* and *III*, each having one fixed point (points *1* and O_1, respectively) and hinge-connected at point *2*, and the second by plates *III* and *IV* fixed at points O_1 and *4* and hinge-connected at point *3*. Each of these two systems is in every respect similar to the system of Fig. 13.6*a*.

As fixing the point O_1 does not prevent infinitesimal displacements of plates *II*, *III* and *IV* with respect to plate *I* reckoned immovable, it is clear that this point is in effect the instantaneous centre of rotation of plate *III*. It follows that the projection of point O_1 on the axis of the graph will provide the extra point required and therefore line *2'-O'-3'* will constitute the displacement graph for the plate *III*. Repeating the same reasoning for plate *IV* we shall find that line *4'-3'-O'* forms the displacement graph for the latter.

Thus the broken line *1'-2'-3'-4'* constitutes the entire displacement graph of the system formed by four hinge-connected plates.

If, for instance, plate *IV* were regarded as the reference one, line *3'-4'* would constitute the axis of the graph from which all the displacements should be measured. The instantaneous centre of rotation of plate *II* would be located at point O_2 formed by the intersection of lines *1-4* and *2-3*. On the displacement graph the corresponding point should lie on the axis of the graph (line *3'-4'*), its displacement being nil. Points *1'*, *2'* and O_2' must also lie on one and the same straight line, for all the three belong to plate *II*.

5.6. *Determination of the Scale Factor*

As already known (Art. 2.6), the displacement in the direction of the force X equals

$$\delta_x = rd\varphi$$

whereas the displacement of any point of the plate along the direction parallel to the load P amounts to

$$\delta_p = xd\varphi$$

It follows that for $x = r$

$$\delta_x = \delta_p$$

Thus, *the scale factor may be obtained by measuring the ordinate to the displacement graph at a distance r from the projection of the centre of rotation, where r is the lever arm of the force X about this centre.* On the graph distance r must be always measured along a normal to the direction of the mobile load, regardless of the angle formed by the axis of the graph and the direction of the load (see Fig. 12.6*c*).

Examples of scale factor determination, when the system consists of one or two plates, were given in Art. 4.6.

There are several ways of obtaining the value of the scale factor when three or more plates are involved. It is obvious that regardless of the procedure adopted we must always obtain the same results, but nevertheless for the sake of clarity we shall denote by δ_{1x}, δ_{2x}, etc., the values of the scale factors determined in different ways.

Assume that in Fig. 15.6*a* plate *I* is fixed in which case line *1′-4′* in Fig. 15.6*b* will constitute the axis of the graph. The displacement δ_{1x} will be conditioned solely by the motion of hinge *2*, point *5* belonging to plate *I* which is regarded as immovable.

Assume that the entire force X is applied to plate *II* in which case the corresponding lever arm will equal r_1. The scale factor δ_{1x} will then equal to ordinate to line *1′-2′* measured a distance r_1 either to the left or to the right of point *1′*.

If, on the other hand, the force X were applied to plate *III*, the lever arm r_2 should be measured from point O_1' and the scale factor δ_{2x} would be given by the ordinate to the line *2′-3′* measured a distance r_2 from point O_1'.

Let us determine the scale factor δ_{3x} assuming that plate *IV* is rendered immovable. In this case both points *2* and *5* will acquire a certain displacement and therefore the required scale factor will be represented by the algebraic sum

$$\delta_{3x} = \delta_{3x}' + \delta_{3x}''$$

where δ'_{3x} is the displacement of point *2* and δ''_{3x} that of point *5*.

In order to determine δ'_{3x} let force X act solely on plate *III*. In that case the lever arm equals r_3 and the insert between the axis of the graph (line *3'-4'* in this case) and the line corresponding to the displacement of plate *III* (line *2'-3'*) measured a distance

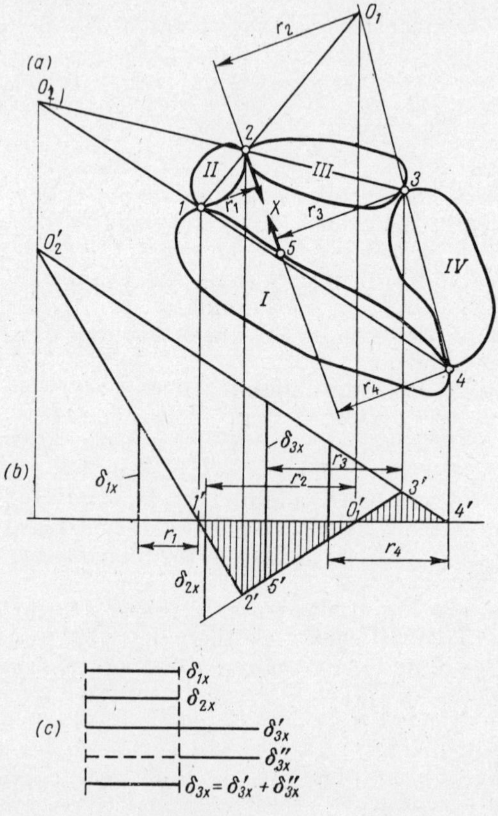

Fig. 15.6

r_3 from point *3'* will represent δ'_{3x}. It must be reckoned positive, for point *2* moves along the direction of the force X. As for δ''_{3x} its value will be found by applying force X to plate *I*, the lever arm in that case equalling r_4.

The insert between the graph axis and line *1'-4'* (representing the displacement of plate *I* with reference to plate *IV*) measured a distance r_4 from point *4'* will yield the value required. This displacement is negative and therefore the value of δ'_{3x} will be found by subtracting the length of δ''_{3x} from that of δ_{3x}. A comparison of the three scale factors obtained appears in Fig. 15.6*b*. If all the operations were carried out correctly all the scale factors obtained will be in strict coincidence.

6.6. *The Sign Convention*

The correct determination of signs will be greatly simplified if the rotations of the plates were such as to ensure in every case a positive displacement along the line of action of the force X, for in this case the scale factor δ_x will be always positive.

This will be fulfilled if the motion imparted to the plates coincides with the direction of X. All the ordinates to the influence line in that case will be opposite in sign as compared with the ordinates to the displacement graph, since X is equal in amount and opposite in sign to $\frac{\delta_p}{\delta_x}$ [see expression (1.6)].

If the load P is directed downwards (in which case positive displacements δ_p are laid off below the graph axis and the negative ones above the axis), those of the ordinates to the influence line which are above the x-axis will be positive, and those below the same axis negative. Vice versa, when the load is directed upwards, positive influence line ordinates will be below the x-axis and negative ones above it.

7.6. *Examples of Influence Line Construction*

Problem 1. Required the influence line for reaction at B of a multispan statically determinate beam shown in Fig. 16.6*a*.

Solution. Eliminate the constraint at the support B and replace it by a force X. Impart an upward motion to point *1* coinciding in direction and sign with force X and construct the displacement graph for plate *I* which will be represented by the line *0'-1'-2'* of Fig. 16.6*b*. Line *2'-3'-4'* will correspond to the displacements of plate *II* and line *4'-5'-6'* to those of plate *III*. The lever arm of X about point *0* is equal to *1*. The value of scale factor δ_x will be given by the ordinate to the displacement graph at point *1*. If the latter is adopted for unity the influence line will merge with the displacement graph.

The ordinates at points *2*, *4* and *6* will be found from the similitude of triangles

$$y_2=+\frac{3}{2};\ y_4=-\frac{3}{2};\ y_6=+\frac{3}{4}$$

Positive ordinates are above the x-axis, negative ones below. The same influence line was obtained previously using statics (see Fig. 57.2*c*).

Problem 2. Required the influence line for the shear in cross section *m-n* of the beam represented in Fig. 17.6*a*.

Solution. Introduce a movable connection as shown in Fig. 17.6*b* between the two parts of the beam separated by section *m-n* and two forces

$X = Q_{mn}$ replacing the vertical constraint at this cross section. Select a graph axis, say, line *1'-11'* and mark on it all the fixed points of the beam (points *1*, *3*, *9* and *11*).

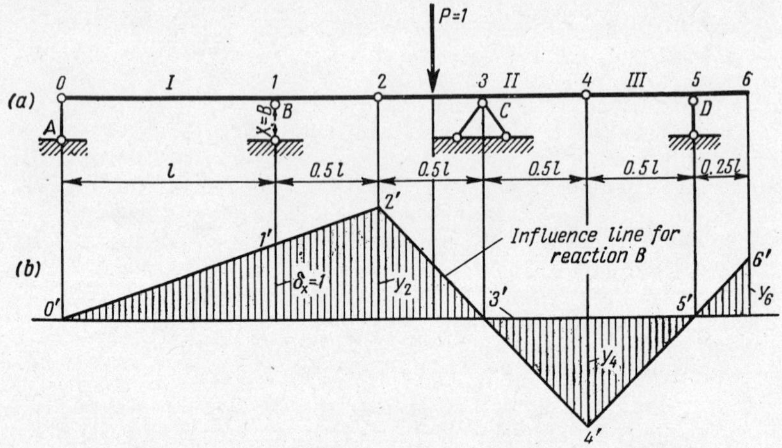

Fig. 16.6

Impart a clockwise rotation to plate *I* about point *A* and a similar rotation to plate *II* about point *B*. The two displacements will be represented

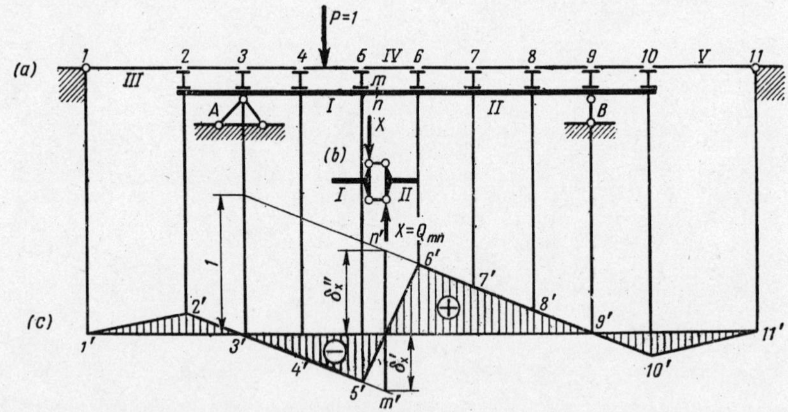

Fig. 17.6

in the graph by the lines *2'-3'-m'* and *n'-9'-10'*, respectively, these two lines being parallel as both parts of the beam are rotated through the same infinitesimal angle $d\varphi$. Mark points *5'* and *6'* on the corresponding lines of the graph.

Line *1'-2'* will constitute the graph for plate *III*, line *5'-6'* for plate *IV* and line *10'-11'* that for plate *V*, the whole graph consisting of the broken line *1'-2'-5'-6'-10'-11'*.

In order to determine the scale factor assume that plate *I* is fixed. Then δ_x will be equal to $m'n'$ which will be regarded as unity

$$\delta_x = m'n' = 1$$

It is readily seen that δ_x is the sum of δ_x' and δ_x''. The same influence line had been obtained previously using statics (see Fig. 55.2*c*).

Problem 3. Required the influence line for the stress U_{46} of a deck-bridge truss of Fig. 18.6a.

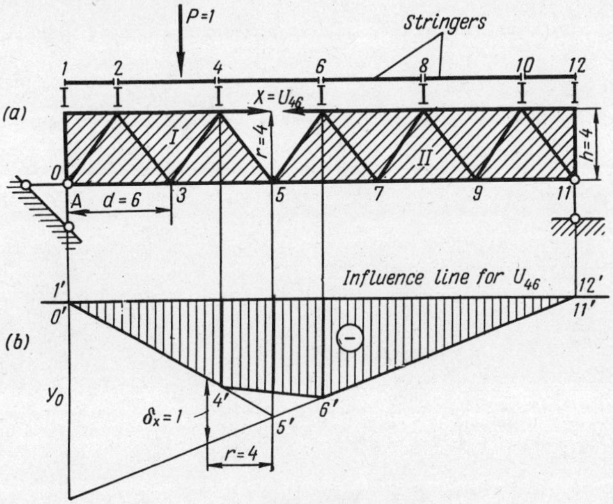

Fig. 18.6

Solution. Replace upper chord member *4-6* by the stress $X = U_{46}$. It should be noted that the elimination of bar *4-6* does not entail that of the corresponding stringer.

The force X will cause a clockwise rotation of plate *I* and counter-clockwise one of plate *II*. The corresponding displacement graphs will be represented by lines *1'-5'* and *5'-12'* (Fig. 18.6*b*). Projecting on these lines points *4'* and *6'* we obtain the displacement of joints *4* and *6* of the upper chord.

The scale factor is found assuming that plate *I* is rendered immobile, plate *II* rotating about point *5*. The lever arm r of stress X about this point equals $h = 4$ m and, accordingly, the scale factor will be given by the insert between lines *1'-5'* and *5'-12'* measured vertically a distance of 4 m from point *5'* (assuming plate *I* fixed, line *1'-5'* becomes the graph axis). Knowing the value of this insert and reckoning it equal to unity, it is easy to determine the influence line ordinate y_0 at the abutment A. Indeed, from the similitude of triangles, $y_0 = 3$. The same influence line had been obtained previously (see Fig. 71.4).

Problem 4. Required the influence line for the stress in diagonal *5-6* of a through bridge truss shown in Fig. 19.6.

Solution. Eliminate the diagonal under consideration and replace it by two forces $X = D_{56}$. The system will be thus transformed into two plates *I* and *II* (hatched on the drawing) connected to one another by two bars *5-7* and *4-6* the directions of which intersect at point K.

Let plate *II* be fixed. Its displacements being in that case nil, the corresponding displacement graph *6'-12'* will merge with the x-axis (Fig. 19.6*b*) and the instantaneous centre of rotation of plate *I* will be at point K. The force X will impart a clockwise rotation to plate *I* about this centre, line

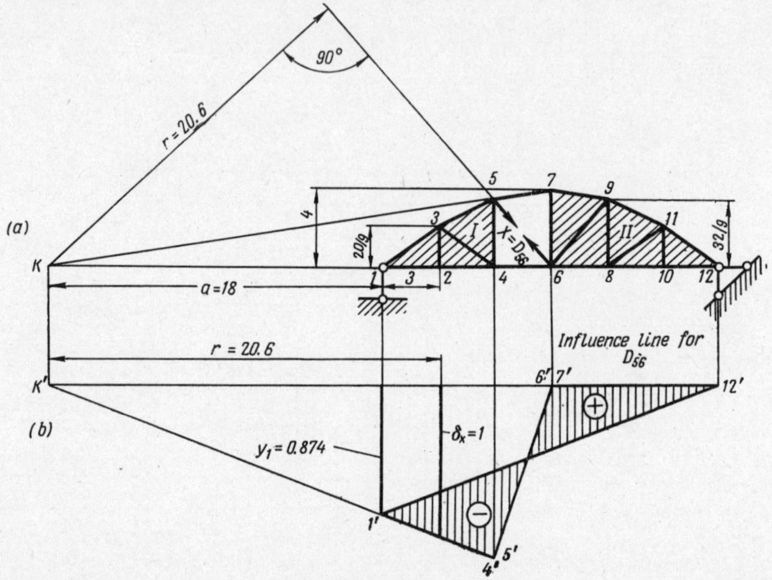

Fig. 19.6

K'-4' representing its displacement graph while the lines *4'-6'* and *5'-7'* will represent that of bars *4-6* and *5-7*.

Actually it is not the plate *II* but the ground, i.e., points *1* and *12*, that should be regarded as fixed. Therefore trace line *1'-12'* and adopt it as the final axis of the graph.

The entire displacement graph will then be represented by the broken line *1'-4'-6'-12'*, the sign convention stipulated in the previous section remaining in force.

Thereafter proceed with the determination of the scale factor. Assuming that plate *II* is immobile and that force X acts on plate *I*, the lever arm r of this force about point K will equal 20.6 metres (see Problem 2 in Art. 7.4). The insert between the graph axis *6'-12'* and the line representing the displacement of plate *I* (line *1'-4'*) measured at a distance of 20.6 m from point K' will provide the value of $\delta_x = 1$.

Using the similitude of triangles obtain ordinate y_1 under the left-hand abutment

$$\frac{1}{y_1}=\frac{20.6}{18}$$

wherefrom

$$y_1=\frac{18}{20.6}=0.874$$

The influence line thus obtained coincides fully with that of Fig. 69.4 constructed using statics.

Problem 5. Required the influence line for the bending moment acting over cross section K of a parabolic three-hinged spandrel arch of Fig. 20.6*a*.

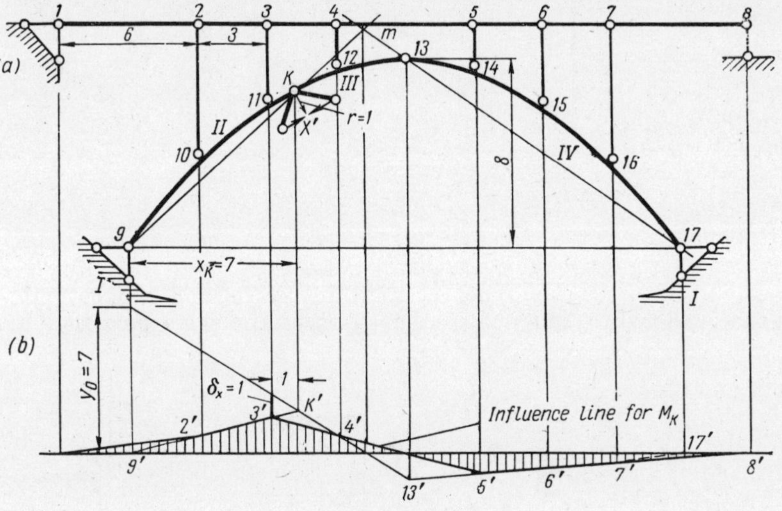

Fig. 20.6

Solution. Introduce an extra hinge at cross section K which leads to the formation of four plates *I*, *II*, *III* and *IV* connected together by means of four hinges *9*, K, *13* and *17*.

Construct the displacement graph of this system of plates using the instantaneous centre of rotation of plate *III* with reference to plate *I* (point m) which will be represented by the broken line *9′-K′-13′-17′* (Fig. 20.6*b*).

Points *1* and *8* are plotted on the graph axis. Points *2′* and *3′* are then marked on the displacement graph for plate *II*, point *4′* on that for plate *III*, and points *5′*, *6′* and *7′* on that for plate *IV*. Connecting all these points together the displacement graph of all the panel points of the deck will be obtained.

In order to determine the scale factor fix plate *II* and let force X act on plate *III* causing it to rotate with reference to plate *II* about the hinge K. The lever arm of force X may be taken equal to 1 metre.

The scale factor δ_x will be given by the length of the segment between the graph axis *9′-K′* and the line representing the displacement of plate *III* (line *K′-13′*) measured one metre away from point K'.

Knowing the value of this segment the ordinate to the influence line for the bending moment at the abutment hinge *9* will be found from

$$\frac{y_0}{\delta_x}=\frac{x_k}{1}$$

wherefrom

$$y_0=\delta_x x_k=1x_k=7$$

All the other ordinates to the influence line will be readily found thereafter.

7. RETAINING WALLS AND EARTH PRESSURE COMPUTATION

1.7. *General*

Retaining walls are structures intended to prevent the sliding down of slopes too steep to remain standing on their own. Fig. 1.7 shows two different types of retaining walls and a sheet pile-wall which in numerous cases may serve the same purpose.

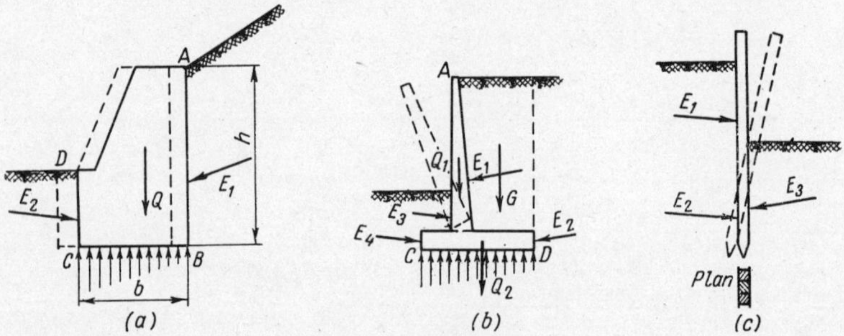

Fig. 1.7

The retaining wall shown in Fig. 1.7*a* is a massive construction, its main dimensions b and h being of the same order. Walls of this type are usually built of rubble or mass concrete. They are subjected to their dead weight Q, the active and passive pressure of the earth E_1 and E_2 developed over the rear and front faces AB and CD and the reaction CB acting over the foundation. Retaining walls of much lighter construction shown in Fig. 1.7*b* are usually built of reinforced concrete and consist of a foundation slab CD and a vertical wall AB. The forces acting on a wall of this type consist of the dead weight Q_1, Q_2, . . ., etc., of the weight G of the column of earth resting on the foundation slab, the active and passive pressure of the earth E_1, E_2, . . ., and the reactive forces distributed over the lower surface of the foundation slab. The reduced weight of these walls renders it possible to make use of prefabrication techniques.

Sheet pile-walls are built up of separate wooden, reinforced concrete, or steel sheet piles which are sunk into the ground side by side using special equipment. The dead weight of sheet piles and the vertical reaction applied to their points are so small that they are always neglected. Accordingly, the only forces that must be considered are the active and the passive pressures of the earth E_1, E_2, E_3, ..., etc., which must balance each other.

In all computations pertaining to retaining walls the depth of the structure in the direction normal to the surface of the drawing will be always taken equal to one metre. The design of retaining walls and of sheet piling must be always preceded by the determination of the loads and forces acting on these structures including the earth pressure E. Without committing any serious error, both the active and the passive earth pressures may be computed on the assumption that the earth constitutes a granular material.

2.7. *Physical Properties of Granular Materials*

Granular materials consist of very small solid rounded particles and therefore the only internal stresses that can develop in such materials are friction and compression. Dry sand and grains of cereals in large quantities constitute granular materials which are as close as possible to the definition given above. In the actual design of retaining walls cohesive soils are frequently met with but the forces of cohesion are usually neglected and the soil is regarded as a granular mass.

In order to determine the pressure exerted by a granular material on a retaining wall the following physical properties of this material must be known:

1. Its weight per cubic metre γ usually given in tons. This weight varies from 1.6 tons per cubic metre for dry sand to 2.0 tons per cubic metre for water saturated materials.

2. Its porosity η given in per cent and representing the ratio of all the intergranular voids to the total volume of the material. For compacted sand $\eta = 30$ per cent, for loose sand it is close to 50 per cent and for dry clay it may vary from 25 to 40 per cent.

3. The weight of the material suspended in water γ_0 also given in tons per cubic metre. As one cubic metre of the material contains η per cent of voids, the loss in weight due to its immersion will be equal to the weight of the water displaced or, in other words, to

$$\left(1-\frac{\eta}{100}\right)\gamma_w$$

where γ_w is the density of the water.

Consequently

$$\gamma_0 = \gamma - \gamma_w \left(1 - \frac{\eta}{100}\right) \tag{1.7}$$

4. The angle of repose φ which is the steepest angle to the horizontal at which a heap of this material will stand on its own (Fig. 2.7). This angle is characteristic of the friction developed

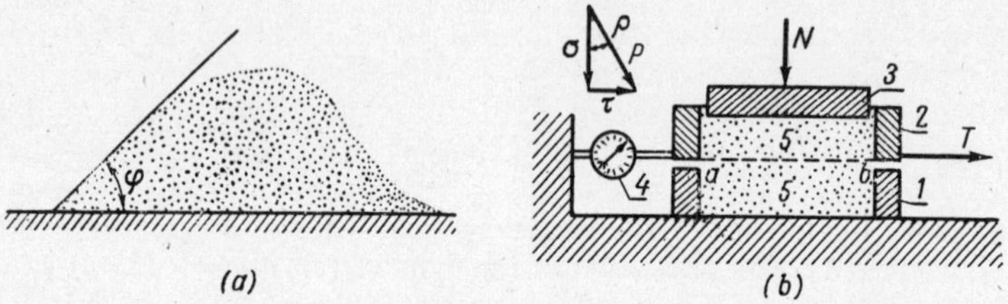

Fig. 2.7

between the particles at the surface of the granular material (all forces of cohesion being neglected).

The magnitude of the angle of repose φ depends greatly on the degree of humidity of the material. Thus

for dry sand	$\varphi = 30\text{-}35°$
for humid sand	$\varphi = 40°$
for wet sand	$\varphi = 25°$
for dry clay	$\varphi = 40\text{-}45°$
for wet clay	$\varphi = 20\text{-}25°$

5. The angle of internal friction ϱ characterizing the friction between the inner particles of a large volume of the material. The magnitude of this angle can be determined experimentally using a device schematically represented in Fig. 2.7*b*. This device consists of a metal cylinder separated horizontally in two parts (*1* and *2*), a plunger die *3* and a dial indicator *4*. The lower part of the cylinder (part *1*) is fixed whilst the upper one (part *2*) can move horizontally under the action of a force T. The specimen of the granular material *5* contained in the cylinder is subjected to a constant vertical pressure N, transmitted through the plunger die *4*, and to a gradually increasing shearing force T. The magnitude of this force is registered at the precise moment when the state of limit equilibrium is reached, in other words, at the moment when the first sign of sliding of the upper part of the cylinder along the plane *a-b* is detected by the dial indicator *4*.

At this moment the compressive stress σ acting across section *a-b* is equal to $\frac{N}{F}$ while the shearing stress τ equals $\frac{T}{F}$, F being the area of the cross section *a-b*.

When the state of limit equilibrium is reached, the resulting stress p is deviated from the normal to the plane along which the sliding occurs by an angle equal to the angle of internal friction given by

$$\tan \varrho = \frac{\tau}{\sigma}$$

wherefrom

$$\tau = \sigma \tan \varrho$$

The U.S.S.R. Building Codes usually stipulate the following values for the angle of the internal friction:

for fine sand	$\varrho = 20\text{-}30°$
for medium sand	$\varrho = 30\text{-}40°$
for course sand, gravel and rounded pebbles	$\varrho = 40\text{-}45°$
for sandy loam	$\varrho = 15\text{-}30°$
for ordinary loam	$\varrho = 10\text{-}30°$

The value of the angle of internal friction in sandy soils may be considered approximately equal to its angle of repose φ i. e., $\varrho = \varphi$.

6. The angle of friction between the material and the face of the wall δ, which depends mainly on the condition of the surface along which the contact occurs. When the surface is very smooth δ almost equals 0, and for very course surfaces δ may approach the angle of internal friction ϱ. In actual design work δ is frequently taken equal to zero. Otherwise it may be expressed as a fraction of the angle of internal friction

$$\delta \approx \frac{1}{2} \varrho \text{ up to } \frac{3}{4} \varrho$$

7. The cohesion C which is usually expressed in kg per sq cm or in tons per sq m. In dry granular materials, such as sand or grain, C is practically nil. In other usual soils the cohesion will amount only to a fraction of a ton per square metre and therefore it may be safely neglected. A device similar to the one described above can be used for the determination of the cohesion C which is related to the normal and shearing stresses by Coulomb's formula

$$\tau = C + \sigma \tan \varrho$$

3.7. *Active Pressure of Granular Materials*

The active pressure of a granular material is the force which it will develop on some surface when the latter moves over a very small distance away.

As the surface AB of Fig. 3.7a shifts to a new position A_1B_1 a part of the granular material contained in the wedge ABC starts

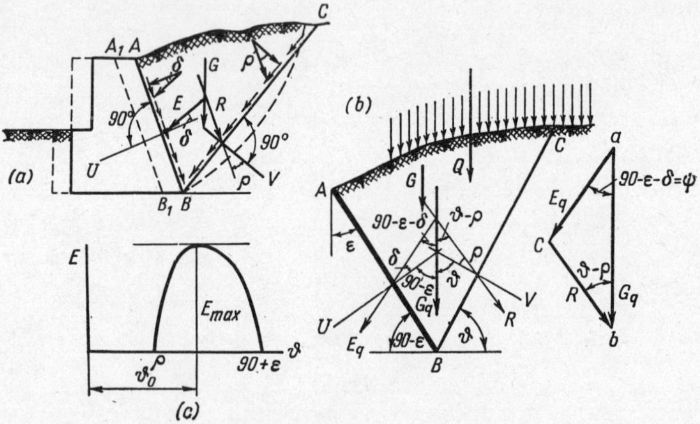

Fig. 3.7

moving downwards. The surface which separates the moving part from the one remaining immovable is called the *cleavage* or *slip plane* (sufrace) and its projection on the plane of the drawing—the *cleavage* or *slip line*. The paths of the particles contained in the wedge ABC are very intricate and depend both on the character and the magnitude of the displacement of the surface AB. The directions of the pressures E and R exerted by the granular material cannot be determined with certainty, for the state of limit equilibrium will never be reached simultaneously at all points along the surfaces concerned and therefore the stress will not be deviated everywhere from the normal by an angle equal to the angle of friction.

The correct determination of the pressure developed by the earth against some surface is therefore extremely complicated and has as yet not found a comprehensive solution. The simplified wedge theory given by Coulomb (1736-1806) is based on the following assumptions:

1. The curved cleavage surface is replaced by a plane whereby its projection on the plane of the drawing becomes a straight line BC.

2. The granular materials contained within the wedge are considered solid.

3. The wedge itself is in a state of unstable equilibrium, i. e., in a state preceding immediately its sliding down. The latter assumption permits to determine the directions of the resultant pressures E and R. When the surface AB moves away, the wedge ABC starts sliding down and the forces of friction which develop along the surfaces AB and BC within the material will be also directed downwards. If the limit equilibrium is reached simultaneously at every point along the surface AB the resultant stress will be deviated everywhere from the normal to this surface by an angle equal to the angle of friction δ and therefore the resultant pressure will also make an angle δ with the normal U. Similarly the pressure R will be deviated from the normal V by an angle equal to the angle of internal friction ϱ.

Let us determine the pressure E_q developed against the surface AB (Fig. 3.7*b*) when an arbitrary surcharge is applied to the surface of the earth.

Assume that $G =$ dead weight of the wedge ABC ($G =$ area $ABC\gamma$)
$Q =$ resultant of the surcharge acting on the wedge
$G_q =$ resultant of the forces G and Q; $G_q = G + Q$

Knowing the magnitude of G_q and the directions of the pressures E_q and R we may construct the triangle of forces *abc*.

The angles of this triangle are

$$\angle abc = \vartheta - \varrho; \quad \angle cab = 90° - \varepsilon - \delta = \psi$$
$$\angle acb = 180° - (\vartheta - \varrho + \psi)$$

From this triangle we obtain

$$\frac{E_q}{\sin(\vartheta - \varrho)} = \frac{G_q}{\sin[180° - (\vartheta - \varrho + \psi)]}$$

wherefrom

$$E_q = G_q \frac{\sin(\vartheta - \varrho)}{\sin(\vartheta + \psi - \varrho)} \tag{2.7}$$

This expression cannot be used as yet for the determination of the active pressure E_q for it contains the angle ϑ made by the cleavage plane with the horizontal which remains unknown as well as the dead weight of the wedge G and the magnitude of the surcharge Q, both depending on the angle just mentioned.

When the angle ϑ made by the cleavage plane with the horizontal varies it entails a corresponding variation in the value of the pressure E_q, this variation, if represented graphically, having the shape of a curve shown in Fig. 3.7*c*. When $\vartheta = \varrho$ $\sin(\vartheta - \varrho) = 0$ and $E_q = 0$; for $\vartheta = 90° + \varepsilon$ the cleavage plane BC will coincide with the back of the wall AB and both E_q and the resultant G_q will also reduce to zero.

It is obvious that the maximum value of the active pressure E_q will correspond to some intermediate value of $\vartheta = \vartheta_0$.

When designing a retaining wall this maximum value of the active pressure should be taken into consideration, for if the strength and stability of the wall are insured under these most adverse conditions, the wall will remain standing for any other direction of the cleavage plane. The value of the angle ϑ_0 corresponding to the maximum of E_q may be determined from the equation

$$\frac{dE_q}{d\vartheta} = 0$$

The sign of the second derivative shows that the pressure thus obtained by Coulomb's wedge theory is indeed the maximum one. In actual practice the maximum active pressure developed by the earth against the back of a retaining wall may be somewhat smaller than $E_{q\,max}$ determined as above. However, in certain cases when the displacement of the wall becomes extremely small (for instance, when the wall is founded on solid rock) the pressure it will sustain may exceed substantially the maximum pressure computed on the basis of the aforesaid theory.

When the surface of the earth is of irregular shape, the equation $\frac{dE_q}{d\vartheta} = 0$ may be solved only by graphical methods. If the surface is plane, direct computation becomes possible.

4.7. *Graphical Determination of Maximum Active Pressure*

Let us determine the direction of the cleavage plane corresponding to the maximum pressure developed against the back of a retaining wall AB when the surface of the earth is irregular in shape but no surcharge is applied thereto. Adopting an oblique system of coordinates HBD we shall first construct the graph of the variation of the active pressure E_n in terms of the direction of the cleavage plane (Fig. 4.7). For this purpose let us measure to some scale the dead weight of the wedges along the axis BD and the pressures E_n along the axis BH. It may be shown that the length of the line K_nF_n will represent the amount of the pressure corresponding to the direction of cleavage line BC_n. Indeed, the weight G_n of the corresponding ABC_n will equal

$$G_n = \gamma \times (\text{area of triangle } ABC_n)$$

Assume that BF_n represents to scale this weight. The line F_nK_n parallel to the axis of coordinates BH will meet the line BC_n at point K_n.

The angles of the triangle BF_nK_n are equal to

$$\angle K_nBF_n = \vartheta_n - \varrho$$
$$\angle K_nF_nB = \alpha = (90° - \varepsilon) + \varrho - (\varrho + \delta) = 90° - \varepsilon - \delta = \psi$$

Let us now construct the triangle of forces abc in which the ray $ab = G_n = BF_n$ and the ray $ac = E_n$. Comparing the triangles

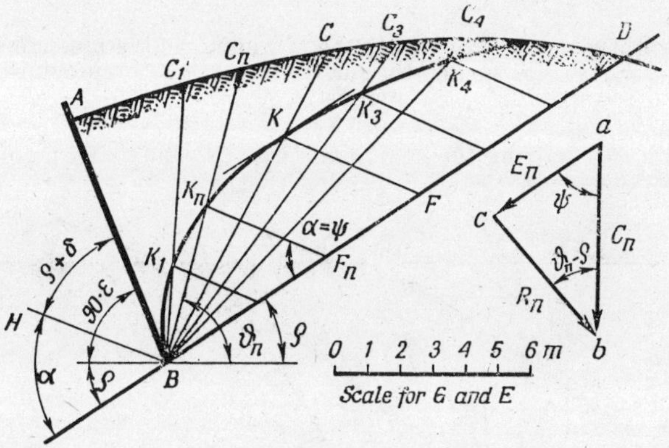

Fig. 4.7

abc and F_nBK_n we remark immediately that they are identical and therefore

$$E_n = K_nF_n$$

Thus, in order to determine the pressure developed by a granular material against the face AB for any given direction of the cleavage plane BC_n we must lay off along the axis BD the dead weight of the wedge ABC_n (represented by the length BF_n) and then trace through the point F_n a line parallel to the other axis BH until its intersection with the corresponding cleavage line BC_n at point K_n. The length of the line K_nF_n measured to scale will represent the magnitude of the pressure E_n.

If we repeat the construction just described for a number of conveniently chosen directions of the cleavage planes BC_1, BC_3, etc., we shall find a series of points K_1, K_3, etc. Connecting these points by a smooth curve we shall obtain the required graph showing the variation of E_n in terms of the angle ϑ exactly in the same way as the graph schematically represented in Fig. 3.7c. In order to find the maximum of E_n we may now trace a tangent to the curve parallel to the axis BD and through the point of tangency K we must trace the line KF parallel to the other coordinate axis BH. The length of this line (always measured to scale) will give us the maximum value of the active pressure E which will be developed against the back of the wall AB while the line BKC will indicate the inclination of the cleavage plane.

The graphical method described above remains valid when a surcharge is applied to the surface of the earth. In that case the dead weight of each wedge should be increased by the amount of the load which it carries.

Problem. Determine graphically the maximum active pressure developed by a granular material against the surface AB (Fig. 5.7), provided $\varrho = 40°$, $\delta = 5°$ and $\gamma = 1.6$ tons per cubic metre.

Solution.

1. Start with tracing the coordinate axes BD and BH.

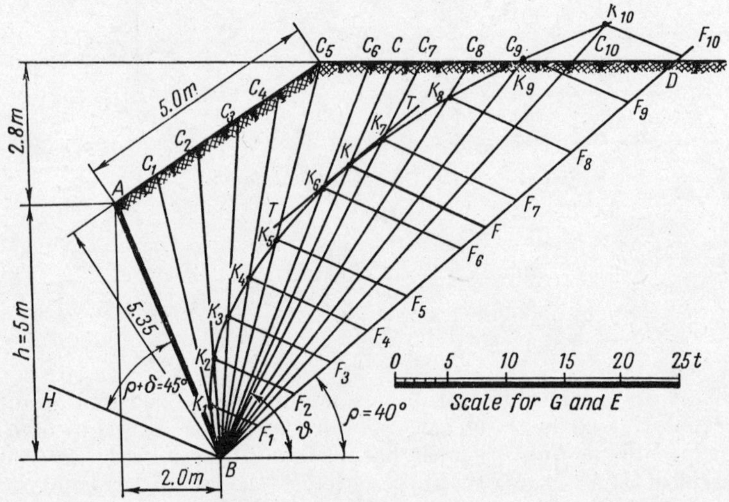

Fig. 5.7

2. Adopt a number of cleavage plane directions given by BC_1, BC_2, etc. For this purpose divide the line AC_5 into five segments each one metre long and select points C_6, C_7, . . ., C_{10} equally at one metre intervals.

3. Compute the dead weight of the wedges. For the wedge ABC_1 this weight equals

$$G_1 = \frac{1}{2} \times 5.35 \times 1 \times 1.6 = 4.28 \text{ tons}$$

The weights of the other wedges abutting to the line AC_5 will be exactly the same.

The weight of the wedge C_5BC_6 and of all the other wedges abutting to the horizontal C_5C_{10} will be equal to

$$G_6 - G_5 = \frac{1}{2} \times 7.8 \times 1 \times 1.6 = 6.24 \text{ tons}$$

4. Set out to scale along the axis BD the dead weights of the wedges G_1, G_2, . . ., G_{10} which are as follows

$BF_1 = G_1 = 4.28$ tons	$BF_6 = G_6 = 27.64$ tons
$BF_2 = G_2 = 8.56$ tons	$BF_7 = G_7 = 33.88$ tons
$BF_3 = G_3 = 12.84$ tons	$BF_8 = G_8 = 40.12$ tons
$BF_4 = G_4 = 17.12$ tons	$BF_9 = G_9 = 46.36$ tons
$BF_5 = G_5 = 21.40$ tons	$BF_{10} = G_{10} = 52.60$ tons

5. Through the points F_1, F_2, . . ., F_{10} trace the lines F_1K_1, F_2K_2, . . ., $F_{10}K_{10}$ parallel to the axis BH.

6. Connect the points B, K_1, . . ., K_{10} by a smooth curve thus obtaining the graph showing the variation of the pressure E developed against the surface AB.

7. Trace the line TT tangent to the graph and parallel to the axis BD.

8. Connect the point of tangency K and the foot of the wall B by a straight line BKC which will constitute the cleavage line.

9. Through the same point of tangency trace a line KF parallel to the axis BH and measure to scale the length of this line which will represent the maximum active pressure developed against the surface AB

$$E = KF = 13 \text{ tons}$$

5.7. *Poncelet's Method*

In all cases when the surface of the granular material and the surface AB are plane, the determination of the maximum active pressure may be carried out by a graphical method devised by Poncelet.

Without entering into the theoretical demonstration of this method (based equally on Coulomb's wedge theory) we shall describe hereunder the procedure to be followed when a uniformly distributed surcharge q acts on the surface of the earth.

Start with replacing this surcharge by an equivalent layer of earth, the thickness of which is given by

$$h_0 = \frac{q}{\gamma}$$

This being done, the position of the cleavage plane corresponding to the maximum of the active pressure E_q is determined as follows.

The line AB is continued until its intersection at point A_1 with the upper surface of the equivalent layer (Fig. 6.7). Thereafter:

(1) through the point B trace a line BL_1 making an angle ϱ with the horizontal and meeting the upper surface of the equivalent layer at L_1;

(2) through the point A_1 trace the line A_1M making an angle $(\varrho + \delta)$ with the surface A_1B until its intersection with the line BL_1 at point M;

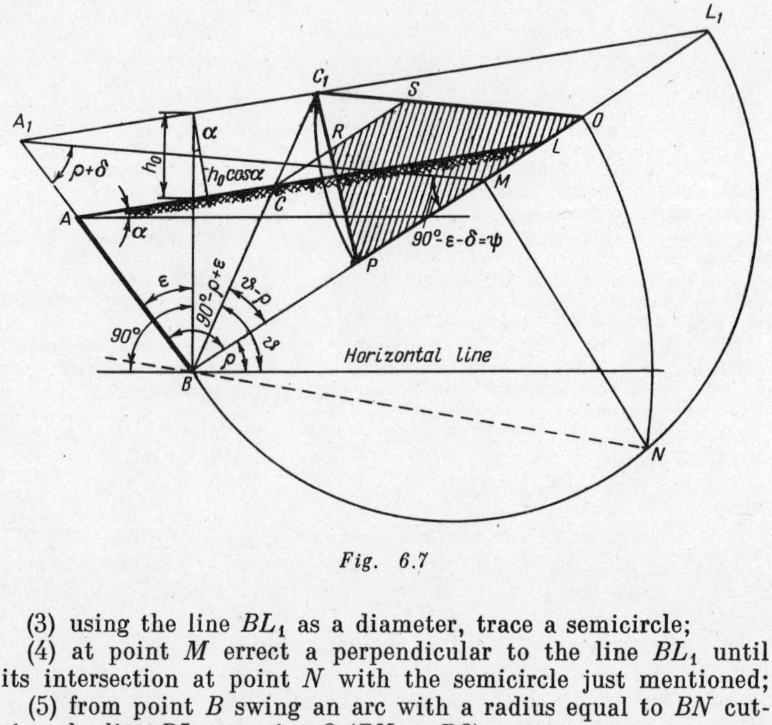

Fig. 6.7

(3) using the line BL_1 as a diameter, trace a semicircle;

(4) at point M errect a perpendicular to the line BL_1 until its intersection at point N with the semicircle just mentioned;

(5) from point B swing an arc with a radius equal to BN cutting the line BL_1 at point O ($BN = BO$);

(6) from point O trace line OC_1 parallel to A_1M until its intersection at point C_1 with A_1L_1;

(7) the line BC_1 connecting the foot of the wall with point C_1 constitutes the projection on the paper of the cleavage plane.

This being done, proceed with the determination of the magnitude of the pressure E_q developed against the surface AB:

(8) from point O trace an arc using OC_1 as radius until its intersection with line BL_1 at point P;

(9) connect points C_1 and P by a straight line thus obtaining the triangle OC_1P;

(10) through point C draw the line CRS parallel to BL_1.

The area of the trapezoid PRSO multiplied by 1 and by the density of the granular material γ *will give the magnitude of* $E_{q\,max}$ *developed against the surface AB.*

If the surcharge q were nil, it would suffice to take $h_0 = 0$ in which case the line $A_1C_1L_1$ of Fig. 6.7 would coincide with the line ACL and the point C_1 would coincide with the point C. The

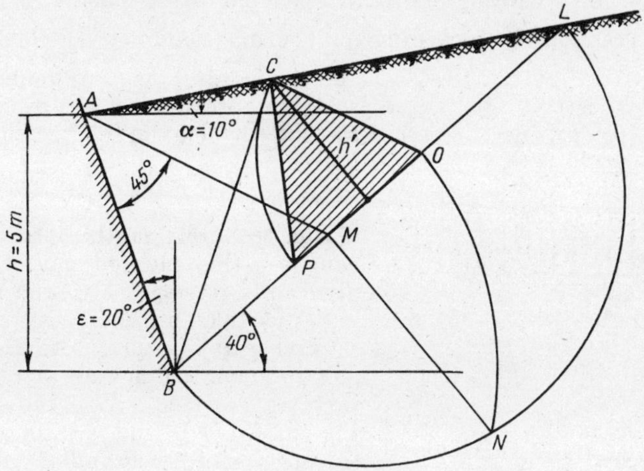

Fig. 7.7

direction of the cleavage plane will remain unchanged, as for the magnitude of the maximum pressure it will be given by the area of triangle POC (see Fig. 7.7) multiplied by 1 and by γ

$$E = \gamma \times (\text{area of triangle } POC)$$

Problem. Required to determine graphically the active pressure developed against the back of the wall AB (Fig. 7.7), provided $h = 5$ metres, $\varrho = 40°$, $\delta = 5°$, $\varepsilon = 20°$, $\alpha = 10°$, $\gamma = 1.6$ tons per cubic metre.

Solution. 1. Determine as just explained the position of the points A, M, N, O, C, P.

2. Determine the position of the cleavage plane BC.

3. Measure to scale the base and the height of the triangle POC equal, respectively, to 3.4 metres and 3.08 metres.

4. Compute the area of the triangle POC

$$F = \frac{1}{2}\, 3.4 \times 3.08 = 5.24 \text{ square metres}$$

5. Determine the magnitude of the active pressure required

$$E = 1.6 \times 5.24 = 8.38 \text{ tons}$$

6.7. *Method of Direct Computation of the Earth Pressure*

In order to compute directly the earth pressure we must first reduce Coulomb's formula (2.7) to the following form

$$E_q = Cf(\vartheta)$$

where C is a certain factor independent of ϑ. Thereafter using the expression $\frac{dE_q}{d\vartheta}$ providing for the maximum of E_q determine the position of the cleavage plane (angle ϑ_0). Having found this angle, substitute its value in the expression of E_q thus obtaining the maximum pressure required

$$E_q = Cf(\vartheta_0)$$

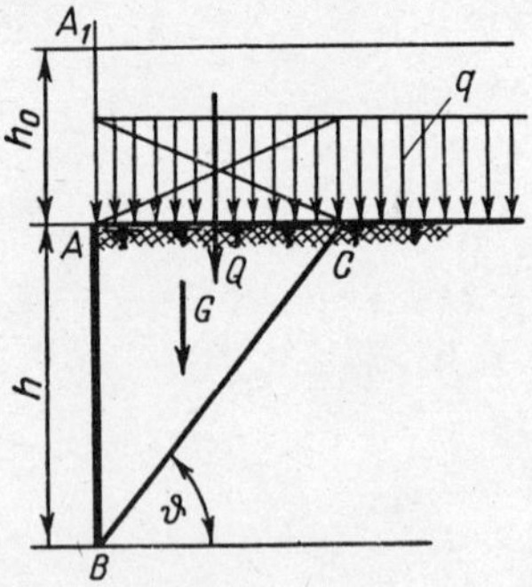

Fig. 8.7

Due to purely mathematical difficulties, this method may be applied only in some particular cases.

Let us take up the most simple case when it is required to find the pressure exerted against a smooth vertical surface AB shown in Fig. 8.7 ($\delta = 0$ and $\varepsilon = 0$), when the surface of the granular mass is horizontal and loaded with a uniformly distributed surcharge q tons per square metre.

Let BC represent the direction of some cleavage plane. In that case the dead weight of the wedge ABC will be given by $G = 0.5 \cdot AB \cdot AC \cdot 1 \cdot \gamma$ and the resultant of the surcharge acting on this wedge by $Q = AC \cdot 1 \cdot q$. The resultant of G and Q will be

$$G_q = G + Q = \frac{1}{2} AB \cdot AC \cdot \gamma + ACq = \frac{1}{2} AC \cdot \gamma \left(AB + 2\frac{q}{\gamma} \right)$$

Substituting in this expression h for AB, by $h \cot \vartheta$ for AC and h_0 for $\frac{q}{\gamma}$ (where h_0 is the thickness of the equivalent layer) we obtain

$$G_q = \frac{1}{2} \gamma h (h + 2h_0) \cot \vartheta$$

Reverting to the expression 2.7 and replacing $\psi = 90° - \varepsilon - \delta$ by 90°, the fractional part of this expression becomes equal to

$$\frac{\sin(\vartheta - \varrho)}{\sin(\vartheta + \psi - \varrho)} = \frac{\sin(\vartheta - \varrho)}{\cos(\vartheta - \varrho)} = \tan(\vartheta - \varrho)$$

After the above transformations, the expression (2.7) becomes

$$E_q = \frac{1}{2}\gamma h(h+2h_0)\cot\vartheta\cdot\tan(\vartheta-\varrho) = C\cdot f(\vartheta)$$

where

$$C = \frac{1}{2}\gamma h(h+2h_0);\; f(\vartheta) = \cot\vartheta\cdot\tan(\vartheta-\varrho)$$

The angle of the cleavage plane and the horizontal will be determined using the equation

$$\frac{dE_q}{d\vartheta} = 0 \text{ or } \frac{d}{d\vartheta}[Cf(\vartheta)] = C\left[-\frac{1}{\sin^2\vartheta}\tan(\vartheta-\varrho) + \right.$$

$$\left. +\cot\vartheta\frac{1}{\cos^2(\vartheta-\varrho)}\right] = 0$$

Reducing both terms in brackets to the same denominator and dividing the equation by

$$\frac{C}{\sin^2\vartheta\cos^2(\vartheta-\varrho)}$$

we obtain

$$\sin\vartheta\cdot\cos\vartheta = \sin(\vartheta-\varrho)\cos(\vartheta-\varrho)$$

or

$$\sin 2\vartheta = \sin 2(\vartheta-\varrho)$$

The roots of this equation are

$$\vartheta = n90^\circ + (-1)^n(\vartheta-\varrho)$$

where $n = 0, 1, 2, 3, \ldots$

If $n = 0$, we obtain $\vartheta = \vartheta - \varrho$ leading to $\varrho = 0$. This solution is incompatible with the physical properties of the granular materials for which we always have $\varrho \neq 0$.

When $n = 1$, we obtain $\vartheta = 90^\circ - (\vartheta - \varrho)$ leading to $\vartheta_0 = 45^\circ + \frac{\varrho}{2}$.

For values of n greater than one we obtain again a series of solutions incompatible with the terms of the problem. Therefore, the only root of the equation to be retained corresponds to $n = 1$ in which case the angle formed by he cleavage plane with the horizon equals

$$\vartheta_0 = 45^\circ + \frac{\varrho}{2} \tag{3.7}$$

Substituting this value of ϑ_0 in the expression of the pressure we obtain

$$E_q = C\cdot f(\vartheta_0) = C\cot(\vartheta_0)\tan(\vartheta_0-\varrho) = C\cot\left(45^\circ+\frac{\varrho}{2}\right)\times$$

$$\times\tan\left(45^\circ-\frac{\varrho}{2}\right)$$

Replacing in this expression cot $\left(45° + \frac{\varrho}{2}\right)$ by tan $\left(45° - \frac{\varrho}{2}\right)$ and substituting its value for C we finally obtain

$$E_q = \frac{1}{2}\gamma h(h + 2h_0)\tan^2\left(45° - \frac{\varrho}{2}\right) \tag{4.7}$$

If the surface AB had a batter ($\varepsilon \neq 0$) and were rough ($\delta \neq 0$) and the surface of the earth sloped towards the wall (Fig. 9.7), the magnitude of the active pressure would be given by the following formula

$$E_q = \frac{1}{2}\gamma h(h + 2h_0K_q)K \tag{5.7}$$

where

$$\left.\begin{aligned} h_0 &= \frac{q}{\gamma} \\ K_q &= \frac{\cos\varepsilon\cos\alpha}{\cos(\varepsilon - \alpha)} \\ K_1 &= \frac{\sin(\varrho - \alpha)}{\cos(\varepsilon - \alpha)} \\ K_0 &= \sqrt{\frac{\sin(\varrho + \delta)\cos(\varepsilon - \alpha)}{\cos(\varepsilon + \delta)\sin(\varrho - \alpha)}} \\ K &= \left[\frac{\cos(\varrho - \varepsilon)}{(1 + K_0K_1)\cos\varepsilon}\right]^2 \frac{1}{\cos(\varepsilon + \delta)} \end{aligned}\right\} \tag{6.7}$$

The position of the cleavage plane would be determined by

$$x_0 = K_0h_1 \tag{7.7}$$

where $h_1 = \frac{h}{\cos\varepsilon}$ (see Fig. 9.7).

The determination of the point of application of the active pressure requires that the distribution of the unit pressures along the surface of the wall be known.

In order to obtain this distribution let us first consider the variation of the active pressure E_q in terms of the depth y (Fig. 10.7*a*). For this purpose we may use expression (5.7) replacing in the latter h by the ordinate y, thus obtaining

$$E_{qy} = \frac{1}{2}\gamma y(y + 2h_0K_q)K$$

This expression permits to construct the graph just mentioned (shown in Fig. 10.7*b*) which represents the increase of the pressure E_{qy} with the increase of the depth of the foot of the wall. It is easily seen that this graph is a conic parabola.

When the depth y is increased by dy the active pressure E_{qy} is increased by dE_{qy}. This increment dE_{qy} is distributed over an elementary area, the vertical projection of which is equal to dy multiplied by 1 (as the depth of the structure in the direction normal to the surface of the drawing is considered equal to unity). Thus, the unit pressure referred to the vertical projection of the surface it acts upon equals

$$P_{qy}=\frac{dE_{qy}}{dy}$$

or, in other words, it equals the first derivative of the resultant pressure in terms of y.

Differentiating E_{qy} as indicated we obtain

$$P_{qy}=\gamma(y+h_0K_q)K \quad (8.7)$$

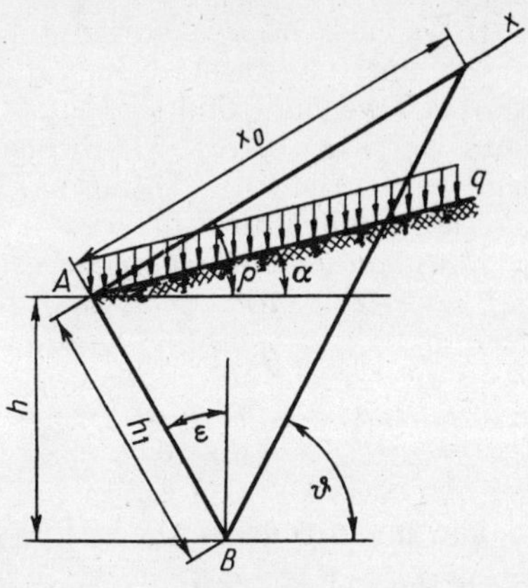

Fig. 9.7

This expression shows that the unit pressure varies along the surface AB linearly. In order to construct the corresponding graph it will suffice therefore to determine the unit pressures at any two points, say, at A and at B (Fig. 10.7c)

$$P_A=\gamma h_0K_qK; \quad P_B=\gamma(h+h_0K_q)K$$

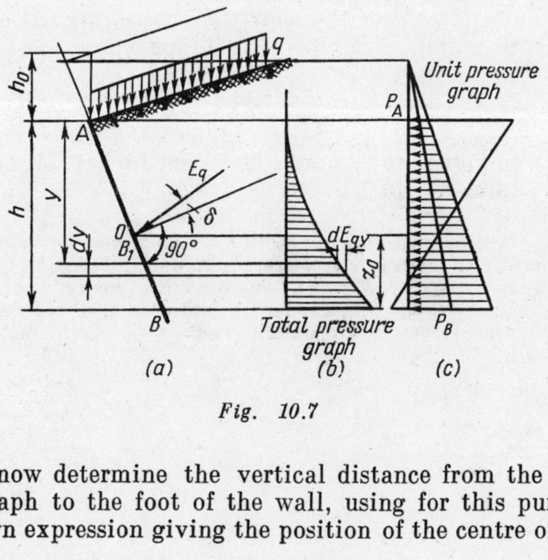

Fig. 10.7

Let us now determine the vertical distance from the centroid of this graph to the foot of the wall, using for this purpose the well-known expression giving the position of the centre of gravity of a trapezoid

$$z_0 = \frac{h}{3} \cdot \frac{2P_A + P_B}{P_A + P_B} \tag{9.7}$$

If we now trace a horizontal line through the centroid of the graph until its intersection at point O with the rear face of the wall AB we shall find the point of application of the active pressure E_q (Fig. 10.7*a*). The direction along which the pressure E_q acts will form with the normal to the surface AB an angle equal to the angle of friction δ.

Thus, *the magnitude of the active pressure developed by a granular material against some surface may be calculated using expression* (5.7); *its point of application will be situated at the same level as the centroid of the unit pressure graph, the position of the point may be calculated using expression* (9.7), *and the direction of the active pressure will form an angle* δ *with the normal to the surface under consideration.*

The magnitude of the active pressure may also be determined with the aid of the unit pressure graph. Indeed, from $P_{qy} = \frac{dE_{qy}}{dy}$ it follows that $dE_{qy} = P_{qy} \cdot dy$. Upon integration of both parts of this equation we obtain

$$E_q = \int_0^h P_{qy} \cdot dy$$

The right-hand part of this equation represents the area of the unit pressure graph (Fig. 10.7*c*). In other words

$$E_q = \frac{1}{2}(P_A + P_B)\,h \tag{10.7}$$

The latter expression is more convenient for actual computation than the expression (5.7).

Problem. It is required to compute the active pressure developed against the lower part BC of the rear face of a retaining wall AB (Fig. 11.7) if $\varrho = 35°$, $\delta = 8°$, $\alpha = 20°$, $\varepsilon = 10°$, $\gamma = 1.6$ tons per cubic metre and $q = 0.8$ ton per square metre. All the dimensions are indicated in the figure.

Solution. Using formulas (6.7) determine h_0 as well as the factors of the K group

$$h_0 = \frac{q}{\gamma} = 0.5 \text{ m}$$

$$K_q = \frac{\cos 10° \cdot \cos 20°}{\cos 10°} = 0.94$$

$$K_1 = \frac{\sin 15°}{\cos 10°} = \frac{0.259}{0.985} = 0.263$$

$$K_0=\sqrt{\frac{\sin 43^\circ}{\cos 18^\circ \times 0.263}}=\sqrt{\frac{0.682}{0.951\times 0.263}}=\sqrt{2.72}=1.65$$

$$K=\left[\frac{\cos 25^\circ}{(1+1.65\times 0.263)\cos 10^\circ}\right]^2\frac{1}{\cos 18^\circ}=\left(\frac{0.906}{1.434\times 0.984}\right)^2\times$$

$$\times\frac{1}{0.951}=0.434$$

This being done, determine the values of the unit pressures at points B and C using formula (8.7)

$$P_B=1.6\,(2.5+0.5\times 0.94)\,0.434=2.06 \text{ tons per sq m}$$

$$P_C=1.6\,(6.5+0.5\times 0.94)\,0.434=4.84 \text{ tons per sq m}$$

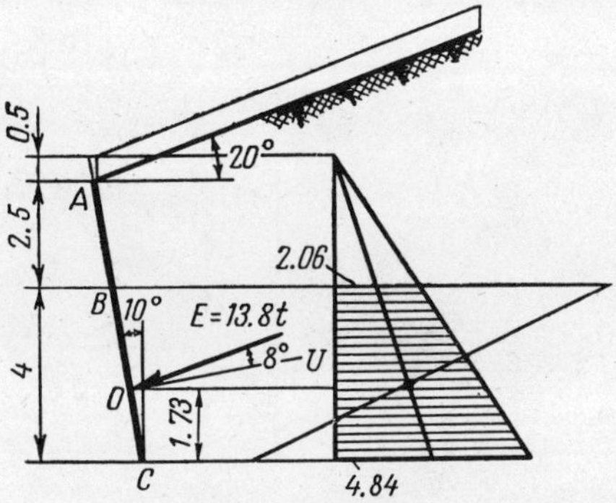

Fig. 11.7

Thereafter compute the area of the graph corresponding to the lower portion of the wall face BC, this area representing the magnitude of the active pressure required

$$E=\frac{1}{2}(2.06+4.84)4=13.8 \text{ tons}$$

The ordinate of the centroid of the graph will be given by the expression (9.7)

$$z_0=\frac{4}{3}\times\frac{2\times 2.06+4.84}{2.06+4.84}=1.73 \text{ metres}$$

The point of application of E will lie at the intersection of the horizontal passing through this centroid and the face of the wall AC. The direction of E will form an angle $\delta = 8^\circ$ with the normal U to this surface.

7.7. *Particular Cases of Pressure Computation*

(a) *Pressure developed by an unsurcharged granular material* (Fig. 12.7). Substituting $h_0 = 0$ in the expressions (5.7), (6.7),

(8.7) and (9.7) we obtain

$$E = \frac{1}{2}\gamma h^2 K \tag{11.7}$$

$$P_y = \gamma y K;\; P_A = 0;\; P_B = \gamma h K;\; z_0 = \frac{h}{3} \tag{12.7}$$

The direction of the cleavage plane will remain unaltered as the factor K_0 is independent of the intensity of the surcharge q acting on the surface of the earth.

(b) *Pressure developed against a vertical smooth surface by a uniformly surcharged granular material having a horizontal surface.* This case was already considered above [see expression (4.7)].

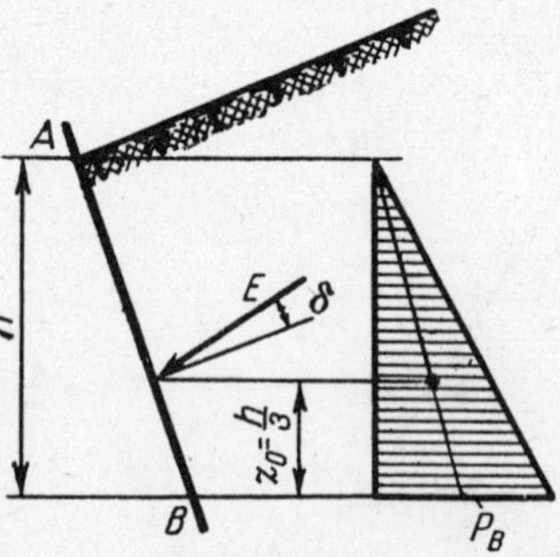

Fig. 12.7

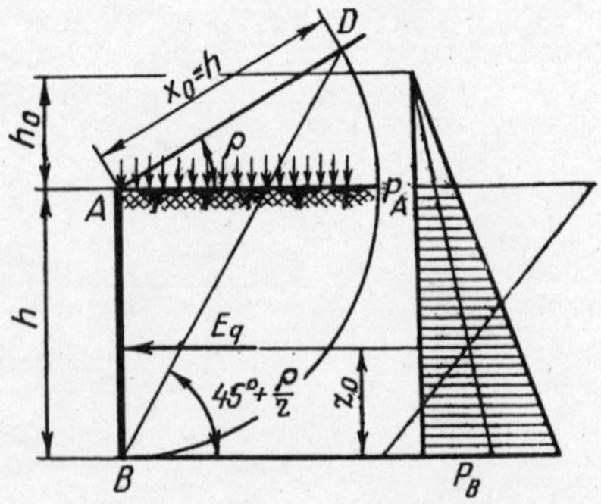

Fig. 13.7

Hereunder we shall use the more general expressions (5.7) and (6.7) for the same case. Putting $\varepsilon = \delta = \alpha = 0$ (Fig. 13.7) in the expressions (5.7) and (6.7) we obtain

$$K_1 = \frac{\sin(\varrho-\alpha)}{\cos(\varepsilon-\alpha)} = \sin\varrho;\; K_q = \frac{\cos\varepsilon\cos\alpha}{\cos(\varepsilon-\alpha)} = 1$$

$$K_0 = \sqrt{\frac{\sin(\varrho+\delta)\cos(\varepsilon-\alpha)}{\cos(\varepsilon+\delta)\sin(\varrho-\alpha)}} = 1$$

$$K = \left[\frac{\cos(\varrho-\varepsilon)}{(1+K_0K_1)\cos\varepsilon}\right]^2 \frac{1}{\cos(\varepsilon+\delta)} = \frac{\cos^2\varrho}{(1+\sin\varrho)^2} = \frac{1-\sin^2\varrho}{(1+\sin\varrho)^2} =$$

$$= \frac{1-\sin\varrho}{1+\sin\varrho} = \tan^2\left(45° - \frac{\varrho}{2}\right)$$

With these values of the K factors, the expressions (5.7), (8.7) and (9.7) become

$$E_q = \frac{1}{2}\gamma h(h+2h_0)\tan^2\left(45° - \frac{\varrho}{2}\right) \tag{4.7}$$

$$\left.\begin{aligned} P_{yq} &= \gamma (y + h_0) \tan^2 \left(45^\circ - \frac{\varrho}{2}\right) \\ P_A &= \gamma h_0 \tan^2 \left(45^\circ - \frac{\varrho}{2}\right) \\ P_B &= \gamma (h + h_0) \tan^2 \left(45^\circ - \frac{\varrho}{2}\right) \end{aligned}\right\} \tag{13.7}$$

$$z_0 = \frac{h}{3} \cdot \frac{h + 3h_0}{h + 2h_0} \tag{14.7}$$

The position of the cleavage plane will be determined by laying off as heretofore the length x_0 along the x-axis [see expression (7.7)] where

$$x_0 = K_0 h_1 = 1 \times h$$

The line connecting point O with the foot of the wall represents the cleavage line required. The angle ϑ_0 made by this line with the horizontal is equal to

$$\vartheta_0 = 90^\circ - \angle ABD = 90^\circ - \frac{180^\circ - \angle BAD}{2} =$$

$$= 90^\circ - \frac{180^\circ - (90^\circ + \varrho)}{2} = 45^\circ + \frac{\varrho}{2}$$

which coincides with its value given by the expression (3.7) found previously.

(c) *The same case as in (b) but without surcharge* (Fig. 14.7). Substituting in the expressions obtained above $h_0 = 0$ we obtain

$$E = \frac{1}{2} \gamma h^2 \tan^2 \left(45^\circ - \frac{\varrho}{2}\right) \tag{15.7}$$

$$P_y = \gamma y \tan^2 \left(45^\circ - \frac{\varrho}{2}\right); \ P_A = 0; \ P_B = \gamma h \tan^2 \left(45^\circ - \frac{\varrho}{2}\right) \tag{16.7}$$

$$z_0 = \frac{h}{3}$$

The position of the cleavage plane remains the same as in case (b) when a uniform load was acting on the surface of the earth, i. e.,

$$\vartheta_0 = 45^\circ + \frac{\varrho}{2}$$

(d) *Pressure developed against a polygonally shaped surface* (Fig. 15.7).

The pressure E_q developed against the upper portion AB of the polygonal surface ABB_1 will be determined as heretofore using formulas (5.7) through (10.7).

The pressure developed against the lower portion BB_1 may be computed approximately assuming that this pressure will be the same as that acting on an equivalent portion of a plane surface A_2BB_1. In order to compute this pressure, trace through point B the line BD parallel to the surface of the earth and consider the weight of the overlaying portion of the material as a uniformly distributed surcharge of intensity $q = \gamma h_0'$. The depth of this layer h_0' will be taken equal to the sum of the thickness of the layer h_0, the vertical projection of AA' equal to h_a and the vertical projection of AB equal to h. Computing as usual the factors

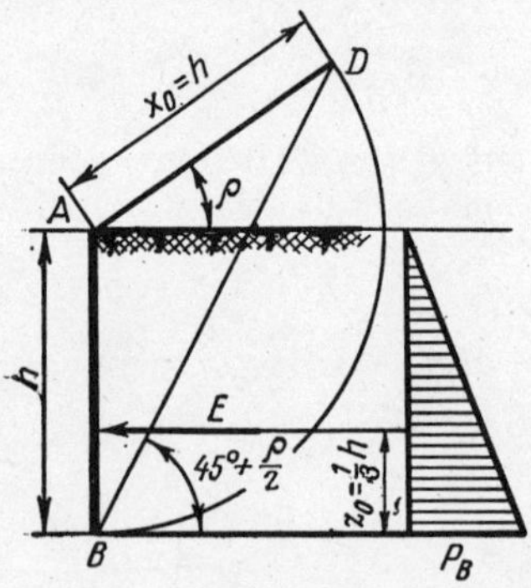

Fig. 14.7

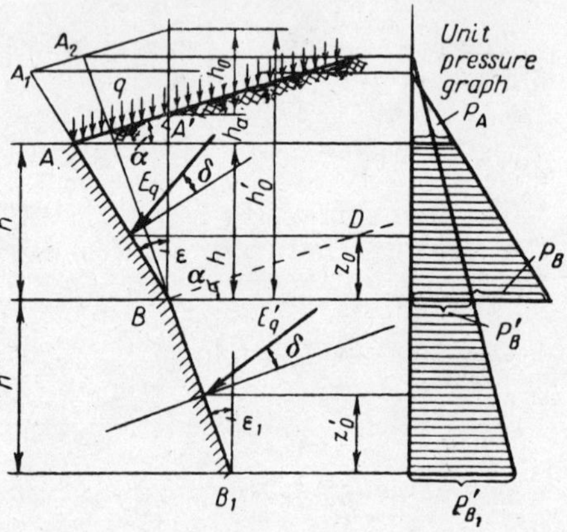

Fig. 15.7

of the K group and substituting them in the usual formulas in which the batter of the wall is taken equal to ε_1 we obtain

$$E_q' = \frac{1}{2}\gamma h'(h' + 2h_0'K_q)K$$

$$P_{qy}' = \gamma(y + h_0'K_q)K$$

$$P_B' = \gamma h_0'K_qK$$

$$P_{B_1}' = \gamma(h' + h_0'K_q)K$$

$$z_0' = \frac{h'}{3}\,\frac{h' + 3h_0'K_q}{h' + 2h_0'K_q}$$

The unit pressure graph for the case under consideration is represented in Fig. 15.7*b*.

Fig. 16.7*a* represents a more complicated case which may be met with in the design of reinforced concrete pertaining walls provided with a spur.

The pressure developed against a wall of this type will be determined separately for each of the plane surfaces constituting its rear face. Thus, the pressures exerted against the portions AB

and CD will be computed using expressions (12.7) in which the ordinates y_B and y_C (corresponding to points B and C, respectively) will be taken equal to h and the ordinate y_D of the point D equal to $(h + h_1)$. The factors of the K group will be computed

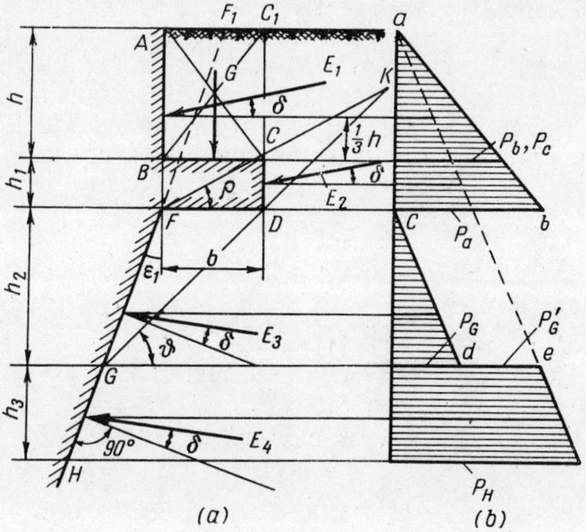

Fig. 16.7

using formulas (6.7) in which $\alpha = \varepsilon = 0$; the unit pressure graph for both parts will be given by one common straight line ab (Fig. 16.7*b*).

The magnitude of the pressures developed against AB and CD will be provided by the corresponding areas of the above graph

$$E_1 = \frac{1}{2} hP_b$$

$$E_2 = \frac{1}{2} h_1 (P_c + P_a)$$

The horizontal surface BC is subjected to the action of a vertical load G equal to the dead weight of the material contained in the prism $ABCC_1$ and equal to

$$G = hb\gamma$$

There is no load applied to the horizontal surface FD.

As for the portion FH it should be subdivided into two parts FG and GH, point G being determined by the intersection with the face of the wall of a line parallel to the cleavage plane and passing through point D. The position of the cleavage plane is determined using formula (7.7)

$$x_0 = FK = GFK_0$$

The factor K_0 entering this expression will be computed using formulas (6.7) putting $\alpha = 0$ and $\varepsilon = -\ \varepsilon_1, \varrho, \delta$. The portion FD will be subjected to the pressure developed by the layer h_2 (Fig. 16.7) alone, this pressure being independent of the weight of the overlaying material. The corresponding unit pressure graph will be represented by a straight line cd, the slope of which is steeper than that of line ab. At point G the unit pressure will be computed

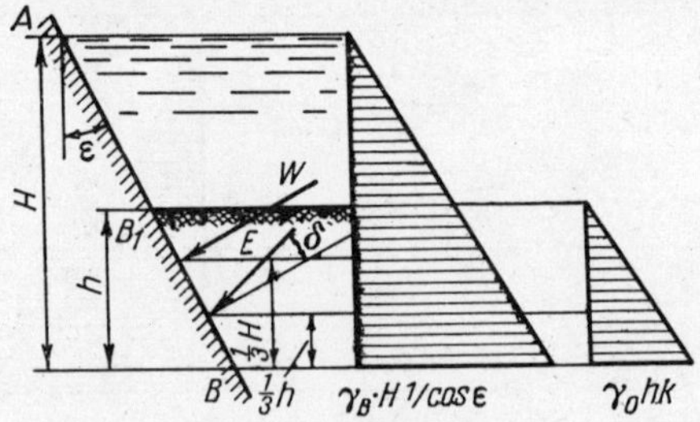

Fig. 17.7

using formula (12.7) for $y_G = h_2$. The pressure acting on the portion of the wall considered will equal

$$E_3 = \frac{1}{2} h_2 P_G$$

As for the pressure sustained by the wall below G it is already dependent on the weight of the whole granular mass. Therefore, when unit pressures are determined at points G and H we must adopt

$$y'_G = h + h_1 + h_2;\ y_H = h + h_1 + h_2 + h_3$$

The corresponding graph will consist of a straight line ef parallel to cd and intersecting the graph axis at point a. The resultant pressure sustained by portion GH will therefore be

$$E_4 = \frac{1}{2} h_3 (P'_G + P_H)$$

(e) *Pressure developed by water saturated earth* (Fig. 17.7). In the case under consideration the rear face of the wall AB may be regarded as subjected separately to the hydrostatic pressure W and to the pressure of the earth whose weight is reduced by the amount of water expelled.

The hydrostatic pressure W can be found using expressions (6.7) through (10.7) on the assumption that

$$\varrho=\alpha=\delta=0 \qquad \text{and} \qquad h_0=0$$

We obtain

$$K_1=0;\ K_0=\frac{0}{0};\ K=\frac{1}{\cos\varepsilon}$$

$$W=\frac{1}{2}\gamma_B H^2\frac{1}{\cos\varepsilon}$$

The indeterminate value of factor K_0 indicates that the hydrostatic pressure is completely independent of the position of the cleavage surface.

Referred to the vertical projection of the rear face of the wall, the hydrostatic pressure at point B will equal

$$w=\gamma_B H\frac{1}{\cos\varepsilon}$$

The point of application of W will be at the same level with the centroid of the unit pressure graph

$$z_0'=\frac{H}{3}$$

In the computations relative to the active pressure of the earth itself, its weight per cubic metre must be taken equal to γ_0 instead of γ [see expression (1.7)] owing to the presence of water. In that case expressions (11.7) and (12.7) will give

$$E=\frac{1}{2}\gamma_0 h^2 K$$

$$P_y=\gamma_0 yK$$

$$P_{B_1}=0$$

$$P_B=\gamma_0 hK$$

$$z_0=\frac{h}{3}$$

The total pressure sustained by the wall will be thus composed by the hydrostatic pressure W and the earth pressure E computed with due regard to the alteration of its weight per cubic metre caused by the water.

(f) *Pressure exerted by a layer of impervious soil surmounted by water* (Fig. 18.7).

Pressure computations are very approximate in this case and are carried out assuming that the water acts on the upper part of the wall situated above the surface of the soil alone, while the lower part of the wall is subjected to the pressure of the earth on which the water acts as a surcharge.

The hydrostatic pressure W will be computed as heretofore and will amount to

$$W = \frac{1}{2} \gamma_B (H-h)^2 \frac{1}{\cos \varepsilon}$$

Its point of application being given by

$$z_0' = \frac{H-h}{3}$$

The vertical pressure developed by the layer of water on the surface of the earth amounts to

$$q = \gamma_B (H-h)$$

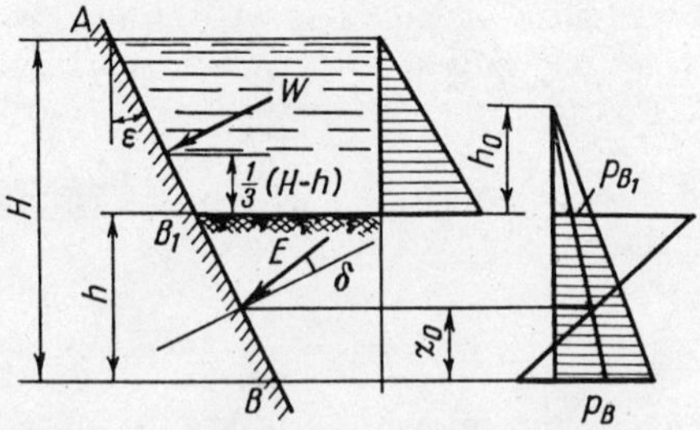

Fig. 18.7

the depth h_0 of the equivalent layer of earth being

$$h_0 = \frac{\gamma_B (H-h)}{\gamma}$$

As the surface of the earth is assumed horizontal ($\alpha = 0$)

$$K_q = \frac{\cos \varepsilon \cos \alpha}{\cos (\varepsilon - \alpha)} = 1$$

The value of the active pressure E_q, the values of the unit pressures and the point of application of E_q may be now found using expressions (5.7) to (9.7)

$$E_q = \frac{1}{2} \gamma h (h + 2h_0) K$$

$$P_{qy} = \gamma (y + h_0) K$$

$$P_{B_1} = \gamma h_0 K$$

$$P_B = \gamma (h + h_0) K$$

$$z_0 = \frac{h}{3} \cdot \frac{2P_{B_1} + P_B}{P_{B_1} + P_B}$$

8.7. *Passive Pressure of Granular Materials*

The term *passive pressure* refers to the resultant pressure developed by a granular material against some surface when the latter shifts over a very small distance towards this material.

The magnitude of the passive pressure may be determined using the same wedge theory of Coulomb (see Art. 3.7), all the assumptions made in the development of this theory remaining valid.

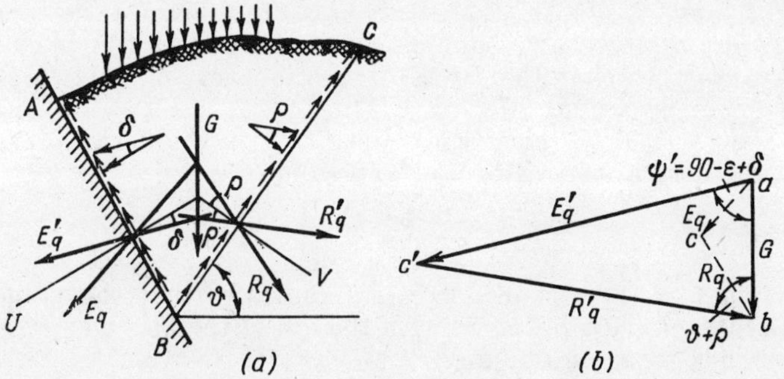

Fig. 19.7

When the surface *AB* of Fig. 19.7*a* is forced towards the granular material a wedge *ABC* is formed again, this wedge behaving as a solid body and sliding upwards along the surface *AB* and the cleavage plane *BC*. The forces of friction which develop within the wedge along the two surfaces just mentioned are directed upwards. It will be remembered that in the case of the active pressure the wedge moves downwards, and the forces of friction act in the same direction. When a state of limit equilibrium is reached the passive pressure E'_q (which is the resultant of the normal earth pressure and of the forces of friction) will be deviated clockwise from the normal *U* to the surface of the wall by an angle δ. Similarly, the resultant pressure R'_q is deviated from the normal *V* to the cleavage plane *BC* counterclockwise through an angle ϱ equal to the angle of internal friction of the material. The resultant of the two pressures E'_q and R'_q will be equal to the dead weight G_q of the wedge *ABC*.

The triangle of forces *abc'* for the case of the passive pressure is represented in Fig. 19.7*b*. For comparison the triangle of forces corresponding to the case of active pressure is represented in the same figure in dash lines. It is clearly seen that for one and the same position of the cleavage plane the passive pressure is considerably greater than the active pressure.

From the triangle abc' we can determine the magnitude of the passive pressure E_q'

$$E_q' = G_q \frac{\sin(\vartheta + \varrho)}{\sin(\vartheta + \psi' + \varrho)} \qquad (17.7)$$

where

$$\psi' = 90° - \varepsilon + \delta$$

Comparing expressions (17.7) and (2.7), we come to the conclusion that the magnitude of the passive pressure can be computed using the expression for the active pressure, provided the angles ϱ and δ are replaced by $(-\varrho)$ and $(-\delta)$, which is easily understood if we remember that the forces of friction act in the two cases in opposite directions.

The angle ϑ of the cleavage plane BC with the horizontal will be again determined using the expression

$$\frac{dE_q'}{d\vartheta} = 0$$

The sign of the second derivative indicates that the value of the passive pressure obtained with the aid of the above expression corresponds to a minimum.

The general expressions for the computation of the passive pressure obtained by replacing ϱ and δ in expressions (5.7) and (6.7) by $(-\varrho)$ and $(-\delta)$ are

$$E_q' = \frac{1}{2} \gamma h (h + 2h_0 K_q) K' \qquad (18.7)$$

$$P_{yq}' = \gamma (h + h_0 K_q) K' \qquad (19.7)$$

The factors of the K group entering these expressions are

$$\left.\begin{aligned} K_1' &= \frac{\sin(-\varrho - \alpha)}{\cos(\varepsilon - \alpha)} \\ K_q &= \frac{\cos\varepsilon \cos\alpha}{\cos(\varepsilon - \alpha)} \\ K_0' &= \sqrt{\frac{\sin(\varrho + \delta)\cos(\varepsilon - \alpha)}{\cos(\varepsilon - \delta)\sin(\varrho + \alpha)}} \\ K' &= \left[\frac{\cos(\varrho + \varepsilon)}{(1 + K_0' K_1')\cos\varepsilon}\right]^2 \frac{1}{\cos(\varepsilon - \delta)} \end{aligned}\right\} \qquad (20.7)$$

The ordinate of the point of application of the passive pressure will be derived from

$$z_0' = \frac{h}{3} \cdot \frac{2P_A' + P_B'}{P_A' + P_B'} \qquad (21.7)$$

In case the rear face of the wall is vertical and smooth and the surface of the earth is horizontal, the magnitude of the passive pressure can be calculated using formulas (4.7), (13.7) and (14.7) after replacing in these formulas ϱ by $(-\varrho)$

$$E'_q = \frac{1}{2}\gamma h(h+2h_0)\tan^2\left(45^\circ + \frac{\varrho}{2}\right) \tag{22.7}$$

$$\left.\begin{aligned} P'_{yq} &= \gamma(y+h_0)\tan^2\left(45^\circ+\frac{\varrho}{2}\right) \\ P'_A &= yh_0\tan^2\left(45^\circ+\frac{\varrho}{2}\right) \\ P'_B &= \gamma(h+h_0)\tan^2\left(45^\circ+\frac{\varrho}{2}\right) \end{aligned}\right\} \tag{23.7}$$

$$z'_0 = \frac{h}{3}\cdot\frac{h+3h_0}{h+2h_0} \tag{24.7}$$

In the above expressions h_0 is as usual the thickness of the layer of earth equivalent to the surcharge of intensity q.

The corresponding position of the cleavage plane will be obtained by tracing through the top of the wall A (Fig. 20.7) the axis x' making an angle $-\varrho$ with the horizontal and by laying off along this line a length $AD' = x'_0 = K'_0 h = h$. The line BD' will represent the projection of the cleavage or slip plane on the plane of the drawing. For the sake of comparison we give again in the same figure the position of the cleavage plane BD corresponding to the case of active pressure.

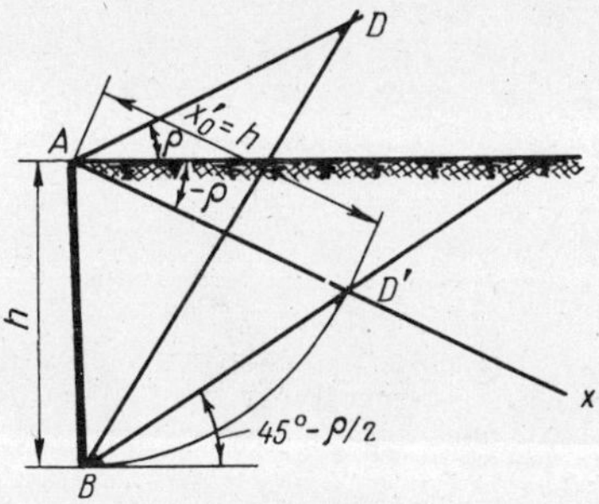

Fig. 20.7

The triangle ABD' permits the determination of the angle ϑ' formed by the cleavage plane BD' with the horizon

$$\vartheta' = 45^\circ - \frac{\varrho}{2} \tag{25.7}$$

In conclusion it is worth mentioning that all the graphical methods permitting the determination of the active pressure remain applicable to the case of the passive pressure, provided the angles ϱ and δ are replaced everywhere by $(-\varrho)$ and $(-\delta)$.

Problem. It is required to determine graphically the passive pressure developed against the surface AB of Fig. 21.7, if $h = 5$ m, $\varrho = 40^\circ$, $\delta = 5^\circ$, $\varepsilon = 20^\circ$, $\alpha = 10^\circ$, $\gamma = 1.6$ tons per cubic metre.

Solution. 1. Start with determining the position of points A, M, N, O, D, C and P in a way exactly similar to the one used above (see Fig. 6.7) but replacing everywhere the angles δ and ϱ by $-\delta$ and $-\varrho$.

2. Determine the position of the cleavage plane BDC.

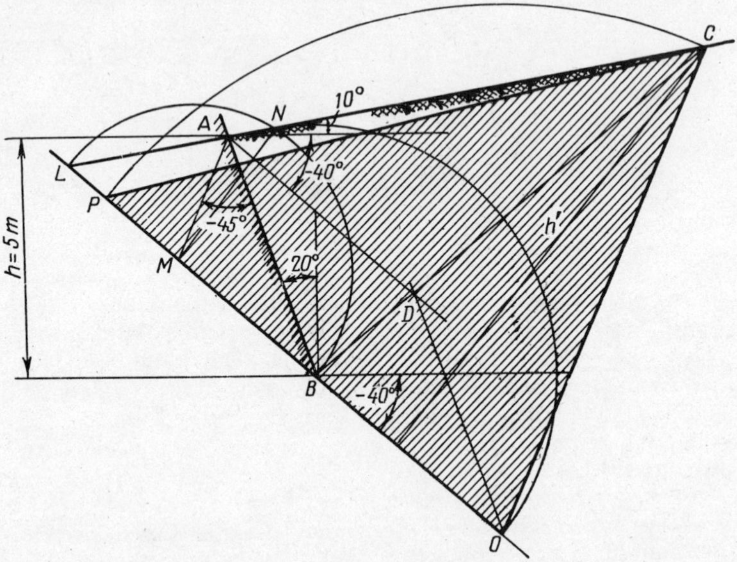

Fig. 21.7

3. Measure to scale the base and the height of the triangle OPC, which are equal, respectively, to 11.0 m and 10.7 m.

4. Compute the area of triangle OPC

$$F=\frac{1}{2}\,11.0\times 10.7=58.85 \text{ sq m}$$

5. Determine the passive pressure exerted against the surface AB

$$E'=1.6\times 58.85=94.2 \text{ tons}$$

6. Compare the value of the passive pressure thus obtained with that of the active pressure computed for an identical case in Art 5.7.

$$\frac{E'}{E}=\frac{94.2}{8.38}=11.2$$

8. STRAIN ENERGY THEORY AND GENERAL METHODS OF DISPLACEMENT COMPUTATION

1.8. *General*

The stress analysis of redundant structures requires that use should be made of displacement equations in addition to the usual equilibrium equations. It becomes therefore necessary to determine the deformations and strains in different parts of the structure. Moreover, the deflections of statically determinate structures must be also frequently determined, such structures having to fulfil certain requirements concerning both their strength and their rigidity, in order to avoid excessive deformations under service loads. For this reason the study of various methods of strain and deflection computation for elastic systems acquires the greatest importance in the theory of structures.

This chapter will be devoted to the study of general methods permitting the determination of the strains and deflections of various framed structures, arches, rigid frames, etc. We shall start with reviewing certain questions concerning the work accomplished by the external forces and the potential or strain energy accumulated in various elastic systems during their deformation.

2.8. *Work of External Forces*

During the loading of any system its elements are put into motion, acquiring certain velocities and accelerations. It is clear that the rate of growth of the deformations will increase proportionally to the rate of loading, and if the latter becomes very small, the momentum acquired by the system when passing from one state to another will become quite negligible. Hereafter this latter type of loading will be referred to as *statical* loading.

In order to determine the work of any external load P applied gradually to any elastic system (Fig. 1.8) we shall make use of Maxwell's principle of superposition, provided the material follows Hooke's law. Consequently, the displacements suffered by different points of an elastic system will be in direct proportion to the loads which have caused them. In its most general form this may be expressed by the following equation

$$\Delta = \alpha P \tag{1.8}$$

In this expression Δ is the deformation sustained by the system along the line of action of force P, and α is a factor depending on the material itself, on the pattern and the dimensions of the structure and on the point of application of the load P.

Let force P increase by dP; this will immediately cause a corresponding increase of Δ by $d\Delta$. The work performed by the load

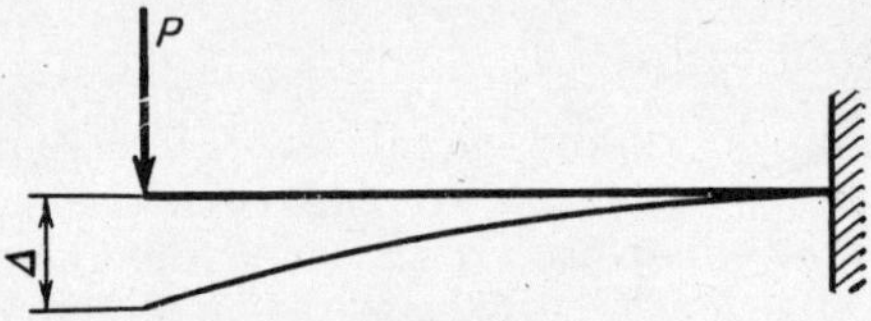

Fig. 1.8

P along the displacement $d\Delta$, neglecting as usual the infinitesimals of the higher orders, will be

$$dA = (P + dP)\, d\Delta = P\, d\Delta$$

Replacing $d\Delta$ by its value αdP (1.8) we obtain

$$dA = P\, d\Delta = \alpha P\, dP$$

Integrating this expression from zero to the final value of the external load, we obtain the expression of the work accomplished by this load during its statical application

$$A = \alpha \int_0^P P\, dP = \frac{\alpha P^2}{2}$$

As $\Delta = \alpha P$, this may be equally written

$$A = \frac{1}{2} P\Delta$$

It should be noted that the direction of the displacement caused by a load P may differ from that of the load. As the work accomplished by a load is always expressed by the product of a force by the length of the displacement measured along the line of action of this force, *the displacement Δ will always represent the projection of the total displacement of the load point on the direction of the load.* Thus, for instance, if a load P acts at an angle β to the axis of a beam (Fig. 2.8), the displacement Δ will be given by the length of the line ab, this length being equal to the projection of the total deflection aa_1 on the line of action of load P.

The work accomplished by a couple or moment $\mathfrak{M}$ can be found in the same way provided the displacement Δ corresponds to that type of loading. It will be readily seen that in this case Δ must represent the angular rotation of the cross section to which the aforesaid moment is applied.

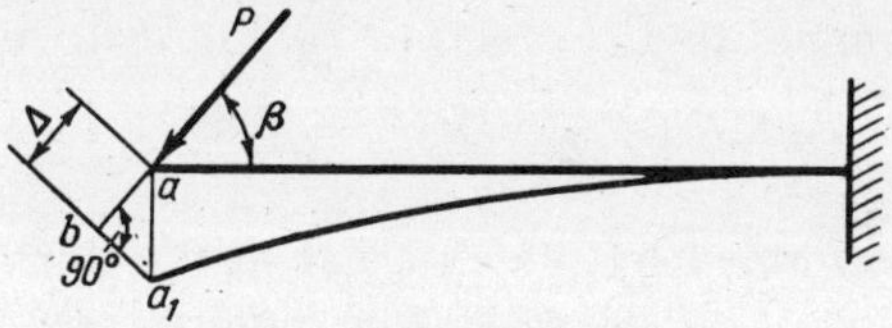

Fig. 2.8

Thus, the work accomplished by a moment $\mathfrak{M}$ applied statically to the beam of Fig. 3.8 will be given by

$$A = \frac{1}{2}\,\mathfrak{M}\vartheta$$

where ϑ is the angular rotation (in radians) of the cross section to which the moment $\mathfrak{M}$ is directly applied. Thus, *the work accomplished by any external force applied gradually to an elastic system*

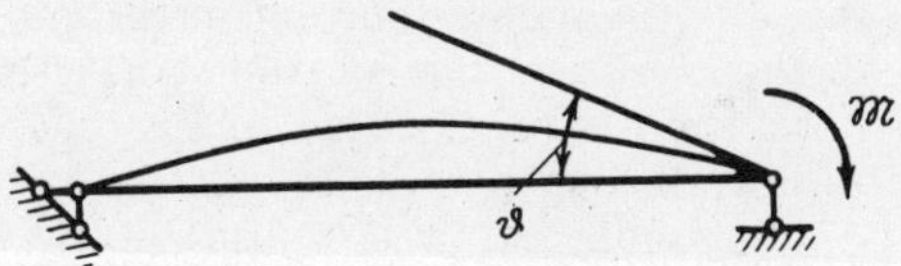

Fig. 3.8

will be always given by half the product of this force by the length of the displacement measured in the direction of this force. The term *force* applies in this case to any external action including moments, distributed loads, etc.

As for the term *displacement*, it will mean the deformation corresponding to the type of action whose work is being studied. Thus, a linear displacement will correspond to a concentrated load P, an angular rotation to a moment $\mathfrak{M}$ and the area of the displacement graph of a loaded stretch to distributed loads.

When a system of loads is gradually applied to a structure, the work accomplished by each of these loads will equal half the product of its magnitude by the displacement corresponding to this load but caused by all the loads in question. Thus, in the case of the beam of Fig. 4.8 which carries two concentrated loads P_1 and P_2 and which is subjected at the same time to the action of two moments $\mathfrak{M}_1$ and $\mathfrak{M}_2$ the work of the external forces will equal

$$A=\frac{P_1\Delta_1}{2}+\frac{P_2\Delta_2}{2}+\frac{\mathfrak{M}_1\vartheta_1}{2}-\frac{\mathfrak{M}_2\vartheta_2}{2}$$

The negative sign of the last term of this equation indicates that the angular rotation of the cross section to which moment $\mathfrak{M}_2$ is applied is opposite in direction to the said moment. Thus,

$$A=\Sigma\frac{P_i\Delta_i}{2}+\Sigma\frac{\mathfrak{M}_i\vartheta_i}{2} \tag{2.8}$$

The work performed by the external forces along the displacements caused by these forces can be equally expressed in terms

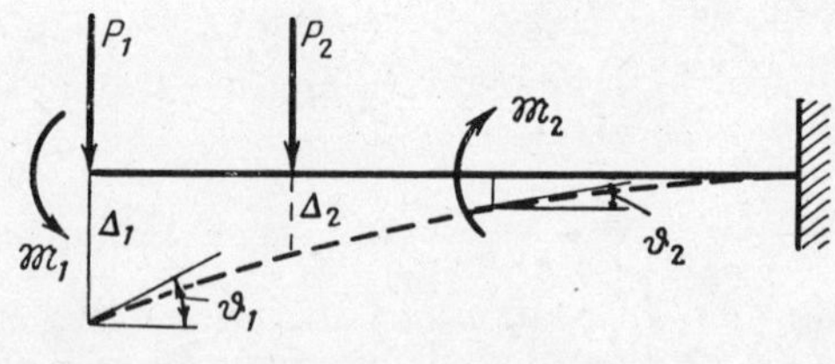

Fig. 4.8

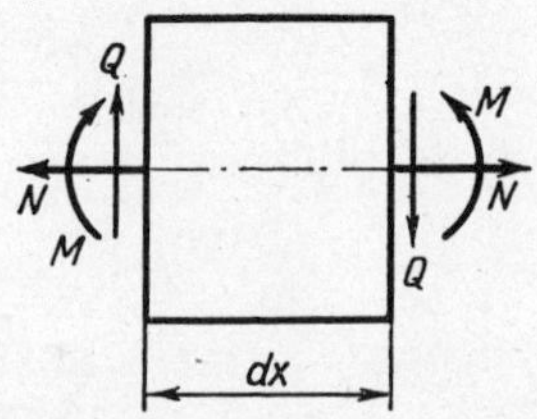

Fig. 5.8

of the stresses (bending moments, normal forces and shears) which are developed in the cross sections of the structure under consideration. Let us take the bar represented in Fig. 5.8 and let us consider an infinitely small length dx bounded by two planes normal to the bar axis. The whole bar will comprise an infinite number of such sections. If all the loads act in the plane of the bar axis, the element dx will be subjected to a normal force N, a bending moment M and a shearing force Q.

For a bar as a whole these actions constitute internal forces while for the element dx they may be regarded as external loads whose work will then be expressed by the products of N, M and Q by the corresponding displacements sustained by the said element.

Hereunder let us study separately the work performed by each of these actions.

An element dx subjected solely to a normal force N appears in Fig. 6.8. If we admit that its left extremity is held fast, the right-hand one will move along the direction of force N towards the right over a length equal to

$$\Delta_x=\frac{N\,dx}{EF}$$

where EF is the tensile or compressive rigidity of the bar under consideration.

The work performed by the stress N along the displacement Δ_x will be therefore expressed by

$$dA_N = \frac{1}{2} N\Delta_x = \frac{1}{2} N \frac{N\,dx}{EF}$$

An element dx acted upon solely by a bending moment is represented in Fig. 7.8. Once again let us assume that its left-hand

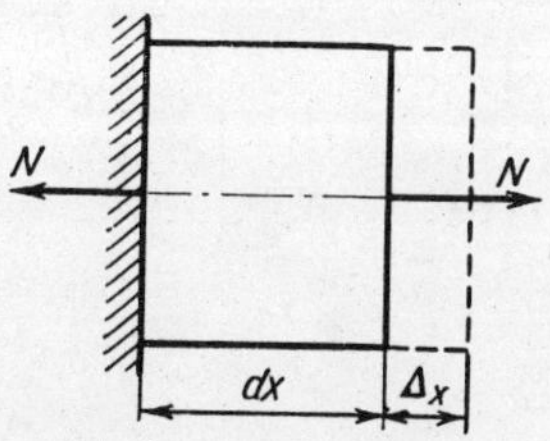

Fig. 6.8

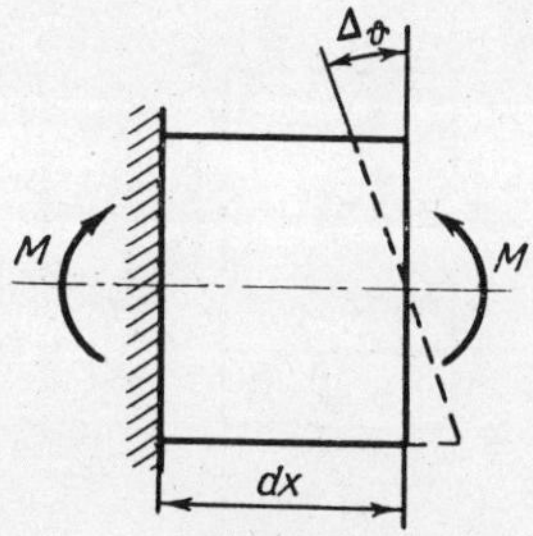

Fig. 7.8

extremity remains fixed in which case the angular rotation of the right-hand one will be given by

$$\Delta_\vartheta = \frac{M\,dx}{EJ}$$

EJ being the flexural rigidity of the bar section under consideration. During its statical application the bending moment will therefore accomplish the work given by

$$dA_M = \frac{1}{2} M\Delta_\vartheta = \frac{1}{2} M \frac{M\,dx}{EJ}$$

Let us further examine the element dx of Fig. 8.8*a* acted upon by a shearing force Q. If we fix again the left end face (Fig. 8.8*b*) we must apply to the right-hand face transversal stresses $\tau\,dF$ of which the shearing force Q is the resultant. In the case of pure bending these transversal stresses will be given by Zhurawsky's formula

$$\tau\,dF = \frac{QS}{Jb}\,dF$$

where dF is the area of a horizontal elementary strip situated a distance y from the neutral axis, while S is the statical moment of that part of the cross section above (or below) this strip about the same axis (Fig. 8.8c). The magnitude of the mutual displacement of two identical strips, one belonging to the left end face and the other to the right one, will be equal to the displacement $\gamma\,dx$ of the right end (the left one being assumed fixed) and will therefore be given by the expression

$$\gamma\,dx = \frac{\tau}{G}\,dx$$

where γ is the angle of shear.

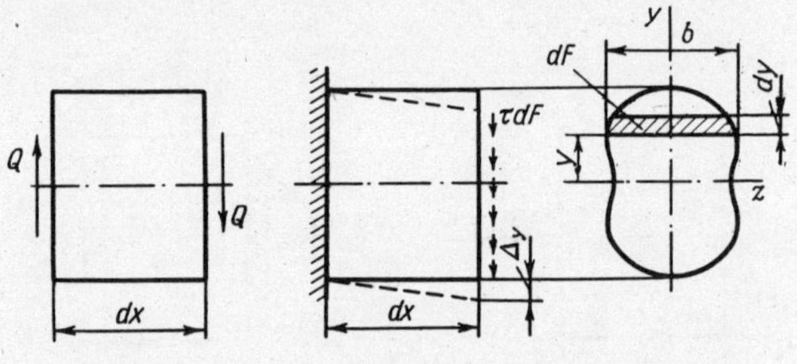

(a) (b) (c)

Fig. 8.8

Hence, the work of an elementary transversal stress $\tau\,dF$ along the displacement $\gamma\,dx$ will be given by

$$\frac{1}{2}\tau\,dF\cdot\gamma\,dx$$

Integrating this expression over the whole area of the cross section F we obtain the work of all the shearing stresses acting across this section

$$dA_Q = \int\limits_F \frac{1}{2}\tau\gamma\,dx\,dF = \int\limits_F \frac{\tau^2\,dx}{2G}\,dF = \int\limits_F \frac{Q^2S^2}{J^2b^2}\,\frac{dx}{2G}\,dF =$$

$$= \frac{Q^2\,dx}{2GJ^2}\int\limits_F \frac{S^2}{b^2}\,dF = \eta\,\frac{Q^2\,dx}{2GF}$$

In this expression GF is the transversal rigidity of the cross section considered, while $\eta = \frac{F}{J^2}\int\limits_F \frac{S^2}{b^2}\,dF$ is nondimensional factor depending solely on the shape and size of the cross section.

Denoting $\eta\,\frac{Qdx}{GF}$ by Δ_y, the elementary work dA_Q will be expressed by

$$dA_Q = \frac{1}{2}\,Q\Delta_y = \frac{1}{2}\,Q\,\frac{Q\,dx}{GF}\,\eta$$

In this expression Δ_y may be regarded as the mutual vertical displacement of the two cross sections bounding the element dx (see Fig. 8.8*b*). For rectangular cross sections the value of factor η will be obtained replacing in the corresponding expression F by bh, J by $\frac{bh^3}{12}$, and S by $\frac{b}{2}\left(\frac{h^2}{4}-y^2\right)$ which leads to $\eta = 1.2$. For a circular section the same procedure will yield $\eta=\frac{10}{9}$ whilst for H- or for I-shaped section the approximate value of $\eta=\frac{F}{F_w}$ may be adopted, F_w being the cross section of the web. If the elements under consideration are acted upon simultaneously by a normal stress N, a bending moment M and a shear Q, the work accomplished by each of these actions along the displacements caused by the two other ones will remain nil. Consequently, the total work will be expressed by

$$dA=dA_N+dA_M+dA_Q=\frac{1}{2}\left(N\frac{N\,dx}{EF}+M\frac{M\,dx}{EJ}+Q\frac{Q\,dx}{GF}\eta\right)$$

Integrating the expression of dA over the whole length l of each bar constituting the structure and summing up the results, we obtain the following expression which permits the computation of the work of external forces expressed in terms of the internal ones for the whole structure

$$A=\frac{1}{2}\left(\Sigma\int_0^l M\frac{M\,dx}{EJ}+\Sigma\int_0^l N\frac{N\,dx}{EF}+\Sigma\int_0^l Q\frac{Q\,dx}{GF}\eta\right) \tag{3.8}$$

which may be written as follows

$$A=\Sigma\int_0^l \frac{M^2\,dx}{2EJ}+\Sigma\int_0^l \frac{N^2\,dx}{2EF}+\Sigma\int_0^l \frac{Q^2\,dx}{2GF}\eta \tag{4.8}$$

In the expression (3.8) the letters M, N and Q represent the internal forces acting over a cross section situated a distance x from the origin of coordinates, while the terms $\frac{Mdx}{EJ}$, $\frac{Ndx}{EF}$ and $\frac{Qdx}{EG}\eta$ are the corresponding displacements of the element dx of the bar. The above two expressions permit the computation of the work accomplished by the loads in terms of the internal stresses developed under the action of these loads. Expression (4.8) shows that the work of the external loads will be always positive.

3.8. *Strain Energy*

During the loading of a body the external forces accomplish a certain amount of work part of which may be used to overcome the internal friction, to alter the temperature or the magnetic properties of the material, etc. In the materials usually considered as elastic this part of the work is negligible and therefore we may admit that all the work of external forces is transformed in that case into potential or strain energy. The latter is accumulated in the body under consideration during the period of increasing strains and deformations caused by these forces. When the body is unloaded, this energy is restituted as work accomplished by the internal stresses. As no energy is ever lost, we may say that all the work A accomplished by the external forces is transformed into strain energy W or, in other words, that

$$A = W$$

Substituting in this equation the value of A given by the expression (4.8) we obtain

$$W = \Sigma \int_0^l \frac{M^2\,dx}{2EJ} + \Sigma \int_0^l \frac{N^2\,dx}{2EF} + \Sigma \int_0^l \frac{Q^2\,dx}{2GF} \eta \qquad (5.8)$$

The analysis of this expression leads to the following conclusions:

1. The strain energy is always positive, for the above expression contains the values of the internal forces M, N and Q in the second power.

2. The strain energy is expressed by a homogeneous function of the stresses or strains in the second power, the strains being directly proportional to the stresses.

3. The strain energy accumulated under the action of a certain system of forces is not equal to the sum of strain energies due to each of these forces separately and therefore the principle of superposition is no longer valid. This follows from the fact that the strain energy is a function of the second power of the stresses M, N and Q and that the square of a sum is never equal to the sum of the squares.

4. The strain energy accumulated in a body is independent of the sequence in which the external forces are applied, the final values of the stresses M, N and Q being independent of this sequence. Consequently, the strain energy depends only on the final state of an elastic body.

Statement 3 can be confirmed by the following example. Let us consider three different ways of load application to the elastic bar shown in Fig. 9.8*a*, viz.:

(1) loading by a single force P_1 (Fig. 9.8*b*),
(2) loading by a single force P_2 (Fig. 9.8*c*),
(3) simultaneous loading by both forces P_1 and P_2 (Fig. 9.8*d*).
The strain energy accumulated in the first two cases as given by expression (5.8) amounts to

$$W_1 = \frac{P_1^2 l}{2EF} \qquad W_2 = \frac{P_2^2 l}{2EF}$$

In the third case it will be given by

$$W_3 = \frac{(P_1 + P_2)^2 l}{2EF} = \frac{P_1^2 l}{2EF} + \frac{P_2^2 l}{2EF} + \frac{P_1 P_2 l}{EF}$$

Comparing W_3 with the sum $(W_1 + W_2)$ we note that the sum of strain energies due to each of the forces separately is not equal

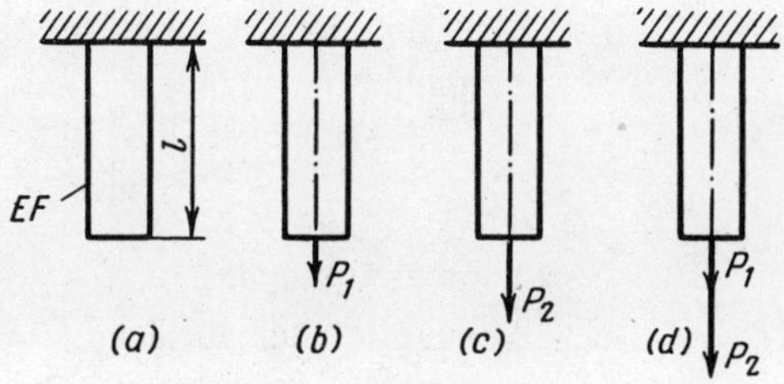

Fig. 9.8

to the strain energy due to the simultaneous action of these same forces. Indeed

$$W_3 = W_1 + W_2 + \frac{P_1 P_2 l}{EF}$$

For a better understanding of the above equation let us imagine that at first load P_1 is increased gradually from zero to its final value and then remains constant while load P_2 slowly reaches its full value in the same way. It is clear that the application of load P_2 will cause the end of the bar to move downwards an amount $\frac{P^2 l}{EF}$, and that during that time the load P_1 (assumed constant) will perform the work equal to $\frac{P_1 P_2 l}{EF}$.

Thus, the last term of the expression for W_3 gives the value of the work performed by the load P_1 when its point of application is shifted by force P_2 (or vice versa, if the sequence of loading is inverted).

The above example shows clearly that the principle of superposition does not apply to the computation of the strain energy accumulated in an elastic body for otherwise the terms of the equation, taking care of the work accomplished by one part of the loads along the displacement caused by the other part of the loads, would be completely lost.

Problem 1. Required to determine the strain energy accumulated by an end-supported beam of rectangular cross section (its width and depth equalling b and h, respectively), the beam being loaded by a couple $\mathfrak{M}$ acting at its right-hand extremity (Fig. 10.8*a*).

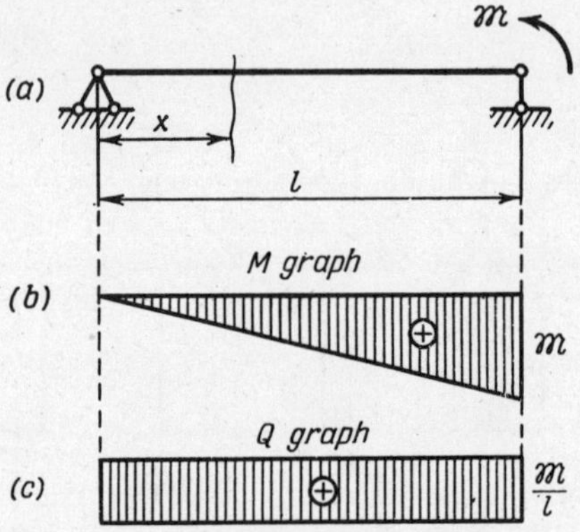

Fig. 10.8

Solution. Draw the bending moment and the shearing force diagrams as shown in Fig. 10.8*b* and *c* (normal stresses in this particular case being nil). The magnitude of these stresses in any cross section will be given by

$$M=\frac{\mathfrak{M}}{l}x \quad \text{and} \quad Q_x=\frac{\mathfrak{M}}{l}$$

Introducing these values in the expression for the strain energy (5.8) we obtain

$$W=\int_0^l \frac{M^2\,dx}{2EJ}+\int_0^l \frac{Q^2\,dx}{2GF}\eta=\frac{\mathfrak{M}^2}{2l^2EJ}\int_0^l x^2\,dx+$$

$$+\frac{\mathfrak{M}^2}{2l^2GF}\eta\int_0^l dx=\frac{\mathfrak{M}^2}{2l^2}\left(\frac{l^3}{3EJ}+\frac{\eta l}{GF}\right)=\frac{\mathfrak{M}^2}{2l}\left(\frac{l^2}{3EJ}+\frac{\eta}{GF}\right)$$

Let us compare now the magnitudes of the strain energies due, on the one hand, to the shearing forces and, on the other, to the bending moments. For this purpose let us replace G, F, J and η by their values corresponding to a cross section of rectangular shape

$$G=0.4E, \quad F=bh, \quad J=\frac{bh^3}{12} \quad \text{and} \quad \eta=1.2$$

This leads to

$$W=\frac{\mathfrak{M}^2}{2l}\left(\frac{l^2}{\frac{3Ebh^3}{12}}+\frac{1.2}{0.4Ebh}\right)=\frac{2\mathfrak{M}^2l}{Ebh^3}\left[1+\frac{3}{4}\left(\frac{h}{l}\right)^2\right]$$

The second term in brackets represents the relative value of the strain energy due to the shearing forces. This term is directly proportional to the ratio $\frac{h}{l}$ where h is the depth of the cross section and l is the span of the beam.

Hence, the influence of the shearing forces will drop rapidly with the decrease of this ratio. When this ratio is equal to $\frac{1}{5}$ (beams with a greater ratio are seldomly met)

$$W=\frac{2\mathfrak{M}^2l}{Ebh^3}(1+0.03)$$

It follows that in the case under consideration the strain energy due to the shearing forces constitutes about 3 per cent of the total energy accumulated. In the case of beams met with in actual practice for which the ratio $\frac{h}{l}$ is usually much smaller, the influence of the shearing forces becomes quite negligible.

Problem 2. Required to determine the strain energy accumulated in the truss of Fig. 11.8, all the elements of this truss having the same cross section F.

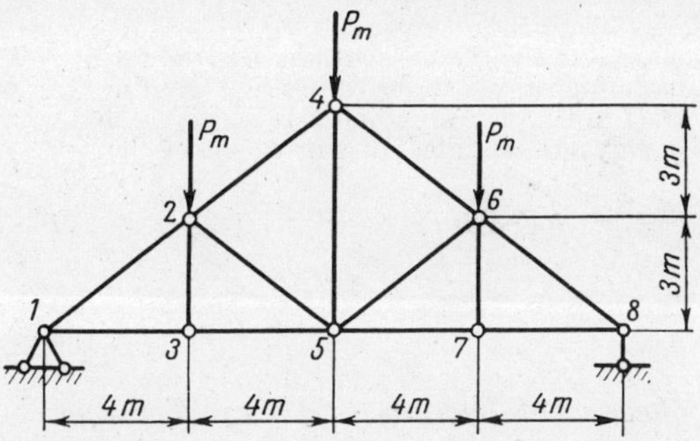

Fig. 11.8

Solution. As the bending moments and shearing forces remain nil in all the bars of the truss and as the normal stresses N and the rigidities EF remain constant over the whole length of each bar, expression (5.8) giving the amount of the strain energy accumulated becomes

$$W=\Sigma\frac{N^2}{2EF}\int_0^l dx=\Sigma\frac{N^2l}{2EF} \tag{6.8}$$

In this expression $N=$ total direct stress in each of the bars caused by the system of loads applied

$l=$ length of the bar.

The sign Σ shows that the summation of the energies must be carried over all the bars of the truss. Of course, those bars which remain idle may be neglected, the product N^2l remaining always nil when $N = 0$.

In the case of trusses and similar structures strain energy computations should be carried out in tabular form as indicated hereunder.

Table 1.8

Bar No.	Total stress N^*, tons	N^2	l, m	N^2l
1-2; 6-8	$-\frac{5}{2}P$	$\frac{25}{4}P^2$	5	$\frac{125}{4}P^2$
2-4; 4-6	$-\frac{5}{3}P$	$\frac{25}{9}P^2$	5	$\frac{125}{9}P^2$
1-3; 3-5 *5-7; 7-8*	$2P$	$4P^2$	4	$16P^2$
2-3; 6-7	0	0	3	0
2-5; 5-6	$-\frac{5}{6}P$	$\frac{25}{36}P^2$	5	$\frac{125}{36}P^2$
4-5	P	P^2	6	$6P^2$

The last column of the Table contains the values of N^2l for each bar of the truss. Summing up all these values and dividing the result by $2EF$ (E being expressed in tons per sq m and F in sq m) we shall obtain the value of the strain energy accumulated in the whole of the truss

$$W=\left(2\,\frac{125}{4}P^2+2\,\frac{125}{9}P^2+16P^2\times 4+0\times 2+2\,\frac{125}{36}P^2+6P^2\right)\frac{1}{2EF}=$$

$$=83\,\frac{11}{18}\,\frac{P^2}{EF}$$

4.8. *Theorem of Reciprocal Works (Theorem of Betty)*

Let us consider two different states of the same elastic system in equilibrium and let us assume that in the first state the system is acted upon by a single statically applied load P_1 and in the second by a statically applied load P_2 (Fig. 12.8).

We shall denote by Δ_{mn} the deflection sustained by any point of the system, the first of the index letters m indicating the direction of the deflection and the second n the number of the load which

◆

* The computation of stresses N has not been included in the above Table.

has caused this deflection. Thus, Δ_{mn} will indicate *the deflection along the line of action of load m caused by the load n.* When the body is acted upon by a moment, Δ_{mn} will represent an angular rotation expressed in radians. The action n may consist also of several concentrated loads, moments or combinations of distributed loads.

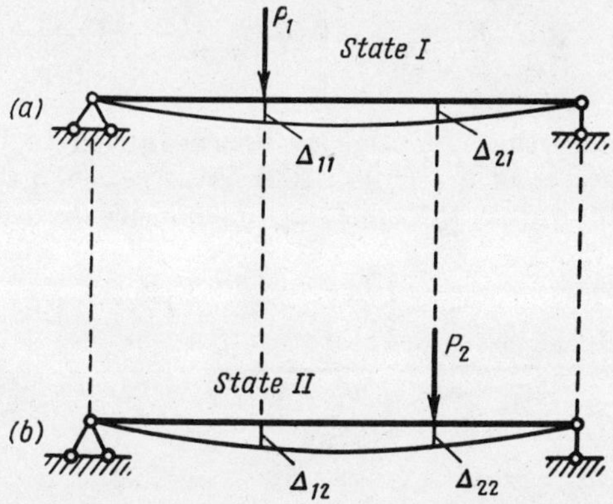

Fig. 12.8

In the case under consideration the various displacements Δ_{mn} are:

Δ_{11} = deflection along the direction of load P_1 due to the same load

Δ_{12} = deflection along the direction of load P_1 due to the load P_2

Δ_{21} = deflection along the direction of load P_2 due to the load P_1

Δ_{22} = deflection along the direction of load P_2 due to the same load.

These four deflections are clearly shown in Fig. 12.8.

Let A_{11} be the work performed by load P_1 along the direction of this same load (in other words, the work corresponding to state 1). Let also A_{22} be the work performed by the load P_2 along the deflections corresponding to state 2.

Expression (2.8) leads to the following values of the work corresponding to each of these states, provided the loads are applied gradually

$$A = \frac{P_1\Delta_{11}}{2}; \qquad A_{22} = \frac{P_2\Delta_{22}}{2}$$

This work could also be expressed in terms of the internal forces acting in the beam [see expression (4.8)]

$$\left.\begin{aligned} A_{11} &= \Sigma\int_0^l \frac{M_1^2\,dx}{2EJ} + \Sigma\int_0^l \frac{N_1^2\,dx}{2EF} + \Sigma\int_0^l \frac{Q_1^2\,dx}{2GF}\eta \\ A_{22} &= \Sigma\int_0^l \frac{M_2^2\,dx}{2EJ} + \Sigma\int_0^l \frac{N_2^2\,dx}{2EF} + \Sigma\int_0^l \frac{Q_2^2\,dx}{2GF}\eta \end{aligned}\right\} \qquad (7.8)$$

Let us assume that the same system is loaded in the following sequence: *first*, load P_1 (Fig. 13.8) is increased gradually from zero to its final value; the deflections sustained by the system and

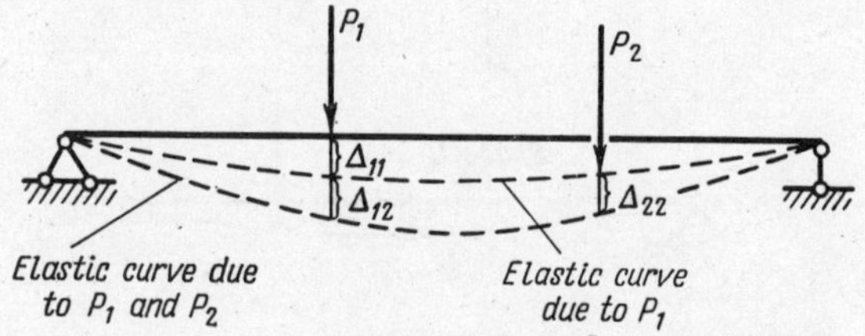

Fig. 13.8

the stresses developed in that case will be exactly the same as those corresponding to state 1 of Fig. 12.8*a*. In particular, the deflection under load P_1 will equal Δ_{11} and the work performed by this load during its application will amount to $A_{11} = \frac{P_1\Delta_{11}}{2}$. *After that* let load P_2 increase in the same way. This will entail the development of additional stresses and deflections, these stresses and deflections being equal to those sustained by the system in state 2 of Fig. 12.8*b*; thus the additional deflection at the point of application of load P_1 will equal Δ_{12}. As the load P_1 did not vary during the application of load P_2 it will travel downwards a distance equal to the additional deflection Δ_{12} performing the work $A_{12} = P_1\Delta_{12}$; at the same time load P_2 will perform the work $A_{22} = \frac{P_2\Delta_{22}}{2}$. It follows that the total work accomplished during the loading of the system first by load P_1 and next by load P_2 will equal

$$A = A_{11} + A_{12} + A_{22} = \frac{P_1\Delta_{11}}{2} + P_1\Delta_{12} + \frac{P_2\Delta_{22}}{2} \qquad (8.8)$$

At the same time the work performed by loads P_1 and P_2 may be expressed (see Eq. 2.8) by half the product of each of these loads by the total deflection along the direction of this load (Fig. 14.8).

$$A=\frac{P_1(\Delta_{11}+\Delta_{12})}{2}+\frac{P_2(\Delta_{21}+\Delta_{22})}{2}$$

Equating the above two expressions we obtain

$$\frac{P_1\Delta_{11}}{2}+P_1\Delta_{12}+\frac{P_2\Delta_{22}}{2}=\frac{P_1(\Delta_{11}+\Delta_{12})}{2}+\frac{P_2(\Delta_{21}+\Delta_{22})}{2}$$

wherefrom

$$P_1\Delta_{12}=P_2\Delta_{21}$$

The product $P_1\Delta_{12}$ represents the work A_{12} performed by load P_1 (corresponding to state 1 of Fig. 12.8*a*) along the deflection

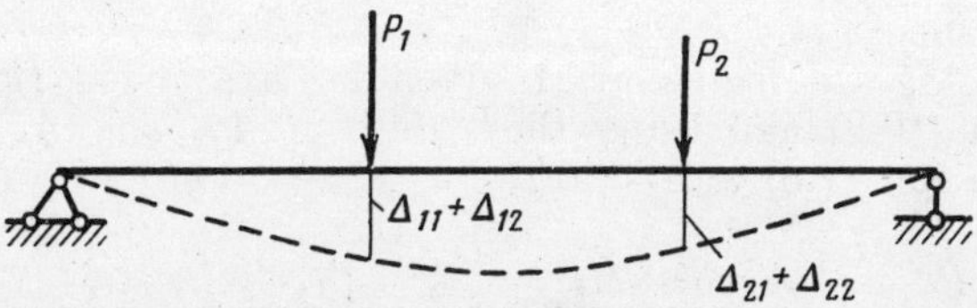

Fig. 14.8

caused by load P_2 following the direction of load P_1 (state 2 of Fig. 12.8*b*). In the same way $P_2\Delta_{21}$ represents the work A_{21} performed by load P_2 of state 2 along the deflection following the line of action of this load caused by load P_1 of state 1.

Consequently

$$A_{12}=A_{21} \tag{9.8}$$

The same result would be obtained if the body under consideration were acted upon by any number of concentrated or distributed loads or moments.

Thus, *the work performed by the actions of state 1 along the deflections caused by the actions corresponding to state 2 is equal to the work performed by the actions of state 2 along the deflections due to the actions of state 1, all the deflections being measured in the direction of the said actions.*

Let us express now the work A_{12} in terms of the bending moments, normal stresses and shears developed in the first and in the second state.

From expression (8.8) we obtain

$$A_{12}=A-A_{11}-A_{22} \tag{10.8}$$

Here A represents the total work produced by loads P_1 and P_2 along the displacements due to these same loads. Using expression (4.8), this work may be expressed by

$$A = \Sigma \int_0^l \frac{(M_1+M_2)^2\,dx}{2EJ} + \Sigma \int_0^l \frac{(N_1+N_2)^2\,dx}{2EF} + \Sigma \int_0^l \frac{(Q_1+Q_2)^2\,dx}{2GF}\eta \tag{11.8}$$

In this expression M_1, N_1 and Q_1 are respectively the bending moments, the normal stresses and the shears developed in the members of the system under consideration due to the application of load P_1, while M_2, N_2 and Q_2 are those due to the application of load P_2.

The sums $(M_1 + M_2)$, $(N_1 + N_2)$ and $(Q_1 + Q_2)$ represent the total resultant stresses in cross sections due to the combined action of both loads P_1 and P_2.

Introducing the value of A given by expression (11.8) into expression (10.8) and using the values of A_{11} and A_{22} derived from equation (7.8) we obtain

$$A_{12} = \Sigma \int_0^l \frac{(M_1+M_2)^2 - M_1^2 - M_2^2}{2EJ}\,dx + \Sigma \int_0^l \frac{(N_1+N_2)^2 - N_1^2 - N_2^2}{2EF}\,dx +$$

$$+ \Sigma \int_0^l \frac{(Q_1+Q_2)^2 - Q_1^2 - Q_2^2}{2GF}\,\eta\,dx$$

wherefrom

$$A_{12} = \Sigma \int_0^l M_1 \frac{M_2\,dx}{EJ} + \Sigma \int_0^l N_1 \frac{N_2\,dx}{EF} + \Sigma \int_0^l Q_1 \frac{Q_2\,dx}{GF}\eta \tag{12.8}$$

In this expression each of the terms preceded by the integral sign may be considered as the product of a total stress (say, the bending moment M_1) due to the actions of state 1 and the total strains of the element dx, say, $\frac{M_2 dx}{EJ}$, due to the actions of state 2.

5.8. *Theorem of Reciprocal Displacements (Theorem of Maxwell)*

Let us take up once again two different states of one and the same system, the first state corresponding to the application of a a unit load P_1 and the second to that of a unit load P_2 (Fig. 15.8). Hereafter we shall use the sign δ to indicate the displacements (strains, angular rotations or deflections) caused by unit loads $P = 1$ or unit moments $\mathfrak{M} = 1$, in order to distinguish them from those due to loads or moments of arbitrary magnitudes which shall be denoted by Δ. Thus, δ_{21} will indicate the displacement due to the unit load P_1 along the direction of load P_2 whilst δ_{12} will indicate the displacement along the line of action of load P_1 due to the application of load unity P_2.

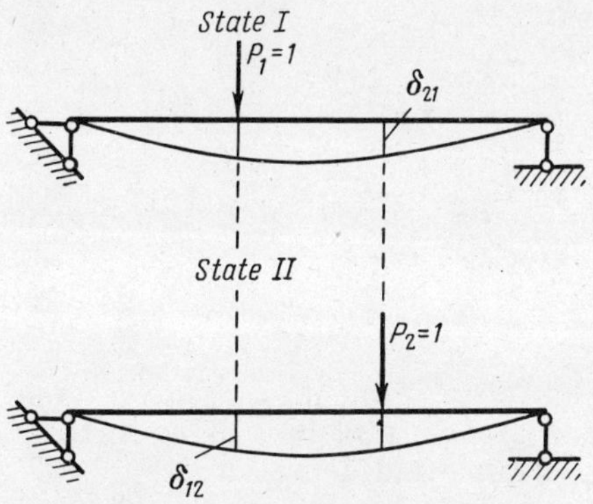

Fig. 15.8

In the preceding article we have shown that

$$P_1\delta_{12} = P_2\delta_{21}$$

As $P_1 = P_2 = 1$, this expression becomes

$$\delta_{12} = \delta_{21}$$

Generalizing we may write for any unity actions

$$\delta_{mn} = \delta_{nm} \tag{13.8}$$

The expression thus obtained is the algebraic expression of *Maxwell's theorem* which runs as follows: *in any elastic system the displacements caused by a load unity along the line of action of another load unity are always equal to the displacement due to this second load unity along the line of action of the first one.*

Equation (13.8) will obviously remain true even when loads P_1 and P_2 have arbitrary but equal values. In this case the said expression will become

$$\Delta_{12} = \Delta_{21} \tag{14.8}$$

An illustration of Maxwell's theorem is afforded by the example of Fig. 16.8. In state 1 the beam with a built-in end is acted upon by a unit load P_1 while in the second one by a unit moment $\mathfrak{M}$.

The rotation ϑ_a due to the unit load P_1 must be numerically equal to the deflection y_l due to the moment $\mathfrak{M}$, i.e., $\vartheta_a = y_l$.

Let us compute now the values of ϑ_a and of y_l using one of the procedures developed in the treatises on the strength of materials.

In state 1 (Fig. 16.8*a*)

$$\vartheta_a = \frac{1}{EJ}\left(-Pla + \frac{Pa^2}{2}\right) = -\frac{Pa}{EJ}\left(l - \frac{a}{2}\right)$$

and in state 2 (Fig. 16.8*b*)

$$y_l = \frac{1}{EJ}\left[-\mathfrak{M}\frac{l^2}{2} + \mathfrak{M}\frac{(l-a)^2}{2}\right] = -\frac{\mathfrak{M}a}{EJ}\left(l - \frac{a}{2}\right)$$

Since $\mathfrak{M} = P = 1$

$$\vartheta_a = -\frac{a}{EJ}\left(l - \frac{a}{2}\right) \quad \text{and} \quad y_l = -\frac{a}{EJ}\left(l - \frac{a}{2}\right)$$

which confirms that

$$\vartheta_a = y_l$$

The strains and displacements caused by nondimensional unit loads $P = 1$ and moments $\mathfrak{M} = 1$ differ in their dimensionality from the usual strains and deflections.

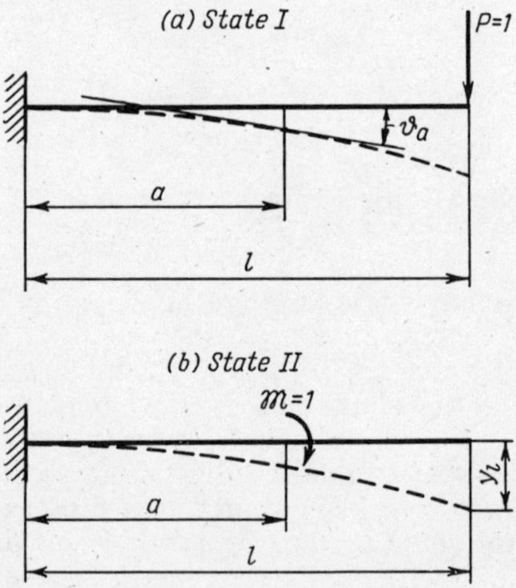

Fig. 16.8

Indeed, the dimensionality of a displacement caused by a unit load is given by the ratio of a displacement to the action which has caused it. Thus, in the previous example the angle of rotation ϑ_a produced by a nondimensional load unity $P = 1$ (which is entirely different from a load equal to 1 ton or 1 kg) will be expressed in kg^{-1}. Similarly the deflection produced by the unit moment

$\mathfrak{M} = 1$ will also be expressed in cm per kg·cm or kg^{-1}; in other words, its dimensionality will be exactly the same as that of an angular rotation due to a unit load.

6.8. *Methods of Displacement Computation*

Let us consider two different states of one and the same system. In its first state the system is acted upon by any number of loads and moments whose values may be chosen at will (Fig. 17.8*a*) and in the second state by one single load unity P_2 (Fig. 17.8*b*).

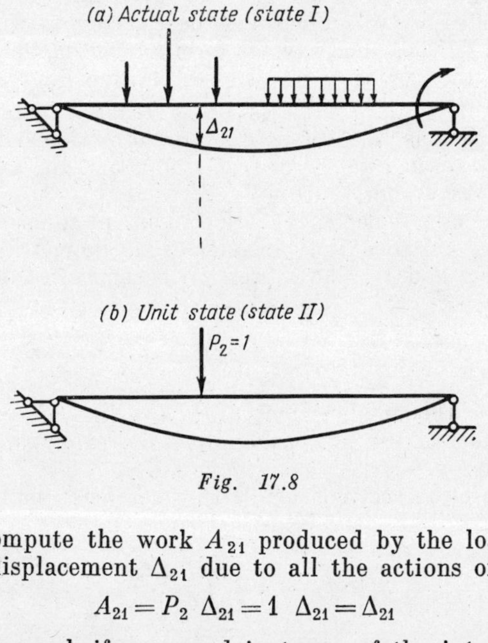

Fig. 17.8

Let us compute the work A_{21} produced by the load unity P_2 along the displacement Δ_{21} due to all the actions of state 1

$$A_{21} = P_2 \ \Delta_{21} = 1 \ \Delta_{21} = \Delta_{21}$$

This same work if expressed in terms of the internal stresses using formulas (9.8) and (12.8) becomes

$$A_{21} = \Sigma \int_0^l \overline{M}_2 \frac{M_1 \, dx}{EJ} + \Sigma \int_0^l \overline{N}_2 \frac{N_1 \, dx}{EF} + \Sigma \int_0^l \overline{Q}_2 \frac{Q_1 \, dx}{GF} \eta \qquad (15.8)$$

(The dashes placed over M_2, N_2 and Q_2 indicate that these stresses are due to a load unity.)

Thus the displacement caused by any combination of loads may be expressed in terms of the stresses developed by the said combination and by those due to a load unity. The line of action of this load unity must coincide with the direction of the displacement under consideration.

When a linear displacement is required (say, a deflection at any point of the system) the load unity must be a concentrated nondimensional load acting at this particular point. If, on the other hand, it is required to find the angular rotation of a certain cross section, the unit action must be a nondimensional concentrated moment applied to this section.

Hereafter we shall refer to that state when a single load unity is applied to the system as the *imaginary* or *unity state*, while the case when the same system is acted upon by the combination of actions effectively applied will be referred to as the *real* or *actual* state. In the same way the term *unit graph* or *unit diagram* will refer to the graphs of the stresses developed under the action of a load unity, these stresses being denoted by $\overline{M}$, $\overline{N}$ and $\overline{Q}$, while the diagrams of stresses due to the actions effectively applied will be termed *actual* or *real graphs* and the corresponding stresses will be designated by M_p, N_p and Q_p.

In certain cases it becomes more convenient to use alphabetical indices in expression (15.8) instead of the numerical ones (say, m and n, or p and k). The expression becomes in this case

$$\Delta_{mn} = \Sigma \int_0^l \overline{M}_m \frac{M_n\,dx}{EJ} + \Sigma \int_0^l \overline{N}_m \frac{N_n\,dx}{EF} + \Sigma \int_0^l \overline{Q}_m \frac{Q_n\,dx}{GF}\,\eta \qquad (16.8)$$

where Δ_{mn} is the displacement along the line of action of load unity P_m due to the actions applied in reality and belonging to the group n.

When the cross sections of all the members remain constant, the expression 16.8 may be rewritten as follows

$$\Delta_{mn} = \Sigma \frac{1}{EJ} \int_0^l \overline{M}_m M_n\,dx + \Sigma \frac{1}{EF} \int_0^l \overline{N}_m N_n\,dx + \Sigma \frac{\eta}{GF} \times$$

$$\times \int_0^l \overline{Q}_m Q_n\,dx \qquad (17.8)$$

The three expressions (15.8), (16.8) and (17.8) are frequently referred to as the *general displacement equations or Mohr's equations.*

For the computation of displacements with the help of these expressions the following sequence will be adopted:

1. In the first place determine the stresses M_n, N_n and Q_n due to the applied loads for an arbitrary cross section in terms of its abscissa x.

2. Apply a unity action at the cross section whose deflection or angular rotation is required, a concentrated load corresponding to a deflection or any other translation and a moment to an angular rotation.

3. Compute the stresses $\overline{M}_m$, $\overline{N}_m$ and $\overline{Q}_m$ due to this unit action for the same cross section situated a distance x from the origin of coordinates.

4. Introduce the values of the stresses M_n, N_n and Q_n as well as those of $\overline{M}_m$, $\overline{N}_m$ and $\overline{Q}_m$ in one of the three expressions (15.8),

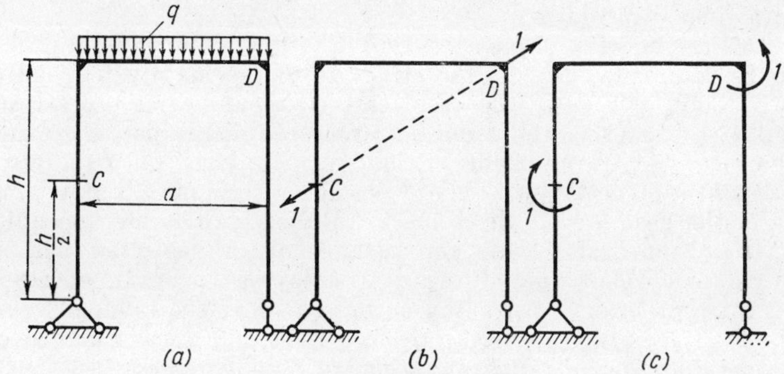

Fig 18.8

(16.8) or (17.8) and integrate along all the elements of the entire structure. When the displacement Δ_{mn} thus obtained is positive, its direction coincides with that adopted for the unit action and when it is negative, it is directly opposite to the one adopted for the unit action.

In the design of the redundant structure it is sometimes required to find the mutual displacement of two preselected points. In that case a system of two unit loads of opposite direction should be applied along the direction of the displacement required, these unit loads being replaced by unit moments when the displacement in question is an angular rotation. Thus, if it were required, for instance, to find the increase in the distance between point C and D of a portal frame appearing in Fig. 18.8*a*, unit loads acting along the line CD should be applied to both of these points as indicated in Fig. 18.8*b*. All the computations will be carried out thereafter using Mohr's formula in the sequence mentioned above, keeping in mind that the stresses $\overline{M}_m$, $\overline{N}_m$ and $\overline{Q}_m$ will be those developed under the simultaneous action of both load unities just mentioned.

If the displacement obtained is positive, its direction coincides with the one adopted for the load unities, in other words, the distance between points C and D will increase. If, on the contrary, the displacement obtained were negative, it would mean that points C and D are brought closer together.

The relative angular rotation of two cross sections of some structure may be calculated in exactly the same way. In the example just mentioned two unit moments should be applied in that case to points C and D, these moments acting in opposite directions as shown in Fig. 18.8*c*. As for the computations themselves, they will not differ in any respect from those just described for the case of a linear translation.

In the majority of cases, only one term of Mohr's formula has to be retained. Thus, if the structure works mainly in bending (this being generally the case for rectilinear beams, portal and building frames as well as for flat arches) it will suffice to use only the term containing bending moments. Similarly when the members of the structure work mainly in direct tension or compression, as is the case for all the hinged systems, both terms depending on the bending moments and shears may be neglected without any appreciable reduction in the accuracy of the results obtained. In all that follows, with the exception of a few specified cases, *we shall always neglect the influence of normal stresses and shears on the deflection of rectilinear beams and rigid frames.*

If the element under consideration is a curved bar whose radius of curvature is at least ten times as great as the depth of its cross section, Mohr's formula for rectilinear bars may be used, provided the length of the straight element dx is replaced by the length of the arch ds. The influence of normal stresses and shears may be usually neglected.

Problem 1. Determine the deflection of an end-supported beam of constant cross section acted upon by a concentrated load P_n at midspan (Fig. 19.8*a*). All the three terms of Mohr's formula should be used.

Solution. The imaginary state of this beam will correspond to the application of a concentrated load unity in the direction of the deflection required (Fig. 19.8*b*). The normal stresses will remain constantly nil and therefore Mohr's formula will become

$$\Delta_{mn}=\Sigma\frac{1}{EJ}\int_0^l \overline{M}_m M_n\,dx+\Sigma\frac{\eta}{GF}\int_0^l \overline{Q}_m Q_n\,dx=\Delta_{mn}^M+\Delta_{mn}^Q$$

where

$$\Delta_{mn}^M=\Sigma\frac{1}{EJ}\int_0^l \overline{M}_m M_n\,dx$$

is the deflection in pure bending (i. e., due solely to the bending moments) whilst

$$\Delta^Q_{mn}=\Sigma \frac{\eta}{GF}\int\limits_0^l \overline{Q}_m Q_n\,dx$$

is that part of the total deflection caused solely by the shearing forces.

For all the cross sections of the beam to the left of point C the bending moments M_n and $\overline{M}_m$ and the shearing forces Q_n and $\overline{Q}_m$ are given by

$$M_n=\frac{P_n}{2}x \qquad \overline{M}_m=\frac{1}{2}x$$

$$Q_n=\frac{P_n}{2} \qquad \overline{Q}_m=\frac{1}{2}$$

The corresponding graphs are given in Fig. 19.8*c*, *d*, *e* and *f*. Introducing these values into the expressions giving the two different parts of the total deflection, we obtain

$$\Delta^M_{mn}=\frac{2}{EJ}\int\limits_0^{l/2}\frac{x}{2}\cdot\frac{P_n}{2}x\,dx=$$

$$=\frac{P_nl^3}{48EJ}$$

$$\Delta^Q_{mn}=\frac{2\eta}{GF}\int\limits_0^{l/2}\frac{1}{2}\cdot\frac{P_n}{2}\,dx=$$

$$=\frac{P_n\eta l}{4GF}$$

The beam being symmetrical about a vertical axis, we may integrate only along one half of its length, say, the left one. The total deflection will be given by

$$\Delta_{mn}=\Delta^M_{mn}+\Delta^Q_{mn}=$$

$$=\frac{P_nl^3}{48EJ}+\frac{P_n\eta l}{4GF}$$

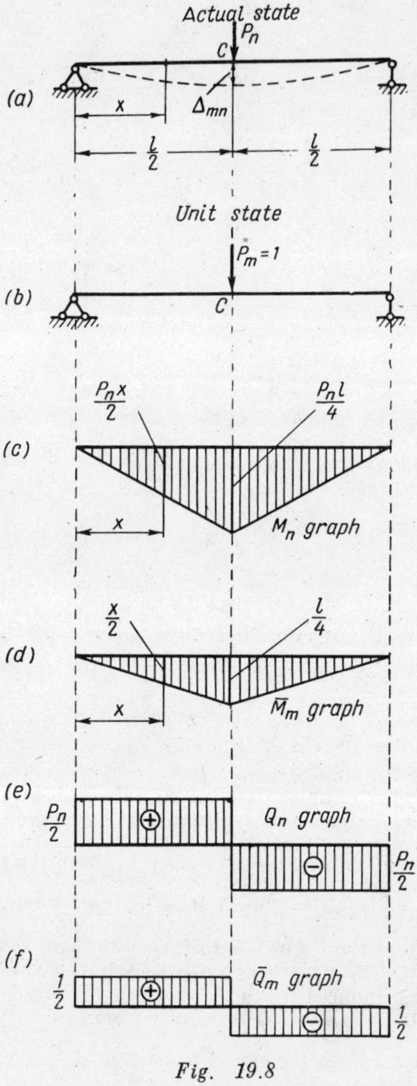

Fig. 19.8

When the value of this deflection is positive, its direction will coincide with that of the load unity (if a negative value were obtained, it would indicate that the beam is deflected in the opposite direction).

Let us now determine the relative importance of both parts of the total deflection, the one due to the bending moment and the other due to the shears.

Let the cross-sectional dimensions of the beam be b and h with $h = 0.1l$

$$\frac{\Delta_{mn}^{Q}}{\Delta_{mn}^{M}}=\frac{P_n\eta l 48EJ}{P_n l^3 4FG}=\frac{12\eta EJ}{l^2FG}$$

Replacing in the above expression J, F, η and G by the following values

$$J=\frac{bh^3}{12}=\frac{bl^3}{12{,}000}, \qquad F=bh=\frac{bl}{10}, \qquad \eta=1.2 \text{ and } G=0.4E$$

we obtain

$$\frac{\Delta_{mn}^{Q}}{\Delta_{mn}^{M}}=\frac{12\times 1.2\times Ebl^3\times 10}{12{,}000\times l^2\times 0.4Ebl}=\frac{3}{100}$$

It follows that the deflection produced by the shears amounts to 3 per cent only of that part of the deflection produced by the bending moments.

The unfluence of the shears will decrease together with the ratio $\frac{h}{l}$, and for $h = \frac{l}{20}$ we have already

$$\frac{\Delta_{mn}^{Q}}{\Delta_{mn}^{M}}=\frac{3}{400}$$

It is obvious that in the great majority of cases the term Δ_{mn}^{Q} may be completely neglected by comparison with the term Δ_{mn}^{M}. Thus we obtain the well-known expression

$$\Delta_{mn}=\Delta_{mn}^{M}=\frac{P_n l^3}{48EJ}$$

Problem 2. Compute the vertical deflection Δ_C of point C of a uniformly loaded beam built-in at its left end (Fig. 20.8*a*).

Solution. The bending moment curve due to the uniformly distributed load is represented in Fig. 20.8*b*. The magnitude of the bending moment at any cross section a distance x from the right-hand end of the beam equals $-\frac{qx^2}{2}$. The imaginary state will correspond to the application of a concentrated load unity at point C, its direction coinciding with that of the deflection required, i. e., being vertical (Fig. 20.8*c*). The diagram of the bending moments $\overline{M}_m$ induced by the load unity P_m is represented in Fig. 20.8*d*. It is clear that this moment will differ from zero only at the cross sections of the beam situated to the left of point C $\left(\text{at } \frac{l}{2}\leqslant x\leqslant l\right)$ where its amount will be given by $\overline{M}_m=-\left(x-\frac{l}{2}\right)$.

Neglecting the shears and integrating the term depending on the bending moments from $\frac{l}{2}$ to l ($\overline{M}_m$ remaining constantly nil to the right of section C)

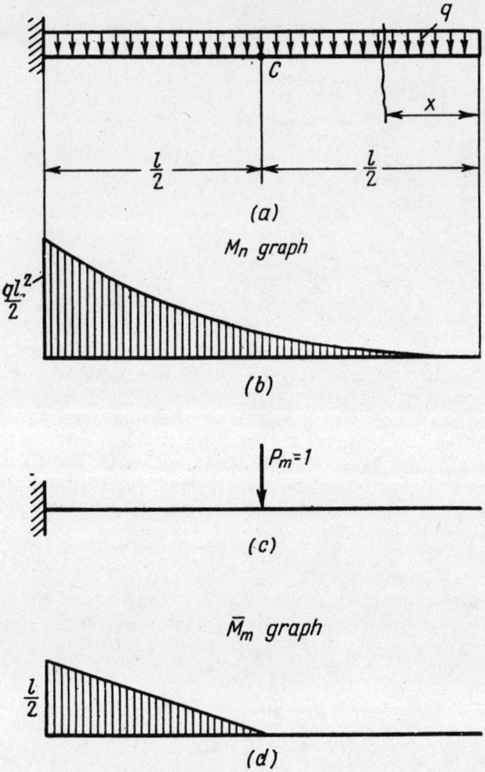

Fig. 20.8

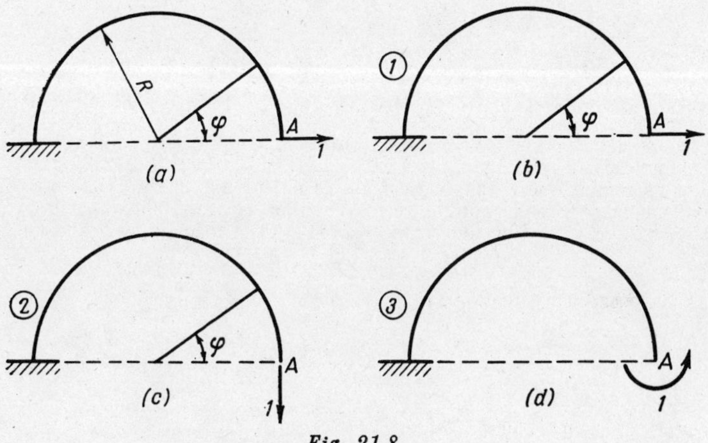

Fig. 21.8

we obtain, using expression (17.8),

$$\Delta_C=\frac{1}{EJ}\int\limits_{l/2}^{l}\overline{M}_mM_n\,dx=\frac{1}{EJ}\int\limits_{l/2}^{l}\left(x-\frac{l}{2}\right)\frac{qx^2}{2}\,dx=$$

$$=\frac{q}{2EJ}\int\limits_{l/2}^{l}\left(x^3-\frac{x^2l}{2}\right)dx=\frac{q}{2EJ}\left(\frac{x^4}{4}-\frac{x^3l}{6}\right)_{l/2}^{l}=$$

$$=\frac{q}{2EJ}\left[\frac{l^4}{4}-\frac{l^4}{6}-\frac{(l/2)^4}{4}+\frac{(l/2)^3\,l}{6}\right]=\frac{17ql^4}{384EJ}$$

Problem 3. Determine the maximum translation of point A belonging to the neutral axis of a curved beam as well as the angular rotation of the cross section passing through this same point (Fig. 21.8*a*).

Solution. The influence of the shears and normal forces being negligible and the direction of the translation remaining unknown, we shall compute separately its horizontal and vertical components induced by the bending moments. The total value of the displacement will be found thereafter in the usual way. The magnitude of the bending moment induced at any section by the load P is given by

$$M_p=PR\sin\varphi$$

In order to determine the horizontal component Δ_{1p} of the total displacement let us apply at point A a horizontal load unity as indicated in Fig. 21.8*b*. The value of the bending moment induced in this case by the load unity will constitute for any section

$$\overline{M}_1=1\cdot R\sin\varphi=R\sin\varphi$$

Remembering that $ds = Rd\varphi$, Mohr's formula (16.8) becomes

$$\Delta_{1p}=\int\limits_{\varphi=0}^{\varphi=\pi}\frac{M_p\overline{M}_1\,ds}{EJ}=\int\limits_{0}^{\pi}\frac{PR\sin\varphi R\sin\varphi R\,d\varphi}{EJ}=$$

$$=\frac{PR^3}{EJ}\int\limits_{0}^{\pi}\sin^2\varphi\,d\varphi=\frac{PR^3}{EJ}\left(\frac{\varphi}{2}-\frac{\sin 2\varphi}{4}\right)_0^{\pi}=\frac{\pi PR^3}{2EJ}$$

The value of the displacement thus obtained being positive, its direction will coincide with that adopted for the load unity in Fig. 21.8*b*.

Let us compute now the vertical displacement Δ_{2p} of this same point A. For this purpose we shall apply a vertical load unity as indicated in Fig. 21.8*c*. In that case the bending moments induced by the load unity become

$$\overline{M}_2=-1\cdot R\,(1-\cos\varphi)=-R\,(1-\cos\varphi)$$

Using once again expression (16.8) we obtain

$$\Delta_{2p}=\int\limits_{\varphi=0}^{\varphi=\pi}\frac{M_p\overline{M}_2\,ds}{EJ}=\int\limits_{0}^{\pi}\frac{RP\sin\varphi\,[-R\,(1-\cos\varphi)]\,R\,d\varphi}{EJ}=$$

$$=\frac{PR^3}{EJ}\left(\int\limits_{0}^{\pi}-\sin\varphi\,d\varphi+\int\limits_{0}^{\pi}\sin\varphi\cos\varphi\,d\varphi\right)=-\frac{2PR^3}{EJ}$$

This displacement being negative, point A must travel in a direction opposite to the one selected for the load unity, i.e., upwards. The total displacement of point A will be given by

$$\Delta_A=\sqrt{\Delta_{1p}^2+\Delta_{2p}^2}=\sqrt{\left(\frac{\pi PR^3}{2EJ}\right)^2+\left(-\frac{2PR^3}{EJ}\right)^2}=\frac{PR^3}{EJ}\sqrt{\frac{\pi^2}{4}+4}$$

The angular rotation Δ_{3p} of cross section A will be obtained applying to the said cross section an imaginary unit moment $\overline{M}_3$ (Fig. 21.8d). In that case the bending moments induced in the beam will remain constant and equal to unity. Consequently, the angular rotation of cross section A will be given by

$$\Delta_{3p}=\int_{\varphi=0}^{\varphi=\pi}\frac{M_p\overline{M}_3\,ds}{EJ}=$$

$$=\int_0^{\pi}\frac{PR\sin\varphi\cdot 1\cdot R\,d\varphi}{EJ}=\frac{2PR^2}{EJ}$$

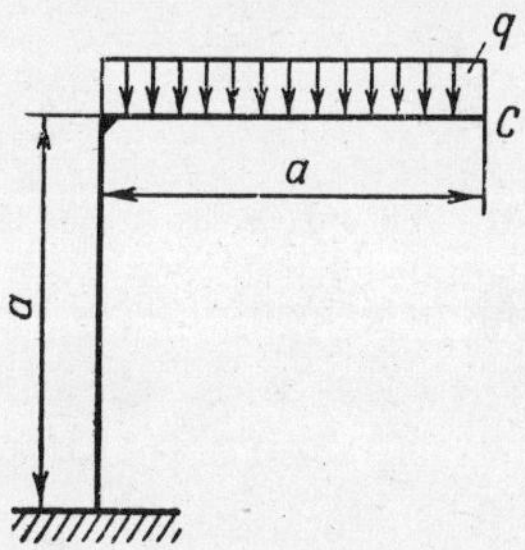

Fig. 22.8

The direction of this rotation will coincide with that of the unit moment, which means that the cross section will turn counterclockwise.

Problem 4. Determine the angular rotation of the free end C of a knee frame appearing in Fig. 22.8.

Solution. Apply to section C of the knee frame a unit moment M tending to turn this section in the direction of the rotation required. In that case

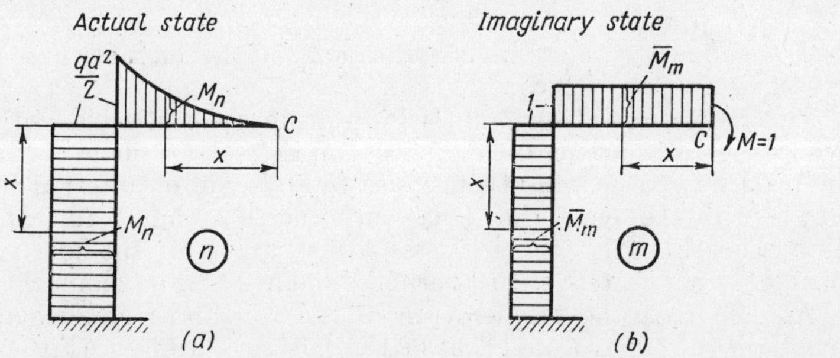

Fig. 23.8

the normal forces and the shears throughout the structure will remain constantly nil and Mohr's formula will comprise one term only depending on the bending moments even if it were desired to account for all the stresses induced in the structure.

Draw the two bending moment curves, one for the distributed loads q effectively applied and one for the imaginary unit moment M. These two curves are given in Fig. 23.8a and b, respectively. Analytically the values of the bending moments for both cases will be given by:

for the upright

$$M_n = -\frac{qa^2}{2} \text{ and } \overline{M}_m = -1$$

for the horizontal beam

$$M_n = -\frac{qx^2}{2} \text{ and } \overline{M}_m = -1$$

Introducing these values in expression (16.8) we obtain

$$\varphi = \Delta_{mn} = \int_0^a \frac{1}{2} \cdot \frac{qx^2}{EJ} dx + \int_0^a \frac{1}{2} \cdot \frac{qa^2}{EJ} dx = \frac{2qa^3}{3EJ}$$

7.8. *Temperature Strains*

Mohr's formula (16.8) may be written as follows

$$\Delta_{mn} = \Sigma \int_0^l \overline{M}_m \Delta_{\varphi n} + \Sigma \int_0^l \overline{N}_m \Delta_{xn} + \Sigma \int_0^l \overline{Q}_m \Delta_{yn} \qquad (18.8)$$

where $\Delta_{\varphi n} = \frac{M_n dx}{EJ}$ = mutual angular rotation of the two end faces of element dx induced by the applied loads

$\Delta_{xn} = \frac{N_n dx}{EF}$ = mutual linear displacement of the same faces along the axis of the element

$\Delta_{yn} = \frac{Q_n dx}{GF} \eta$ = their mutual displacement in the direction normal to the axis of the member (see Art. 2.8).

In this transcription Mohr's formula may be utilized not only when the displacements $\Delta_{\varphi n}$, Δ_{xn} and Δ_{yn} of an element dx are induced by stresses due themselves to a system of external loads, but also in the event the strains are due to a change in temperature. Consequently this expression may serve for the solution of problems connected with thermal expansions and contractions.

Assume that the temperature of the top fibres of element dx has been raised by t_1° and that of the bottom ones by t_2° (Fig. 24.8). Assume also that within the body itself the temperature varies linearly.

The expansion of the top fibres of the elment dx will equal $\alpha t_1 dx$ and that of the bottom fibres $\alpha t_2 dx$ where α is the coefficient of thermal expansion. In the case of a symmetrical cross section the expansion at midheight will equal half the sum of the expansions of the extreme fibres

$$\Delta_{xn} = \frac{\alpha (t_1 + t_2)}{2} dx$$

The mutual angular rotation of the two cross sections bounding this element will be given by

$$\Delta_{\varphi n} = \Delta_{\varphi t} = \frac{\alpha (t_1 - t_2)}{h} dx$$

As the rise of temperature will lead to no vertical displacements of the element dx the term Δ_{yn} will remain nil.

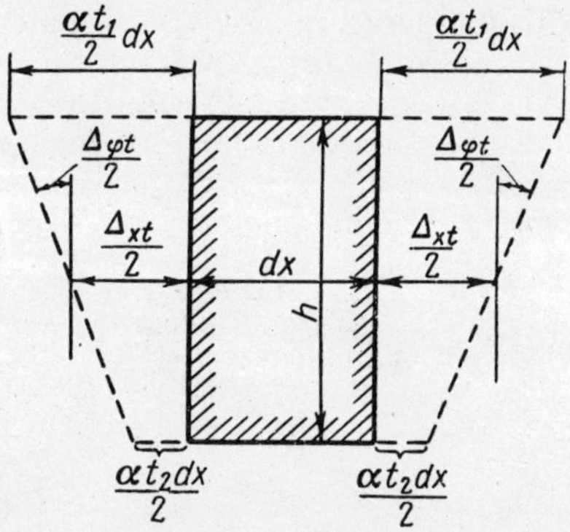

Fig. 24.8

Introducing the above values into formula (18.8) we shall obtain the expression permitting direct computation of strains and deflections arising from temperature changes*

$$\Delta_{mt} = \Sigma\alpha \frac{t_1 - t_2}{h} \int_0^l \overline{M}_m \, dx + \Sigma\alpha \frac{t_1 + t_2}{2} \int_0^l \overline{N}_m \, dx \qquad (19.8)$$

In this expression the sign Σ indicates that the summation must be carried over all the members of the system.

It is obvious that only those members which have been submitted to a temperature change must be taken into consideration. For rectilinear or polygonal bars of constant cross section the corresponding integrals may be computed as the areas bounded by the diagrams of unit stresses, which permits to reduce the above expression to the following form which is extremely convenient for practical design

$$\Delta_{mt} = \Sigma\alpha \frac{t_1 - t_2}{h} \Omega_{\overline{M}} + \Sigma\alpha \frac{t_1 + t_2}{2} \Omega_{\overline{N}} \qquad (20.8)$$

* This expression will be valid only if the change in temperature and the height of the cross section do not vary within the length of each particular member forming the structure.

Here $\Omega_{\overline{M}}$ and $\Omega_{\overline{N}}$ indicate the areas bounded by the $\overline{M}$ and $\overline{N}$ curves. When the cross section is nonsymmetrical about its neutral axis the term $\frac{t_1+t_2}{2}$ must be replaced by $t_2+\frac{t_1-t_2}{h}y$ where y is the distance of the lower fibre to the gravity axis.

The signs of all the terms appearing in the above formula will be obtained as follows: when the strains of element dx induced both by the variation in temperature and by the load unity are of the same direction the corresponding term of the equation will be positive; if it were otherwise, the term would be negative.

In the computation of thermal displacements *the strains and deflections produced by the shearing forces may no longer be neglected for their relative value may be quite appreciable.*

Problem. Required the vertical displacement of point C of the knee frame appearing in Fig. 22.8, when the indoor temperature rises by 10°C, the outdoor temperature remaining constant (Fig. 25.8*a*).

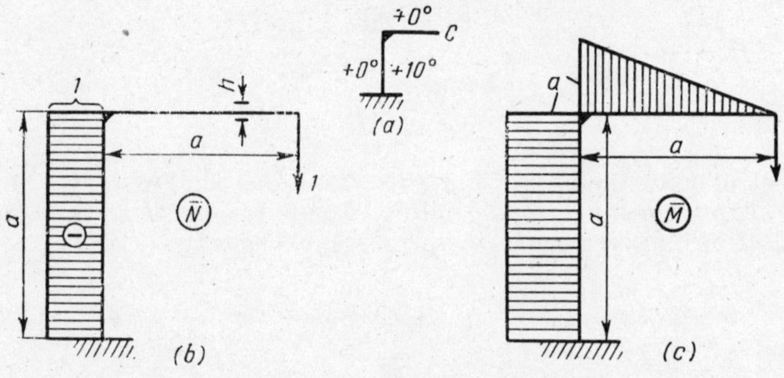

Fig. 25.8

Solution. Apply a load unity along the direction of the displacement required and draw the corresponding $\overline{M}$ and $\overline{N}$ curves (Fig. 25.8*b* and *c*). The areas bounded by these curves will amount to

$$\Omega_{\overline{N}}=1\cdot a=a$$

$$\Omega_{\overline{M}}=\frac{aa}{2}+aa=1.5a^2$$

Let us also compute the terms depending directly on the temperature change

$$\frac{t_1+t_2}{2}=\frac{0+10}{2}=5$$

$$|t_1-t_2|=|0-10|=10$$

It should be observed that the last term representing the total change in temperature must be always taken in absolute value regardless of its sign.

It should also be noted that an increase in the indoor temperature leads to an extension of the inner fibres of the knee frame while the unit load shown in Fig. 25.8 causes their contraction. Consequently, that term of the equation which accounts for the bending moments will be negative. The same will apply to the term accounting for the normal stresses, for an increase in temperature leads to an extension of the upright while the load unity adopted entails its contraction. Consequently

$$\Delta_{mt} = -5\alpha a - 15\alpha \frac{a^2}{h}$$

8.8. *Displacement Computation Techniques*

In a number of cases displacement computations may be simplified very considerably by the introduction of a special technique for the calculation of the integrals belonging to the type $\int_0^l \overline{M}_m M_n dx$*.

We shall name this technique the *graph multiplication method* for it is based on the fact that the expression preceded by the sign of the integral contains the product of two ordinates to the $\overline{M}_m$ and M_n curves. This technique will apply, provided at least one of the curves (say, that of $\overline{M}_m$) reduces to a straight line. The other diagram may be bounded by any curve or broken line. The ordinate to any straight line may be always expressed by $\overline{M}_m = x \tan\alpha$; the meaning of x and α being clearly shown in Fig. 26.8.

Introducing this value of $\overline{M}_m$ into the integral under consideration, we obtain

$$\int_0^l \overline{M}_m M_n \, dx = \tan\alpha \int_0^l x M_n \, dx = \tan\alpha \int_0^l x d\Omega_n$$

where $M_n dx = d\Omega_n$ represents the differential of the area Ω_n bounded by the M_n curve (Fig. 26.8).

Consequently, the expression $\int_0^l x d\Omega_n$ represents the statical moment of the graph area about the O-O' axis (Fig. 26.8).

◆

* The same technique will apply to similar integrals

$$\int_0^l \overline{N}_m N_n \, dx \text{ and } \int_0^l \overline{Q}_m Q_n \, dx$$

It is well known that this statical moment may be expressed by

$$\int_0^l x d\Omega_n = \Omega_n x_c$$

x_c representing the abscissa of the graph centroid. It follows that $\int_0^l \overline{M}_m M_n dx = x_c \tan\alpha\,\Omega_n$, but since $x_c \tan\alpha = y_c$ we obtain finally

$$\int_0^l \overline{M}_m M_n \, dx = \Omega_n y_c \tag{21.8}$$

Hence, *the product of the multiplication of two graphs, one of which at least is bounded by a straight line, equals the area bounded*

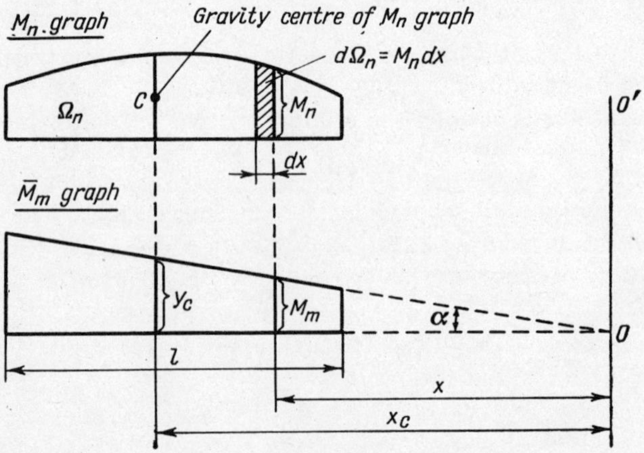

Fig. 26.8

by the graph of arbitrary outline multiplied by the ordinate y_c to the first graph measured along the vertical passing through the centroid of the second one. This product will be reckoned positive when the graph of arbitrary outline and the ordinate to the rectilinear graph are both of the same sign and negative when the two are of opposite sign. This procedure has been suggested in 1925 by Prof. A. Vereshchagin when he was still a student of the Moscow Railway Transport Institute and therefore in the U.S.S.R. this method is known also as *Vereshchagin's method.*

It should be noted that the left part of expression (21.8) differs from Mohr's integral by the absence of the factor $\frac{1}{EJ}$. Hence the result of the graph multiplication carried out by Vereshchagin's method must be later divided by EJ.

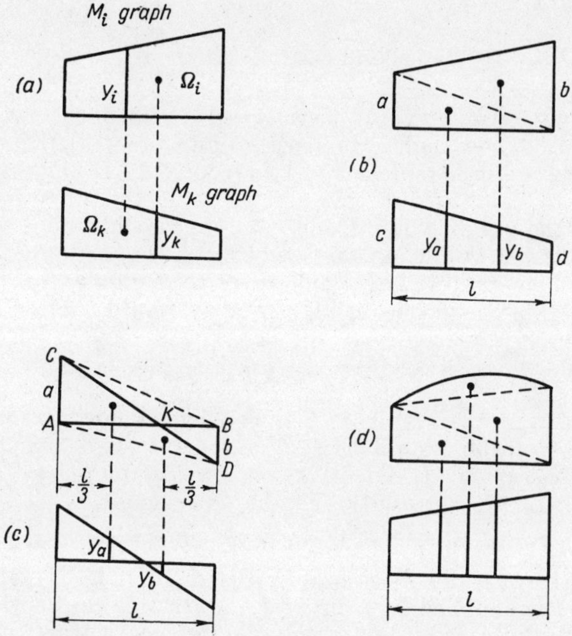

Fig. 27.8

It should be always kept in mind that *the ordinate y_c must be measured on the graph bounded by a straight line*. If both of the graphs were bounded by straight lines the ordinate y_c could be measured on any one of the two. Thus, if it were required to find the product of the graphs for M_i and M_k of Fig. 27.8*a* one could either multiply the area Ω_i, bounded by the M_i curve, by the ordinate y_k measured along the vertical passing through the centroid of this area to the M_k curve, or else one could multiply the area Ω_k, bounded by the M_k curve, by the ordinate y_i measured to the M_i curve along the vertical passing through the Ω_k centroid.

When a trapezoidal graph has to be multiplied by another graph of the same shape, it is convenient to subdivide one of the two into two triangles as indicated in Fig. 27.8*b* and to multiply thereafter the area of each of these triangles by the ordinate to the other graph measured along the vertical passing through the centroid of each of these triangles.

Thus, in the case just mentioned we would obtain

$$\frac{al}{2}y_a+\frac{bl}{2}y_b=\frac{al}{2}\left(\frac{2c}{3}+\frac{d}{3}\right)+\frac{bl}{2}\left(\frac{c}{3}+\frac{2d}{3}\right)=$$

$$=\frac{l}{6}(2ac+2bd+ad+bc)$$

The same procedure could be followed if each of the two graphs consisted of two triangles of opposite sign (Fig. 27.8c). One of the two graphs could be replaced by two triangles ABC and ABD the ordinates to which would remain of the same sign along the whole length of the graph. The introduction of two additional triangles CBK and AKD would have no effect on the final results for their ordinates are equal in value and opposite in sign.

Multiplying the graphs of Fig. 27.8*c* we would obtain

$$\frac{al}{2}y_a+\left(-\frac{bl}{2}\right)(-y_b)=\frac{al}{2}y_a+\frac{bl}{2}y_b$$

When one of the graphs is bounded by a conic parabola, the area of this graph should be subdivided into two triangles and a parabolic segment. It will be remembered that a parabolic graph is peculiar to the uniformly distributed loads and that the ordinate to the centre of the parabolic segment is always equal to $\frac{ql^2}{8}$.

It may happen that both graphs are irregular in shape but one of the two is bounded by a broken line. In this case both graphs should be subdivided into a number of portions so that in each of them at least one of the graphs should be bounded by a straight line. Thus, if it were necessary to multiply the two graphs represented in Fig. 28.8*a* and *b*, both should be subdivided into two parts, the result of their multiplication being given by the sum $\Omega_1 y_1+\Omega_2 y_2$.

One could also subdivide these graphs in three portions as indicated in Fig. 28.8*c* and *d*. In that case the result of their multiplication would be given by $\Omega_1 y_1+\Omega_2 y_2+\Omega_3 y_3$.

Vereshchagin's method requires the rapid evaluation of graph areas of different shape and the determination of the position of their centroid. Table 2.8 represents the areas and the centroid positions for various graphs and is intended to facilitate these computations.

Table 3.8 gives the values of Mohr's integrals $\int M_i M_k dx$ computed for various graphs of different outline. This table will be of considerable help in the computation of displacements.

Vereshchagin's method is particularly well fit for the computation of the deflections of beams and framed structures, the different

Table 2.8

No.	Shape of the graph	Area Ω	Position of the centroid	
			z_1	z_2
1	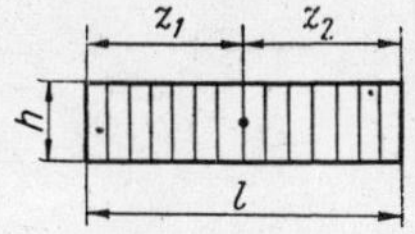	hl	$\frac{l}{2}$	$\frac{l}{2}$
2	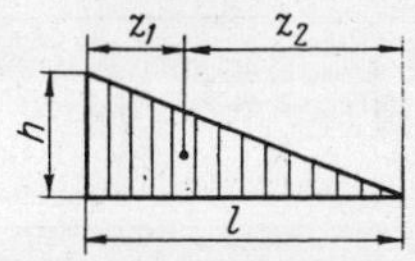	$\frac{hl}{2}$	$\frac{l}{3}$	$\frac{2l}{3}$
3	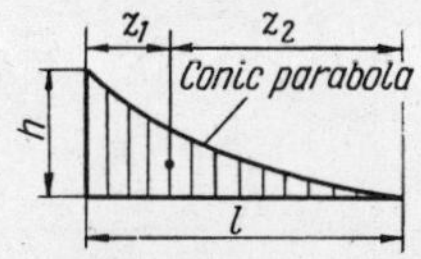	$\frac{hl}{3}$	$\frac{l}{4}$	$\frac{3l}{4}$
4	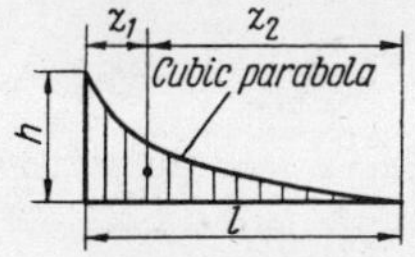	$\frac{hl}{4}$	$\frac{l}{5}$	$\frac{4l}{5}$
5	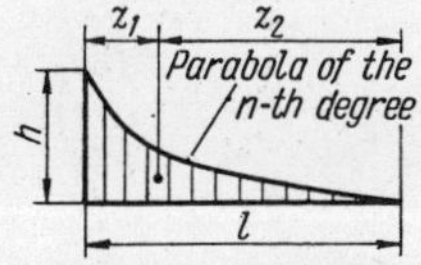	$\frac{hl}{n+1}$	$\frac{l}{n+2}$	$\frac{(n+1)\,l}{n+2}$
6	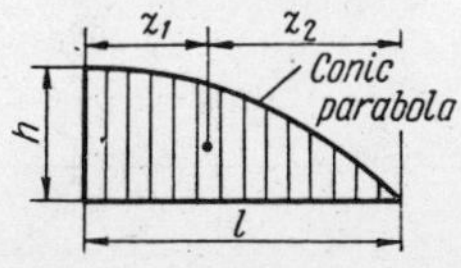	$\frac{2hl}{3}$	$\frac{3l}{8}$	$\frac{5l}{8}$

spans or members of which do not vary in their rigidity. Should this rigidity vary along an element, the product EJ would have to remain under the integral sign which would make Vereshchagin's method inapplicable. It would then become necessary to calculate analytically Mohr's integrals or, if an approximate solution were deemed sufficient, all the members of the structure could be fictitiously replaced by other ones whose rigidity would vary by increments.

Value of Mohr's Integrals $\int M_i M_k ds$ (the

M_i graph M_k graph	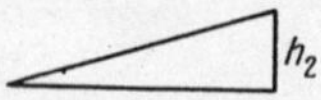	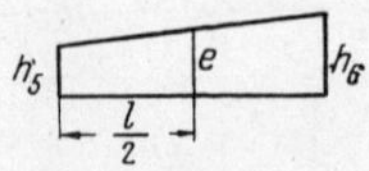
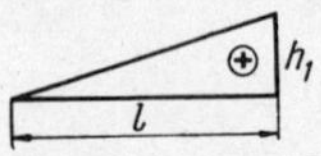	$\frac{lh_1h_2}{3}$	$\frac{lh_1}{6}(2h_6+h_5)$
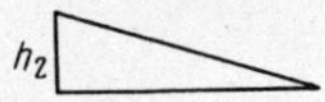	$\frac{lh_2^2}{6}$	$\frac{l}{6}h_2(2h_5+h_6)$
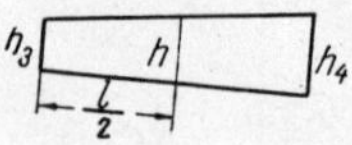	$\frac{l}{6}h_2(2h_4+h_3)$	$\frac{l}{6}[2(h_3h_5+h_4h_6)+$ $+h_3h_6+h_4h_5]$
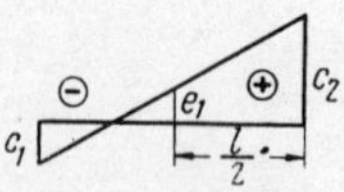	$\frac{l}{6}h_2(2c_2-c_1)$	$\frac{l}{6}[2(-c_1h_5+c_2h_6-$ $-c_1h_6+c_2h_5)]$
Conic parabola 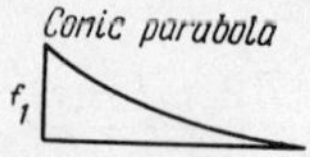	$\frac{lf_1h_2}{12}$	$\frac{lf_1}{12}(3h_5+h_6)$
Conic parabola 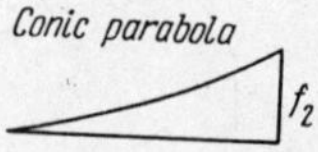	$\frac{lf_2h_2}{4}$	$\frac{lf_2}{12}(3h_6+h_5)$
Conic parabola 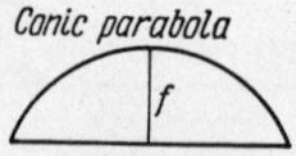	$\frac{lfh_2}{3}$	$\frac{lf}{3}(h_5+h_6)$
Conic parabola 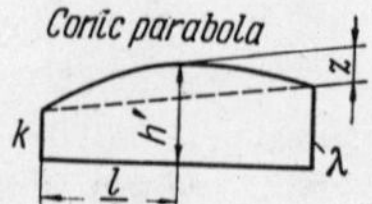	$\frac{l}{6}h_2(2h'+\lambda)$	$\frac{l}{6}(h_5k+4eh'+h_6\lambda)$

base length of each graph being equal to l)

Table 3.8

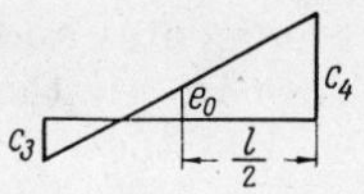	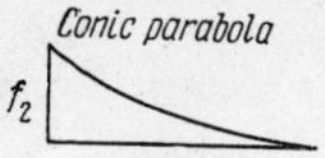	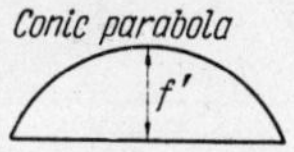	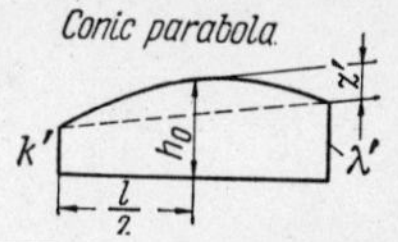
$\frac{lh_1}{6}(2c_4-c_3)$	$\frac{lf_2h_1}{12}$	$\frac{lf'h_1}{3}$	$\frac{lh_1}{6}(2h_0+\lambda')$
$\frac{l}{6}h_2(-2c_3+c_4)$	$\frac{lf_2h_2}{4}$	$\frac{lf'h_2}{3}$	$\frac{l}{6}h_2(k'+2h_0)$
$\frac{l}{6}[2(-h_3c_3+$ $+h_4c_4)+h_3c_4-$ $-h_4c_3]$	$\frac{lf_2}{12}(3h_3+h_4)$	$\frac{lf'}{3}(h_3+h_4)$	$\frac{l}{6}(h_3k'+4hh_0+$ $+h_4\lambda')$
$\frac{l}{6}[2(c_1c_3+c_2c_4-$ $-c_2c_3-c_1c_4)]$	$\frac{lf_2}{12}(-3c_1+c_2)$	$\frac{lf'}{3}(-c_1+c_2)$	$\frac{l}{6}(-c_1k'+$ $+4e_1h_0+c_2\lambda')$
$\frac{lf_1}{12}(-3c_3+c_4)$	$\frac{lf_1f_2}{5}$	$\frac{lf_1f'}{5}$	$\frac{lf_1}{60}[5(3k'+$ $+\lambda')+12z']$
$\frac{lf_2}{12}(3c_4-c_2)$	$\frac{lf_2^2}{30}$	$\frac{lf_2f'}{5}$	$\frac{lf_2}{60}[5(3\lambda'+k')+$ $+12z']$
$\frac{lf}{3}(-c_3+c_4)$	$\frac{lff_2}{5}$	$\frac{8}{15}lff'$	$\frac{lf}{15}[5(k'+\lambda')+$ $+8z']$
$\frac{l}{6}(-c_3k+4e_0h'+$ $+c_4\lambda)$	$\frac{lf_2}{60}[5(3k+\lambda)+$ $+12z]$	$\frac{lf'}{15}[5(k+\lambda)+$ $+8z]$	$\frac{l}{6}[2kk'+2\lambda\lambda'+$ $+k\lambda'+\lambda k'+2z\times$ $\times(k'+\lambda')+$ $+2z'(k+\lambda)+$ $+3.2zz']$

The deflections of hinged structures are computed using only that term of Mohr's equation which takes into account the normal stresses. It will be remembered that bending moments and shears remain nil in all the members of this class of structures. Consequently, Mohr's formula reduces to

$$\Delta_{mn} = \Sigma \int_0^i \overline{N}_m \frac{N_n\,dx}{EF}$$

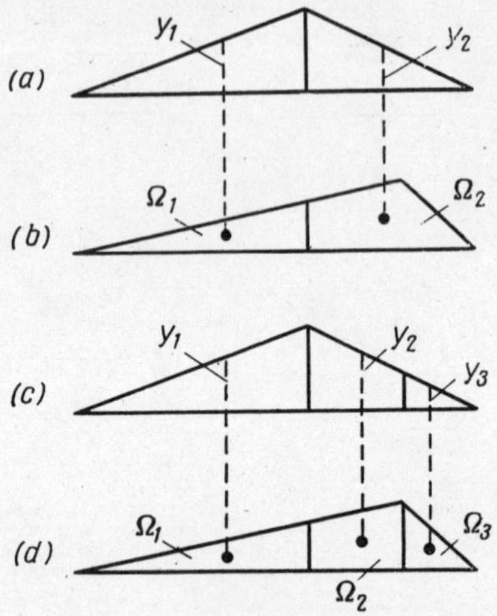

Fig. 28.8

The integrals must be calculated separately for the whole length of each member of the structure whereafter all of the values of these integrals must be summed up. In the great majority of cases the normal stresses $\overline{N}_m$, N_n, the cross-sectional area F and Young's modulus E will remain constant within the limits of each member in which case the above expression becomes

$$\Delta_{mn} = \Sigma \frac{\overline{N}_m N_n}{EF} \int_0^i dx = \Sigma \frac{\overline{N}_m N_n}{EF}\, l \qquad (22.8)$$

Thus the computation of the deflections of trusses and similar structures reduces to the summing up of the products $\frac{\overline{N}_m N_n}{EF}\, l$ calculated separately for each bar. These computations should be carried out in tabular form.

9.8. *Examples of Displacement Computation Using Vereshchagin's Method*

Problem 1. Required the deflection of point C of the beam appearing in Fig. 19.8*a*. The effect of both bending moments and shears should be accounted for.

Solution. The imaginary state of the beam as well as the graphs of the stresses induced by the load P_n and by the unit load P_m are represented in Fig. 19.8*b*, *c*, *d*, *e* and *f*. Using Vereshchagin's method we find

$$\Delta_{mn} = \frac{2}{EJ}\cdot\frac{P_n l}{4}\cdot\frac{l}{2}\cdot\frac{1}{2}\cdot\frac{2}{3}\cdot\frac{l}{4} + \frac{2\eta}{GF}\cdot\frac{P_n}{2}\cdot\frac{l}{2}\cdot\frac{1}{2} = \frac{P_n l^3}{48EJ} + \frac{P_n \eta l}{4GF}$$

This result coincides exactly with the one obtained in Art. 6.8 by direct integration.

Problem 2. Required the horizontal displacement of point C of the portal frame shown in Fig. 29.8a. The moments of inertia of all the members of the frame are indicated in the same figure, Young's modulus E remaining constant throughout.

Solution. The bending moment diagram corresponding to the actual loading is given in Fig. 29.8b. The imaginary loading consists of one horizontal load unity acting at point C. The corresponding bending moment graph

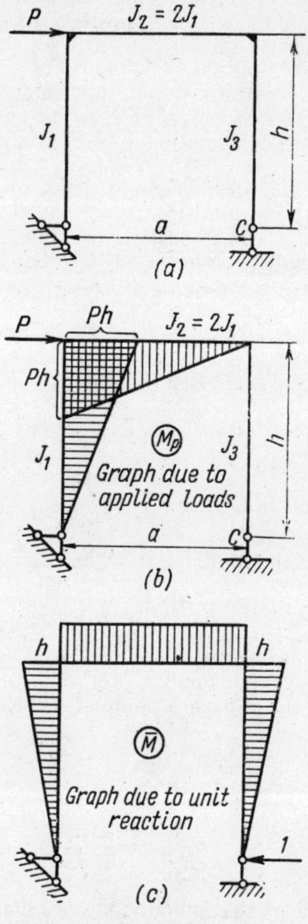

Fig. 29.8

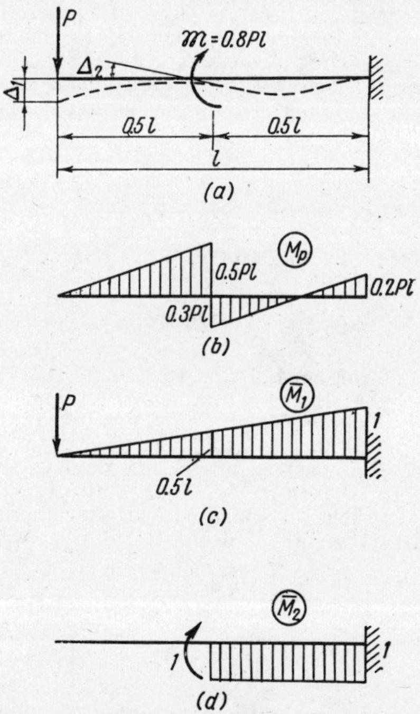

Fig. 30.8

is given in Fig. 29.8c. The signs of the bending moments appearing in these graphs may be omitted if desired for these graphs are always drawn on the side of the extended fibres. The displacement of point C will be obtained by multiplying all the ordinates to the bending moment graph corresponding to the actual loading by the ordinates to the graph due to the fictitious load unity. Using Vereshchagin's method and taking into account the different rigidities of the column and of the cross beam we find

$$\Delta_C = -Phh\cdot\frac{1}{2}\cdot\frac{2}{3}\cdot h\cdot\frac{1}{EJ_1} - Pha\cdot\frac{1}{2}\cdot h\cdot\frac{1}{EJ_2} =$$

$$= -\frac{Ph_3}{3EJ_1} - \frac{Pah^2}{2EJ_2} = -\frac{Ph^2}{EJ_1}\left(\frac{h}{3}+\frac{a}{4}\right)$$

The displacement thus found will be negative for the M_p and $\overline{M}$ graphs are situated on different sides of each member of the frame thus indicating that the bending moments M_p and $\overline{M}$ are of opposite signs. Hence the actual displacement of point C will occur in a direction opposite to the one adopted for the load unity, i.e., towards the right.

Problem 3. Required the deflection Δ_1 and the angular rotation Δ_2 of the beam with a built-in end appearing in Fig. 30.8*a*, these two displacements being due respectively to the application of a concentrated load P and of a moment $\mathfrak{M}$. The bending moment graphs corresponding to the actual loading are indicated in Fig. 30.8*b*.

Solution. Apply along the directions of the displacements required a unit load (Fig. 30.8*c*) and a unit moment (Fig. 30.8*d*) and trace the corresponding bending moment diagrams (Fig. 30.8*c* and *d*). The deflections and angular rotations will be computed using Mohr's formula together with Vereshchagin's method. The M_p graph will be first multiplied by the $\overline{M}_1$ graph and then by the $\overline{M}_2$ graph.

$$\Delta_1=\frac{1}{EJ}\left[0.5Pl\times 0.5l\times\frac{1}{2}\times\frac{2}{3}\times 0.5l+\frac{0.5l}{6}\times\right.$$

$$\left.\times(-2\times 0.3Pl\times 0.5l+2\times 0.2Pl\times l-0.3Pl\times l+0.2Pl\times 0.5l)\right]=\frac{Pl^3}{30EJ}$$

$$\Delta_2=\frac{1}{EJ}\left(0.3Pl\times 0.5l\times\frac{1}{2}\times 1-0.2Pl\times 0.5l\times\frac{1}{2}\times 1\right)=\frac{Pl^2}{40EJ}$$

Problem 4. Required the horizontal displacement of point D belonging to the structure represented in Fig. 31.8*a*.

Solution. The bending moment graph corresponding to the actual loading appears in Fig. 31.8*b*. Let us apply a load unity along the direction of the displacement required. The graph of the bending moments induced by this load is given in Fig. 31.8*c*. Multiplying the two graphs one by another using Vereshchagin's method and remembering that the separate members of the structure are of different rigidity, we obtain

$$\Delta_1=\frac{qa^2}{2}\cdot a\cdot\frac{1}{3}\cdot\frac{3a}{4}\cdot\frac{1}{EJ_1}+\frac{qa^2}{2}\cdot a\cdot a\cdot\frac{1}{EJ_1}+\frac{2a}{6}\times$$

$$\times\left(2\,\frac{qa^2}{2}\cdot a+2\cdot\frac{3qa^2}{2}\cdot a-\frac{qa^2}{2}\cdot a-\frac{3qa^2}{2}\cdot a\right)\frac{1}{EJ_2}=\frac{23qa^4}{24EJ_1}$$

The area of the M_p graph pertaining to element CD is bounded by a concave parabolic curve and therefore its area is equal to one third of the product of its base length by the maximum ordinate, i.e.,

$$\Omega_{CD}=\frac{1}{3}\cdot\frac{qa^2}{2}\cdot a=\frac{qa^3}{6}$$

The centroid of this graph is situated a distance $\frac{a}{4}$ as measured from point C (see Table 2.8). Hence the corresponding ordinate to the M graph equals $\frac{3a}{4}$.

Problem 5. Required the deflection Δ_C of a beam built in at its left extremity and carrying a uniformly distributed load (Fig. 32.8*a*).

Solution. Having drawn the M_p graph (Fig. 32.8*b*) apply a vertical load unity at point C and draw the corresponding bending moment diagram

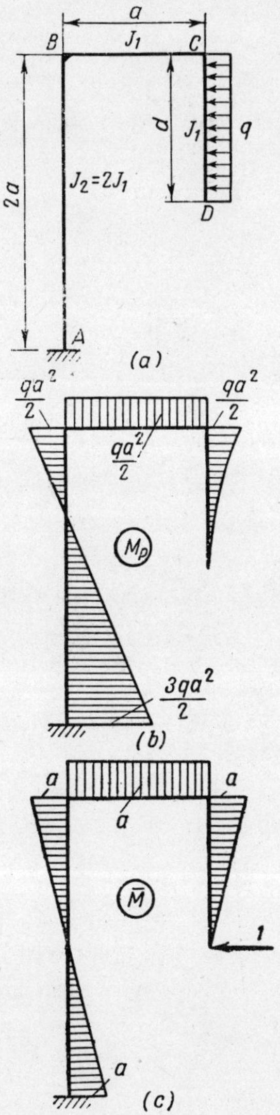

Fig. 31.8

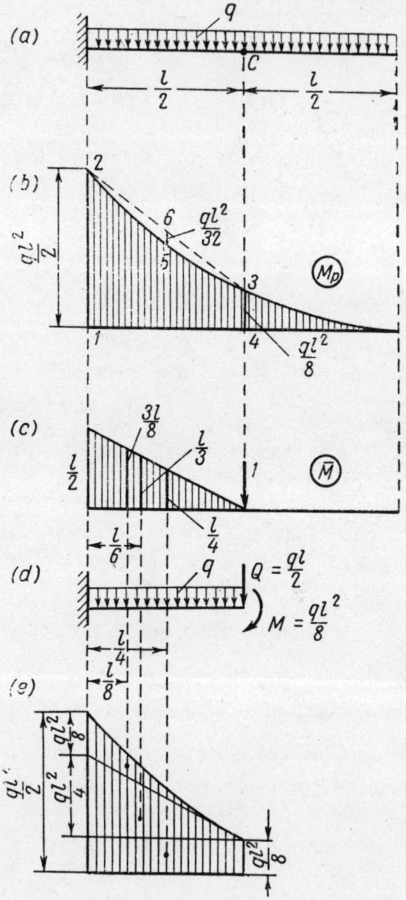

Fig. 32.8

(Fig. 32.8*c*). The value of the deflection Δ_C will be obtained multiplying the ordinates to the parabolic graph by these of the one bounded by a straight line. This operation may be carried out in two different ways:

1. The M_p graph for the left-hand part of the beam will be regarded as consisting of a trapezoid *1-2-6-3-4-1* with negative ordinates and a parabolic segment *2-6-3-5-2* with positive ordinates (Fig. 32.8*b*).

The maximum ordinate to the parabolic graph will equal

$$\frac{q\left(\frac{l}{2}\right)^2}{8}=\frac{ql^2}{32}$$

Multiplying the M_p graph by the $\overline{M}$ graph we obtain

$$\Delta_C=\frac{1}{EJ}\left[\frac{l}{2}\cdot\frac{l}{2}\cdot\frac{1}{2}\left(\frac{2}{3}\cdot\frac{ql^2}{2}+\frac{1}{3}\cdot\frac{ql^2}{8}\right)-\right.$$
$$\left.-\frac{ql^2}{32}\cdot\frac{l}{2}\cdot\frac{2}{3}\cdot\frac{1}{2}\cdot\frac{l}{2}\right]=\frac{17ql^4}{384EJ}$$

2. Isolate the left half of the beam replacing the action exerted by the right-hand part by a bending moment $M=\frac{ql^2}{8}$ and a shearing force $Q=\frac{ql}{2}$ (Fig. 32.8*d*). The graph of the bending moment M_p acting across the sections of the left-hand portion of the beam is shown in Fig. 32.8*e*.

As will be readily seen this graph and the graph *1-2-5-3-4-1* of Fig. 32.8*b* are absolutely identical. It can be subdivided into three parts as indicated in Fig. 32.8*e*, each of these parts representing respectively:

(a) a rectangular graph having for ordinate $\frac{ql^2}{8}$ due to the bending moment $M=\frac{ql^2}{8}$ applied to section C;

(b) a triangular graph due to the shearing force $Q=\frac{ql}{2}$ equally applied to section C. The maximum ordinate to this graph equals $\frac{ql^2}{4}$;

(c) a parabolic graph due to the uniformly distributed loads applied to the left half of the beam. The maximum ordinate to this graph totals $\frac{ql^2}{8}$.

Multiplying each of these three graphs by the bending moment graph due to the load unity we obtain

$$\Delta_C=\frac{1}{EJ}\left(\frac{ql^2}{8}\cdot\frac{l}{2}\cdot\frac{l}{4}+\frac{ql^2}{4}\cdot\frac{l}{2}\cdot\frac{1}{2}\cdot\frac{l}{3}+\frac{ql^2}{8}\cdot\frac{l}{2}\cdot\frac{1}{3}\cdot\frac{3l}{8}\right)=\frac{17ql^4}{384EJ}$$

Problem 6. Required the deflection of the structure appearing in Fig. 33.8*a* at load point. The left end of the horizontal bar is hinge-supported while its right extremity is suspended to a flexible wire. The moment of inertia of the beam equals J, the cross section of the wire F, Young's modulus of the material of both being E.

Solution. In the structure under consideration the beam will work only in bending and the wire in direct tension. Hence Mohr's formula for the beam will reduce to one term which contains the product of bending moments. For the wire Mohr's formula will consist of the term containing normal stresses. The total displacement of the load point will be therefore given by

$$\Delta=\int_0^l \frac{M_p\overline{M}\,dx}{EJ}+\int_0^a \frac{N_p\overline{N}\,dx}{EF}$$

Let us apply to the beam a unit load acting along the direction of the displacement required. The corresponding bending moment graph appears

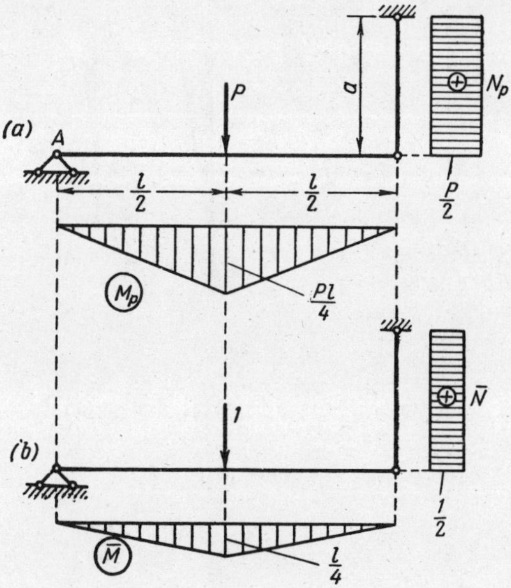

Fig. 33.8

in Fig. 33.8*b*. Computing the first term of the expression given above by Vereshchagin's method we obtain

$$\int_0^l \frac{M_p\overline{M}\,dx}{EJ}=\frac{1}{EJ}\cdot 2\left(\frac{1}{2}\cdot\frac{Pl}{4}\cdot\frac{l}{2}\right)\frac{2}{3}\cdot\frac{l}{4}=\frac{Pl^3}{48EJ}$$

The total tensile stresses N_p and $\overline{N}$ in this hanger due to load P and to the unit load will amount to $\frac{P}{2}$ and $\frac{1}{2}$, respectively. The corresponding graphs are given in Fig. 33.8*a* and *b*. The multiplication of these graphs gives

$$\int_0^a \frac{N_p\overline{N}\,dx}{EF}=\frac{P}{2}\cdot a\cdot\frac{1}{2}\cdot\frac{1}{EF}=\frac{Pa}{4EF}$$

The total deflection will be given by

$$\Delta = \frac{Pl^3}{48EJ} + \frac{Pa}{4EF}$$

Problem 7. Required the angular rotation of cross section Δ_B of the three-hinged frame appearing in Fig. 34.8*a*. The flexural rigidity EJ of all

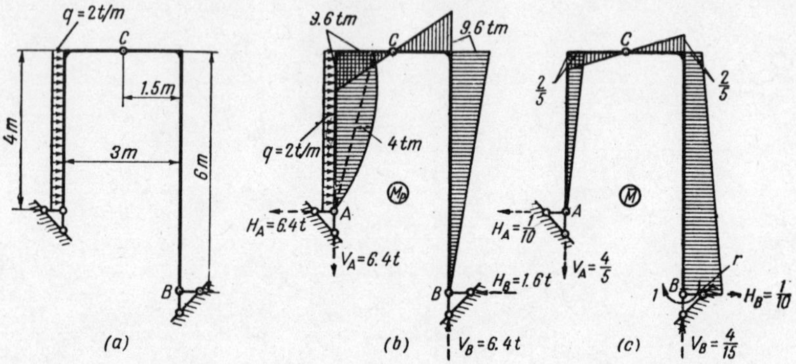

Fig. 34.8

the elements of this frame remains constant and equal to 25 × 10⁴ tons sq m.

Solution. The reactions at the supports of this frame will be derived from the usual equilibrium equations (Fig. 34.8*b*)

$$\sum_L M_C = H_A \times 4 - V_A \times 1.5 - q \times 4 \times \frac{4}{2} = 4H_A - 1.5V_A - 2 \times 4 \times \frac{4}{2} = 0$$

wherefrom

$$4H_A - 1.5V_A = 16 \tag{a}$$

$$\Sigma M_B = -H_A \times 2 - V_A \times 3 + q \times 4\left(\frac{4}{2} + 6 - 4\right) =$$

$$= -2H_A - 3V_A + 2 \times 4 \times 4 = 0$$

wherefrom

$$2H_A + 3V_A = 32 \tag{b}$$

$$\Sigma Y = -V_A + V_B = 0 \tag{c}$$

$$\Sigma X = -H_A - H_B + q \times 4 = -H_A - H_B + 2 \times 4 = 0$$

wherefrom

$$H_B = 8 - H_A \tag{d}$$

From equations (a) and (b) we obtain $V_A = 6.4$ tons and $H_A = 6.4$ tons. Equation (c) gives $V_B = 6.4$ tons and equation (d) gives $H_B = 1.6$ tons.

Knowing these reactions we can draw the bending moment diagram appearing in Fig. 34.8*b*. This being done, let us apply a unit moment at section B which will turn this section in the direction of the angular rotation required.

The reactions at the supports induced by this action will be obtained as follows (Fig. 34.8c):

$$\sum_L M_C = H_A \times 4 - V_A \times 1.5 = 0 \tag{e}$$

$$\Sigma M_B = -H_A \times 2 - V_A \times 3 + 1 = 0 \tag{f}$$

$$\Sigma Y = -V_A + V_B = 0 \tag{g}$$

$$\Sigma X = -H_A + H_B = 0 \tag{h}$$

Equations (e) and (f) yield $V_A = \frac{4}{15}$ and $H_A = \frac{1}{10}$ while equations (g) and (h) give $V_B = \frac{4}{15}$ and $H_B = \frac{1}{10}$.

Knowing these reactions we may draw the bending moment graph due to the unit moment (Fig. 34.8c).

The value of the angular rotation will be obtained multiplying the first of the graphs obtained as described above by the second one

$$\Delta_B = \frac{1}{EJ}\left[\frac{9.6\times 4}{2}\times\frac{2}{3}\times\frac{2}{5}+\frac{4\times 4\times 2}{3}\times\frac{1}{2}\times\frac{2}{5}+\right.$$

$$\left.+\frac{9.6\times 1.5}{2}\times\frac{2}{3}\times\frac{2}{5}\times 2+\frac{9.6\times 6}{2}\left(\frac{2}{3}\times\frac{2}{5}+\frac{1}{3}\times 1\right)\right]=$$

$$=\frac{1}{25\times 10^4}\left(\frac{76.8}{15}+\frac{32}{15}+\frac{57.6}{15}+\frac{259.2}{15}\right)=1.13\times 10^{-4} \text{ radians}$$

Problem 8. Required the vertical deflection Δ_5 of joint *5* of a steel truss represented in Fig. 35.8*a*. The cross sections of all the members of this truss

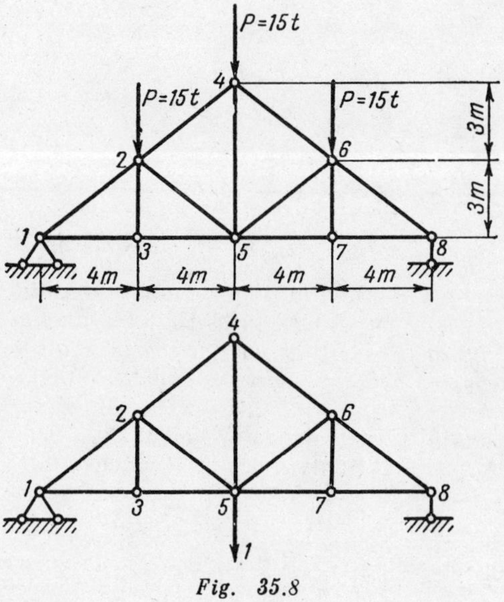

Fig. 35.8

remain constant and equal to $F = 125$ sq cm and Young's modulus $E = 2 \times 10^6$ kg/sq cm.

Solution. The diflection Δ_5 will be given by the expression

$$\Delta_5 = \Sigma \frac{N_p \bar{N}}{EF} l$$

All the necessary computations are tabulated hereunder (see Table 4.8). The entries in the last column of this table give the values of the product $N_p \bar{N} l$ for each bar in ton metres. Summing up all these entries and dividing the total by EF we shall obtain the displacement required

$$\Delta_5 = \frac{1}{2 \times 10^7 \times 0.0125} (156.25 \times 2 + 104.17 \times 2 + 80 \times 4 + 90 \times 1) = \\ = 3.72 \times 10^{-3} \text{ m} = 3.72 \text{ mm}$$

Table 4.8

Bar No.		Bar length l, m	Stresses due to actual loading N_p, tons	Stresses due to load unity $\bar{N}$	$N_p \bar{N} l$, ton metres
1-2;	*6-8*	5	—37.5	—0.833	156.25
2-4;	*4-6*	5	—25	—0.833	104.17
1-3;	*3-5*	4	30	0.667	80
5-7;	*7-8*				
2-3;	*6-7*	3	0	0	0
2-5;	*5-6*	5	—12.5	0	0
4-5		6	15	1	90

10.8. *Strain Energy Method of Displacement Computation*

The strain energy method of displacement computation is based on *Castilliano's theorem* which states that *the partial derivative of the strain energy in terms of the unit action is equal to the displacement induced by the actual loading along the direction of the said unit action.*

In order to simplify the demonstration of this theorem we shall retain in Mohr's formula only one term which contains the bending moments (see Art. 3.8)

$$W = \Sigma \int_0^l \frac{M^2\, dx}{2EJ}$$

Let us express the bending moment as follows

$$M=\overline{M}_1P_1+\overline{M}_2P_2+\ldots+\overline{M}_kP_k+\ldots+\overline{M}_nP_n$$

In this expression $\overline{M}_1$, $\overline{M}_2$, . . ., $\overline{M}_k$, $\overline{M}_n$ are the bending moments due to unit forces $P_1=1$, $P_2=1$, . . ., $P_k=1$, $P_n=1$ whose lines of action coincide with those of the corresponding applied loads.

The partial derivative of W in terms of P_k will be

$$\frac{\partial W}{\partial P_k}=\frac{\partial}{\partial P_k}\left(\Sigma\int\limits_0^l \frac{M^2\,dx}{2EJ}\right)=\Sigma\int\limits_0^l \frac{M\frac{\partial M}{\partial P_k}\,dx}{EJ}$$

but $\frac{\partial M}{\partial P_k}=\overline{M}_k$ and consequently $\frac{\partial W}{\partial P_k}=\Sigma\int\limits_0^l \frac{\overline{M}_k M\,dx}{EJ}$

As shown in Art. 6.8, the right-hand part of this expression represents Δ_{kp} and therefore

$$\frac{\partial W}{\partial P_k}=\Delta_{kp}$$

In actual design practice Castilliano's theorem is seldomly used, its interest being mainly theoretic. If it were desired however to use it the sequence of operations would be as follows:

1. A certain action should be applied to the system under consideration along the direction of the displacement required.
2. The strain energy due to the actual loading and to this unit action should be calculated.
3. The expression of the strain energy should be differentiated in terms of the said unit action. The magnitude of the displacement will be then obtained from this expression reducing to zero the magnitude of the unit action for in reality this action is nonexistent.

In the particular case when one of the load points coincides with the cross section the deflection of which is sought, it will be the value of this particular load that will be introduced into the partial derivative.

Problem. Required the angular rotation of the free end of a uniformly loaded beam with a built-in end (Fig. 36.8*a*).

Solution. Let us apply to the free end of the beam a couple $\mathfrak{M}$ as shown in Fig. 36.8*b*. The value of the bending moment at a cross section situated a distance x from the free end will be given by

$$M=-\left(q\frac{x^2}{2}+\mathfrak{M}\right)$$

The corresponding strain energy will amount to

$$W=\int_0^l \frac{M^2\,dx}{2EJ}=\int_0^l \frac{\left(q\frac{x^2}{2}+\mathfrak{M}\right)^2}{2EJ}\,dx$$

which gives after integration

$$W=\frac{1}{EJ}\left(\frac{q^2l^5}{40}+\frac{ql^3\mathfrak{M}}{6}+\frac{\mathfrak{M}^2l}{2}\right)$$

Differentiating W in terms of $\mathfrak{M}$ and reducing thereafter its value to zero we obtain

$$\varphi=\left(\frac{\partial W}{\partial \mathfrak{M}}\right)_{\mathfrak{M}=0}=\frac{1}{EJ}\left(\frac{ql^3}{6}+\mathfrak{M}l\right)=\frac{ql^3}{6EJ}$$

As already stated, when one of the loads is actually applied at the point the displacement of which is desired to determine there is no need to apply

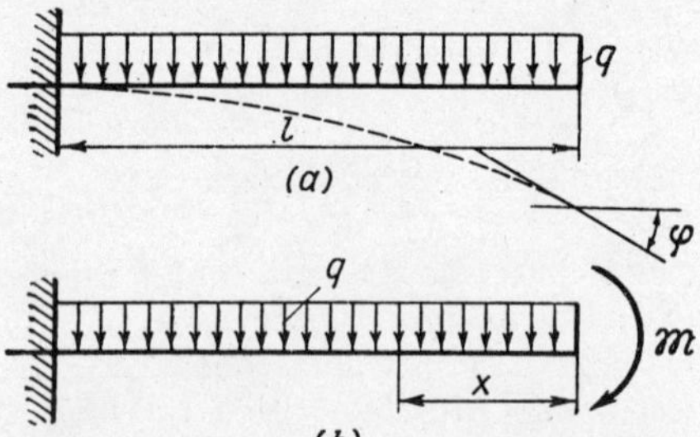

Fig. 36.8

imaginary loads. Assume, for instance, that it is required to determine the maximum deflection of a beam built in at one end and carrying a concentrated load at the other. In that case

$$M=-Px$$

$$W=\int_0^l \frac{P^2x^2}{2EJ}\,dx=\frac{P^2l^3}{6EJ}$$

$$\Delta=\frac{\partial W}{\partial P}=\frac{Pl^3}{3EJ}$$

11.8. *Elastic Loads Method*

The method described hereunder permits the determination of deflections and angular rotations at a certain number of isolated points of the structure. Increasing their number, the elastic curve of the deflected system will be obtained with a precision increasing in direct proportion to the number of points considered. The curve so obtained might be also termed the displacement graph of the system.

Indeed, if the values of the deflections determined at a certain number of points were set out along the verticals passing through these points, the broken line connecting the ordinates so obtained would constitute an approximation of the elastic curve. The deflections of all the intermediate points might be obtained with a certain degree of approximation by measuring the ordinates to this broken line, for in reality the elastic line of a member will be a smooth curve. The above does not apply to hinged structures

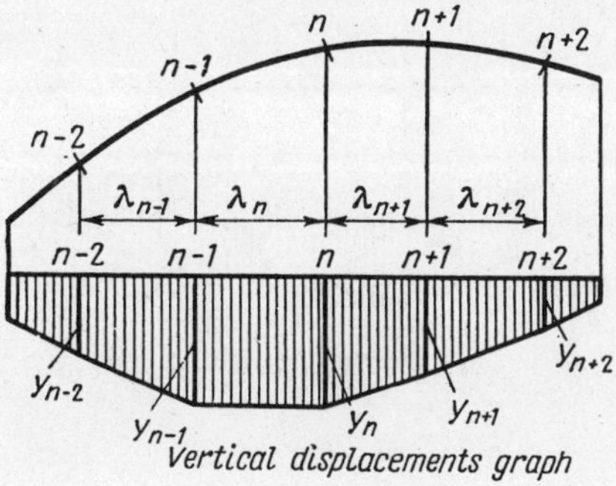

Fig. 37.8

such as trusses, for as long as the loads act at the joints all the bars remain straight and consequently the deformed axis of the lower (or upper) chord will follow a broken line connecting the diflected joints. If the axis of the displacement graph is normal to the direction of the deflections, such a graph will resemble very closely a bending moment curve of an end-supported beam acted upon by several concentrated loads. It is this resemblance which forms the basis of the method described hereunder.

Fig. 37.8 represents a part of some structure for which it is required to find the deflections at a certain number of points. Assume that the broken line of Fig. 38.8*a* represents the bending moment graph. Each apex of this graph will lie in the vertical passing through one of the load points. Let us find the magnitudes of these loads. For this purpose we shall compute the shearing forces Q_n and Q_{n+1} acting at the extremities of the stretches λ_n and λ_{n+1}. Using the theorem of Zhuravsky we may write

$$Q_n = \frac{M_n - M_{n-1}}{\lambda_n}; \quad Q_{n+1} = \frac{M_{n+1} - M_n}{\lambda_{n+1}}$$

Let us pass two sections through the beam one immediately to the right and the other immediately to the left of point n. The element isolated by these two sections is represented in Fig. 38.8*b*. The end faces of this element are acted upon by the shearing forces Q_n and Q_{n+1} reckoned positive. Projecting all the forces applied to this element on a vertical we obtain

$$\Sigma Y = Q_n - P_n - Q_{n+1} = 0$$

wherefrom

$$P_n = Q_n - Q_{n+1}$$

The latter expression shows that the load acting at point n is equal to the difference between the shearing forces Q_n and Q_{n+1}.

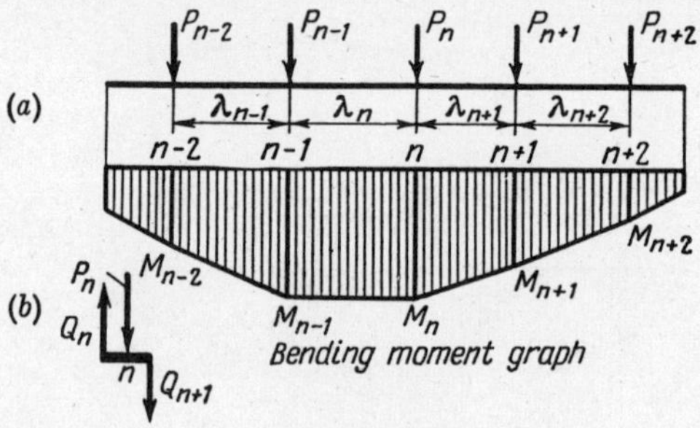

Fig. 38.8

It is clear therefore that the concentrated loads acting at points $(n+1)$ and $(n-1)$ will also amount to

$$P_{n-1} = Q_{n-1} - Q_n$$
$$P_{n+1} = Q_{n+1} - Q_{n+2}$$

Introducing into these expressions the values of the shearing forces in terms of the bending moments we obtain

$$P_n = \frac{M_n - M_{n-1}}{\lambda_n} - \frac{M_{n+1} - M_n}{\lambda_{n+1}} = -M_{n-1}\frac{1}{\lambda_n} +$$
$$+ M_n\left(\frac{1}{\lambda_n} + \frac{1}{\lambda_{n+1}}\right) - M_{n+1}\frac{1}{\lambda_{n+1}} \qquad (23.8)$$

It follows that if the beam is subjected to the action of the concentrated loads P_n calculated as above, its bending moment graph will coincide exactly with the deflection graph of the structure under consideration. Since the two curves given in Figs. 37.8 and 38.8*a* are identical, let us replace in the expression of P_n the values of M_{n-1}, M_n and M_{n+1} by the corresponding deflections y_{n-1}, y_n and y_{n+1}.

We shall thus obtain the expression of the so-called elastic loads W_n which will induce in an end-supported beam a bending moment curve coinciding with the approximate deflection graph of the actual structure. The value of these elastic loads will be thus given by

$$W_n = -y_{n-1}\frac{1}{\lambda_n} + y_n\left(\frac{1}{\lambda_n} + \frac{1}{\lambda_{n+1}}\right) - y_{n+1}\frac{1}{\lambda_{n+1}} \qquad (24.8)$$

At first sight this expression appears thoroughly unfit for practical use. Indeed the values of the elastic loads given by this expression depend themselves on the unknown deflections.

However, it is possible to obtain an expression of the elastic loads in terms of the external forces acting on the structure. The procedure is as follows. Let us designate by the term *actual state* the state of the structure characterized dy the actual or existing loading and let us apply at points $(n-1)$, n and $(n+1)$ two couples both equal to unity but of opposite sign. Let us assume also that each of these couples is constituted by two vertical forces amounting to $\frac{1}{\lambda_n}$ for the first one and to $\frac{1}{\lambda_{n+1}}$ for the second.

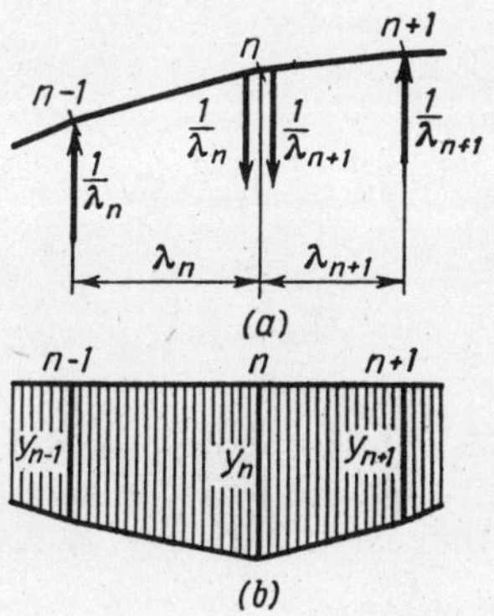

Fig. 39.8

The corresponding state of the structure (Fig. 39.8*a*) will be designated by the term *imaginary state*. Taking up expression (24.8) we note that its right-hand part represents the work performed by the imaginary loads along the deflections caused by the actual ones.

Indeed, the product of the vertical load $\frac{1}{\lambda_n}$ acting at point $(n-1)$ by the deflection y_{n-1} (Fig. 39.8*a* and *b*) represents the work accomplished by this load along the deflection of this point caused by the actual loading. The negative value of this term is due to the fact that the force $\frac{1}{\lambda_n}$ is directed upwards whilst the deflection of point $(n-1)$ is directed downwards. Similarly, the product $y_n\left(\frac{1}{\lambda_n} + \frac{1}{\lambda_{n+1}}\right)$ represents the work produced by the forces $\frac{1}{\lambda_n}$ and $\frac{1}{\lambda_{n+1}}$ of the imaginary state acting at point n along the deflection y_n due to the actual loads. The third term of expression (24.8) is again equal to the work performed by the force $\frac{1}{\lambda_{n+1}}$ along the deflection y_{n+1}, the minus sign showing that the directions of the force and of this deflection are directly opposed.

Let us now express the work accomplished by the forces which form the imaginary couples along the actual displacements in terms of the stresses $\overline{M}$, $\overline{N}$ and $\overline{Q}$ induced dy these unit couples and of the stresses M_p, N_p and Q_p due to the applied loads. Using expression (12.8) we may write

$$-\frac{1}{\lambda_n} y_{n-1} + \left(\frac{1}{\lambda_n} + \frac{1}{\lambda_{n+1}}\right) y_n - \frac{1}{\lambda_{n+1}} \cdot y_{n+1} =$$

$$= \Sigma \int_0^l \overline{M} \frac{M_p\, dx}{EJ} + \Sigma \int_0^l \overline{N} \frac{N_p\, dx}{EF} + \eta \Sigma \int_0^l \overline{Q} \frac{Q_p\, dx}{GF}$$

and consequently

$$W_n = \Sigma \int_0^l \overline{M} \frac{M_p\, dx}{EJ} + \Sigma \int_0^l \overline{N} \frac{N_p\, dx}{EF} + \eta \Sigma \int_0^l \overline{Q} \frac{Q_p\, dx}{GF} \qquad (25.8)$$

This latter expression constitutes the *general equation* giving the elastic loads in terms of the internal stresses. When used for the computation of deflections of beams and rigid frames, this expression is considerably simplified as only the term containing the bending moments must be retained. In the case of flat arches, the normal stresses must be also accounted for, while the shearing forces are taken into consideration only in a few particular cases. The deflections of trusses and other hinge-connected structures are computed using only the term containing direct stresses.

The values of the elastic loads corresponding to different points of the structure are obtained through the application of imaginary unit couples successively to two neighbouring elements of the structure. Once the values of these elastic loads are known, the deflections are readily calculated using the following procedure. The loads just mentioned are applied to an imaginary beam of appropriate length and rigidity and the bending moment curve is drawn in the usual way. The ordinates to this curve will be numerically equal to the deflections required. The choice of the beam mentioned above is governed by the following considerations:

1. To each point of the real structure which remains fixed there must correspond a point in the beam where the bending moment induced by the elastic loads is nil. On the other hand, to any deflected point of the structure there must correspond a cross section of the beam where the bending moment differs from zero.

2. Wherever the slope of the deflected axis of the real structure varies or, in other words, wherever two adjacent cross sections of

the beam rotate one with reference to the other, the corresponding cross sections of the imaginary beam must be acted upon by shearing forces induced by the elastic loads.

Thus, in the case of a beam built in at one of its ends ($y_a = 0$ and $\varphi_a = 0$) and free at the other end ($y_b \neq 0$, $\varphi_b \neq 0$) we must have in the imaginary beam

$$M_a^c = 0 \text{ and } Q_a^c = 0$$

At the other end of the imaginary beam the values of the bending moment M_b^c and of the shearing force Q_b^c must be on the contrary different from zero.

Table 5.8 contains the schematic drawings of conjugate imaginary beams corresponding to the structures represented in the left-hand column.

Table 5.8

No.	Real structure	Imaginary beam
1		
2		
3		
4		
5		
6	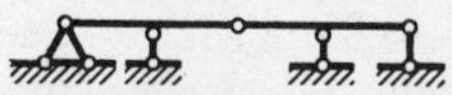	

The following sequence should be adopted for the construction of the displacement graphs using the method under consideration:

1. Begin with the determination of the M_p, N_p and Q_p graphs induced in the real structure by the actual loading.
2. Choose such points of the structure whose deflections will be characteristic for the structure as a whole.
3. Apply successively to the adjacent points chosen as above two unit couples, the direction of forces constituting these couples being parallel to those of the deflections required.
4. Draw the $\overline{M}$, $\overline{N}$ and $\overline{Q}$ graphs induced by the said unit couples (Fig. 39.8).

5. Compute the values of the elastic loads either by direct integration or using Vereshchagin's method described previously.

6. Choose an imaginary beam in conformity with the character of the deformations of the real structure.

7. Trace the diagrams of the bending moments induced in this beam by the elastic loads. These loads will be reckoned positive when they are of the same direction as the adjacent forces forming two neighbouring unit couples and the bending moment graph will be always traced on the side of the extended fibres.

The ordinates to the bending moment diagram thus obtained will be equal both in amount and in direction to the deflections of the real structure.

The elastic loads acting on the imaginary beam at its supports have no influence on the corresponding bending moment graph and therefore their computation becomes unnecessary.

12.8. *Simplified Expression of Elastic Loads for Beams and Rigid Frames*

The determination of the deflection line for solid web structural members is carried out by subdividing the total length of such members in a series of short stretches, for which it may be admitted that the unit stresses remain constant.

Let us consider two adjacent stretches meeting at point n (Fig. 40.8*a*). The bending moment curves due to the applied loads are as usual drawn on the side of the extended fibres. Normal stresses are considered constant and positive within the boundaries of each stretch.

In order to find the elastic load W_n let us apply to the system two couples consisting of vertical forces $\frac{1}{\lambda_n}$ and $\frac{1}{\lambda_{n+1}}$ (Fig. 40.8*b*). Incidentally this means that the elastic load will be also directed vertically. The direction in which each of the two couples tends to rotate the corresponding stretch must produce an extension in the member on the same side as produced by the actual loading. The forces constituting these unit couples will lead to the appearance in each stretch of normal forces equal to:

(1) within the stretch between points $(n—1)$ and n

$$\overline{N}_n = -\frac{1}{\lambda_n}\sin\beta_n$$

where

$$\lambda_n = S_n\cos\beta_n$$

and consequently

$$\overline{N}_n = -\frac{\sin\beta_n}{S_n\cos\beta_n} = -\frac{\tan\beta_n}{S_n}$$

(2) within the stretch between points n and $(n+1)$

$$\overline{N}_{n+1}=\frac{1}{\lambda_{n+1}}\sin\beta_{n+1}$$

where

$$\lambda_{n+1}=S_{n+1}\cos\beta_{n+1}$$

wherefrom

$$\overline{N}_{n+1}=\frac{\sin\beta_{n+1}}{S_{n+1}\cos\beta_{n+1}}=\frac{\tan\beta_{n+1}}{S_{n+1}}$$

The multiplication of the bending moment graph due to the actual loading (Fig. 40.8*a*) by the bending moment graph due to

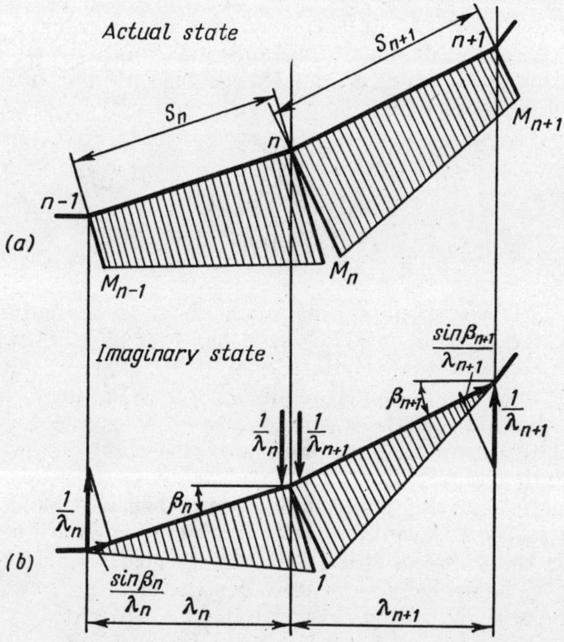

Fig. 40.8

the unit couples (Fig. 40.8*b*) carried out by Vereshchagin's method provides the following expression for the elastic loads

$$W_n^M=\Sigma\int_0^s\overline{M}M\frac{dx}{EJ}=\frac{S_n}{6EJ_n}(M_{n-1}+2M_n)+\frac{S_{n+1}}{6EJ_{n+1}}(2M_n+M_{n+1}) \tag{26.8}$$

The above expression accounts only for bending moments. In order to take care of the normal stresses let us compute the value of the integral

$$W_n^N = \Sigma \int_0^s \overline{N} N \frac{dx}{EF} = -\frac{\tan \beta_n}{S_n} \frac{N_n S_n}{EF_n} + \frac{\tan \beta_{n+1}}{S_{n+1}} \times$$

$$\times \frac{N_{n+1} S_{n+1}}{EF_{n+1}} = -\frac{N_n}{EF_n} \tan \beta_n + \frac{N_{n+1}}{EF_{n+1}} \tan \beta_{n+1} =$$

$$= -\varepsilon_n \tan \beta_n + \varepsilon_{n+1} \tan \beta_{n+1}$$

In this expression ε_n and ε_{n+1} are the unit strains of elements S_n and S_{n+1} caused dy the normal forces N_n and N_{n+1}.

Thus, the total value of the elastic load taking into account both bending moments and normal stresses will be given by

$$W_n = \frac{S_n}{6EJ_n} (M_{n-1} + 2M_n) + \frac{S_{n+1}}{6EJ_{n+1}} (2M_n + M_{n+1}) - \frac{N_n}{EF_n} \tan \beta_n +$$

$$+ \frac{N_{n+1}}{EF_{n+1}} \tan \beta_{n+1} \tag{27.8}$$

It will be observed that it is much easier to compute the elastic load using expression (27.8) as it becomes possible to dispense with a number of intermediate operations. Thus, there will be no longer any need to apply to the structure the unit couples, to trace the diagrams of the corresponding stresses and to carry out the multiplication of the graphs due to the actual loading and to the said unit couples.

The elastic load computed as just described will have the same direction as the adjacent forces of two neighbouring unit couples as long as the value of this load remains positive. If the normal stresses may be neglected and provided the bending moment graph due to the actual loading does not change sign within the length of elements S_n and S_{n+1}, the elastic load W_n will be directed towards the bending moment curve.

Problem. Required the deflection line of the cantilever beam of Fig. 41.8 supporting at its free end a concentrated load P.

Solution. Subdivide the beam in two equal parts choosing points *0*, *1* and *2* at the ends of these parts. Trace the bending moment graph due to the actual loading on the side of the extended fibres. Using expression (26.8) determine the magnitude of the elastic loads at points *0* and *1*. It will serve no useful purpose to determine the elastic load at point *2*, this load having no influence on the stresses in the imaginary beam. In computing the magnitude of the elastic load at point *0* it is assumed that the built-in end is replaced by a stretch of infinite rigidity.

$$W_0=\frac{\lambda_0}{6EJ_0}(M_{-1}+2M_0)+\frac{\lambda_1}{6EJ_1}(2M_0+M_1)=\frac{\lambda_0}{\infty}(M_{-1}+2M_0)+$$

$$+\frac{l}{12EJ}\left(2Pl+\frac{Pl}{2}\right)=\frac{5Pl^2}{24EJ}$$

The elastic load corresponding to point *1* will be

$$W_1=\frac{\lambda_1}{6EJ_1}(M_0+2M_1)+\frac{\lambda_2}{6EJ_2}(2M_1+M_2)=\frac{l}{12EJ}\left(Pl+2\frac{Pl}{2}\right)+$$

$$+\frac{l}{12EJ}\left(2\frac{Pl}{2}+0\right)=\frac{Pl^2}{4EJ}$$

The values of the elastic loads being calculated, apply these two loads at points *0* and *1* of the imaginary beam built in at its right-hand extremity (Fig. 42.8) and construct the corresponding bending moment diagram. The

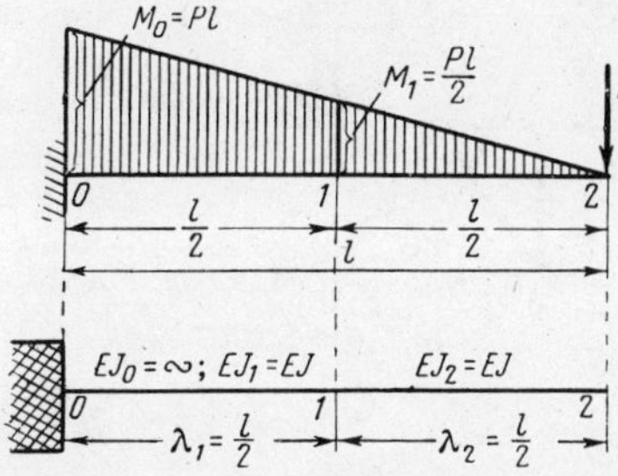

Fig. 41.8

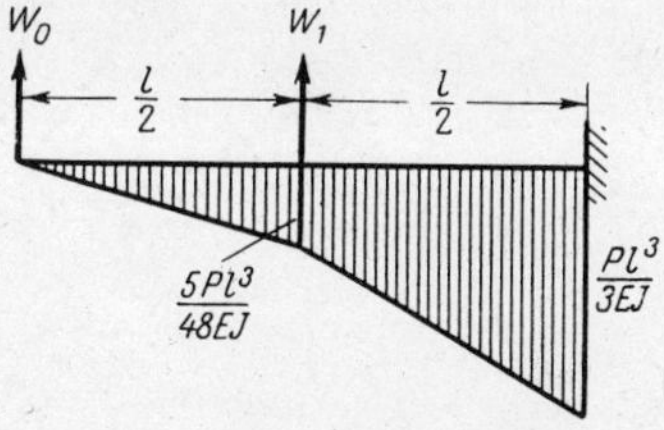

Fig. 42.8

ordinates of this diagram will be reckoned positive when situated on the side of the extended fibres. The elastic loads are directed upwards, in other words, towards the bending moment curve due to the actual loading, and therefore the bending moments at points *0*, *1* and *2* of the imaginary beam will have the following values

$$M_0^c=0;\quad M_1^c=W_0\lambda_1=\frac{5Pl^3}{48EJ};\quad M_2^c=W_0l+W_1\frac{l}{2}+W_2\times 0=\frac{Pl^3}{3EJ}$$

The graph of the bending moments induced in the imaginary beam by the elastic loads is given in Fig. 42.8. It represents at the same time the deflection graph of the real beam. At points *0*, *1* and *2* the deflections of the real beam will coincide exactly with the deflections represented by the ordinates to the above graph, whilst at intermediate points there will be a slight difference between the two. If the real beam were subdivided into a greater number of parts, the deflection curve of this beam would have been obtained with greater precision.

13.8. *Simplified Expression of Elastic Loads for Hinge-Connected Structures*

When applied to hinge-connected structures expression (25.8) becomes

$$W_n = \Sigma \overline{N} \frac{N_p l}{EF} = \Sigma \overline{N} \Delta l_p \tag{28.8}$$

In this expression $\overline{N}$ represents the normal stress due to the unit couples applied to the bars meeting at the joint n for which the value of the elastic load is sought, while Δl_p represents the total

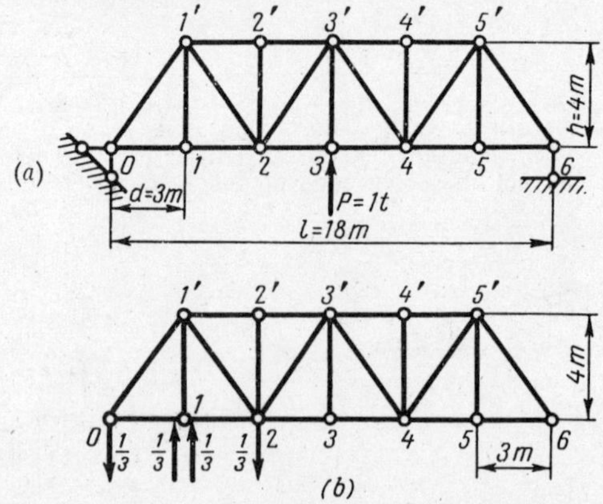

Fig. 43.8

strain of these bars caused by the actual loading. The application of the elastic loads method to the deflection computation for a truss is illustrated in the following example.

Let us assume that it is required to determine the deflection line of the lower chord of a truss represented in Fig. 43.8*a*. The truss is acted upon by a single vertical load $P = 1$ ton acting at joint *3* and directed upwards. The cross sections of all the members of the truss are the same. Let us compute the values of the elastic loads which must be applied to the imaginary beam at points corresponding to joints *1*, *2*, *3*, *4* and *5* of the lower chord. For this purpose let us apply unit couples successively to each two bars of the lower chord meeting at a joint.

If it were desired to find the deflection line of the upper chord the unit couples should be applied to the bars of this chord. Similarly, the construction of a displacement graph for points situated along the broken line *0-1'-2-3'-4-5'-6* would necessitate the determination of the elastic loads acting at joints *1'*, *2*, *3'*, *4* and *5'*.

Let us proceed with the determination of elastic load W_1. Incidentally, this load represents the angular rotation of bar *0-1* with reference to bar *1-2*. In order to find the magnitude of this load let us apply to bars *0-1* and *1-2* unit couples consisting of forces $\frac{1}{d}=\frac{1}{3}$ (Fig. 43.8*b*) and compute the normal stresses induced in the bars by these couples. It is readily seen that all the bars excepting bars *0-1*, *0-1'*, *1'-2*, *1-2* and *1-1'* will remain idle. The reactions at the ends of the truss will also remain nil. Stresses in the loaded bars are given in Table 6.8.

Table 6.8

Bar No.	Total stress	Bar No.	Total stress
0-1	$-\frac{1}{4}$	*1'-2*	$+\frac{5}{12}$
1-2	$-\frac{1}{4}$	*1-1'*	$-\frac{2}{3}$
0-1'	$+\frac{5}{12}$		

Stresses in all the members of the truss due to the application of the load P are given in Table 7.8. This table contains also all the necessary information regarding the length and the cross-sectional areas of the bars.

Table 7.8

Bar No.	Length of bar, m	Cross section, sq m	Stress, tons	Bar No.	Length of bar, m	Cross section, sq m	Stress, tons
0-1; 5-6	3	F	$-\frac{3}{8}$	*2'-3'; 3'-4'*	3	F	$+\frac{3}{4}$
1-2; 4-5	3	F	$-\frac{3}{8}$	*2-3'; 3'-4*	5	F	$+\frac{5}{8}$
0-1'; 5'-6	5	F	$+\frac{5}{8}$	*2-3; 3-4*	3	F	$-\frac{9}{8}$
1'-2; 4-5'	5	F	$-\frac{5}{8}$	*3-3';*	4	F	-1
				1-1'; 5-5'	4	F	0
1'-2'; 4'-5'	3	F	$+\frac{3}{4}$	*2-2'; 4-4'*	4	F	0

The data contained in these two tables permits the computation of the elastic load W_1

$$W_1=\Sigma\overline{N}N_p\frac{l}{EF}=\frac{1}{EF}(\overline{N}_{01}N_{01}l_{01}+\overline{N}_{12}N_{12}l_{12}+\overline{N}_{01'}N_{01'}N_{01'}+$$

$$+\overline{N}_{1'2}N_{1'2}l_{1'2}+\overline{N}_{11'}N_{11'}l_{11'}=\frac{1}{EF}\left[-\frac{1}{4}\left(-\frac{3}{8}\right)\times 3+\left(-\frac{1}{4}\right)\times\right.$$

$$\times\left(-\frac{3}{8}\right)\times 3+\frac{5}{12}\times\frac{5}{8}\times 5+\frac{5}{12}\left(-\frac{5}{8}\right)\times 5+\left(-\frac{2}{3}\right)\times$$

$$\left.\times 0\times 4\right]=+\frac{9}{16EF}$$

Since the system is completely symmetrical, elastic load W_5 will have the same value

$$W_5=W_1=+\frac{9}{16EF}$$

The value of the elastic load acting at joint *2* will be obtained applying the unit couples to bars *1-2* and *2-3* (Fig. 44.8*a*).

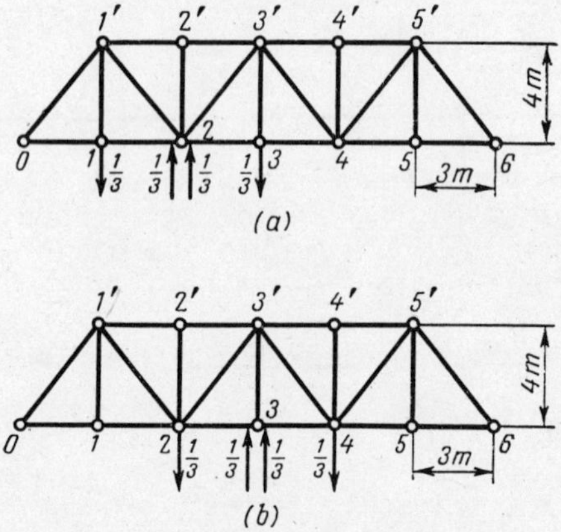

Fig. 44.8

In this case stresses will be developed in bars *1′-2′*, *2′-3′*, *1′-2*, *2-3′*, *1-1′* and *3-3′*, all the other bars remaining idle.

The values of these stresses are given in Table 8.8.

All the computations relative to elastic load W_2 are carried out in Table 9.8 using data contained in Tables 7.8 and 8.8.

The final value of W_2 will be obtained by summing up all the entries of the last column of Table 9.8

$$W_2=-\frac{5}{24EF}$$

Table 8.8.

Bar No.	Stress	Bar No.	Stress	Bar No.	Stress
1'-2	$-\frac{5}{12}$	*2'-3'*	$+\frac{1}{4}$	*1-1'*	$+\frac{1}{3}$
1'-2'	$+\frac{1}{4}$	*2-3'*	$-\frac{5}{12}$	*3-3'*	$+\frac{1}{3}$

Table 9.8.

Bar No.	$\frac{l}{EF}$	Stresses $\overline{N}$ induced by unit couples	Stresses N_p induced by actual loading	$\frac{\overline{N}N_p l}{EF}$
1'-2	$\frac{5}{EF}$	$-\frac{5}{12}$	$-\frac{5}{8}$	$+\frac{125}{96EF}$
1'-2'	$\frac{3}{EF}$	$+\frac{1}{4}$	$+\frac{3}{4}$	$+\frac{9}{16EF}$
2'-3'	$\frac{3}{EF}$	$+\frac{1}{4}$	$+\frac{3}{4}$	$+\frac{9}{16EF}$
2-3'	$\frac{5}{EF}$	$-\frac{5}{12}$	$+\frac{5}{8}$	$-\frac{125}{96EF}$
3-3'	$\frac{4}{EF}$	$+\frac{1}{3}$	-1	$-\frac{4}{3EF}$

Owing to the symmetry of the system, the elastic load W_4 will have the same value. Negative values of these two loads indicate that the mutual rotation of bars *1-2* and *2-3* occurs in a direction opposite to the one of the unit couples. In other words, bar *1-2* will rotate clockwise with respect to bar *2-3*.

It remains to find the value of the last elastic load W_3. To this end let us apply unit couples to bars *2-3* and *3-4* (Fig. 44.8*b*) repeating all the computations in the same order as heretofore. These computations appear in the appropriate columns of Table 10.8.

Adding up all the entries of the last column of this table we obtain

$$W_3 = \frac{167}{24EF}$$

Table 10.8

Bar No.	$\frac{l}{EF}$	Stresses $\overline{N}$ induced by unit couples	Stresses N_p induced by actual loading	$\frac{\overline{N}N_p l}{EF}$
2-3′	$\frac{5}{EF}$	$+\frac{5}{12}$	$+\frac{5}{8}$	$+\frac{125}{96EF}$
4-3′	$\frac{5}{EF}$	$+\frac{5}{12}$	$+\frac{5}{8}$	$+\frac{125}{96EF}$
2-3	$\frac{3}{EF}$	$-\frac{1}{4}$	$-\frac{9}{8}$	$+\frac{27}{32EF}$
3-4	$\frac{3}{EF}$	$-\frac{1}{4}$	$-\frac{9}{8}$	$+\frac{27}{32EF}$
3-3′	$\frac{4}{EF}$	$-\frac{2}{3}$	-1	$+\frac{8}{3EF}$

Elastic loads W_1, W_3 and W_5 being positive, these loads will be directed upwards, that is in the same direction as the adjacent

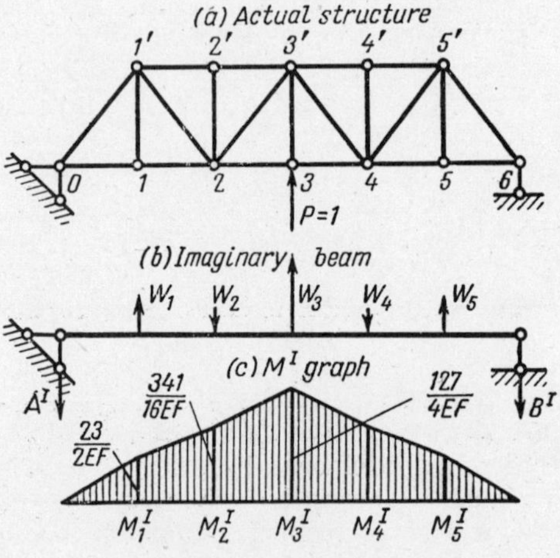

Fig. 45.8

forces of neighbouring unit couples. The negative elastic loads W_2 and W_4 will be directed downwards.

The conjugate imaginary beam corresponding to the truss under consideration appears in Fig. 45.8*b*. It represents a horizontal end-supported beam carrying 5 symmetrical loads. The abutment reactions produced by these loads will be directed downwards and will amount to

$$A^I = B^I = \frac{1}{2} \times \frac{1}{EF} \left(\frac{9}{16} - \frac{5}{24} + \frac{167}{24} - \frac{5}{24} + \frac{9}{16} \right) = \frac{23}{6EF}$$

The deflection line of the lower chord will be given dy the values of the bending moments at the point of application of the elastic loads to this imaginary beam

$$M_1^I = y_1 = M_5^I = y_5 = -\frac{23}{6EF} \times 3 = -\frac{23}{2EF}$$

$$M_2^I = y_2 = M_4^I = y_4 = -\frac{23}{6EF} \times 6 + \frac{9}{16EF} \times 3 = -\frac{341}{16EF}$$

$$M_3^I = y_3 = -\frac{23}{6EF} \times 9 + \frac{9}{16EF} \times 6 - \frac{5}{24EF} \times 3 = -\frac{127}{4EF}$$

These data being obtained, we may trace the diagram ot the bending moment produced by the elastic loads. The ordinates to this diagram plotted on the side of extended fibres will correspond exactly both in amount and direction to those of the lower chord deflection line of the truss. The load being directed upwards, all the joints of the lower chord will deflect in the same direction. All the above computations may be checked determining with the aid of Mohr's formula the deflection of joint *3* induced by the application of the vertical load $P = 1$ ton. The value of this deflection will be given by

$$\Delta_{pp} = \frac{\overline{N} N_p}{EF} = \frac{N^2 pl}{EF}$$

$\overline{N}$ being numerically equal in this particular case to N_p, for the load P itself equals 1 ton. All the data necessary for these computations will be found in Table 7.8 given above

$$\Delta_{pp} = \delta_{pp} = \frac{1}{EF} \left[\left(\frac{3}{8} \right)^2 \times 3 \times 4 + \left(\frac{5}{8} \right)^2 \times 5 \times 6 + \left(\frac{3}{4} \right)^2 \times \right.$$

$$\left. \times 3 \times 4 + \left(\frac{9}{8} \right)^2 \times 3 \times 2 + 1^2 \times 4 \right] = \frac{1}{EF} \times$$

$$\times \frac{108 + 750 + 432 + 486 + 256}{64} = \frac{127}{4EF}$$

It is seen that the deflection of point *3* computed by the method of elastic loads is exactly the same as that computed using Mohr's formula.

14.8. *Deformations of Statically Determinate Structures Caused by the Movement of Supports*

No stresses result from a displacement of one or more supports of a statically determinate structure provided the supports travel along the direction of the corresponding reactions.

Let us examine, for instance, the frame of Fig. 46.8. Assume that the right-hand support settles vertically an amount Δ due to undermining or any other cause. Such a settlement will produce no bending moments or normal stresses in the members of the structure. In order to determine the displacement of point k along the direc-

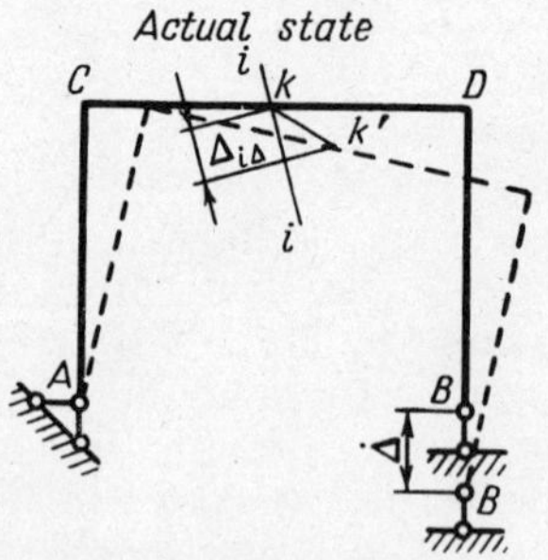

Fig. 46.8

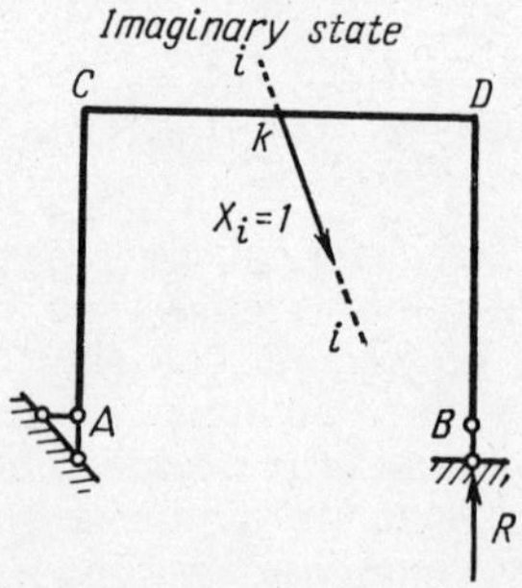

Fig. 47.8

tion i-i imagine that a unit load $X_i = 1$ acts on the crossbeam at point k along the direction of the displacement required (Fig. 47.8). Let R be the reaction caused by this unit load at the support which has settled.

For the two states represented in Figs. 46.8 and 47.8 in one of which (the actual state) the structure carries no load at all, we may write on the basis of Maxwell's theorem of reciprocal works that

$$X_i\Delta_{i\Delta} - R\Delta = 0$$

and, since $X = 1$

$$1\Delta_{i\Delta} - R\Delta = 0$$

and, consequently

$$\Delta_{i\Delta} = R\Delta$$

meaning that the displacement at any point of the statically determinate system caused by the settlement (or any other movement) of a support is equal to the product of the amount of this settlement by the reaction R at the corresponding support induced by a unit load acting along the direction of the displacement studied. This displacement will be reckoned positive when the directions of reaction R and of the displacement Δ are opposed and negative when their directions coincide.

The same result could be obtained from the strain energy equation

$$\frac{1}{2}X_i\delta_{ii}+X_i\Delta_{i\Delta}-R\Delta=\frac{1}{2}X_i\delta_{ii}$$

The left part of this equation represents the work accomplished by all the external forces (reactions included) acting on the statically determinate system of Fig. 48.8*a* in case the settlement of support B takes place after the application of the load unity X_i, and the right part of the same equation represents the work produced by these same forces in case the settlement would reach its final value before the application of this load

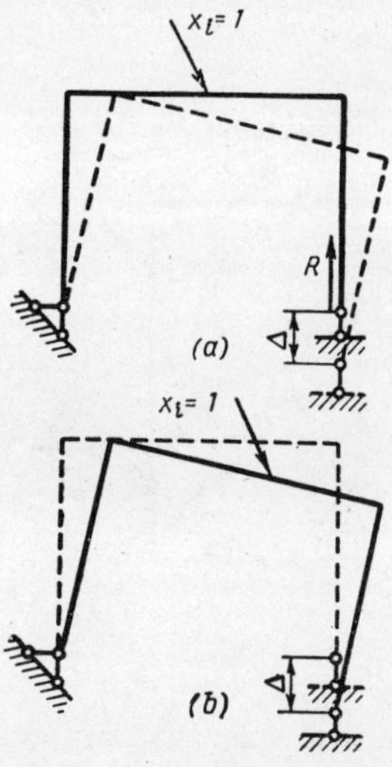

Fig. 48.8

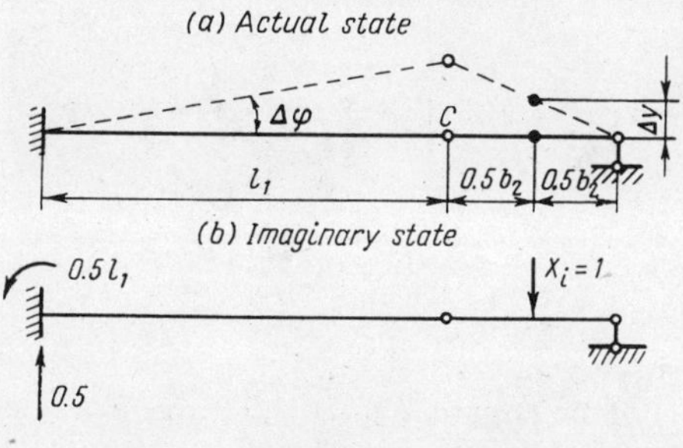

Fig. 49.8

(Fig. 48.8*b*). The two parts of this equation must have exactly the same value because in both cases the total deformation of the system remains the same. It follows that the strain energy accumulated in the first case (represented by the left part of the equation) must be exactly the same as the strain energy acquired by the structure in the second case and represented by the right-hand part of the equation. As $X_i = 1$, this equation leads immediately to

$$\Delta_{i\Delta}=R\Delta$$

which coincides with the result obtained on the basis of the theorem of reciprocal works.

Let us take up a beam provided with an intermediate hinge as represented in Fig. 49.8*a*. It is required to determine the vertical displacement Δ_y of point C of this beam when the fixed end is rotated through the angle Δ_φ. This may be done applying a unit load $X_i = 1$ at point C (Fig. 49.8*b*). On the basis of Maxwell's theorem we may write

$$X_i\Delta_y + 0.5l_1\Delta_\varphi = 0$$

wherefrom

$$1\Delta_y + 0.5l_1\Delta_\varphi = 0$$

$$\Delta_y = -0.5l_1\Delta_\varphi$$

The negative value obtained for the displacement indicates that point C will shift upwards in a direction opposite to the one adopted for the unit load X_i.

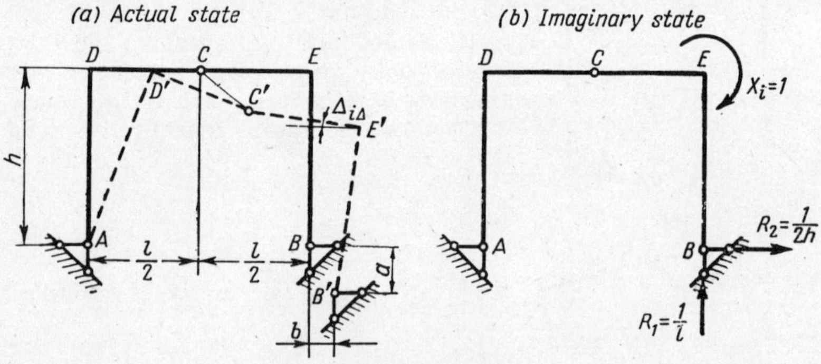

Fig. 50.8

Let us consider now the more general case when several support constraints of a statically determinate structure yield simultaneously. As an example, we shall study the frame appearing in Fig. 50.8*a*. The deformations of this frame are due to a horizontal displacement and a vertical settlement of the right-hand support, at the outcome of which the system will occupy the position indicated in dotted lines.

In order to find the angular rotation of joint E let us apply at this point a unit moment X_i acting in the direction of the rotation required (Fig. 50.8*b*) At the right-hand support this unit moment will give rise to a reaction whose vertical component R_1 will be equal to $\frac{1}{l}$ and the horizontal one R_2 to $\frac{1}{2h}$. Equating the work accomplished by the external forces in the case of the actual displacement (Fig. 50.8*a*) and in the imaginary one (Fig. 50.8*b*) we obtain

$$X_i\Delta_{i\Delta} - R_1a + R_2b = 0$$

and, since $X_i = 1$

$$\Delta_{i\Delta} = R_1a - R_2b$$

Introducing in this expression the values of reactions R_1 and R_2 we find

$$\Delta_{i\Delta} = \frac{a}{l} - \frac{b}{2h}$$

Thus, in order to determine the displacements induced at any point of a statically determinate structure by the movement of its supports (these supports being shifted along the directions of the existing constraints) we must:

1. Choose an imaginary state of the structure for which *the support in question remains fixed.*
2. Apply to the structure a unit action $X_i = 1$ coinciding in direction with the displacement required.
3. Determine the reactions produced by the said unit action along those of the constraints which remaining stationary in the imaginary state yield in the actual one.
4. Form an equation expressing that the work accomplished by the loads and reactions of the imaginary state along the displacements of the real one equals zero.
5. Obtain the value of the displacement required solving the aforesaid equation.

15.8. *Deformations of a Kinematic Chain Caused by the Mutual Rotation of Two Neighbouring Links*

Hereunder the term *kinematic chain* shall apply to any system consisting of a number of hinge-connected rectilinear elements, forming a broken line.

Let us examine the displacement of any point C of such a system along the direction *i-i* when the angle formed by two neighbouring

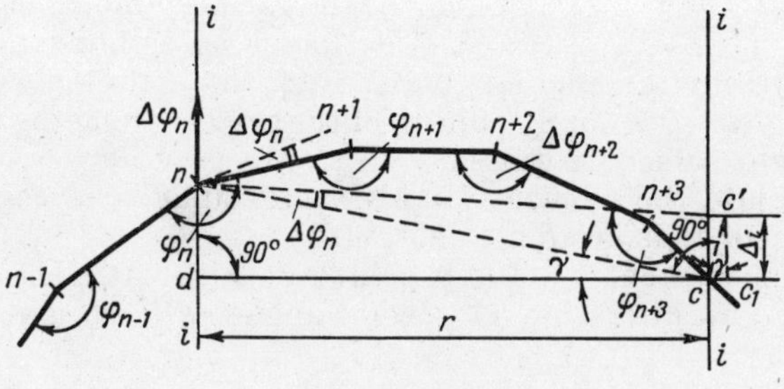

Fig. 51.8

links $n—1$, n and n, $n + 1$ is modified (Fig. 51.8). Assume that this angle has changed an amount $\Delta\varphi_n$ and that the part of the system situated to the left of joint n remains fixed. The angular rotation $\Delta\varphi_n$ will cause a displacement of point C which will occupy a new position C'. It is clear that the angle $C'nC$ will be equal to $\Delta\varphi_n$ and since the rotation is supposed to be very small, the circular arc CC' may be replaced by the normal to nC.

Let Δ_i be the projection of CC' on the direction i-i and γ the angle formed by line nC with Cd normal to i-i (Fig. 51.8). The similarity of triangles $CC'C_1$ and CnC' yields

$$C'C_1 = \Delta_i = C'C \cos\gamma = nC\,\Delta\varphi_n \cos\gamma$$

As

$$nC \cos\gamma = Cd = r$$

we obtain finally

$$\Delta_i = \Delta\varphi_n r$$

Thus, the displacement Δ_i of any point of the system produced by a change of angle φ_n by an amount $\Delta\varphi_n$ is equal to the product of $\Delta\varphi_n$ by the length r, which represents the projection of the nC segment on a normal to the displacement required.

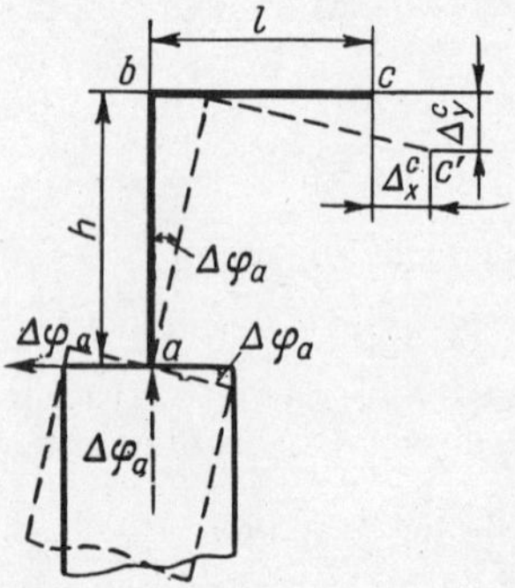

Fig. 52.8

Should we represent the angular rotation $\Delta\varphi_n$ by a vector applied at point n and directed parallel to the displacement Δ_i required (Fig. 51.8), this displacement will be numerically equal to the moment of this vector about the point C whose displacement is studied. Thus, the displacement Δ_i of some point of the kinematic chain along a direction i-i caused by a change of the angle φ_n formed by two neighbouring links and amounting to $\Delta\varphi_n$ may be found as follows:

1. Represent the angular rotation $\Delta\varphi_n$ by a vector.
2. Apply this vector at point n of the system along the direction of the displacement required.
3. Compute the moment of this vector about the point whose displacement it is desired to obtain.

The displacement Δ_i of any point of a kinematic chain resulting from the alteration of several angles will be given by the expression

$$\Delta_i = \Delta\varphi_1 r_1 + \Delta\varphi_2 r_2 + \ldots + \Delta\varphi_m r_m = \sum_1^m \Delta\varphi r \qquad (29.8)$$

Problem. Determine the horizontal and vertical displacements of point C belonging to the knee frame of Fig. 52.8 when the foundation of this frame is rotated about point a clockwise through an angle $\Delta\varphi_a$.

Solution. Apply at point a a horizontal vector $\Delta\varphi_a$. Its moment about point C will give immediately the horizontal displacement required

$$\Delta_x^c = \Delta\varphi_a h$$

The vertical displacement of point C will be obtained in exactly the same way

$$\Delta_y^c = \Delta\varphi_a l$$

The total displacement of point C (the distance CC') will be given by

$$CC' = \sqrt{(\Delta_x^c)^2 + (\Delta_y^c)^2} = \Delta\varphi_a \sqrt{h^2 + l^2}$$

The same result could be obtained following the procedure outlined in the previous article.

16.8. *Deflections of Three-Dimensional Framed Structures*

In the most general case three different stresses M, N and Q act across a section passed through a member of any plane system, and therefore the general expression giving the deformations of such systems will contain three terms, each of which characterizes the displacement due to one of the three stresses mentioned.

In three-dimensional framed structure the cross sections of any member will be acted upon dy six stresses: two bending moments M_y and M_z about the principal axes of inertia y and z of the section under consideration, one torque moment M_t about the longitudinal axis x of the bar, one normal stress N_x and two shearing forces Q_y and Q_z parallel to the aforementioned axes y and z. Consequently, in this case the general expression of the displacements will consist of six terms, each of which will represent the displacement due to one of the aforesaid stresses.

Following exactly the same procedure as in Art. 6.8, we shall obtain the expression given hereunder permitting the displacement computation for three-dimensional framed structures

$$\Delta_{mn} = \Sigma \int_0^l \overline{M}_{ym} \frac{M_{yn}\,dx}{EJ_y} + \Sigma \int_0^l \overline{M}_{zm} \frac{M_{zn}\,dx}{EJ_z} + \Sigma \int_0^l \overline{M}_{tm} \frac{M_{tn}\,dx}{GJ_t} + \Sigma \int_0^l \overline{N}_{xm} \frac{N_{xn}\,dx}{EF} + \Sigma \int_0^l \overline{Q}_{ym} \times \frac{Q_{yn}\,dx}{GF} \eta_y + \Sigma \int_0^l \overline{Q}_{zm} \frac{Q_{zn}\,dx}{GF} \eta_z \qquad (30.8)$$

In this expression $\overline{M}_{ym}$ and $\overline{M}_{zm}$ represent the bending moments due to a unit action (concentrated load, when linear displacements are studied, and unit moments in the case of angular rotations) whose direction coincides with that of the displacement. In the same way $\overline{M}_{tm}$ represents the torque produced by the same unit action, and $\overline{N}_{xm}$, $\overline{Q}_{ym}$ and $\overline{Q}_{zm}$ are the normal stress and the shears produced thereby. At the same time M_{yn}, M_{zn}, M_{tn}, M_{xn}, Q_{yn} and Q_{zn} will indicate the stresses induced by the actual loading.

Coefficients η_y and η_z will be determined in relation with the shape of the cross section (see Art. 2.8). The magnitude of J_t appearing in the expression of the torque rigidity may be approximately taken equal to:

for a square cross section

$$J_t = 0.143a^4$$

for an elongated rectangular cross section (at $a > b$)

$$J_t = \frac{b^3}{3}(a - 0.63b)$$

for cross sections consisting of several rectangles of small width (such as the cross sections of T-beams, H-beams, etc.)

$$J_t = \frac{1}{3}\Sigma d^3 l$$

(l being the length and d the width of the rectangle)

for a circular cross section

$$J_t = J_p = \frac{\pi d^4}{32} = \frac{\pi r^4}{2}$$

and for an annular cross section

$$J_t = \frac{\pi}{32}(D^4 - d^4) = \frac{\pi}{2}(R^4 - r^4)$$

(where D and R indicate the external and d and r the internal diameters and radii of the ring).

When the cross sections of all the bars remain constant, the rigidities EJ_y, EJ_z, GJ_t, EF and GF as well as the coefficients η_y and η_z appearing in expression (30.8) may be placed in front of the integral signs.

The computation of displacements is carried out with the aid of expression (30.8) in exactly the same way as in the case of plane structures described in Art. 6.8. When computing the displacements of three-dimensional structures with rigid joints only the first three terms of the expression (30.8) will be retained, while the influence of the normal and shearing stresses may be neglected. On the contrary, if it were desired to determine the deflections of a three-dimensional hinge-connected structure one should take into consideration solely the normal stresses.

Problem 1. Required the vertical deflection of the free end C of a horizontal knee frame appearing in Fig. 53.8*a*. The frame is loaded with one

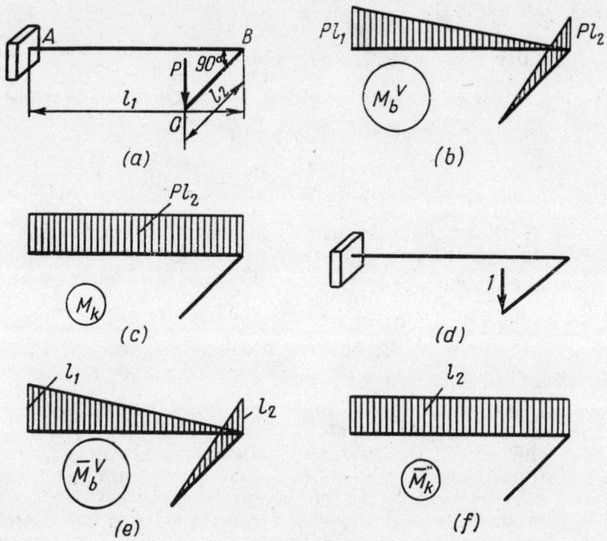

Fig. 53.8

vertical force P, its cross section is circular in shape and remains constant throughout. The value of G shall be taken equal to 0.4 E.

Solution. Fig. 53.8*b* represents the diagram for the bending moments M_b^V acting in a vertical plane normally to the axes of the frame members, these moments being induced by the actual loads, and Fig. 53.8*c* represents the torque curve $\overline{M}_t$*. No horizontal bending moments will be induced by the vertical load P acting at point C.

Apply a vertical unit load as indicated in Fig. 53.8*d* and trace the graphs of the bending moment $\overline{M}_b^V$ and of the torque $\overline{M_t}$ induced by this unit load as shown in Fig. 53.8*e* and *f*. The deflection Δ of point C will be obtained applying Vereshchagin's method to expression (30.8)

$$\Delta = \frac{Pl_1 \frac{l_1}{2} \times \frac{2}{3} l_1}{EJ} + \frac{Pl_2 \frac{l_2}{2} \times \frac{2}{3} l_2}{EJ} + \frac{Pl_2 l_1 l_2}{GJ_p}$$

Each term of right-hand part of this expression represents one of the components of the total vertical displacements of point C. Thus, the first term $\frac{Pl_1^3}{3EJ}$ is the vertical deflection of point B (see Fig. 53.8*a*) caused by the bending of member AB. This deflection entails an identical deflection of point C. The second term $\frac{Pl_2^3}{3EJ}$ is the vertical deflection of point C which

* In Figs. 53.8 and 54.8 M_t is designated by M_k—*Tr*.

results from the bending of the element BC. The torque $M_t = Pl_2$ caused in element AB by the load P rotates cross section B about the horizontal axis through an angle $\varphi = \frac{M_t l_1}{GJ_p} = \frac{Pl_2 l_1}{GJ_p}$. This rotation will cause point C to travel vertically over a stretch $\varphi l_2 = \frac{Pl_1 l_2^2}{GJ_p}$.

Introducing in the above expression for Δ the values of J, J_p and G equal to $\frac{\pi d^4}{64}$, $\frac{\pi d^4}{32}$ and $0.4\,E$, respectively, we obtain finally

$$\Delta = \frac{64P}{\pi d^4 E}\left(\frac{l_1^3}{3} + \frac{l_2^3}{3} + \frac{l_1 l_2^2}{0.8}\right)$$

Problem 2. Required the horizontal displacement Δ along axis BC of cross section K of a polygonal beam appearing in Fig. 54.8*a* as well as the angular rotation of the same cross section in the plane BCD. The beam is built in at point A and is of circular cross section which remains constant for all of its elements.

Solution. The graphs of the bending moments M_b^V (acting in a vertical plane) and of M_b^H (acting in a horizontal plane) as well as the graph of the torque M_t induced by the actual loads are shown in Fig. 54.8*b*, *c* and *d*. These graphs are drawn on the side of the extended fibres of each member of the beam. The sign of the torque is indicated in the graph, this torque being reckoned positive when seen from that part of the structure which has been cut off, it tends to rotate the remaining part clockwise.

In order to determine the horizontal displacement Δ a unit load parallel to element BC must be applied at point K (Fig. 54.8*e*). The corresponding graphs of the bending moments $\overline{M}_b^V$, $\overline{M}_b^H$ and of the torque $\overline{M}_t$ are represented in Fig. 54.8*f*. Using once again expression (30.8) and carrying out the graph multiplication by the method proposed by Vereshchagin, we obtain

$$\Delta = \frac{1\times 1}{2}\left(\frac{2}{2}\times 4 + \frac{1}{3}\times 2\right)\frac{1}{EJ} + \frac{1\times 1\times 4}{EJ} + \frac{2\times 2}{2}\times\frac{2}{3}\times 4\times\frac{1}{EJ} + \frac{1\times 2\times 4}{EJ_p}$$

The two first terms of this expression account for the bending moments acting in the vertical planes, the third for those acting in horizontal planes and the last one takes care of the torque. All the products are positive because the graphs of the bending moments which are being multiplied one by the other remain all the time on one and the same side of the corresponding members, and the torques are also of the same sign. Replacing in the above expression J_p by $2J$ $\left(\text{where } J = \frac{\pi d^4}{64}\right)$ we obtain finally

$$\Delta = \frac{1}{EJ}\left(\frac{4}{3} + \frac{1}{3} + 4 + \frac{16}{3} + 4\right) = \frac{15}{EJ}$$

The angular rotation φ of the cross section K will be determined applying at this point a unit bending moment acting in the plane BCD (Fig. 54.8*g*). The corresponding graphs of the bending moments $\overline{M}_b^V$ and $\overline{M}_b^H$ and of the torque $\overline{M}_t$ are shown in Fig. 54.8*h*. It will be noted that the bending moments acting in the horizontal plane will remain constantly nil.

Expression (30.8) gives

$$\varphi = \frac{2+4}{2}\times 1\times 1\times\frac{1}{EJ} + \frac{4\times 1\times 1}{EJ} + \frac{4\times 2\times 1}{EJ_p} = \frac{1}{EJ}(3+4+4) = \frac{11}{EJ}$$

The magnitude of the loads and the length of the beam members being expressed in tons and in metres, respectively, the value of Young's modu-

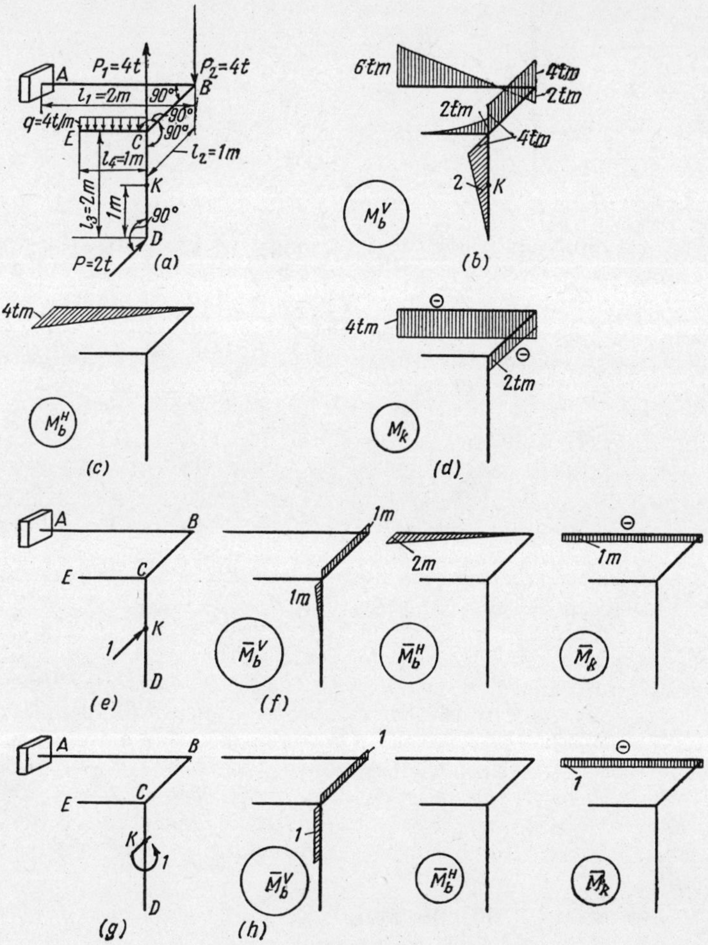

Fig. 54.8

lus E must be expressed in tons per square metre and that of J in metres in the fourth power. The value of the horizontal displacement Δ will be then obtained in metres and that of the angular rotation φ in radians. Both these values being positive, the directions of Δ and φ will coincide with those chosen for the unit load (see Fig. 54.8*e*) and for the unit moment (see Fig. 54.8*g*).

9. ANALYSIS OF THE SIMPLER STATICALLY INDETERMINATE STRUCTURES BY THE METHOD OF FORCES*

1.9. *General*

While taking his course in the strength of materials, the reader has already met with structures which cannot be analyzed using solely equilibrium equations. The computation of stresses set up in these structures requires the use of additional equations, namely deformation equations. Such structures are called *statically indeterminate* or *redundant.*

The main difference between the redundant structures and the statically determinate ones resides in the fact that the stress distribution depends for the first ones not only on the loading but also on the relative dimensions of their members. If these members are made of different materials the stress distribution will equally depend on the elastic properties of these materials. Statically indeterminate structures are also very sensible to such factors as the settlement of their supports, temperature variation, manufacturing and erection defects, etc., which give rise to additional stresses, while the same factors would have no influence whatsoever on statically determinate structures. At present redundant structures are widely used in numerous branches of engineering activities. Their analysis must always start with a close examination of arrangement of their members, the primary goal of this examination being the determination of the *degree of redundancy.*

*This degree of redundancy is equal to the number of redundant constraints*** *whose elimination would transform the given system into*

* The method of analysis described in the present Chapter is referred to by various authors either as the method of deflections or the method of least work, depending on the procedure adopted for the determination of the coefficients to the unknowns.

We prefer to translate literally its name from Russian and to call it *method of forces.* Indeed, in that way we are sure to avoid confuses with the *slope and deflection method* (see Chapter 13) and moreover both methods will be consistently named in conformity with the nature of the unknowns.

** Hereunder the term constraint will signify everything capable of preventing the mutual displacement of different points or cross sections of a structure. The adjective *redundant* should never be regarded as synonymous to *superfluous* or *useless.*

a statically determinate one without impeding its geometrical stability.

In the previous articles it has been already explained that *geometrically stable systems are such systems whose shape cannot be altered without a deformation of their elements.*

There are no redundant constraints in a statically determinate system and the elimination of a single constraint will always transform these systems in mechanisms whose elements are endowed with a certain freedom of movement.

The beam appearing in Fig. 1.9*a* constitutes a structure, whose degree of redundancy is equal to one, for one of the supporting

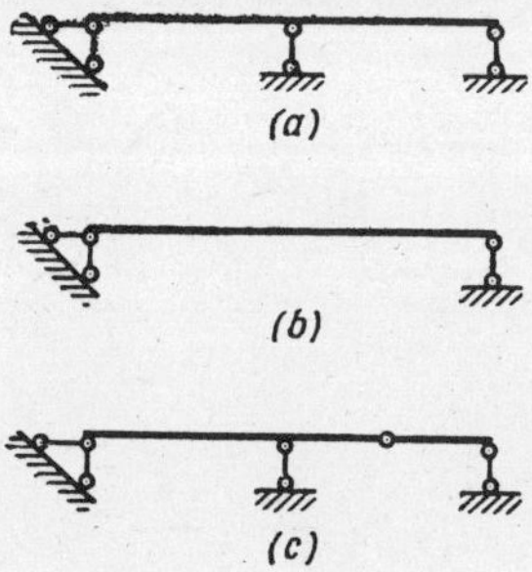

Fig. 1.9

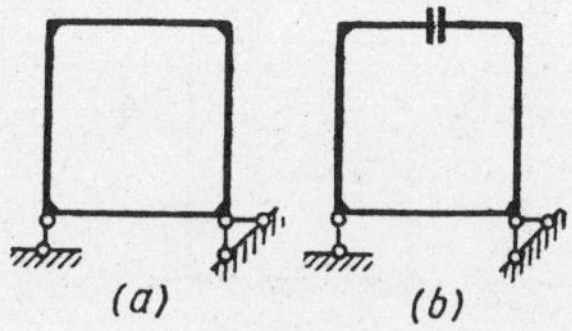

Fig. 2.9

bars constitutes a redundant connection with the ground. This beam will become statically determinate as soon as one of these bars is eliminated (as in Fig. 1.9*b*) or through the introduction of an intermediate hinge (as in Fig. 1.9*c*).

The frame appearing in Fig. 2.9*a* constitutes a structure redundant to the third degree, its transformation into a statically determinate system requiring that at least one of its members should be cut in two (Fig. 2.9*b*). We have seen previously that this operation is equivalent to the elimination of three internal constraints corresponding to three internal forces acting across the section, namely, the bending moment, the shear and the normal stress. The equilibrium equations alone do not permit the determination of these internal forces.

Any other closed frame with rigid joints lying in one plane will also form a system with a degree of redundancy equal to three.

The two framed bents appearing in Fig. 3.9 are typical examples of similar structures. In the frame in Fig. 3.9*b* whose uprights are rigidly fixed in the ground, the latter may be regarded as constituting an additional member of infinite rigidity.

The structure appearing in Fig. 4.9*a* is provided with a hinge at midspan of the top girder. If we pass a section through this hinge, this section would be acted upon by two stresses N and Q only (Fig. 4.9*b*). Consequently, the upper frame has two degrees of redundancy while the whole structure is redundant to the fifth degree, for the lower frame is completely closed and therefore its degree of redundancy equals three. The elimination of all the redundant constraints could reduce this system to two columns built in at their lower ends and provided at their upper parts with two horizontal brackets as indicated in Fig. 4.9*b*.

The total number of redundant constraints could also be obtained in the following way. The top frame being provided with a hinge is redundant in the second degree; in addition a built-in end is always equivalent to three constraints and therefore two

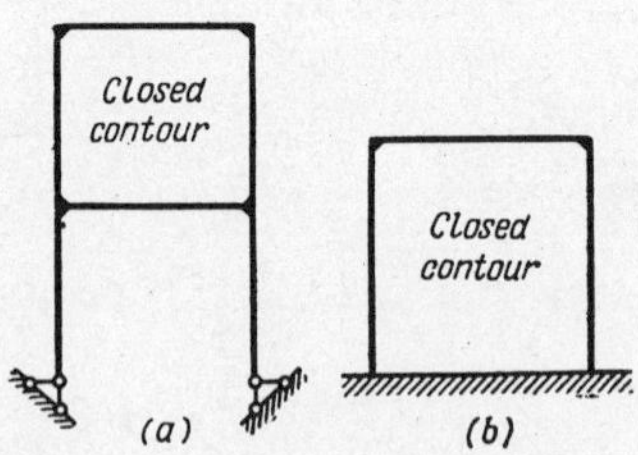

Fig. 3.9

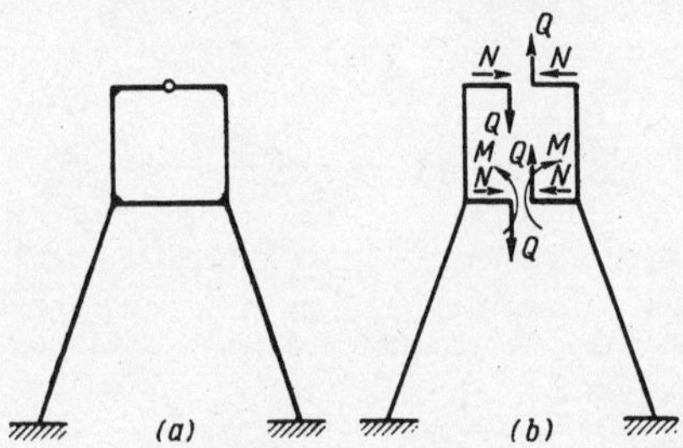

Fig. 4.9

fixed supports of the frame represent a total of six constraints. As the equilibrium equations will permit the determination of three reactions only, the other three constraints are redundant. Therefore the whole system will have a degree of redundancy equal to five.

It should be noted that there are usually several ways of eliminating the redundant constraints in order to convert the given structure into a statically determinate one, but the number of eliminated constraints will always remain the same. Thus, the simple statically determinate structures appearing in Fig. 1.9*b* and *c* have been derived from one and the same redundant structure of Fig. 1.9*a*, the first one by the elimination of the intermediate support and the second one by the introduction of a hinge. The latter eliminates the constraint preventing mutual rotation of two cross sections, one situated to its right and one to its left.

The introduction of a hinge into one of the members of a redundant structure or the replacement of a rigid joint formed by the meeting of two bars by a hinge is always equivalent to the elimination of one constraint and will therefore lower by one degree the redundancy of the whole structure.

Hereunder hinges of this type shall be referred to as ordinary hinges.

In eliminating the redundant constraints of some structure care should be taken not to disturb its stability. From this point of view the elimination of one of the vertical supporting bars of the framed bent shown in Fig. 5.9*b* would be unacceptable, for the

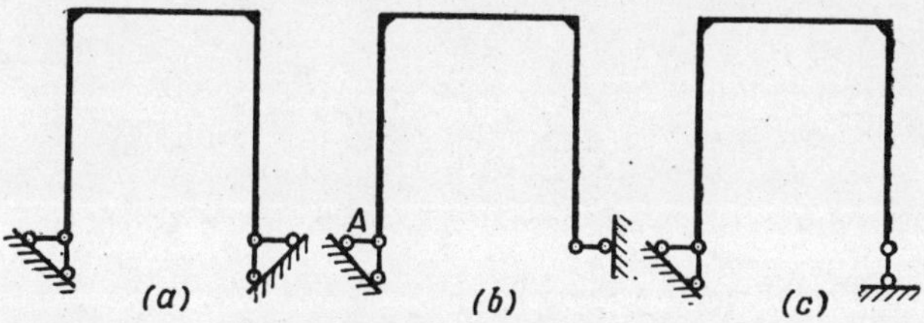

Fig. 5.9

three remaining bars would concur at point A and, consequently, these bars would be uncapable of preventing the rotation of the whole system about this point. The correct way of eliminating the redundant constraint of this structure is shown in Fig. 5.9*c*.

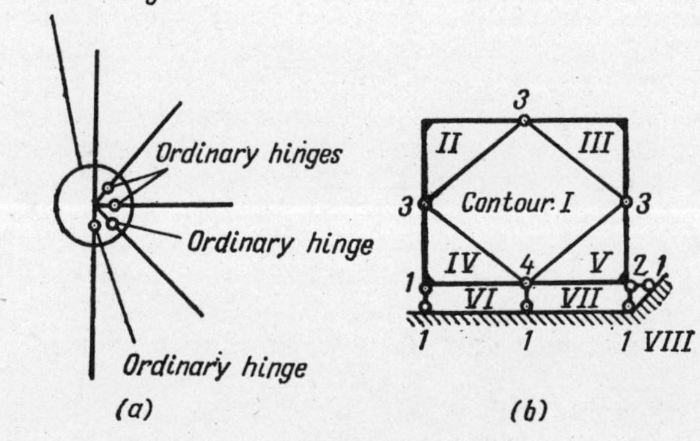

Fig. 6.9

The degree of redundancy of complicated structures may be determined remembering that each hinge introduced instead of a rigid joint formed by the meeting of K bars reduces the degree of redundancy of the system by (K—1), for such a hinge replaces (K—1) ordinary hinges (Fig. 6.9*a*). Hence the degree of redundancy of a structure can be obtained multiplying by three the number of closed contours forming this structure (regardless of any hinges, whether within the structure itself or at the supports) and then reducing the number so obtained by the number of all the ordinary hinges existing in the system.

Hinges common to K bars meeting at one point should be regarded as equivalent to $(K-1)$ ordinary hinges.

Mathematically this rule may be expressed by the following formula

$$n = 3m - H \tag{1.9}$$

In this expression n is the degree of redundancy, m is the number of closed contours which form the structure, and H is the number of ordinary hinges. It will be remembered that we have agreed to use the term ordinary hinge for a hinge placed at the meeting of two bars, the term double hinge meaning a hinge introduced at the meeting of three bars and so forth.

The structure appearing in Fig. 6.9*b* consists of eight closed contours (marked with Roman figures) and against each joint we have entered the equivalent number of ordinary hinges. The horizontal and vertical bars meeting at the outer joints of the system are regarded as a single knee shaped member, these bars being rigidly connected together. Consequently, $m = 8$; $H = 3 + 3 + 3 + 4 + 1 + 2 + 1 + 1 + 1 + 1 = 20$ and $n = 3 \times 8 - 20 = 4$, meaning that the structure is statically indeterminate in the fourth degree.

As already mentioned, the elimination of any one of the constraints of a statically determinate system transforms immediately this system into a mechanism, showing thereby that the number of constraints in such systems constitutes the *absolute minimum* required to ensure their stability. Any additional constraint in excess of this minimum transforms the system into a redundant one.

It is clear that in such a system there is at least one constraint that can be eliminated without prejudice to its stability. However, there may exist such constraints which cannot be excluded without interfering with the stability of the structure. Hereunder we shall designate such constraints by the term *necessary constraints*. It is interesting to note that the stresses corresponding to the necessary constraints can be always determined with the aid of statics alone. An example of a necessary constraint is afforded by the vertical supporting bars of the framed bent represented in Fig. 5.9*a*.

Neither of these two bars can be removed without rendering the system unstable.

The constraints which can be eliminated without prejudicing the stability of the system form the *ordinary redundant constraints*.

The stresses corresponding to these constraints cannot be derived from the equations of equilibrium alone. The horizontal supporting bars of the portal frame just mentioned (see Fig. 5.9*a*) constitute an example of the latter type of constraints. We know that for each system of coplanar forces in equilibrium statics provides three independent equations. Hence, if some system is connected to the ground by means of three supporting bars, the stresses in these bars may be computed using equilibrium equations alone irrespectively of the degree of redundancy of the whole system. A similar structure may be therefore regarded as *internally redundant.*

Externally this structure is statically determinate for the abutment reactions and the external loads constitute a balanced system of forces all of which can be completely determined with the aid of statics alone. For such systems all the external (support) constraints belong to the category of *necessary* ones. If, on the contrary, a structure is endowed with more than three external constraints such a structure can usually be considered both as externally or internally redundant. Indeed, one can choose at will those of the constraints which will be regarded as the redundant ones. Thus, the frame of Fig. 5.9*a* may be regarded as externally redundant if one decides to eliminate one of the horizontal supports in order to transform it into the statically determinate system shown in Fig. 5.9*c*. On the other hand, if one decides to consider as redundant the constraint which prevents the rotation of one part of the crossbar about the other, in other words, if one decides to transform the frame into a statically determinate structure by the introduction of a hinge as shown in Fig. 7.9, this same frame should be considered as an internally redundant one. The frame of Fig. 8.9*a* whose degree of redundancy equals six may be considered:

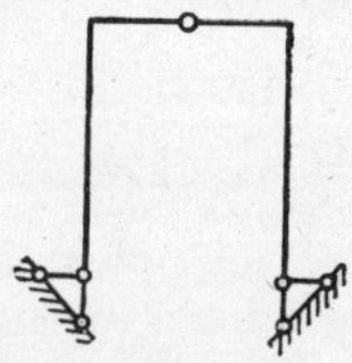

Fig. 7.9

(a) as being three times internally and three times externally redundant, if its conversion into a simple statically determinate structure appearing in Fig. 8.9*b* is carried out by the removal of three external constraints (for which purpose one built-in end is set free) and of three internal constraints;

(b) as being four times internally and twice externally redundant, if it is decided to transform the given system into a statically determinate one as indicated in Fig. 8.9*c*;

(c) finally the same frame may be regarded as being six times internally redundant, if its conversion into two separate statically determinate parts is carried out as shown in Fig. 8.9*d*.

The same frame cannot be regarded as statically indeterminate only from the point of view of its external constraints. Indeed, the system is redundant in the sixth degree while a maximum of

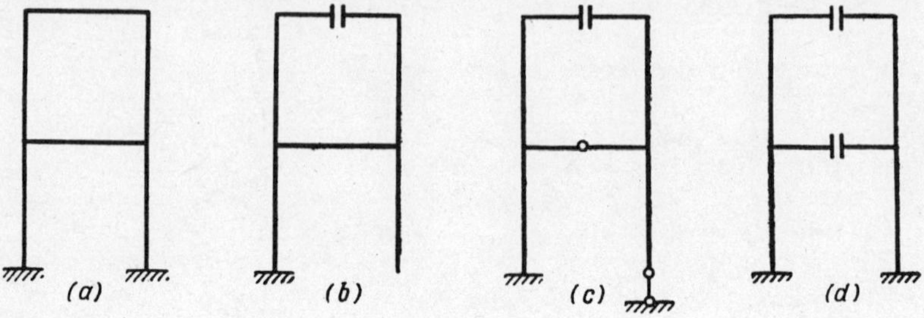

Fig. 8.9

three external constraints can be removed without disrupting its stability. It is clear therefore that this system cannot be converted into a statically determinate one by the elimination of external constraints alone.

2.9. *Canonical Equations Deduced by the Method of Forces*

In the previous article it was shown that the stress analysis of redundant structures requires the use of additional equations based on the strains and deflections suffered by these structures. In the method of forces these equations are obtained through the transformation of *the given structure redundant to the nth degree into a simple statically determinate one.*

The elimination of any constraints will introduce alterations neither in the stress distribution nor in the strains and deflections if in the place of constraints so removed we introduce forces * equivalent to the reactions developed by these constraints. Consequently, if the simple structure is acted upon both by the actual loads and the additional actions which replace the eliminated constraints, the strains and deflections of such a system as well as the stresses induced therein will be exactly the same as in the original one and therefore the two become equivalent.

Since no displacement of the given redundant structure along the reactions at the supports is possible, the displacements of the

* As previously (see Art. 2,8), the term force will apply equally to moments.

conjugate simple statically determinate structure along the same directions must also equal zero, even if some of these supports were eliminated when converting the former to the latter. This means that *the reactions developed by the redundant constraints will be such as is necessary to render nil the deformations of the simple statically determinate structure along the direction of these reactions.*

The equation translating the above statement into mathematical symbols will be as follows

$$\Delta_i = \Delta_{i1} + \Delta_{i2} + \ldots + \Delta_{i,\,n-1} + \Delta_{in} + \Delta_{ip} = 0 \qquad (2.9)$$

In this expression the first of the two indices following the letter Δ shows the direction of the displacement (the latter coinciding with the eliminated constraint) and the second one the action causing this displacement. Thus Δ_{ik} indicates a displacement along the direction i caused by the reaction of the constraint k. In the same way, Δ_{ip} will indicate the displacement along the direction of constraint i caused by the applied loads.

Let us indicate by X_k the magnitude of the reaction developed by the constraint k (this reaction being either a moment or a direct stress). At the same time let us designate by δ_{ik} the displacement caused by a unit action. In that case we can replace Δ_{ik} by $X_k\delta_{ik}$ and the expression (2.9) will become

$$\Delta_i = X_1\delta_{i1} + X_2\delta_{i2} + \ldots + X_{n-1}\delta_{i,\,n-1} + X_n\delta_{in} + \Delta_{ip} = 0 \quad (3.9)$$

In this way the equivalence of the original structure and of the simple statically determinate one will be mathematically interpreted by a system of n linear equations

$$\left.\begin{array}{l} X_1\delta_{11} + X_2\delta_{12} + \ldots \ldots + X_n\delta_{1n} + \Delta_{1p} = 0 \\ X_1\delta_{21} + X_2\delta_{22} + \ldots \ldots + X_n\delta_{2n} + \Delta_{2p} = 0 \\ \ldots \quad \ldots \quad \ldots \quad \ldots \\ X_1\delta_{n1} + X_2\delta_{n2} + \ldots \ldots + X_n\delta_{nn} + \Delta_{np} = 0 \end{array}\right\} \quad (4.9)$$

Main diagonal

Secondary diagonals

Equations (4.9) constitute the additional expressions based on the deformations of the system which permit complete determination of all the support reactions and of all the stresses induced by the given system of loads in the original redundant structure. The first of these equations expresses the idea that the displacement of the simple structure along the direction of the first eliminated constraint (that is along the direction of force or moment X_1) is equal to zero, the second, that the displacement of this same structure along the direction of the second constraint which has been removed is also equal to zero, and so forth.

The system of simultaneous linear expressions such as (4.9) form the so-called *canonical equations of the method of forces*, this name indicating that these equations are of standard form and that the unknowns are the reactive forces developed by the eliminated constraints. The number of these equations is always equal to the number of the constraints removed, which means that it corresponds to the degree of redundancy of the given structure. It is important to note that both the number of terms in each of the separate equations as well as the total number of these equations depend solely on the degree of redundancy of the structure and are in no respect influenced by any of its other peculiarities.

The coefficients to the unknowns of equations (4.9) represent the deflections of the simple structure obtained by elimination of the redundant members, these deflections being due to unit loads and moments acting along the direction of the eliminated constraints. Numerically the values of these coefficients depend on the layout of the structure and on the cross-sectional dimensions of its members. Should these members be made of different materials, these coefficients will also depend on the elastic properties of the latter.

Thus coefficient δ_{ik} entering the above equations will represent the deflection along the direction i induced by a unit action (moment or load) acting along the direction k. The unit displacement δ_{ii} situated in the main diagonal of the canonical equations and characterized by two identical indices will be termed hereafter *principal deflection* whereas the deflections such as δ_{ik} standing in the secondary diagonals of the aforesaid equations will be termed *secondary deflections*. On the basis of Maxwell's theorem of reciprocal displacements, the secondary deflections situated symmetrically about the main diagonal will be always equal between themselves

$$\delta_{ik} = \delta_{ki} \, ^*$$

* It should be remembered that the dimensionality of a unit deflection is that of a ratio of a deflection to the action which has caused it. Consequently, a unit translation due to a concentrated load will be given in cm/kg while that due to unit couple in cm/kg·cm or in kg^{-1}. In the same way a unit angular rotation due to a unit load will be given in kg^{-1} and an angular rotation due to a unit couple in $kg \cdot cm^{-1}$.

This reduces considerably the volume of work necessary to determine the coefficients to the unknowns. These are usually obtained by computing the deflections of the simple structure produced by unit actions applied along the directions of the eliminated constraints. It is recommended to carry out these computations using the procedures developed in the preceding chapter.

The diagrams of bending moments induced in the conjugate simple structure by each of the unit actions ($X_i = 1$) will be traced separately, each of these graphs bearing the number of the eliminated constraints, the same applying to the actual loading (the M_P graph).

The unit deflections δ_{ik} will be obtained through the multiplication of the corresponding unit graph $\overline{M}_i$ by the unit graph $\overline{M}_k$ whereas the deflection due to the applied loads Δ_{iP} through the multiplication of the unit graph $\overline{M}_i$ by the graph of the actual bending moment M_P. *

The *main* or *principal deflections will be always positive* whilst the secondary ones as well as those due to the applied loads might be both positive or negative. When all the coefficients to the unknowns entering the system of simultaneous equations (unit displacements) as well as the deflections due to the applied loads are known, one may proceed with the solution of the said equations. The roots of these equations will furnish the values of the unknown stresses acting in the redundant members. These will permit the construction of the bending moment diagrams induced by $X_1, X_2, \ldots, X_i$, etc., in all the necessary members of the structure. It is convenient to use for this purpose the unit graphs traced previously. The operation consists in the multiplication of all the ordinates to each of these graphs by a constant factor equal to the magnitude of the action just obtained. The pertinent ordinates to the diagram of the bending moments acting in the redundant structure will be obtained through the summation of the ordinates to the graphs induced by the stresses X and by the actual loading in the aforementioned simple statically determinate structure.

The same result will be achieved if the simple structure obtained by eliminating all the redundant members were subjected simultaneously to the applied loads and to all the stresses acting in the eliminated members determined as described above.

The bending moment graph due to the combination of all these actions may be constructed using any of the well-known procedures.

◆

* For simplicity we have neglected the influence of normal and shearing forces. If it were desired to account for these, one should trace the corresponding diagrams and compute the corresponding products.

It is worth noting that several different simple structures may be used for the computation of the same redundant structure, these different structures being obtained by the elimination of different members regarded as redundant. It is very important to choose the one leading to the greatest possible simplification of the computation through the reduction to zero of a maximum number of secondary deflections. One should also endevour to obtain bending moment graphs of the simplest possible configuration for the members of the simple structure.

To make clear the above statement, let us take as an example the portal frame appearing in Fig. 9.9*a* and let us examine the var-

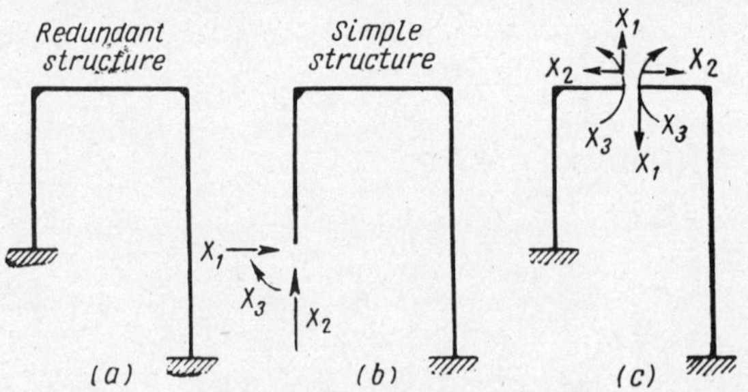

Fig. 9.9

ious simple structures which may be derived therefrom. To begin with let us eliminate the three constraints which prevent both the horizontal and the vertical movements and the angular rotation of the lower extremity of the left-hand column. The simple structure obtained in that case appears in Fig. 9.9*b*. The unknowns X_1, X_2 and X_3 will represent the reactions developed by the eliminated constraints and the simultaneous equations will express the idea that the deflections and rotations along the directions of the eliminated constraints remain nil.

Let us choose another way of rendering the redundant structures statically determinate, namely, by cutting in two the top bar as indicated in Fig. 9.9*c*. This is equivalent to the elimination of three constraints preventing mutual displacement of the two faces of the crossbar situated to the right and to the left of the cut.

Hence, each of the unknowns X_1, X_2 and X_3 will represent in this case a group of two opposite forces or couples acting over the two cross sections just mentioned.

As to the system of canonical equations, it will always remain of the same form regardless of the way in which the simple statically determinate structure has been obtained.

In the first of the two cases considered above these equations would express the idea that the movements of the lower end of the left-hand column remain nil. In the second case the same equations would mean that the two adjacent sections through the crossbar remain motionless with reference to each other. However, these equations do not exclude the possibility of the two sections moving or rotating together.

In the case of the simple structure of Fig. 9.9*b* the coefficient δ_{12} represents the horizontal motion of the lower end of the left column caused by the vertical unit load $X_2 = 1$. As for the simple structure of Fig. 9.9*c*, the coefficient δ_{12} represents alteration of the vertical distance between two adjacent cross sections of the top beam induced by two horizontal unit forces $X_2 = 1$.

3.9. *Analysis of the Simpler Redundant Structures*

Let us examine the sequence of operations leading to the determination of all stresses in redundant structures taking as an example a beam built in at one end and freely supported at the other (Fig. 10.9*a*). The simple statically determinate structure can be derived from the above by eliminating the right-hand support thus obtaining the beam appearing in Fig. 10.9*b*. A single constraint has to be eliminated for this purpose (that corresponding to a roller support) and therefore the given structure is statically indeterminate in the first degree. Apply now the unknown reaction X_1 to the cantilever beam at its free end together with the uniform load of q kg per unit length as shown in Fig. 10.9*b*. The equation expressing that the deflections of the simple statically determinate structure and those of the given redundant beam are identical becomes

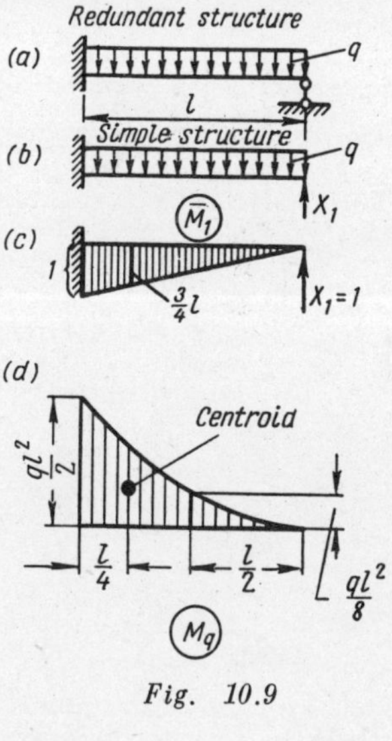

Fig. 10.9

$$X_1\delta_{11} + \Delta_{1q} = 0 \qquad (5.9)$$

More precisely this equation shows that the deflection along the direction of the eliminated reaction is nil. The determination of X_1 requires that the values of the coefficient δ_{11} and of the term Δ_{1q} should be previously calculated, the first of the two representing the deflection of the right-hand extremity of the cantilever beam along the reaction X_1 caused by a unit load acting in the same direction (Fig. 10.9*c*), and the second—the deflection along the same direction due to the loads applied. The coefficient δ_{11} will be found raising to the second power the unit bending moment graph $\overline{M}_1$ (Fig. 10.9*c*). As for the term Δ_{1q} it will be obtained by multiplying the area of the same unit bending moment graph $\overline{M}_1$ by the M_q diagram due to the actual loading (Fig. 10.9*d*).

Hence

$$\delta_{11} = l \cdot \frac{l}{2} \cdot \frac{2}{3} \cdot \frac{l}{EJ} = \frac{l^3}{3EJ}$$

$$\Delta_{1q} = -\frac{1}{3} \cdot \frac{ql^2}{2} \cdot l \cdot \frac{3}{4} \cdot \frac{l}{EJ} = -\frac{ql^4}{8EJ}$$

Substituting these values in equation (5.9) and solwing this equation with respect to X_1 we obtain

$$X_1 = -\frac{\Delta_{1q}}{\delta_{11}} = \frac{3}{8} ql$$

The diagram of the resulting bending moments acting at the cross sections of the given redundant beam will be found summing

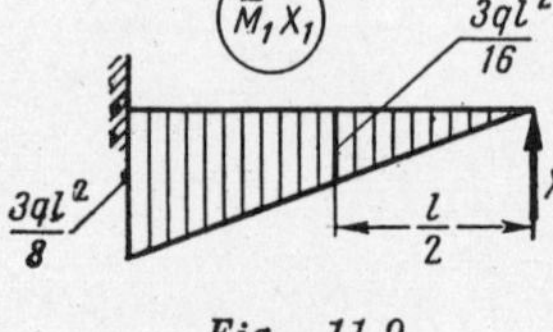

Fig. 11.9

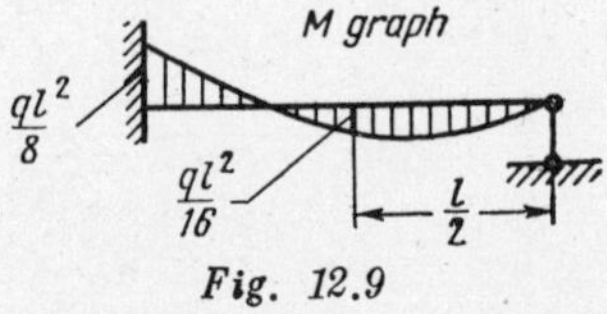

Fig. 12.9

up the ordinates to the M_q (Fig. 10.9*d*) graph with those to the $\overline{M}_1$ graph all the ordinates to which have been previously multiplied by the magnitude of X_1 (Fig. 11.9). The diagram so obtained appears in Fig. 12.9. Thus the ordinate to the resulting bending moment curve will equal

at midspan

$$M = \overline{M}_1 X_1 + M_q = \frac{3}{16} ql^2 - \frac{ql^2}{8} = \frac{ql^2}{16}$$

and at the wall

$$M = \overline{M}_1 X_1 + M_q = \frac{3}{8} ql^2 - \frac{ql^2}{2} = -\frac{ql^2}{8}$$

The maximum and minimum values of the resulting bending moments can be easily derived from the diagram of the resulting shears Q for, as it is well known, the zero ordinate points of this diagram always correspond to the extremal values of the bending moments.

This same beam (Figs. 10.9*a* and 13.9*a*) could be analyzed using for simple statically determinate structure the one obtained eliminating the constraint which prevent the rotation of the built-in end. The simple end-supported beam obtained in this way appears in Fig. 13.9*b*. The graph of the bending moments produced in this new system by a unit couple acting across the section at the wall appears in Fig. 13.9*c*, while the graph of the bending moments due to the applied loads is given in Fig. 13.9*d*.

Raising to the second power the area of the $\overline{M}_1$ graph we obtain

$$\delta_{11} = 1 \cdot l \cdot \frac{1}{2} \cdot \frac{2}{3EJ} = \frac{l}{3EJ}$$

Multiplying the same graph by the area of the M_q one we get

$$\Delta_{1q} = \frac{ql^2}{8} \cdot l \cdot \frac{2}{3} \cdot \frac{1}{2EJ} = \frac{ql^3}{24EJ}$$

The introduction of these values in expression (5.9) gives

$$X_1 = -\frac{\Delta_{1q}}{\delta_{11}} = -\frac{ql^2}{8}$$

This shows that the simple statically determinate system is acted upon by a moment $X_1 = -\frac{ql^2}{8}$ applied to the left end of the beam and by a uniformly distributed load of q kg per metre (Fig. 13.9*e*). The resulting bending moment diagram induced in the simple structure by these two actions will represent the bending moment diagram for the given redundant structure (Fig. 13.9*f*). It is readily seen that this diagram coincides exactly with that of Fig. 12.9 obtained previously using a different simple structure (see Fig. 10.9*b*).

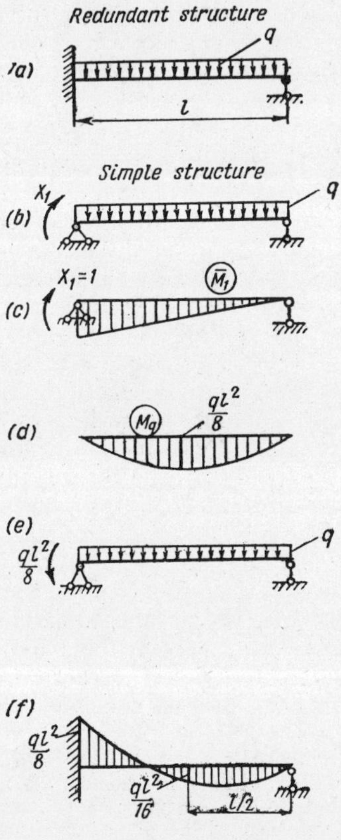

Fig. 13.9

The above example shows that the following sequence of operations may be conveniently adopted for the stress analysis of redundant structures by the method of forces:

1. Choose a simple statically determinate structure obtained by eliminating all the redundant constraints of the given one.

2. Replace the eliminated constraints by unknown forces acting in the same direction.

3. Form the canonical equations (4.9) expressing that the displacements of the simple structure along the directions of the eliminated constraints under the combined action of the loads applied and of the unknown moments and forces replacing these constraints are equal to zero.

4. Apply successively to the simple structure the unit actions $X_1 = 1$, $X_2 = 1$, $X_3 = 1$, . . ., $X_n = 1$ and trace the diagrams of the corresponding bending moments $\overline{M}_i$. Trace equally the diagram of the bending moments M_p due to the applied loads. *

5. Calculate all the coefficients δ_{ik} to the unknowns multiplying one by the other the unit graphs mentioned in item 4.

6. Calculate by the same procedure the free terms Δ_{ip}. For this purpose the unit graphs must be multiplied by the M_p graph due to the applied loads.

7. Solve the system of simultaneous equations with reference to the unknown actions X_1, X_2, . . ., X_n.

8. Compute the ordinates to the resulting bending moment curve by summing up the ordinates to the unit graphs multiplied previously by the magnitude of the corresponding action ** with the ordinates to the bending moment curve due to the actual loading.

One may also apply to the simple statically determinate structure all the redundant reactions and stresses just determined together with the actual loads, tracing thereafter the combined bending moment diagram. This diagram will coincide with that of the given redundant structure.

Let us proceed now with the solution of a few problems.

Problem 1. Trace the bending moment diagram for the portal frame of Fig. 14.9*a*. The moment of inertia of the crossbeam is twice as large as that of the uprights.

Solution. The portal frame under consideration being redundant to the first degree, the simple statically determinate structure may be obtained eliminating the horizontal constraint at the right-hand support (Fig. 14.9*b*).

◆

* All the above refers to structures, deformations of which remain practically uneffected by direct and shearing forces. If it were otherwise, it would be necessary to trace equally the diagrams for the shears and normal stresses due both to the unit actions ($\overline{Q}_i$ and $\overline{N}_i$) and to the applied loads (Q_p and N_p).

** It is strongly advised to trace new bending moment diagrams induced by the redundant reactions and not to alter the scale of the unit graphs traced previously, for the latter procedure is a source of frequent errors.

The bending moment graphs due to the applied loads and to a unit load acting along the direction of the eliminated constraint have been given in Fig. 29.8*b* and *c* of Art. 9.8. The equation expressing that the horizontal deflection of the lower end of the simple structure is nil becomes

$$X_1\delta_{11}+\Delta_{1p}=0$$

The coefficient δ_{11} will be given by

$$\delta_{11}=2\frac{h^2}{2}\cdot\frac{2}{3}h\cdot\frac{1}{EJ_1}+\frac{hah}{2EJ_1}=\frac{h^2}{6EJ_1}(4h+3a)$$

The deflection due to the applied load has been computed in Problem 2 of Art. 9.8

$$\Delta_{1p}=-\frac{Ph^2}{EJ_1}\left(\frac{h}{3}+\frac{a}{4}\right)=-\frac{Ph^2}{12EJ_1}(4h+3a)$$

Hence

$$X_1=-\frac{\Delta_{1p}}{\delta_{11}}=\frac{P}{2}$$

The resulting bending moment graph will be obtained by multiplying all the ordinates to the unit graph by $\frac{P}{2}$ and by adding them thereafter to the

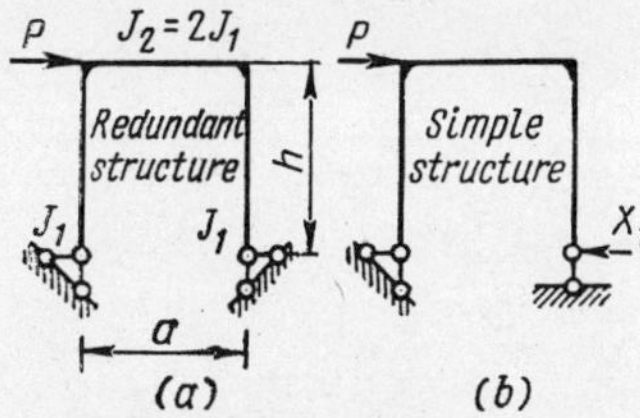

Fig. 14.9

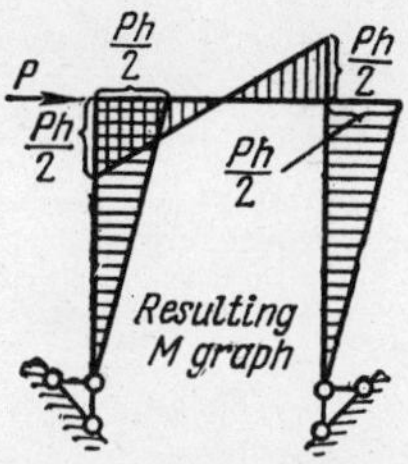

Fig. 15.9

ordinates to the bending moment diagram due to the applied loads. This graph appears in Fig. 15.9.

Problem 2. Trace the bending moment diagram for the redundant knee frame represented in Fig. 16.9*a*.

Solution. This structure is statically indeterminate to the second degree. Let us eliminate the constraints at the lower support obtaining thereby the simple statically determinate structure of Fig. 16.9*b*. The corresponding system of equations (4.9) becomes

$$X_1\delta_{11}+X_2\delta_{12}+\Delta_{1q}=0$$
$$X_1\delta_{21}+X_2\delta_{22}+\Delta_{2q}=0$$

The graphs of the bending moments induced by unit loads acting along the eliminated constraints as well as by the loads applied appear in Fig. 17.9. The coefficient δ_{11} will be obtained raising to the second power the $\overline{M}_1$ graph

$$\delta_{11}=\frac{1}{EJ}\left(\frac{a^2}{2}\cdot\frac{2}{3}a+a^2a\right)=\frac{4a^3}{3EJ}$$

The coefficient δ_{12} is given by the product of the $\overline{M}_1$ and $\overline{M}_2$ graphs

$$\delta_{12}=\delta_{21}=-\frac{1}{EJ}a\cdot a\cdot\frac{a}{2}=-\frac{a^3}{2EJ}$$

Raising to the second power the $\overline{M}_2$ graph, we obtain

$$\delta_{22}=\frac{1}{EJ}\cdot\frac{a^2}{2}\cdot\frac{2}{3}a=\frac{a^3}{3EJ}$$

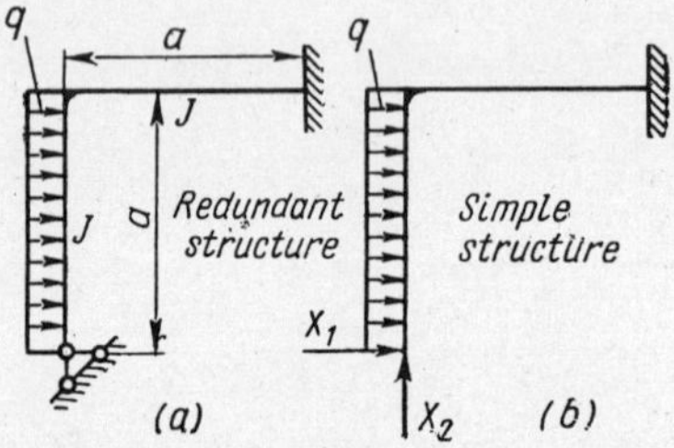

Fig. 16.9

The free terms of both equations will be obtained multiplying the $\overline{M}_1$ and the $\overline{M}_2$ graphs by the M_q graph

$$\Delta_{1q}=\frac{1}{EJ}\left(\frac{qa^2}{2}\cdot\frac{a}{3}\cdot\frac{3}{4}a+\frac{qa^2}{2}a\cdot a\right)=\frac{5qa^4}{8EJ}$$

$$\Delta_{2q}=-\frac{1}{EJ}\cdot\frac{qa^2}{2}\cdot a\frac{a}{2}=-\frac{qa^4}{4EJ}$$

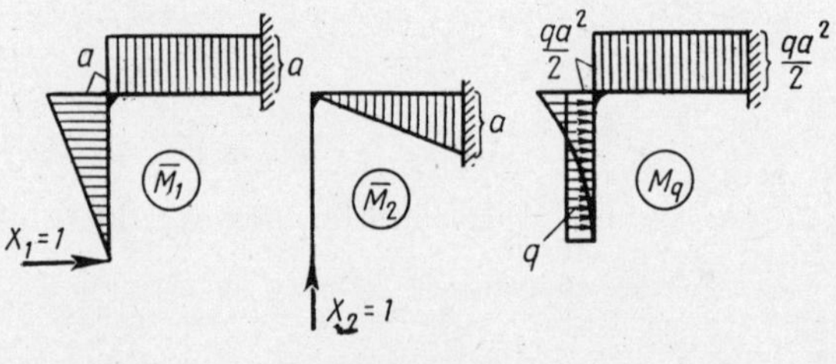

Fig. 17.9

Introducing the values so obtained into the system of equations and dividing both of these equations by $\frac{a^3}{EJ}$ we obtain

$$\frac{4}{3}X_1-\frac{1}{2}X_2+\frac{5}{8}qa=0$$

$$-\frac{1}{2}X_1+\frac{1}{3}X_2-\frac{1}{4}qa=0$$

The solution of these two equations yields

$$X_1 = -\frac{3}{7}\,qa; \qquad X_2 = \frac{3}{28}\,qa$$

In order to obtain the bending moment diagram for the redundant structure apply simultaneously to the simple statically determinate one both the actual loads and the unknown reactions just determined. Reaction X_1 must be directed towards the left, its value being negative (Fig. 18.9a). The expres-

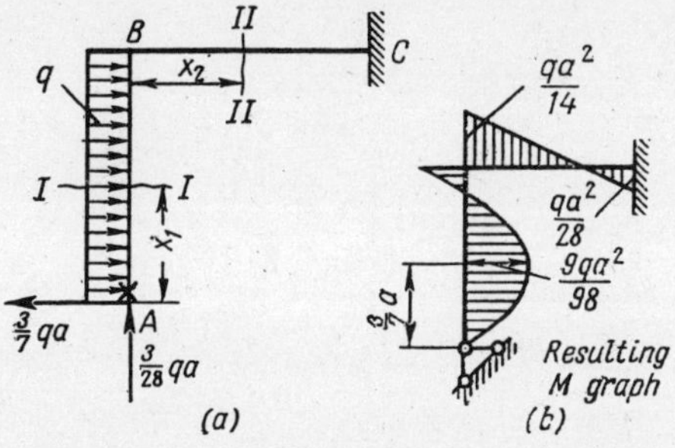

Fig. 18.9

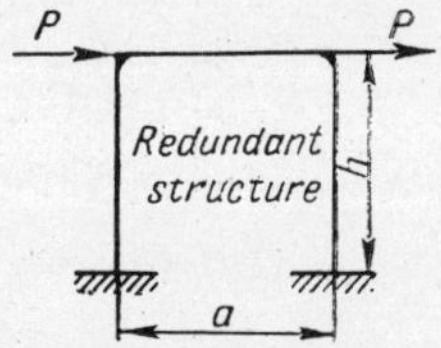

Fig. 19.9

sions of the bending moments acting in each member of the structure will be obtained as usual considering the lower end of the column as its left-hand extremity and marking this end with an asterisk.

Section I-I

$$M^I = \frac{3}{7}\,qax_1 - \frac{qx_1^2}{2}$$

$$\text{for } x_1 = 0 \qquad M^I = 0$$

$$\text{for } x_1 = \frac{a}{2} \qquad M^I = \frac{3}{7}\,qa\,\frac{a}{2} - \frac{qa^2}{8} = \frac{5}{56}\,qa^2$$

$$\text{for } x_1 = a \qquad M^I = \frac{3}{7}\,qaa - \frac{qa^2}{2} = -\frac{qa^2}{14}$$

The maximum value of M^I will be found equating to zero the first derivative of the above expression with reference to x_1

$$\frac{dM^I}{dx_1} = \frac{3}{7}\,qa - qx_1 = 0$$

wherefrom

$$x_1 = x_0 = \frac{3}{7}\,a$$

$$M^I_{max} = \frac{3}{7}\,qa\,\frac{3}{7}\,a - \frac{q}{2}\cdot\frac{9}{49}\,a^2 = \frac{9}{98}\,qa^2$$

Section II-II

$$M^{II} = \frac{3}{28}\,qax_2 + \frac{3}{7}\,qa^2 - \frac{qa^2}{2}$$

$$\text{for } x_2 = 0 \qquad M^{II} = -\frac{qa^2}{14}$$

$$\text{for } x_2 = a \qquad M^{II} = \frac{3}{28}\,qaa + \frac{3}{7}\,qa^2 - \frac{qa^2}{2} = \frac{qa^2}{28}$$

The resulting bending moment graph for the redundant structure is shown in Fig. 18.9*b*.

Problem 3. Trace the bending moment diagram for the portal frame of Fig. 19.9.

Solution. This structure is redundant to the third degree. Let us compare the three simple statically determinate structures shown in Fig. 20.9 with a view of choosing the one which will reduce to a minimum the amount

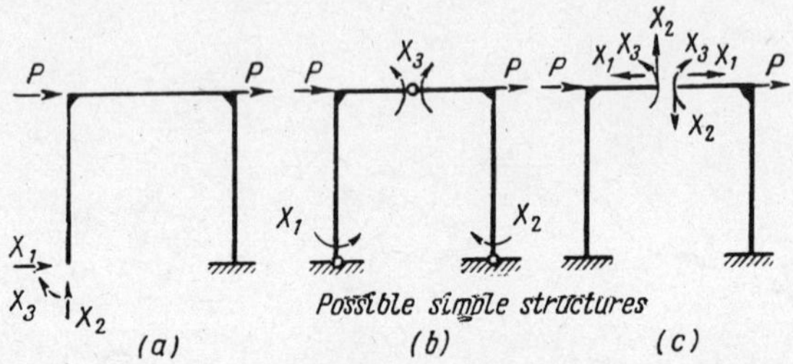

Fig. 20.9

of computation. From this view-point preference should be given to symmetrical systems, for in that case it becomes possible to trace unit bending moment graphs and to compute their products only for one half of the structure.

On these grounds the simple structure appearing in Fig. 20.9*a* should be rejected forthwith. Both structures of Fig. 20.9*b* and *c* are symmetrical

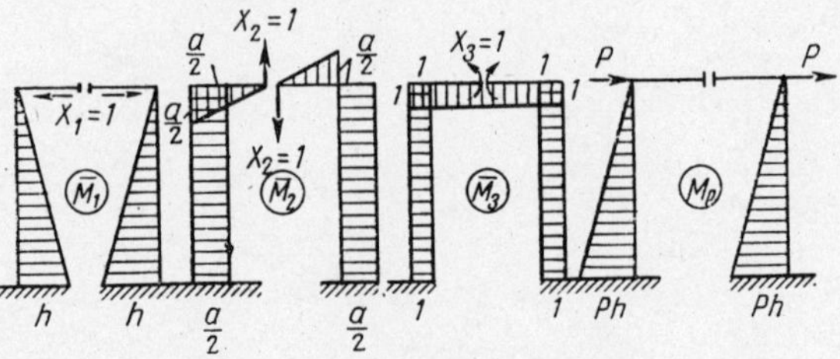

Fig. 21.9

but it will be easier to trace all the necessary bending moment graphs for the one appearing in Fig. 20.9*c*. Hence our choice will fall on the latter. The corresponding bending moment graphs are represented in Fig. 21.9.

Let us form the equations expressing that the mutual displacements of the two faces of the crossbeam on both sides of the cut are nil.

$$X_1\delta_{11}+X_2\delta_{12}+X_3\delta_{13}+\Delta_{1p}=0$$
$$X_1\delta_{21}+X_2\delta_{22}+X_3\delta_{23}+\Delta_{2p}=0$$
$$X_1\delta_{31}+X_2\delta_{32}+X_3\delta_{33}+\Delta_{3p}=0$$

Before proceeding with the computation of all the coefficients attention should be drawn to the fact that all the graphs can be subdivided into symmetrical ones ($\overline{M}_1$ and $\overline{M}_3$) and antisymmetrical ones ($\overline{M}_2$ and M_p). The ordinates to the left-hand and to the right-hand halves of the latter are equal in amount but opposite in sign, being situated on different sides of the corresponding members of the portal frame. It can be easily proved that *all the deflections computed multiplying symmetrical graphs by antisymmetrical ones will be always nil.* For this reason the following deflections appearing in the above equations will reduce to zero

$$\delta_{12},\quad \delta_{21},\quad \delta_{23},\quad \delta_{32},\quad \Delta_{1p} \text{ and } \Delta_{3p}$$

Consequently, the equations themselves become

$$X_1\delta_{11}+X_3\delta_{13}=0$$
$$X_1\delta_{31}+X_3\delta_{33}=0$$
$$X_2\delta_{22}+\Delta_{2p}=0$$

the first two leading immediately to

$$X_1=0 \text{ and } X_3=0$$

and the third one yielding

$$X_2=-\frac{\Delta_{2p}}{\delta_{22}}$$

Hence we need calculate only the deflections Δ_{2p} and δ_{22}

$$\Delta_{2p}=-\frac{2Ph^2}{2}\cdot\frac{a}{2}\cdot\frac{1}{EJ}=-\frac{Ph^2a}{2EJ}$$

$$\delta_{22}=\left[\left(\frac{a}{2}\right)^2\cdot\frac{1}{2}\cdot\frac{2}{3}\cdot\frac{a}{2}+\frac{a}{2}\cdot h\cdot\frac{a}{2}\right]\frac{2}{EJ}=$$

$$=\frac{a^2}{12EJ}(a+6h)$$

Consequently

$$X_2=\frac{6Ph^2}{a(a+6h)}$$

The final diagram of the bending moments acting in the redundant structure may be now obtained multiplying the ordinates to the $\overline{M}_2$ graph by the magnitude of X_2 and summing them up with the ordinates to the M_p graph due to the actual loading. This bending moment diagram is represented in Fig. 22.9.

Problem 4. Determine the thrust developed at the abutments of the two-hinged arch appearing in Fig. 23.9*a*. The neutral line of this arch follows a parabolic curve given by

$$y=\frac{4f}{l^2}(l-x)x$$

The rise of this arch is less than one fifth of its span and the stiffness of its cross sections remains constant and equal to EJ.

Solution. Let us regard as redundant the horizontal component (thrust) of the reaction developed at the left-hand support (Fig. 23.9*b*). This leads to

$$X_1\delta_{11}+\Delta_{1p}=0 \qquad (6.9)$$

Seeing that the neutral line of the arch is a curve, Vereshchagin's method becomes inoperative and therefore Mohr's integrals will have to be computed analytically. For flat arches this problem is not very complicated for without appreciable error ds may be replaced by dx and $\cos\varphi$ may be taken equal to 1. The integration will be carried out between $x = 0$ and $x = l$. The angle φ just mentioned is the angle formed by the tangent to the neutral line and the x-axis. It should be remembered that normal stresses must be taken into

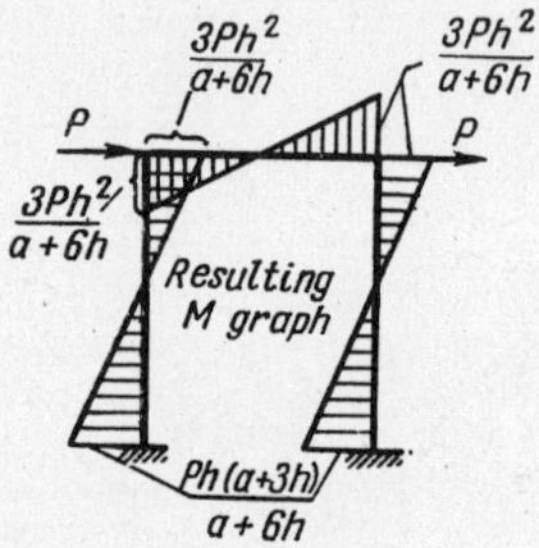

Fig. 22.9

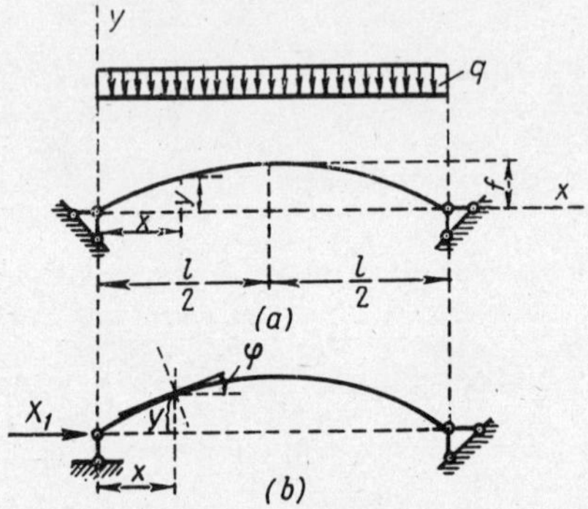

Fig. 23.9

consideration when computing the horizontal displacement of flat arches due to horizontal loads. Hence the coefficient δ_{11} will be calculated using the expression

$$\delta_{11}=\int_0^s \frac{\overline{M}_1^2\,ds}{EJ}+\int_0^s \frac{\overline{N}_1^2\,ds}{EF}$$

where

$$\overline{M}_1=-1y=-\frac{4f}{l^2}(l-x)\,x$$

and

$$\overline{N}_1=-1\cos\varphi=-\cos\varphi$$

Replacing ds by dx and putting $\cos\varphi = 1$ we obtain

$$\delta_{11}=\frac{16f^2}{EJl^4}\int_0^l x^2(l-x)^2\,dx+\frac{1}{EF}\int_0^l dx=\frac{8}{15}\,\frac{f^2l}{EJ}+\frac{l}{EF}$$

The displacement Δ_{1p} due to vertical loads will be determined using the formula

$$\Delta_{1p}=\int_0^s \frac{\overline{M}_1 M_p\,ds}{EJ}$$

where

$$M_p=\frac{ql}{2}\,x-q\,\frac{x^2}{2}=q\,\frac{x}{2}\,(l-x)$$

This leads to

$$\Delta_{1p}=\int_0^s -\frac{1}{EJ}\frac{4f}{l^2}(l-x)\,x\,\frac{qx}{2}(l-x)\,ds=-\frac{2qf}{l^2EJ}\times$$

$$\times\int_0^l (l-x)^2x^2\,dx=-\frac{2qf}{l^2EJ}\int_0^l (l^2x^2-2lx^3+x^4)\,dx=$$

$$=-\frac{2qf}{l^2EJ}\left(l^2\frac{l^3}{3}-2l\frac{l^4}{4}+\frac{l^5}{5}\right)=-\frac{qfl^3}{15EJ}$$

The solution of equation (6.9) yields immediately the value of the desired thrust X_1

$$X_1=\frac{ql^2}{8f+\frac{15}{f}\cdot\frac{J}{F}}$$

Problem 5. Required the stresses in all the elements of the framed structure appearing in Fig. 24.9*a*. All the members of this structure are of the same cross section. Bars *5* and *6* have no common hinge at midlength.

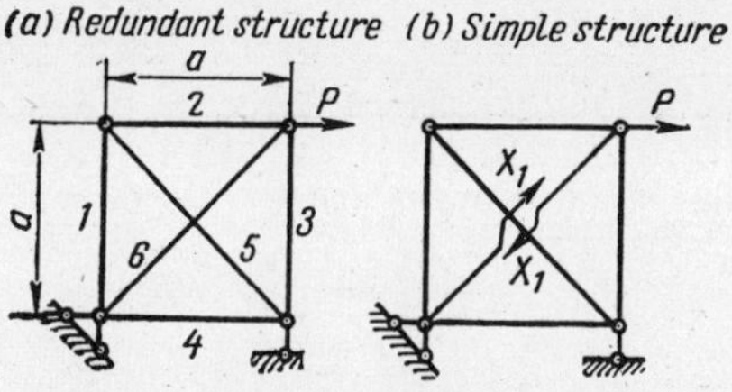

Fig. 24.9

Solution. Since the structure is redundant to the first degree we may obtain the simple statically determinate one by cutting diagonal *6* (Fig. 24.9*b*). The corresponding equation will be of the standard form

$$X_1\delta_{11}+\Delta_{p1}=0$$

The deflections δ_{11} and Δ_{1p} may be obtained using the expressions developed previously for through structures

$$\delta_{11}=\Sigma\frac{\overline{N}_i^2 l}{EF}=\frac{1}{EF}\Sigma N_i^2 l$$

$$\Delta_{1p}=\Sigma\frac{\overline{N}_i N_p l}{EF}=\frac{1}{EF}\Sigma\overline{N}_i N_p l$$

In these expressions $\overline{N}_i$ are the stresses induced in the different bars by the unit load $X_1=1$, and N_p are the stresses due to the applied loads.

All the necessary calculations are given in Table 1.9. The column which should normally contain the cross-sectional areas of all the bars has been omitted, these areas remaining constant throughout the structure.

Table 1.9

Bar No.	l	$\overline{N}_i$	N_p	$\overline{N}_i N_p l$	$\overline{N}_i^2 l$
1	a	$-\frac{1}{\sqrt{2}}$	P	$-\frac{Pa}{\sqrt{2}}$	$\frac{a}{2}$
2	a	$-\frac{1}{\sqrt{2}}$	P	$-\frac{Pa}{\sqrt{2}}$	$\frac{a}{2}$
3	a	$-\frac{1}{\sqrt{2}}$	0	0	$\frac{a}{2}$
4	a	$-\frac{1}{\sqrt{2}}$	P	$-\frac{Pa}{\sqrt{2}}$	$\frac{a}{2}$
5	$a\sqrt{2}$	1	$-P\sqrt{2}$	$-2Pa$	$a\sqrt{2}$
6	$a\sqrt{2}$	1	0	0	$a\sqrt{2}$
Total				$-\frac{Pa}{\sqrt{2}}(3+2\sqrt{2})$	$(2+2\sqrt{2})\,a$

Using the data contained in the above table we may easily compute the values of Δ_{1p} and δ_{11}

$$\Delta_{1p}=-\frac{Pa}{EF\sqrt{2}}(3+2\sqrt{2})$$

$$\delta_{11}=\frac{a}{EF}(2+2\sqrt{2})$$

Introducing these values into the standard equation we obtain

$$X_1=P\frac{3+2\sqrt{2}}{2\sqrt{2}+4}$$

Stresses in all the different members of the given structure will be readily obtained using expression

$$N_i=N_{ip}+\overline{N}_i X_1$$

The first term of the right-hand part of this expression represents the stress induced in the corresponding member of the simple structure by the applied loads, and the second the stress induced in the same member by the reaction of the redundant constraint. This expression constitutes thus one of the applications of the principle of superposition. For bar 2, the total stress will equal

$$N_2=N_{2p}+\overline{N}_2 X_1=P+\left(-\frac{1}{\sqrt{2}}\right)P\frac{3+2\sqrt{2}}{2\sqrt{2}+4}=P\frac{1+2\sqrt{2}}{4(1+\sqrt{2})}$$

Problem 6. Required the stresses in a trussed beam appearing in Fig. 25.9*a*. The *main beam* whose length is equal to the total span of the structure works in bending, while the *reinforcing members* work in direct tension or compression just as those of an ordinary truss. We shall assume that the cross sections of all the reinforcing members remain constant. The displacement of the two different parts of the structure will be calculated using expressions peculiar to the type of stress developed in each of these parts.

Solution. The simple statically determinate structure will be obtained by cutting bar *1-2* as indicated in Fig. 25.9*b*. The equation will be of the standard form

$$X_1\delta_{11}+\Delta_{1p}=0$$

The values of the coefficient δ_{11} to the unknown X_1 and of the free term Δ_{1p} will be obtained using bending moment graph for the main beam (Fig. 26.9)

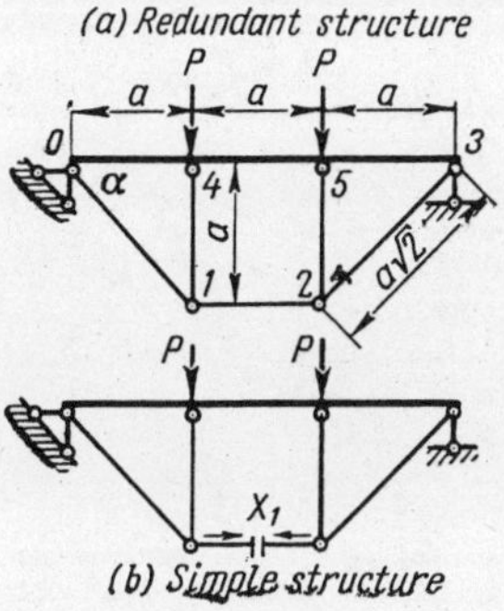

Fig. 25.9

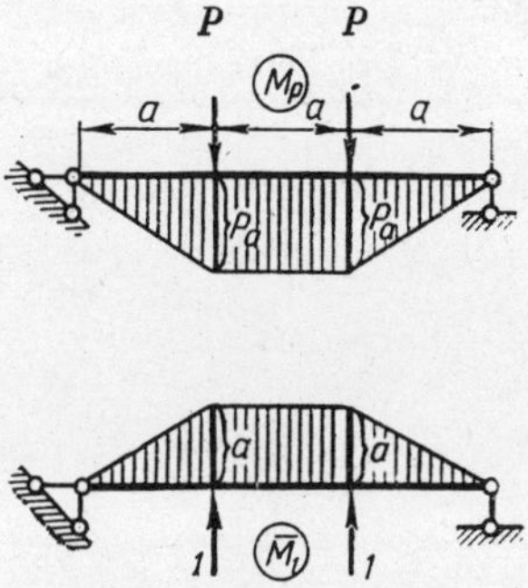

Fig. 26.9

and the values of the normal stresses developed in the reinforcing members. The unit stress X_1 acting along the horizontal bar *1-2* will produce a compression in both the queen posts, the magnitude of this compression in this particular case being also equal to unity. These stresses will be transmitted directy to the main beam. The following table gives the amounts of stresses in all the bars of the auxiliary system.

Table 2.9

Bar No.	l	$\overline{N}_i$	N_p	$\overline{N}_i N_p l$	$\overline{N}_i^2 l$
0-1	$a\sqrt{2}$	$\sqrt{2}$	0	0	$2a\sqrt{2}$
1-2	a	1	0	0	a
2-3	$a\sqrt{2}$	$\sqrt{2}$	0	0	$2a\sqrt{2}$
1-4	a	—1	0	0	a
2-5	a	—1	0	0	a
Total	—	—	—	0	$a(3+4\sqrt{2})$

The values of displacements δ_{11} and Δ_{1p} are obtained as follows

$$\delta_{11}=\frac{1}{EJ}\left(\frac{a^2}{2}\cdot\frac{2}{3}a\cdot 2+a^2a\right)+\frac{1}{EF}a(3+4\sqrt{2})=$$

$$=\frac{5a^3}{3EJ}+\frac{a}{EF}(3+4\sqrt{2})$$

$$\Delta_{1p}=\frac{1}{EJ}\left[\frac{Pa^2}{2}\left(-\frac{2}{3}\right)a\cdot 2-Pa^2a\right]=-\frac{5Pa^3}{3EJ}$$

Introducing the latter values into the standard equation and solving it for X_1 we obtain

$$X_1=\frac{P}{1+\frac{3}{5}\cdot\frac{EJ(3+4\sqrt{2})}{EFa^2}}$$

The diagrams of the bending moments acting in the main beam and of the normal stresses in the auxiliary members of the redundant structure will

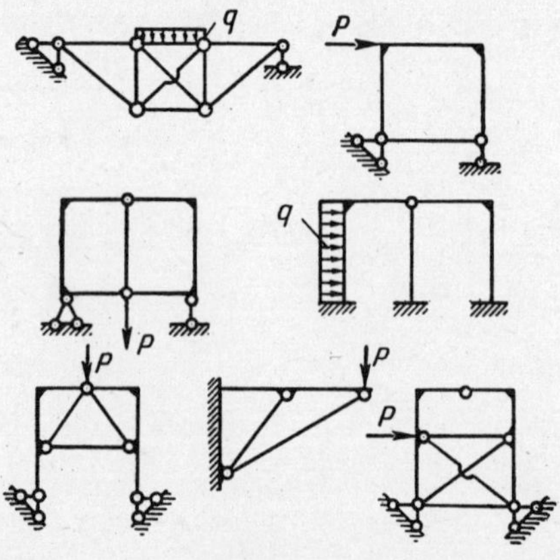

Fig. 27.9

be obtained as usual by the summation of the ordinates to the stress curve due to the applied loads with the ordinates to the unit curves multiplied by the magnitude of X_1.

Referring to the different structures appearing in Fig. 27.9 the student will

(1) determine their degree of redundancy;

(2) find appropriate simple statically determinate structures and choose the better ones;

(3) form the corresponding systems of canonical equations;

(4) trace the diagram of the unit stresses;

(5) compute one of the unit displacements δ for each of those structures.

4.9. *Stresses in Redundant Structures due to Temperature Changes*

In the event of temperature changes the standard equations used in the method of forces become

$$\left.\begin{array}{l} X_1\delta_{11}+X_2\delta_{12}+X_3\delta_{13}+\ldots+X_n\delta_{1n}+\Delta_{1t}=0 \\ X_1\delta_{21}+X_2\delta_{22}+X_3\delta_{23}+\ldots+X_n\delta_{2n}+\Delta_{2t}=0 \\ \ldots\ldots\ldots\ldots\ldots\ldots\ldots\ldots \\ \ldots\ldots\ldots\ldots\ldots\ldots\ldots\ldots \\ X_1\delta_{n1}+X_2\delta_{n2}+X_3\delta_{n3}+\ldots+X_n\delta_{nn}+\Delta_{nt}=0 \end{array}\right\} \quad (7.9)$$

In these equations the coefficients δ_{11}, δ_{12}, . . ., δ_{nn} have the same meaning as heretofore, while the terms Δ_{1t}, Δ_{2t}, . . ., Δ_{nt} are the deflections of the conjugate simple structure along the redundant constraints X_1, X_2, . . ., X_n caused by the thermal gradient.

These deflections may be obtained using expressions (19.8) and (20.8) developed in Art. 7.8. *

$$\Delta_{mt}=\Sigma\alpha\frac{t_1-t_2}{h}\int_0^l \overline{M}_m\,dx+\Sigma\alpha\frac{t_1+t_2}{2}\int_0^l \overline{N}_m\,dx \quad (19.8)$$

or

$$\Delta_{mt}=\Sigma\alpha\frac{t_1-t_2}{h}\Omega_{\overline{M}}+\Sigma\alpha\frac{t_1+t_2}{2}\Omega_{\overline{N}} \quad (20.8)$$

Equations (7.9) express as usual the idea that the deflections of the simple determinate structure along the directions of the eliminated constraints remain nil.

Problem. Determine the stresses induced in a one-time redundant structure appearing in Fig. 28.9 and trace the corresponding bending moment curves assuming that the indoor temperature rises by 10°C while the outdoor one remains unchanged.

Solution. Adopt the simple structure appearing in Fig. 29.9 for which the standard equation becomes

$$X_1\delta_{11}+\Delta_{1t}=0 \quad (8.9)$$

◆

* We shall admit that the cross sections of all the elements involved are symmetrical about the horizontal gravity axis. Were it otherwise $\frac{t_1+t_2}{2}$ should be replaced by $t_2+\frac{t_1-t_2}{h}\,y$ where y is the distance from the fibre heated so t_2 to the centroid of the cross sections.

Using expression (20.8) just cited we obtain

$$\Delta_{1t} = -\alpha\,\frac{10-0}{h}\left(aa+\frac{aa}{2}\right) - \alpha\,\frac{10+0}{2}\,a = -15\alpha\,\frac{a^2}{h} - 5\alpha a$$

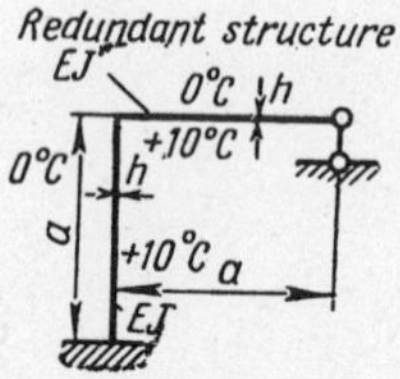

Fig. 28.9

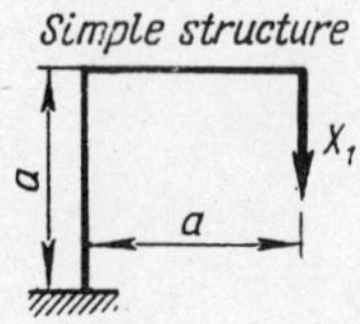

Fig. 29.9

As for δ_{11} its value will be found raising to the second power the area of the $\overline{M}_1$ graph (Fig. 30.9)

$$\delta_{11} = \frac{1}{EJ}\cdot\frac{aa}{2}\cdot\frac{2}{3}\,a + a^3\,\frac{1}{EJ} = \frac{4a^3}{3EJ}$$

Introducing these values into equation (8.9) and solving the same we obtain

$$X_1 = -\frac{\Delta_{1t}}{\delta_{11}} = \frac{5\alpha a\left(\frac{3a}{h}+1\right)3EJ}{4a^3} = \frac{15\alpha EJ\left(\frac{3a}{h}+1\right)}{4a^2}$$

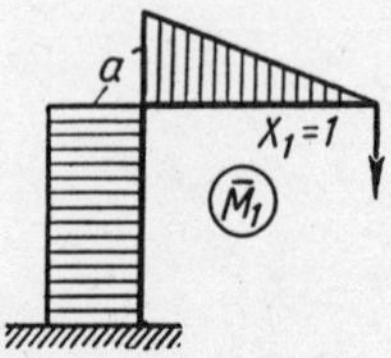

Fig. 30.9

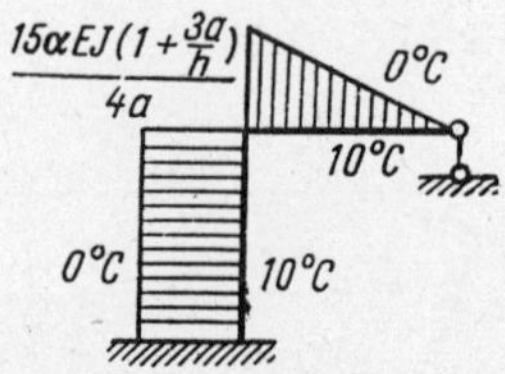

Fig. 31.9

The bending moment diagram induced in the given redundant structure by the given temperature change can now be obtained multiplying all the ordinates to the $\overline{M}_1$ curve by X_1. This diagram is represented in Fig. 31.9.

5.9. *Stresses in Redundant Structures Caused by the Movement of Supports*

As already mentioned, statically indeterminate structures may become severely stressed not only due to the application of external loads or due to temperature changes but also in the event when one or more of their supports suffer a linear translation, an angular rotation or both.

Let us study this problem using as an example the portal frame of Fig. 32.9*a*. The shape taken by this frame after the right-hand support has shifted from B to B' is indicated in the same figure by dash lines. The horizontal and vertical displacements of the support will be taken equal to a and b respectively and its angular rotation to φ.

The influence exercised by the simple structure adopted on the formation of the standard equations will be investigated using the examples of Fig. 32.9*b*, *c* and *d*. In the first case (Fig. 32.9*b*) the directions of the redundant constraints coincide exactly with

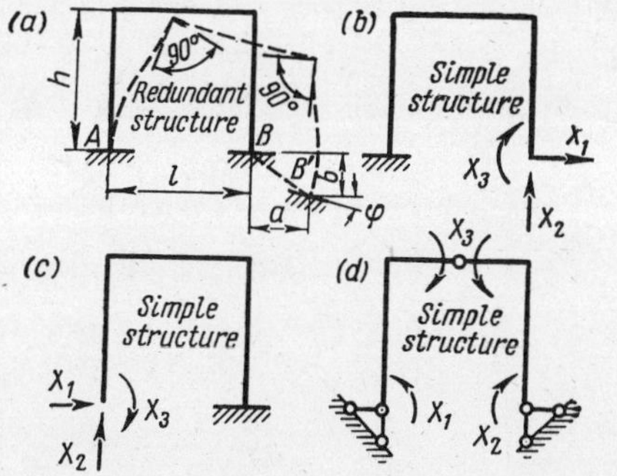

Fig. 32.9

those of the support displacements. Thus, the unknown reaction X_1 follows the direction of the horizontal displacement, the reaction X_2 that of the vertical one (though being opposite in sign) while the moment X_3 acts along the direction of the rotation suffered by the cross section at the support. The magnitudes of these reactions must be such as to render the displacements of the simple statically determinate structure exactly equal to those stipulated in the problem. Hence the canonical equations expressing this idea will be of the following form

$$\left.\begin{aligned} X_1\delta_{11}+X_2\delta_{12}+X_3\delta_{13}&=a\\ X_1\delta_{21}+X_2\delta_{22}+X_3\delta_{23}&=-b\\ X_1\delta_{31}+X_2\delta_{32}+X_3\delta_{33}&=\varphi \end{aligned}\right\} \tag{9.9}$$

The negative value of the last term of the second equation is due to the fact that reaction X_2 is directed upwards while the support moves downwards.

On the other hand, if the simple structure of Fig. 32.9*c* were adopted it would become necessary to regard the displacements of the support B as a system of external loads. This would be reflected by the introduction into the canonical equations of free terms corresponding to the said loads, these terms being designated as usual by $\Delta_{1\Delta}$, $\Delta_{2\Delta}$ and $\Delta_{3\Delta}$. It is clear that these terms will have the following values (see Art. 14.8)

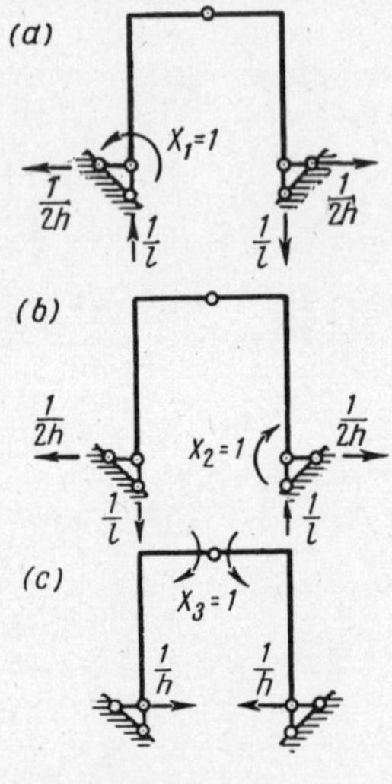

Fig. 33.9

$$\Delta_{1\Delta}=a;\quad \Delta_{2\Delta}=-b+l\varphi;\quad \Delta_{3\Delta}=\varphi$$

Consequently, the canonical equations will take the following shape

$$\left.\begin{aligned} X_1\delta_{11}+X_2\delta_{12}+X_3\delta_{13}+a=0\\ X_1\delta_{21}+X_2\delta_{22}+X_3\delta_{23}-b+l\varphi=0\\ X_1\delta_{31}+X_2\delta_{32}+X_3\delta_{33}+\varphi=0 \end{aligned}\right\} \quad (10.9)$$

For the simple structure of Fig. 32.9*d* these same equations would become

$$\left.\begin{aligned} X_1\delta_{11}+X_2\delta_{12}+X_3\delta_{13}+\Delta_{1\Delta}=0\\ X_1\delta_{21}+X_2\delta_{22}+X_3\delta_{23}+\Delta_{2\Delta}=\varphi\\ X_1\delta_{31}+X_2\delta_{32}+X_3\delta_{33}+\Delta_{3\Delta}=0 \end{aligned}\right\} \quad (11.9)$$

Here $\Delta_{1\Delta}$, $\Delta_{2\Delta}$ and $\Delta_{3\Delta}$ are the displacements of the conjugate simple structure along the directions of X_1, X_2 and X_3 due to the vertical and horizontal movements of the right-hand support.*

It was shown in Art. 14.8 that these displacements are readily computed using expression

$$X_i\Delta_{i\Delta}+\Sigma R\Delta=0$$

in which the left part represents the work accomplished by the forces of the imaginary state along the displacements of the simple structure due to the motion of the supports. In this case the imaginary state of the simple structure permitting the determination of the angular rotation along X_1 due to the displacements of the right-hand support is that of Fig. 33.9*a*. Hence the work accomplished by the forces of the imaginary state along the displacements of the simple structure when its right-hand support is moved both vertically and horizontally will be expressed by

$$X_1\Delta_{1\Delta}+\frac{1}{2h}a+\frac{1}{l}b=0$$

wherefrom

$$\Delta_{1\Delta}=-\frac{a}{2h}-\frac{b}{l}$$

* The angular rotation of the right-hand support will produce no displacement of the simple structure along the directions of X_1, X_2 and X_3.

Similarly, the work accomplished by the forces of the second imaginary state shown in Fig. 33.9*b* along the displacements of the simple structure due to the movements of the same support will be given by

$$X_2\Delta_{2\Delta}+\frac{1}{2h}a-\frac{1}{l}b=0$$

wherefrom

$$\Delta_{2\Delta}=-\frac{a}{2h}+\frac{b}{l}$$

As for the displacement $\Delta_{3\Delta}$ it will be obtained from the equation corresponding to Fig. 33.9*c*.

$$X_3\Delta_{3\Delta}-\frac{1}{h}a=0$$

wherefrom

$$\Delta_{3\Delta}=\frac{a}{h}$$

Introducing these values into equations (11.9) we obtain

$$\left.\begin{aligned} X_1\delta_{11}+X_2\delta_{12}+X_3\delta_{13}-\left(\frac{a}{2h}+\frac{b}{l}\right)&=0\\ X_1\delta_{21}+X_2\delta_{22}+X_3\delta_{23}-\left(\frac{a}{2h}-\frac{b}{l}\right)&=\varphi\\ X_1\delta_{31}+X_2\delta_{32}+X_3\delta_{33}+\frac{a}{h}&=0 \end{aligned}\right\} \qquad (12.9)$$

It should be remembered that each term of the left part of these equations represents the deflection of the simple statically determinate structure along the direction of a redundant reaction induced either by this same reaction or by the movement of the support at B.

All the equations of the present article have been thus based on the principle of superposition. It may be easily shown that these same equations may be based on the theorem of reciprocal works. Indeed let us consider two different states of the same simple statically determinate structure, namely those represented in Fig. 34.9*a* and *b*. Using the above theorem we obtain immediately

$$X_1\delta_{11}+X_2\delta_{21}+X_3\delta_{31}=1a \qquad (13.9)$$

The left-hand part of this equation represents the work accomplished by the applied loads of Fig. 34.9*a* along the deflections of the imaginary state of Fig. 34.9*b* while the right-hand part—that done by the unit load $X_1 = 1$ of the first imaginary state along the actual displacement equal to a.

Exactly the same reasoning will lead to the formation of the two following equations (Fig. 35.9*a* and *b*)

$$\left.\begin{aligned} X_1\delta_{12}+X_2\delta_{22}+X_3\delta_{32}&=-1b\\ X_1\delta_{13}+X_2\delta_{23}+X_3\delta_{33}&=1\varphi \end{aligned}\right\} \tag{14.9}$$

The left parts of these equations represent the work accomplished by the applied loads along the imaginary displacements induced by the unit loads of the second and third imaginary states of Fig. 35.9*a* and *b* while the right-hand parts—those accomplished by the imaginary unit loads of the two latter states along the given displacements of the support.

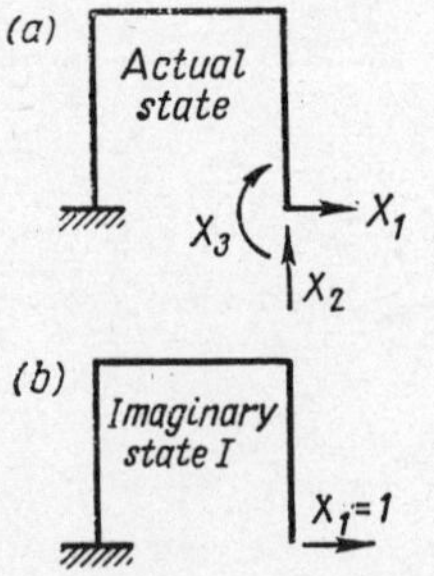

Fig. 34.9

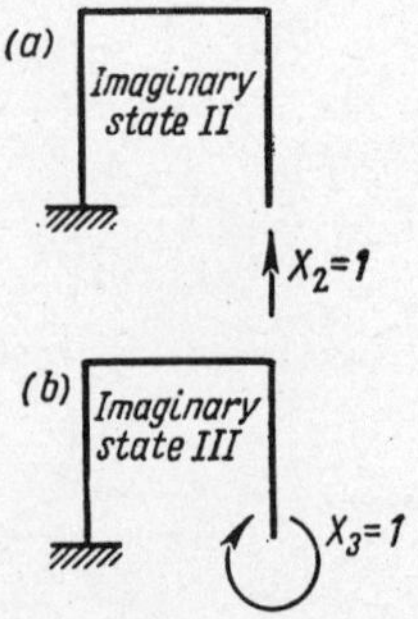

Fig. 35.9

Comparing equations (9.9) obtained previously using the simple structure of Fig. 32.9*b* with those based on the theorem of reciprocal works [equations (13.9) and (14.9)], it becomes immediately apparent that the two systems are absolutely identical for $\delta_{12} = \delta_{21}$, $\delta_{13} = \delta_{31}$ and $\delta_{23} = \delta_{32}$.

Nevertheless the basic ideas conveyed by these two systems of equations are entirely different. Indeed, the equations based on the principle of superposition express that the sum of displacements along the directions of the redundant constraints are either nil or equal to predetermined amounts; as for those based on the theorem of reciprocal works, they express that the work on the simple statically determinate structure accomplished by the applied loads along the displacements of this same structure caused by any one of the imaginary unit loads is equal to the work produced by the said unit load (together with the support reactions due to this load) along the displacement caused by the actual loading.

For exercise let us use once again the theorem of reciprocal works for the determination of stresses in the same portal frame, adopting for conjugate simple structure the one appearing in Fig. 36.9*a*.

The corresponding imaginary states are given in Fig. 36.9*b*, *c* and *d*.

The standard equations based on the principle of reciprocal works become

$$\left.\begin{aligned} X_1\delta_{11}+X_2\delta_{21}+X_3\delta_{31}&=\frac{b}{l}+\frac{a}{2h}\\ X_1\delta_{12}+X_2\delta_{22}+X_3\delta_{32}&=\varphi+\frac{a}{2h}-\frac{b}{l}\\ X_1\delta_{13}+X_2\delta_{23}+X_3\delta_{33}&=-\frac{a}{h}\end{aligned}\right\} \qquad (15.9)$$

Comparing the latter set of equations with those derived from the principle of superposition [equations (12.9)] we see once again that they are absolutely identical.

In actual design practice it is more convenient to base the equations on the principle of reciprocal works when solving problems connected with the settlement of supports, the equations so obtaind affording a clearer picture of the physical reality. The same method could be used for stress analysis of redundant structures subjected to a system of external loads, but it would lose the advantage just mentioned, for in the latter case the principle of superposition gives a better representation of the phenomenon.

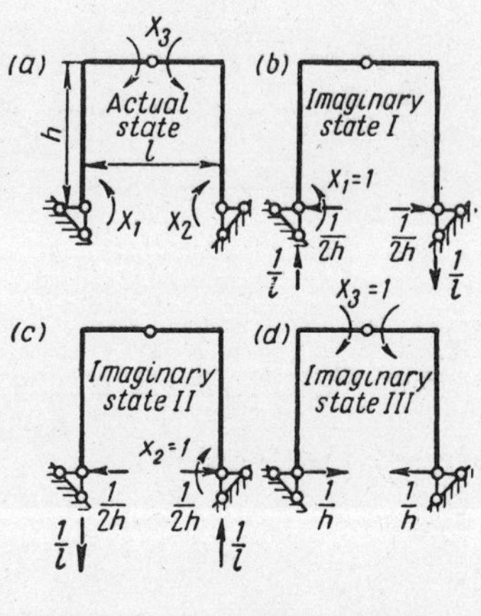

g. 36.9

6.9. *Diagrams for Shearing and Direct Stresses. Checking of Diagrams*

Once all the redundant stresses and reactions X_1, X_2, . . . , X_n have been found, one may proceed with the determination of shearing and normal forces acting in the structure under consideration. These will be exactly the same as those arising in the simple statically determinate structure under the combined action of the applied loads and of the said redundant stresses and reactions.

The same results may be achieved using the bending moment curves obtained for the given redundant structure as described in the previous articles. Indeed let us isolate from the rest of this structure a rectilinear bar AB and let l be its length (Fig. 37.9*a*). In the most general case this bar will be acted upon:

(a) by the loads actually applied within its limits;

(b) by the bending moments M_{AB} and M_{BA} at the end sections, the magnitudes of these bending moments may be scaled off directly from the corresponding diagram;

(c) by the shearing forces Q_{AB} and Q_{BA} as well as by the normal stresses N_{AB} and N_{BA} developed at the same cross sections.

Here and after the first of the two indices will indicate the position of the cross section, while both of these indices together will designate the member containing this section. Thus, M_{AB} will mean the bending moment acting at section A of bar AB.

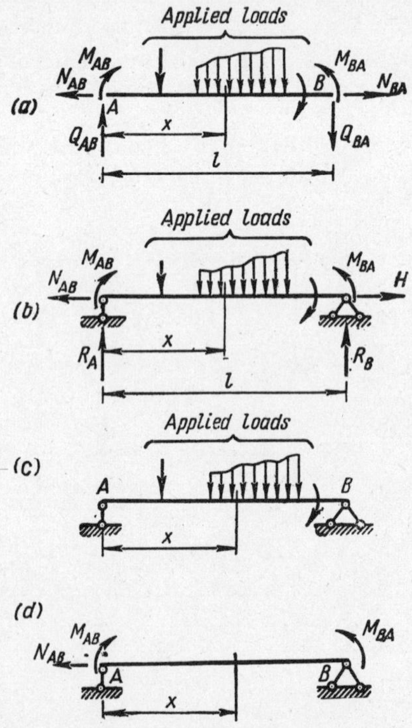

Fig. 37.9

Since the bar AB is in equilibrium the stresses Q_{AB}, Q_{BA} and N_{BA} may be regarded as the vertical and horizontal reactions of an end-supported beam appearing in Fig. 37.9*b*, which we shall call as before the reference beam. It follows that the stresses acting at any cross section of the said reference beam and those existing in the corresponding section of the given structure will be absolutely identical. Hence the bending moment at any cross section x of the bar AB will equal the sum of the bending moments induced in the corresponding section of the reference beam by all the actions shown in Fig. 37.9*c* and *d*

$$M = M+{}^0M_{AB}+\frac{M_{BA}-M_{AB}}{l}x$$

where M^0 represents the bending moment produced in the reference beam by the external loads of Fig. 37.9*c* while $M_{AB}+\frac{M_{BA}-M_{AB}}{l}x$ is the moment arising from the application of moments M_{AB} and M_{BA} to its end sections (Fig. 37.9*d*).

The theorem of Zhuravsky stating that the first derivative of the bending moment represents the shear in the same cross section we may write

$$Q=\frac{dM}{dx}=\frac{dM^0}{dx}+\frac{M_{BA}-M_{AB}}{l}=Q^0+\frac{M_{BA}-M_{AB}}{l} \qquad (16.9)$$

Here Q^0 is the shear induced in the corresponding cross section of the reference beam by the loads directly applied thereto (Fig. 37.9*c*).

The above expression permits the determination of the bending moments and shears in any section of a rectilinear member belonging to a redundant framed structure provided the loads directly applied to this member and the bending moments acting at the end sections are known.

When the bending moment curves are traced on the side of the extended fibres, the sign of the shearing forces may be ascertained as follows: *the shear will be reckoned positive if the axis of the member must be rotated clockwise in order to come in coincidence with the tangent to the bending moment curve, provided the angle of rotation is smaller than 90°. Numerically the shear is directly proportional to the value of the natural tangent of this angle.* This rule presented in Art. 1.2 permits the immediate determination of the shear sign for any cross section of bar *AB*.

The direction of the shearing force will be obtained remembering that a positive shear will always tend to rotate clockwise the section it is acting upon about the far end of that same part of the member.

Normal stresses will be determined isolating in succession each joint of the structure and applying thereto both the actual loads and the shearing forces obtained as described above. One could also use the procedure outlined at the beginning of this article.

Problem. Trace the Q and the N curves for the portal frame appearing in Fig. 38.9 together with the diagrams of bending moments acting in all of its members.

Solution. First trace the shear diagram for column *1-2*. No external load being applied to this member, the bending moment diagram forms a straight

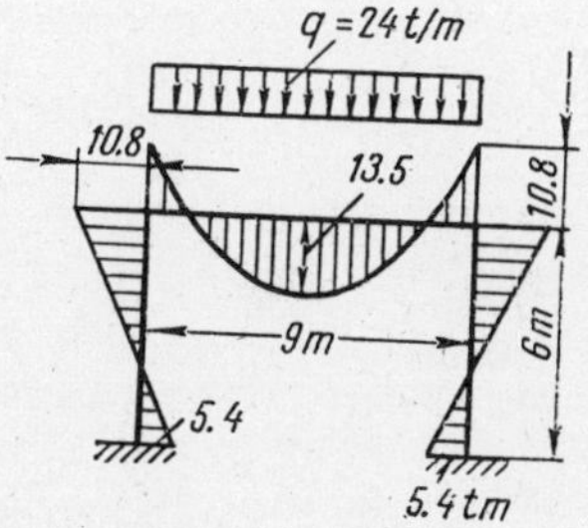

Fig. 38.9

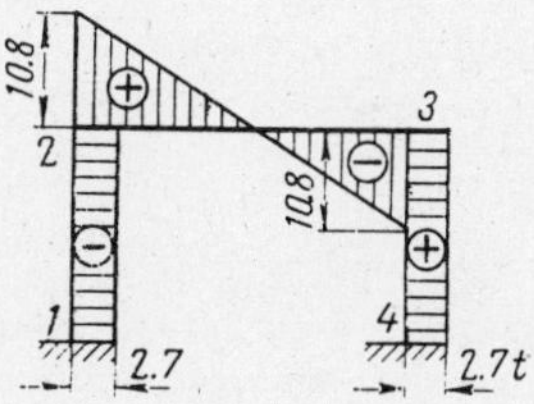

Fig. 39.9

line and therefore the shear will remain constant. It will be reckoned negative for the column axis must be rotated counterclockwise to come in coincidence with the tangent to the bending moment diagram (the two coinciding in that particular case). Numerically the shearing force will equal the natural tangent of the aforesaid angular rotation, viz.,

$$Q_{12}=Q_{21}=-\frac{10.8+5.4}{6}=-2.7 \text{ tons}$$

The same results would have been obtained through the application of the formula given at the beginning of the present article

$$Q_{12}=Q_{21}=Q_{12}^{0}+\frac{M_{21}-M_{12}}{l_{12}}$$

Seeing that no load is directly applied to the column in question, the shear Q_{12}^{0} will reduce to zero and therefore

$$Q_{12}=Q_{21}=\frac{M_{21}-M_{12}}{l_{12}}=\frac{-10.8-(+5.4)}{6}=-2.7 \text{ tons}$$

When computing the shearing forces each of the members should be placed mentally in a horizontal position; the bending moments reckoned positive

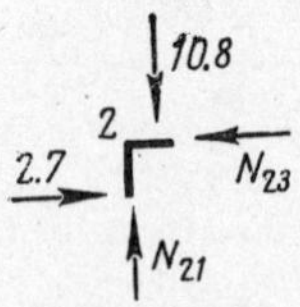

Fig. 40.9

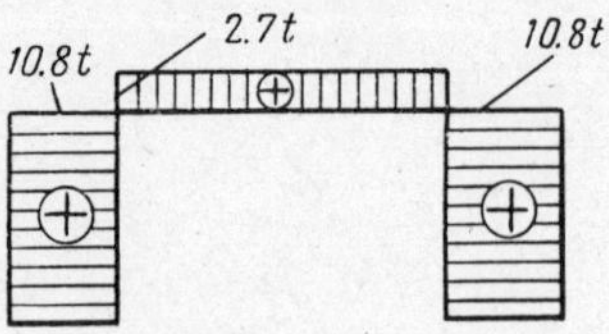

Fig. 41.9

will then produce an extension of the lower fibres of this member and those reckoned negative—an extension of its upper fibres.

The shearing force in the right-hand column will be determined in exactly the same way and will equal +2.7 tons.

As for the shear in the crossbeam, its value at any section situated a distance x from joint *2* will be given by

$$Q=Q^{0}+\frac{M_{32}-M_{23}}{l_{23}}=+\frac{2.4\times 9}{2}-2.4x+\frac{-10.8-(-10.8)}{9}=$$
$$=+10.8-2.4x$$

When $x=0$ (that is, immediately to the right of joint *2*)

$$Q_{23}=+10.8 \text{ tons}$$

and when $x=9$ metres (that is, immediately to the left of joint *3*)

$$Q_{32}=+10.8-2.4\times 9=-10.8 \text{ tons}$$

The diagram of shearing forces thus obtained is represented in Fig. 39.9.

The diagram of the normal stresses can be derived either from that for the shears or alternatively its ordinates may be calculated knowing the reactions of all the redundant constraints.

Let us determine the normal stresses using the equilibrium of joints. At first we may isolate joint *2* (Fig. 40.9) acted upon by the shear $Q_{23}=10.8$ tons developed at the left extremity of the crossbeam and directed downwards, the shear $Q_{21}=-2.7$ tons developed at the top of the column and directed from left to right and by the normal stresses N_{23} and N_{21} (both reckoned positive if entailing compression) and acting along the crossbeam and the column, respectively.

Equilibrium considerations yield immediately

$$N_{23}=+2.7 \text{ tons and } N_{21}=+10.8 \text{ tons}$$

The normal stress acting in the right-hand column will be obtained isolating joint *3* and will amount to +10.8 tons.

The complete diagram of normal stresses is given in Fig. 41.9.

A convenient method of checking the M, Q and N diagrams consists in the successive isolation of different parts or joints of the structure which must always remain in equilibrium. Thus, the projection on the vertical of all the support reactions of any framed structure must always equal the vertical component of the resultant of all the applied loads. Similarly, the sum of moments of all the reactions about any point of the structure must always equal the moment about the same point of the resultant of the applied loads, and so forth.

A rapid check of the diagram of the shearing forces may be taken comparing this diagram with that of the bending moments: indeed when the moment curve becomes parallel to the beam axis, the shear must equal zero; when the tangent to the bending moment curve remains inclined towards the same side, the shear may not change sign; its magnitude will be greater for that section for which the slope of the tangent to the bending moment curve is the steeper.

When two bars form a joint, the ordinates to their bending moment curves at this joint must always have numerically the same values (provided no outside moments act at this joint) since the bending moments must always balance. In the same case direct and shearing forces considered separately will not balance, but considered together they must form a system of concurrent forces in equilibrium

However, the control of stress curves based on statics alone does not provide complete guarantee of the exactitude of all the computations for equilibrium conditions may be satisfied even if errors were committed when calculating the redundant reactions. Indeed, the bending moment curve for any redundant structure always results from the summation of the ordinates to the curve induced in the simple statically determinate structure by the applied loads with those to the curves due to the redundant reactions and stresses. If all of these curves were constructed correctly, equilibrium conditions will remain satisfied even if the values of these reactions and stresses are completely wrong.

In the majority of cases any errors committed when computing the reactions of the redundant constraints will be detected checking that the *deflections* of certain points are consistent with the stipulations of the problem. The following example will serve to illustrate the above.

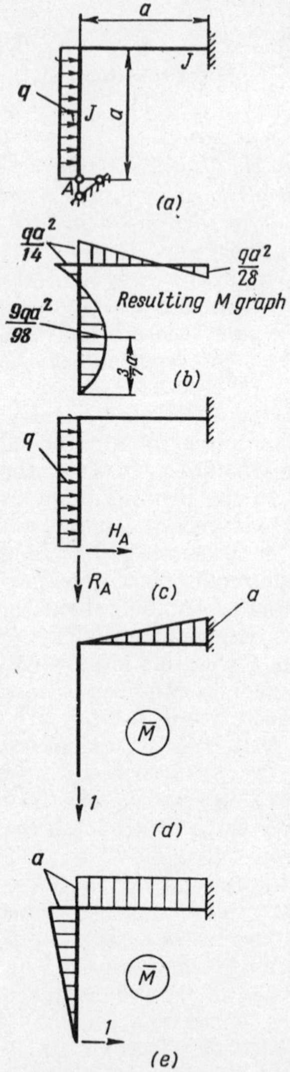

Fig. 42.9

Fig. 42.9a represents a knee frame statically indeterminate to the second degree. The computed bending moment diagram is shown in Fig. 42.9b. This diagram will remain unaltered should we transform the given structure into a statically determinate one, say, by elimination of the two support constraints at the lower end of the column (Fig. 42.9c) provided these constraints are replaced by their reactions.

Let us now compute the vertical deflection Δ^V of the lower end of the column in order to make sure that this deflection remains nil. For this purpose we shall first trace the diagram of the bending moment induced by a vertical load unity acting at point A (Fig. 42.9d) whereafter we shall multiply this diagram by the M diagram pertaining to the given structure (Fig. 42.9b). The result is

$$\Delta^V = \frac{1}{EJ}\, a\cdot a\,\frac{1}{2}\left(\frac{qa^2}{14}\cdot\frac{1}{3}-\frac{qa^2}{28}\cdot\frac{2}{3}\right)=0$$

Let us check also that the horizontal displacement of the same point remains equally nil. For that purpose we may multiply the bending moment graph due to a horizontal load unity applied at this point by the area of the same bending moment diagram as in the preceding paragraph (Fig. 42.9b).

$$\Delta^H = \frac{1}{EJ}\left(\frac{qa^2}{14}\,a\frac{1}{2}\cdot\frac{2a}{3}-\frac{qa^2}{8}\,a\cdot\frac{2}{3}\cdot\frac{a}{2}+\right.$$
$$\left.+\frac{qa^2}{14}\,a\cdot\frac{1}{2}\,a-\frac{qa^2}{28}\,a\frac{1}{2}\,a\right)=$$
$$=\frac{qa^4}{EJ}\left(\frac{1}{42}-\frac{1}{24}+\frac{1}{28}-\frac{1}{56}\right)=0$$

Thus, the above method of checking the computed stresses acting in members of redundant structures consists in the following:

1. Transform the given redundant structure into a simple statically determinate one.

2. Replace successively each of the eliminated constraints by a unit load or a unit moment as the case may be.

3. Trace for each of these unit actions a bending moment diagram.

4. Compute the deflection of the simple structure along the direction of each of these unit actions. The amount of this deflection will be given by the product of the ordinates to the bending moment

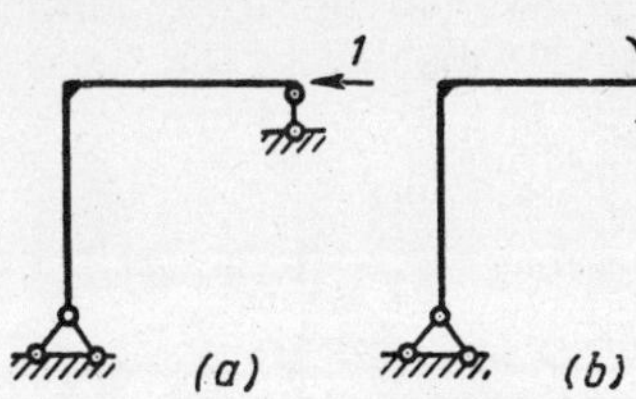

Fig. 43.9

curve due to the unit action by those to the diagram induced in the given redundant structure by the applied loads.

5. If these deflections are consistent with the stipulations of the problem (nil in the majority of cases) one may be reasonably sure that all the computations were carried out correctly.

The simple statically determinate structure used in that case need not coincide necessarily with the one used for the determination of the redundant stresses and reactions. Different simple structures may be used for the computation of different deflections of one and the same redundant structure. Thus, for instance, the resulting bending moment diagram of Fig. 42.9*b* could be checked using the simple structure of Fig. 43.9*a* for the computation of the horizontal deflection of the right end of the crossbeam and that of Fig. 43.9*b* for the computation of the angular rotation at the same point.

One can also use for the same purpose the graphs of the bending moments due to imaginary unit actions utilized in the original computations. In the latter case all that need be done to control the accuracy of the resulting diagram is to multiply this diagram by the former graphs and to make sure that their product remains equal to zero.

The control just described is particularly simplified for structures forming closed contours or those with built-in ends (which theoretically is one and the same).

Assume that it is required to control the accuracy of all the computations pertaining to the multispan frame with built-in columns (Fig. 44.9*a*) . Let us isolate a single bent applying at the cuts external moments and forces equivalent to the internal ones acting at these cross sections. Obviously the bending moment diagram relating to the isolated part of the frame will undergo no change whatsoever.

Now let us pass any arbitrary section through one of the members of the isolated bent, applying once again at the cut external actions equivalent to the stresses which existed at this section (Fig. 44.9*b*). There will be again no change in the bending moment curves pertaining to the two portions of the frame. It may be easily shown that the sections adjacent to the cut will undergo no mutual rotation. Indeed, let us multiply the resulting bending moment

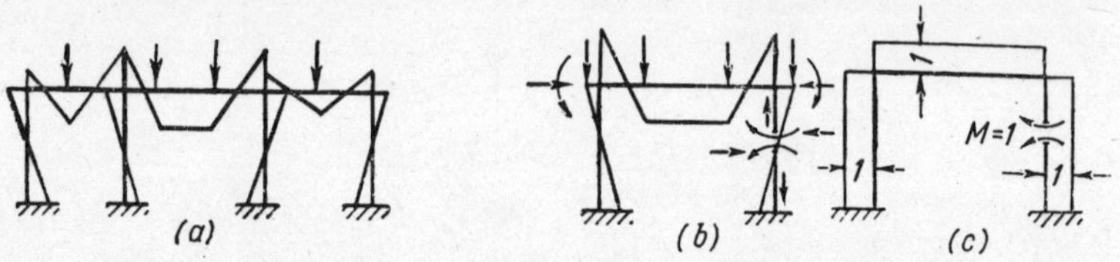

Fig. 44.9

graph by the graph due to a unit moment acting at the cut (Fig. 44.9*c*). As the ordinates to the latter graph will be constant and equal to unity, the above mentioned multiplication will reduce to a simple summing up of graph areas bounded by the resulting bending moment curve. These must be naturally taken with due consideration to their signs and the sum so obtained must equal zero.

If the different members of the structure differ in stiffness, the areas of each graph must be previously divided by the stiffness of the corresponding member. *Thus, for any structure forming a closed contour the algebraic sum of bending moment graph surfaces must reduce to zero, these surfaces being previously divided by EJ when necessary*. As for the sign of the graph areas, those situated within the contour will be taken with one sign and those situated outside with the opposite one.

This method of controlling the accuracy of computations is the simplest. If the results obtained are satisfactory, one may be reasonably sure that all the computations were carried out correctly.

It should be noted however that this method is inapplicable to framed structures with hinged joints or parts thereof.

Complete certitude that no error has been committed in any of the computations can be gained only if the number of control operations carried out is equal to the number of redundant constraints, provided these operations do not repeat one another. * Thus, for

* These operations may consist either in the multiplication of graphs or in the summation of their areas.

instance, if the graph areas for two contiguous parts of a structure have been summed up, the same procedure may not be applied to the same two parts taken as a whole, for this would simply repeat the controls already carried out and could therefore furnish no new data.

7. 9. *Strains and Deflections of Statically Indeterminate Structures*

Expressions (15.8) through (17.8) developed in Art. 6.8 were based on the assumptions that the material of the structure follows Hooke's law and that the strains and deflections of the structure are very small compared to its dimensions. Hence, these expressions as well as the corresponding computation techniques can be applied to all framed structures regardless of whether they are statically determinate or not.

Let us therefore use one of these expressions for the determination of the vertical deflection Δ_C of point C located along the neutral axis of a knee frame subjected to a uniformly distributed horizontal load of q kg per unit length as shown in Fig. 45.9*a*. This frame was analyzed in Problem 2 of Art. 3.9 (see Fig. 16.9*a*). The resulting bending moment graph is represented in Fig. 45.9*b* (see also Fig. 18.9*b*). In order to find the desired deflection let us apply at point C a vertical unit load, which will give rise to the bending moment curve of Fig. 45.9*c*. *

Multiplying the two graphs appearing in Fig. 45.9*b* and *c* one by the other we obtain

$$\Delta_C = \Big[\frac{qa^2}{14}\cdot\frac{a}{2}\cdot\frac{2}{3}\cdot\frac{3a}{56} - \frac{qa^2}{8}\cdot\frac{2}{3}a\cdot\frac{3a}{56}\cdot\frac{1}{2} + \frac{a}{2}\cdot\frac{1}{6}\times$$
$$\times\left(\frac{3a}{56}\cdot\frac{qa^2}{14}\cdot 2 - \frac{a}{7}\cdot\frac{qa^2}{56}\cdot 2 + \frac{3a}{56}\cdot\frac{qa^2}{56} - \frac{a}{7}\cdot\frac{qa^2}{14}\right) + \frac{a}{2}\cdot\frac{1}{6}\times$$
$$\times\left(-\frac{a}{7}\cdot\frac{qa^2}{56}2 - \frac{9a}{56}\cdot\frac{qa^2}{28}2 + \frac{a}{7}\cdot\frac{qa^2}{28} + \frac{9a}{56}\cdot\frac{qa^2}{56}\right)\Big]\frac{1}{EJ} = -\frac{qa^4}{448EJ}$$

The negative value found for the deflection Δ_C indicates that point C moves upwards, for the load unity was directed downwards.

The procedure described remains rather complicated since it requires that all the stresses in the redundant structure should be computed twice: once for the case of applied loads and once for the case of the imaginary load unity This procedure will be greatly simplified if we remember that the deformations of the simple statically determinate structure acted upon both by the applied

◆

* The corresponding calculations are omitted here.

loads and the redundant stresses and reactions will be exactly the same as those of the given indeterminate structure. Hence, in the case under consideration the deflection Δ_C may be computed

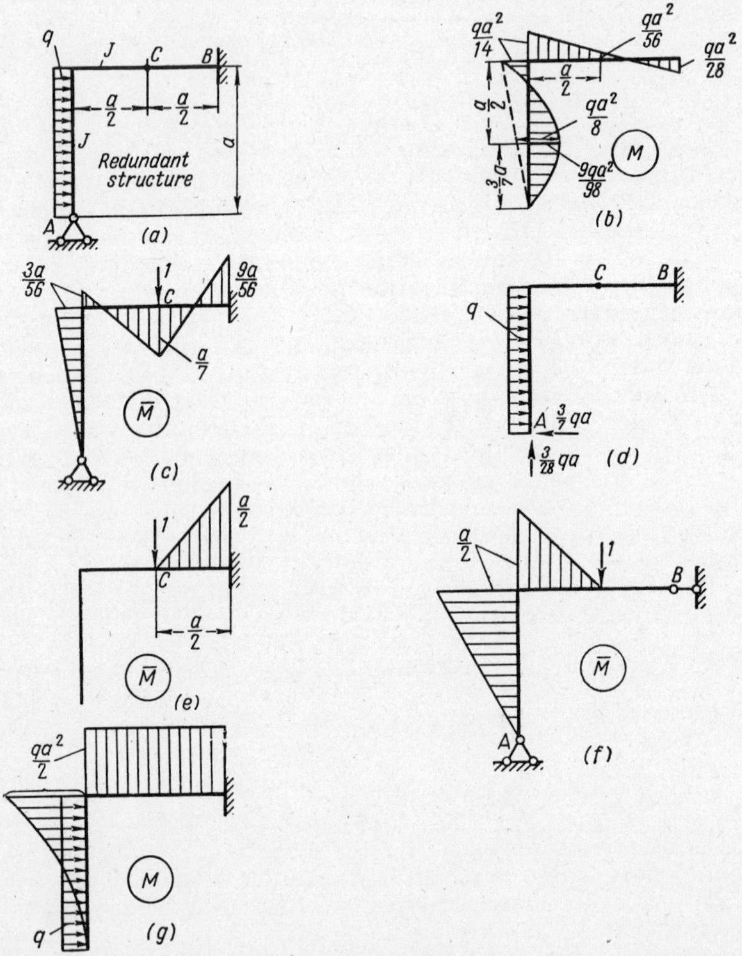

Fig. 45.9

with equal success either for the redundant structure of Fig. 45.9*a* or for the simple statically determinate one of Fig. 45.9*d*.

Let us apply at point C of the latter structure a unit load following the direction of the required deflection and let us trace the corresponding bending moment diagram (Fig. 45.9*e*). Multiplying this diagram by the resulting bending moment graph given in Fig. 45.9*b* we obtain

$$\Delta_C = \frac{a}{2}\cdot\frac{a}{2}\cdot\frac{1}{2}\left(\frac{qa^2}{56}\cdot\frac{1}{3}-\frac{qa^2}{28}\cdot\frac{2}{3}\right)\frac{1}{EJ} = -\frac{qa^4}{448EJ}$$

Any statically determinate structure derived from the given only by the elimination of redundant constraints can be used for deflection computation. It is in no way necessary that this simple structure should be the same as the one used for stress analysis. Thus, the deflection of point C of the knee frame could be obtained just as well using for auxiliary simple structure the one shown in Fig. 45.9*f*. The application to this structure of a vertical load unity at point C would lead to a bending moment diagram shown in the same figure. The subsequent multiplication of this diagram by that bounded by the curve of the resulting bending moments given in Fig. 45.9*b* leads to

$$\Delta_C = \left[\frac{qa^2}{14}\cdot\frac{a}{2}\cdot\frac{2}{3}\cdot\frac{a}{2}-\frac{qa^2}{8}\cdot\frac{2}{3}a\cdot\frac{a}{2}\cdot\frac{1}{2}+\frac{a}{2}\cdot\frac{a}{2}\times\right.$$
$$\left.\times\frac{1}{2}\left(\frac{qa^2}{14}\cdot\frac{2}{3}+\frac{qa^2}{56}\cdot\frac{1}{3}\right)\right]\frac{1}{EJ} = -\frac{qa^4}{448EJ}$$

The choice of the auxiliary simple structure should be governed by the following considerations: the bending moment curve due to the load unity must be as simple as possible, this curve must be obtained with the minimum of computations and the ordinates to this curve should reduce to zero wherever the outline of the resulting bending moment diagram for the given redundant structure becomes too complicated. Hence, in the example dealt with preference should be given to the simple structure of Fig. 45.9*e* as compared to that of Fig. 45.9*f*.

The necessity might arise to determine the deflection sustained by a redundant structure under a given set of loads without being interested in the corresponding stresses. In that case one may compute only the stresses induced in the structure by a unit load acting in the direction of the desired deflection, disregarding entirely those due to the applied loads. The deflection will be then obtained multiplying the bending moment graph due to the unit load and pertaining to the given redundant structure by the diagram of bending moments induced in the auxiliary simple structure by the actual loading. In the previous example the deflection Δ_C could be thus obtained multiplying the $\overline{M}$ graph of

Fig. 45.9c by that appearing in Fig. 45.9g

$$\Delta_C = \left[\frac{qa^2}{2}\cdot\frac{a}{3}\cdot\frac{3a}{56}\cdot\frac{3}{4}+\frac{qa^2}{2}\cdot\frac{a}{2}\left(\frac{3a}{56}-\frac{a}{7}\cdot 2+\frac{9a}{56}\right)\frac{1}{2}\right]\times$$
$$\times\frac{1}{EJ} = -\frac{qa^4}{448EJ}$$

Thus, it may be stated that *the deflections of a redundant structure may be determined using only one bending moment diagram pertaining to the given structure, either that induced by the applied loads or else that due to a load unity acting along the desired deflection. The second graph may be traced for any simple structure derived from the first one by elimination of redundant constraints.*

The deflections and distortions of statically indeterminate trusses and other hinge-connected structures will be obtained in exactly the same way with the only difference that the bending moment curves and graph areas must be in that case replaced by those relating to the normal stresses.

8.9. *The Elastic Centre Method*

As a rule, stress analysis of redundant structures requires the simultaneous solution of several equations with several unknowns. The higher the degree of indeterminancy the greater the number of these equations, the harder their solution and the lower the accuracy of the final results. It is quite natural therefore that attempts are frequently made to form the above equations in such a way that each of them should contain one unknown only, or, alternatively, that the system of these equations should fall into separate groups each containing a reduced number of unknowns not entering the other groups.

The system of equations pertaining to a structure redundant in the second degree is so simple that it is obviously senseless to seek any further simplifications even if such were possible.

Passing to structures indeterminate in the third degree let us investigate the possibilities of simplifying the analysis of a closed contour of arbitrary configuration such as shown, for instance, in Fig. 46.9*a*. Having eliminated the constraints at the left-hand abutment and having replaced them with three unknown reactions X_1, X_2 and X_3 as indicated in Fig. 46.9*b*, we should normally form three standard equations with three unknowns.

In order to avoid the simultaneous solution of these equations let us fix to the free end of the simple structure an infinitely stiff bracket *ab*, both the length and the direction of which remain as yet unknown (Fig. 47.9).

At the free end b of this bracket let us apply at right angles two forces Z_1 and Z_2 (the direction of these forces coinciding with that of a new set of coordinate axes u and v) and a moment Z_3.

If the magnitude of these actions were such that they would immobilize completely the end b of the bracket, preventing both its rotation and translation, the left extremity of the simple structure (point a) would also become fixed and thus from the viewpoint of their deflections the given redundant structure and the simple structure of Fig. 47.9 would be in the same conditions.

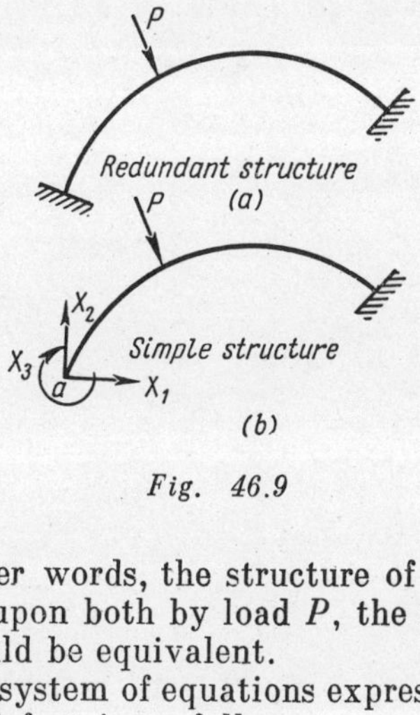

Fig. 46.9

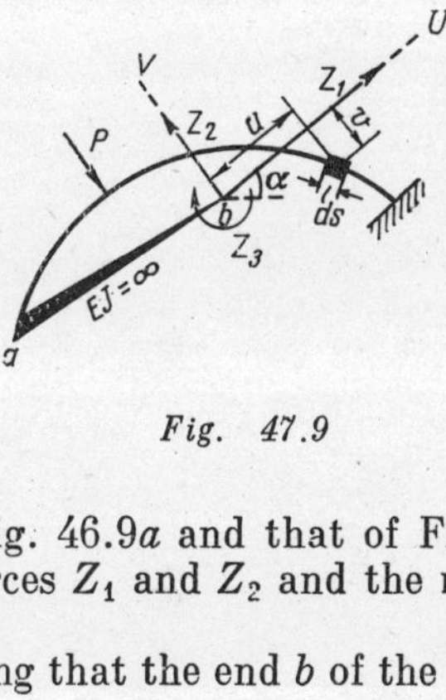

Fig. 47.9

In other words, the structure of Fig. 46.9a and that of Fig. 47.9 acted upon both by load P, the forces Z_1 and Z_2 and the moment Z_3 would be equivalent.

The system of equations expressing that the end b of the bracket is held fast is as follows

$$\left.\begin{aligned} Z_1\delta_{11}+Z_2\delta_{12}+Z_3\delta_{13}+\Delta_{1p}=0 \\ Z_1\delta_{21}+Z_2\delta_{22}+Z_3\delta_{23}+\Delta_{2p}=0 \\ Z_1\delta_{31}+Z_2\delta_{32}+Z_3\delta_{33}+\Delta_{3p}=0 \end{aligned}\right\} \qquad (17.9)$$

It will be remembered that the coefficients δ_{ik} entering these equations represent the displacements of the free end of bracket ab induced by load unities following the direction of the unknowns Z_1, Z_2 and Z_3.

The magnitude of these displacements depends, of course, both on the size of the bracket and on the direction of the axes u and v. Let us choose these parameters in such a way as to render nil all the secondary displacements of point b. In that case each of the above three equations will contain only one unknown, these equations reducing to

$$\begin{aligned} Z_1 &= -\frac{\Delta_{1p}}{\delta_{11}} \\ Z_2 &= -\frac{\Delta_{2p}}{\delta_{22}} \\ Z_3 &= -\frac{\Delta_{3p}}{\delta_{33}} \end{aligned} \tag{18.9}$$

Let us express mathematically the conditions governing our choice of the aforesaid parameters. Denoting by u and v the coordinates of an element ds of the given structure (see Fig. 47.9) we may write

$$\delta_{12} = \int_0^l \frac{\overline{M}_1\overline{M}_2}{EJ}\, ds = \int_0^l \frac{v\cdot u ds}{EJ} = 0$$

$$\delta_{13} = \int_0^l \frac{\overline{M}_1\overline{M}_3}{EJ}\, ds = \int_0^l \frac{v\cdot 1 ds}{EJ} = 0$$

$$\delta_{23} = \int_0^l \frac{\overline{M}_2\overline{M}_3}{EJ}\, ds = \int_0^l \frac{u\cdot 1 ds}{EJ} = 0$$

Dividing each of these three equations by E we obtain

$$\int_0^l u\,\frac{ds}{J} = 0$$

$$\int_0^l v\,\frac{ds}{J} = 0$$

$$\int_0^l vu\,\frac{ds}{J} = 0$$

It is fairly easy to find a geometrical interpretation to the conditions thus expressed. Indeed, let us subdivide the whole of our structure into elements ds and let us apply at the centre of gravity of each of these elements an imaginary load $\frac{ds}{J}$. In that case the first two of the above integrals will mean that the statical moments of these imaginary loads about the coordinate axes u and v having for origin point b are nil. That will happen only when point b coincides with the centre of gravity of loads $\frac{ds}{J}$. The third of the integrals means that the polar moment of these same loads about point b is equally nil, this becoming possible only when the coordinate axes coincide with the principal axes of inertia of the system.

It follows that *all the secondary displacements of a structure redundant in the third degree and forming a closed contour will reduce to zero provided the new origin of coordinate axes is brought in coincidence with the centre of gravity of the imaginary loads* $\frac{ds}{J}$, *and the direction of these axes—with that of the principal axes of inertia of these same loads. If the given structure is symmetrical, the principal axes of inertia will coincide with the axes of symmetry.*

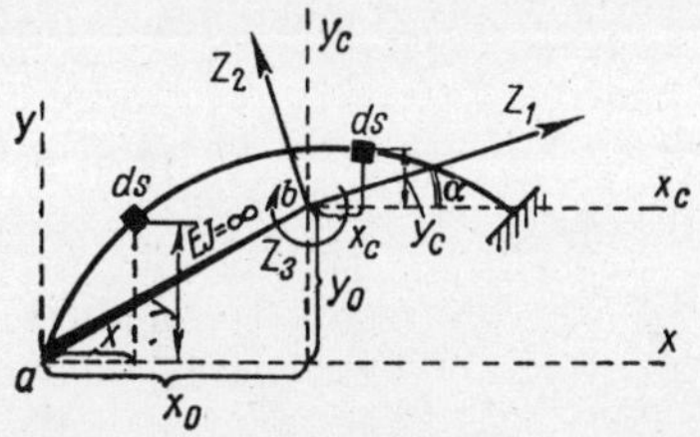

Fig. 48.9

The origin of the new coordinate axes may by determined using formulas provided by theoretical mechanics: thus, using the notations of Fig. 48.9 the position of the centre of gravity of the loads $\frac{ds}{J}$ will be given by

$$x_0=\frac{\int_0^l x\frac{ds}{J}}{\int_0^l \frac{ds}{J}};\quad y_0=\frac{\int_0^l y\frac{ds}{J}}{\int_0^l \frac{ds}{J}} \tag{19.9}$$

In each case the numerator represents the sum of the statical moments of the imaginary loads $\frac{ds}{J}$ about some axis of coordinates (the summation being carried over the whole length of the contour), and the denominator equals the total of these same loads.

The angle formed by the principal axes of inertia with the x-axis is given by the following formula developed in the treatises on the strength of materials

$$\tan\,2\alpha=-\frac{2J_{x_c y_c}}{J_{x_c}-J_{y_c}} \tag{20.9}$$

where

$$\left.\begin{aligned} J_{x_c y_c} &= \int_0^l x_c y_c \frac{ds}{J} \\ J_{x_c} &= \int_0^l y_c^2 \frac{ds}{J} \\ J_{y_c} &= \int_0^l x_c^2 \frac{ds}{J} \end{aligned}\right\} \qquad (21.9)$$

The imaginary loads $\frac{ds}{J}$ are sometimes called the *elastic loads* or *masses* and the centre of gravity of these elastic masses—the *elastic centre of gravity* or simply the *elastic centre*.

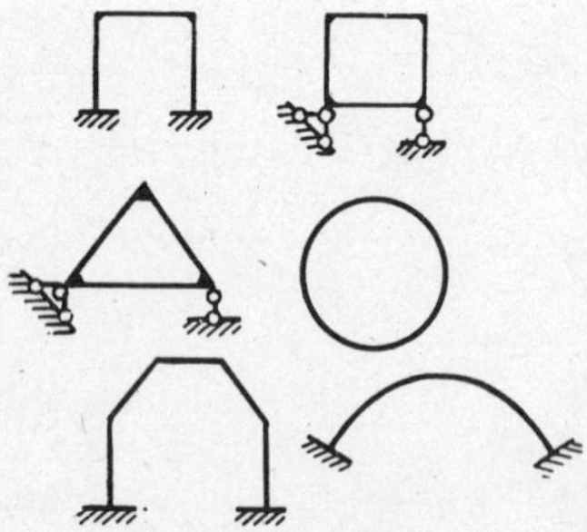

Fig. 49.9

It is not always advisable to resort to the method just described. Thus, for unsymmetrical systems the amount of work required to determine the position of the new coordinate axes may become so important that it will outweight completely the advantages gained from the simplification of equations.

However, if the structure is endowed with at least one axis of symmetry it may be stated off hand that the method of the elastic centre will yield tangible results. Some examples of such structures are given in Fig. 49.9.

Problem. Determine the position of the elastic centre for the fixed end arch of Fig. 50.9. The neutral line of this arch follows a conic parabola given

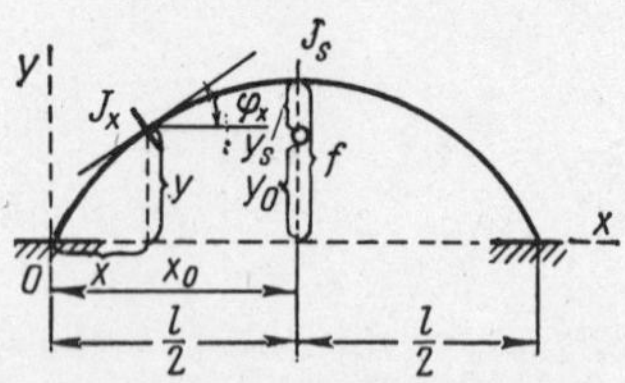

Fig. 50.9

by $y = \frac{4f}{l^2}(l - x)x$ and its cross-sectional moments of inertia vary inversely to cos φ_x

$$J_x = \frac{J_s}{\cos \varphi_x}$$

where J_x = moment of inertia of an arbitrary cross section whose abscissa equals x

J_s = moment of inertia of the cross section at the crown of the arch

φ_x = angle formed by the tangent to the neutral line of the arch at point x with the axis of abscissa.

Solution. Start with determining the position of the elastic centre with reference to the coordinate axes xoy (Fig. 50.9). Owing to the symmetry of the structure both as regards its dimensions and the stiffness of its cross sections, one of the coordinate axes is known off hand: it is vertical in direction and passes through the crown. Therefore, the abscissa of the elastic centre x_0 equals $\frac{l}{2}$. As for the ordinate it will be found using expression (19.9) and remembering that

$$ds = \frac{dx}{\cos \varphi_x} \quad \text{and} \quad J_x = \frac{J_s}{\cos \varphi_x}$$

hence

$$y_0 = \frac{\int_0^l \frac{4f}{l^2}(l-x)\,x\,\frac{dx}{J_s}}{\int_0^l \frac{dx}{J_s}} = \frac{2}{3}f$$

9.9. *Influence Lines for the Simpler Redundant Structures*

The construction of influence lines for redundant structures may be carried out using both the *statical* and the *kinematic* methods.

Let us compare both these methods using as an example the beam of constant cross section appearing in Fig. 51.9*a*. The influence line for the right-hand

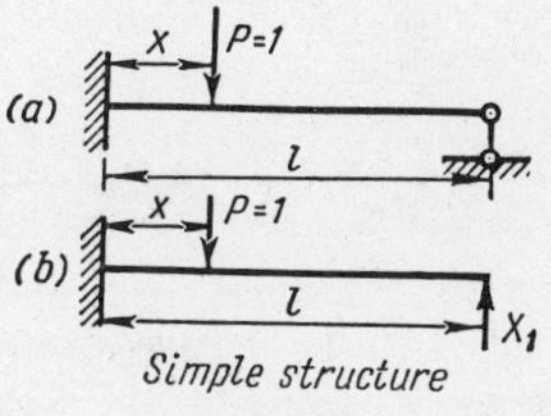

Fig. 51.9

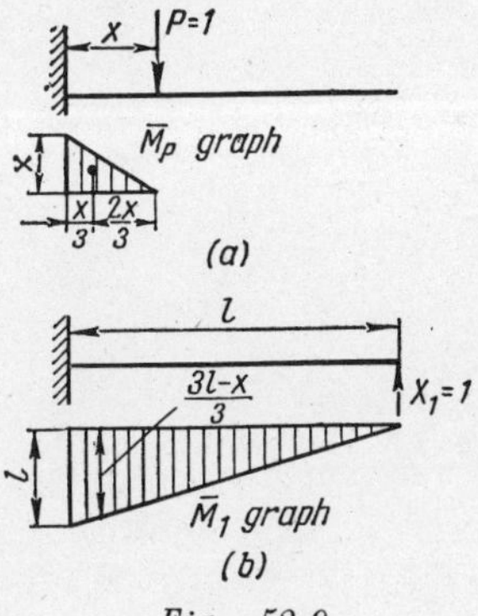

Fig. 52.9

abutment reaction will be obtained using as simple statically determinate structure the one shown in Fig. 51.9*b*. The standard equation expressing that the deflection of the latter structure along X_1 is nil becomes

$$X_1\delta_{11}+\delta_{1p}=0$$

wherefrom

$$X_1=-\frac{\delta_{1p}}{\delta_{11}} \tag{22.9}$$

The deflections δ_{11} and δ_{1p} will be derived from the $\overline{M}_1$ and the $\overline{M}_p$ diagrams, the first of these diagrams being due to the application of a load unity along the direction of X_1 and the second—to the application of load $P = 1$ a distance x from the wall.

These diagrams appear in Fig. 52.9*a* and *b*.

Hence

$$\delta_{1p}=-\frac{xx}{2}\cdot\frac{3l-x}{3}\cdot\frac{1}{EJ}=-\frac{x^2(3l-x)}{6EJ}$$

and

$$\delta_{11}=\frac{ll}{2}\cdot\frac{2l}{3}\cdot\frac{1}{EJ}=\frac{l^3}{3EJ}$$

Introducing these values in expression (22.9) we obtain

$$X_1=-\frac{\delta_{1p}}{\delta_{11}}=\frac{x^2(3l-x)\,3EJ}{6EJl^3}=\frac{x^2(3l-x)}{2l^3}$$

The above expression gives the value of reaction X_1 for any position of the load unity P along the beam and therefore the

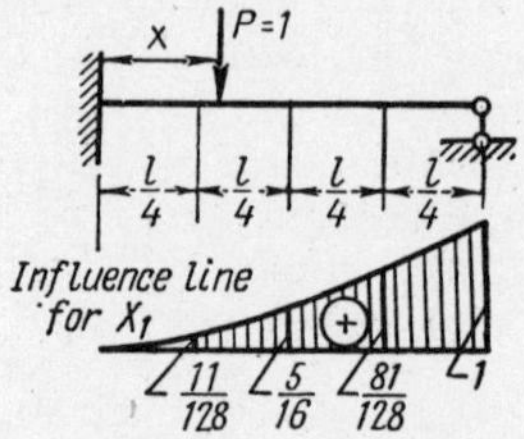

Fig. 53.9

graphical representation of this expression will constitute the influence line for the said reaction.

Table 3.9 gives the ordinates to this influence line at quarter-span increments computed using the above expression for X_1. The influence line itself is represented in Fig. 53.9.

Next, let us take up the influence line for the shear Q_C acting at midspan of the same beam. Based on the principle of superposition the magnitude of this shear will be

$$Q_C=Q_C^0+\overline{Q}_CX_1 \tag{23.9}$$

Table 3.9

x	x^2	$3l-x$	X_1
0	0	$3l$	0
$\frac{l}{4}$	$\frac{l^2}{16}$	$\frac{11}{4}l$	$\frac{11}{128}$
$\frac{l}{2}$	$\frac{l^2}{4}$	$\frac{5}{2}l$	$\frac{5}{16}$
$\frac{3}{4}l$	$\frac{9l^2}{16}$	$\frac{9}{4}l$	$\frac{81}{128}$
l	l^2	$2l$	1

where

Q_C^0 = shear at midspan of the simple cantilever beam due to load unity P

$\overline{Q}_C$ = shear at the same cross section due to the abutment reaction $X_1 = 1$. The value of $\overline{Q}_C$ remains constant and equal to -1 ($\overline{Q}_C = -1$).

As for Q_C^0 when the load unity P is to the right of section C

$$Q_C^0 = 1$$

and when it has shifted to the left of this section

$$Q_C^0 = 0$$

The influence line for Q_C^0 is shown in Fig. 54.9.

The value of the last term of expression (23.9) for any position of load unity P is equal to the abutment reaction X_1 multiplied

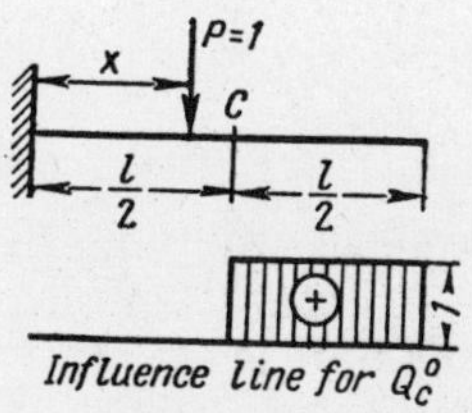

Fig. 54.9

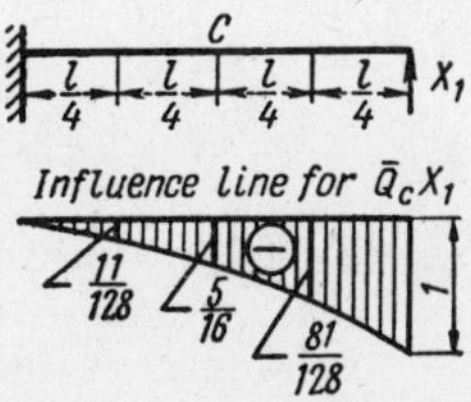

Fig. 55.9

by (—1). The graphical representation of this relation is given in Fig. 55.9. Adding the ordinates to the influence lines for Q_C^0 and for $\overline{Q}_C X_1$ we obtain the influence line for the full shearing force Q_C acting across section C of the redundant beam (Fig. 56.9).

The influence line for the bending moment at midspan of the same beam will be obtained in a similar way. Indeed, this bending moment is equal to

$$M_C = M_C^0 + \overline{M}_C X_1$$

where

M_C^0 = bending moment at cross section C of the simple cantilever beam induced by the load unity P

$\overline{M}_C$ = bending moment at the same cross section but induced by a unit load acting along the direction of X_1; the value of this bending moment is constantly equal to $\frac{l}{2}$.

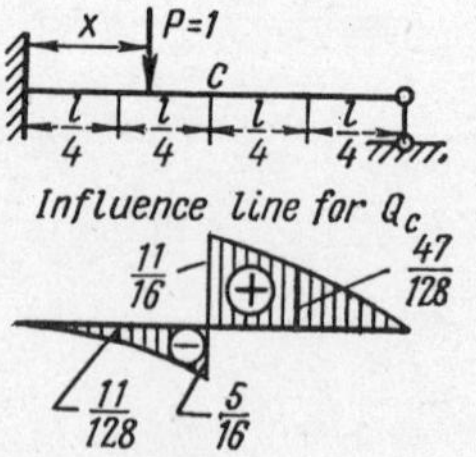

Fig. 56.9

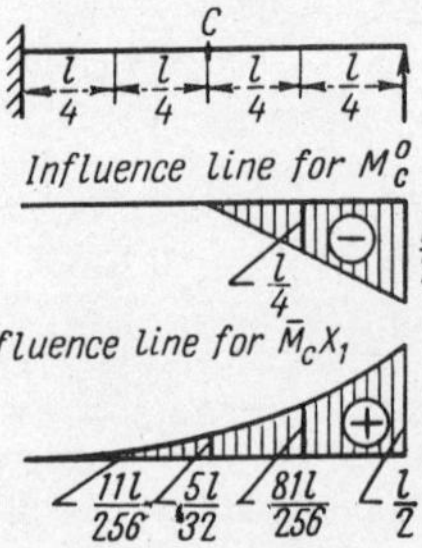

Fig. 57.9

The influence lines for M_C^0 as well as for $\overline{M}_C X_1$ are represented in Fig. 57.9. Adding their ordinates together we obtain the influence line for M_C (Fig. 58.9).

The method just described is based on considerations of equilibrium alone and therefore it may be termed *statical*.

Let us replace now the deflection δ_{1p} in expression (22.9) by δ_{p1} as provided for by the theorem of reciprocal deflections. We obtain

$$X_1 = -\frac{\delta_{p1}}{\delta_{11}} \tag{24.9}$$

Though δ_{1p} and δ_{p1} are always numerically equal they convey very different ideas. Indeed, δ_{1p} is the deflection of a fixed point of application of the force X_1 due to a unit load P travelling along the beam; at the same time δ_{p1} is the deflection at the point of application of a load P travelling along the beam, caused by a load unity acting along the direction of X_1. Hence, the variation of δ_{p1} represented graphically will constitute at a certain scale the elastic curve of the beam subjected to the action of a load unity applied along the direction of X_1. However, if we divide all the ordinates to this curve by a constant factor equal to

$(-\delta_{11})$ we shall obtain the ordinates to a graph representing the variation of X_1 when load unity P travels along the beam. By definition this constitutes the influence line for X_1.

It follows that the influence line for X_1 will have the same shape as the elastic curve of the simple statically determinate structure loaded by a unit action following the direction of that unknown.

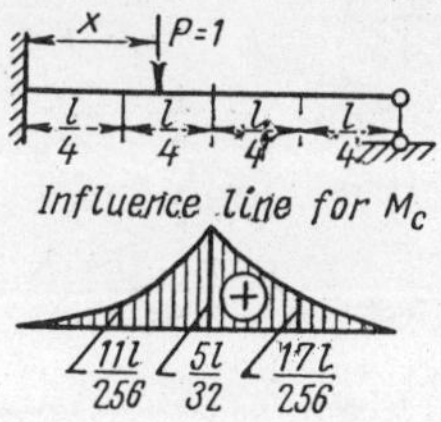

Fig. 58.9

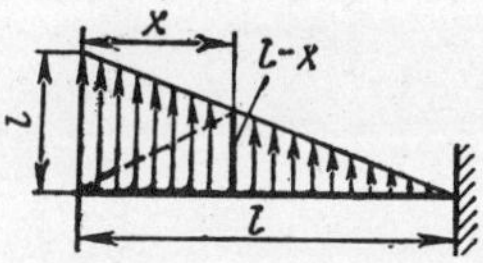

Fig. 59.9

In that case $(-\delta_{11})$ becomes a scale factor permitting the conversion of the ordinates to the deflection curve to those of the desired influence line.

This particular method of influence lines construction shall be termed hereafter the *kinematic* method.

For comparison, let us take up again the beam of Fig. 51.9 and construct the influence line for the abutment reaction X_1, using the kinematic method based on equation (24.9).

The deflection curve δ_{p1} will be obtained by the graph-analytical method of moments, which is usually described in all the treatises on the strength of materials. In essence this method permits to replace deflection computations by those of bending moments arising in an imaginary beam carrying an imaginary load distributed in accordance with the diagram of the bending moments existing in the real beam. The ordinates to the imaginary bending moment curve are then divided by EJ. Such an imaginary beam with a load distribution corresponding to $X_1 = 1$ is represented in Fig. 59.9.

The magnitude of the bending moment (divided by EJ) in any cross section of the imaginary beam situated a distance x from its left end will be

$$\delta_{p1} = -\left(\frac{lx}{2}\cdot\frac{2}{3}x + \frac{(l-x)\,x}{2}\cdot\frac{x}{3}\right)\frac{1}{EJ}$$

wherefrom

$$\delta_{p1} = -\frac{x^2}{6EJ}(3l - x)$$

The value of δ_{p1} obtained by the kinematic method coincides exactly with that of δ_{1p} derived from statical considerations. The negative sign indicates that the force X_1 will cause the beam of Fig. 51.9*b* to deflect upwards, whereas we have convened to reckon the deflections positive when their direction coincides with that of their cause, i.e., load P which is directed downwards.

The value of δ_{11} will be obtained raising to the second power

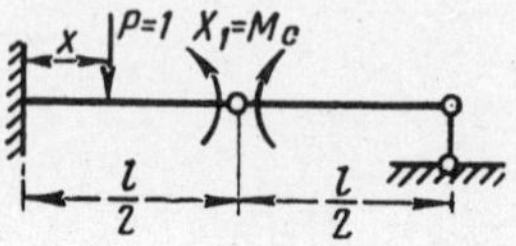

Fig. 60.9

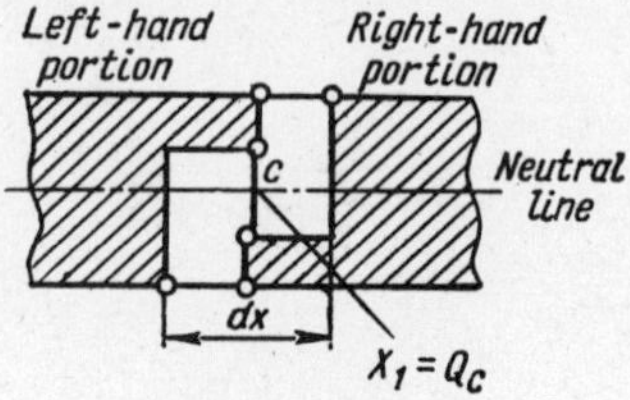

Fig. 61.9

$\overline{M}_1$ graph (see Fig. 52.9*b*) *

$$\delta_{11}=\frac{l^3}{3EJ}$$

Hence the equation of the influence line for reaction X_1 becomes

$$X_1=-\frac{\delta_{p1}}{\delta_{11}}=\frac{x^2(3l-x)}{2l^3}$$

This equation is exactly the same as the one obtained previously by the statical method.

The kinematic method may be used with good results for the construction of influence lines for internal stresses acting in the members of redundant structures and in particular for the construction of influence lines for bending moments and shearing forces.

Thus the influence line for the bending moment could be obtained by the method just described adopting for redundant reaction the bending moment M_C acting at midspan of the simple statically determinate structure of Fig. 60.9. The equation expressing that the mutual rotation of two contiguous sections to the right and to the left of the hinge remains nil becomes

$$X_1\delta_{11}+\delta_{1p}=0 \quad \text{or} \quad X_1\delta_{11}+\delta_{p1}=0$$

wherefrom

$$X_1=-\frac{\delta_{p1}}{\delta_{11}}$$

Consequently, in that case the influence line for the bending moment at midspan will have the same shape as the deflection line

* The value of δ_{11} may be derived from that of δ_{p1} substituting l for x and changing the sign, for the load unities X_1 and P are directly opposed.

of the simple structure adopted when acted upon by a unit moment applied at this same cross section.

The kinematic method permits the easy determination of the shape of the influence line for any action, this shape being the same

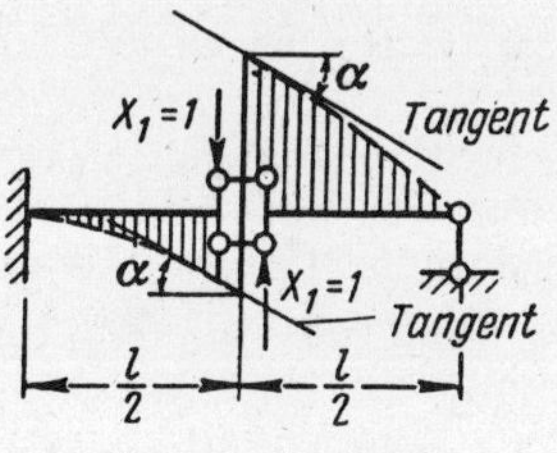

Fig. 62.9

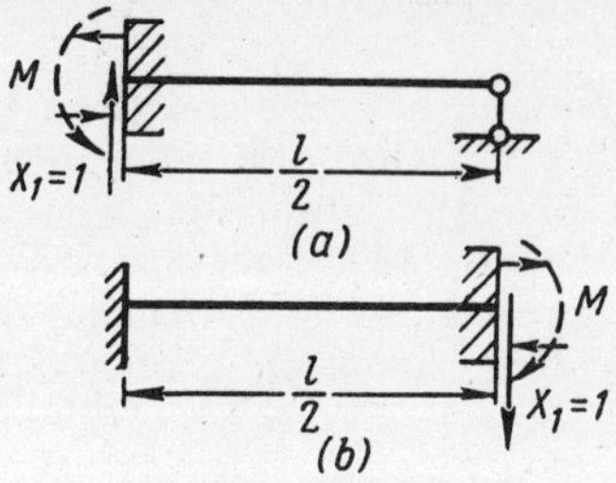

Fig. 63.9

as that of the elastic curve of the corresponding simple structure loaded by a unit force or moment. This analogy can be of considerable value both in checking the accuracy of influence lines obtained by some other method and in seeking those parts of the structure which must be loaded in order to provide for maximum or minimum values of the desired stress.

When using the kinematic method for the construction of shear influence line the connection between the right-hand and the left-hand portions of the beam may be represented by three bars as shown in Fig. 61.9. The stress developed in the vertical bar will have exactly the same value as the shearing force at the same cross section.

The simple statically determinate structure loaded with unit forces X_1 permitting the construction of the Q_C influence line appears in Fig. 62.9, and the equations negating the existence of a mutual displacement of two cross sections contiguous to C along the line of action of X_1 are of the form

$$X_1\delta_{11}+\delta_{1p}=0$$

wherefrom

$$X_1=-\frac{\delta_{1p}}{\delta_{11}} \quad \text{or} \quad X_1=-\frac{\delta_{p1}}{\delta_{11}}$$

Thus, the influence line for the shear at section C will again follow the shape of the deflection curve of the simple structure acted upon by the load unity X_1.

At points immediately to the right and to the left of section C both branches of this curve will have parallel tangents, for the horizontal connection bars preclude the possibility of a mutual rotation of the sections in question.

The right-hand part of the beam (see Fig. 62.9) is subjected to the action of a load unity X_1 tending to rotate it in a clockwise direction (Fig. 63.9*a*). To maintain it in equilibrium the moment developed by the horizontal bars must act in an opposite direction and therefore the top bar will be extended and the lower one compressed (Fig. 63.9*a*). The same reasoning applying to the left-hand portion of the beam (Fig. 63.3*b*), its top fibres will also be extended and the lower compressed.

The deflection graphs δ_{p1} necessary for the construction of influence lines by the kinematic method may be obtained using any of the procedures described above and in particular the one we have called the method of elastic loads (see Arts. 11.8 to 13.8).

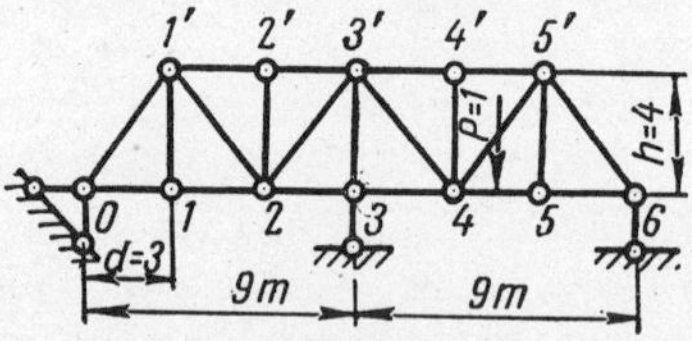

Fig. 64.9

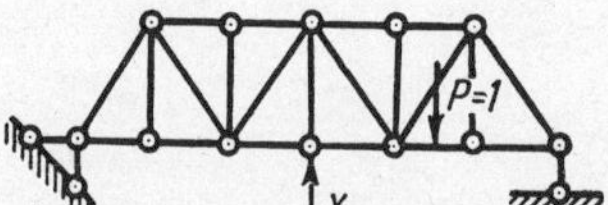

Fig. 65.9

The kinematic method may be conveniently used for the construction of influence lines for stresses or reactions at the supports of statically indeterminate trusses. As an example, let us take the truss redundant in the first degree appearing in Fig. 64.9. The cross-sectional areas of all the members of this truss are the same. The corresponding simple statically determinate structure is given in Fig. 65.9.

The standard equation showing that the displacement along the line of action of the redundant constraint X_1 is nil takes the shape

$$X_1\delta_{11}+\delta_{1p}=0$$

wherefrom

$$X_1=-\frac{\delta_{1p}}{\delta_{11}}=-\frac{\delta_{p1}}{\delta_{11}}$$

The variation of δ_{p1} when load unity P travels along the lower chord of the truss will coincide with the deflection curve of the same chord of the simple structure subjected to the action of a unit load X_1. The ordinates to this curve at all the joints have been computed in Art. 13.8 using the elastic loads method, and the curve itself is represented in Fig. 45.8. All these ordinates are negative for under the action of X_1 the lower chord tends to move upwards while load unity P is directed downwards. The influence line for X_1 obtained by dividing the ordinates to the δ_{p1} curve by $(-\delta_{11})$ is represented in Fig. 66.9. Thus, this influence line will differ from the deflection curve of the simple structure only by a constant factor equal to $\left(-\frac{1}{\delta_{11}}\right)$. As for the value of this factor, it will be easily found by scaling off the value of δ_{p1} at the point of application of the load X_1.

Once the influence line for the reaction at the redundant support has been found, the stresses in all the members of the truss will

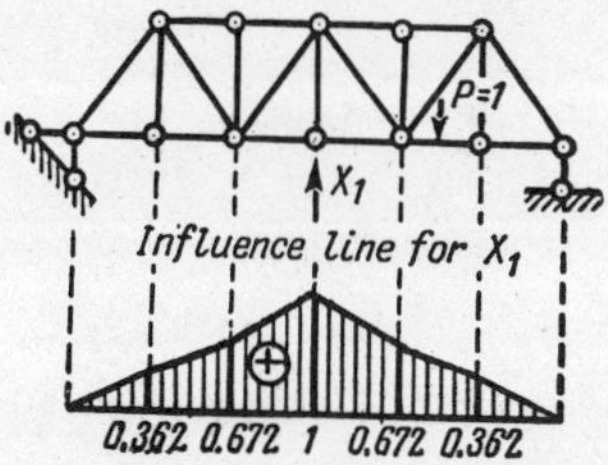

Fig. 66.9

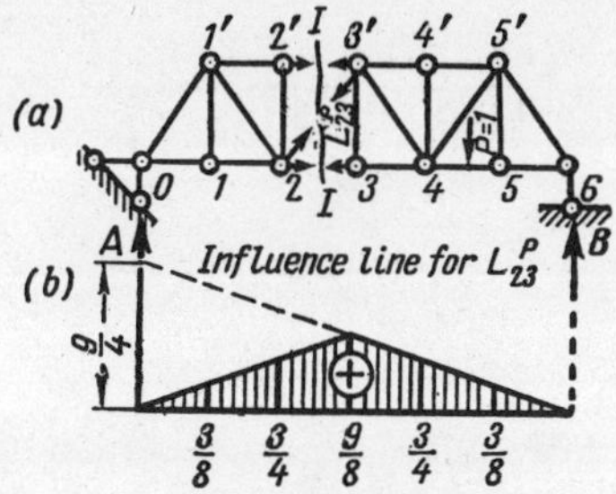

Fig. 67.9

be readily obtained using the well-known expression based on the principle of superposition

$$N_i = N_i^P + \overline{N}_i X_1$$

where N_i^P = stress in bar i resulting from the application to the simple structure of the actual load

$\overline{N}_i$ = stress in the same member resulting from the application to the same structure of the unit load X_1.

It follows that the influence line for the stress N_i may be obtained by the summation of the ordinates to the two other influence lines, that for N_i^P and that for $\overline{N}_i X_1$.

To illustrate the above, let us construct the influence line for stress L_{23} acting in one of the lower chord members of the same truss (see Fig. 64.9).

Using the expression just mentioned we may write

$$L_{23} = L_{23}^P + \overline{L}_{23} X_1$$

In order to obtain the influence line for both terms of the right-hand part of this equation let us pass section *I-I* as in Fig. 67.9. Assuming that load unity P is to the right of this section, the equilibrium of the left portion of the truss requires that

$$\Sigma M_{3'} = A \cdot 3d - L_{23}^P h = 0$$

wherefrom

$$L_{23}^P = \frac{A \cdot 3d}{h} = \frac{9}{4} A$$

The right-hand part of the influence line for L_{23}^{P} will be obtained setting off along the vertical passing through the left support an

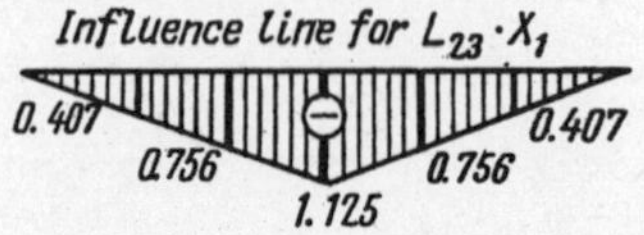

Fig. 68.9

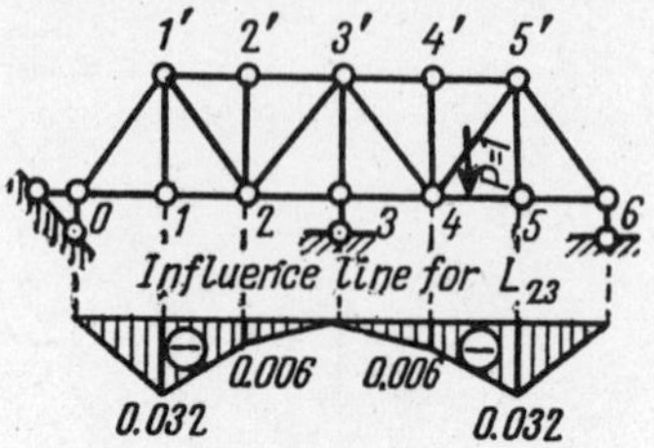

Fig. 69.9

ordinate equal to $\frac{3d}{h} = \frac{9}{4}$ and connecting this point with the point of zero ordinate over the right abutment. The left-hand portion of the same influence line will be drawn remembering that the two always intersect in the vertical passing through the origin of moments.

The completed influence line for L_{23}^{P} pertaining to the simple structure of Fig. 67.9*a* is given in Fig. 67.9*b*.

This same influence line shows that the stress induced in bar *2-3* by the unit load X_1 will equal $-\frac{9}{8}$.

Multiplying by this factor all the ordinates to the influence line for X_1 given in Fig. 66.9, we obtain a graphical interpretation of the variation of $\overline{L}_{23}X_1$ or, in other words, the influence line

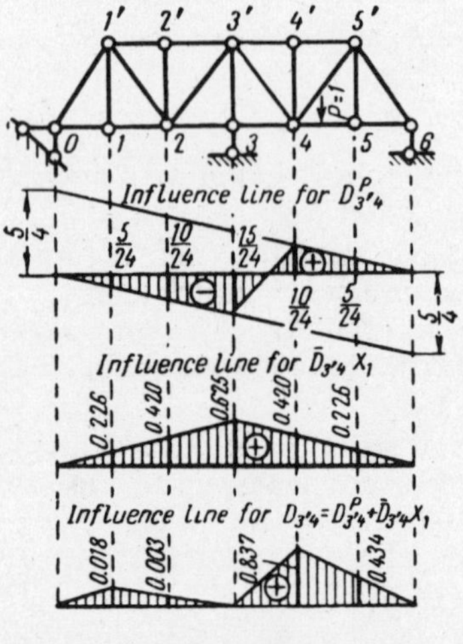

Fig. 70.9

for the second term of our equation (Fig. 68.9). Summing up the ordinates to the influence line for L_{23}^{P} with those to the influence line for $\overline{L}_{23}X_1$ we obtain the influence line for the stress developed in bar *2-3* of the redundant truss. This influence line appears in Fig. 69.9.

The influence line for the stress $D_{3'4}$ obtained by the same procedure is shown in Fig. 70.9.

10. CONTINUOUS BEAMS

1.10. *Theorem of Three Moments*

A continuous beam is a statically indeterminate multispan beam on hinged supports. The end spans may be cantilever, may be freely supported or built in. At least one of the supports of a continuous beam must be able to develop a reaction along the beam axis.

Fig. 1.10*a* represents several spans of a continuous beam carrying an arbitrary system of vertical loads. The supports will be numbered from left to right *0*, *1*, *2*, *3*, . . ., $n-1$, n, $n+1$, etc.. and the included spans will be designated by l_1, l_2, l_3, . . ., l_{n-1}, l_n, l_{n+1}, etc., the index of each span coinciding with that of its right-hand support. It will be assumed that the moment of inertia remains constant within each span, but may vary from one span to the other.

A conjugate statically determinate system for a continuous beam may be obtained by elimination of constraints considered as redundant which prevent mutual rotation of two contiguous sections over the supports or putting it otherwise, by the introduction of a hinge at each of these supports as indicated in Fig. 1.10*b*.

The equation, expressing mathematically that the angles of rotation of the aforementioned sections over the supports one with respect to the other remain nil, will be of the following form

$$\begin{aligned} \ldots X_{n-2}\delta_{n,\,n-2}+X_{n-1}\delta_{n,\,n-1}+X_n\delta_{n,\,n}+X_{n+1}\delta_{n,\,n+1}+ \\ +X_{n+2}\delta_{n,\,n+2}+\ldots+\Delta_{np}=0 \end{aligned} \qquad (1.10)$$

The coefficients to all the unknowns as well as the free term will be calculated with the aid of diagrams of the bending moments induced by unit couples acting along the direction of each redundant constraint (Fig. 1.10*c*, *d*, *e*, *f*, *g*) and of that due to the actual loads (Fig. 1.10*h*), all referred to the system of end-supported conjugate beams.

Thus, the coefficients $\delta_{n,n-2}$, $\delta_{n,n-1}$, $\delta_{n,n}$, $\delta_{n,n+1}$, $\delta_{n,n+2}$ will be obtained multiplying the $\overline{M}_n$ diagram (Fig. 1.10*e*) by those

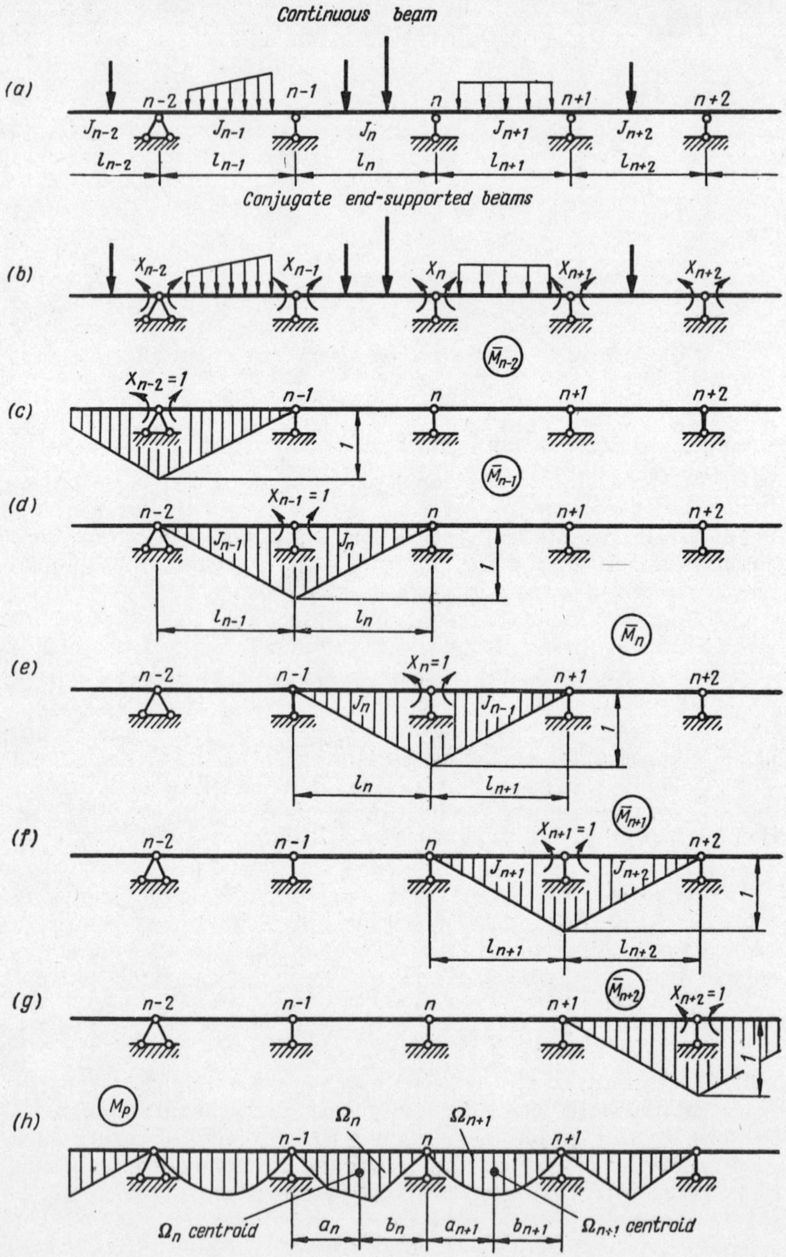

Continuous beam
(a)
n-2
n-1
n
n+1
n+2
J_{n-2} J_{n-1} J_n J_{n+1} J_{n+2}
l_{n-2} l_{n-1} l_n l_{n+1} l_{n+2}
Conjugate end-supported beams
(b)
X_{n-2} X_{n-1} X_n X_{n+1} X_{n+2}
(c)
$\bar{M}_{n-2}$
$X_{n-2}=1$
(d)
$\bar{M}_{n-1}$
$X_{n-1}=1$
(e)
$\bar{M}_n$
$X_n=1$
(f)
$\bar{M}_{n+1}$
$X_{n+1}=1$
(g)
$\bar{M}_{n+2}$
$X_{n+2}=1$
(h)
M_p
Ω_n
Ω_{n+1}
Ω_n centroid
Ω_{n+1} centroid
a_n b_n a_{n+1} b_{n+1}

Fig. 1.10

for $\overline{M}_{n-2}$, $\overline{M}_{n-1}$, $\overline{M}_n$, $\overline{M}_{n+1}$ and $\overline{M}_{n+2}$ (see Fig. 1.10*c*, *d*, *e*, *f*, *g*)

$$\delta_{n,\,n-2}=0$$

$$\delta_{n,\,n-1}=\frac{1}{EJ_n}\cdot 1\cdot\frac{l_n}{2}\cdot\frac{1}{3}\cdot 1=\frac{l_n}{6EJ_n}$$

$$\delta_{n,\,n}=\frac{1}{EJ_n}\left(1\cdot\frac{l_n}{2}\cdot\frac{2}{3}\cdot 1\right)+\frac{1}{EJ_{n+1}}\left(1\cdot\frac{l_{n+1}}{2}\cdot\frac{2}{3}\cdot 1\right)=$$

$$=\frac{2l_n}{6EJ_n}+\frac{2l_{n+1}}{6EJ_{n+1}}$$

$$\delta_{n,\,n+1}=\frac{1}{EJ_{n+1}}\left(1\cdot\frac{l_{n+1}}{2\times 3}\cdot 1\right)=\frac{l_{n+1}}{6EJ_{n+1}}$$

$$\delta_{n,\,n+2}=0$$

Hence, all the coefficients to the unknowns in equation (1.10) with the exception of the coefficients $\delta_{n,n-1}$, $\delta_{n,n}$, and $\delta_{n,n+1}$ reduce to zero.

The multiplication of the $\overline{M}_n$ graph (Fig. 1.10*e*) by the M_p graph (Fig. 1.10*h*) yields the following value for the free term of the above equation

$$\Delta_{np}=\frac{1}{EJ_n}\,\Omega_n y_n+\frac{1}{EJ_{n+1}}\,\Omega_{n+1}y_{n+1}$$

In this expression Ω_n and Ω_{n+1} are the areas of the M_p diagram over the l_n and the l_{n+1} spans (see Fig. 1.10*h*), while y_n and y_{n+1} are the ordinates to the $\overline{M}_n$ graph measured over the centroids of Ω_n and Ω_{n+1}, respectively.

From Fig. 1.10*e* and *h* we draw

$$y_n=\frac{a_n}{l_n}\quad\text{and}\quad y_{n+1}=\frac{b_{n+1}}{l_{n+1}}$$

and accordingly

$$\Delta_{np}=\frac{1}{EJ_n}\cdot\frac{\Omega_n a_n}{l_n}+\frac{1}{EJ_{n+1}}\cdot\frac{\Omega_{n+1}b_{n+1}}{l_{n+1}}$$

Designating $\Omega_n a_n$ by $S^M_{n,n-1}$ and $\Omega_{n+1}b_{n+1}$ by $S^M_{n+1,\,n+1}$ we obtain finally

$$\Delta_{np}=\frac{S^M_{n,\,n-1}}{EJ_n l_n}+\frac{S^M_{n+1,\,n+1}}{EJ_{n+1}l_{n+1}}$$

where $S^M_{n,n-1}$ = static moment of Ω_n about the vertical passing through the left-hand support of span l_n

$S^M_{n+1,n+1}$ = static moment of Ω_{n+1} about the vertical passing through the right-hand support of span l_{n+1}. The signs of these static moments will be the same as those of the corresponding graph areas.

Substituting the values thus found in equation (1.10) and collecting the terms we find

$$X_{n-1}\cdot\frac{l_n}{J_n}+2X_n\left(\frac{l_n}{J_n}+\frac{l_{n+1}}{J_{n+1}}\right)+X_{n+1}\frac{l_{n+1}}{J_{n+1}}=$$

$$=-\frac{6S^M_{n,\,n-1}}{l_nJ_n}-\frac{6S^M_{n+1,\,n+1}}{l_{n+1}J_{n+1}}$$

As the unknowns X_{n-1}, X_n and X_{n+1} represent the bending moments at the supports of the redundant beam M_{n-1}, M_n and M_{n+1}, respectively, the above equation, universally known as the equation of three moments will take the following form

$$M_{n-1}\frac{l_n}{J_n}+2M_n\left(\frac{l_n}{J_n}+\frac{l_{n+1}}{J_{n+1}}\right)+M_{n+1}\frac{l_{n+1}}{J_{n+1}}=$$

$$=-\frac{6S^M_{n,\,n-1}}{l_nJ_n}-\frac{6S^M_{n+1,\,n+1}}{l_{n+1,\,n+1}} \tag{2.10}$$

If the cross-sectional areas of all the spans of the beam remain constant, i.e., if $J_{n-2}=J_{n-1}=J_n=J_{n+1}=J_{n+2}$, etc., the equation of three moments becomes

$$M_{n-1}l_n+2M_n(l_n+l_{n+1})+M_{n+1}l_{n+1}=$$

$$=-\frac{6S^M_{n,\,n-1}}{l_n}-\frac{6S^M_{n+1,n+1}}{l_{n+1}} \tag{3.10}$$

The right-hand part of equation (3.10) is equal to six times the imaginary reaction R_n which would arise at the nth support of the conjugate system of end-supported beams if the spans contiguous to that support were loaded with the areas of the bending moment diagrams due to the actual loads acting over these same spans (see Fig. 1.10*h*). This reaction will be reckoned positive when the loads just mentioned cause an extension of the lower fibres of the beams. Equation (3.10) becomes then

$$M_Ll_L+2M_C(l_L+l_R)+M_Rl_R=-6R^I_C \tag{4.10}$$

where M_L, M_C, and M_R stand for the moments over the left-hand, the central and the right-hand supports, l_L and l_R for the length of the spans to the left and to the right of the central support, and R^I_C for the imaginary reaction of the central support. All the three equations (2.10), (3.10) and (4.10) are known under the name of *equations of three moments*. Each of them expresses the idea that the mutual angular rotation of two adjacent cross sections over the nth support is nil. Therefore the equations relate particularly to this nth support.

At the same time each of these equations establishes the relation between the bending moments acting over three consecutive supports of the continuous beam. The number of such equations that can be written for a beam, all the supports of which are hinged, will be exactly equal to the number of the intermediate supports. The simultaneous solution of these equations will yield all the values of the unknown bending moments at these supports.

The system formed by all the equations of three moments constitutes a particular form of the system of canonical equations. Its great advantage resides in the fact that it necessitates neither the construction of unit moment graphs nor the computation of the displacements caused by the unit actions and the actual loads, simplifying thereby very considerably the analysis of continuous beams.

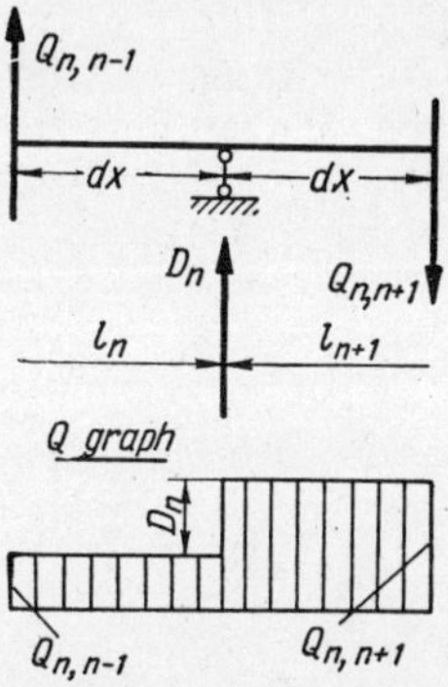

Fig. 2.10

When all the moments at the supports are known one may proceed with the determination of bending moments within the spans, of the shears and of the reactions developed at each support. These computations will be carried out assuming that each span is simply supported at its ends and is acted upon both by the applied loads and the moments at the supports just determined. The following expressions will be used

$$M = M^0 + M_{n-1} + \frac{M_n - M_{n-1}}{l_n} x \tag{5.10}$$

$$Q = Q_x^0 + \frac{M_n - M_{n-1}}{l_n} \tag{6.10}$$

where x is the distance from the left-hand support of this span. These same expressions may be used for the construction of the corresponding diagrams.

The support reactions D_n will be determined as follows: isolating an infinitely small element over the support under consideration from the rest of the beam we see that the left face of this element is acted upon by the shear $Q_{n,n-1}$ while its right face by the shear $Q_{n,n+1}$. Such an element is represented in Fig. 2.10*a*. Projecting on the vertical all the forces acting on this element we obtain

$$Q_{n,\,n-1} + D_n - Q_{n,\,n+1} = 0$$

wherefrom

$$D_n = Q_{n,\,n+1} - Q_{n,\,n-1} \tag{7.10}$$

In the above expressions reactions D_n are reckoned positive when directed upwards.

Thus, *in a continuous beam the reaction at any support is equal to the difference between the shears acting over two contiguous cross sections located both sides of the support under consideration. Hence the numerical value of this reaction will be equal to the rise or to the fall in the shear diagram over the corresponding support* (Fig. 2.10*b*).

The same rection D_n may be obtained if both spans meeting over the support in question are regarded as simple end-supported beams (Fig. 3.10), these beams being acted upon both by the actual loads and the moments at the supports already determined. In

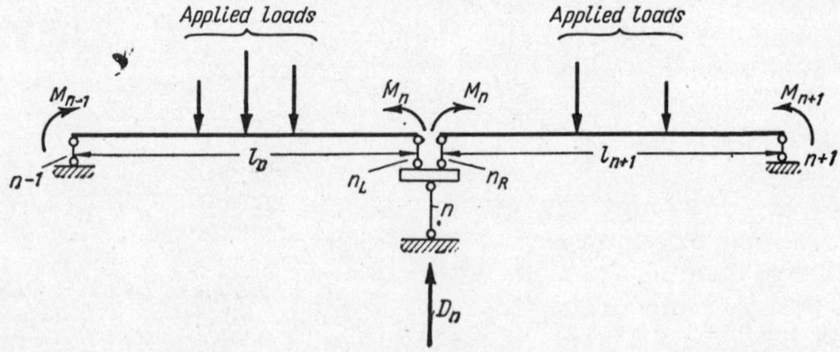

Fig. 3.10

this case reaction D_n will be given by the sum of D_{nL} developed at the right end of the left-hand beam and of D_{nR} developed at the left end of the right-hand beam or, in other words

$$D_n = D_{nL} + D_{nR} \tag{8.10}$$

in this expression

$$\left.\begin{aligned} D_{nL} &= D^0_{nL} + \frac{M_{n-1} - M_n}{l_n} \\ D_{nR} &= D^0_{nR} + \frac{M_{n+1} - M_n}{l_{n+1}} \end{aligned}\right\} \tag{9.10}$$

where D^0_n, D^0_{nL} and D^0_{nR} are the reactions at the common supports of both beams due solely to the action of actual loads, the bending moments at the supports being disregarded. *

Introducing the values given by expressions (9.10) in equation (8.10) we obtain

$$D_n = D^0_{nL} + D^0_{nR} + \frac{M_{n-1} - M_n}{l_n} + \frac{M_{n+1} - M_n}{l_{n+1}} \tag{10.10}$$

* The magnitude of D_{nL} and D_{nR} could also be deduced from the equilibrium of the end-supported beams just mentioned.

Both formulas (7.10) and (10.10) permit the computation of all the support reactions of the continuous beam when the moments at the supports are known. The reaction at the support of a cantilever-end span will be determined using expression (7.10) or, in other words, computing the difference between the shears acting on both sides of this support. It will be remembered that the shearing force acting at the support of a cantilever beam is always equal to the projection of all forces acting on this beam in the direction normal to its axis.

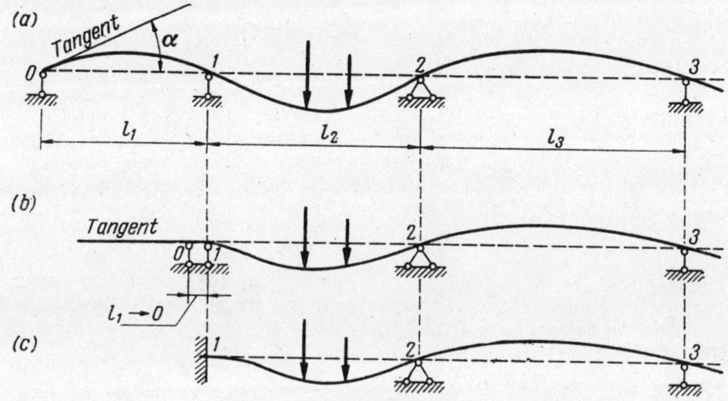

Fig. 4.10

Let us now examine the continuous beam appearing in Fig. 4.10*a* when the length of the left-end span of this beam reduces gradually to zero. In this case the tangent to the elastic curve at the left-hand support will also tend towards the axis of the beam (Fig. 4.10*b*), which indicates that the cross section over support *1* will suffer no rotation whatsoever. This means that the end of the beam appearing in Fig. 4.10*a* becomes fixed when the length l_1 of its end span tends towards zero (Fig. 4.10*c*).

It follows that *the analysis of continuous beams with fixed ends may be carried out using the same equation of three moments, provided a built-in end is replaced by an additional freely supported span of zero length.*

Concluding we may recommend the following sequence of operations for the analysis of continuous beams:

1. Trace a schematic drawing of the beam indicating all the applied loads. If one of the end spans of the beam is built in, replace the built-in end by an additional simply supported span of zero length.

2. Number from left to right all the supports as well as all the spans.

3. Write for each intermediate support of the beam an equation of three moments.

4. Proceed with the simultaneous solution of all these equations obtaining thus the magnitude of all the bending moments at the supports (except the end ones which remain zero).

5. Determine the bending moments and the shears along the spans using expression (5.10) and (6.10) thus obtaining all the data necessary for the construction of stress diagrams.

6. Compute all the support reactions using expressions (7.10) or (10.10).

7. The accuracy of the diagrams obtained will be checked using one of the methods described in Art. 6.9.

Problem 1. Trace the M and Q diagrams for a continuous beam of constant cross section represented in Fig. 5.10*a*.

Solution. Construct the M_p diagram regarding each span as a separate end-supported beam (Fig. 5.10*b*) and form for support *1* the equation of three moments. Expression (4.10) leads to

$$M_0 l_1 + 2M(l_1 + l_2) + M_2 l_2 = -6R_1^I$$

The beam being symmetrical and symmetrically loaded, the moments over supports *1* and *2* will have the same values, which means that $M_1 = M_2$. In addition we have $M_0 = 0$ and $l_1 = l_2 = l$.

Hence

$$R_1^I = \frac{Pl}{4} \cdot \frac{l}{2} \cdot \frac{1}{2} = \frac{Pl^2}{16}$$

and consequently

$$2M_1 \cdot 2l + M_1 l = -\frac{3Pl^2}{8}$$

wherefrom

$$M_1 = M_2 = -\frac{3Pl}{40}$$

Both bending moments M_1 and M_2 thus obtained being negative, the top fibres over the supports will be extended and the bottom ones compressed. In order to construct the bending moment diagram let us determine now the moment at load point P using formula (5.10)

$$M_{\frac{l}{2}} = \frac{Pl}{4} - \frac{3Pl}{40} + \frac{-\frac{3Pl}{40} - \left(-\frac{3Pl}{40}\right)}{l} \cdot \frac{l}{2} = \frac{7Pl}{40}$$

Knowing the values of the bending moments at the supports as well as that of the moment at the middle of the central span we can proceed with the construction of the bending moment diagram for the whole beam (Fig. 5.10*c*).

At the same time expression (6.10) permits the calculation of shearing forces at all the cross sections of the beam.

Thus, for the span *0-1* we find

$$Q_x = 0 + \frac{-\frac{3Pl}{40} - 0}{l} = -\frac{3P}{40}$$

and for the span *1-2* we have between support *1* and the load point

$$Q=\frac{P}{2}+\frac{\frac{-3Pl}{40}-\left(-\frac{3Pl}{40}\right)}{l}=\frac{P}{2}$$

The above data are sufficient for the construction of the left half of the shear diagram appearing in Fig. 5.10*d*. This diagram being symmetrical, its right half will be obtained immediately.

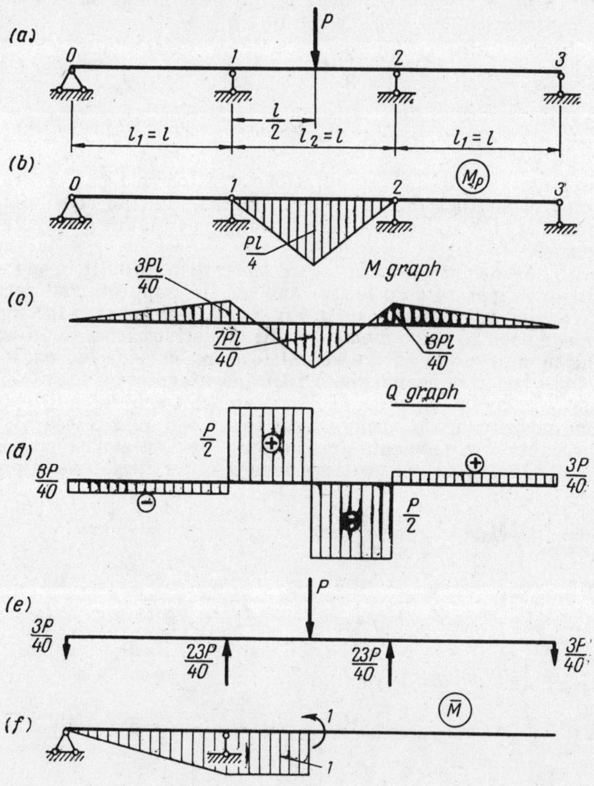

Fig. 5.10

The reactions at the supports will be given by expression (7.10)

$$D_0=D_3=-\frac{3P}{40}-0=-\frac{3P}{40}$$

$$D_1=D_2=\frac{P}{2}-\left(-\frac{3P}{40}\right)=\frac{23P}{40}$$

Fig. 5.10e represents the given beam together with all the loads and reactions acting thereupon. It is obvious that the whole system is in equilibrium.

Let us check the accuracy of the M diagram which must provide for deflections consistent with the stipulations of the problem. Since the beam and the load distribution are symmetrical, the angular rotation v_p of the cross section at load point must equal zero. The value of this rotation can be obtained eliminating supports 2 and 3 and applying to the simple statically determinate system thus obtained a unit moment at load point P. The bending moment diagram due to this unit moment appears in Fig. 5.10f. Multiplying this graph by the bending moment graph due to the actual loads and reactions (Fig. 5.10c) we obtain

$$v_p=\frac{1}{EJ}\left[-\frac{3Pl}{40}\cdot l\frac{1}{2}\cdot\frac{2}{3}1+\left(\frac{7Pl}{40}-\frac{3Pl}{40}\right)\frac{1}{2}\cdot\frac{l}{2}\cdot 1\right]=$$

$$=\frac{Pl^2}{40EJ}\left(-1+\frac{7}{4}-\frac{3}{4}\right)=0$$

which indicates that all the computations were carried out accurately.

Problem 2. Construct the M and Q diagrams for the continuous beam appearing in Fig. 6.10a; the cross-sectional areas of this beam vary from span to span.

Solution. The schematic drawing of the beam with its built-in end replaced by an additional span of zero length and all the supports and spans duly numbered from left to right appears in Fig. 6.10b. The terms of the right-hand part of the equation of three moments will be derived from the diagrams of the bending moments due to the actual loading, considering each span as a separate end-supported beam (Fig. 6.10c). The diagram for the cantilevering end will be constructed in the same way as for a simple cantilever beam.

The bending moment M_3 being equal to —5 ton metres (see Fig. 6.10c), the only two unknown moments are those at the first and second supports (M_1 and M_2). The equations of three moments for these two supports are

$$\frac{M_0 l_1}{J_1}+2M_1\left(\frac{l_1}{J_1}+\frac{l_2}{J_2}\right)+M_2\frac{l_2}{J_2}=-\frac{6S^M_{1;\,0}}{l_1J_1}-\frac{6S^M_{2;\,2}}{l_2J_2}$$

$$\frac{M_1 l_2}{J_2}+2M_2\left(\frac{l_2}{J_2}+\frac{l_3}{J_3}\right)+M_3\frac{l_3}{J_3}=-\frac{6S^M_{2;\,1}}{l_2J_2}-\frac{6S^M_{3;\,3}}{l_3J_3}$$

Seeing that $M_0=0$; $M_3=-5$ ton metres; $l_1=0$; $l_2=6\text{m}$; $l_3=4\text{m}$; $J_2=2J$; $J_3=J$ and $S^M_{1;0}=0$ we find

$$S^M_{2;\,2}=-\frac{5\times 3}{2}\left(3+\frac{3}{3}\right)+\frac{5\times 3}{2}\cdot\frac{2}{3}\cdot 3=-15$$

$$S^M_{2;\,1}=-\frac{5\times 3}{2}\cdot\frac{2}{3}\cdot 3+\frac{5\times 3}{2}\left(3+\frac{3}{3}\right)=15$$

$$S^M_{3;\,3}=\frac{3\times 4}{2}\cdot\frac{4}{2}+\frac{3}{2}\cdot\frac{4}{2}\cdot\frac{2}{3}\cdot 4\cdot\frac{3}{4}=18$$

Consequently

$$\frac{2M_1 6}{2J}+M_2\frac{6}{2J}=-\frac{6\,(-15)}{6\times 2J}$$

$$\frac{M_1 6}{2J}+2M_2\left(\frac{6}{2J}+\frac{4}{J}\right)-\frac{5\times 4}{J}=-\frac{6\times 15}{6\cdot 2J}-\frac{6\times 18}{4J}$$

Upon collection of terms and other simplifications these two equations become

$$4M_1+2M_2=5 \text{ and } 6M_1+28M_2=-29$$

these two equations yield $M_1 = 1.98$ ton metres and $M_2 = -1.46$ ton metres.

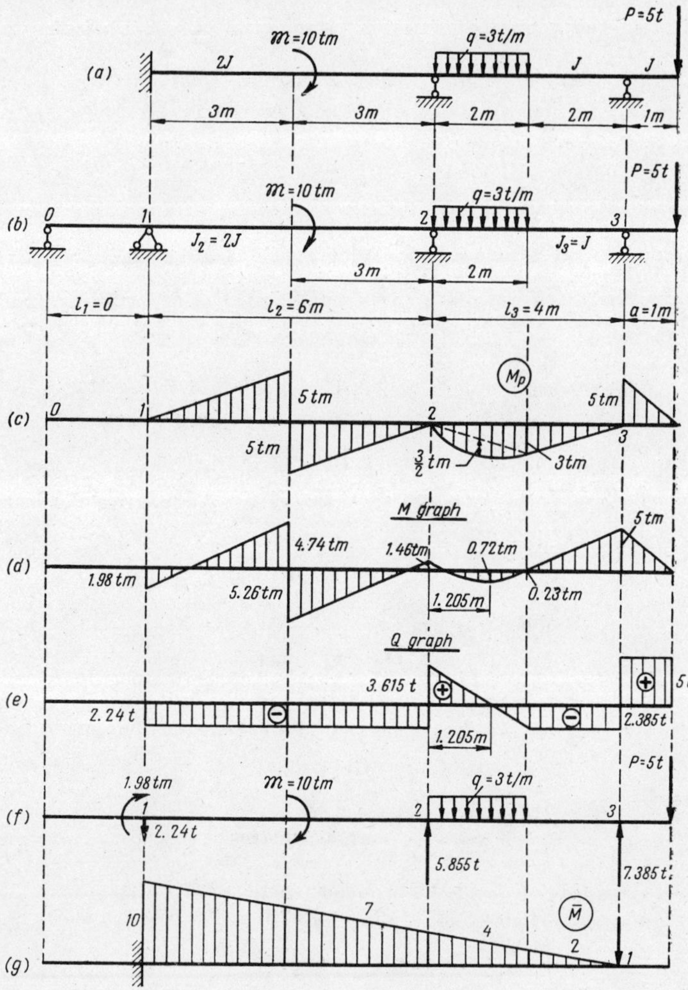

Fig. 6.10

The bending moments acting along the spans may now be easily found using expression (5.10), those acting along the cantilever span are already known.

Span 1-2

for $0 \leqslant x \leqslant 3$ m $\quad M=-\frac{5}{3}x+1.98+\frac{-1.46-1.98}{6}x=1.98-2.24x$

for $x=3$ m $\quad M=1.98-2.24\times 3=-4.74$ ton metres

for $3 \text{ m} \leqslant x \leqslant 6$ m $\quad M=10-\frac{5}{3}x+1.98+\frac{-1.46-1.98}{6}x=11.98-2.24x$

for $x=3$ m $\quad M=11.98-2.24\times 3=5.26$ ton metres

for $x=6$ m $\quad M=11.98-2.24\times 6=-1.46$ ton metres

Span 2-3

for $0 \leqslant x \leqslant 2$ m $\quad M=4.5x-\frac{3x^2}{2}-1.46+\frac{-5+1.46}{4}x=$

$$=-1.46+3.615x-1.5x^2$$

for $x=2$ m $\quad M=-1.46+3.615\times 2-1.5\times 2^2=-0.23$ ton metre

for $2 \text{ m} \leqslant x \leqslant 4$ m $\quad M=4.5x-3\times 2(x-1)-1.46+$

$$+\frac{-5+1.46}{4}x=4.54-2.385x$$

for $x=2$ m $\quad M=4.54-2.385\times 2=-0.23$ ton metre

for $x=4$ m $\quad M=4.54-2.385\times 4=-5.00$ ton metres

The completed bending moment diagram is represented in Fig. 6.10*d*.

The magnitude of the shearing forces acting within the spans will be given by the first derivative of the expressions obtained for the bending moment.

Span 1-2

$$Q=-2.24 \text{ tons}$$

Span 2-3

for $0 \leqslant x \leqslant 2$ m $\quad Q=3.615-3x$

for $x=0$ $\quad Q=3.615$ tons

for $x=2$ m $\quad Q=3.615-3\times 2=-2.385$ tons

The shearing force will become zero at the cross section determined by

$$Q=3.615-3x=0$$

wherefrom

$$x=\frac{3.615}{3}=1.205 \text{ metres}$$

At this cross section the bending moment will reach its maximum *

$$M_{max}=-1.46+3.615\times 1.205-1.5\times 1.205^2=0.72 \text{ ton metre}$$

$$2 \text{ m} \leqslant x \leqslant 4 \text{ m} \qquad Q=-2.385 \text{ tons}$$

The shearing force diagram constructed using the above data is represented in Fig. 6.10*e*.

◆

* By maximum we mean here any point corresponding to a horizontal tangent to the bending moment curve.

The reactions at the support will be obtained as previously using formula (7.10)

$$D_1=-2.24 \text{ tons}; \quad D_2=3.615+2.24=5.855 \text{ tons}$$

$$D_3=5-(-2.385)=7.385 \text{ tons}$$

Fig. 6.10*f* represents the beam with all the loads and reactions acting thereon. Let us check whether the equilibrium conditions are satisfied

$$\Sigma Y=-2.24+5.855+7.385-3\times 2-5=13.24-13.24=0$$

$$\Sigma M_1=1.98+10-5.855\times 6-7.385\times 10+3\times 2\times 7+$$
$$+5\times 11=108.98-108.98=0$$

Let us control also the accuracy of the M diagram using the method based on the consistency of deflections. For this purpose let us compute the deflection of the section situated directly over support *3*. This deflection must be necessarily nil, the support precluding any vertical movement of the beam. The bending moment diagram induced by a vertical unit load acting at the corresponding section of a simple statically determinate beam (obtained by elimination of supports *2* and *3*) is represented in Fig. 6.10*g*. Multiplying the ordinates to the curve of the resultant moments M by the diagram of Fig. 6.10*g* we obtain

$$\Delta_3=\frac{3}{6}(-2\times 10\times 1.98+2\times 7\times 4.74+10\times 4.74-7\times 1.98)\times$$
$$\times\frac{1}{2EJ}+\frac{3}{6}(-2\times 7\times 5.26+2\times 4\times 1.46+7\times 1.46-4\times 5.26)\times$$
$$\times\frac{1}{2EJ}+\frac{2}{6}(2\times 4\times 1.46+2\times 2\times 0.23+4\times 0.23+2\times 1.46)\frac{1}{EJ}-$$
$$-1\times 5\times 2\times\frac{2}{3}\times\frac{4+2}{2}\times\frac{1}{EJ}+\frac{2\times 2}{2}\times\left(0.23\times\frac{2}{3}+5\times\frac{1}{3}\right)\frac{1}{EJ}=$$
$$=\frac{1}{EJ}(15.075-18.195+5.48-6+3.64)=0$$

which indicates that all the computations were carried out correctly.

2.10. *The Focal Points Method*

The method described in the present article becomes particularly interesting when continuous beams with a very large number of spans must be dealt with or when only a few of the spans carry the loads.

Let us examine the beam of constant cross section appearing in Fig. 7.10*a* assuming that only the *n*th span is loaded. We may eliminate all the redundant constraints opposing mutual rotation of adjacent cross sections at the supports, provided we replace these constraints by the corresponding bending moments which are as yet unknown. In order to find these moments let us write consecutively the equations of three moments for each of the supports.

(a) *Support 1*

The equation of three moments becomes

$$M_0 l_1 + 2M_1 (l_1 + l_2) + M_2 l_2 = 0$$

but since $M_0 = 0$

$$2M_1 (l_1 + l_2) + M_2 l_2 = 0$$

hence

$$\frac{M_2}{M_1} = -\frac{2(l_1 + l_2)}{l_2} = -K_2$$

The latter relation shows that

(1) the bending moments M_1 and M_2 at two neighbouring supports are of opposite signs;

(2) the ratio $\frac{M_2}{M_1}$ depends solely on span lengths l_1 and l_2, but is completely uninfluenced by the magnitude of the moments and loads acting along the spans further to the right.

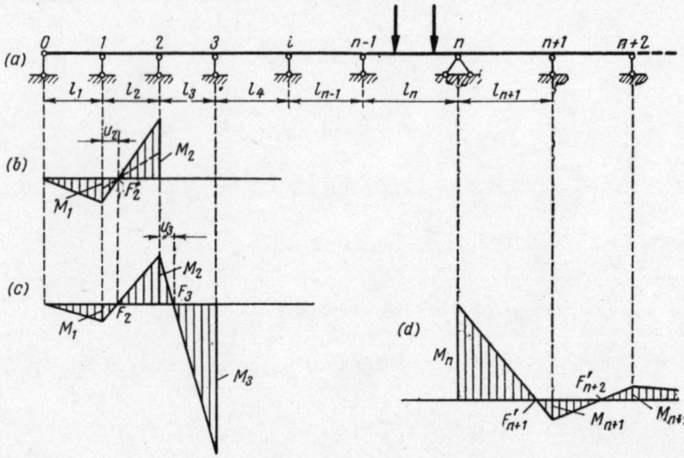

Fig. 7.10

In the particular case when $l_1 = l_2$ the factor K_2 becomes equal to 4, or, in other words, the bending moment M_2 is in absolute value four times as large as the bending moment M_1.

The bending moment diagram for span l_2 has the shape indicated in Fig. 7.10*b*. This diagram shows that in the left half of the span there is a point where the bending moment becomes zero. This point is known as the *left-hand focal point* of the second span and will be hereafter designated by F_2. The location of this point along the span depends on the value of K_2 which we shall call the *left-hand focal factor* for the second span. The location of the focal point remains uninfluenced by the lengths of the spans further to the right, nor by the loads these spans may carry. The bending moments at the first and second supports may vary in terms both of the amount and of the distribution of the loads, but the ratio between these moments will remain constant as long as the spans l_1 and l_2 remain unloaded. Consequently, the *bending moment at the focal point F_2 will remain always nil when the latter condition is fulfilled.*

The distance u_2 between the focal point F_2 and the nearest support to the left is given by

$$u_2 = \frac{l_2}{1+K_2}$$

and in the particular case when $l_1 = l_2 = l$

$$u_2 = \frac{1}{5}\, l$$

(b) *Support 2*

The equation of three moments becomes

$$M_1 l_2 + 2M_2 (l_2 + l_3) + M_3 l_3 = 0$$

substituting in this equation $M_1 = -\frac{M_2}{K_2}$ we find

$$-\frac{M_2}{K_2}\, l_2 + 2M_2 (l_2 + l_3) + M_3 l_3 = 0$$

wherefrom

$$\frac{M_3}{M_2} = -\left[2 + \frac{l_2}{l_3}\left(2 - \frac{1}{K_2}\right)\right] = -K_3$$

It is clear that the ratio $\frac{M_3}{M_2}$ is again independent of the length of the spans *further* to the right as well as of the loads applied to these spans. The shape of the bending moment diagram along the third span will be as shown in Fig. 7.10*c*. Within the left half of this span there will again exist a focal point where the bending moment will remain zero as long as three spans l_1, l_2 and l_3 remain unloaded.

(c) *Support 3*

The equation of three moments will be in this case

$$M_2 l_3 + 2M_3 (l_3 + l_4) + M_4 l_4 = 0$$

Substituting in this equation $-\frac{M_3}{K_3}$ for M_2 we obtain as previously

$$\frac{M_4}{M_3} = -\left[2 + \frac{l_3}{l_4}\left(2 - \frac{1}{K_3}\right)\right] = -K_4$$

Thus, the focal factor for span *4* will be given by exactly the same expression as the one for span *3* with the only difference that all the intervening indices are increased by one. Consequently, the general expression for the focal factor K_n relative to the left-hand focal point of span n will be

$$K_n = \left[2 + \frac{l_{n-1}}{l_n}\left(2 - \frac{1}{K_{n-1}}\right)\right] = -\frac{M_n}{M_{n-1}} \qquad (11.10)$$

The above expression permits the computation of all the focal factors one after the other. Thus, for the first span of a simply supported continuous beam we have

$$M_0 = 0 \quad \text{and} \quad K_1 = -\frac{M_1}{M_0} = -\infty$$

which indicates that the left-hand focal point of the first span coincides with the left-end support. For the second span we obtain as already mentioned

$$K_2 = 2 + \frac{l_1}{l_2}\left(2 - \frac{1}{\infty}\right) = \frac{2(l_1 + l_2)}{l_2}$$

It should be always remembered that *the left-hand focal point is a point situated along the axis of a continuous beam at which the bending moment remains nil as long as the span under consideration and all the other spans to its left remain unloaded.*

Let us now investigate those spans of the same continuous beam which are located to the right from the loaded ones. Reasoning in exactly the same way, we shall obtain an expression giving the value of the right-hand focal factor. The bending moment diagram for the unloaded right-hand spans will have the shape indicated in Fig. 7.10*d*. This diagram shows clearly that in the right half of each span there exists equally a certain point where the bending moment remains nil as long as the span under consideration and the spans located further to the right carry no loads. These points are the *right-hand focal points* and the expression giving the value of the right-hand focal factor which we shall indicate by K'_n will be derived from that for the left-hand focal factor keeping in mind that the numbers allotted to the supports and spans decrease from right to left

$$K'_n = 2 + \frac{l_{n+1}}{l_n}\left(2 - \frac{1}{K'_{n+1}}\right) = -\frac{M_{n-1}}{M_n} \qquad (12.10)$$

When the right-hand extremity of a continuous beam is hinge-supported the focal factor becomes infinitely great just as in the case of a hinge-supported left end. All the other right-hand focal factors will be determined in succession with the aid of equation (12.10).

The expressions (11.10) and (12.10) may be used for the determination of focal factors pertaining to continuous beams with fixed ends if these ends are replaced by additional spans of zero length. Thus, for a continuous beam with a built-in left end the value of the focal factor K_1 will be given by

$$K_1 = 2 + \frac{l_0}{l_1}\left(2 - \frac{1}{K_0}\right)$$

where $K_0 = \infty$ for the left end if the additional span is hinge-supported. As $l_0 = 0$ we obtain $K_1 = 2$ which means that the focal point will be situated one third of the span to the right

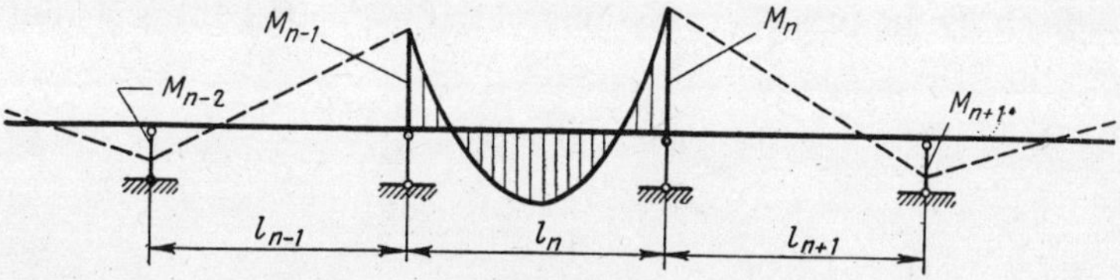

Fig. 8.10

from the wall. It is worth mentioning that this is the maximum distance which can separate the focal point from the corresponding support. As for the minimum distance, it is equal to zero as we have already seen.

Let us apply the focal points method to the determination of the bending moments at the supports of a continuous beam appearing in Fig. 8.10, assuming that only one span of this beam (say, span n) is loaded. The equations of three moments for each of the two supports limiting the loaded span become

for support $n-1$

$$M_{n-2}l_{n-1} + 2M_{n-1}(l_{n-1} + l_n) + M_n l_n = -6R^I_{n-1} = -6A^I_n$$

for support n

$$M_{n-1}l_n + 2M_n(l_n + l_{n+1}) + M_{n+1}l_{n+1} = -6R^I_n - 6B^I_n$$

In these expressions A^I_n and B^I_n are the imaginary reactions of the left- and of the right-hand supports of the nth span respectively. Replacing the bending moments at supports M_{n-2} and M_{n+1} by their values expressed in terms of the focal factors

$$M_{n-2} = -\frac{M_{n-1}}{K_{n-1}} \text{ and } M_{n+1} = -\frac{M_n}{K'_{n+1}}$$

we obtain

$$M_{n-1}\left[2(l_{n-1}+l_n)-\frac{l_{n-1}}{K_{n-1}}\right]+M_n l_n=-6A_n^I$$

$$M_{n-1}l_n+M_n\left[2(l_n+l_{n+1})-\frac{l_{n+1}}{K'_{n+1}}\right]=-6B_n^I$$

Dividing both parts of these expressions by l_n

$$M_{n-1}\left[2+\frac{l_{n-1}}{l_n}\left(2-\frac{1}{K_{n-1}}\right)\right]+M_n=-\frac{6A_n^I}{l_n}$$

$$M_{n-1}+M_n\left[2+\frac{l_{n+1}}{l_n}\left(2-\frac{1}{K'_{n+1}}\right)\right]=-\frac{6B_n^I}{l_n}$$

and keeping in mind expressions (11.10) and (12.10) we finally obtain

$$M_{n-1}K_n+M_n=-\frac{6A_n^I}{l_n}$$

$$M_{n-1}+M_nK'_n=-\frac{6B_n^I}{l_n}$$

Solving these equations for M_{n-1} and M_n we find

$$\left.\begin{aligned}M_{n-1}&=-\frac{6\,(A_n^IK'_n-B_n^I}{l_n\,(K_nK'_{n-1})}\\ M_n&=-\frac{6\,(B_n^IK_n-A_n^I)}{l_n\,(K_nK'_n-1)}\end{aligned}\right\}\qquad(13.10)$$

If the loads are located over one of the end spans, say, the left one, and provided the left-end support is hinged, the value of the focal factor for this span will become infinitely great. The bending moment at the left-end support becomes nil and the value of the bending moment at the other support of the span becomes indeterminate. This indetermination will be eliminated dividing both the numerator and the denominator of the expression by K_n. When K_n increases indefinitely we obtain

$$M_n=-\frac{6\left(B_n^I-\dfrac{A_n^I}{\infty}\right)}{l_n\left(K'_n-\dfrac{1}{\infty}\right)}=-\frac{6B_n^I}{l_nK'_n}$$

Having thus determined the bending moments at the supports of the loaded span, the bending moments at all the other supports will be found easily using the expressions for the focal factors.

When several spans of a continuous beam are loaded directly the problem will be solved using the principle of superposition.

Problem 1. Determine the bending moments M_1 and M_2 at the supports of span *1-2* carrying a load unity P situated a distance x from support *1* (Fig. 9.10*a*).

Solution. Using expression (11.10) determine the left-hand focal factors for spans l_1 and l_2

$$K_1=-\frac{M_1}{M_0}=-\infty, \quad K_2=2+\frac{l_1}{l_2}\left(2-\frac{1}{\infty}\right)=4$$

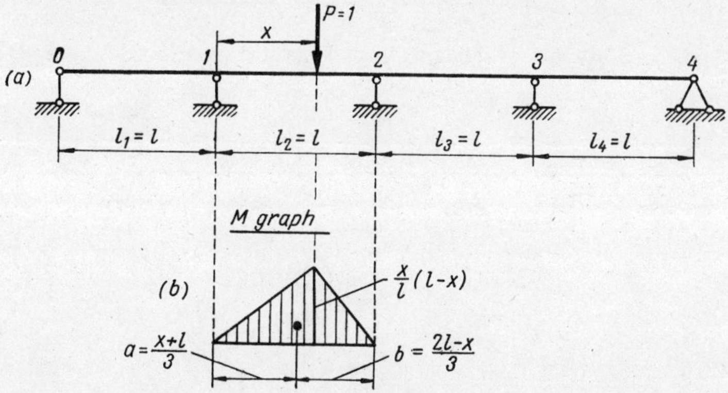

Fig. 9.10

Determine in the same way the right-hand focal factors for spans l_4, l_3 and l_2 using formula (12.10)

$$K_4'=-\frac{M_3}{M_4}=-\infty$$

$$K_3'=2+\frac{l_4}{l_3}\left(2-\frac{1}{\infty}\right)=4$$

$$K_2'=2+\frac{l_3}{l_2}\left(2-\frac{1}{4}\right)=3.75$$

The bending moments at the supports of the loaded span will be found using expression (13.10)

$$M_1=-\frac{6\,(A_2^IK_2'-B_2^I)}{l_2\,(K_2K_2'-1)}; \quad M_2=-\frac{6\,(B_2^IK_2-A_2^I)}{l_2\,(K_2K_2'-1)}$$

in which

$$A_2^I=\frac{x\,(l-x)}{l}\cdot\frac{l}{2}\cdot\frac{(2l-x)}{3l}=\frac{(l-x)\,(2l-x)\,x}{6l}$$

$$B_2^I=\frac{x\,(l-x)}{l}\cdot\frac{l}{2}-\frac{(x+l)}{3l}=\frac{x\,(l^2-x^2)}{6l}$$

(see the bending moment graph represented in Fig. 9.10*b*). Using these expressions we find

$$M_1=-\frac{6\left[\frac{(l-x)\,(2l-x)\,x}{6l}\times 3.75-\frac{x\,(l^2-x^2)}{6l}\right]}{l\,(4\times 3.75-1)}=$$

$$=-\frac{(l-x)(2l-x)x\times 3.75-x(l^2-x^2)}{14l^2}$$

$$M_2=-\frac{6\left[\frac{x(l^2-x^2)}{6l}\times 4-\frac{(l-x)(2l-x)x}{6l}\right]}{14l}=$$

$$=\frac{4x(l^2-x^2)-(l-x)(2l-x)x}{14l^2}$$

Problem 2. Required the complete analysis of a six-span continuous beam uniformly loaded over the whole length of span *4* (Fig. 10.10*a*).

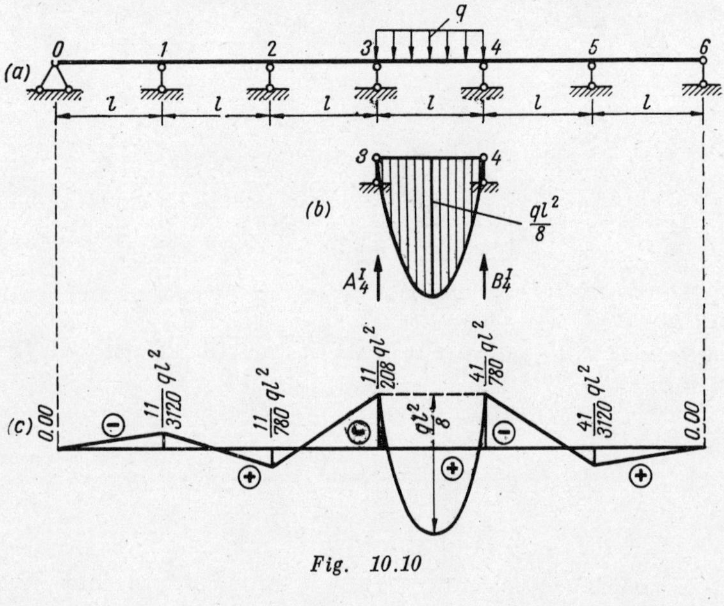

Fig. 10.10

Solution. Begin with computing the left-hand focal factors using expression (11.10)

$$K_1=-\frac{M_1}{M_0}=-\infty$$

$$K_2=2+\frac{l}{l}\left(2-\frac{1}{\infty}\right)=4$$

$$K_3=2+\frac{l}{l}\left(2-\frac{1}{4}\right)=\frac{15}{4}$$

$$K_4=2+\frac{l}{l}\left(2-\frac{1\times 4}{15}\right)=\frac{56}{15}$$

There is no need to compute the focal factors for the following spans. The right-hand focal factors will be computed beginning with the right-end span of the beam. All the spans of the beam being of the same length, we have

$$K_6' = K_1 = -\infty$$
$$K_5' = K_2 = 4$$
$$K_4' = K_3 = 3.75$$

The bending moment diagram induced in the conjugate statically determinate beam by the given loading is represented in Fig. 10.10*b*.

Using this diagram as the imaginary load diagram we shall find the following values for the support reactions

$$A_4^I = B_4^I = \frac{ql^2}{8} \cdot \frac{2}{3} l \cdot \frac{1}{2} = \frac{ql^3}{24}$$

Formula (13.10) yields immediately the bending moments at the supports *3* and *4*

$$M_3 = -\frac{6\left(\frac{ql^3}{24} \times \frac{15}{4} - \frac{ql^3}{24}\right)}{l\left(\frac{56}{15} \times \frac{15}{4} - 1\right)} = -\frac{11}{208} ql^2$$

$$M_4 = -\frac{6\left(\frac{ql^3}{24} \times \frac{56}{15} - \frac{ql^3}{24}\right)}{l\left(\frac{56}{15} \times \frac{15}{4} - 1\right)} = -\frac{41}{780} ql^2$$

Knowing the magnitude of these moments and the values of the focal factors all the other bending moments at the supports of the beam are obtained with no difficulty

$$M_2 = -\frac{M_3}{K_3} = \frac{11}{208} ql^2 \frac{4}{15} = \frac{11}{780} ql^2$$

$$M_1 = -\frac{M_2}{K_2} = -\frac{11}{780} ql^2 \frac{1}{4} = -\frac{11}{3,120} ql^2$$

$$M_0 = 0$$

$$M_5 = -\frac{M_4}{K_5'} = \frac{41}{780} ql^2 \frac{1}{4} = \frac{41}{3,120} ql^2$$

$$M_6 = 0$$

The data so obtained have permitted the construction of the diagram appearing in Fig. 10.10*c*.

3.10. *Bending Moment Envelope Curves*

The control of fibre stresses in continuous beams and the choice of their cross-sectional dimensions will frequently require the knowledge of the extreme values the bending moments may attain at different points under different loading conditions. The dead loads will be usually considered uniformly distributed but the position of the live loads may vary quite considerably. If at every cross section of the beam we set off two ordinates—one representing the maximum value of the bending moment (M_{max})

and the other its minimum value (M_{min}) and if we connect these ordinates by two smooth curves we shall obtain what is usually referred to as the *bending moment envelope curves.*

The construction of such curves can be best explained using as an example the three-span continuous beam represented in Fig. 11.10*a*. Let q be the uniformly distributed dead load per unit length of the beam, and p the live load equally uniformly distributed, which occupies either the whole length of the beam or is spread over certain span lengths only, or is completely absent. Suppose $q = 2$ tons per metre and $p = 3.75$ tons per metre.

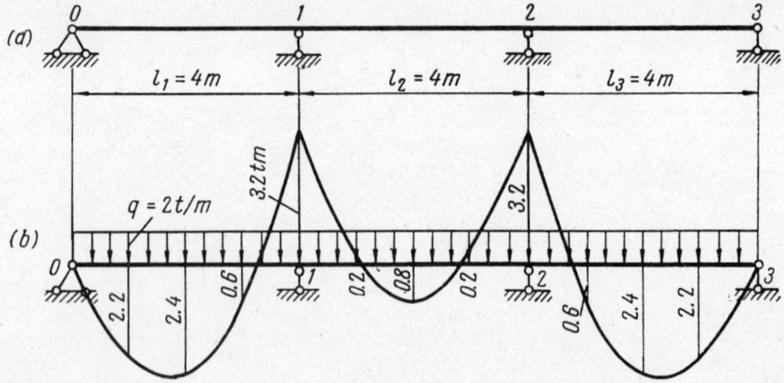

Fig. 11.10

The bending moments at the supports may be determined using either the equations of three moments alternatively or can be deduced from the position of focal points.

The diagram of the bending moments due to the dead load is given in Fig. 11.10*b*. Fig. 12.10*a*, *b* and *c* represents the bending moment diagrams due to the live load occupying successively the first, the second and the third spans.

Let us proceed now with the construction of the envelope curve. For this purpose we shall first take at each section the sum of all positive ordinates due to the live loads and add it to the ordinate at this same cross section due to the dead load. We shall thus obtain for each cross section the ordinate representing the maximum bending moment M_{max} that can be produced by the given loads. Thereafter we must pick out for each of the cross sections under consideration all the negative ordinates that may arise under the effect of the live loads, sum them up and then add them algebraically to the ordinate induced at the same section by the dead load. The resulting ordinate will represent the minimum bending moment M_{min} possible under the given loading conditions.

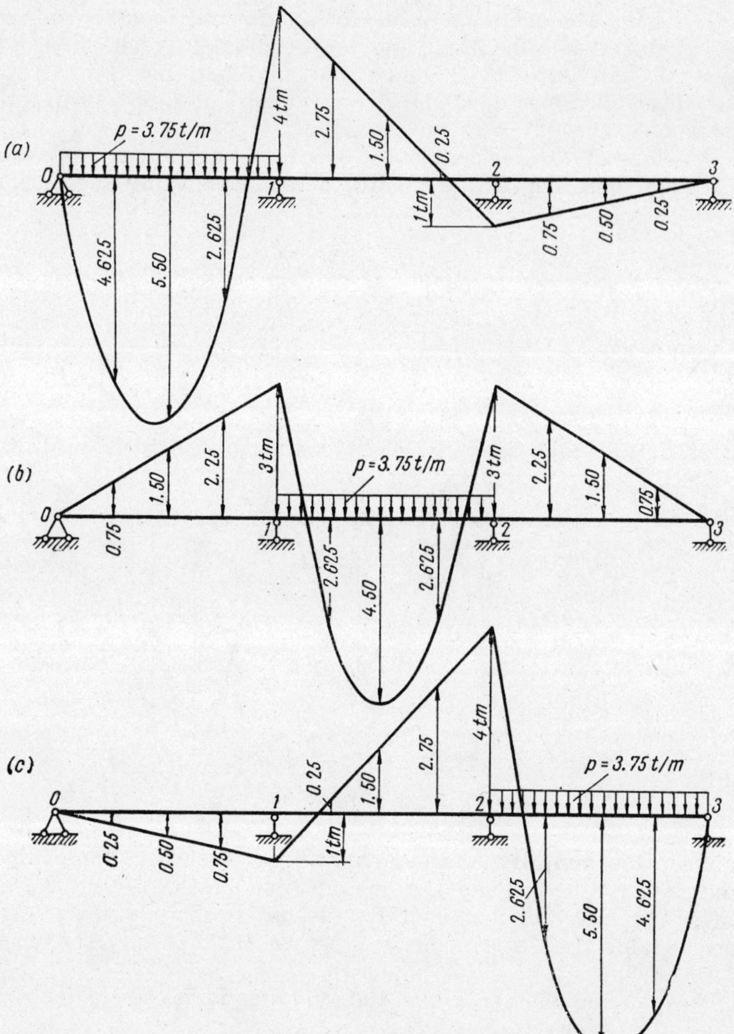

Fig. 12.10

Thus, for instance, the M_{max} and M_{min} ordinates for section *1* over the first support will be

$$M_{max} = 1.0 + (-3.2) = -2.2 \text{ ton metres}$$

$$M_{min} = -4 + (-3) + (-3.2) = -10.2 \text{ ton metres}$$

Repeating the same operation for a sufficient number of sections we shall find the M_{max} and M_{min} ordinates which, connected together, will form the two envelope curves desired (Fig. 13.10). The shearing forces envelope curves can be obtained in exactly the same way.

Envelope curves for continuous beams of constant cross section and even span lengths are usually constructed using appropriate

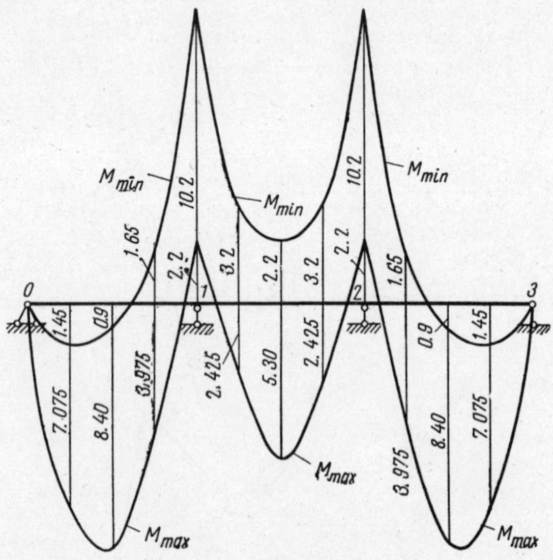

Fig. 13.10

tables which simplify the operation very considerably. These tables contain data permitting the computation of the M and Q ordinates due both to dead and live loads. Hereunder we present such a table (Table 1.10) for a two-span beam simply supported at its ends.

The bending moments and the shearing forces are calculated using the following relations

$$\left.\begin{aligned} M &= (\alpha q + \beta p)\, l^2 \\ Q &= (\gamma q + \delta p)\, l \end{aligned}\right\} \qquad (14.10)$$

where q = dead load per unit length
p = live load per unit length
α, β, γ and δ = coefficients whose values are drawn from the aforesaid tables.

Table 1.10

$\frac{x}{l}$	Bending moments M			Shearing forces Q		
	D. L.	L. L.		D. L.	L. L.	
	α	$+\beta$	$-\beta$	γ	$+\delta$	$-\delta$
0.0	0	0	0	+0.375	0.4375	0.0625
0.1	+0.0325	0.03875	0.00625	+0.275	0.3437	0.0687
0.2	+0.0550	0.06750	0.01250	+0.175	0.2624	0.0874
0.3	+0.0675	0.08625	0.01875	+0.075	0.1932	0.1182
0.375	+0.0703	0.09375	0.02344	0	0.1491	0.1491
0.4	+0.0700	0.09500	0.02500	—0.025	0.1359	0.1609
0.5	+0.0625	0.09375	0.03125	—0.125	0.0898	0.2148
0.6	+0.0450	0.08250	0.03750	—0.225	0.0544	0.2794
0.7	+0.0175	0.06125	0.04375	—0.325	0.0287	0.3537
0.75	0	0.04688	0.04688	—0.375	0.0193	0.3943
0.8	—0.0200	0.03000	0.05000	—0.425	0.0119	0.4369
0.85	—0.0425	0.01523	0.05773	—0.475	0.0064	0.4814
0.9	—0.0675	0.00611	0.07361	—0.525	0.0027	0.5277
0.95	—0.0950	0.00138	0.09638	—0.575	0.0007	0.5757
1.0	—0.1250	0	0.12500	—0.625	0	0.6250
Support reaction $A_1=$				1.250	1.2500	0

Maximum bending moments are obtained using the values of coefficient β contained in the $+\beta$ column, minimum bending moments using those in the $-\beta$ column. In exactly the same way if it is desired to find the maximum shear, the coefficient δ should be selected in the $+\delta$ column, and if it is the minimum value of the shear that is needed this same coefficient shall be taken from the $-\delta$ column.

The same table permits also the determination of the support reactions due to the application of both dead and live loads. The formula to be used is the same as for Q.

Tables such as Table 1.10 contained in handbooks usually take care of partial loadings of different spans when such partial loadings may lead to greater (or smaller) values of bending moments, shears and support reactions as compared to those due to the loading of complete span lengths. For this reason the envelope curves obtained with the use of such tables are even more accurate than those constructed as explained above.

Problem. A reinforced concrete double-span ceiling beam carries the weight of the ceiling itself amounting to 600 kg per metre of beam length. The ceiling may be eventually loaded with a layer of insulating material at the rate of 200 kg per metre of the beam. This insulation may be applied to any part of the ceiling (Fig. 14.10). Required the most unfavourable values of the bending moment and of the shear at a section situated a distance $x = 0.4l$ from the left-end support and of the reaction A_1 at the intermediate support.

Solution. Determine the bending moment using the first of the expressions (14.10). This expression may be rewritten as follows

$$M = \alpha q l^2 + \beta p l^2 = M_q + M_p$$

Here $\alpha q l^2$ represents the bending moment induced by the dead load of 600 kg/m alone. From Table 1.10 the coefficient α corresponding to $x = 0.4\ l$ equals 0.0700 and therefore

$$M_q = \alpha q l^2 = 0.0700 \times 600 \times 10^2 = 4{,}200 \text{ kg}\cdot\text{m}$$

The term $M_p = \beta p l^2$ represents the bending moment induced by the load p whose situation along the beam is such that it will provide either for a maximum or for a minimum value of the moment.

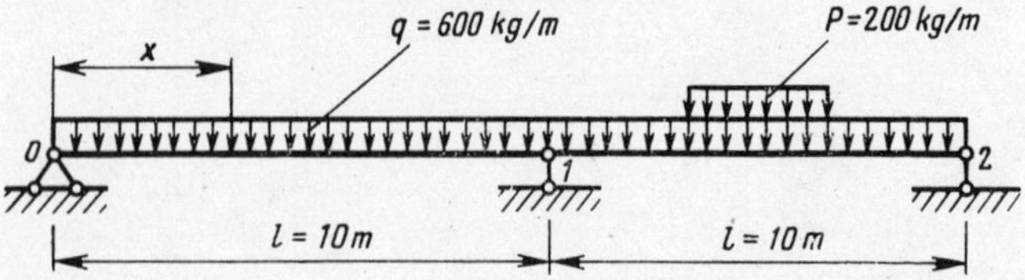

Fig. 14.10

In the present case for the section under consideration the value of coefficient β corresponding to M_{max} equals $+\ 0.09500$ and to M_{min} -0.02500. Therefore

$$M_{p\,max} = \beta p l^2 = 0.09500 \times 200 \times 10^2 = 1{,}900 \text{ kg}\cdot\text{m}$$

and

$$M_{p\,min} = \beta p l^2 = -0.02500 \times 200 \times 10^2 = -500 \text{ kg}\cdot\text{m}$$

Hence the most unfavourable value of the bending moment at a section distant $0.4\ l$ from the left-end support amounts to

$$M_{max} = 4{,}200 + 1{,}900 = 6{,}100 \text{ kg}\cdot\text{m}$$

If we use the value of $M_{p\,min}$ the resulting bending moment at the given cross section will be considerably smaller

$$M_{min} = 4{,}200 - 500 = 3{,}700 \text{ kg}\cdot\text{m}$$

and therefore, since the two values obtained are of the same sign, the first one alone will be retained for further computations.

As for the shear, the use of the second one of the expressions (14.10) together with Table 1.10 yields

$$Q_{max} = (-0.025 \times 600 + 0.1359 \times 200) \times 10 = -150 + 271 = 121 \text{ kg}$$

$$Q_{min} = (-0.025 \times 600 - 0.1609 \times 200) \times 10 = -150 - 321 = -471 \text{ kg}$$

The greatest value of the reaction at the intermediate support will be given by

$$A_1 = (1.250q + 1.250p)\, l = (1.250 \times 600 + 1.250 \times 200)\, 10 = \\ = 10{,}000 \text{ kg} = 10 \text{ tons}$$

4.10. *Influence Lines for Continuous Beams*

Consider a continuous beam acted upon by a moving unit load P travelling along the span l_n (Fig. 15.10a) and assume that the distance of this load to support $(n - 1)$ is given by $x = \eta l_n$

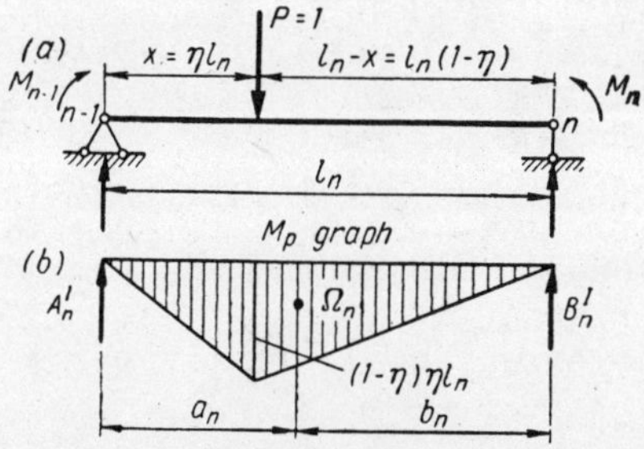

Fig. 15.10

(Fig. 15.10b). In order to find the values of the bending moments M_n and M_{n-1} at the supports let us first determine the imaginary reactions at these same supports

$$A_n^I = \frac{(1-\eta)\,\eta l_n \cdot \eta l_n \cdot \frac{1}{2}\left(l_n - \frac{2}{3}\eta l_n\right) + (1-\eta)\,\eta l_n \cdot l_n}{l_n} \rightarrow$$

$$\rightarrow \frac{(1-\eta)\frac{1}{2}\cdot\frac{2}{3}\, l_n\,(1-\eta)}{l_n} = (1-\eta)\,\eta l_n^2\,\frac{2-\eta}{6}$$

$$B_n^I = (1-\eta)\,\eta l_n l_n\,\frac{1}{2} - A_n^I = (1-\eta)\,\eta l_n^2\,\frac{1+\eta}{6}$$

Using expression (13.10) we obtain then

$$\left.\begin{aligned} M_{n-1} &= -\frac{6\,(A_n^I K_n' - B_n^I)}{l_n\,(K_n K_n' - 1)} = -\frac{(1-\eta)\,\eta l_n}{K_n K_n' - 1}\,[(2-\eta)\,K_n' - (1+\eta)] = \\ &= -c\,(1-\eta)\,\eta\,[(2-\eta)\,K_n'' - (1+\eta)] \\ M_n &= -\frac{6\,(B_n^I K_n - A_n^I)}{l_n\,(K_n K_n' - 1)} = -\frac{(1-\eta)\,\eta l_n}{K_n K_n' - 1}\,[(1+\eta)\,K_n - (2-\eta)] = \\ &= -c\,(1-\eta)\,\eta\,[(1+\eta)\,K_n - (2-\eta)] \end{aligned}\right\} \quad (15.10)$$

where

$$c = \frac{l_n}{K_n K_n' - 1}$$

With the aid of the latter expression let us prepare a table giving the values of M_{n-1} and M_n for different positions of the moving load P (at 0.1 l_n increments).

Table 2.10

η	Formulas to be determined	
	M_{n-1}	M_n
0	0	0
0.1	$-c\,(0.171K'_n-0.099)$	$-c\,(0.099K_n-0.171)$
0.2	$-c\,(0.288K'_n-0.192)$	$-c\,(0.192K_n-0.288)$
0.3	$-c\,(0.357K'_n-0.273)$	$-c\,(0.273K_n-0.357)$
0.4	$-c\,(0.384K'_n-0.336)$	$-c\,(0.336K_n-0.384)$
0.5	$-c\,(0.375K'_n-0.375)$	$-c\,(0.375K_n-0.375)$
0.6	$-c\,(0.336K'_n-0.384)$	$-c\,(0.384K_n-0.336)$
0.7	$-c\,(0.273K'_n-0.357)$	$-c\,(0.357K_n-0.273)$
0.8	$-c\,(0.192K'_n-0.288)$	$-c\,(0.288K_n-0.192)$
0.9	$-c\,(0.099K'_n-0.171)$	$-c\,(0.171K_n-0.099)$
1	0	0

Knowing the values of the bending moments at the supports (see Table 2.10) and the values of the focal factors K_n and K'_n we may easily obtain the moments at all the other supports of the

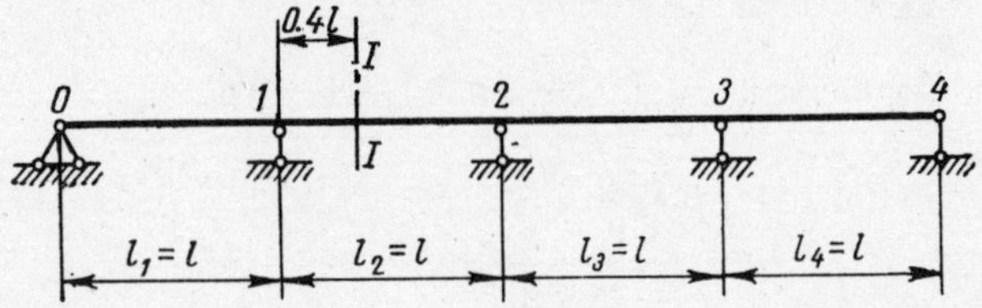

Fig. 16.10

beam for any position of the moving load along any of the spans. This being done, we may proceed with the construction of the influence lines either for the bending moments or for the shearing forces acting at any section of span l_n as well as of the influence lines for any of the support reactions.

Let us take up, for instance, the continuous beam appearing in Fig. 16.10 and let us construct the influence lines for the bending moments at all the supports as well as the influence lines for the bending moment and shear at section $I=I$ of the second span and for the reaction D_0.

At first we shall calculate all the focal distances whereafter we shall construct the influence lines for the bending moments at the supports of the span under consideration on the assumption that the unit load P travels along this span.

Hereafter we shall denote the bending moments at the support by M with two indices, the first giving the number of the support and the second the number of the span along which the mobile load is travelling. Thus, for instance, M_{22} will mean the bending moment at support *2* due to a load travelling along the second span; in the same way M_{12} will indicate the bending moment produced by the same load travelling along the same span but acting over support *1*.

The values of all focal factors obtained with the aid of expressions (11.10) and (12.10) are given in Table 3.10.

Table 3.10

Left-hand focal factors	Right-hand focal factors
$K_1=\infty$	$K'_1=3.733$
$K_2=4$	$K'_2=3.75$
$K_3=\frac{15}{4}=3.75$	$K'_3=4$
$K_4=\frac{56}{15}=3.733$	$K'_4=\infty$

Table 4.10 contains the values of M_{11}, M_{12} and M_{22} computed using data given in Tables 2.10 and 3.10. For a symmetrical beam the values of M_{34}, M_{33} and M_{23} appearing in the same table require no calculations.

The expressions for the bending moments at the inner supports of the end spans (in the case under consideration M_{11} and M_{34}) contain both in the numerator and in the denominator focal values which become infinitely great when the end supports are hinged. In order to overcome this difficulty both the numerator and the denominator should be divided by the said focal factor. Thus, the bending moment M_{11} for load point given by $x=0.1l_1=$ $=0.1l$ becomes

$$M_{11}=-\frac{l_1\left(\frac{0.099K_1}{K_1}-\frac{0.171}{K_1}\right)}{\frac{K_1K'_1-1}{K_1}}=-\frac{l_1\left(0.099-\frac{0.171}{\infty}\right)}{K'_1-\frac{1}{\infty}}=$$

$$=-\frac{0.099l_1}{3.733}=-0.02652l_1$$

Table 4.10

Load point	Bending moments at the supports		
	M_{11}	M_{12}	M_{22}
$x=0$	0	0	0
$x=0.1l_n$	$-0.02652l_1$	$-0.03874l_2$	$-0.01608l_2$
$x=0.2l_n$	$-0.05144l_1$	$-0.06342l_2$	$-0.03428l_2$
$x=0.3l_n$	$-0.07312l_1$	$-0.07612l_2$	$-0.05250l_2$
$x=0.4l_n$	$-0.09000l_1$	$-0.07886l_2$	$-0.06858l_2$
$x=0.5l_n$	$-0.10046l_1$	$-0.07366l_2$	$-0.08036l_2$
$x=0.6l_n$	$-0.10286l_1$	$-0.06286l_2$	$-0.08572l_2$
$x=0.7l_n$	$-0.09564l_1$	$-0.04762l_2$	$-0.08250l_2$
$x=0.8l_n$	$-0.07714l_1$	$-0.03086l_2$	$-0.06858l_2$
$x=0.9l_n$	$-0.04580l_1$	$-0.01430l_2$	$-0.04178l_2$
$x=l_n$	0	0	0

Table 4.10 (continued)

Load point	Bending moments at the supports		
	M_{34}	M_{33}	M_{23}
$x=0$	0	0	0
$x=0.1l_n$	$-0.04580l_4$	$-0.01430l_3$	$-0.04178l_3$
$x=0.2l_n$	$-0.07714l_4$	$-0.03086l_3$	$-0.06858l_3$
$x=0.3l_n$	$-0.09564l_4$	$-0.04762l_3$	$-0.08250l_3$
$x=0.4l_n$	$-0.10286l_4$	$-0.06286l_3$	$-0.08572l_3$
$x=0.5l_n$	$-0.10046l_4$	$-0.07366l_3$	$-0.08036l_3$
$x=0.6l_n$	$-0.09000l_4$	$-0.07886l_3$	$-0.06858l_3$
$x=0.7l_n$	$-0.07312l_4$	$-0.07612l_3$	$-0.05250l_3$
$x=0.8l_n$	$-0.05144l_4$	$-0.06342l_3$	$-0.03428l_3$
$x=0.9l_n$	$-0.02652l_4$	$-0.03874l_3$	$-0.01608l_3$
$x=l_n$	0	0	0

The magnitude of M_{11} for any other position of the load point within the limits of the first span will be obtained in exactly the same way. The values of the bending moments M_{12} and M_{22} for a load point situated at $x = 0.1l_2$ will be obtained using the following expressions (see Table 2.10)

$$M_{12} = -\frac{l_2}{K_2K_2'-1}(0.171K_2'-0.099) = -\frac{l_2}{4\times 3.75-1}\times$$

$$\times(0.171\times 3.75-0.099) = -0.03874l$$

$$M_{22} = -\frac{l_2}{K_2K_2'-1}(0.099K_2-0.171) = -\frac{l_2}{4\times 3.75-1}\times$$

$$\times(0.099\times 4-0.171) = -0.01608l$$

The magnitude of these moments for other positions of the load along the same span will be computed in the same way.

Table 5.10

Load point	Bending moments at the supports			
	$M_{13}=-\frac{M_{23}}{4}$	$M_{14}=\frac{M_{34}}{15}$	$M_{21}=-\frac{M_{11}}{3.75}$	$M_{24}=-\frac{M_{34}}{3.75}$
$x=0$	0	0	0	0
$x=0.1l_n$	$+0.010445l$	$-0.003053l$	$+0.007072l$	$+0.012213l$
$x=0.2l_n$	$+0.017145l$	$-0.005143l$	$+0.013717l$	$+0.020571l$
$x=0.3l_n$	$+0.020625l$	$-0.006376l$	$+0.019499l$	$+0.025504l$
$x=0.4l_n$	$+0.021430l$	$-0.006857l$	$+0.024000l$	$+0.027429l$
$x=0.5l_n$	$+0.020090l$	$-0.006697l$	$+0.026789l$	$+0.026789l$
$x=0.6l_n$	$+0.017145l$	$-0.006000l$	$+0.027429l$	$+0.024000l$
$x=0.7l_n$	$+0.013125l$	$-0.004875l$	$+0.025504l$	$+0.019499l$
$x=0.8l_n$	$+0.008570l$	$-0.002429l$	$+0.020571l$	$+0.013717l$
$x=0.9l_n$	$+0.004020l$	$-0.001768l$	$+0.012213l$	$+0.007072l$
$x=l_n$	0	0	0	0

Note: The value of index n appearing in the first column must be taken equal to the second index allotted to M.

The graphical interpretation of M_{11}, M_{12}, M_{22}, M_{34}, M_{33} and M_{23} obtained with the aid of Table 4.10 is given in Fig. 17.10. Each pair of contiguous curves, those for M_{11} and M_{12}, for M_{22} and M_{23}, for M_{33} and M_{34} constitute the influence line for bending moments induced at the support separating the two spans by a unit load situated along one of these two spans.

Thus, for instance, the two curves for M_{11} and M_{12} constitute the influence line for the bending moment induced at the cross section over support *1* by a unit load travelling along the first two spans of the beam. In the event the unit load travels along more distant spans the ordinates to the influence line for the same bending moments at the supports may be obtained using the focal point method.

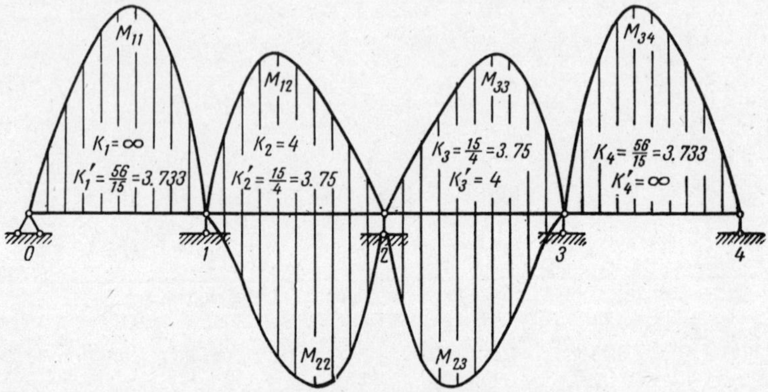

Fig. 17.10

Thus, the ordinates to the influence line for the bending moment over support due to a unit load located over span *3* will be easily obtained using the following expression

$$M_{13} = -\frac{M_{23}}{K_2} = -\frac{M_{23}}{4}$$

The magnitude of the bending moment acting at the same support when the moving load has passed to span *4* will be given by

$$M_{14} = -\frac{M_{24}}{K_2}$$

but since

$$M_{24} = -\frac{M_{34}}{K_3}$$

we obtain finally

$$M_{14} = +\frac{M_{34}}{K_3K_2} = \frac{M_{34}}{3.75\times 4} = \frac{M_{34}}{15}$$

The ordinates to the influence lines for M_{13} and M_{14} are given in Table 5.10; this table contains the ordinates to the influence lines for M_{21} and M_{24} deduced from

$$M_{21} = -\frac{M_{11}}{K_2'} = -\frac{M_{11}}{3.75}$$

$$M_{24} = -\frac{M_{34}}{K_3} = -\frac{M_{34}}{3.75}$$

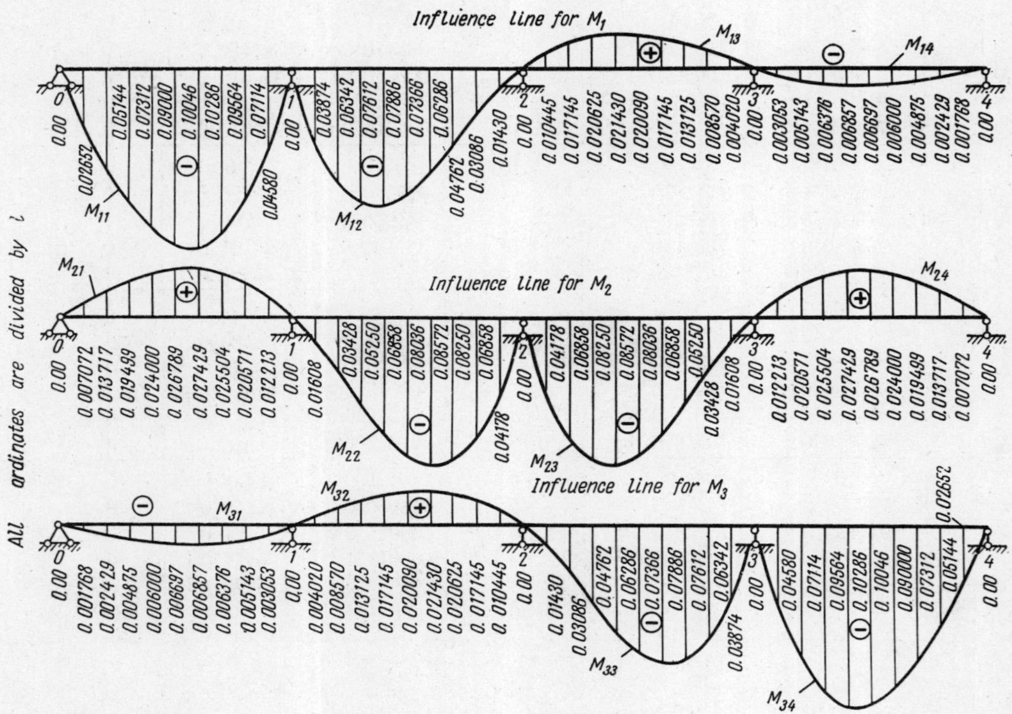

All ordinates are divided by l
Influence line for M_1
M_{11} M_{12} M_{13} M_{14}
0 1 2 3 4
0.00 0.02652 0.05144 0.07312 0.09000 0.10046 0.10286 0.09564 0.07114 0.04580 0.00 0.03874 0.06342 0.07612 0.07886 0.07366 0.06286 0.04762 0.03086 0.01430 0.00 0.010445 0.017145 0.020625 0.021430 0.020090 0.017145 0.013125 0.008570 0.004020 0.00 0.003053 0.005143 0.006376 0.006857 0.006697 0.006000 0.004875 0.002429 0.001768 0.00
Influence line for M_2
M_{21} M_{22} M_{23} M_{24}
0 1 2 3 4
0.00 0.007072 0.013717 0.019499 0.024000 0.026789 0.027429 0.025504 0.020571 0.012213 0.00 0.01608 0.03428 0.05250 0.06858 0.08036 0.08572 0.08250 0.06858 0.04178 0.00 0.04178 0.06858 0.08250 0.08572 0.08036 0.06858 0.05250 0.03428 0.01608 0.00 0.012213 0.020571 0.025504 0.027429 0.026789 0.024000 0.019499 0.013717 0.007072 0.00
Influence line for M_3
M_{31} M_{32} M_{33} M_{34}
0 1 2 3 4
0.00 0.001768 0.002429 0.004875 0.006000 0.006697 0.006857 0.006376 0.005143 0.003053 0.00 0.004020 0.008570 0.013125 0.017145 0.020090 0.021430 0.020625 0.017145 0.010445 0.00 0.01430 0.03086 0.04762 0.06286 0.07366 0.07886 0.07612 0.06342 0.03874 0.00 0.04580 0.07114 0.09564 0.10286 0.10046 0.090000 0.07312 0.05144 0.02652 0.00

Fig. 18.10

Table 6.10

	Load point	Q^0	$\frac{1}{l}(M_2 - M_1)$	Ordinate to the influence line for $Q\,(x = 0.4l_2)$	M^0	$0.4\,M_2$	$0.6\,M_1$	Ordinate to the influence line for $M\,(x = 0.4l_2)$
Span 1	$\eta=0$	0	0	0	0	0	0	0
	$\eta=0.2$	0	+0.055157	+0.055157	0	$+0.005487l$	$-0.030864l$	$-0.025377l$
	$\eta=0.4$	0	+0.114000	+0.114000	0	$+0.009600l$	$-0.054000l$	$-0.044400l$
	$\eta=0.6$	0	+0.130289	+0.130289	0	$+0.010972l$	$-0.061716l$	$-0.050744l$
	$\eta=0.8$	0	+0.097711	+0.097711	0	$+0.008228l$	$-0.046284l$	$-0.038056l$
	$\eta=1.0$	0	0	0	0	0	0	0
Span 2	$\eta=0.2$	−0.2	+0.02914	−0.17086	$+0.12l$	$-0.013712l$	$-0.038052l$	$+0.068236l$
	$\eta=0.4-0$	−0.4	+0.01028	−0.38972	$+0.24l$	$-0.027432l$	$-0.047316l$	$+0.165252l$
	$\eta=0.4+0$	+0.6	+0.01028	+0.61028	$+0.24l$	$-0.027432l$	$-0.047316l$	$+0.165252l$
	$\eta=0.6$	+0.4	−0.02286	+0.37714	$+0.16l$	$-0.034288l$	$-0.037716l$	$+0.087996l$
	$\eta=0.8$	+0.2	−0.03772	+0.16228	$+0.08l$	$-0.027432l$	$-0.018516l$	$+0.034052l$
	$\eta=1.0$	0	0	0	0	0	0	0
Span 3	$\eta=0.2$	0	−0.085725	−0.085725	0	$-0.027432l$	$+0.010287l$	$-0.017145l$
	$\eta=0.4$	0	−0.107150	−0.107150	0	$-0.034288l$	$+0.012858l$	$-0.021430l$
	$\eta=0.6$	0	−0.085725	−0.085725	0	$-0.027432l$	$+0.010287l$	$-0.017145l$
	$\eta=0.8$	0	−0.042850	−0.042850	0	$-0.013712l$	$+0.004342l$	$-0.009370l$
	$\eta=1.0$	0	0	0	0	0	0	0
Span 4	$\eta=0.2$	0	+0.015428	+0.015428	0	$+0.008228l$	$-0.03086l$	$+0.005142l$
	$\eta=0.4$	0	+0.020572	+0.020572	0	$+0.010972l$	$-0.004114l$	$+0.006858l$
	$\eta=0.6$	0	+0.018000	+0.018000	0	$+0.009600l$	$-0.003600l$	$+0.006000l$
	$\eta=0.8$	0	+0.011288	+0.011288	0	$+0.005487l$	$-0.001457l$	$+0.004030l$
	$\eta=1.0$	0	0	0	0	0	0	0

The influence lines for the bending moments M_1, M_2 and M_3 are represented in Fig. 18.10. These influence lines may be used for the design of continuous beams consisting of four spans of

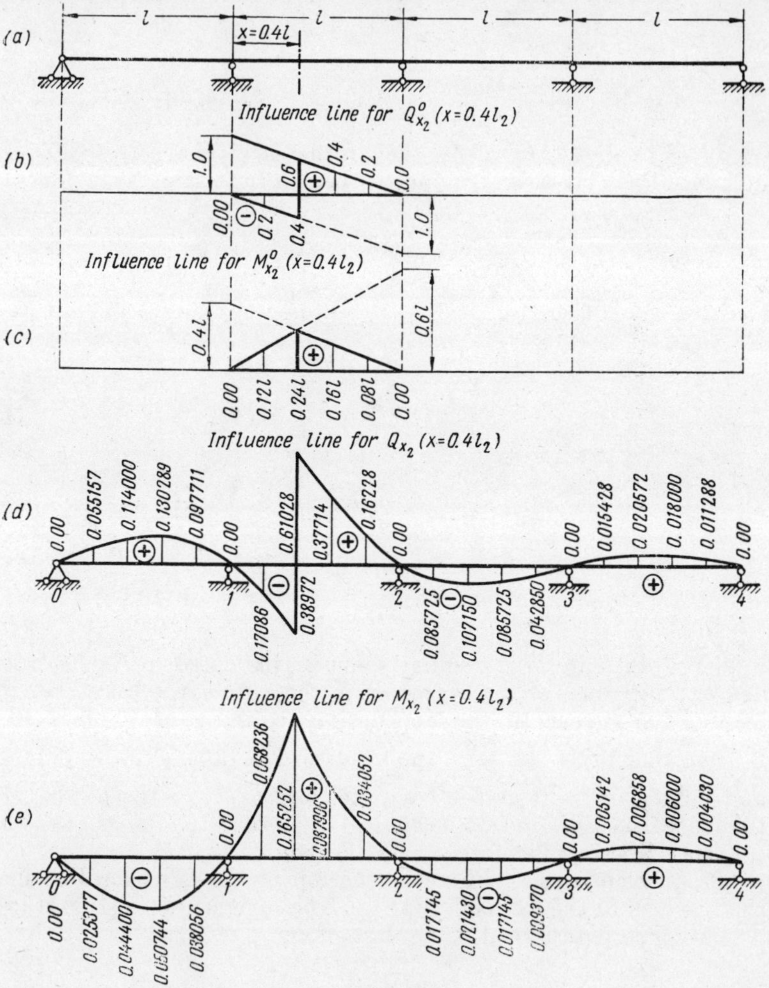

Fig. 19.10

equal length with freely supported ends. Influence lines for bending moments at the supports of the continuous beam with uneven spans and any arbitrary number of supports can be obtained in exactly the same way.

Let us consider now the influence lines for the bending moments and shearing forces at a cross section of the second span located a distance $x = 0.04l$ to the right of support l (Fig. 19.10a). For this purpose we shall use expressions (5.10) and (6.10)

$$Q = Q_2^0 + \frac{M_2 - M_1}{l}$$

$$M = M^0 + \frac{M_2 - M_1}{l} 0.4l + M_1 = M^0 + 0.4M_2 + 0.6M_1$$

where Q^0 and M^0 represent the ordinates to the corresponding influence lines for an end-supported beam of the same span. These

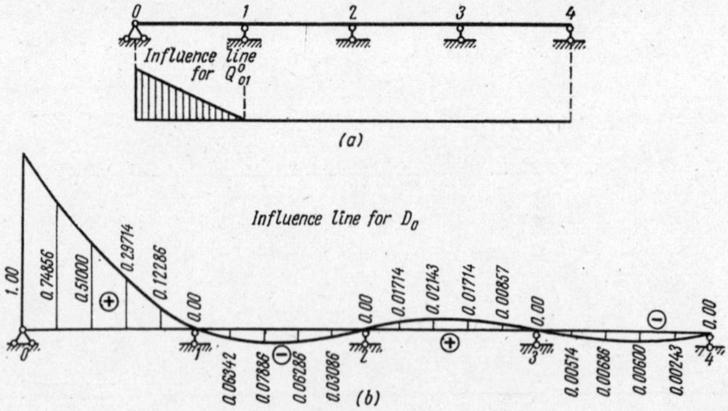

Fig. 20.10

influence lines are of triangular shape as indicated in Fig. 19.10b and c. The values of the ordinates to these influence lines at $0.2l$ increments appear in the appropriate columns of Table 6.10.

The same table contains the values of $\frac{1}{l}(M_2 - M_1)$, $0.4M_2$ and $0.6M_1$ as well as the computed ordinates to the Q and M influence lines desired. A graphical representation of these two influence lines appears in Fig. 19.10d and e.

Let us examine next the construction of the influence line for the reaction of the left-end support. The magnitude of this reaction may be determined using expression

$$D_0 = Q_{01}^0 + \frac{M_1}{l}$$

where Q_{01}^0 represents the left-end reaction of a simply supported beam corresponding to the first span. The influence line for Q_{01}^0 appears in Fig. 20.10a. All the calculations necessary to obtain the ordinates to the influence line for D_0 are tabulated hereunder.

The completed influence line for D_0 is given in Fig. 20.10*b*. The influence lines for all the other support reactions of a continuous beam can be obtained using exactly the same procedure.

Table 7.10

Load point		Q^0_{01}	$\frac{M_1}{l}$	Ordinates to the influence line for D^0
Span 1	$\eta=0$	+1.0	0	+1.0
	$\eta=0.2$	+0.8	−0.05144	+0.74856
	$\eta=0.4$	+0.6	−0.09000	+0.51000
	$\eta=0.6$	+0.4	−0.10286	+0.29714
	$\eta=0.8$	+0.2	−0.07714	+0.12286
	$\eta=1$	0	0	0
Span 2	$\eta=0.2$	0	−0.06342	−0.06342
	$\eta=0.4$	0	−0.07886	−0.07886
	$\eta=0.6$	0	−0.06286	−0.06286
	$\eta=0.8$	0	−0.03086	−0.03086
	$\eta=1$	0	0	0
Span 3	$\eta=0.2$	0	+0.01714	+0.01714
	$\eta=0.4$	0	+0.02143	+0.02143
	$\eta=0.6$	0	+0.01714	+0.01714
	$\eta=0.8$	0	+0.00857	+0.00857
	$\eta=1$	0	0	0
Span 4	$\eta=0.2$	0	−0.00514	−0.00514
	$\eta=0.4$	0	−0.00686	−0.00686
	$\eta=0.6$	0	−0.00600	−0.00600
	$\eta=0.8$	0	−0.00243	−0.00243
	$\eta=1$	0	0	0

Influence lines permit easy and rapid determination of maximum and minimum values of support reactions, shears and bending moments due to the combined action of moving loads and dead loads (it is frequently assumed that moving loads are uniformly distributed over whole span lengths). Thus, for instance, reaction D_0 will attain its design value when in addition to the dead loads of q kg per metre the beam will be acted upon by mobile loads of p kg per metre distributed along the whole length of the first and third spans. The ordinates to the corresponding influence line over these spans being positive, the value of reaction $D_{0\,max}$ will be given by

$$D_{0\,max} = q\,(\omega_1 + \omega_2 + \omega_3 + \omega_4) + p\,(\omega_1 + \omega_3)$$

In the latter expression ω_1, ω_2, ω_3 and ω_4 represent the areas bounded by the influence line for D_0 over the corresponding spans. These areas may be easily calculated using the numerical values of the ordinates given in Fig. 20.10*c* and assuming that the segments of the curve between two neighbouring ordinates can be replaced by straight lines. These areas will be reckoned positive or negative depending on the sign of the ordinates.

The minimum value of reaction D_0 will be given by the following expression

$$D_{0\,min} = q\,(\omega_1 + \omega_2 + \omega_3 + \omega_4) + p\,(\omega_2 + \omega_4)$$

11. REDUNDANT ARCHES

1.11. *Definitions. Choice of the Neutral Line*

Arches as distinguished from simply supported systems are thrust developing structures whose general form is that of a curve. Classified with reference to the number of hinges arches fall into the following categories: *three-hinged* arches (Fig. 1.11*a*), *two-hinged* arches (Fig. 1.11*b*), arches of *one hinge* (Fig. 1.11*c*) and *hingeless* or *fixed end* arches (Fig. 1.11*d*). All the arches with the exception of the three-hinged ones are statically indeterminate. In bridge construction, especially in railroad bridges, the more frequently used arches are the two-hinged and the fixed end ones.

In all calculations in solid masonry or reinforced concrete arches it is customary to consider strips of unit width separated in imagination from the rest by two parallel planes as shown in Fig. 2.11. The whole structure becomes thus replaced by a series of parallel arches, the deflections of which remain strictly identical as long as each one of them carries the same load. In this way the stress analysis remains exactly the same for all the arched structures of the same type irrespective of their depth. Arch ribs are frequently loaded at certain points only, as in the case of trusses or plate girders supporting floor beams.

When designing an arch great care should be taken to reach as close coincidence as possible between the outline of the axis (also called the neutral or the centre line) of the arch and the pressure line (or equilibrium polygon). As previously stated, such a coincidence would provide an arch of *maximum economy*. However, complete coincidence can be achieved only in the case of three-hinged arches. As for the statically indeterminate arches, it is impossible to obtain full coincidence of the arch axis with the pressure line, for bending moments in such arches are absolutely unavoidable. Therefore the most economical design of an arch will be the one providing for minimum fibre stresses in the arch.

The pressure line for statically indeterminate arches can be obtained only if all the redundant reactions are already determined. However, these reactions depend themselves on the deformations of arch and for this reason it becomes extremely difficult to find the most economical configuration of an arch of this type even when dead loads alone are involved. The problem can be solved only by a series of successive approximations. This may be done by selecting first some arbitrary curve (usually a parabola) for the arch neutral line which is then corrected on the grounds of comparison with the pressure line obtained for that particular arch. Alternatively, the centre line of the arch may, in the first approximation, follow the pressure line of a three-hinged arch of the same span and rise. Upon correction of the arch neutral line a new

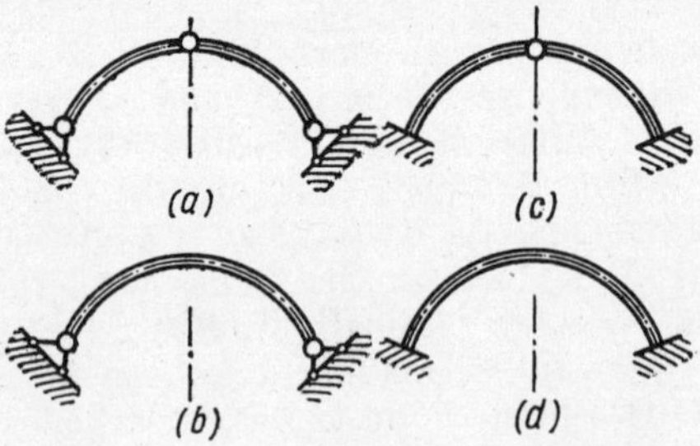

Fig. 1.11

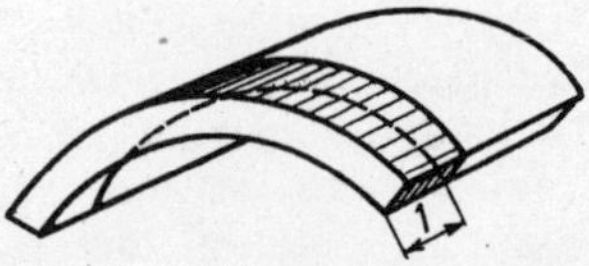

Fig. 2.11

pressure polygon is constructed for the corrected arch. The operation is repeated as many times as necessary to obtain a satisfactory coincidence of the two lines.

For arches carrying moving loads the choice of the neutral line becomes even more complicated. In actual practice this choice is most frequently based on the simple comparison of several arches differing both in outline and in cross-sectional dimensions.

The configuration of masonry arches must be selected with the view of maintaining the pressure line for all possible load combinations as close as possible to the central core boundaries, masonry being uncapable of resisting tensile stresses of any appreciable magnitude.

2.11. *Arches with Variable Cross-Sectional Dimensions*

The coefficients to the unknowns and the free terms of the simultaneous equations used for purposes of stress analysis of redundant arches depend on the cross-sectional dimensions and the moments of inertia of the structure. As a rule, neither of these two remain constant through the whole length of the arch. Thus, for instance, in fixed end arches the height of the section and consequently its moment of inertia increase very frequently from crown to abutments because the bending moments are as a rule much smaller at midspan than in the immediate neighbourhood of the supports. On the other hand, the thickness of two-hinged arches decreases usually from the crown to the abutments following the bending moment diagram.

Direct computation of the coefficients to the unknowns and of the free terms of the simultaneous equations requires the integration of expressions containing the values of F and J and therefore it becomes necessary to express mathematically the variation of these quantities along the arch. The following equation has been found very useful in practical design

$$J_x = \frac{J_c}{\left[1-(1-n)\frac{x}{l_1}\right]\cos\varphi_x}$$

where x = abscissa of the neutral line referred to a coordinate origin coinciding with the centroid of the crown section

J_c = moment of inertia of the same section

J_x = moment of inertia of a section situated a distance x from the origin of coordinates

φ_x = angle the tangent forms with the neutral line of the arch and the horizontal

l_1 = one half of the arch span.

As for n its value is given by

$$n = \frac{J_c}{J_0 \cos\varphi_0}$$

J_0 and φ_0 correspond to the section at the support.

Modifying the value of n we modify at the same time the law governing the variation of cross-sectional dimensions along the arch. Frequently n is taken equal to unity in which case the expression for J_x becomes

$$J_x = \frac{J_c}{\cos\varphi_x}$$

For a rectangular arch of constant width b, J_x may be replaced by $\frac{d_x^3 b}{12}$ and J_c by $\frac{d_c^3 b}{12}$, where d_x and d_c represent the thickness of the arch at the crown and at an arbitrary section a distance x from the coordinate origin. When $n = 1$

$$\frac{d_x^3 b}{12} = \frac{d_c^3 b}{12\cos\varphi_x}$$

wherefrom

$$d_x^3 = \frac{d_c^3}{\cos\varphi_x}$$

The cross-sectional areas F_c and F_x become in this case

$$F_x = bd_x \quad \text{and} \quad F_c = bd_c$$

It follows that

$$F_x^3 = \frac{F_c^3}{\cos \varphi_x}$$

wherefrom

$$F_x = \frac{F_c}{\sqrt[3]{\cos \varphi_x}}$$

For simplicity this expression is very frequently replaced by

$$F_x = \frac{F_c}{\cos \varphi_x}$$

It has been proved that this simplification entails an error in the bending moment and thrust values which does not exceed 1 per cent.

When the rise of an arch is less than 1/8 of its span (flat arches) the value of $\cos \varphi_x$ for all the cross sections will remain very close to unity, thus permitting to adopt a constant thickness of the arch throughout and therefore

$$F_x = F_c = \text{const}$$

In the design of flat arches the length of the elementary segment *ds* is also usually replaced by the length of its horizontal projection *dx*.

3.11. *Conjugate Statically Determinate Structures Used for Stress Analysis of Fixed End Arches*

A fixed end arch (Fig. 3.11*a*) constitutes always a closed contour and is therefore redundant in the third degree. It follows that the simple statically determinate structure can be obtained by elimination of three redundant constraints which must be replaced by three unknown actions X_1, X_2 and X_3. Several of such statically determinate structures are shown in Fig. 3.11*b*, *c*, *d*, *e* and *f*. The statically determinate structure of Fig. 3.11*b* is formed by a curved bar built in at its left end. The three unknowns represent in this case the reactive forces developed by the right-hand abutment. The structure of Fig. 3.11*c* consisting of two curved bars fixed at one of their ends has been obtained by cutting the arch in two. In this case the unknowns will represent the bending moment, the shear and the normal stress acting across the cut. If the simple structure is constituted by a three-hinged arch, the unknowns will represent the bending moments at the crown and at the abutments.

The simultaneous equations used for stress computation of a fixed end arch will take the following shape

$$\left.\begin{aligned} X_1\delta_{11} + X_2\delta_{12} + X_3\delta_{13} + \Delta_{1p} = 0 \\ X_1\delta_{21} + X_2\delta_{22} + X_3\delta_{23} + \Delta_{2p} = 0 \\ X_1\delta_{31} + X_2\delta_{32} + X_3\delta_{33} + \Delta_{3p} = 0 \end{aligned}\right\} \tag{1.11}$$

provided that in these computations resort is made to one of the simple structures appearing in Fig. 3.11*b*, *c* or *d*.

In Art. 8.9 it has been shown that all the secondary coefficients of the simultaneous equations may be reduced to zero by an appropriate choice of the simple structure. In this case the simultaneous equations mentioned above become

$$\left.\begin{aligned} X_1\delta_{11} + \Delta_{1p} = 0 \\ X_2\delta_{22} + \Delta_{2p} = 0 \\ X_3\delta_{33} + \Delta_{3p} = 0 \end{aligned}\right\} \tag{2.11}$$

leading immediately to the following values of the redundant reactions

$$X_1 = -\frac{\Delta_{1p}}{\delta_{11}}; \quad X_2 = -\frac{\Delta_{2p}}{\delta_{22}}; \quad X_3 = -\frac{\Delta_{3p}}{\delta_{33}} \tag{3.11}$$

The simple statically determinate structures will be obtained in that case by the addition to the free ends of the curved built-in bars of one or two infinitely stiff brackets as indicated in Fig. 3.11*e* and *f*. The unknown actions will be applied to the free ends of the said brackets, these ends coinciding with the elastic centre of the structure. The directions of these actions will coincide with those of the principal axes of inertia of the elastic loads $\frac{ds}{J}$. When a fixed end arch is symmetrical about a vertical axis the elastic centre of the structure will always lie in this vertical, one of the principal axes of inertia being horizontal and the other vertical. In that case the conjugate structures of Fig. 3.11*e* and *f* will be replaced by those appearing in Fig. 4.11*a* and *b*.

Since all the secondary displacements due to the unknown forces acting at the elastic centre are nil, it becomes very easy to determine the nature of the displacements of the brackets produced by these unknown actions.

Let us examine first the simple structure of Fig. 5.11*a* subjected to two unit horizontal forces $X_1 = 1$. Since $\delta_{31} = 0$, both brackets will remain vertical and parallel to one another. Their deflection in the vertical direction will depend on the direction of the unit forces X_1 but the amount of this deflection will be exactly the same for both brackets since otherwise δ_{21} would

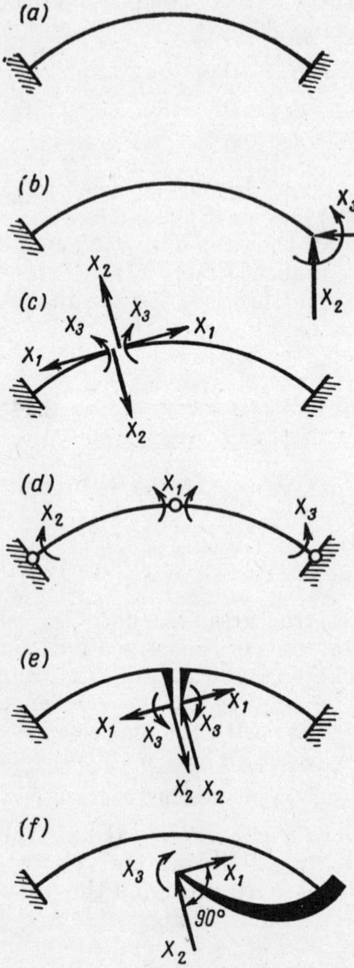

Fig. 3.11

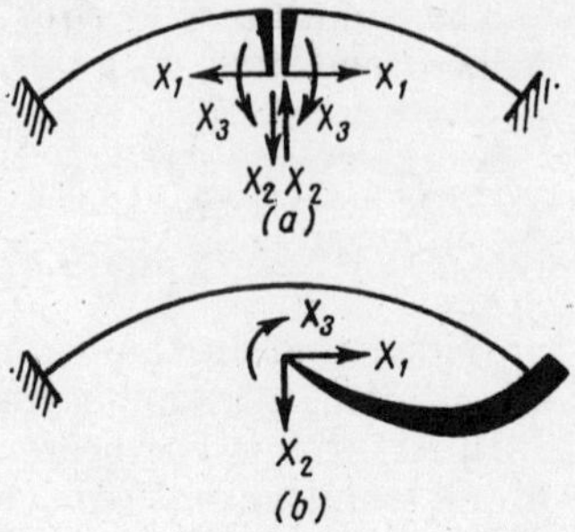

Fig. 4.11

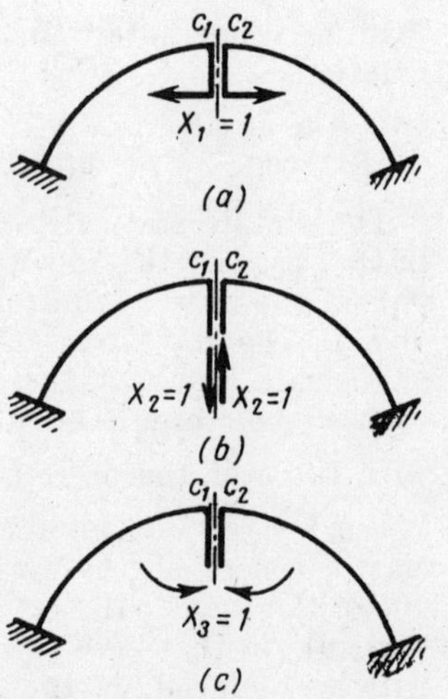

Fig. 5.11

be different from zero. In the horizontal direction the mutual displacement of two brackets will equal δ_{11}.

When the same system is subjected to two vertical unit forces X_2 (Fig. 5.11*b*) the two brackets will rotate together remaining parallel to one another (for otherwise the displacement δ_{32} would be different from zero), will shift together along the horizontal and sustain a mutual vertical displacement equal to δ_{22}.

The unit couples X_3 shown in Fig. 5.11*c* will entail a mutual rotation of the two brackets, each bracket becoming inclined to the vertical at an angle equal to $\frac{1}{2}\delta_{33}$. The total mutual displacement of the two brackets will equal δ_{33}. The free ends of these brackets will remain at the same distance from one another, δ_{13} being nil. They may shift vertically upwards or downwards depending on the direction of the unit couples but both must shift the same amount, for otherwise δ_{23} will differ from zero.

4.11. *Approximate Methods of Design and Analysis of Fixed End Arches*

The designer is frequently called upon to deal with arches whose neutral line and law of cross-sectional variation cannot be expressed by analytical equations fit for practical use. In such

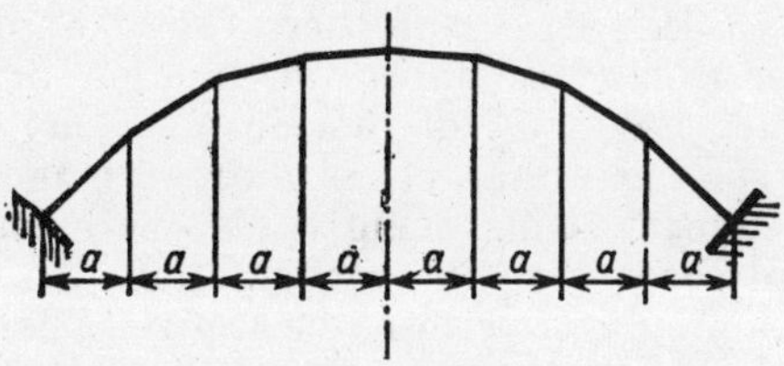

Fig. 6.11

cases the exact analysis of the deflections of the conjugate simple structure becomes impossible for this analysis is based on integral calculus. Resort must be then made to approximate methods, two of which will be described hereunder.

In the first of these methods the neutral line of the arch is replaced by a polygon of from 8 to 20 sides (Fig. 6.11). In addition it is assumed that cross-sectional areas remain constant along each of these sides, their dimensions being equal to those of the given arch as measured over the centre of that particular side. All the loads applied to the arch are replaced by concentrated loads acting at the apices of the polygon. The magnitude of these loads is taken equal to the combined support reaction of two contiguous simple beams carrying the same loads and having the same length as the corresponding portion of the arch. The polygon of structure obtained in this way can be analyzed by one of the methods described in Art. 3.9. For the analysis, any of the structures shown in Figs. 3.11 and 4.11 may be adopted to be conjugate with or without transfer of the redundant constraints to the elastic centre. Any other statically determinate system could be equally used if that were found expedient. All the deflections and rotations can be easily calculated using Vereshchagin's method of graph multiplication. The stresses obtained by this procedure

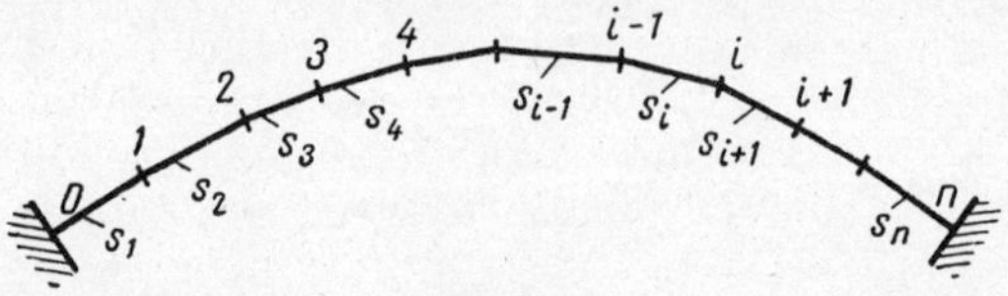

Fig. 7.11

are practically equal to those induced in the curved arch. Hereunder in Problem 1 of the present article we shall give an example of stress analysis by the method just described.

The second of the approximate methods consists in the subdivision of the arch into a number of segments generally comprised between 8 and 20. Having chosen thereafter an appropriate simple structure either from those appearing in Figs. 3.11 and 4.11 or any other deemed better fit for this purpose, one should proceed with the construction of the stress diagrams due to the actual and to the unit loads. This being done, the calculations of the coefficients and free terms of the simultaneous equations are carried out assuming that within the limits of each segment the expressions under the integral sign vary linearly. Consequently, the corresponding integral will be equal to the length of the segment multiplied by half the sum of the values of the expression under the integral sign calculated for the sections limiting this particular segment. Thus, for instance, the displacement δ_{12} will be taken equal to

$$\delta_{12}=\int_0^s \frac{\overline{M}_1\overline{M}_2}{EJ}\,ds \approx \sum_{i=1}^{i=n} \frac{s_i}{2}\left(\frac{\overline{M}_{1,i-1}\,\overline{M}_{2,\,i-1}}{EJ_{i-1}}+\right.$$

$$\left.+\frac{\overline{M}_{1,\,i}\;\overline{M}_{2,\,i}}{EJ_i}\right)=\sum_{i=0}^{i=n}\frac{\overline{M}_{1,\,i}\;\overline{M}_{2,\,i}}{EJ_i}\left(\frac{s_i}{2}+\frac{s_{i+1}}{2}\right)$$

where s_i = length of the segment i limited by sections $(i-1)$ and i (Fig. 7.11)

$\overline{M}_{1,i}$ and $\overline{M}_{2,i}$ = bending moments induced at section i by the unknowns X_1 and X_2;
J_i = moment of inertia of the same section
n = number of segments into which the arch has been subdivided.

The above expression can also be written as follows

$$\delta_{12} = \sum_{i=0}^{i=n} \overline{M}_{1,i} \overline{M}_{2,i} \frac{\bar{s}_i}{EJ_i} \tag{4.11}$$

where s_i is half the sum of the segment lengths contiguous to section i

$$s_i = \frac{s_i + s_{i+1}}{2}$$

Thus, in order to determine one of the displacements δ or Δ proceed as follows:

1. Compute the values of the expressions under the integral sign corresponding to each section situated at the boundaries of the segments forming the arch.

2. Multiply each of the values obtained in this way by half the sum of the contiguous segment lengths.

3. Calculate the sum of all the values obtained as explained above.

All subsequent computations (solution of simultaneous equations, construction of stress diagrams, etc.) will be exactly the same as for any other statically indeterminate structure.

Problem 2 presented at the end of this article will give an example of stress analysis of a fixed end arch using the latter method.

Regardless of the method of analysis selected the displacements δ and Δ may be obtained using the method of elastic loads described in Art. 11.8. Recourse to this method is strongly advised when it is desired to obtain the influence lines for internal stresses induced in the arch, for in this case it becomes necessary to construct whole deflection graphs and not only to determine the deflection of particular points. The construction of influence lines for redundant arches is shown in detail in Problem 3 of the present article.

When the rise of the arch is greater than one fifth of its span, the calculation of the deflections and angular rotations may be carried out neglecting the influence of shears and normal stresses. This does not apply to flat arches whose rise is smaller than $\frac{1}{5}$ of the span. For these unit displacements δ_{ii} along the direction of the thrust must be carried out taking due account of the corresponding unknown $X_i = 1$ as well as the normal stresses resulting from the thrust. However, the other displacements due both to the unit actions and to the applied loads may be again calculated neglecting normal stresses and shears.

Problem 1. Using the first of the methods described above compute the stresses induced in the arch of Fig. 8.11*a*. The neutral line of this arch follows a conic parabola, and the cross-sectional moments of inertia vary in

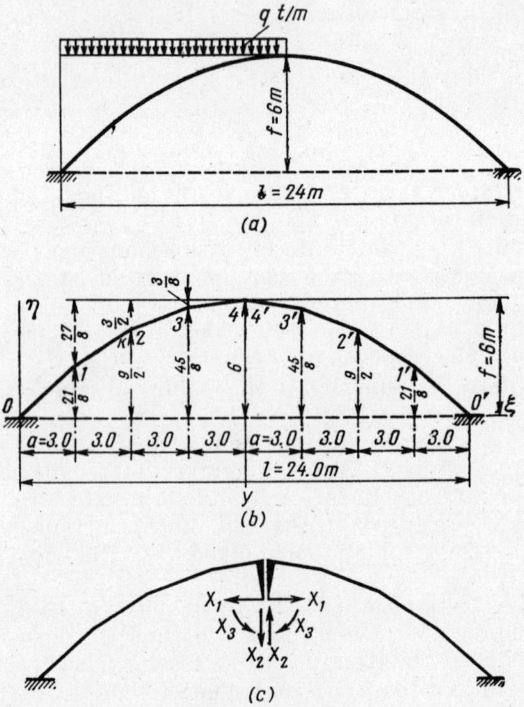

Fig. 8.11

accordance with the expression $J_x = \frac{J_c}{\cos \varphi_x}$. The M, Q and N diagrams will be constructed assuming that the left semiarch carries a uniformly distributed load $q = 2$ tons per metre.

Solution. Subdivide the arch span into 8 equal parts thus adopting $n = 8$ and $a = 3$ metres. Inscribe into the given arch a polygonal one as indicated in Fig. 8.11*b*. The equation of a conic parabola in the coordinate system $\eta 0 \xi$ whose origin coincides with the centroid of the section at the left-hand support becomes

$$\eta = \frac{4f}{l^2}(l-\xi)\,\xi = \frac{4 \times 6}{24 \times 24}(24-\xi)\,\xi = \frac{24-\xi}{24}\,\xi$$

The values of η at the boundaries of all the different segments are indicated in the same figure. Assume that the conjugate simple structure is obtained by cutting the arch at the crown as indicated in Fig. 8.11c with transfer of the redundant constraints to the elastic centre.

Compute the ordinate y_s of this elastic centre in the coordinate system xy

$$y_s=\frac{\Sigma\int y\frac{ds}{J}}{\Sigma\int\frac{ds}{J}}$$

Since J_i is assumed constant for each side of the polygonal arch the expression for y_s becomes

$$y_s=\frac{\sum\limits_{i=1}^{i=n}\frac{y_{i-1}+y_i}{2}\cdot\frac{s_i}{J_i}}{\sum\limits_{i=1}^{i=n}\frac{s_i}{J_i}}$$

where s_i represents the length of side i and $\frac{y_{i-1}+y_i}{2}$ is the ordinate to the centre of this side.

In the case under consideration we have $J_i=\frac{J_c}{\cos\varphi_i}$ where φ_i is the inclination of side i of the polygon to the horizontal and therefore

$$\frac{s_i}{J_i}=\frac{s_i\cos\varphi_i}{J_c}=\frac{a}{J_c}$$

Hence

$$y_s=\frac{\frac{a}{2J_c}\sum\limits_{i=1}^{i=n}(y_{i-1}+y_i)}{\frac{a}{J_c}n}=\frac{\sum\limits_{i=1}^{i=n}(y_{i-1}+y_i)}{2n}$$

or with due regard to the symmetry of the arch

$$y_s=\frac{2\sum\limits_{i=1}^{i=n/2}(y_{i-1}+y_i)}{2n}=\frac{1}{n}\sum_{i=1}^{i=n}(y_{i-1}+y_i)$$

wherefrom

$$y_s=\frac{1}{8}\left(6+\frac{27}{8}\times2+\frac{3}{2}\times2+\frac{3}{8}\times2+0\right)=2.0625\text{ metres}$$

The exact value of y_s for a parabolic arch whose moments of inertia vary in accordance with the $J_x=\frac{J_c}{\cos\varphi_x}$ is equal to $\frac{1}{3}f$ which in the present case is smaller than the value obtained above by 6.25 cm. This difference can be further reduced increasing the number of sides of the inscribed polygonal arch.

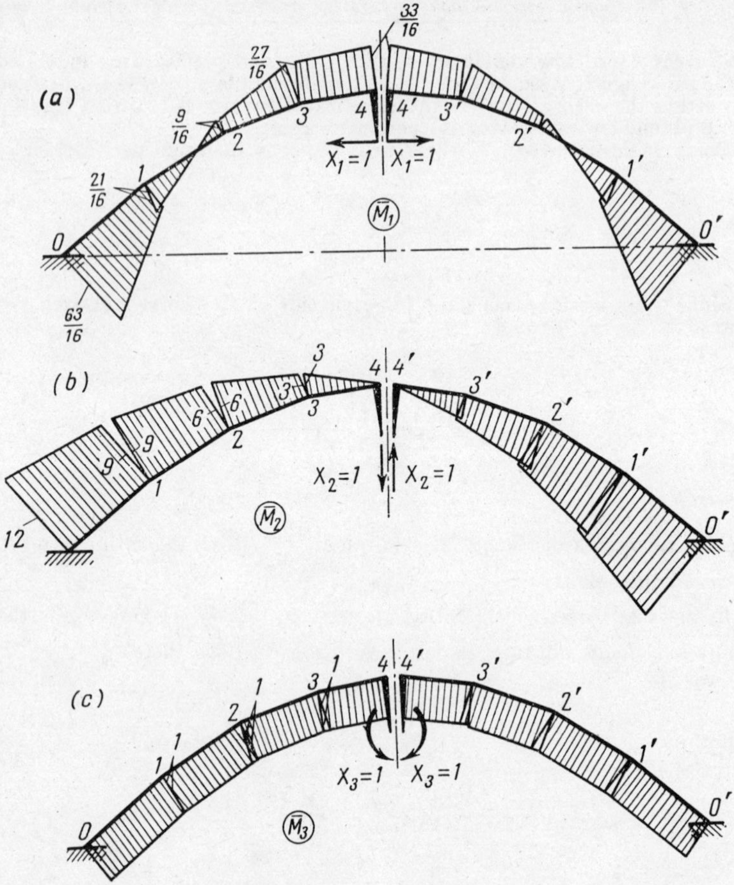

Fig. 9.11

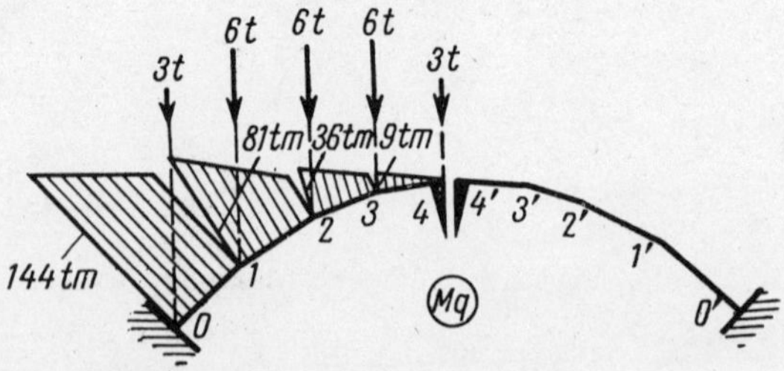

Fig. 10.11

Fig. 9.11 represents the bending moment diagrams induced in the simple structure by unit actions applied at the elastic centre, and Fig. 10.11 those due to the actual loads concentrated at the apices of the left half of the polygonal arch. At both ends of this arch (points *0* and *4*) these loads are $\frac{qa}{2} = 3$ tons, and at the intermediate points (points *1*, *2* and *3*) these loads amount to $qa = 6$ tons. The reactions of the unknown constraints will be given by

$$X_1 = -\frac{\Delta_{1q}}{\delta_{11}}; \qquad X_2 = -\frac{\Delta_{2q}}{\delta_{22}}; \qquad X_3 = -\frac{\Delta_{3q}}{\delta_{33}} \tag{3.11}$$

Displacements δ and Δ will be obtained multiplying the graphs of Fig. 9.11 by that of Fig. 10.11 using Vereshchagin's method. It should be remembered that the ratio between the length of each side of the polygon and the moment of inertia of the corresponding cross section remains constant and equal to

$$\frac{s_i}{J_i} = \frac{s_i \cos \varphi_i}{J_i} = \frac{a}{J_c}$$

The unit displacements will be determined using formulas peculiar to trapezoidal stress diagrams (see Art 8.8). Thus, raising to the second power the $\overline{M}_1$ graph (Fig. 9.11*a*) we obtain

$$\delta_{11} = \frac{2a}{6EJ_c}\left[\left(2\cdot\frac{63}{16}\cdot\frac{63}{16} + 2\cdot\frac{21}{16}\cdot\frac{21}{16} + 2\cdot\frac{63}{16}\cdot\frac{21}{16}\right) + \right.$$
$$+\left(2\cdot\frac{21}{16}\cdot\frac{21}{16} + 2\cdot\frac{9}{16}\cdot\frac{9}{16} - 2\cdot\frac{21}{16}\cdot\frac{9}{16}\right) + \left(2\cdot\frac{9}{16}\cdot\frac{9}{16} + 2\times\right.$$
$$\left.\times\frac{27}{16}\cdot\frac{27}{16} + 2\cdot\frac{9}{16}\cdot\frac{27}{16}\right) + \left.\left(2\cdot\frac{27}{16}\cdot\frac{27}{16} + 2\cdot\frac{33}{16}\cdot\frac{33}{16} + 2\cdot\frac{27}{16}\cdot\frac{33}{16}\right)\right] = \frac{2,457}{32EJ_c}$$

The values of δ_{22} and δ_{33} will be obtained in the same way, i. e., raising to the second power the $\overline{M}_2$ and $\overline{M}_3$ graphs (Fig. 9.11*b* and *c*)

$$\delta_{22} = \frac{2a}{6EJ_c}\left[(2\times 12\times 12 + 2\times 9\times 9 + 2\times 12\times 9) + \right.$$
$$+ (2\times 9\times 9 + 2\times 6\times 6 + 2\times 9\times 6) +$$
$$\left.+ (2\times 6\times 6 + 2\times 3\times 3 + 2\times 6\times 3) + 2\times 3\times 3\right] = \frac{1,152}{EJ_c}$$

$$\delta_{33} = \frac{a}{EJ_c}\times 8 = \frac{24}{EJ_c}$$

The displacements Δ_{1q}, Δ_{2q} and Δ_{3q} due to the actual loading will be given by the product of the $\overline{M}_1$, $\overline{M}_2$ and $\overline{M}_3$ graphs by the M_q graph (Fig. 10.11)

$$\Delta_{1q} = \frac{a}{6EJ_c}\left[\left(-2\times\frac{63}{16}\times 144 - 2\times\frac{21}{16}\times 81 - \frac{63}{16}\times 81 - \frac{21}{16}\times 144\right) + \right.$$
$$+\left(-2\times\frac{21}{16}\times 81 + 2\times\frac{9}{16}\times 36 - \frac{21}{16}\times 36 + \frac{9}{16}\times 81\right) +$$
$$+\left(2\times\frac{9}{16}\times 36 + 2\times\frac{27}{16}\times 9 + \frac{9}{16}\times 9 + \frac{27}{16}\times 36\right) +$$
$$\left.+\left(2\times\frac{27}{16}\times 9 + \frac{33}{16}\times 9\right)\right] = -\frac{7,371}{8EJ_c}$$

$$\Delta_{2q}=\frac{a}{6EJ_c}\,[(2\times 12\times 144+2\times 9\times 81+12\times 81+9\times 144)+$$
$$+(2\times 9\times 81+2\times 6\times 36+9\times 36+6\times 81)+(2\times 6\times 36+$$
$$+2\times 3\times 9+6\times 9+3\times 36)+2\times 3\times 9]=\frac{5{,}292}{EJ_c}$$

$$\Delta_{3q}=\frac{a}{EJ_c}\left(\frac{144}{2}+81+36+9\right)=-\frac{594}{EJ_c}$$

Introducing the values of these displacements in equations (3.11) we obtain

$$X_1=-\frac{\Delta_{1q}}{\delta_{11}}=\frac{7{,}371\times 32}{8\times 2{,}457}=12 \text{ tons}$$

$$X_2=-\frac{\Delta_{2q}}{\delta_{22}}=\frac{-5{,}292}{1{,}152}=-4.594 \text{ tons}$$

$$X_3=-\frac{\Delta_{3q}}{\delta_{33}}=\frac{594}{24}=24.75 \text{ ton metres}$$

The M, Q and N diagrams may now be obtained applying to the elastic centre of the conjugate structure two forces $X_1 = 12$ tons and $X_2 = -4.594$ tons and a couple $X_3 = 24.75$ ton metres together with the uniform loads distributed over the left semiarch.

The following formulas may be used provided the ordinates pass through apices of the polygonal arch:

(a) for the left semiarch (Fig. 11.11*a*)

$$M=X_1\,(y-y_s)+X_2x+X_3-q\,\frac{x^2}{2}$$

$$Q=-X_1 \sin\varphi+X_2\cos\varphi-qx\cos\varphi$$
$$N=X_1\cos\varphi+X_2\sin\varphi-qx\sin\varphi$$

(b) for the right semiarch (Fig. 11.11*b*)

$$M=X_1\,(y-y_s)+X_2x+X_3$$
$$Q=-X_1\sin\varphi+X_2\cos\varphi$$
$$N=X_1\cos\varphi+X_2\sin\varphi$$

In the above expressions x and y are the coordinates of the neutral line of the arch, x becoming negative to the left of the axis of symmetry (see Fig. 8.11*b*), while φ is the angle between the tangent to this neutral line and the horizontal. The values of this angle are positive for the left half of the arch and negative for the right one. *

Normal stresses will be reckoned positive when they cause a compression of the arch. For the bending moments and shearing forces the usual sign convention given in Art. 1.2 will be maintained. It is easily seen that all the expressions for the stresses induced in the left half of the arch differ from those for its right half only by the presence of terms due to the distributed

◆

* The stress diagrams must be constructed for the real arch and not for the imaginary polygonal one adopted solely with the view of simplifying the computations.

loads q. The values of the angle φ will be deduced from the equation of the neutral line

$$\eta = \frac{4f}{l^2}(l-\xi)\,\xi$$

Table 1.11

Section No.	ξ m	tan φ	φ	sin φ	cos φ	x m	y m
0	0	1	45°	0.707	0.707	—12	6
1	3	0.75	36°52′	0.600	0.800	—9	3.375
2	6	0.5	26°34′	0.447	0.894	—6	1.5
3	9	0.25	14°02′	0.2425	0.970	—3	0.375
4; 4′	12	0	0°	0	1	0	0
3′	15	—0.25	—14°02′	—0.2425	0.970	3	0.375
2′	18	—0.5	—26°34′	—0.447	0.894	6	1.5
1′	21	—0.75	—36°52′	—0.600	0.800	9	3.375
0′	24	—1	—45°	—0.707	0.707	12	6

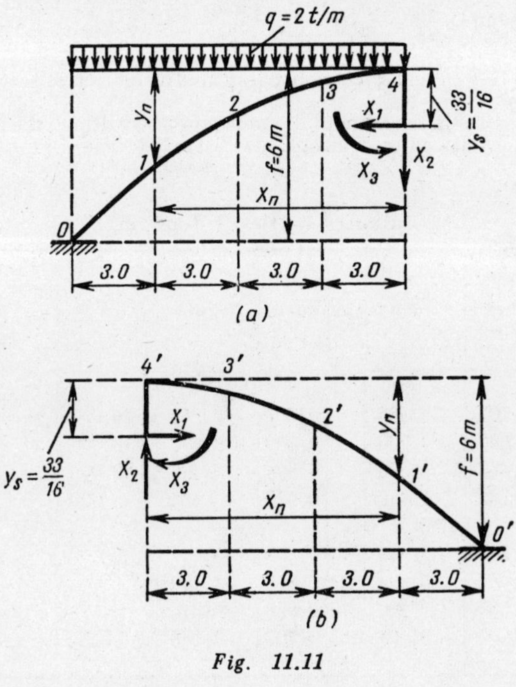

Fig. 11.11

Table 2.11

Ordinates to the M diagram

Section No.	X_1	$y-y_s$	$X_1(y-y_s)$	X_2	x	X_2x	X_3	$\frac{-qx^2}{2}=-x^2$ for the left semiarch	Ordinates to the M diagram, ton metres
0		3.9375	47.25		—12.00	55.13		—144.00	—16.87
1		1.3125	15.75		—9.00	41.35		—81.00	0.85
2		—0.5625	—6.75		—6.00	27.56		—36.00	9.56
3		—1.6875	—20.25		—3.00	13.78		—9.00	9.28
4; 4'	12.000	—2.0625	—24.75	—4.594	0	0	24.75	0	0
3'		—1.6875	—20.25		3.00	—13.78		—	—9.28
2'		—0.5625	—6.75		6.00	—27.56		—	—9.56
1'		1.3125	15.75		9.00	—41.35		—	—0.85
0'		3.9375	47.25		12.00	—55.13		—	16.87

wherefrom

$$\tan\varphi=\frac{d\eta}{d\xi}=\frac{4f}{l^2}(l-2\xi)=\frac{4\times 6}{24^2}(24-2\xi)=1-\frac{\xi}{12}$$

The values of x, y and φ corresponding to different cross sections of the arch appear in Table 1.11

The ordinates to the M, Q and N diagrams together with all the corresponding computations are entered into Tables 2.11, 3.11 and 4.11.

Table 3.11

Ordinates to the Q diagram

Section No.	X_1	$\sin\varphi$	$-X_1\sin\varphi$	X_2	$\cos\varphi$	$X_2\cos\varphi$	qx	$-qx\cos\varphi$ for the left semiarch	Ordinates to the Q diagram, tons
0		0.707	—8.484		0.707	—3.248	—24.000	16.968	5.24
1		0.600	—7.200		0.800	—3.675	—18.000	14.400	3.53
2		0.447	—5.364		0.894	—4.109	—12.000	10.728	1.26
3		0.2425	—2.910		0.970	—4.457	—6.000	5.820	—1.55
4; 4'	12.000	0	0	—4.594	1.0	—4.594	0	0	—4.59
3'		—0.2425	2.910		0.970	—4.457	—	—	—1.55
2'		—0.447	5.364		0.894	—4.109	—	—	1.26
1'		—0.600	7.200		0.800	—3.675	—	—	3.53
0'		0.707	8.484		0.707	—3.248	—	—	5.24

Table 4.11

Ordinates to the N diagram

Section No.	X_1	$\cos\varphi$	$\frac{X_1}{\cos\varphi}$	X_2	$\sin\varphi$	$X_2\sin\varphi$	qx	$-qx\sin\varphi$ for the left semi-arch	Ordinates to the N diagrams, tons
0		0.707	8.484		0.707	—3.248	—24.000	16.968	22.21
1		0.800	9.600		0.600	—2.756	—18.000	10.800	17.64
2		0.894	10.728		0.447	—2.054	—12.000	5.364	14.04
3		0.970	11.640		0.2425	—1.114	—6.000	1.455	11.98
4; 4'	12.000	1.0	12.000	—4.594	0	0	0	0	12.00
3'		0.970	11.640		—0.2425	1.114	—	—	12.76
2'		0.894	10.728		—0.447	2.054	—	—	12.79
1'		0.800	9.600		—0.600	2.756	—	—	12.36
0'		0.707	8.484		—0.707	3.248	—	—	11.73

The diagrams given in Fig. 12.11 have been constructed using the data contained in the above tables.

Let us check the M diagram of Fig. 12.11 using the method based on the consistency of deflections (See Art. 6.9). This can be done multiplying the said diagram by, say, the unit moment diagram $\overline{M}_2$ (Fig. 9.11*b*).

$$\frac{2a}{6EJ_c}[(2\times12\times16.87-2\times9\times0.85-12\times0.85+9\times16.87)-$$
$$-(2\times9\times0.85+2\times6\times9.56+9.56\times9+6\times0.85)-$$
$$-(2\times6\times9.56+2\times3\times9.28+6\times9.28+3\times9.56)-$$
$$-2\times3\times9.28]=\frac{1}{EJ_c}(556.7-557.1)\approx0$$

Problem 2. Required the M, Q and N diagrams for the parabolic arch carrying a horizontal load of $q = 2$ tons per sq metre (Fig. 13.11). Compressive stresses arising in the arch will be accounted for *. The span of the arch $l = 36$ m, its rise $f = 18$ m. At the crown the arch is 1.2 m thick and at the abutments 1.8 m thick. At intermediate sections the thickness of the arch is given by

$$d_x=d_c+\frac{d_a-d_c}{0.5l}\,|\,x\,|$$

where $|\,x\,|$ is the horizontal distance of the cross section under consideration to the crown (Fig. 15.11). The width of the arch (in the direction normal to the plane of the drawing) will be assumed equal to 1.0 m.

◆

* In this case $f = 0.5\,l$ (i. e., $f > \frac{1}{5}l$) and consequently direct stresses should be neglected. In this problem they will be considered solely in order to acquaint the reader with the corresponding computation techniques.

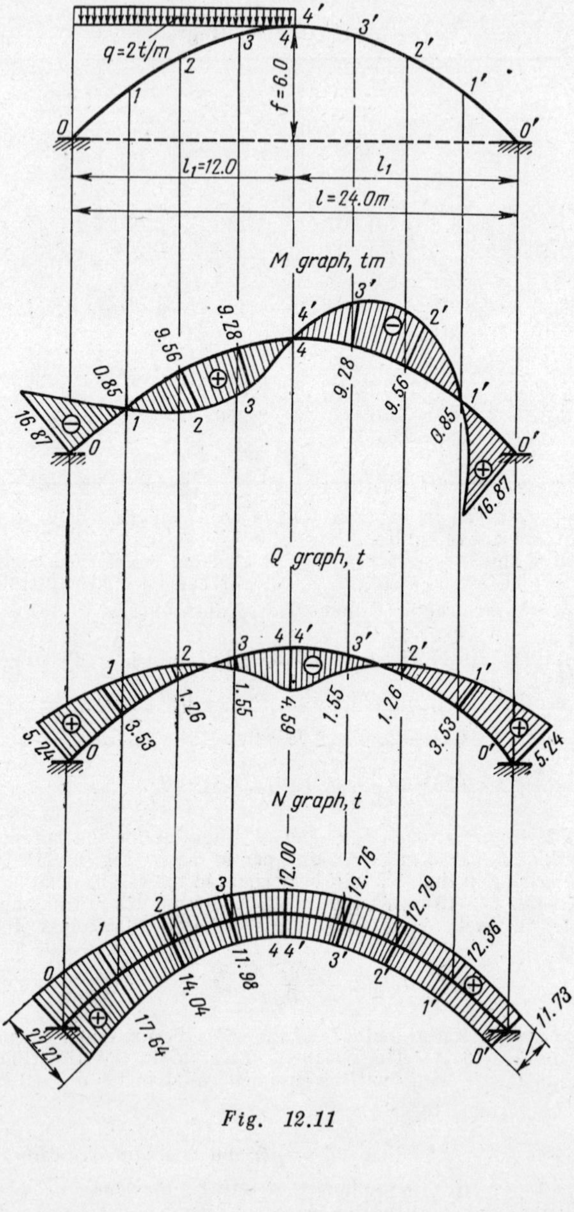
q=2t/m
f=6.0
l_1=12.0
l_1
l=24.0m
M graph, tm
16.87
0.85
9.56
9.28
Q graph, t
5.24
3.53
1.26
1.55
4.59
N graph, t
22.21
17.64
14.04
11.98
12.00
12.76
12.79
12.36
11.73

Fig. 12.11

Solution. The conjugate simple structure with all the redundant reactions transferred to the elastic centre appears in Fig. 14.11. The corresponding canonical equations become

$$\left.\begin{aligned} X_1\delta_{11}+\Delta_{1q}=0 \\ X_2\delta_{22}+\Delta_{2q}=0 \\ X_3\delta_{33}+\Delta_{3q}=0 \end{aligned}\right\} \quad (2.11)$$

whence

$$\left.\begin{aligned} X_1=-\frac{\Delta_{1q}}{\delta_{11}} \\ X_2=-\frac{\Delta_{2q}}{\delta_{22}} \\ X_3=-\frac{\Delta_{3q}}{\delta_{33}} \end{aligned}\right\} \quad (3.11)$$

In order to determine the displacements δ_{11}, δ_{22}, δ_{33}, Δ_{1q}, Δ_{2q} and Δ_{3q}, subdivide the arch into twelve segments having equal horizontal projections (Fig. 15.11).

$$a=\frac{l}{12}=\frac{36}{12}=3.0 \text{ m}$$

The ordinates to the neutral line of the arch represented in Fig. 15.11 are computed using expression

$$y=\frac{4f}{l^2}x^2=\frac{4\times 18}{36^2}x^2=\frac{x^2}{18}$$

All the data necessary for further calculations are given in Table 5.11. The angle φ between the tangent to the neutral line of the arch and the horizontal has been computed using expression

$$\tan\varphi=\frac{dy}{dx}=\frac{2x}{18}=\frac{x}{9}$$

For the right semiarch $\tan\varphi$ and consequently the angles φ themselves are positive and for the left semiarch they are negative. The mean values of segment lengths contiguous to section subdividing the arch have been calculated using the approximate relation $\bar{s}_i=\frac{a}{\cos\varphi_i}$ with the exception of sections *0* and *6* for which

$$\bar{s}_0=\frac{a}{2\cos\varphi_0}=1.5 \text{ m}$$

and

$$\bar{s}_6=\frac{a}{2\cos\varphi_6}=\frac{3}{2\times 0.4472}=3.35 \text{ m}$$

The arch thickness at different cross sections has been calculated using the relation specified above

$$d=1.20+\frac{1.80-1.20}{18}|x|=1.20+\frac{|x|}{30}$$

while the moments of inertia are given by $J=\frac{d^3}{12}$.

Table 5.11

Section No.	x, m		y, m	tan φ right semiarch	cos φ	sin φ		$\bar{s}$, m	d, m	J m⁴	$\frac{\bar{s}}{J}$	$y\frac{\bar{s}}{J}$
	left semi-arch	right semi-arch				left semiarch	right semiarch					
0	0	0	0	0	1	0	0	1.50	1.20	0.144	10.4	0
1	—3	3	0.5	0.3333	0.9487	—0.3162	0.3162	3.16	1.30	0.183	17.3	8.7
2	—6	6	2	0.6667	0.8321	—0.5547	0.5547	3.61	1.40	0.229	15.8	31.6
3	—9	9	4.5	1.0000	0.7071	—0.7071	0.7071	4.24	1.50	0.281	15.1	67.9
4	—12	12	8	1.3333	0.6000	—0.8000	0.8000	5.00	1.60	0.341	14.6	116.8
5	—15	15	12.5	1.6667	0.5145	—0.8575	0.8575	5.83	1.70	0.409	11.8	147.5
6	—18	18	18	2.0000	0.4472	—0.8944	0.8944	3.35	1.80	0.486	6.9	124.2
								Total			91.9	496.7

Table 6.11

Section No.	y, m	$y-y_s$	$(y-y_s)^2$	$\frac{\bar{s}}{J}$	$(y-y_s)^2 \frac{\bar{s}}{J}$	$\cos\varphi$	$\cos^2\varphi$	$\bar{s}$, m	F, m²	$\frac{\bar{s}}{F}$	$\cos^2\varphi \frac{\bar{s}}{F}$
0	0	—5.40	29.16	10.4	303.3	1	1	1.50	1.20	1.25	1.25
1	0.5	—4.90	24.01	17.3	415.4	0.949	0.901	3.16	1.30	2.43	2.18
2	2	—3.40	11.58	15.8	182.6	0.832	0.692	3.61	1.40	2.58	1.79
3	4.5	—0.90	0.81	15.1	12.2	0.707	0.500	4.24	1.50	2.83	1.41
4	8	2.60	6.76	14.6	98.7	0.600	0.360	5.00	1.60	3.12	1.12
5	12.5	7.10	50.41	11.8	594.8	0.514	0.264	5.83	1.70	3.43	0.90
6	18	12.60	158.76	6.9	1095.4	0.447	0.200	3.35	1.80	1.86	0.37
		Total			2702.4	—	—	—	—	—	9.02

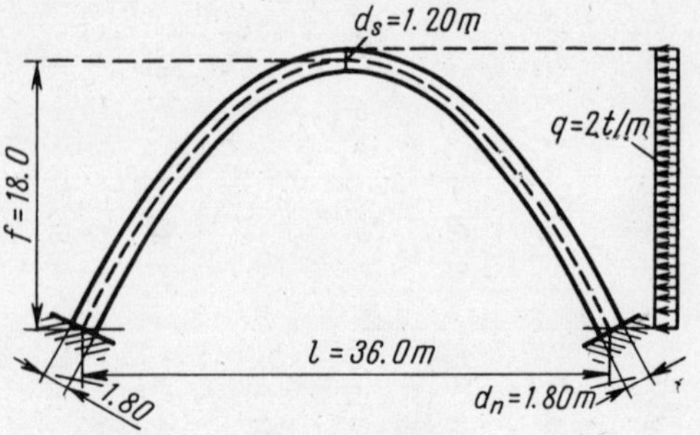

Fig. 13.11

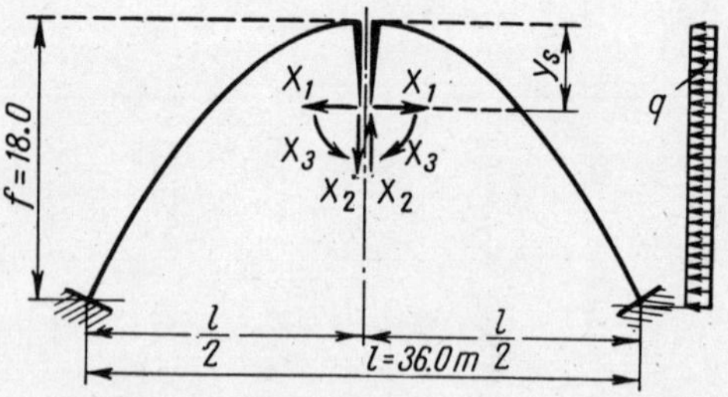

Fig. 14.11

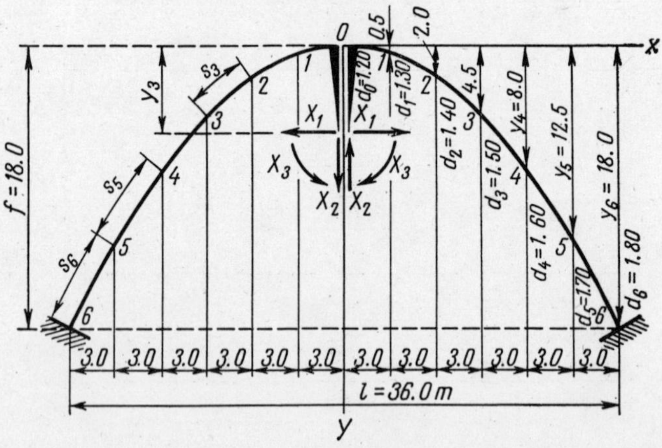

Fig. 15.11

The coordinates of the elastic centre with reference to the adopted axes (see Fig. 15.11) will be $x_s = 0$ and

$$y_s = \frac{\Sigma y \frac{ds}{J}}{\Sigma \frac{ds}{J}} \approx \frac{\Sigma y \frac{\bar{s}}{J}}{\Sigma \frac{\bar{s}}{J}} = \frac{496.7}{91.9} = 5.40 \text{ m}$$

The values of the numerator and the denominator of the latter expression have been taken from Table 5.11.

Let us now compute the *unit displacements*. Displacement δ_{11} will be obtained using the relation

$$E\delta_{11} = 2\left(\Sigma \overline{M}_1^2 \frac{\bar{s}}{J} + \Sigma \overline{N}_1^2 \frac{\bar{s}}{F}\right)$$

which takes due care of the normal stresses. In the above expression

$$\overline{M}_1 = 1\,(y - y_s);\ \overline{N}_1 = 1\cos\varphi;\ F = 1d$$

and therefore

$$E\delta_{11} = 2\left[\Sigma\,(y - y_s)^2 \frac{\bar{s}}{J} + \Sigma\cos^2\varphi \frac{\bar{s}}{F}\right]$$

In the latter expression the term in brackets is multiplied by 2 for the summation is carried along half the arch only. All calculations relative to $E\delta_{11}$ are entered into Table 6.11. Using the data thus obtained we find

$$E\delta_{11} = 2\,(2702.4 + 9.02) = 5422.8$$

The displacement δ_{22} will be computed neglecting the influence of the normal stresses, consequently

$$E\delta_{22} = 2\Sigma \overline{M}_2^2 \frac{\bar{s}}{J}$$

Since $\overline{M}_2 = 1x$, this expression reduces to

$$E\delta_{22} = 2\Sigma x^2 \frac{\bar{s}}{J}$$

The corresponding calculations are entered into Table 7.11

Table 7.11

Section No.	x, m	x^2	$\frac{\bar{s}}{J}$	$x^2 \frac{\bar{s}}{J}$
0	0	0	10.4	0
1	3	9	17.3	156
2	6	36	15.8	569
3	9	81	15.1	1223
4	12	144	14.6	2102
5	15	225	11.8	2655
6	18	324	6.9	2236
			Total	8941

Using the total shown at the foot of the last column we obtain

$$E\delta_{22}=2\times 8{,}941=17{,}882$$

Displacement δ_{33} will be obtained in exactly the same way

$$E\delta_{33}=2\Sigma\overline{M}_3^2\frac{\overline{s}}{J}$$

where $\overline{M}_3 = 1$, leading to $E\delta_{33} = 2\Sigma\frac{\overline{s}}{J}$

Using the value of the total of Table 5.11 we get

$$E\delta_{33}=2\times 91.9=183.8$$

The displacements of the simple structure due to the applied loads will be obtained using the following expressions

$$E\Delta_{1q}=\Sigma\overline{M}_1 M_q\frac{\overline{s}}{J}$$

$$E\Delta_{2q}=\Sigma\overline{M}_2 M_q\frac{\overline{s}}{J}$$

$$E\Delta_{3q}=\Sigma\overline{M}_3 M_q\frac{\overline{s}}{J}$$

where $\overline{M}_1 = y - y_s$; $\overline{M}_2=x$; $\overline{M}_3 = 1$; $M_q = -\frac{qy^2}{2}=-\frac{2y^2}{2}=-y^2$, and, consequently,

$$E\Delta_{1q}=-\Sigma(y-y_s)\,y^2\frac{\overline{s}}{J}$$

$$E\Delta_{2q}=-\Sigma xy^2\frac{\overline{s}}{J}$$

$$E\Delta_{3q}=-\Sigma y^2\frac{\overline{s}}{J}$$

In all the three of these expressions the summation will be carried over the right semiarch only, the bending moments in the left semiarch remaining constantly nil. Further calculations are carried out in tabular form (see Table 8.11).

Using the results of these calculations we get

$$E\Delta_{1q}=-43{,}178;\quad E\Delta_{2q}=-82{,}255;\quad E\Delta_{3q}=-5387.1$$

Introducing these values into expressions (3.11) we obtain the magnitudes of the unknown redundant reactions

$$X_1=-\frac{\Delta_{1q}}{\delta_{11}}=\frac{43{,}178}{5422.8}=7.96 \text{ tons}$$

$$X_2=-\frac{\Delta_{2q}}{\delta_{22}}=\frac{82{,}255}{17{,}882}=4.60 \text{ tons}$$

$$X_3=-\frac{\Delta_{3q}}{\delta_{33}}=\frac{5387.1}{183.8}=29.31 \text{ ton metres}$$

Table 8.11

Section No.	x	y	$\frac{\bar{s}}{J}$	$y-y_s$	y^2	$y^2\frac{\bar{s}}{J}$	$(y-y_s)x$ $xy^2\frac{\bar{s}}{J}$	$xy^2\frac{\bar{s}}{J_i}$
0	0	0	10.4	−5.40	0	0	0	0
1	3	0.5	17.3	−4.90	0.25	4.3	−21	13
2	6	2	15.8	−3.40	4	63.2	−215	379
3	9	4.5	15.1	−0.90	20.25	305.8	−275	2752
4	12	8	14.6	2.60	64	934.4	2429	11213
5	15	12.5	11.8	7.10	156.25	1843.8	13091	27657
6	18	18	6.9	12.60	324	2235.6	28169	40241
					Total	5387.1	43178	82255

These forces are applied at the elastic centre of the statically determinate conjugate structure. Together with the uniform loads applied to the arch, they constitute the complete system of loads permitting the computation of all the stresses and reactions (Fig. 14.11).

Table 9.11

Ordinates to the M diagram

	Section No.	X_1	$y-y_s$	$X_1(y-y_s)$	X_2	x	X_2x	X_3	$\frac{-qy^2}{2}=-y^2$ (for the right semiarch)	Ordinates to the M diagram, ton metres
Left semiarch	*0*		−5.40	−42.96		0	0		—	−13.67
	1		−4.90	−39.00		−3	−13.80		—	−23.49
	2		−3.40	−27.06		−6	−27.60		—	−25.35
	3		−0.90	−7.16		−9	−41.40		—	−19.25
	4		2.60	20.70		−12	−55.20		—	−5.19
	5		7.10	56.52		−15	−69.00		—	16.83
	6	7.96	12.60	100.30	4.60	−18	−82.80	29.31	—	46.81
Right semiarch	*0*		−5.40	−42.98		0	0		0	−13.67
	1		−4.90	−39.00		3	13.80		−0.25	3.86
	2		−3.40	−27.06		6	27.60		−4	25.85
	3		−0.90	−7.16		9	41.40		−20.25	43.30
	4		2.60	20.70		12	55.20		−64	41.21
	5		7.10	56.52		15	69.00		−156.25	−1.42
	6		12.60	100.30		18	82.80		−324	−111.59

The ordinates to the M, Q and N graphs will be obtained using the following equations:

(a) for the left semiarch

$$M = X_1(y - y_s) + X_2 x + X_3$$

$$Q = X_1 \sin\varphi + X_2 \cos\varphi$$

$$N = X_1 \cos\varphi - X_2 \sin\varphi$$

(b) for the right semiarch

$$M = X_1(y - y_s) + X_2 x + X_3 - \frac{qy^2}{2}$$

$$Q = X_1 \sin\varphi + X_2 \cos\varphi - qy \sin\varphi$$

$$N = X_1 \cos\varphi - X_2 \sin\varphi - qy \cos\varphi$$

It will be noted that the expressions for the left semiarch differ from those for the right one solely by the absence of the term accounting for the uniformly distributed load q.

All the calculations relative to the ordinates to the M, N and Q diagrams are carried out in Tables 9.11, 10.11 and 11.11.

Table 10.11

Ordinates to the Q diagram

	Section No.	X_1	$\sin\varphi$	$X_1\sin\varphi$	X_2	$\cos\varphi$	$X_2\cos\varphi$	$-qy=-2y$	$-qy\sin\varphi$ (for the right semiarch)	Ordinates to the Q diagram, tons
Left semiarch	0	7.96	0	0	4.60	1	4.60	—	—	4.60
	1		−0.316	−2.52		0.949	4.37	—	—	1.85
	2		−0.555	−4.42		0.832	3.83	—	—	−0.59
	3		−0.707	−5.63		0.707	3.25	—	—	−2.38
	4		−0.800	−6.37		0.600	2.76	—	—	−3.61
	5		−0.858	−6.83		0.514	2.36	—	—	−4.47
	6		−0.894	−7.11		0.447	2.06	—	—	−5.05
Right semiarch	0		0	0		1	4.60	0	0	4.60
	1		0.316	2.52		0.949	4.37	−1	−0.32	6.57
	2		0.555	4.42		0.832	3.83	−4	−2.22	6.03
	3		0.707	5.63		0.707	3.25	−9	−6.36	2.52
	4		0.800	6.37		0.600	2.76	−16	−12.80	−3.67
	5		0.858	6.83		0.514	2.36	−25	−21.45	−12.26
	6		0.894	7.11		0.447	2.06	−36	−32.18	−23.01

The diagrams shown in Fig. 16.11 have been plotted using the ordinates calculated in the above tables.

To check the accuracy of the M diagram let us multiply this diagram by the $\overline{M}_1$, $\overline{M}_2$ and $\overline{M}_3$ graphs. In other words, we shall obtain the values

Table 11.11

Ordinates to the *N* diagram

	Section No.	X_1	cos φ	X_1 cos φ	$-X_2$	sin φ	$-X_2$ sin φ	$-qy=-2y$	$-qy$ cos φ (for the right semiarch)	Ordinates to the *N* diagram, tons
Left semiarch	*0*	7.96	1	7.96	−4.60	0	0	—	—	7.96
	1		0.949	7.55		−0.316	1.45	—	—	9.00
	2		0.832	6.63		−0.555	2.55	—	—	9.18
	3		0.707	5.63		−0.707	3.25	—	—	8.88
	4		0.600	4.78		−0.800	3.68	—	—	8.46
	5		0.514	4.09		−0.858	3.95	—	—	8.04
	6		0.447	3.56		−0.894	4.11	—	—	7.67
Right semiarch	*0*		1	7.96		0	0	0	0	7.96
	1		0.949	7.55		0.316	−1.45	−1	−0.95	5.15
	2		0.832	6.63		0.555	−2.55	−4	−3.33	0.75
	3		0.707	5.63		0.707	−3.25	−9	−6.36	−3.98
	4		0.600	4.78		0.800	−3.68	−16	−9.60	−8.60
	5		0.514	4.09		0.858	−3.95	−25	−12.85	−12.71
	6		0.447	3.56		0.894	−4.11	−36	−16.09	−16.64

of $\Sigma \overline{M}_1 M \frac{\bar{s}}{J}$, $\Sigma \overline{M}_2 M \frac{\bar{s}}{J}$ and $\Sigma \overline{M}_3 M \frac{\bar{s}}{J}$ both for the left- and the right-hand semiarches. Remembering that $\overline{M}_1 = y - y_s$, $\overline{M}_2 = x$ and $\overline{M}_3 = 1$, these expressions simplify and become

$$\Sigma (y - y_s) M \frac{\bar{s}}{J}; \quad \Sigma x M \frac{\bar{s}}{J} \quad \text{and} \quad \Sigma M \frac{\bar{s}}{J}$$

The necessary calculations are entered into Table 12.11.

It will be observed that the totals of the entries in the last three columns of Table 12.11 differ very little from zero, which confirms the accuracy of the diagram. The slight discrepancies, which remain below 1 per cent, are due to the fact that we neglected the normal stresses when checking this diagram whereas in computing δ_{11} these stresses were taken into consideration.

Problem 3. Required the influence lines for the redundant reactions X_1, X_2 and X_3 as well as for the stresses M_k, N_k and Q_k acting at section *K* of the arch represented in Fig. 8.11 of Problem 1. The abscissa of this section will be taken equal to the quarter span of the arch, i. e., to 6 m. The construction of influence lines will be carried out using the method of elastic loads. The first of the approximate methods described in the present article will be used throughout and in addition the influence line for the unknown X_2 will be determined using the second of the approximate methods.

Solution. The coordinates of the inscribed polygon apices as well as the ordinate y_s of the elastic centre of the arch were calculated in Problem 1. These coordinates are given in Figs. 8.11 and 9.11 which represent equally the simple structure used in this problem and the bending moment diagrams

Table 12.11

	Section No.	$y-y_s$	x	$\frac{\bar{s}}{J}$	$(y-y_s)\frac{\bar{s}}{J}$	$x\frac{\bar{s}}{J}$	M	$M(y-y_s)\frac{\bar{s}}{J}$	$Mx\frac{\bar{s}}{J}$	$M\frac{\bar{s}}{J}$
Left semiarch	0	—5.40	0	10.4	—56.16	0	—13.67	768	1	—142
	1	—4.90	—3	17.3	—84.77	—51.9	—23.49	1991	1219	—406
	2	—3.40	—6	15.8	—53.72	—94.8	—25.35	1362	2403	—401
	3	—0.90	—9	15.1	—13.59	—135.9	—19.25	262	2616	—291
	4	2.60	—12	14.6	37.96	—175.2	—5.19	—197	909	—76
	5	7.10	—15	11.8	83.78	—177.0	16.83	1410	—2979	199
	6	12.60	—18	6.9	86.94	—124.2	46.81	4070	—5814	323
Right semiarch	0	—5.40	0	10.4	—56.16	0	—13.67	768	0	—142
	1	—4.90	3	17.3	—84.77	51.9	3.86	—327	200	67
	2	—3.40	6	15.8	—53.72	94.8	25.85	—1389	2451	408
	3	—0.90	9	15.1	—13.59	135.9	43.30	—588	5884	654
	4	2.60	12	14.6	37.96	175.2	41.21	1564	7220	602
	5	7.10	15	11.8	83.78	177.0	—1.42	—119	—251	—17
	6	12.60	18	6.9	86.94	124.2	—111.59	—9802	—13959	—770
		Total						+12195 —12422	+22903 —23003	+2253 —2245
								—227	—100	+8

induced by unit actions X_1, X_2 and X_3 applied along the redundant constraints. The simplified expression for the elastic loads was given in Art. 12.8

$$W_n = \frac{S_n}{6EJ_n}(M_{n-1}+2M_n) + \frac{S_{n+1}}{6EJ_{n+1}}(2M_n+M_{n+1}) - \\ -\varepsilon_n \tan\beta_n + \varepsilon_{n+1}\tan\beta_{n+1}$$

Neglecting the longitudinal strains of the arch and remembering that the $\frac{s_i}{J_i}$ ratios remain constant and equal to $\frac{a}{J_c}$, this expression is further simplified and becomes

$$W_n = \frac{a}{6EJ_c}(M_{n-1}+4M_n+M_{n+1}) \tag{5.11}$$

At the crown and at the abutments $S_n = 0$ and $S_{n+1} = 0$, and therefore the elastic loads will equal

$$W_0 = \frac{a}{6EJ_c}(2M_0+M_1) \tag{6.11}$$

$$W_4 = \frac{a}{6EJ_c}(M_3+2M_4) \tag{7.11}$$

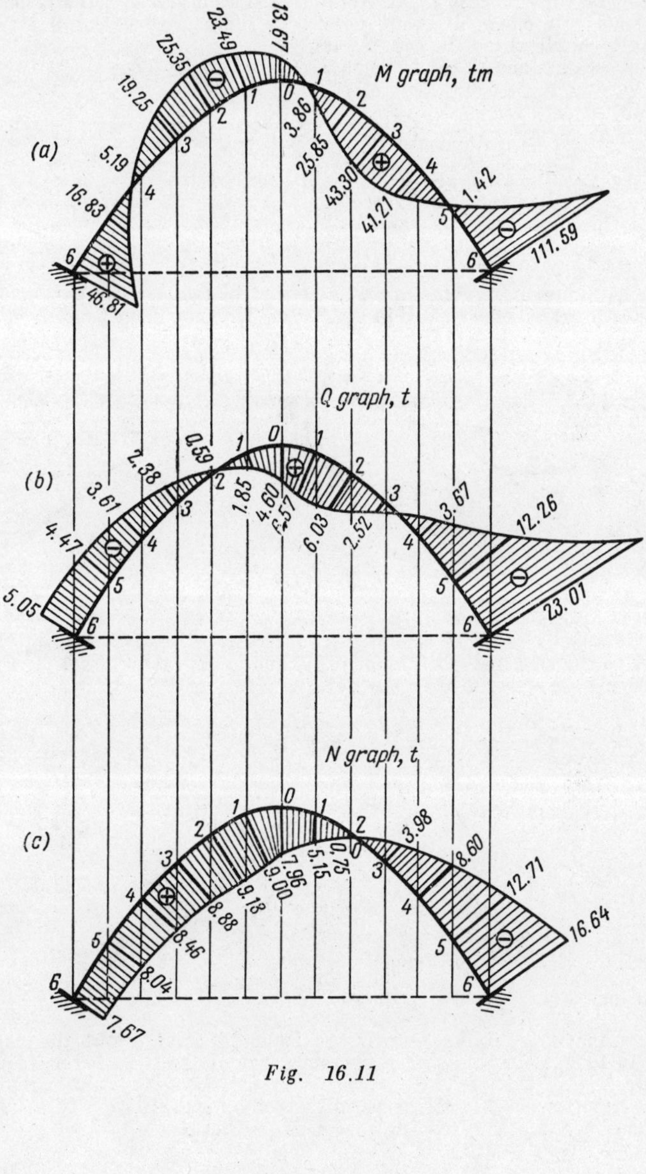
M graph, tm
(a)
13.67
23.49
25.35
19.25
5.19
16.83
46.81
3.86
25.85
43.30
41.21
1.42
111.59
Q graph, t
(b)
0.59
2.38
3.61
4.47
5.05
1.85
4.60
6.57
6.03
2.52
3.67
12.26
23.01
N graph, t
(c)
7.96
9.00
9.18
8.88
8.46
8.04
7.67
5.15
0.75
3.98
8.60
12.71
16.64

Fig. 16.11

Let us compute the ordinate y_s of the elastic centre using the equation $\delta_{13}=0$. For this purpose let us determine the elastic loads corresponding to the bending moment diagram induced in the conjugate simple structure by a unit couple $X_3=1$ (Fig. 9.11*c*).

Using expressions (5.11) through (7.11) we find

$$W_0=\frac{a}{6EJ_c}(2\times 1+1)=\frac{a}{2EJ_c}$$

$$W_1=W_2=W_3=\frac{a}{6EJ_c}(1+4\times 1+1)=\frac{a}{EJ_c}$$

$$W_4=\frac{a}{6EJ_c}(1+2\times 1)=\frac{a}{2EJ_c}$$

The new imaginary structure will consist of two semiarches held fast at the elastic centre as indicated in Fig. 17.11. The imaginary bending moment

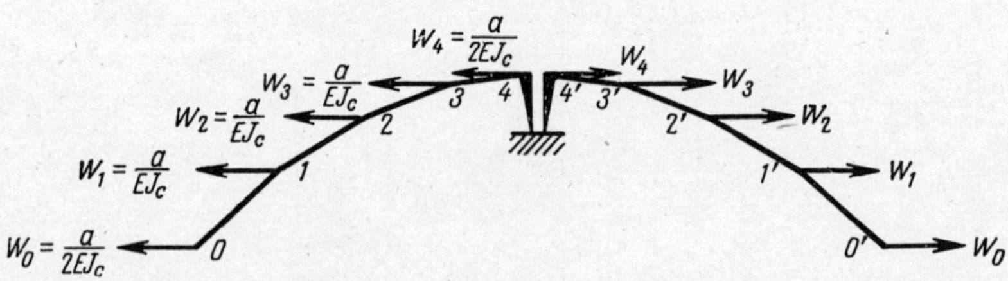

Fig. 17.11

at the fixed ends of these semiarches loaded by a system of elastic loads parallel to the required displacement δ_{13} and acting at points *0*, *1*, *2*, etc., must be nil, thus permitting the determination of y_s

$$W_0(y_s-6)+W_1\left(y_s-\frac{27}{8}\right)+W_2\left(y_s-\frac{3}{2}\right)+W_3\left(y_s-\frac{3}{8}\right)+W_4y_s=0$$

or

$$\frac{a}{2EJ_c}\left[(y_s-6)+2\left(y_s-\frac{27}{8}\right)+2\left(y_s-\frac{3}{2}\right)+2\left(y_s-\frac{3}{8}\right)+y_s\right]=0$$

wherefrom

$$8y_s-\frac{33}{2}=0$$

and

$$y_s=\frac{33}{16}=2.0625 \text{ metres}$$

The value of y_s obtained in this way coincides exactly with the one mentioned in Problem 1 though the two were calculated by entirely different methods.

The simultaneous equations permitting the determination of the redundant reactions due to a moving load unity P become

$$X_1\delta_{11}+\delta_{1p}=0;\quad X_2\delta_{22}+\delta_{2p}=0$$

$$X_3\delta_{33}+\delta_{3p}=0$$

wherefrom

$$X_1=-\frac{\delta_{1p}}{\delta_{11}};\quad X_2=-\frac{\delta_{2p}}{\delta_{22}}$$

$$X_3=-\frac{\delta_{3p}}{\delta_{33}}$$

or

$$X_1=-\frac{\delta_{p1}}{\delta_{11}};\quad X_2=-\frac{\delta_{p2}}{\delta_{22}};\quad X_3=-\frac{\delta_{p3}}{\delta_{33}} \tag{8.11}$$

The graphs of δ_{p1}, δ_{p2} and δ_{p3} for various positions of the load unity P constitute the deflection graphs of the simple structure due to the application of the redundant reactions $X_1 = 1$, $X_2 = 1$ and $X_3 = 1$, respectively. Divided by $-\delta_{11}$, $-\delta_{22}$ and $-\delta_{33}$ the ordinates to these graphs will represent the ordinates to the influence lines for X_1, X_2 and X_3.

Previously we have agreed to call this method of construction of the deflection graphs induced by unit loads by the term *kinematic method*. The same deflection graphs could be also obtained by the elastic loads method. If that were desired, one should start by constructing the $\overline{M}_1$, $\overline{M}_2$ and $\overline{M}_3$ graphs due to unit loads X_1, X_2 and X_3 calculating thereafter the corresponding elastic loads with the aid of expressions (5.11) through (7.11) (see Fig. 9.11). The values of these elastic loads multiplied by EJ_c are given in Table 13.11. Thus, the elastic load W_2 (at point *2*) when the simple structure is loaded by $X_1 = 1$ (see Fig. 9.11*a*) is given by

$$W_2=\frac{a}{6EJ_c}(M_1+4M_2+M_3)=\frac{3}{6EJ_c}\left(\frac{21}{16}-4\times\frac{9}{16}-\frac{27}{16}\right)=-\frac{21}{16EJ_c}$$

and load W_4 (at point *4*) due to the application of forces $X_2 = 1$ to the simple structure (see Fig. 9.11*b* and Eq. 7.11) will amount to

$$W_4=\frac{a}{6EJ_c}(M_3+2M_4)=\frac{3}{6EJ_c}(-3+2\times 0)=-\frac{3}{2EJ_c}$$

Table 13.11

Values of elastic loads (multiplied by EJ_c)

Load point	Unit forces		
	X_1	X_2	X_3
0	$\frac{73.5}{16}$	$\frac{-33}{2}$	$\frac{3}{2}$
1	$\frac{69}{16}$	$\frac{-54}{2}$	$\frac{6}{2}$
2	$\frac{-21}{16}$	$\frac{-36}{2}$	$\frac{6}{2}$
3	$\frac{-75}{16}$	$\frac{-18}{2}$	$\frac{6}{2}$
4	$\frac{-46.5}{16}$	$\frac{-3}{2}$	$\frac{3}{2}$

In order to obtain the deflection diagram δ_{p1} we must apply to the left-hand part of the imaginary structure the elastic loads corresponding to the unit action X_1. These elastic loads must be parallel to the displacement in question and must be directed towards the more extended fibres of the simple structure (Fig. 18.11*a*).

Since the rigid bracket fixed to the simple structure may not be rotated by the unit load X_1, the algebraic sum of all the elastic loads applied to the system of Fig. 18.11*a* must equal zero

$$\Sigma W=\frac{73.5}{16EJ_c}+\frac{69}{16EJ_c}-\frac{21}{16EJ_c}-\frac{75}{16EJ_c}-\frac{46.5}{16EJ_c}=0$$

This relation will be utilized for a check on the accuracy of the computed elastic loads.

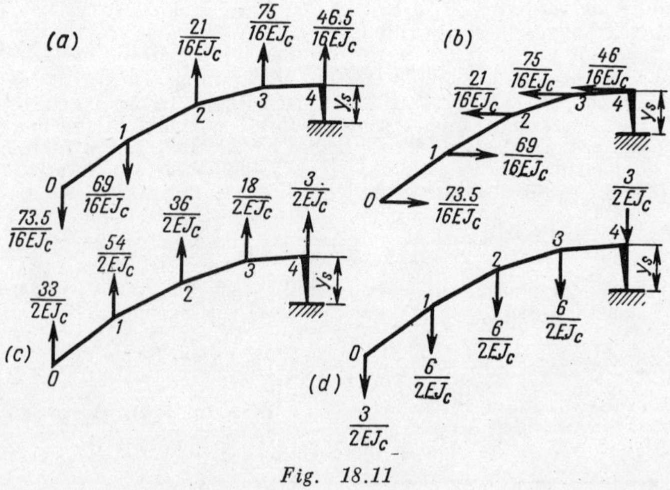

Fig. 18.11

The vertical displacements δ_{p1} of the simple structure due to the unit action X_1 will equal the ordinates to the diagram of bending moments induced in the imaginary structure by the elastic loads. Let us determine the values of these ordinates at different cross sections, the position of these cross sections being indicated by the upper index in parentheses

$$\delta_{p1}^{(1)}=\frac{73.5\times 3}{16EJ_c}=\frac{220.5}{16EJ_c}$$

$$\delta_{p1}^{(2)}=\frac{73.5\times 6-69\times 3}{16EJ_c}=\frac{648}{16EJ_c}$$

$$\delta_{p1}^{(3)}=\frac{73.5\times 9+69\times 6-21\times 3}{16EJ_c}=\frac{1012.5}{16EJ_c}$$

$$\delta_{p1}^{(4)}=\frac{73.5\times 12+69\times 9-21\times 6-75\times 3}{16EJ_c}=\frac{1,152}{16EJ_c}$$

The δ_{p1} diagram appearing in Fig. 19.11*b* has been obtained setting off the ordinates calculated as just described on the side of the more extended fibres of the simple structure (Fig. 18.11*a*). *

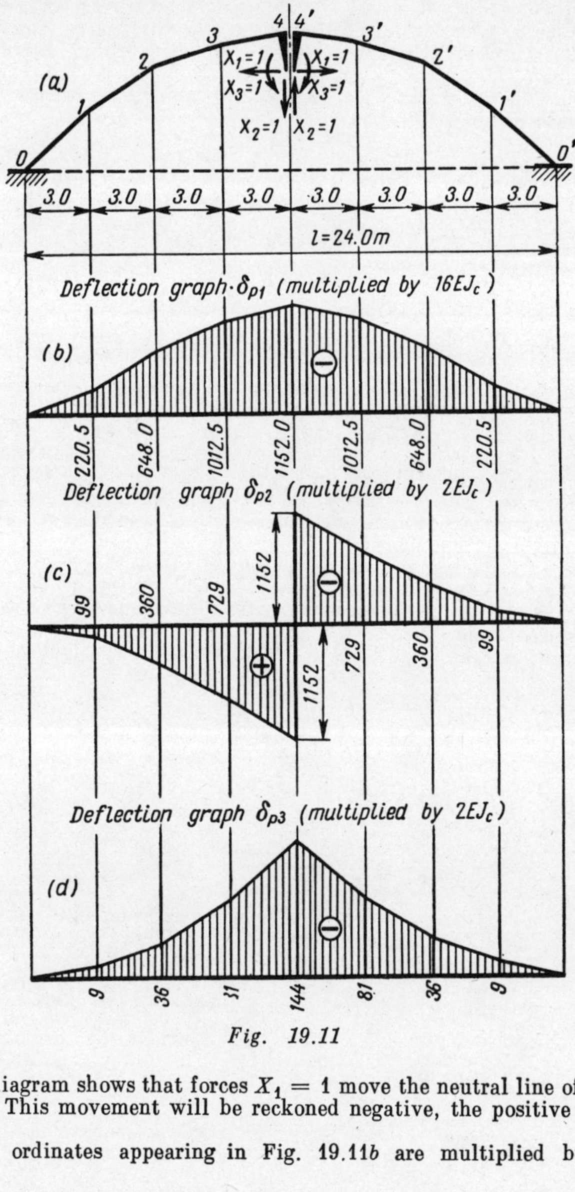

Fig. 19.11

This diagram shows that forces $X_1 = 1$ move the neutral line of the arch upwards. This movement will be reckoned negative, the positive direction

* The ordinates appearing in Fig. 19.11*b* are multiplied by $16EJ_c$.

coinciding as convened with the direction of the unit load P, that means downwards. Consequently, the whole area of the diagram will also be reckoned negative.

The displacement δ_{11} will be obtained rotating the imaginary loads through an angle of 90° until they become horizontal as indicated in Fig. 18.11*b*. This being done, calculate the moment of these loads about the elastic centre and double its value, for the displacement δ_{11} represents the total change in the distance between the lower ends of the brackets along the direction of the horizontal unit loads X_1

$$\delta_{11}=2\left[\frac{73.5}{16EJ_c}\left(\frac{48}{8}-\frac{16.5}{8}\right)+\frac{69}{16EJ_c}\left(\frac{27}{8}-\frac{16.5}{8}\right)+\right.$$
$$+\frac{21}{16EJ_c}\left(\frac{16.5}{8}-\frac{12}{8}\right)+\frac{75}{16EJ_c}\left(\frac{16.5}{8}-\frac{3}{8}\right)+\frac{46.5}{16EJ_c}\times$$
$$\left.\times\frac{16.5}{8}\right]=\frac{3}{8\times 16EJ_c}(73.5\times 31.5+69\times 10.5+21\times 4.5+$$
$$+75\times 13.5+46.5\times 16.5)=\frac{1228.5}{16EJ_c}$$

The value of δ_{11} obtained above coincides exactly with the one computed in Problem 1 using an entirely different procedure. Dividing all the ordinates to the δ_{p1} graph by $(-\delta_{11})$ we obtain the ordinates to the influence line for X_1. This influence line will show the variation of X_1 when the unit load P travels along the arch (Fig. 20.11*b*).

Following the same procedure, we shall find the displacement δ_{p2}. The elastic loads will be applied once more to the imaginary structure and the corresponding bending moment will be determined. In this case the extended fibres of the simple structure acted upon by the force X_2 will be situated at the extrados (see Fig. 9.11*b*) and therefore the elastic loads must be directed upwards (Fig. 18.11*c*). The values of the bending moments induced by these loads in the imaginary structure will furnish the values of the ordinates to the deflection graph for the simple structure under consideration

$$\delta_{p2}^{(1)}=\frac{33\times 3}{2EJ_c}=\frac{99}{2EJ_c}$$
$$\delta_{p2}^{(2)}=\frac{33\times 6+54\times 3}{2EJ_c}=\frac{360}{2EJ_c}$$
$$\delta_{p2}^{(3)}=\frac{33\times 9+54\times 6+36\times 3}{2EJ_c}=\frac{729}{2EJ_c}$$
$$\delta_{p2}^{(4)}=\frac{33\times 12+54\times 9+36\times 6+18\times 3}{2EJ_c}=\frac{1{,}152}{2EJ_c}$$

Setting off these ordinates on the side of the extended fibres of the simple structure we obtain the diagram represented in Fig. 19.11*c*. It will be observed that the diagram thus obtained is antisymmetrical. *

The displacement δ_{22} representing the mutual translation of the free ends of the brackets along the direction of X_2, this displacement must be the double of $\delta_{p2}^{(4)}$

$$\delta_{22}=2\delta_{p2}^{(4)}=2\frac{1{,}152}{2EJ_c}=\frac{1{,}152}{EJ_c}$$

* All the ordinates indicated in Fig. 19.11*c* have been multiplied by $2EJ_c$.

This value coincides again with that obtained in Problem 1, multiplying the bending moment diagrams. The ordinates to the influence line for X_2

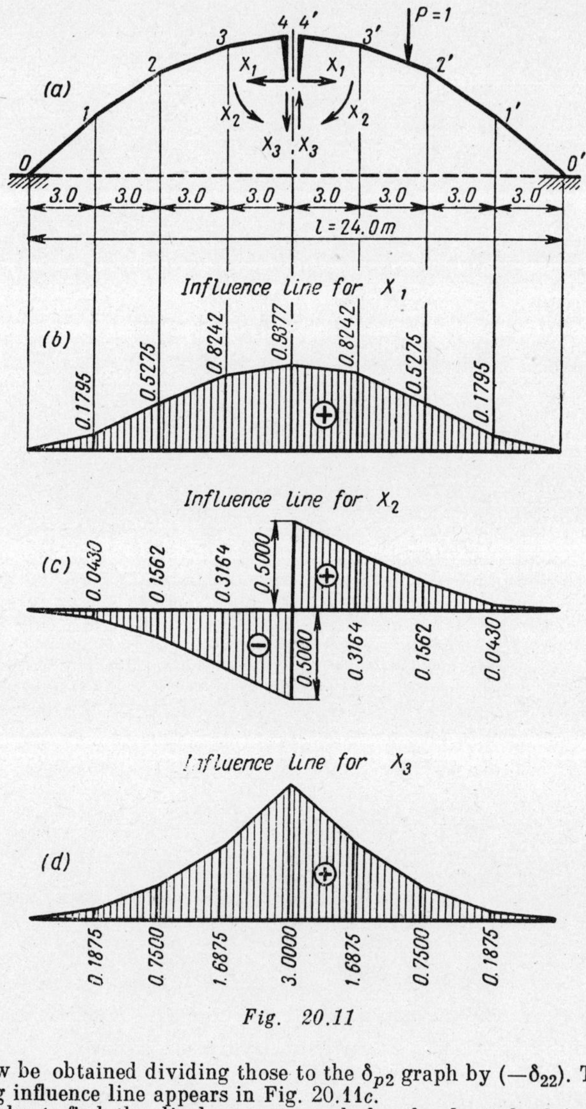

Fig. 20.11

may now be obtained dividing those to the δ_{p2} graph by $(-\delta_{22})$. The corresponding influence line appears in Fig. 20.11*c*.

In order to find the displacement graph δ_3 the elastic loads must be applied once again to the imaginary structure (Fig. 18.11*d*), these loads being directed as usual towards the extended fibres, that means downwards. The desired displacements will be furnished by the values of the bending moments acting at the corresponding sections of the imaginary structure

$$\delta_{p3}^{(1)}=\frac{3\times3}{2EJ_c}=\frac{9}{2EJ_c}$$

$$\delta_{p3}^{(2)}=\frac{3\times6+6\times3}{2EJ_c}=\frac{36}{2EJ_c}$$

$$\delta_{p3}^{(3)}=\frac{3\times9+6\times6+6\times3}{2EJ_c}=\frac{81}{2EJ_c}$$

$$\delta_{p3}^{(4)}=\frac{3\times12+6\times9+6\times6+6\times3}{2EJ_c}=\frac{144}{2EJ_c}$$

These ordinates must be set off above the abscissa axis for the extended fibres of the simple structure are at the extrados. The displacement graph for δ_{p3} thus obtained is represented in Fig. 19.11*d*. *

The displacement δ_{33} representing the mutual angular rotation of the brackets (in other words, that of faces *4* and *4'* of the semiarches), its value will be equal to the sum of the elastic loads, i. e.,

$$\delta_{33}=2\left(\frac{3\times2}{2EJ_c}+\frac{6\times3}{2EJ_c}\right)=\frac{24}{EJ_c}$$

The same value has been obtained previously by the method of graph multiplication. The ordinates to the influence line for X_3 (Fig. 20.11*d*) were obtained dividing the ordinates to the δ_{p3} graph by $(-\delta_{33})$.

Once the influence lines for all the redundant reactions X_1, X_2 and X_3 have been found, we may proceed with the construction of the influence lines for bending moments, shearing forces and normal stresses at any cross section of the arch.

Let us construct these three influence lines for section K situated 6 metres to the right from the left-hand abutment (section *2* of Fig. 8.11*b*).

The angle φ_2 between the tangent to the neutral line of the arch and the horizontal equals for this section 26°34′, $\sin\varphi_2 = 0.447$ and $\cos\varphi_2 = 0.894$ (see Table 1.11 of Problem 1). The lever arm of the force X_1 with reference to section K equals

$$y_s-y_2=\frac{33}{16}-\frac{3}{2}=\frac{9}{16}\text{ metre}$$

and the lever arm of force X_2 with reference to the same section equals 6 metres. The ordinates to the influence lines for M_k, Q_k and N_k will be calculated using the following expressions: **

(a) when the unit load P is between the left-hand abutment and section K or when it has shifted to the right-hand semiarch

$$M_k=-X_1\frac{9}{16}-X_2 6+X_3$$

$$Q_k=-X_1\sin\varphi_2+X_2\cos\varphi_2=-X_1 0.447+X_2 0.894$$

$$N_k=X_1\cos\varphi_2+X_2\sin\varphi_2=X_1 0.894+X_2 0.447$$

◆

* The ordinates to this graph have been once again multiplied by $2EJ_c$.

** Normal stresses will be reckoned positive when they cause compressive stresses in the arch, the usual sign convention being maintained for bending moments and shears.

Table 14.11

Ordinates to the M_k influence line

Load point	X_1	$-X_1 \frac{9}{16}$	X_2	$-X_2\,6$	X_3	$-r$	Ordinates to the M_k influence line, metres
0	0	0	0	0	0	—	—
1	0.1795	—0.101	—0.0430	0.258	0.187	—	0.344
2	0.5275	—0.297	—0.1562	0.937	0.750	0	1.390
3	0.8242	—0.464	—0.3164	1.898	1.687	—3	0.124
4	0.9377	—0.528	—0.5000	3.000	3.000	—6	—0.528
4'	0.9377	—0.528	0.5000	—3.000	3.000	—	—0.528
3'	0.8242	—0.464	0.3164	—1.898	1.687	—	—0.675
2'	0.5275	—0.297	0.1562	—0.937	0.750	—	—0.484
1'	0.1795	—0.101	0.0430	—0.258	0.187	—	—0.172
0'	0	0	0	0	0	—	0

Table 15.11

Ordinates to the Q_k influence line

Load point	X_1	$-X_1\,0.447$	X_2	$X_2\,0.894$	0.894	Ordinates to the Q_k influence line
0	0	0	0	0	—	0
1	0.1795	—0.080	—0.0430	—0.039	—	—0.119
Immediately to the left of section *2*	0.5275	—0.236	—0.1562	—0.140	—	—0.376
Immediately to the right of section *2*	0.5275	—0.236	—0.1562	—0.140	0.894	0.518
3	0.8242	—0.368	—0.3164	—0.283	0.894	0.243
4	0.9377	—0.419	—0.5000	—0.447	0.894	0.028
4'	0.9377	—0.419	0.5000	0.447	—	0.028
3'	0.8242	—0.368	0.3164	0.283	—	—0.085
2'	0.5275	—0.236	0.1562	0.140	—	—0.096
1'	0.1795	—0.080	0.0430	0.039	—	—0.041
0'	0	0	0	0	—	0

(b) when the unit load P is between section K and the crown

$$M_k = -X_1 \frac{9}{16} - X_2 6 + X_3 - 1r$$

where r is the lever arm of the unit load P about section K and

$$Q_k = -X_1 0.447 + X_2 0.894 + 1 \times 0.894$$
$$N_k = -X_1 0.894 + X_2 0.447 + 1 \times 0.447$$

It will be observed that the expressions obtained for case (b) differ from those for case (a) only by the presence of a term accounting for the uniformly distributed load. The ordinates to the influence lines for M_k, Q_k and N_k will be obtained introducing in the above equations the values of X_1, X_2 and X_3 scaled off the corresponding influence lines given in Fig. 20.11.

It is advised to carry out all the calculations in tabular form as indicated in Tables 14.11, 15.11 and 16.11. The entries in the last column of each of these three tables have been used for the construction of influence lines appearing in Fig. 21.11.

Table 16.11

Ordinates to the N_k influence line

Load point	X_1	X_1 0.894	X_2	X_2 0.447	0.447	Ordinates to the N_k influence line
0	0	0	0	0	—	0
1	0.1795	0.160	—0.0430	—0.019	—	0.141
Immediately to the left of section *2*	0.5275	0.471	—0.1562	—0.070	—	0.401
Immediately to the right of section *2*	0.5275	0.471	—0.1562	—0.070	0.447	0.848
3	0.8242	0.737	—0.3164	—0.141	0.447	1.043
4	0.9377	0.838	—0.5000	—0.223	0.447	1.061
4'	0.9377	0.838	0.5000	0.223	—	1.061
3'	0.8242	0.737	0.3164	0.141	—	0.878
2'	0.5275	0.471	0.1562	0.070	—	0.541
1'	0.1795	0.160	0.0430	0.019	—	0.179
0'	0	0	0	0	—	0

Using these influence lines *let us find the magnitudes of M_k, Q_k and N_k induced by uniformly distributed loads acting over the whole of the left semiarch, the intensity of this loading being equal to 2 tons per metre* (see Fig. 8.11*a*). For this purpose we shall replace the uniformly distributed loads by concentrated ones acting at points *0*, *1*, *2*, *3* and *4* of the polygonal arch. These concentrated loads will amount to

$$P_0=P_4=q\frac{a}{2}=\frac{2\times 3}{2}=3 \text{ tons}$$

$$P_1=P_2=P_3=qa=2\times 3=6 \text{ tons}$$

The desired values of M_k, Q_k and N_k will be obtained multiplying the magnitudes of these concentrated loads by the corresponding ordinates to the influence lines, which are then summed up as indicated below

$$M_k=3\times 0+6\times 0.344+6\times 1.390+$$
$$+6\times 0.124-3\times 0.528=9.564 \text{ ton metres}$$

$$Q_k=3\times 0-6\times 0.119+6\times$$
$$\times\frac{-0.376+0.518}{2}+6.0\times 0.243+$$
$$+3\times 0.028=1.254 \text{ tons}$$

$$N_k=3\times 0+6\times 0.114+6\times$$
$$\times\frac{0.401+0.848}{2}+6\times 1.043+$$
$$+3\times 1.061=14.034 \text{ tons}$$

These values are practically the same as those given in Tables 2.11, 3.11 and 4.11 and in Fig. 12.11 of Problem 1.

The influence lines for the redundant constraints X_1, X_2 and X_3 permit equally the construction of stress diagrams induced in the arch by vertical loads. For that purpose the actual loading will be replaced by a system of equivalent concentrated loads acting at the apices of the polygonal arch. The magnitudes of X_1, X_2 and X_3 corresponding to each of the concentrated loads will be easily found using the said influence lines. The diagrams of the stresses induced in the different sections of the arch will then be obtained in the usual way applying to the conjugate simple structure both actual loads and redundant reactions.

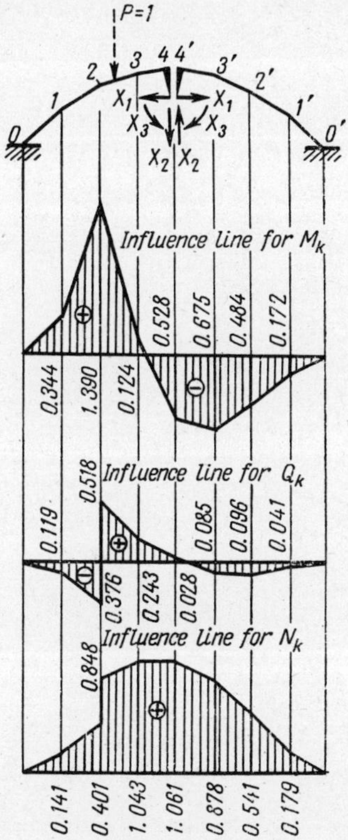

Fig. 21.11

Let us consider an example of the *construction of the influence line for* X_2 using the second of the approximate methods described in the first part of the present article. *

Neglecting compression strains, the expression for the elastic loads will become in this case

$$W_n=M_n\frac{\overline{S}_n}{EJ_n}$$

* The same method was used in Problem 2.

where $\overline{S}_n = \dfrac{S_n + S_{n+1}}{2}$ is the mean of the segment lengths contiguous to section n. The values of $\overline{S}_i$ may be approximately taken equal to

$$\overline{S}_0 = \frac{a}{2\cos\varphi_0}; \quad \overline{S}_1 = \frac{a}{\cos\varphi_1}; \quad \overline{S}_2 = \frac{a}{\cos\varphi_2}; \quad \overline{S}_3 = \frac{a}{\cos\varphi_3}$$

$$\overline{S}_4 = \frac{a}{2\cos\varphi_4}$$

and since $J_i = \dfrac{J_c}{\cos\varphi_i}$, consequently

$$\frac{\overline{S}_0}{J_0} = \frac{\overline{S}_4}{J_4} = \frac{a}{2J_c} = \frac{3}{2J_c}; \quad \frac{\overline{S}_1}{J_1} = \frac{\overline{S}_2}{J_2} = \frac{\overline{S}_3}{J_3} = \frac{a}{J_c} = \frac{3}{J_c}$$

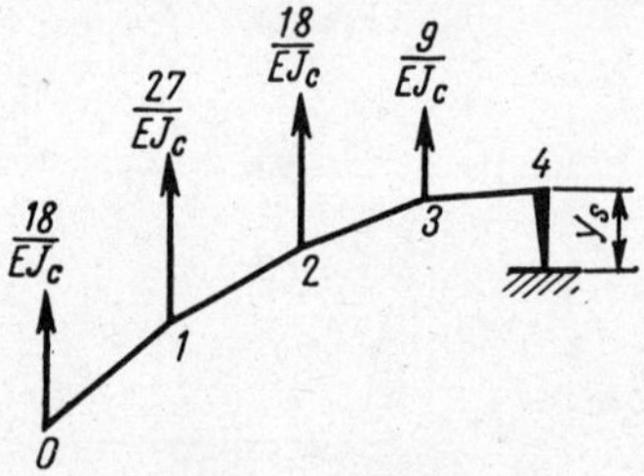

Fig. 22.11

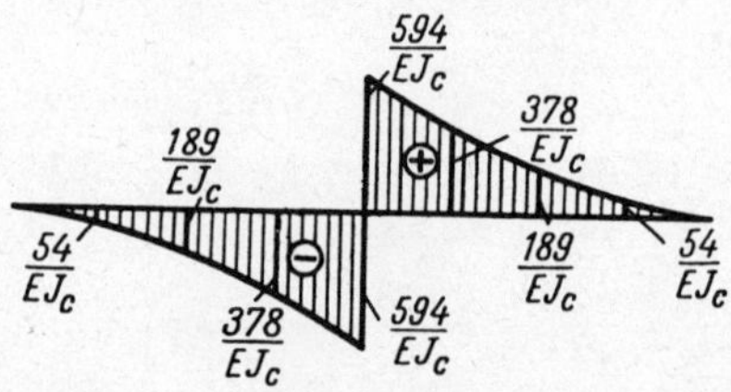

Fig. 23.11

The elastic loads corresponding to the application of unit forces X_2 to the simple structure will be given by (see Fig. 9.11*b*)

$$W_0 = M_0 \frac{\overline{S}_0}{EJ_0} = -12 \cdot \frac{3}{2EJ_c} = -\frac{18}{EJ_c}$$

$$W_1 = M_1 \frac{\overline{S}_1}{EJ_1} = -9 \cdot \frac{3}{EJ_c} = -\frac{27}{EJ_c}$$

$$W_2 = M_2 \frac{\overline{S}_2}{EJ_2} = -6 \cdot \frac{3}{EJ_c} = -\frac{18}{EJ_c}$$

$$W_3 = M_3 \frac{\overline{S}_3}{EJ_3} = -3 \cdot \frac{3}{EJ_c} = -\frac{9}{EJ_c}$$

$$W_4 = M_4 \frac{\overline{S}_4}{EJ_4} = 0 \cdot \frac{3}{2EJ_c} = 0$$

The deflections δ_{p2} will be obtained applying these loads to the imaginary structure of Fig. 22.11 and computing the bending moments induced by these loads

$$\delta_{p2}^{(1)} = \frac{18 \times 3}{EJ_c} = \frac{54}{EJ_c}$$

$$\delta_{p2}^{(2)} = \frac{18 \times 6 + 27 \times 3}{EJ_c} = \frac{189}{EJ_c}$$

$$\delta_{p2}^{(3)} = \frac{18 \times 9 + 27 \times 6 + 18 \times 3}{EJ_c} = \frac{378}{EJ_c}$$

$$\delta_{p2}^{(4)} = \frac{18 \times 12 + 27 \times 9 + 18 \times 6 + 9 \times 3}{EJ_c} = \frac{594}{EJ_c}$$

Setting off the bending moment values thus obtained on the side of the extended fibres of the imaginary structure, we shall obtain the deflection graph for δ_{2p} appearing in Fig. 23.11.

The displacement δ_{22} is the double of $\delta_{p2}^{(4)}$. Thus, δ_{22} equals $2\frac{594}{EJ_c}$ or $\frac{1,188}{EJ_c}$.

Influence line for X_2

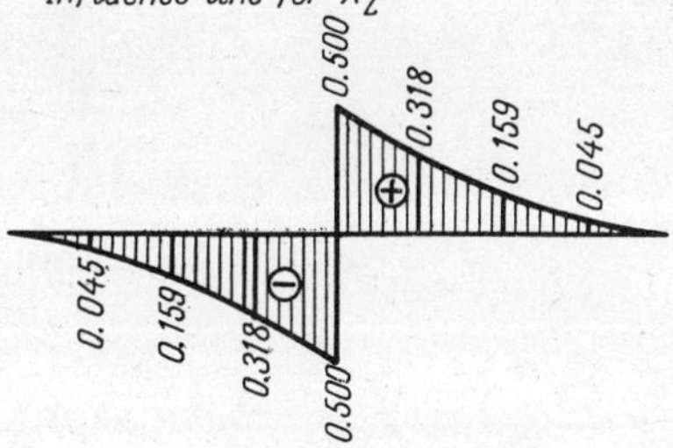

Fig. 24.11

The influence line for X_2 will be obtained dividing all the ordinates to the δ_{p2} graph by $(-\delta_{22})$ because $X_2 = -\frac{\delta_{p2}}{\delta_{22}}$. This influence line appears in Fig. 24.11.

It is practically the same as the one obtained by the first of the methods described in the present article and represented in Fig. 20.11*c*.

5.11. *Effect of Shrinkage and Temperature Changes on Fixed End Reinforced Concrete Arches*

Temperature changes. Every temperature change leads to the appearance of stresses in fixed end arches. Let us establish the expressions permitting to predict these stresses. For this purpose let us assume that the temperature at the extrados has been increased by t_1° while that at the intrados by t_2° (Fig. 25.11). We shall also admit that within the thickness of the arch the temperature varies linearly and therefore the increase in temperature at the neutral line (provided the latter coincides with the centre line of the cross section) will equal $\frac{t_1+t_2}{2}$.

For the simplicity let us denote the difference $(t_1 - t_2)$ by Δt and half the sum of these temperatures by t. The simultaneous equations expressing that the displacements along the redundant constraints transferred to the elastic centre of the arch are nil (Fig. 26.11) become in this case

$$\left.\begin{aligned} X_{1t}\delta_{11}+\Delta_{1t}&=0\\ X_{2t}\delta_{22}+\Delta_{2t}&=0\\ X_{3t}\delta_{33}+\Delta_{3t}&=0 \end{aligned}\right\} \qquad (9.11)$$

Since all the displacements of the arch caused by the said temperature change are symmetrical, the displacement Δ_{t2} must also equal zero and consequently X_{2t} is equally nil. Using

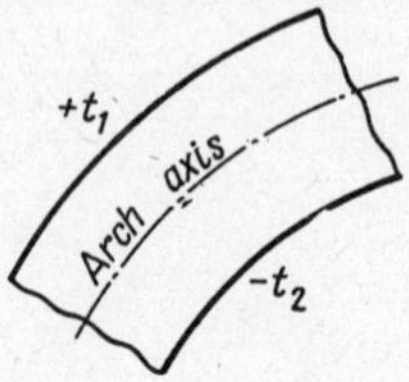

Fig. 25.11

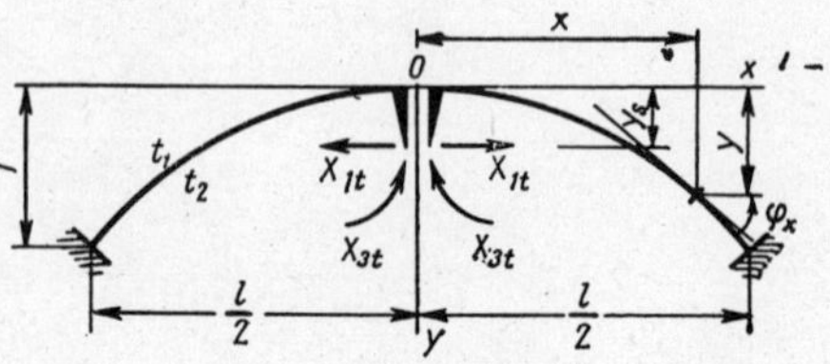

Fig. 26.11

the expressions developed in Art. 7.8 we may determine the deflections due to a temperature change which are given by

$$\Delta_{1t}=\alpha\,(t_1-t_2)\int_s \overline{M}_1\,\frac{ds}{h}+\alpha\,\frac{t_1+t_2}{2}\int_s \overline{N}_1 ds$$

$$\Delta_{3t}=\alpha\,(t_1-t_2)\int_s \overline{M}_3\,\frac{ds}{h}$$

or

$$\Delta_{1t}=-\alpha\Delta t\int_s (y-y_s)\,\frac{ds}{h}-\alpha t\int_s \cos\varphi_x ds$$

$$\Delta_{3t}=-\alpha\Delta t\int_s \frac{ds}{h}$$

In these expressions h represents the thickness of the arch, and α is the coefficient of thermal expansion. The value of δ_{11} (with due regard to the influence of normal stresses) is given by

$$\delta_{11}=\int_s (y-y_s)^2\frac{ds}{EJ}+\int_s \cos^2\varphi_x\,\frac{ds}{EF}$$

while the value of δ_{33} is provided by the expression

$$\delta_{33}=\int\frac{ds}{EJ}$$

Introducing the values of Δ_{1t}, Δ_{3t}, δ_{11} and δ_{33} in equations (9.11) we obtain

$$X_{1t}=-\frac{\Delta_{1t}}{\delta_{11}}=\frac{\alpha\Delta t\int_s (y-y_s)\,\frac{ds}{h}+\alpha t\int_s \cos\varphi_x ds}{\int_s (y-y_s)^2\,\frac{ds}{EJ}+\int_s \cos^2\varphi_x\,\frac{ds}{EF}}$$

$$X_{3t} = -\frac{\Delta_{3t}}{\delta_{33}} = \frac{\alpha \Delta t \int\limits_s \frac{ds}{h}}{\int\limits_s \frac{ds}{EJ}}$$

The fixed end moment due to a temperature change can be computed using the expression

$$M_{At} = X_{3t} + X_{1t}(f - y_s)$$

Let us examine a parabolic arch whose neutral line follows a curve given by $y = \frac{4f}{l^2}x^2$ and whose cross-sectional moments of inertia and thickness vary in accordance with *

$$J = \frac{J_c}{\cos \varphi_x} \qquad h = \frac{h_c}{\cos \varphi_x}$$

In these expressions J_c and h_c represent respectively the moment of inertia and the thickness of the arch at the crown section. Forces X_{1t} and X_{3t} will be given in that case by the following expressions **

$$X_{1t} = \frac{\alpha t l}{(1+\mu)\frac{4}{45} f^2 \frac{1}{EJ_c}} = \frac{45\alpha t EJ_c}{4(1+\mu) f^2}$$

$$X_{3t} = \frac{1}{h_c} \alpha \Delta t EJ_c \qquad (10.11)$$

where $$\mu = \frac{\int\limits_s \cos^2 \varphi_x \frac{ds}{F}}{\int\limits_s (y - y_s)^2 \frac{ds}{J}}$$

The first of the expressions (10.11) indicates that the thrust due to a temperature change increases together with the rigidity of the arch and with the reduction of its rise. On the contrary, a reduction in the cross-sectional dimensions and the use of the materials with a lower modulus of elasticity will reduce the stresses caused by temperature changes.

* For a rectangular arch of constant depth these two expressions are contradictory, the first leading to $h = \frac{h_c}{\sqrt[3]{\cos \varphi_x}}$. Nevertheless the error introduced by the assumption that h equals $\frac{h_c}{\cos \varphi_x}$ is negligible, especially in the case of flat arches.

** It will be remembered that for this type of arches $y_s = \frac{f}{3}$

Shrinkage. Stresses set up in a reinforced concrete arch by the shrinkage of concrete can be calculated in the same way as those due to a temperature change. Indeed if α were the coefficient of thermal expansion of concrete, αt would represent the strain per unit length caused by a change in temperature equal to t°C.

The shrinkage of concrete leads to a shortening of all linear dimensions by approximately 0.025 per cent. If we admit that the coefficient of thermal expansion of concrete is equal to 0.00001, the shrinkage may be regarded as equivalent to a drop in temperature of about 25°C. In actual design practice this is usually reduced to 10 or 15°C, for in reality the arches are cast section by section and therefore only a certain fraction of the total shrinkage must be taken into consideration. Thus, the stresses set up by shrinkage in a fixed end arch may be computed in exactly the same way as the stresses due to a drop in temperature from 10 to 15°C.

It is worth mentioning that there exist means and ways of compensating at least partially the shrinkage effect through artificial variation of internal stresses.

6.11. *Direct Computation of Parabolic Fixed End Arches*

When the neutral line of an arch follows a conic parabola and its cross-sectional moments of inertia vary in accordance with $J_x = \frac{J_c}{\cos \varphi_x}$, where J_c is the moment of inertia at the crown, all the coefficients to the unknowns and all the free terms of the simultaneous equations can be determined by direct computation. Hence, for this particular case it becomes possible to obtain mathematical relations between the redundant reactions acting at the elastic centre of the arch and the applied loads.

Let us find these relations assuming that a vertical load unity P acts a distance a from the left-hand abutment (Fig. 27.11*a*).

The canonical equations relating to the conjugate simple structure of Fig. 27.11*b* become

$$X_1\delta_{11} + \delta_{1p} = 0$$
$$X_2\delta_{22} + \delta_{2p} = 0$$
$$X_3\delta_{33} + \delta_{3p} = 0$$

wherefrom

$$\left.\begin{aligned} X_1 &= -\frac{\delta_{1p}}{\delta_{11}} \\ X_2 &= -\frac{\delta_{2p}}{\delta_{22}} \\ X_3 &= -\frac{\delta_{3p}}{\delta_{33}} \end{aligned}\right\} \tag{11.11}$$

The displacements δ_{1p}, δ_{2p} and δ_{3p} will be calculated neglecting the effect of normal stresses, that means, using the expression

$$\delta_{1p} = \int_0^a \overline{M}_1 \overline{M}_p \frac{ds}{EJ_x}$$

where $\overline{M}_p$ = bending moment induced in the simple structure by the load unity P

$\overline{M}_1$ = bending moment due to the unit reaction X_1.

Substituting the values of $\overline{M}_p$, $\overline{M}_1$, ds and EJ_x in the

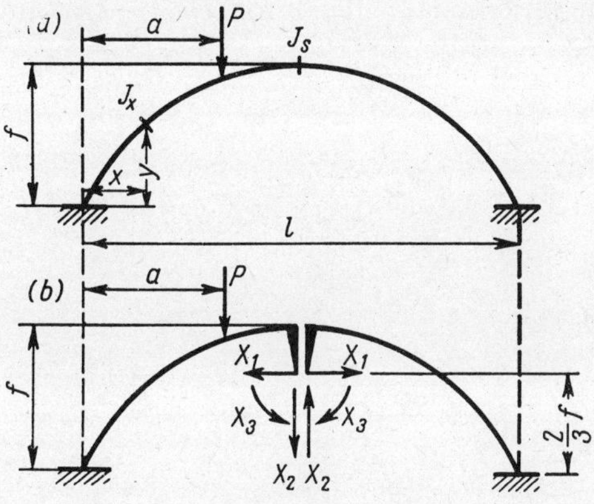

Fig. 27.11

expression for δ_{1p} and remembering that for this particular type of arches the elastic centre ordinate y_s equals $\frac{2f}{3}$ we obtain

$$\delta_{1p} = \int_0^a \overline{M}_1 \overline{M}_p \frac{ds}{EJ_x} = -\int_0^a \left(\frac{2}{3} f - y\right) 1 \times 1 \, (a - x) \frac{dx \cos \varphi_x}{\cos \varphi_x EJ_c} =$$

$$= \int_0^a (x - a) \left[\frac{2}{3} f - \frac{4f}{l^2} (l - x)\, x\right] \frac{dx}{EJ_c} =$$

$$= \frac{fa^2}{3EJ_c} \left(2 \frac{a}{l} - 1 - \frac{a^2}{l^2}\right)$$

since $y = \frac{4f}{l^2} (l - x)\, x$.

Denoting by η the ratio $\frac{a}{l}$ we obtain finally

$$\delta_{1p} = -\frac{fl^2}{3EJ_c} \eta^2 (\eta^2 - 2\eta + 1)$$

The value of δ_{11} is given by

$$\delta_{11} = \int_0^l \overline{M}_1^2 \frac{ds}{EJ_x} = \int_0^l \left(\frac{2}{3}f - y\right)^2 \frac{dx}{EJ_c} = \frac{4}{45} \cdot \frac{f^2 l}{EJ_c}$$

Introducing δ_{1p} and δ_{11} in expression (11.11) for X_1 we obtain immediately

$$X_1 = -\frac{\delta_{1p}}{\delta_{11}} = +\frac{15}{4} \times \frac{l}{f} \eta^2 (\eta^2 - 2\eta + 1) \tag{12.11}$$

This expression can be conveniently used for the construction of the influence line for X_1. The ordinates to this line for the left semiarch will be obtained varying η from 0 to 0.5. For the right semiarch they will be symmetrical to those already found.

Next let us determine the value of displacement δ_{2p}

$$\delta_{2p} = \int_0^a \overline{M}_2 \overline{M}_p \frac{ds}{EJ_x} = + \int_0^a \left(\frac{l}{2} - x\right)(a - x) \frac{dx}{EJ_c} =$$

$$= \frac{l^3}{EJ_c} \eta^2 \left(\frac{\eta}{6} - \frac{1}{4}\right)$$

where η represents as previously the $\frac{a}{l}$ ratio.

The value of δ_{22} is obtained from

$$\delta_{22} = 2 \int_0^{\frac{l}{2}} \overline{M}_2^2 \frac{ds}{EJ_x} = 2 \int_0^{\frac{l}{2}} \left(\frac{l}{2} - x\right)^2 \frac{dx}{EJ_c} = \frac{l^3}{12EJ_c}$$

hence

$$X_2 = -\frac{\delta_{2p}}{\delta_{22}} = -12\eta^2 \left(\frac{\eta}{6} - \frac{1}{4}\right) \tag{13.11}$$

Using the latter expression we may construct the influence line for X_2 varying again η from 0 to 0.5. The ordinates to the same line for $0.5 \leqslant \eta \leqslant 1$ will be equal in value but opposite in sign for the right-hand part of the influence line for X_2 is antisymmetrical with reference to its left-hand part.

Now determine δ_{3p}

$$\delta_{3p} = \int_0^a \overline{M}_3 \overline{M}_p \frac{ds}{EJ_x} = -\int_0 1 (a - x) \frac{dx}{EJ_c} = -\frac{a^2}{2EJ_c}.$$

As for δ_{33} it equals

$$\delta_{33} = \int_0^l \overline{M}_3^2 \frac{ds}{EJ_x} = \int_0^l 1 \frac{dx}{EJ_c} = \frac{l}{EJ_c}$$

which leads to

$$X_3 = -\frac{\delta_{3p}}{\delta_{33}} = \frac{l\eta^2}{2} \tag{14.11}$$

This equation permits the construction of the X_3 influence line for the left-hand semiarch. This influence line will be symmetrical with reference to the vertical passing through the crown.

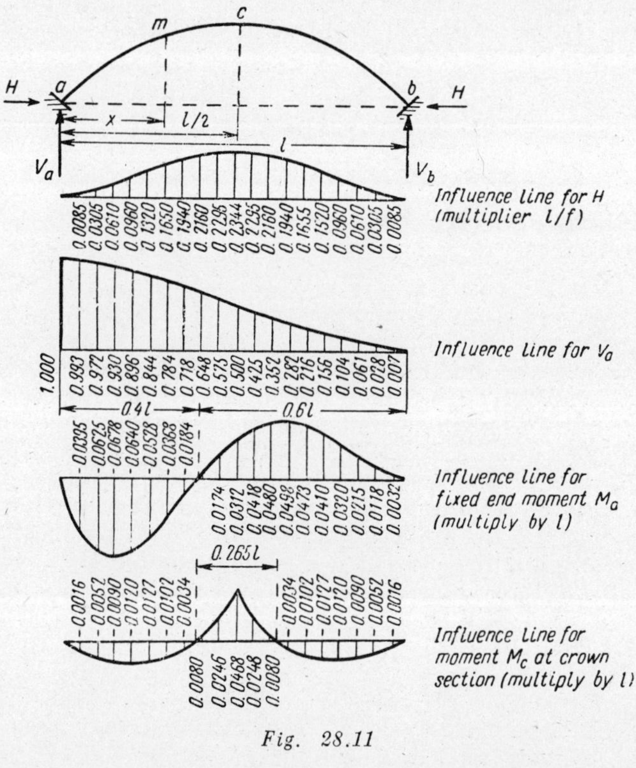

Fig. 28.11

The equations giving the values of the redundant reactions X_1, X_2 and X_3 being known, it becomes possible to find by direct computation the stresses acting at any cross section of the arch or to construct the influence lines for these same stresses provided the applied loads remain vertical.

Fig. 28.11 represents the influence lines for the thrust H, for the vertical reaction V_a and for the fixed end moment M_a acting at the left-hand abutment as well as the influence line for the bending moment M_c acting at the crown section. The ordinates to these influence lines have been found using the above equations for the redundant reactions applied at the elastic centre.

These influence lines (Fig. 28.11) may be used for the design of all fixed end arches whose centre line follows a conic parabola and whose cross-sectional moments of inertia vary in accordance with the relation specified at the beginning of this article. These influence lines permit the determination of stresses acting at any cross section of the arch as long as the loads remain vertical. The following procedure should be adopted for these computations.

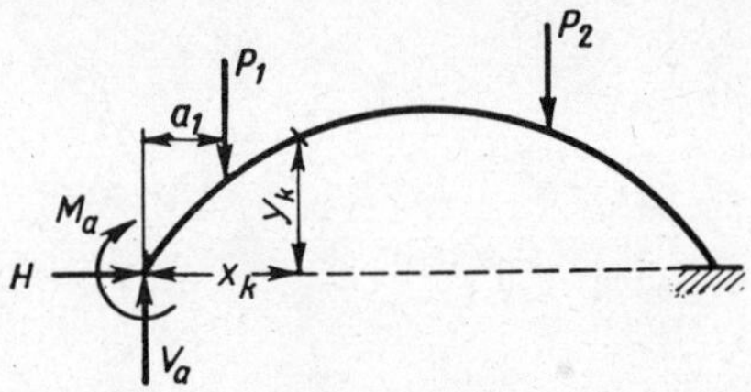

Fig. 29.11

First determine the values of V_a, H and M_a for the given system of loads multiplying each of the latter by the corresponding ordinate to the appropriate influence line. For uniformly distributed loads, their intensity will be multiplied by the areas bounded by the segments of the influence lines. The values of V_a, H and M_a will be then applied to the left end of the arch liberated previously from all the existing constraints. Thereafter the stresses at any section will be easily computed assuming that the arch is a statically determinate curved beam built in at its right end and acted upon both by the applied loads and by the redundant reactions determined as explained previously and applied to the left end.

Thus, for instance, if the fixed end arch were acted upon by two concentrated loads P_1 and P_2 (Fig. 29.11) the bending moment in any arbitrary section K will be given by

$$M_k = V_a x_k - H y_k + M_a - P_1 (x_k - a_1)$$

Problem. A parabolic arch (Fig. 30.11*a*) is acted upon by two vertical loads $P_1 = 10$ tons and $P_2 = 20$ tons as well as by a uniform load of two tons per metre distributed over the quarter span situated immediately to the left of the crown. Required: (1) the thrust H, the vertical reaction V_a and the fixed end moment M_a; (2) the bending moment, the shear and the normal stress acting at the crown section. The neutral line of the arch follows the equation

$$y = \frac{4f}{l_2} (l - x)\, x = 0.2\,(10 - x)\, x$$

and the cross-sectional moments of inertia are given by

$$J_x = \frac{J_c}{\cos \varphi_x}$$

Solution. Start by constructing the influence line for the thrust H. For this purpose multiply all the ordinates to the influence line of Fig. 28.11 by the $\frac{l}{f}$ ratio which in this particular case equals 2 for $l = 10$ metres and $f = 5$ metres. The influence line for the thrust H obtained as just explained is shown in Fig. 30.11*b*.

The influence line for M_a will be obtained by multiplying all the ordinates to the appropriate influence line of Fig. 28.11 by the span length l. The completed influence line is shown in Fig. 30.11*c*.

The ordinates to the influence line for V_a in the event of a parabolic arch are independent of the $\frac{l}{f}$ ratio and therefore the influence line given in Fig. 28.11 may be used without any alterations.

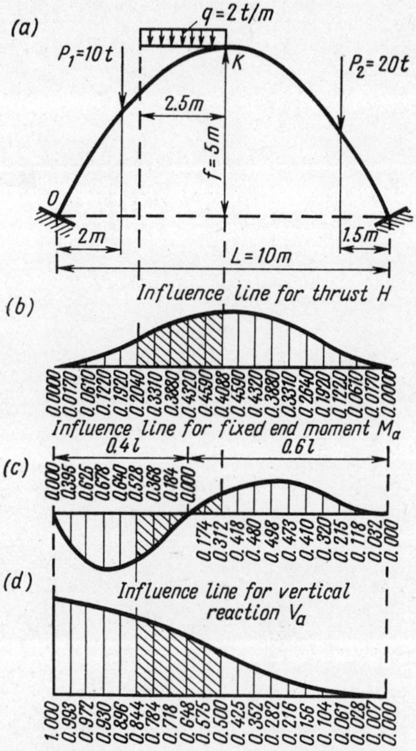

Fig. 30.11

Let us determine now the thrust due to the loads indicated in Fig. 30.11*a*. Its value will be equal to the sum of the product of each concentrated load by the corresponding influence line ordinate with the product of the area bounded by the influence line over that portion of the arch carrying the distributed load by the intensity of the latter. The area mentioned may be calculated approximately replacing curvilinear segments of the influence line between two neighbouring ordinates by straight lines. If the distance a separating these ordinates remains constant, the area bounded by two ordinates h_n and h_m will be given by

$$(\omega)_n^m = a\left(\frac{h_n}{2} + h_{n+1} + h_{n+2} + \ldots + h_{m-1} + \frac{h_m}{2}\right)$$

Hence the thrust H will equal

$$H = P_1 0.1920 + P_2 0.1220 + q\left(\frac{0.2640}{2} + 0.3310 + 0.3880 + \right.$$

$$\left. + 0.4320 + 0.4590 + \frac{0.4688}{2}\right) 0.5 = 6.3364 \text{ tons}$$

The magnitude of the vertical reaction V_a will be obtained in exactly the same way

$$V_a = P_1 0.896 + P_2 0.061 + q\left(\frac{0.844}{2} + 0.784 + 0.718 + \right.$$

$$\left. + 0.648 + 0.575 + \frac{0.500}{2}\right) 0.5 = 12.577 \text{ tons}$$

As for the bending moment M_a it will amount to

$$M_a = P_1(-0.640) + P_2 0.215 + q\left(\frac{-0.528}{2} - 0.368 - 0.184 + \right.$$

$$\left. + 0.000 + 0.174 + \frac{0.312}{2}\right) 0.5 = -2.586 \text{ ton metres}$$

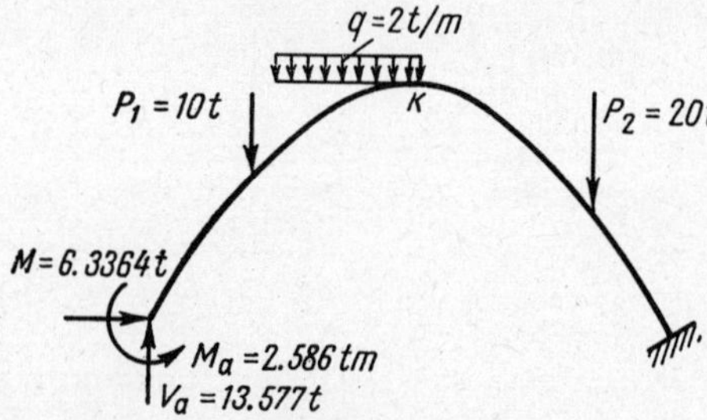

Fig. 31.11

This being done, liberate the left end of the arch from all constraints and replace the latter by the reactive forces just found (Fig. 31.11). The bending moment acting at the crown section will then equal

$$M_k = -M_a + V_a\frac{l}{2} - Hf - P_1\left(\frac{l}{2} - 2\right) - q\frac{l}{2}\cdot\frac{l}{8} =$$

$$= -2.586 + 13.577\times 5 - 6.3364\times 5 - 10\cdot 3 - \frac{2\times 100}{32} = -2.633 \text{ ton metres}$$

The normal stress acting across the same section will be obtained projecting all the forces to the left of this cross section on the horizontal: $N_k = H = 6.3364$ tons (compression).

The shear will be given by the vertical projection of the same forces

$$Q_k = V_a - P_1 - q\frac{l}{4} = 13.577 - 10 - \frac{2\times 10}{4} = -1.423 \text{ tons}$$

In actual practice the design of redundant arches for bridge construction and elsewhere is frequently carried out with the aid of special tables. Such tables have been prepared for widely varying geometrical parameters of arches such as their $\frac{l}{f}$ ratio, the law governing the variation of their cross-sectional dimensions, etc. The use of such tables reduces very considerably the time required for computation work and thereby eliminates in a large measure the risk of errors always present when calculations are long and laborious.

7.11. *Two-Hinged Arches*

In the case of two-hinged arches the stress analysis is usually carried out adopting for simple statically determinate structure

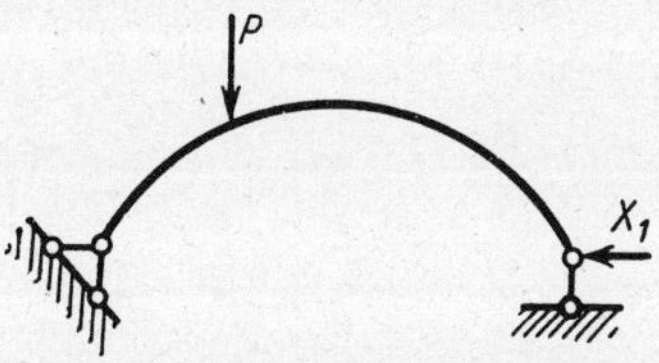

Fig. 32.11

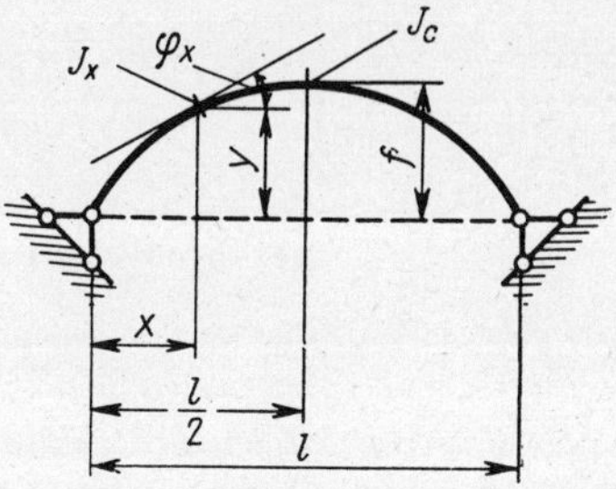

Fig. 33.11

the curved bar shown in Fig. 32.11. The equation expressing that the horizontal displacement along the direction of X_1 is nil becomes

$$X_1\delta_{11}+\Delta_{1p}=0$$

For flat arches the values of δ_{11} and Δ_{1p} will be calculated with due consideration to the effect of normal stresses, i. e., using the expressions

$$\delta_{11}=\Sigma\int_0^s \overline{M}_1^2\frac{ds}{EJ}+\Sigma\int_0^s \overline{N}_1^2\frac{ds}{EF}$$

$$\Delta_{1p}=\Sigma\int_0^s \overline{M}_1M_p\frac{ds}{EJ}+\Sigma\int_0^s \overline{N}_1N_p\frac{ds}{EF}$$

In case direct integration of these expressions becomes too complicated resort should be made to numerical methods or to the method of elastic loads.

Cross-sectional moments of inertia in two-hinged arches remain constant or vary in accordance with

$$J_x=J_c\cos\varphi_x$$

where J_c is the moment of inertia at the crown section (Fig. 33.11). Alternatively the cross-sectional areas of two-hinged arches may vary following the expression

$$F_x=F_c\cos\varphi_x$$

where F_c is again the cross-sectional area at the crown.

12. ANALYSIS OF HIGHLY REDUNDANT STRUCTURES

1.12. *Use of Symmetry*

When analysing the stresses arising in structures with a large number of redundant constraints one is usually called upon to solve a number of simultaneous equations equal to the structure's degree of redundancy, each of these equations containing the same number of unknowns.

Let us investigate, for instance, the frame appearing in Fig. 1.12 which consists of two closed contours and consequently is redundant to the sixth degree. If all the computations pertaining to this frame were carried out, adopting as conjugate statically determinate structure the one given in Fig. 2.12, it would be necessary to form and to solve a system of six simultaneous equations with six unknowns each

$$\begin{aligned}
X_1\delta_{11}+X_2\delta_{12}+X_3\delta_{13}+X_4\delta_{14}+X_5\delta_{15}+X_6\delta_{16}+\Delta_{1p}&=0\\
X_1\delta_{21}+X_2\delta_{22}+X_3\delta_{23}+X_4\delta_{24}+X_5\delta_{25}+X_6\delta_{26}+\Delta_{2p}&=0\\
X_1\delta_{31}+X_2\delta_{32}+X_3\delta_{33}+X_4\delta_{34}+X_5\delta_{35}+X_6\delta_{36}+\Delta_{3p}&=0\\
X_1\delta_{41}+X_2\delta_{42}+X_3\delta_{43}+X_4\delta_{44}+X_5\delta_{45}+X_6\delta_{46}+\Delta_{4p}&=0\\
X_1\delta_{51}+X_2\delta_{52}+X_3\delta_{53}+X_4\delta_{54}+X_5\delta_{55}+X_6\delta_{56}+\Delta_{5p}&=0\\
X_1\delta_{61}+X_2\delta_{62}+X_3\delta_{63}+X_4\delta_{64}+X_5\delta_{65}+X_6\delta_{66}+\Delta_{6p}&=0
\end{aligned} \qquad (1.12)$$

The solution of such a system of equations would be extremely laborious and would require a lot of time. The work can be simplified very considerably due to the *symmetry* of the structure. It should be remembered that *in a symmetrical structure not only the arrangement of its members but also their cross-sectional rigidities are symmetrical about a certain axis.* The simplification is based on the possibility of finding a conjugate statically determinate structure for which the $\overline{M}_i$ diagram for each redundant reaction $X_i = 1$ will be either symmetrical or antisymmetrical.

Thus, for the frame under consideration (see Fig. 1.12) one could adopt for simple structure the one appearing in Fig. 3.12*a*.

In this case the $\overline{M}_2$, $\overline{M}_3$, $\overline{M}_5$ and $\overline{M}_6$ diagrams due to symmetrical unit forces X_2, X_3, X_5 and X_6 would be themselves symmetrical (Fig. 3.12*c*, *d*, *f* and *g*) while the $\overline{M}_1$ and the $\overline{M}_4$ diagrams induced by antisymmetrical unit forces X_1 and X_4 would be also antisymmetrical (Fig. 3.12*b* and *e*).

It is well known that the product of a symmetrical graph by antisymmetrical one is always nil. Thus, for instance, if one were

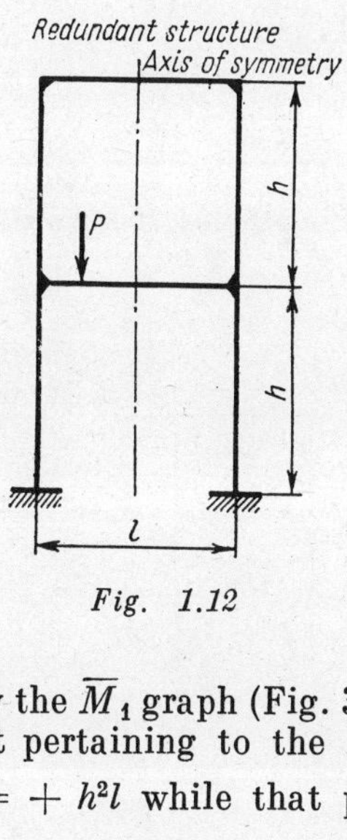

Fig. 1.12

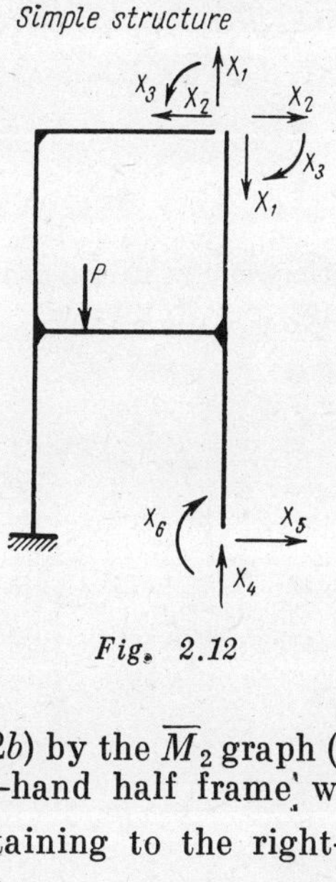

Fig. 2.12

to multiply the $\overline{M}_1$ graph (Fig. 3.12*b*) by the $\overline{M}_2$ graph (Fig. 3.12*c*) the product pertaining to the left-hand half frame would equal $+2h\frac{2h}{2}\frac{l}{2}=+h^2l$ while that pertaining to the right-hand half frame would amount to $-2h\frac{2h}{2}\frac{l}{2}=-h^2l$. Consequently, the displacement δ_{12} equal to the algebraic sum of these two amounts will be nil. For the same reason all the other secondary displacements whose values are obtained multiplying symmetrical graphs by antisymmetrical ones will also reduce to zero.

For the frame under consideration such will be the case for displacements δ_{12}, δ_{13}, δ_{15}, δ_{16}, δ_{21}, δ_{24}, δ_{31}, δ_{34}, δ_{42}, δ_{43}, δ_{45}, δ_{46}, δ_{51}, δ_{54}, δ_{61} and δ_{64}. It follows that the system of simultaneous equations would become

$$X_1\delta_{11}+X_4\delta_{14}+\Delta_{1p}=0$$

$$X_2\delta_{22}+X_3\delta_{23}+X_5\delta_{25}+X_6\delta_{26}+\Delta_{2p}=0$$

$$X_2\delta_{32}+X_3\delta_{33}+X_5\delta_{35}+X_6\delta_{36}+\Delta_{3p}=0$$

$$X_1\delta_{41}+X_4\delta_{44}+\Delta_{4p}=0$$
$$X_2\delta_{52}+X_3\delta_{53}+X_5\delta_{55}+X_6\delta_{56}+\Delta_{5p}=0$$
$$X_2\delta_{62}+X_3\delta_{63}+X_5\delta_{65}+X_6\delta_{66}+\Delta_{6p}=0$$

and would consequently fall into two independent systems

$$\left.\begin{aligned} X_1\delta_{11}+X_4\delta_{14}+\Delta_{1p}=0\\ X_1\delta_{41}+X_4\delta_{44}+\Delta_{4p}=0 \end{aligned}\right\} \quad (2.12)$$

$$\left.\begin{aligned} X_2\delta_{22}+X_3\delta_{23}+X_5\delta_{25}+X_6\delta_{26}+\Delta_{2p}=0\\ X_2\delta_{32}+X_3\delta_{33}+X_5\delta_{35}+X_6\delta_{36}+\Delta_{3p}=0\\ X_2\delta_{52}+X_3\delta_{53}+X_5\delta_{55}+X_6\delta_{56}+\Delta_{5p}=0\\ X_2\delta_{62}+X_3\delta_{63}+X_5\delta_{65}+X_6\delta_{66}+\Delta_{6p}=0 \end{aligned}\right\} \quad (3.12)$$

The first of these systems contains two equations with two antisymmetrical unknowns and the second four equations with four symmetrical unknowns.

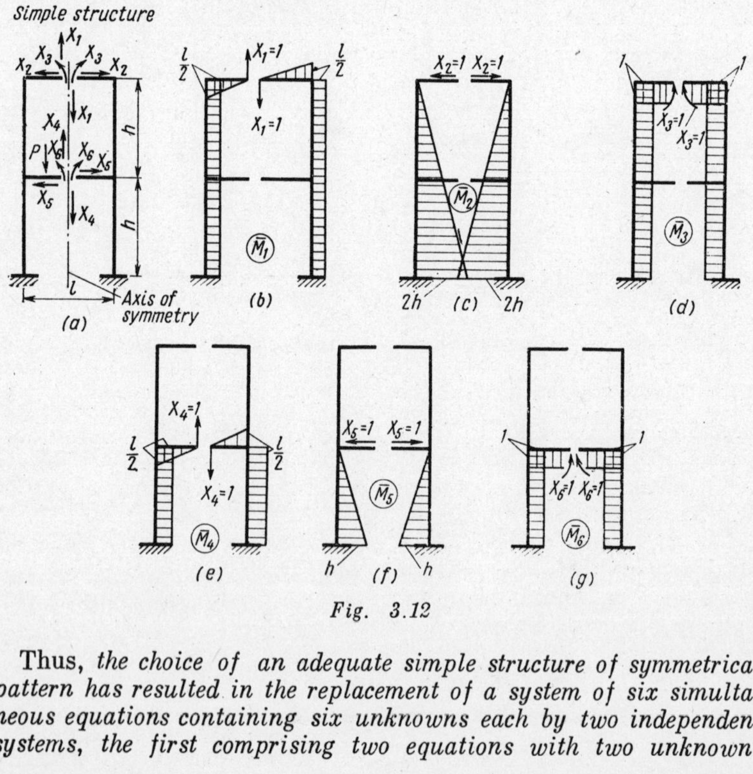

Fig. 3.12

Thus, *the choice of an adequate simple structure of symmetrical pattern has resulted in the replacement of a system of six simultaneous equations containing six unknowns each by two independent systems, the first comprising two equations with two unknowns and the second four equations with four unknowns. This simplifies enormously the computations, enhancing at the same time very considerably the precision of the results obtained.* An additional reduction of computation work has been obtained due to the fact that all the displacements can be calculated for one half of the

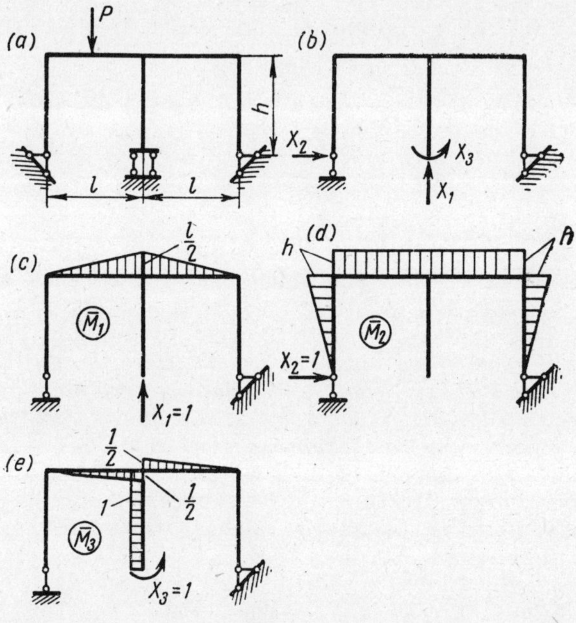

Fig. 4.12

conjugate simple structure only. The total displacement will be the double of that for the half-frame.

If the symmetrical frame contains a central column, the displacement due to the redundant antisymmetrical reactions can be computed multiplying at first the diagrams relative to one half of the frame (without the central column); whereafter the product obtained should be doubled and increased by the product of graph multiplication pertaining to the central column.

Another example of a symmetrical frame is afforded by the frame shown in Fig. 4.12*a*. It is easily seen that this frame is redundant to the third degree. Fig. 4.12*b* represents one of the conjugate simple structures which could be adopted in the present case. This structure is not symmetrical for the lower ends of the extreme columns have different supports. It follows that the redundant reactions X_1, X_2 and X_3 themselves will be also nonsymmetrical. Nevertheless the diagrams of the bending moments induced in this simple structure by unit reactions X_1 and X_2 (Fig. 4.12*c* and *d*) will be symmetrical while the diagram of the bending moment due to $X_3 = 1$ (Fig. 4.12*e*) will be antisymmetrical.

Consequently, for the given simple structures the simultaneous equations will again fall into two different independent groups

$$X_1\delta_{11} + X_2\delta_{12} + \Delta_{1p} = 0$$
$$X_1\delta_{21} + X_2\delta_{22} + \Delta_{2p} = 0 \text{ and } X_3\delta_{33} + \Delta_{3p} = 0$$

just as though all the unknowns were symmetrical or antisymmetrical.

2.12. *Grouping of the Unknowns*

If the structure which is being analyzed consists of several spans it becomes impossible to transfer the points of application of all the redundant reactions to the axis of symmetry (Fig. 5.12*a*). However, *symmetrical and antisymmetrical bending moment diagrams may still be obtained if the unknowns representing a single force or couple are replaced by unknowns representing whole groups of forces.* Let us examine, for example, the six times redundant frame appearing in Fig. 5.12*a*. Should we adopt for conjugate simple structure the one appearing in Fig. 5.12*b* we would have to solve six simultaneous equations with six nonsymmetrical unknowns X_1, X_2, X_3, X_4, X_5 and X_6 given below

$$\left.\begin{aligned}
X_1\delta_{11} + X_2\delta_{12} + X_3\delta_{13} + X_4\delta_{14} + X_5\delta_{15} + X_6\delta_{16} + \Delta_{1p} = 0\\
X_1\delta_{21} + X_2\delta_{22} + X_3\delta_{23} + X_4\delta_{24} + X_5\delta_{25} + X_6\delta_{26} + \Delta_{2p} = 0\\
X_1\delta_{31} + X_2\delta_{32} + X_3\delta_{33} + X_4\delta_{34} + X_5\delta_{35} + X_6\delta_{36} + \Delta_{3p} = 0\\
X_1\delta_{41} + X_2\delta_{42} + X_3\delta_{43} + X_4\delta_{44} + X_5\delta_{45} + X_6\delta_{46} + \Delta_{4p} = 0\\
X_1\delta_{51} + X_2\delta_{52} + X_3\delta_{53} + X_4\delta_{54} + X_5\delta_{55} + X_6\delta_{56} + \Delta_{5p} = 0\\
X_1\delta_{61} + X_2\delta_{62} + X_3\delta_{63} + X_4\delta_{64} + X_5\delta_{65} + X_6\delta_{66} + \Delta_{6p} = 0
\end{aligned}\right\} \quad (4.12)$$

In these equations none of the coefficients δ would normally equal zero.

On the other hand, if the groups of unknown forces Z_1, Z_2, Z_3, Z_4, Z_5 and Z_6 shown in Fig. 5.12*c* were adopted as the unknowns, a very large number of secondary displacements in the simultaneous equations (4.12) would reduce to zero, for these displacements (coefficients) would result from the multiplication of symmetrical graphs by antisymmetrical ones. Here the unknown Z_1 represents two horizontal forces X_1 and X_4 equal in value and opposite in sign, the unknown Z_2—two horizontal forces equal both in value and in sign, Z_3—two vertical forces of equal amount, both directed upwards, Z_4—two vertical forces equal in amount and opposite in direction, Z_5—two couples equal in amount and opposite in direction, and Z_6—two couples of the same magnitude and acting in the same direction. The bending moment diagrams due to the above groups of unit forces are given in Fig. 6.12. It is readily observed that the $\overline{M}_1$, $\overline{M}_3$ and $\overline{M}_5$ diagrams are symmetrical while the $\overline{M}_2$, $\overline{M}_4$ and $\overline{M}_6$ diagrams are antisymmetrical.

Comparing the two simple structures appearing in Fig. 5.12*b* and *c* we realize that the following relations exist between the unknowns X and Z *

$$X_1 = Z_1 + Z_2; \quad X_4 = Z_1 - Z_2$$
$$X_2 = Z_3 + Z_4; \quad X_5 = Z_3 - Z_4$$
$$X_3 = Z_5 + Z_6; \quad X_6 = Z_5 - Z_6$$

The above relations may be rewritten as follows

$$Z_1 = \frac{X_1 + X_4}{2}; \quad Z_4 = \frac{X_2 - X_5}{2}$$
$$Z_2 = \frac{X_1 - X_4}{2}; \quad Z_5 = \frac{X_3 + X_6}{2}$$
$$Z_3 = \frac{X_2 + X_5}{2}; \quad Z_6 = \frac{X_3 - X_6}{2}$$

Grouping the unknowns as indicated above permits the replacement of a single system formed by six simultaneous equations (4.12) by two independent ones (5.12) and (6.12), the first containing only symmetrical unknowns and the second the antisymmetrical ones.

(a) The first system

$$\left.\begin{aligned} Z_1(\delta_{11}) + Z_3(\delta_{13}) + Z_5(\delta_{15}) + (\Delta_{1p}) = 0 \\ Z_1(\delta_{31}) + Z_3(\delta_{33}) + Z_5(\delta_{35}) + (\Delta_{3p}) = 0 \\ Z_1(\delta_{51}) + Z_3(\delta_{53}) + Z_5(\delta_{55}) + (\Delta_{5p}) = 0 \end{aligned}\right\} \qquad (5.12)$$

(b) The second system

$$\left.\begin{aligned} Z_2(\delta_{22}) + Z_4(\delta_{24}) + Z_6(\delta_{26}) + (\Delta_{2p}) = 0 \\ Z_2(\delta_{42}) + Z_4(\delta_{44}) + Z_6(\delta_{46}) + (\Delta_{4p}) = 0 \\ Z_2(\delta_{62}) + Z_4(\delta_{64}) + Z_6(\delta_{66}) + (\Delta_{6p}) = 0 \end{aligned}\right\} \qquad (6.12)$$

◆

* The construction of stress diagrams does not require the determination of the unknowns belonging to the X group.

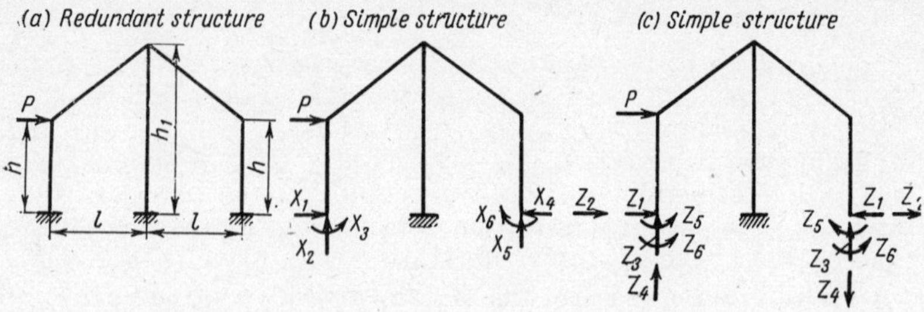

Fig. 5.12

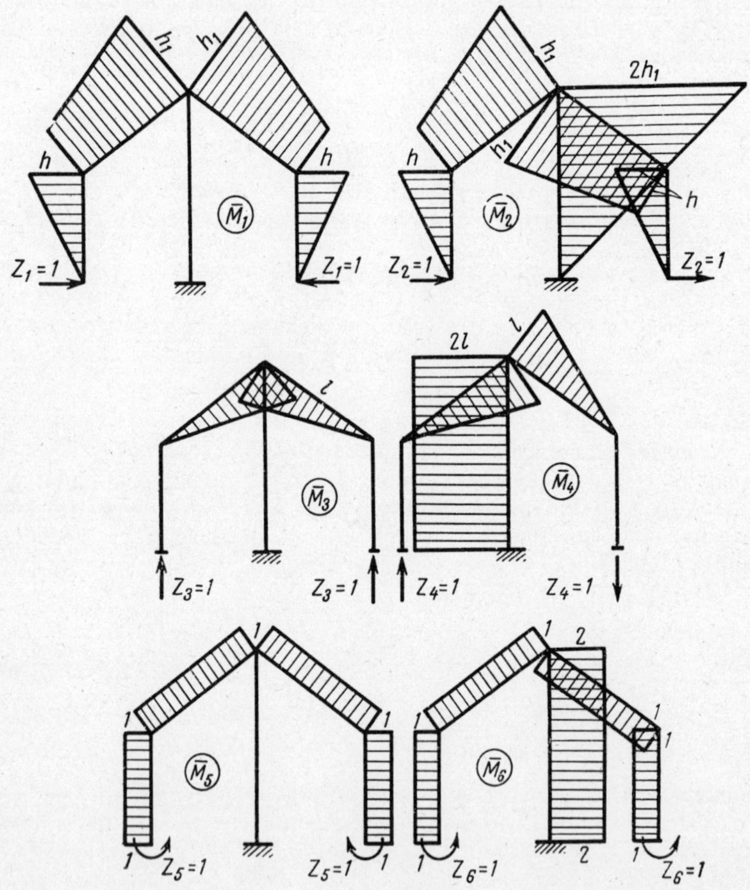

Fig. 6.12

It is clear that in the above expressions the coefficients (δ_{ik}) and the free terms (Δ_{ip}) represent the displacements induced by and along the aforesaid groups of unknown forces.

We have thus succeeded in reducing a system of six equations with six unknowns to two independent systems of three equations with three unknowns. The work required to solve the latter systems will be less important than that needed for the solution of the original one. Hereafter we shall denote by the sign X all the unknowns regardless of whether they represent single forces or whole groups of forces. We shall equally omit the parentheses introduced into equations (5.12) and (6.12) in order to distinguish the displacements due to groups of forces from those due to a single action.

The determination of displacements arising in a statically determinate structure under the action of groups of forces is no more complicated than that of the displacements produced by single actions. These displacements will be computed as usual, multiplying diagrams pertaining to one half of the structure and doubling the result obtained. If the frame contains a central column, graphs pertaining to one half-frame (excluding the central column) will be first multiplied one by the other, the product will then be doubled, and the result so obtained will be added to the product of graphs for the central column.

3.12. *Symmetrical and Antisymmetrical Loading*

If the system of loads acting on a structure is either symmetrical or antisymmetrical all the computations are further simplified, because in this case it becomes possible to find a conjugate simple structure for which the bending moment diagrams due both to the unit actions and to the actual loading become either symmetrical or antisymmetrical. As a result, a number of free terms of the simultaneous equations together with some coefficients to the unknowns will reduce to zero.

Let us examine the frame of Fig. 7.12*a*. This frame is redundant to the sixth degree and is acted upon by a system of symmetrical loads. A symmetrical statically determinate simple structure with symmetrical and antisymmetrical unknowns may be obtained cutting in two the upper crossbar and eliminating three constraints at the supports as indicated in Fig. 7.12*b*. The redundant reactions, which cannot be transferred to the axis of symmetry, such as the horizontal components of the left- and of the right-hand column reactions will be replaced by two groups of forces X_4 and X_5.

The bending moment diagrams for the conjugate structure due both to the unit actions following the direction of the unknowns and to the actual loads are shown in Fig. 7.12*c*, *d*, *e*, *f*, *h* and *i*. All the unknowns being either symmetrical (X_1, X_2, X_3

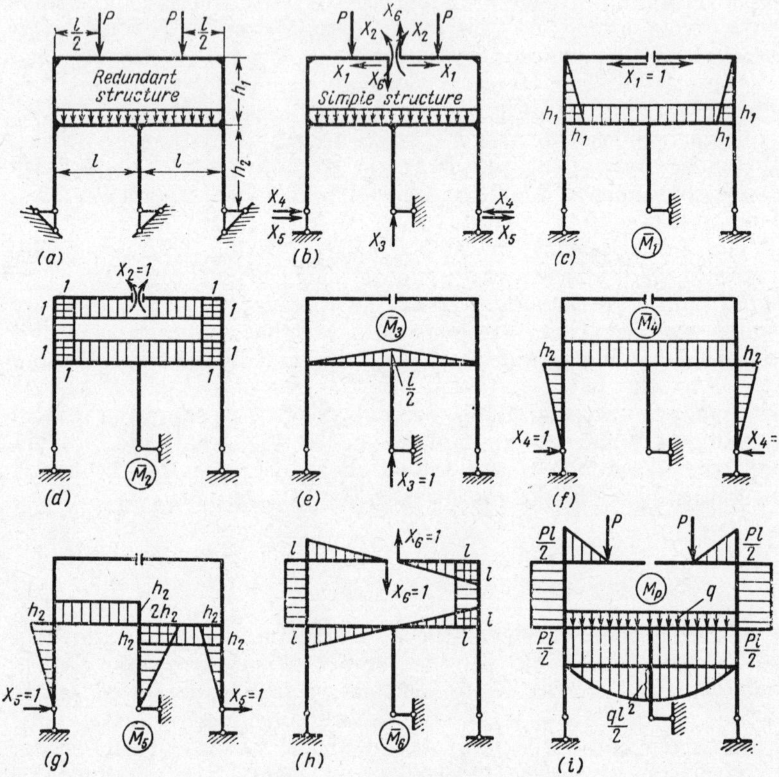

Fig. 7.12

and X_4) or antisymmetrical (X_5 and X_6), the simultaneous equations will form two independent systems given hereunder

$$
\begin{aligned}
X_1\delta_{11}+X_2\delta_{12}+X_3\delta_{13}+X_4\delta_{14}+\Delta_{1p}=0\\
X_1\delta_{21}+X_2\delta_{22}+X_3\delta_{23}+X_4\delta_{24}+\Delta_{2p}=0\\
X_1\delta_{31}+X_2\delta_{32}+X_3\delta_{33}+X_4\delta_{34}+\Delta_{3p}=0\\
X_1\delta_{41}+X_2\delta_{42}+X_3\delta_{43}+X_4\delta_{44}+\Delta_{4p}=0
\end{aligned}
$$

$$
\left.\begin{aligned}
X_5\delta_{55}+X_6\delta_{56}+\Delta_{5p}=0\\
X_5\delta_{65}+X_6\delta_{66}+\Delta_{6p}=0
\end{aligned}\right\} \qquad (7.12)
$$

In the last of the two systems the displacements Δ_{5p} and Δ_{6p} are both nil, their value being obtained multiplying antisymmetrical graphs $\overline{M}_5$ and $\overline{M}_6$ by the symmetrical graph M_p due to the applied loads (Fig. 7.12*i*). Consequently, these two equations become

$$X_5\delta_{55}+X_6\delta_{56}=0$$

$$X_5\delta_{65}+X_6\delta_{66}=0$$

which indicates that both the antisymmetrical unknowns X_5 and X_6 are also nil. Were the frame of Fig. 7.12*a* acted upon by a system of antisymmetrical loads it would be the symmetrical unknowns that would become nil.

Generalizing the above we may formulate the two following rules:

1. *When a symmetrical structure is acted upon by a symmetrical system of loads only those of the unknowns which represent the symmetrical redundant reactions remain different from zero*, while all the unknowns representing antisymmetrical reactions become nil.

2. *When a symmetrical structure is acted upon by an antisymmetrical system of loads, only those of the unknowns which represent the antisymmetrical redundant reactions remain different from zero*, while all the unknowns representing symmetrical unknowns become nil.

4.12. *Load Transformation*

The two rules formulated above are applicable to any symmetrical structure regardless of the actual distribution of loading, for the simple reason that any system of loads can be easily

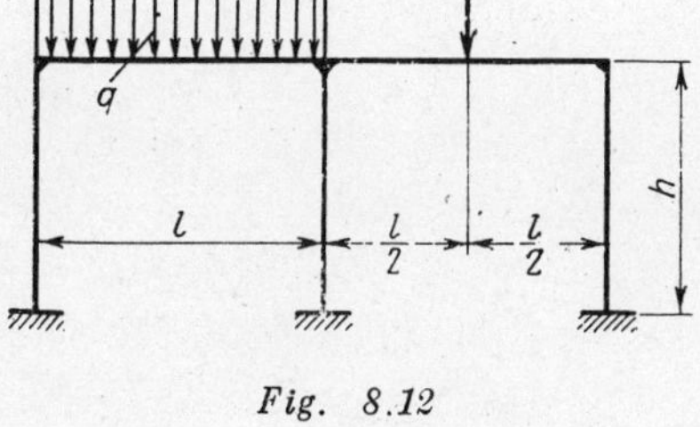

Fig. 8.12

replaced by an equivalent combination of two separate systems, one of which is symmetrical and the other antisymmetrical.

Indeed, let us consider the symmetrical structure of Fig. 8.12 acted upon by one concentrated load P and the uniformly distributed load of q kg per unit length, both are nonsymmetrical. The two loads may be replaced by two groups of components appearing in Fig. 9.12a and b, the first of these groups being symmetrical and the second antisymmetrical. It is clearly seen that the superposition of these two systems of loads leads to a loading absolutely identical to the original one shown in Fig. 8.12.

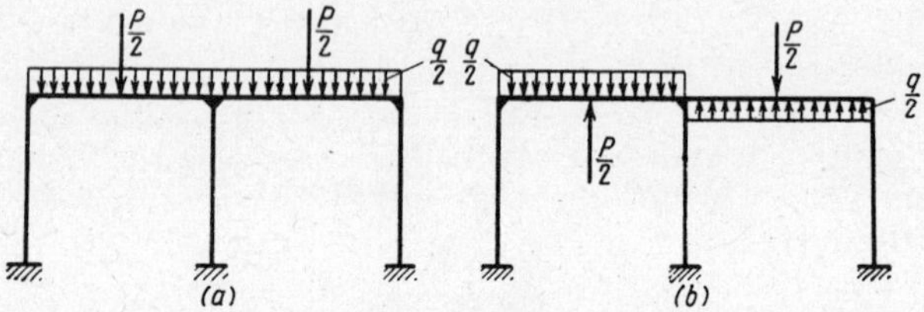

Fig. 9.12

We have seen in the preceding article that when a symmetrical structure is acted upon by symmetrical loads alone the symmetrical unknowns remain different from zero. Consequently, for the simple statically determinate structure (Fig. 10.12a) acted upon

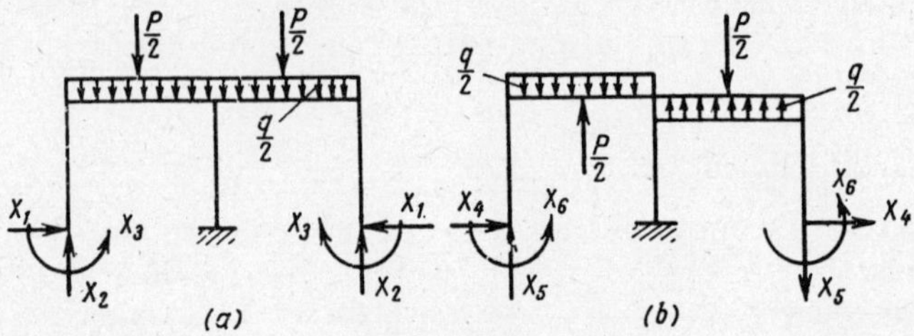

Fig. 10.12

by the loads appearing in Fig. 9.12a only the symmetrical unknowns X_1, X_2 and X_3 must be calculated. The corresponding equations become

$$X_1\delta_{11}+X_2\delta_{12}+X_3\delta_{13}+\Delta_{1p}=0$$
$$X_1\delta_{21}+X_2\delta_{22}+X_3\delta_{23}+\Delta_{2p}=0$$
$$X_1\delta_{31}+X_2\delta_{32}+X_3\delta_{33}+\Delta_{3p}=0$$

For the same reason the system of loads appearing in Fig. 9.12b applied to the simple structure of Fig. 10.12b will provide three antisymmetrical unknowns X_4, X_5 and X_6, all the other unknowns being nil.

Consequently, these equations become

$$X_4\delta_{44}+X_5\delta_{45}+X_6\delta_{46}+\Delta_{4p}=0$$
$$X_4\delta_{54}+X_5\delta_{55}+X_6\delta_{56}+\Delta_{5p}=0$$
$$X_4\delta_{64}+X_5\delta_{65}+X_6\delta_{66}+\Delta_{6p}=0$$

It should be observed, however, that in certain cases the replacement of loads by their symmetrical and antisymmetrical components may complicate the computations instead of simplifying them and consequently the application of this procedure cannot be recommended unconditionally.

Fig. 11.12*a* represents a symmetrical portal frame loaded by one nonsymmetrical force P. If we adopt for simple statically determinate structure the one shown in Fig. 11.12*b*, the simultaneous equations will fall into two independent systems, one containing a single antisymmetrical unknown X_1 and the other two symmetrical unknowns X_2 and X_3.

Nevertheless, it is much easier to construct one diagram of the bending moments due to the single load P (Fig. 11.12*c*) than two bending moment diagrams due to its symmetrical (Fig. 11.12*d*) and antisymmetrical (Fig. 11.12*e*) components. However, in the first case the displacements Δ_{2p} and Δ_{3p} will be computed multiplying the bending moment graphs pertaining to both columns while the use of diagrams of the symmetrical and antisymmetrical components will permit the multiplication of bending moment graphs for one of the columns only. On the other hand, the M_P^c and the M_P^k diagrams are somewhat complicated in outline and for multiplication purposes they must be subdivided into two portions, one rectangular and the other parabolic, which would be unnecessary were the original loading retained.

On the whole, in this particular case the replacement of the load P by its symmetrical and antisymmetrical components will make the calculations rather more complicated instead of simplifying them.

Let us now investigate the frame represented in Fig. 12.12*a* loaded by one horizontal force P. The degree of redundancy of this frame is equal to six.

Fig. 12.12*b* and *c* shows the same structure loaded by the symmetrical and the antisymmetrical components of force P. The symmetrical components will cause no displacement of the top point of the central column and consequently we may admit that this point is held fast as shown in Fig. 12.12*d*. Hence the structure given in the latter figure can be adopted as conjugate simple structure for the case under consideration. The bending moments induced in this structure by the symmetrical components

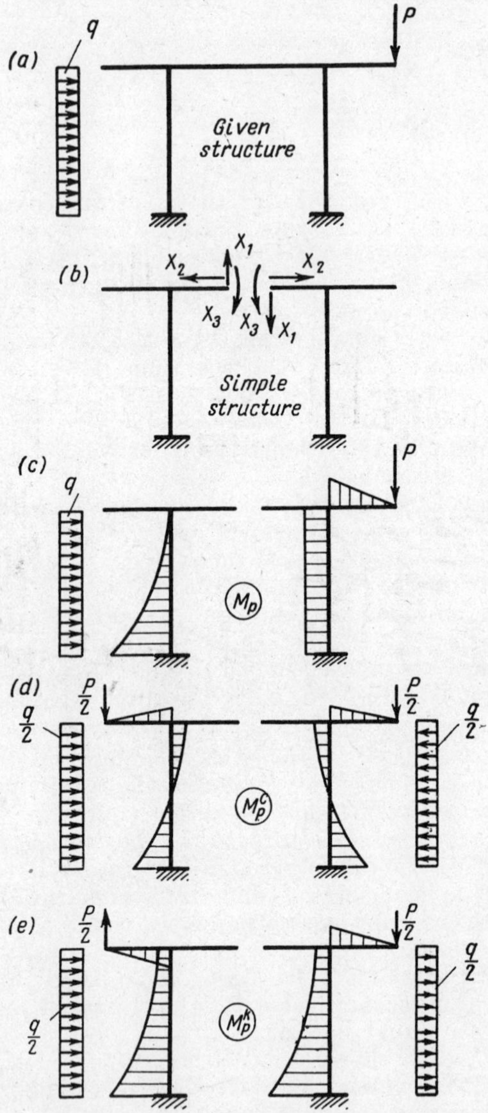

(a)
q
P
Given structure
(b)
X_1
X_2
X_2
X_3
X_3
X_1
Simple structure
(c)
q
P
M_p
(d)
$\frac{P}{2}$
$\frac{q}{2}$
$\frac{P}{2}$
$\frac{q}{2}$
M_p^c
(e)
$\frac{P}{2}$
$\frac{P}{2}$
$\frac{q}{2}$
$\frac{q}{2}$
M_p^k

Fig. 11.12

of load P will remain nil throughout and therefore the displacements Δ produced by these components as well as all the symmetrical unknowns must be equally nil. The same will apply of course to the antisymmetrical unknowns.

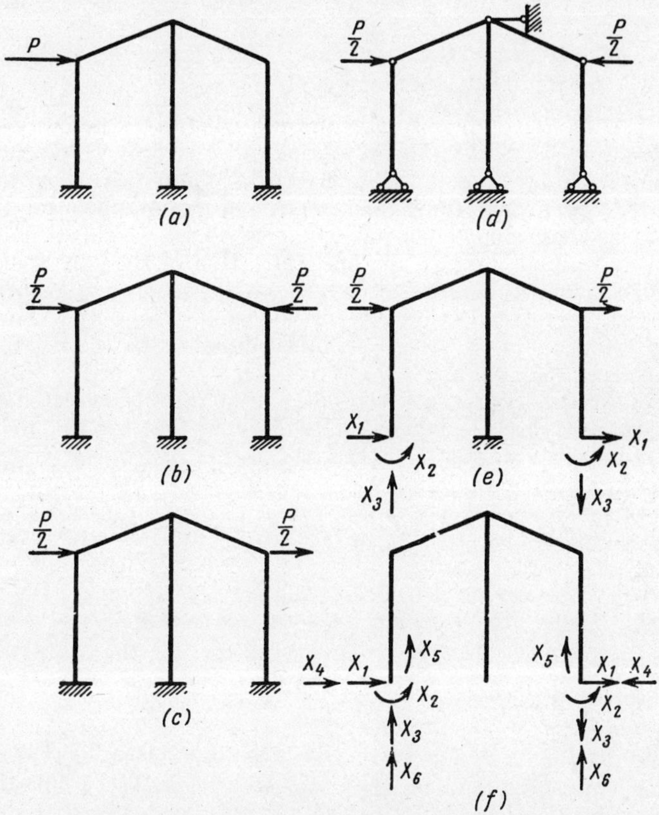

Fig. 12.12

It follows that all the frame members will be acted upon solely by bending moments induced by the antisymmetrical components of load P shown in Fig. 12.12*c*. *

For the latter system of loads we may adopt the simple statically determinate structure represented in Fig. 12.12*e* with unknowns formed by the groups of forces X_1, X_2 and X_3. Thus, for the frame

* Strains due to direct stresses are neglected as usual.

of Fig. 12.12 the transformation of the applied loads into their symmetrical and antisymmetrical components leads to the reduction of the number of the unknowns from six to three. The resulting simplification of all the computations is therefore quite considerable.

5.12. *Accuracy Control of All the Terms Entering the Simultaneous Equations*

It will be remembered that the coefficients to the unknowns and the free terms of the simultaneous equations represent displacements induced in the simple statically determinate structure by unit forces acting along the direction of the unknown reactions, as well as those due to the actual loads. These displacements are usually obtained through the multiplication of the corresponding bending moment diagrams. Errors which occur while carrying out these operations influence further calculations and therefore render erroneous the values of the redundant reactions finally obtained.

Errors in the displacement values can be usually detected using the procedure described hereunder. Suppose that for the analysis of a structure redundant to the nth degree reactions or groups of reactions $X_1, X_2, \ldots, X_i, \ldots, X_n$ have been adopted as the unknowns.

Let us apply to the simple statically determinate structure all the unit forces corresponding to these reactions and let us construct the diagram of the resulting bending moment. We shall call this diagram the *summary unit bending moment diagram* designating its ordinates by $\overline{M}_s$. In any cross section the ordinate $\overline{M}_s$ will thus equal the algebraic sum of all the ordinates $\overline{M}_1, \overline{M}_2, \ldots, \overline{M}_i, \ldots, \overline{M}_n$.

If we multiply successively the $\overline{M}_s$ graph area by the unit bending moment diagrams $\overline{M}_1, \overline{M}_2$, etc., each of the products thus obtained will be equal to the algebraic sum of all the coefficients to the unknowns of the corresponding equation because

$$\sum_{i=1}^{i=n} \delta_{1i} = \delta_{11} + \delta_{12} + \delta_{13} + \ldots + \delta_{1n} = \sum \int \overline{M}_1 \overline{M}_1 \frac{ds}{EJ} +$$

$$+ \sum \int \overline{M}_1 \overline{M}_2 \frac{ds}{EJ} + \ldots + \sum \int \overline{M}_1 \overline{M}_n \frac{ds}{EJ} =$$

$$= \sum \int \overline{M}_1 (\overline{M}_1 + \overline{M}_2 + \ldots + \overline{M}_n) \frac{ds}{EJ} = \sum \int \overline{M}_1 \overline{M}_s \frac{ds}{EJ} = \delta_{1s}$$

For the same reason we may write

$$\delta_{21}+\delta_{22}+\delta_{23}+\ldots+\delta_{2n}=\delta_{2s}$$

It follows that the algebraic sum of all the coefficients to the unknowns contained in equation i must be equal to δ_{is} where

$$\delta_{is}=\Sigma\int \overline{M}_i\overline{M}_s\frac{ds}{EJ} \tag{8.12}$$

Hence the values of all the unit displacements forming the coefficients to the unknowns of the first of the simultaneous equations can be checked comparing their sum with the value of δ_{1s}

$$\sum_{i=1}^{i=n}\delta_{1i}=\delta_{1s} \tag{9.12}$$

The algebraic sums of the coefficients to the unknowns of all the other simultaneous equations may be checked in exactly the same way.

The above procedure permits to verify the coefficients to the unknowns of each equation separately. Let us now sum up all the values of δ_{1s}, δ_{2s}, . . ., δ_{ns} denoting this sum by $\Sigma\delta$. In that case

$$\Sigma\delta=\delta_{1s}+\delta_{2s}+\delta_{3s}+\ldots+\delta_{ns}$$

But since

$$\delta_{1s}=\Sigma\int \overline{M}_1\overline{M}_s\frac{ds}{EJ}$$

$$\delta_{2s}=\Sigma\int \overline{M}_2\overline{M}_s\frac{ds}{EJ}$$

$$\cdots\cdots\cdots$$

$$\delta_{ns}=\Sigma\int \overline{M}_n\overline{M}_s\frac{ds}{EJ}$$

we find that

$$\Sigma\delta=\Sigma\int \overline{M}_s\overline{M}_1\frac{ds}{EJ}+\Sigma\int \overline{M}_s\overline{M}_2\frac{ds}{EJ}+\ldots+$$

$$+\Sigma\int \overline{M}_s\overline{M}_n\frac{ds}{EJ}=\Sigma\int \overline{M}_s(\overline{M}_1+\overline{M}_2+\ldots+\overline{M}_n)\frac{ds}{EJ}=$$

$$=\Sigma\int \overline{M}_s^2\frac{ds}{EJ}=\delta_{ss}$$

Consequently

$$\delta_{ss}=\Sigma\int \overline{M}_s^2\frac{ds}{EJ} \tag{10.12}$$

and

$$\Sigma\delta=\delta_{ss} \tag{11.12}$$

The last expression permits a simultaneous check on the accuracy of all the coefficients to the unknowns contained in all the equations of the given system. This check will be carried out as follows:

a) Find the algebraic sum of all the coefficients to the unknowns (unit displacements) contained in all the equations of the given system

$$\Sigma\delta = (\delta_{11} + \delta_{22} + \delta_{33} + \ldots + \delta_{nn}) + 2(\delta_{12} + \delta_{13} + \ldots + \delta_{1n} + \delta_{23} + \delta_{24} + \ldots + \delta_{2n} + \delta_{34} + \ldots + \delta_{n-1,\,n})$$

In the above expression the first term in parentheses contains all the principal displacements, i. e., those situated on the main diagonal and the other term all the secondary displacements situated on both sides of this diagonal (i. e., on secondary diagonals).

b) Using the method of diagram multiplication find the values of

$$\delta_{ss} = \Sigma \int \overline{M}_s^2 \frac{ds}{EJ}$$

c) Compare the two values obtained as described above.

In a number of cases the following inequality may help in finding the erroneous coefficients to the unknowns

$$\delta_{ii} \times \delta_{kk} \geqslant \delta_{ik}^2$$

The control of the free terms of the canonical equations will be carried out in a similar way:

a) First compute the value of

$$\Delta_{sp} = \Sigma \int \overline{M}_s M_p \frac{ds}{EJ} \qquad (12.12)$$

where M_p represents the bending moments induced in the conjugate simple structure by the actual loads.

b) Check whether

$$\Sigma\Delta = \Delta_{1p} + \Delta_{2p} + \ldots + \Delta_{np} = \Delta_{sp} \qquad (13.12)$$

In general it is quite sufficient to verify simultaneously the coefficients to the unknowns of the whole system of equations. If this check reveals an error, it is recommended to verify these coefficients equation by equation as described at the beginning of the present article.

6.12. *Abridged Solution of Canonical Equations*

Hereunder we shall describe very briefly the abridged method of solving simultaneous canonical equations which was proposed by Gauss. In this method the unknowns are eliminated one by one using a set of certain coefficients. All the operations are carried out in tabular form and checked constantly. *

Let us consider four canonical equations (I), (2), (3) and (4) with four unknowns. **

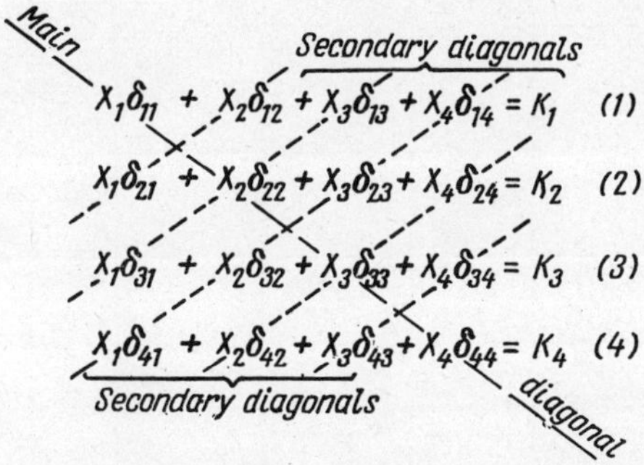

It will be remembered that each pair of coefficients to the unknowns symmetrical about the main diagonal are equal to one another as stipulated by the theorem of reciprocal displacements. The free terms have been transferred to the right-hand part of the equations and are denoted by K_1, K_2, etc. The solution of this system of canonical equations will be carried out as indicated in Table 1.12. All the entries in column 1 can be made off hand. The number of columns and lines of the table is directly dependent on the number of simultaneous equations. Thus, for instance, if our system contained five equations instead of four we should have nine columns instead of the eight contained in Table 1.12. These nine columns would be under the following headings: Equation No.; X_1; X_2; X_3; X_4; X_5; multipliers α_{ik}; S, and K. Similarly instead of 13 lines of Table 1.12 we would have 19, these lines being designated by (I); (2); (I)$\cdot\alpha_{12}$; (II); (3); (I)$\cdot\alpha_{13}$; (II)$\cdot\alpha_{23}$; (III); (4); (I)$\cdot\alpha_{14}$; (II)$\cdot\alpha_{24}$; (III)$\cdot\alpha_{34}$; (IV); (5); (I)$\cdot\alpha_{15}$; (II)$\cdot\alpha_{25}$; (III)$\cdot\alpha_{35}$; (IV)$\cdot\alpha_{45}$; (V).

Having prepared the table, enter the coefficients and the values of K taken directly from the simultaneous equations into lines (I), (2), (3) and (4). No entries are made at this stage in the column for the multipliers α_{ik}. Column S will be filled in with the values

* In this book we shall not consider the "unabridged" method of solving canonical equations, the latter being seldomly used in stress analysis of redundant structures.

** The first of these four equations will be designated by a Roman figure.

Table 1.12

Equation No.	X_1	X_2	X_3	X_4	Multipliers α_{ik}	S	K
(I)	δ_{11}	δ_{12}	δ_{13}	δ_{14}	$\alpha_{12}=-\frac{\delta_{12}}{\delta_{11}}$ $\alpha_{13}=-\frac{\delta_{13}}{\delta_{11}}$ $\alpha_{14}=-\frac{\delta_{14}}{\delta_{11}}$	S_1	K_1
+ { 2 (I)·α_{12}	δ_{21} .	δ_{22} $\delta_{12}\cdot\alpha_{12}$	δ_{23} $\delta_{13}\cdot\alpha_{12}$	δ_{24} $\delta_{14}\cdot\alpha_{12}$		S_2 $S_1\alpha_{12}$	K_2 $K_1\alpha_{12}$
(II)	.	δ'_{22}	δ'_{23}	δ'_{24}	$\alpha_{23}=-\frac{\delta'_{23}}{\delta'_{22}}$ $\alpha_{24}=-\frac{\delta'_{24}}{\delta'_{22}}$	S'_2	K'_2
+ { (3) (I)·α_{13} (II)·α_{23}	δ_{31} . .	δ_{32} . .	δ_{33} $\delta_{13}\cdot\alpha_{13}$ $\delta'_{23}\cdot\alpha_{23}$	δ_{34} $\delta_{14}\cdot\alpha_{13}$ $\delta'_{24}\cdot\alpha_{23}$		S_3 $S_1\alpha_{13}$ $S'_2\alpha_{23}$	K_3 $K_1\alpha_{13}$ $K'_2\alpha_{23}$
(III)	.	.	δ''_{33}	δ''_{34}	$\alpha_{34}=-\frac{\delta''_{34}}{\delta''_{33}}$	S''_3	K''_3
+ { (4) (I)·α_{14} (II)·α_{24} (III)·α_{34}	δ_{41} . . .	δ_{42} . . .	δ_{43} . . .	δ_{44} $\delta_{14}\cdot\alpha_{14}$ $\delta'_{24}\cdot\alpha_{24}$ $\delta''_{34}\cdot\alpha_{34}$		S_4 $S_1\alpha_{14}$ $S'_2\alpha_{24}$ $S''_3\alpha_{34}$	K_4 $K_1\alpha_{14}$ $K'_2\alpha_{24}$ $K''_3\alpha_{34}$
(IV)	.	.	.	δ'''_4		S'''_4	K'''_4

of s_1, s_2, s_3, . . . , s_n which represent

$$s_1 = \delta_{11} + \delta_{12} + \delta_{13} + \ldots + \delta_{1n}$$
$$s_2 = \delta_{21} + \delta_{22} + \delta_{23} + \ldots + \delta_{2n} \text{ etc.}$$

Further operations are carried out in the following sequence:

a) Compute the values of α_{12}, α_{13}, and α_{14} using the expressions indicated in the corresponding lines under the heading "multipliers α_{ik}".

b) Fill line (I)$\cdot\alpha_{12}$ with the values of the products obtained by the multiplication of all the entries of line (I) by α_{12}.

c) Calculate the values of δ'_{22}, δ'_{23}, etc., given in line (II) adding two by two the entries contained in line (2) with those of line (I)$\cdot\alpha_{12}$. No operations are carried out in columns X_1 and "multipliers α_{ik}".

d) Proceeding in the same way determine the values α_{23}, α_{24}, etc., using expressions contained in the column "multipliers α_{ik}".

Gradually the whole of the table will be filled in that way. Whenever it is known beforehand that the value of an entry equals zero, this entry is replaced by a dot. Such is the case of numerous entries in lines (II), (III) and (IV). Each of the lines marked with a Roman figure represents one of the equations to be solved. Thus, for instance, line (III) represents the equation

$$X_3\delta''_{33} + X_4\delta''_{34} = K_3$$

When all the operations are carried out correctly the sum of all the coefficients to the unknowns of each equation will equal the entry in the same line in column S. Thus, for instance, for equation (III) it must be found that

$$\delta''_{33} + \delta''_{34} = S''_3$$

This control should be carried out each time the corresponding line has been completely filled in.

The last of the equations [in our case equation (IV)] will contain only one unknown yielding the value of X_4. Equation (III) containing two unknowns X_3 and X_4 will be easily solved leading thereafter to the solution of equations (II) and (I) thus providing the values of the unknowns X_3, X_2 and X_1 in succession. The final results are checked by the introduction of all the unknowns thus found into the original system of simultaneous equations.

Table 2.12

Equation No.	X_1	X_2	X_3	X_4	Multipliers α_{ik}	S	K
(1)	2	−1	3	−1	$\alpha_{12}=\frac{1}{2}$ $\alpha_{13}=-\frac{3}{2}$ $\alpha_{14}=\frac{1}{2}$	3	5
(2)	−1	3	−2	−5		−5	−21
(1)·α_{12}		$-\frac{1}{2}$	$\frac{3}{2}$	$-\frac{1}{2}$		$\frac{3}{2}$	$\frac{5}{2}$
(II)	·	$\frac{5}{2}$	$-\frac{1}{2}$	$-\frac{11}{2}$	$\alpha_{23}=\frac{1}{5}$ $\alpha_{24}=\frac{11}{5}$	$-\frac{7}{2}$	$-\frac{37}{2}$
+ { (3)	3	−2	−5	4		0	0
(1)·α_{13}	·	·	$-\frac{9}{2}$	$\frac{3}{2}$		$-\frac{9}{2}$	$-\frac{15}{2}$
(II)·α_{23} }	·	·	$-\frac{1}{10}$	$-\frac{11}{10}$		$-\frac{7}{10}$	$-\frac{37}{10}$
(III)	·	·	$-\frac{96}{10}$	$\frac{44}{10}$	$\alpha_{34}=\frac{11}{24}$	$-\frac{52}{10}$	$-\frac{112}{10}$
+ { (4)	−1	−5	4	1		−1	5
(I)·α_{14}	·	·	·	$-\frac{1}{2}$		$\frac{3}{2}$	$\frac{5}{2}$
(II)·α_{24}	·	·	·	$-\frac{121}{10}$		$-\frac{77}{10}$	$-\frac{407}{10}$
(III)·α_{34} }	·	·	·	$\frac{121}{60}$		$-\frac{143}{60}$	$-\frac{154}{30}$
(IV)	·	·	·	$-\frac{575}{60}$		$-\frac{575}{60}$	$-\frac{2,300}{60}$

To illustrate the above let us use the method just described for the solution of the following equations

$$2X_1-X_2+3X_3-X_4=5 \quad (1)\ S_1=2-1+3-1=3$$
$$-X_1+3X_2-2X_3-5X_4=-21 \quad (2)\ S_2=-1+3-2-5=-5$$
$$3X_1-2X_2-5X_3+4X_4=0 \quad (3)\ S_3=3-2-5+4=0$$
$$-X_1-5X_2+4X_3+X_4=5 \quad (4)\ S_4=-1-5+4+1=-1$$

All the coefficients to the unknowns of these equations satisfy the principle of reciprocal displacements. The abbreviated solution of this system of simultaneous equations is contained in Table 2.12.

Having filled in this table, proceed as indicated above with the solution of the equations contained in lines (I), (II), (III) and (IV) starting with the last one

$$-\frac{575}{60}X_4=-\frac{2,300}{60};\ X_4=4$$

$$-\frac{96}{10}X_3+\frac{44}{10}\times 4=-\frac{112}{10};\ X_3=\frac{112+176}{96}=3$$

$$2.5X_2=-18.5+0.5\times 3+5.5\times 4;\ X_2=2$$

$$2X_1=5+2-3\times 3+4;\ X_1=1$$

To verify all the operations enter the values of the unknowns into the original equations.

Table 1.12 shows that an alternation in the values of the free terms will be reflected only in the entries of the last column K.

For this reason the method just described becomes particularly well fit for the stress analysis of redundant structures when these are called upon to carry different loads.

7.12. *Several Problems in Stress Analysis of Redundant Frames*

Problem 1. Construct and check the M, Q and N diagrams for the double-span symmetrical frame shown in Fig. 13.12.

Solution. The frame is three times statically indeterminate. Fig. 14.12 shows one of the simple structures which could be adopted for the solution of the problem. However, that would involve the simultaneous solution of three equations with three unknowns.

The problem will be considerably simplified if the unknowns are grouped as indicated in Fig. 15.12. In the latter case the canonical equations become

$$X_1\delta_{11}+X_2\delta_{12}+\Delta_{1p}=0$$
$$X_1\delta_{21}+X_2\delta_{22}+\Delta_{2p}=0$$
$$X_3\delta_{33}+\Delta_{3p}=0$$

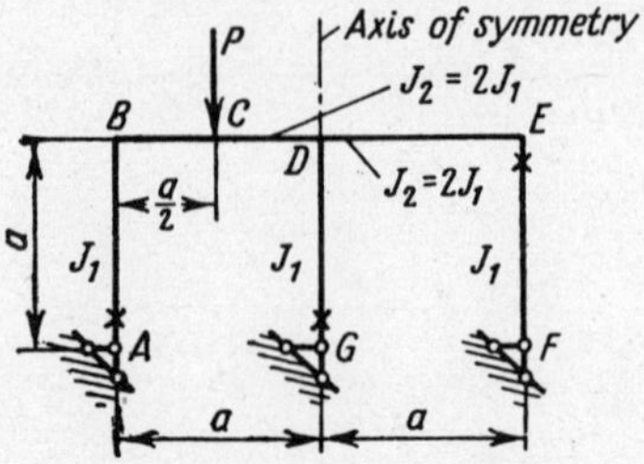

Fig. 13.12

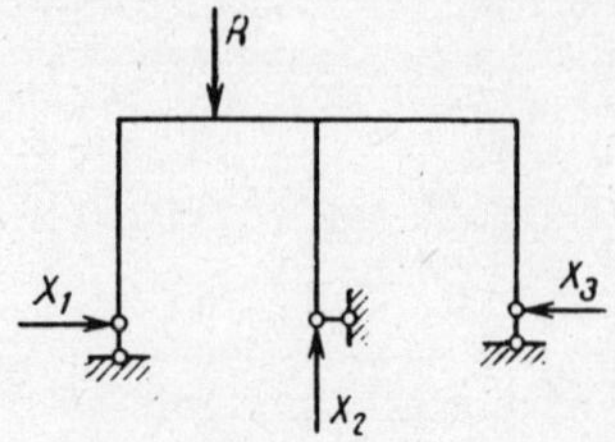

Fig. 14.12

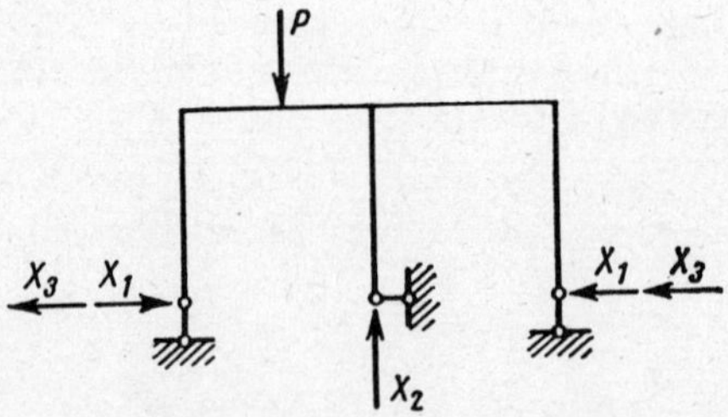

Fig. 15.12

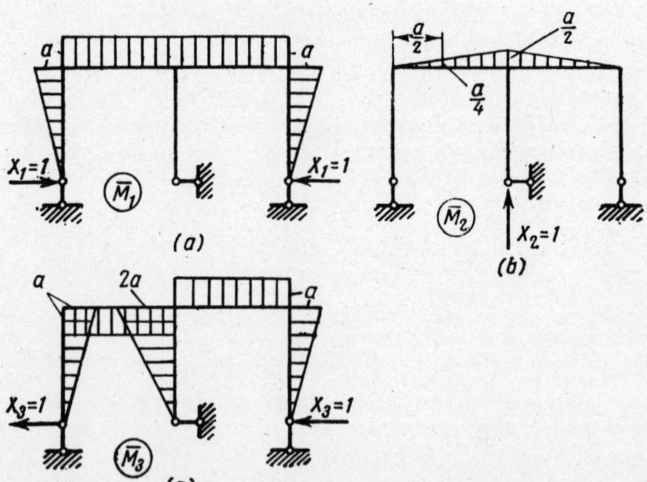

Fig. 16.12

The unit bending moment diagrams $\overline{M}_1$, $\overline{M}_2$ and $\overline{M}_3$ corresponding to the case under consideration are given in Fig. 16.12*a*, *b* and *c* while the diagram due to the applied load is presented in Fig. 17.12.

Multiplying the appropriate graphs one by the other we shall obtain the values both of the coefficients to the unknowns and of the free terms of

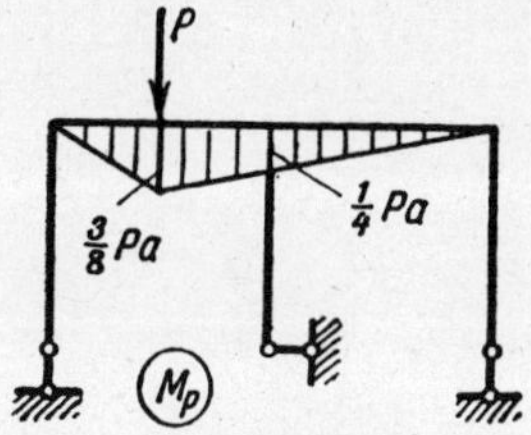

Fig. 17.12

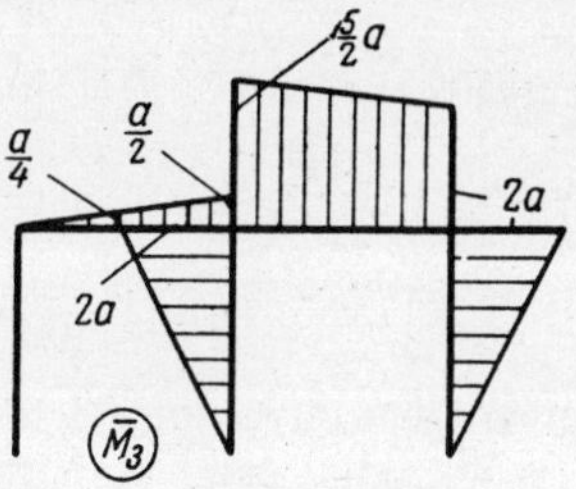

Fig. 18.12

the above equations. It should not be forgotten that the moments of inertia of the columns are only half as great as those of the crossbeams

$$\delta_{11}=2\left(a\cdot\frac{a}{2}\cdot\frac{2}{3}a\frac{1}{EJ_1}+a\cdot a\cdot a\frac{1}{2EJ_1}\right)=\frac{5a^3}{3EJ_1}$$

$$\delta_{12}=\frac{a}{2}\cdot\frac{2a}{2}\cdot\frac{a}{2EJ_1}=\frac{a^3}{4EJ_1}$$

$$\delta_{22}=2\frac{a}{2}\cdot\frac{a}{2}\cdot\frac{2}{3}\cdot\frac{a}{2}\cdot\frac{1}{2EJ_1}=\frac{a^3}{12EJ_1}$$

$$\delta_{33}=2\left(a\frac{a}{2}\cdot\frac{2}{3}a\frac{1}{EJ_1}+a\cdot a\cdot a\frac{1}{2EJ_1}\right)+$$

$$+2a\frac{a}{2}\cdot\frac{2}{3}\cdot 2a\frac{1}{EJ_1}=3\frac{a^3}{EJ_1}$$

$$\Delta_{1p}=-\frac{3Pa}{8}\cdot\frac{2a}{2}\cdot\frac{a}{2EJ_1}=-\frac{3Pa^3}{16EJ_1}$$

$$\Delta_{2p}=-\frac{1}{2EJ_1}\left[\frac{3Pa}{8}\cdot\frac{a}{2}\cdot\frac{1}{2}\cdot\frac{2}{3}\cdot\frac{a}{4}+\frac{a}{2\cdot 6}\left(2\cdot\frac{3}{8}Pa\frac{a}{4}+\right.\right.$$

$$\left.+2\frac{1}{4}Pa\frac{a}{2}+\frac{3}{8}Pa\frac{a}{2}+\frac{1}{4}Pa\frac{a}{4}\right)+$$

$$\left.+\frac{1}{4}Pa\frac{a}{2}\cdot\frac{2}{3}\cdot\frac{a}{2}\right]=-\frac{11Pa^3}{192EJ_1}$$

$$\Delta_{3p}=\left(\frac{3}{8}Pa\frac{a}{2}\cdot\frac{1}{2}+\frac{\frac{3}{8}Pa+\frac{1}{4}Pa}{2}\cdot\frac{a}{2}-\frac{1}{4}Pa\frac{a}{2}\right)\frac{a}{2EJ_1}=\frac{Pa^3}{16EJ_1}$$

The displacements thus computed will be checked using the "summary" unit bending moment diagram $\overline{M}_s$ due to the simultaneous application of all the unit forces acting along the unknowns (Fig. 18.12).

The value of δ_{ss} is obtained raising to the second power the $\overline{M}_s$ diagram:

$$\delta_{ss}=\Sigma\int\overline{M}_s^2\frac{ds}{EJ}=\Big[\frac{a}{2}\cdot\frac{a}{2}\cdot\frac{2}{3}\cdot\frac{a}{2}+\frac{a}{6}\Big(2\cdot\frac{5}{2}a\cdot\frac{5}{2}a+$$

$$+2\cdot 2a\cdot 2a+2\cdot\frac{5}{2}a\cdot 2a\Big)\Big]\frac{1}{2EJ_1}+2\cdot 2a\cdot\frac{a}{2}\cdot\frac{2}{3}\cdot\frac{2a}{EJ_1}=\frac{21a^3}{4EJ_1}$$

Checking that condition (11.12) is satisfied we find

$$\frac{21a^3}{4EJ_1}=\frac{5a^3}{3EJ_1}+\frac{a^3}{12EJ_1}+\frac{3a^3}{EJ_1}+2\cdot\frac{a^3}{4EJ_1}\quad\text{or}$$

$$\frac{21a^3}{4EJ_1}=\frac{21a^3}{4EJ_1}$$

It follows that all the unit displacements are correct.

Let us determine now the value of Δ_{sp} multiplying the $\overline{M}_s$ diagram (Fig. 18.12) by the M_p one (Fig. 17.12)

$$\Delta_{sp}=\Sigma\int\overline{M}_sM_p\frac{ds}{EJ}=\frac{-1}{2EJ_1}\Big[\frac{3}{8}Pa\cdot\frac{a}{2}\cdot\frac{1}{2}\cdot\frac{2}{3}\cdot\frac{a}{4}+$$

$$+\frac{a}{2\cdot 6}\Big(2\cdot\frac{3}{8}Pa\cdot\frac{a}{4}+2\cdot\frac{1}{4}Pa\cdot\frac{a}{2}+\frac{3}{8}Pa\cdot\frac{a}{2}+\frac{1}{4}Pa\cdot\frac{a}{4}\Big)+$$

$$+\frac{1}{4}Pa\cdot\frac{a}{2}\Big(\frac{2}{3}\cdot\frac{5}{2}a+\frac{1}{3}2a\Big)\Big]=-\frac{35Pa^3}{192EJ_1}$$

Check whether condition (13.12) is also satisfied

$$\Delta_{sp}=\Sigma\Delta$$

$$-\frac{35Pa^3}{192EJ_1}=-\frac{3Pa^3}{16EJ_1}-\frac{11Pa^3}{192EJ_1}+\frac{Pa^3}{16EJ_1}\quad\text{or}$$

$$-\frac{35Pa^3}{192EJ_1}=-\frac{35Pa^3}{192EJ_1}$$

This indicates that the displacements due to the applied loads are also correct. We may now introduce the values of the coefficients to the unknowns and of the free terms into our system of canonical equations. Multiplying the first two of these equations by $\frac{192\,EJ_1}{a^3}$ and the last one by $\frac{16\,EJ_1}{a^3}$ we obtain

$$320X_1+48X_2-36P=0$$

$$48X_1+16X_2-11P=0$$

$$48X_3+P=0$$

The solution of these three equations leads to

$$X_1=\frac{3P}{176};\quad X_2=\frac{7P}{11};\quad X_3=-\frac{P}{48}$$

The construction of the resulting bending moment diagram for the given redundant structure will be carried out in the following sequence. First, determine the bending moments induced in the simple structure by the redundant reactions X_1, X_2 and X_3 obtained above. For this purpose multiply all the ordinates to the unit bending moment graphs $\bar{M}_1$, $\bar{M}_2$ and $\bar{M}_3$ by the respective magnitudes of these reactions. The three diagrams thus obtained are represented in Figs. 19.12, 20.12 and 21.12.

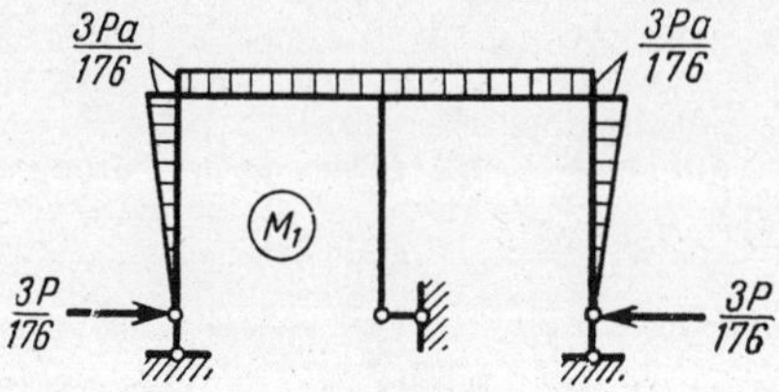

Fig. 19.12

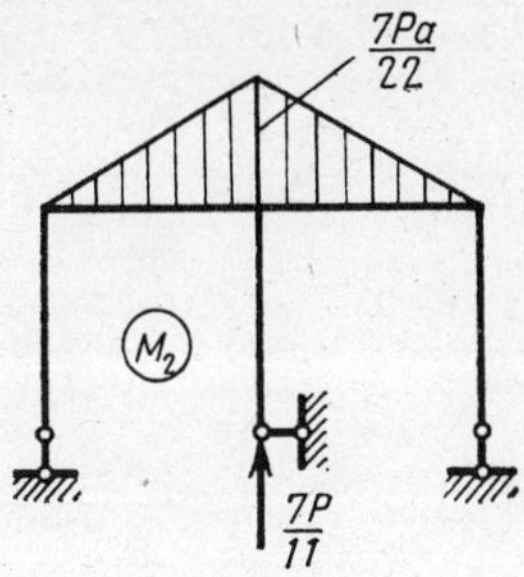

Fig. 20.12

The ordinates to the resulting bending moment diagram can now be obtained summing up the ordinates to these three diagrams with those to the M_p diagram given previously in Fig. 17.12. (In Fig. 13.12 we have marked

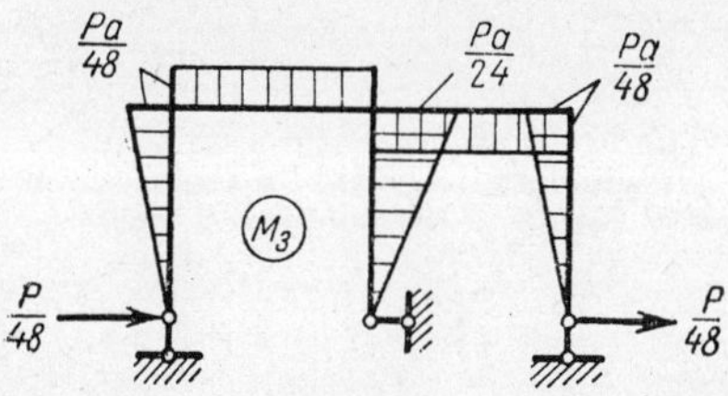

Fig. 21.12

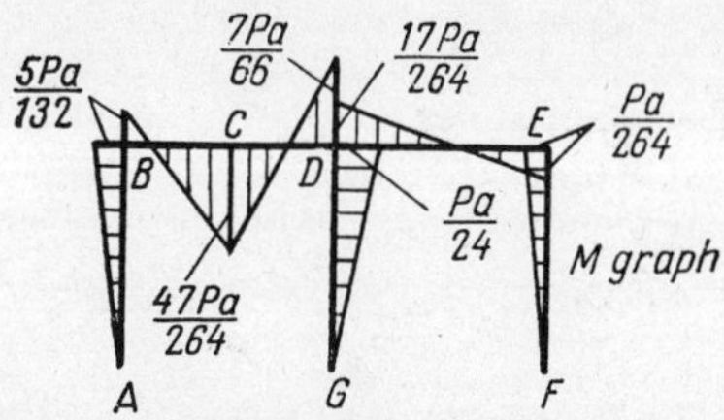

Fig. 22.12

by an asterisk those of the ends of the columns which are considered as being the left-hand ones for the construction of bending moment diagrams.)

$$M_{CB}=M_{CD}=-\frac{3Pa}{176}-\frac{7Pa}{44}-\frac{Pa}{48}+\frac{3Pa}{8}=\frac{47Pa}{264}$$

$$M_{BA}=M_{BC}=-\frac{3Pa}{176}+0-\frac{Pa}{48}+0=-\frac{5Pa}{132}$$

$$M_{DC}=-\frac{3Pa}{176}-\frac{7Pa}{22}-\frac{Pa}{48}+\frac{Pa}{4}=-\frac{7Pa}{66}$$

$$M_{DE}=-\frac{3Pa}{176}-\frac{7Pa}{22}+\frac{Pa}{48}+\frac{Pa}{4}=-\frac{17Pa}{264}$$

$$M_{DG}=0+0+\frac{Pa}{24}+0=\frac{Pa}{24}$$

$$M_{ED}=M_{EF}=-\frac{3Pa}{176}+0+\frac{Pa}{48}+0=\frac{Pa}{264}$$

In all the above expressions M_{BA} denotes the bending moment at cross section B of member BA, M_{BC} denotes the moment at cross section B of member BC and so forth.

The diagram of the resulting bending moments plotted as explained above is represented in Fig. 22.12. In order to check the accuracy of this diagram let us compute the mutual displacement Δ_{FG} of points F and G (see Fig. 13.12). This displacement must be necessarily nil for both points are held fast by the supports of the frame.

In order to find the said displacement we shall eliminate all three constraints at the lower end of the right-hand column, transforming thereby the right half of the structure into a statically determinate polygonal beam shown in Fig. 23.12. Let us now apply two unit forces acting along the

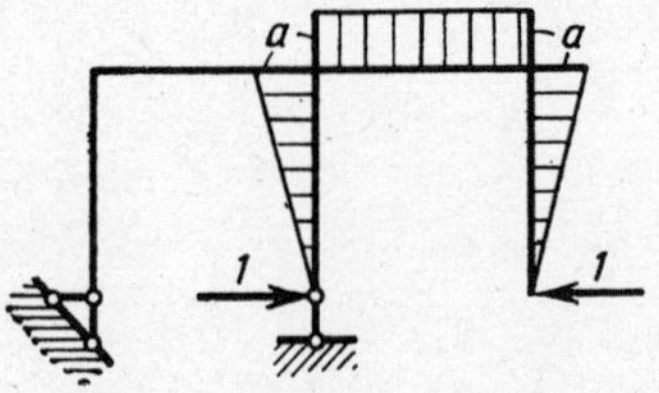

Fig. 23.12

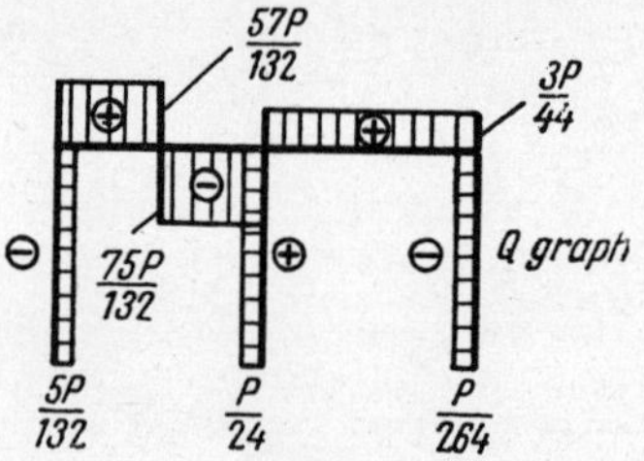

Fig. 24.12

directions of the desired displacement. The bending moment diagram due to these forces is shown in this figure. Multiplying this diagram by the diagram or the resulting bending moments given in Fig. 22.12 we obtain

$$\Delta_{FG} = -a\cdot\frac{a}{2}\cdot\frac{Pa}{264}\cdot\frac{2}{3}\cdot\frac{1}{EJ_1}+a\cdot a\left(\frac{17Pa}{264}\cdot\frac{1}{2}-\frac{Pa}{264}\cdot\frac{1}{2}\right)\times$$
$$\times\frac{1}{2EJ_1}-a\cdot\frac{a}{2}\cdot\frac{2}{3}\cdot\frac{Pa}{24}\cdot\frac{1}{EJ_1}=\frac{Pa^3}{264EJ_1}\left(-\frac{1}{3}+\frac{17}{4}-\frac{1}{4}-\frac{11}{3}\right)=0$$

It will be remembered that the product of the resulting bending moment graph of Fig. 22.12 by the unit diagrams given in Fig. 16.12 must be also nil.

Let us proceed now with the determination of the shearing forces

$$Q_{AB}=Q_{BA}=\frac{M_{BA}-M_{AB}}{a}=\frac{\frac{-5Pa}{132}-0}{a}=\frac{-5P}{132}$$

$$Q_{BC}=Q_{CB}=\left(\frac{47Pa}{264}+\frac{5Pa}{132}\right)\frac{1}{a/2}=\frac{57P}{132}$$

$$Q_{CD}=Q_{DC}=-\left(\frac{7Pa}{66}+\frac{47Pa}{264}\right)\frac{1}{a/2}=-\frac{75p}{132}$$

$$Q_{DE}=Q_{ED}=\left(\frac{Pa}{264}+\frac{17Pa}{264}\right)\frac{1}{a}=\frac{3P}{44}$$

$$Q_{EF}=Q_{FE}=-\frac{Pa}{264}\cdot\frac{1}{a}=-\frac{P}{264}$$

$$Q_{DG}=Q_{GD}=\frac{Pa}{24}\cdot\frac{1}{a}=\frac{P}{24}$$

The signs of the shearing forces can be checked remembering that the shear is reckoned positive when the axis of the element must be rotated clockwise through an angle smaller than 90° in order to come in coincidence with the tangent to the bending moment diagram at the section under consideration.

The magnitudes of the shearing forces Q obtained above have led to the construction of the diagram shown in Fig. 24.12.

The values of the normal forces will be derived from the equilibrium of joints B, D and E isolated in succession (Fig. 25.12*a*, *b* and *c*). In these computations unknown normal stresses will be always reckoned positive, i. e.,

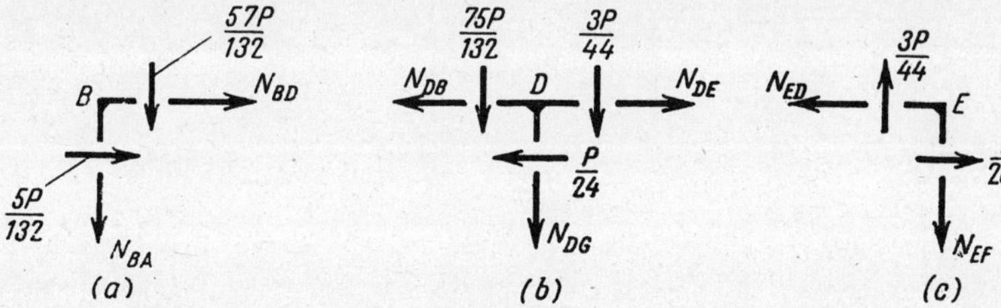

Fig. 25.12

causing the extension of the corresponding member. The values of shearing forces are taken directly from the shear diagram of Fig. 24.12. The equilibrium equations become:

Joint B (Fig. 25.12*a*)

$$\Sigma Y = -N_{BA} - \frac{57P}{132} = 0$$

wherefrom

$$N_{BA} = -\frac{57P}{132} \quad \text{(compression)}$$

$$\Sigma X = \frac{5P}{132} + N_{BD} = 0 \quad \text{therefore } N_{BD} = -\frac{5P}{132} \quad \text{(compression)}$$

Joint D (Fig. 25.12*b*)

$$\Sigma Y = -N_{DG} - \frac{75P}{132} - \frac{3P}{44} = 0$$

wherefrom

$$N_{DG} = -\frac{7P}{11} \quad \text{(compression)}$$

$$\Sigma X = -N_{DB} + N_{DE} - \frac{P}{24} = 0 \quad \text{but since } N_{DB} = N_{BD}$$

we have finally

$$N_{DE} = \frac{P}{24} - \frac{5P}{132} = \frac{P}{264} \quad \text{(extension)}$$

Joint E (Fig. 25.12*c*)

$$\Sigma Y = \frac{3P}{44} = N_{EF} = 0 \quad \text{wherefrom } N_{EF} = \frac{3P}{44} \quad \text{(extension)}$$

$$\Sigma X = -N_{ED} + \frac{P}{264} = 0 \quad \text{therefore } N_{ED} = \frac{P}{264} \quad \text{(extension)}$$

Since $N_{ED} = N_{DE}$ the latter result may be regarded as confirming the accuracy of previous computations. The diagram of the normal stresses given in Fig. 26.12 has been constructed using the data just obtained.

Let us check the accuracy of the M, Q and N diagrams using the method based on equilibrium considerations. Isolating the upper half of the frame we must find that the actual loads applied to that part of the structure are

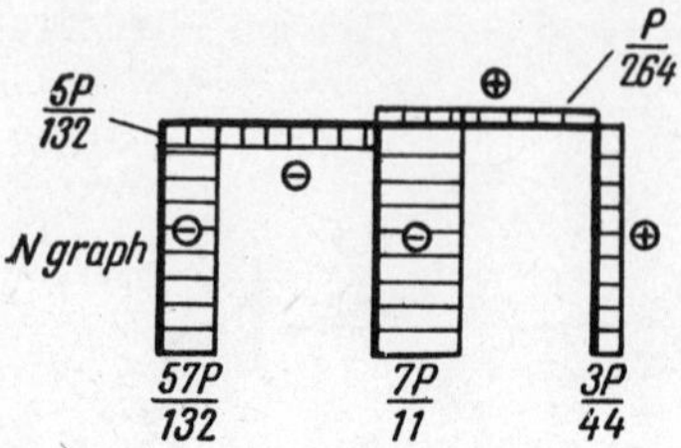

Fig. 26.12

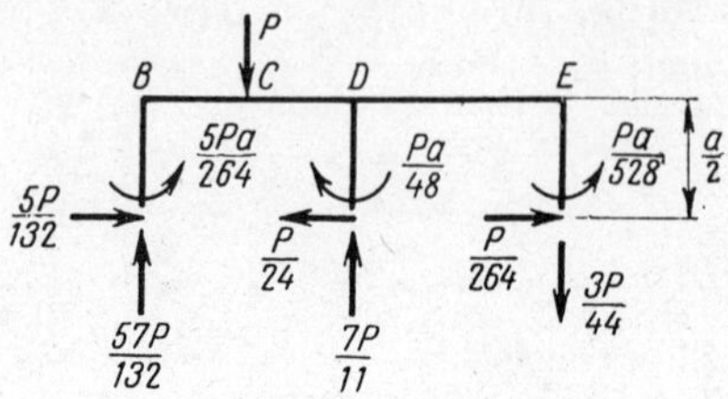

Fig. 27.12

balanced exactly by the stresses acting at the cuts, i. e., at midheight of the columns (Fig. 27.12). The values of these stresses will be scaled off the corresponding diagrams (see Figs. 22.12, 24.12 and 26.12). Indeed we find tha

$$\Sigma X = \frac{5P}{132} - \frac{P}{24} + \frac{P}{264} = \frac{P}{264}(10 - 11 + 1) = 0$$

$$\Sigma Y = -P + \frac{57P}{132} + \frac{7P}{11} - \frac{3P}{44} = \frac{P}{132}(-132 + 57 + 84 - 9) = 0$$

$$\Sigma M_B = \left(-\frac{5P}{132} + \frac{P}{24} - \frac{P}{264}\right)\frac{a}{2} + P\frac{a}{2} -$$

$$-\frac{7P}{11}a + \frac{3P}{44}2a - \frac{5Pa}{264} + \frac{Pa}{48} - \frac{Pa}{528} =$$

$$= \frac{Pa}{528}(-10 + 11 - 1 + 264 - 336 + 72 - 10 + 11 - 1) = 0$$

Hence all the equilibrium requirements are satisfied.

The M, Q and N diagrams could be obtained using a somewhat different procedure. Indeed we could apply to the conjugate statically determinate structure the redundant reactions

$$X_1 = \frac{3P}{176}; \quad X_2 = \frac{7P}{11} \quad \text{and} \quad X_3 = -\frac{P}{48}$$

together with the actual loads as shown in Fig. 28.12*a* and *b*. Thereafter we could calculate the values of the reactions developed at the supports of the latter structure under the simultaneous action of all the forces mentioned above.

The diagrams of the bending moments, shearing forces and normal stresses obtained in that way for the simple statically determinate structure would coincide exactly with the corresponding resulting stress diagrams for the given redundant structure.

Problem 2. Required the complete stress analysis for the two-story frame of a factory building loaded unsymmetrically along the top crossbeam. The different rigidities of all the frame members are indicated in Fig. 29.12.

Solution. The conjugate simple structure will be obtained sectioning both crossbeams at midspan (Fig. 29.12*b*). The bending moment diagrams due to

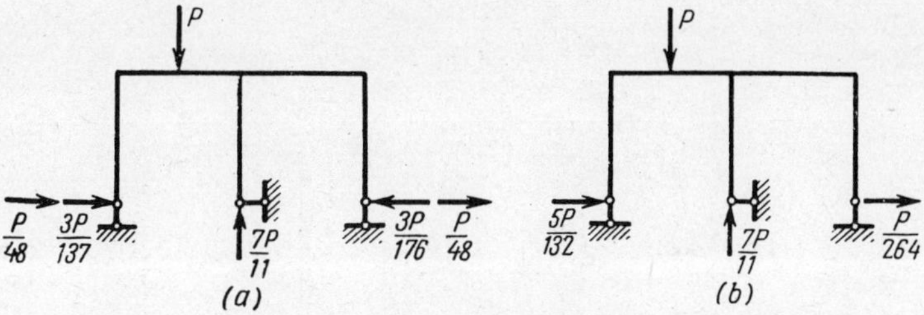

Fig. 28.12

unknown unit forces are given in Fig. 30.12. The diagram of the bending moments induced in the same simple structure by the actual loading is given

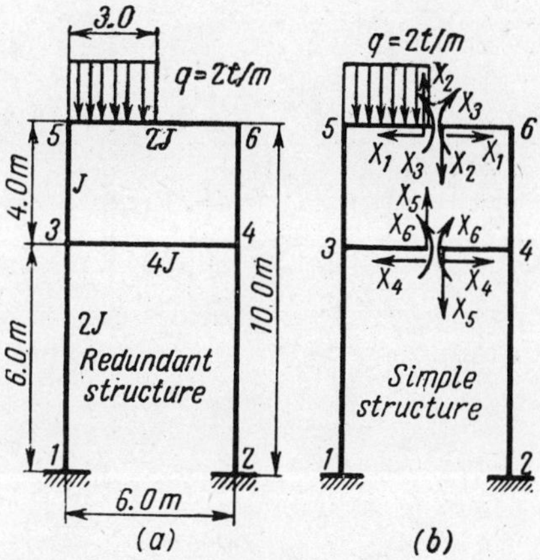

Fig. 29.12

in Fig. 31.12. Owing to the proper choice of the simple structure all the following secondary displacements reduce to zero

$$\delta_{12},\quad \delta_{15},\quad \delta_{23},\quad \delta_{24},\quad \delta_{26},\quad \delta_{35},\quad \delta_{45}\quad \text{and}\quad \delta_{56}$$

In this case the system of simultaneous equations will fall into two separate groups, the first containing four unknowns out of the six and the second two unknowns only:

(a) the first group

$$X_1\delta_{11}+X_3\delta_{13}+X_4\delta_{14}+X_6\delta_{16}=-\Delta_{1p}$$
$$X_1\delta_{31}+X_3\delta_{33}+X_4\delta_{34}+X_6\delta_{36}=-\Delta_{3p}$$
$$X_1\delta_{41}+X_3\delta_{43}+X_4\delta_{44}+X_6\delta_{46}=-\Delta_{4p}$$
$$X_1\delta_{61}+X_3\delta_{63}+X_4\delta_{64}+X_6\delta_{66}=-\Delta_{6p}$$

(b) the second group

$$X_2\delta_{22}+X_5\delta_{25}=-\Delta_{2p}$$
$$X_2\delta_{52}+X_5\delta_{55}=-\Delta_{5p}$$

All the displacements will be calculated assuming that $EJ = 1$ ton sq m

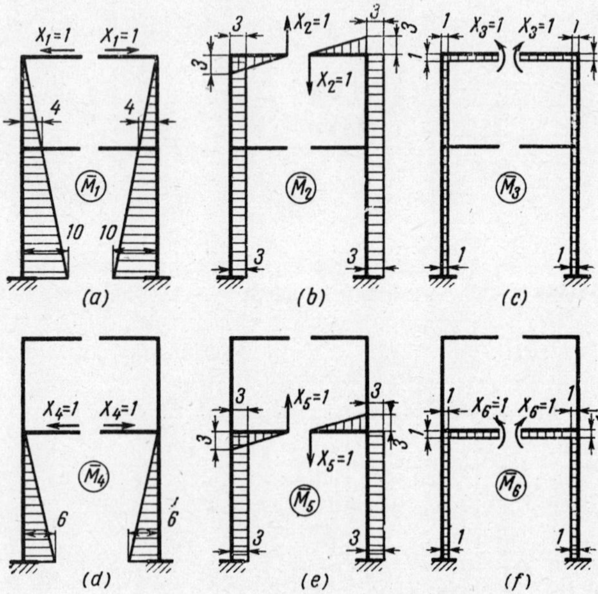

Fig. 30.12

$$\delta_{11}=\frac{4\times 4}{2}\times\frac{2}{3}\times 4\times\frac{2}{1}+\frac{6}{6}(2\times 4^2+2\times 10^2+2\times 4\times 10)\frac{2}{2}=354.7$$

$$\delta_{13}=\frac{4\times 4}{2}\times 1\times\frac{2}{1}+6\times\frac{4+10}{2}\times 1\times\frac{2}{2}=58$$

$$\delta_{14}=\frac{6\times 6}{2}\left(4+\frac{2}{3}\times 6\right)\frac{2}{2}=144$$

$$\delta_{16}=\frac{4+10}{2}\times 6\times 1\times\frac{2}{2}=42$$

$$\delta_{33}=6\times 1\times 1\times\frac{1}{2}+4\times 1\times 1\times 2+6\times 1\times 1\times 1=17$$

$$\delta_{34}=\frac{6\times 6}{2}\times\frac{1}{2}\times 2=18$$

$$\delta_{36}=1\times 6\times\frac{1}{2}\times 2=6$$

$$\delta_{44}=\frac{6\times 6}{2}\times\frac{2}{3}\times 6\times\frac{2}{2}=72$$

$$\delta_{46}=\frac{6\times 6}{2}\times\frac{1}{2}\times 2=18$$

$$\delta_{66}=6\times 1\times 1\times\frac{1}{4}+6\times 1\times 1\times\frac{1}{2}\times 2=7.5$$

$$\delta_{22}=\frac{3\times 3}{2}\times\frac{2}{3}\times 3\times\frac{1}{2}\times 2+4\times 3\times 3\times 2+6\times 3\times 3\times 1=135$$

$$\delta_{25}=3\times 6\times 3\times\frac{1}{2}\times 2=54$$

$$\delta_{55}=\frac{3\times 3}{2}\times\frac{2}{3}\times 3\times\frac{1}{4}\times 2+6\times 3\times 3\frac{1}{2}\times 2=58.5$$

$$\Delta_{1p}=-\frac{4\times 4\times 9}{2\times 1}-\frac{6\,(10+4)\times 9}{2\times 2}=-261$$

$$\Delta_{2p}=-\frac{9\times 3}{3\times 2}\times\frac{3}{4}\times 3-4\times 9\times 3-\frac{6\times 9\times 3}{2}=-199.12$$

$$\Delta_{3p}=-\frac{3\times 9\times 1}{3\times 2}-\frac{4\times 9\times 1}{1}-\frac{6\times 9\times 1}{2}=-67.5$$

$$\Delta_{4p}=-\frac{6\times 6\times 9}{2\times 2}=-81$$

$$\Delta_{5p}=-\frac{6\times 9\times 3}{2}=-81$$

$$\Delta_{6p}=-\frac{6\times 9\times 1}{2}=-27$$

In order to check the values of these displacements construct the summary unit bending moment diagram $\overline{M}_s$ due to the simultaneous action of all the

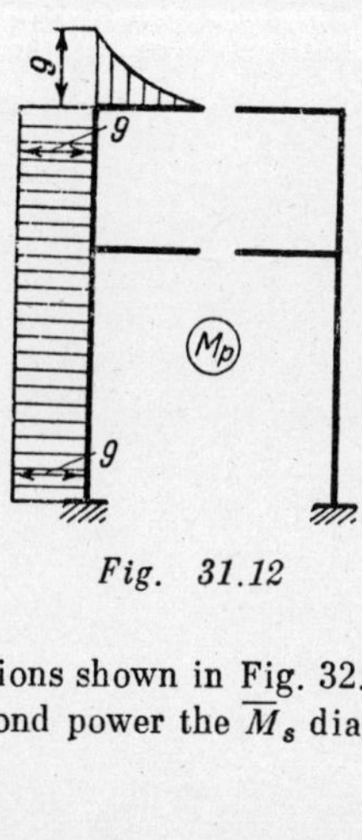

Fig. 31.12

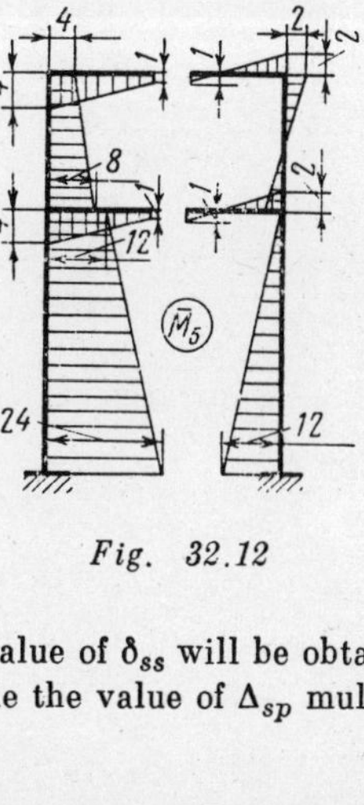

Fig. 32.12

unit reactions shown in Fig. 32.12. The value of δ_{ss} will be obtained raising to the second power the $\overline{M}_s$ diagram while the value of Δ_{sp} multiplying the

$\overline{M}_s$ diagram by the M_p one

$$\delta_{ss}=\Sigma\int\overline{M}_s^2\frac{ds}{EJ}=\frac{6}{6\times 2}(2\times 4^2+2\times 2^2-4\times 2\times 2)+$$

$$+\frac{4}{6\times 1}(2\times 4^2+2\times 8^2+4\times 8\times 2)+\frac{6}{6\times 2}(2\times 12^2+2\times 24^2+12\times 24\times 2)+$$

$$+\frac{6}{6\times 4}(2\times 4^2+2\times 2^2-4\times 2\times 2)+\frac{4}{6\times 1}(2\times 2^2+2\times 2^2-2\times 2\times 2)+$$

$$+\frac{6\times 12\times 2}{2\times 2\times 3}\times 12=1324.7$$

$$\Delta_{sp}=\Sigma\int\overline{M}_sM_p\frac{ds}{EJ}=-\frac{3\times 9}{3\times 2}\left(1+\frac{3}{4}\times 3\right)-\frac{4\times 9}{2}(4+8)-$$

$$-\frac{6\times 9}{2\times 2}(12+24)=-716.62$$

Condition (11.12) requires that $\delta_{ss}=\Sigma\delta$. Indeed

$$1324.7=354.7+17+72+7.5+135+58.5+2\,(58+144+42+$$
$$+18+6+18+54)\quad\text{or}\quad 1324.7=1324.7$$

Condition (13.12) requires that $\Delta_{sp}=\Sigma\Delta$. Indeed

$$-716.62=-261-199.12-67.5-81-81-27=-716.62$$

Both of these conditions being satisfied, we may be sure that no error has been committed in calculating the unit displacements. Substituting the values of these displacements in the two groups of simultaneous equations, we obtain

$$354.7X_1+58X_3+144X_4+42X_6=+261$$
$$58X_1+17X_3+18X_4+6X_6=+67.5$$
$$144X_1+18X_3+72X_4+18X_6=81.0$$
$$42X_1+6X_3+18X_4+7.5X_6=+27.0$$
$$135X_2+54X_5=+199.12$$
$$54X_2+58.5X_5=+81.0$$

The solution of these equations (omitted here) leads to the following values of the redundant reactions

$X_1=+0.607$ ton $\quad X_4=-0.692$ ton

$X_2=+1.460$ tons $\quad X_5=+0.037$ ton

$X_3=+2.753$ ton metres $\quad X_6=-0.340$ ton metre

Applying all the forces to the simple statically determinate structure and calculating the bending moments induced both by the reactions just obtained and the actual loads we find

$$M_{56}=+1.460\times 3+2.753-9=-1.87 \text{ ton metres}$$
$$M_{65}=-1.460\times 3+2.753=-1.63 \text{ ton metres}$$
$$M_{34}=+0.037\times 3-0.340=-0.23 \text{ ton metre}$$
$$M_{43}=-0.037\times 3-0.340=-0.45 \text{ ton metre}$$
$$M_{35}=+0.607\times 4+1.460\times 3+2.753-9=+0.56 \text{ ton metre}$$

$$M_{46} = +0.607\times4-1.460\times3+2.753 = +0.80 \text{ ton metre}$$

$$M_{31} = +0.607\times4+1.460\times3+2.753+0.037\times3-0.340-$$
$$-9 = +0.33 \text{ ton metre}$$

$$M_{42} = +0.607\times4-1.460\times3+2.753-0.037\times3-0.340=$$
$$= +0.36 \text{ ton metre}$$

$$M_{13} = +0.607\times10+1.460\times3+2.753-0.692\times6+$$
$$+0.037\times3-0.340-9 = -0.18 \text{ ton metre}$$

$$M_{24} = +0.607\times10-1.460\times3+2.753-0.692\times6-$$
$$-0.037\times3-0.340 = -0.16 \text{ ton metre}$$

The same bending moments could be obtained as follows: First multiply the unit bending moment graphs of Fig. 30.12 by the corresponding values

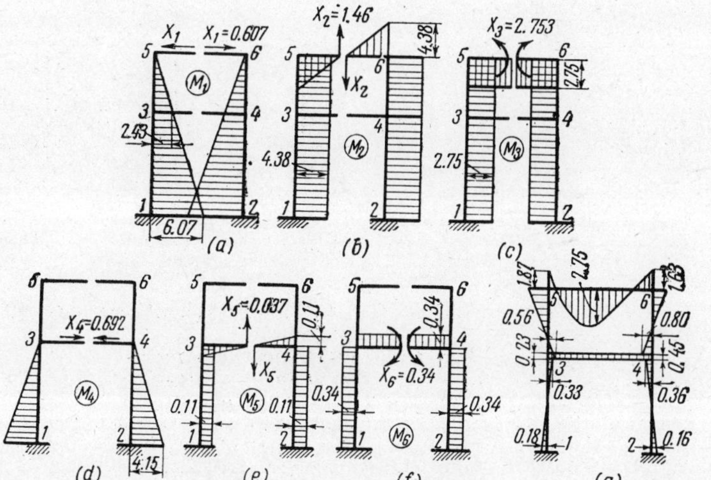

Fig. 33.12

of the unknowns. Thus, all the ordinates to the diagram induced by $X_1 = 1$ (see Fig. 30.12*a*) should be multiplied by $+$ 0.607, the ordinates to the graph induced by $X_2 = 1$ (Fig. 30.12*b*) by $+1.460$, and so forth. This will give the bending moment diagrams shown in Fig. 33.12*a*, *b*, *c*, *d*, *e* and *f* *. This being done, add up the ordinates to all of these diagrams together with those to the diagram due to the actual loads (see Fig. 31.12). The resulting ordinates will represent the ordinates to the bending moment diagram corre-

* It is not advisable to change the scales of the unit diagrams instead of constructing new ones, this procedure constituting a source of frequent errors.

sponding to the given redundant structure. It will be readily observed that the bending moments at the joints are the same as determined previously

$$M_{56}=4.38+2.75-9=-1.87 \text{ ton metres}$$

$$M_{34}=+0.11-0.34=-0.23 \text{ ton metre}$$

$$M_{35}=+2.43+4.38+2.75-9=+0.56 \text{ ton metre}$$

$$M_{31}=+2.43+4.38+2.75+0.11-0.34-9=+0.33 \text{ ton metre}$$

$$M_{13}=+6.07+4.38+2.75-4.15+0.11-0.34-9=-0.18 \text{ ton metre}$$

and so forth. The resulting bending moment diagram for the frame under consideration is given in Fig. 33.12*g*.

Problem 3. Required the complete analysis of a double-span symmetrical frame loaded with one horizontal force (Fig. 34.12*a*).

Solution. The frame under consideration is redundant in the third degree. The simple statically determinate structure adopted is represented in

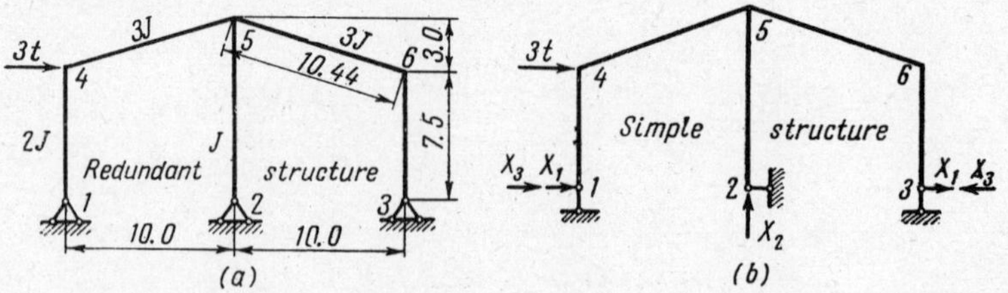

Fig. 34.12

Fig. 34.12*b*. The bending moment diagrams induced by unit reactions will be constructed for the following groups of unknowns:

X_1 consisting of two horizontal antisymmetrical forces,

X_2 consisting of one vertical reaction at the central support,

X_3 consisting of two symmetrical horizontal forces.

The corresponding unit bending moment diagrams are shown in Fig. 35.12. Let us resolve the force acting on the frame also in two groups of components, one symmetrical and the other antisymmetrical as indicated in Fig. 36.12*a* and *b*. The bending moment diagrams due to these components are also given in the same figure.

The displacements Δ induced by the actual loads will be obtained multiplying the bending moment graph due to antisymmetrical unit forces X_1 (Fig. 35.12*a*) by the graph of Fig. 36.12*b* due to the antisymmetrical component of the actual load. The same operation will be repeated with the graphs due to the symmetrical unknowns $X_2 = 1$ and $X_3 = 1$ on one hand, and the graph due to the symmetrical components of force P given in Fig.36. 12*a*, on the other. It is obvious that both secondary displacements δ_{12} and δ_{13} will reduce to zero. Consequently, the three simultaneous equations become

$$X_1\delta_{11}+\Delta_{1p}=0$$

$$X_2\delta_{22}+X_3\delta_{23}+\Delta_{2p}=0$$

$$X_2\delta_{32}+X_3\delta_{33}+\Delta_{3p}=0$$

Let us calculate all the displacements assuming that $EJ = 1$ ton sq m

$$\delta_{11}=2\times\frac{10.44\times 7.5\times 9}{3}+2\times\frac{10.44\times 3\times 9.5}{2\times 3}+2\times$$
$$\times\frac{7.5\times 7.5\times 5}{2\times 2}+\frac{10.5\times 21\times 14}{2\times 1}=+2253.1$$

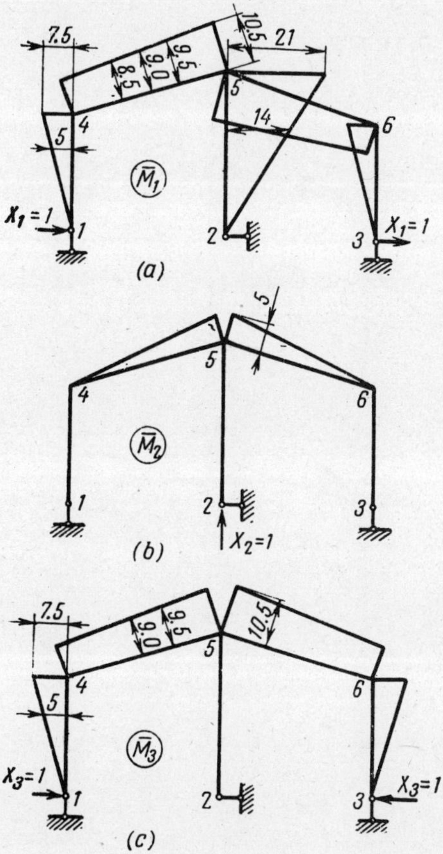

Fig. 35.12

$$\delta_{22}=2\times\frac{10.44\times 5\times\frac{2}{3}\times 5}{2\times 3}=+58$$

$$\delta_{33}=2\times\frac{10.44\times 7.5\times 9}{3}+2\times\frac{10.44\times 3\times 9.5}{2\times 3}+2\times$$
$$\times\frac{7.5\times 7.5\times 5}{2\times 2}=+709.6$$

$$\delta_{23}=2\times\frac{10.44\times5\times9.5}{2\times3}=+165.3$$

$$\Delta_{1p}=2\times\frac{10.44\times15.75\times9.5}{2\times3}+\frac{10.5\times31.5\times14}{2\times1}=+2835.9$$

$$\Delta_{2p}=2\times\frac{10.44\times5\times3}{2\times3}=+52.2$$

$$\Delta_{3p}=2\times\frac{10.44\times4.5\times9.5}{2\times3}=+148.77$$

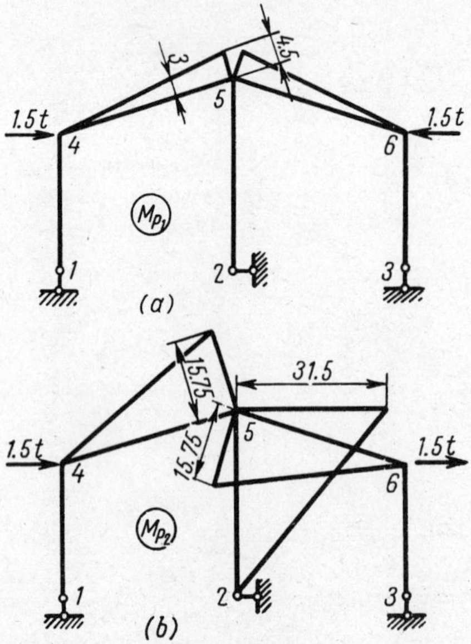

Fig. 36.12

Substituting these values into the above system of equations we obtain

$$2253.1X_1=-2835.9$$

$$58X_2+165.3X_3=-52.2$$

$$165.3X_2+709.6X_3=-148.77$$

The solution of these equations yields

$$X_1=-1.259;\quad X_2=-0.900=-0.9;\quad X_3=0$$

The bending moments at different cross sections of the structure will be

$M_{45}=+1.259\times7.5=+9.44$ ton metres

$M_{54}=+1.259\times10.5+0.9\times5-4.5-15.75=-2.53$ ton metres

$M_{56}=-1.259\times10.5+0.9\times5-4.5+15.75=+2.53$ ton metres

$M_{65}=-1.259\times7{\cdot}5=-9.44$ ton metres

$M_{52}=-1.259\times21+31.5=+5.06$ ton metres

The same bending moments could be obtained using the procedure adopted in the previous problem, that is multiplying the ordinates to the unit graphs by the magnitudes of the appropriate unknowns * and then summing up all these ordinates together with the ordinates to the diagram due to the actual loading.

The resulting bending moment diagram is given in Fig. 37.12.

Certain simplifications could be introduced in the above computations on the following grounds. When the given structure of Fig. 34.12*a* is acted

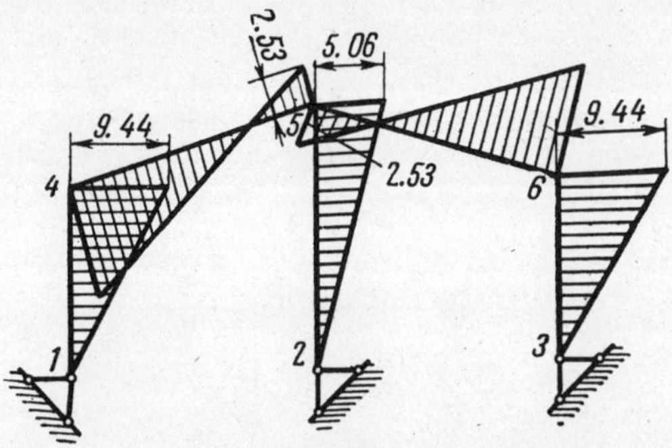

Fig. 37.12

upon by two symmetrical horizontal forces as shown in Fig. 36.12*a* alone the inclined members will work in direct compression, all other members remaining idle. **

For this reason the bending moment diagram due to the single load will be exactly the same as the one produced by the application of its antisymmetrical components. It follows that the horizontal reactions at the supports of the two outside columns must be equal both in value and in direction, which means that $X_3 = 0$. Previously we have arrived at the same conclusion at the outcome of rather laborious calculations.

The symmetrical components of force P will provoke no bending at joint *5*. However, we have found previously that the bending moments in the adjacent cross sections of the inclined members due to the antisymmetrical components equal —4.5 ton metres (Fig. 36.12*a*). These moments must be balanced by the moments induced by the redundant reactions. Of these only X_2 and X_3 can give rise to bending in the inclined members immediately to the left and to the right of joint *5*. Consequently, the bending moments induced by both of these reactions at the cross sections just mentioned must equal 4.5 ton metres. And since $X_3 = 0$ we may write that $-5X_2 = +4.5$

◆

* This means that the ordinates to the $\overline{M}_1$ diagram will be multiplied by (—1.259) and those to the $\overline{M}_2$ diagram by (—0.9).

** This is easily proved by the introduction of imaginary hinges at joints *4* and *6* and by the elimination of the horizontal constraint at joint *2*. The application of the symmetrical components will produce no bending in any member of the latter structure. Therefore, all the free terms of the canonical equations become nil, which requires that all the reactions of the redundunt constraints should be equally nil.

wherefrom

$$X_2=\frac{-4.5}{5}=-0.9 \text{ ton}$$

In the above equation the coefficient (—5) to the unknown X_2 represents the value of the bending moment at the cross section under consideration due to a unit load following the direction of X_2 (see Fig. 35.12*b*). Thus, out of three unknowns two may be determined off hand leaving only one unknown X_1. This will require the computation of two displacements δ_{11} and Δ_{1p} and the solution of one equation with one unknown. The above example shows that in certain cases quite complicated problems can be solved very simply.

Problem 4. Required the stress diagrams for both vertical and horizontal members of a three-span highway bridge schematically represented in Fig. 38.12*a* carrying on the first two spans a uniformly distributed load of three tons per metre.

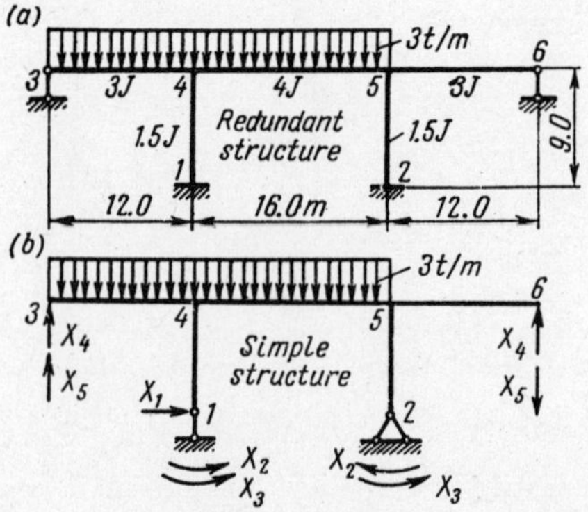

Fig. 38.12

Solution. The simple statically determinate structure to be adopted is indicated in Fig. 38.12*b*. The bending moment diagrams due to unit reactions are given in Fig. 39.12. The diagrams due to the symmetrical and antisymmetrical components of the actual loads are represented in Fig. 40.12. * An examination of the above-mentioned bending moment diagrams leads immediately to the conclusion that

$$\delta_{13}=\delta_{15}=\delta_{23}=\delta_{25}=\delta_{34}=\delta_{45}=0$$

Consequently, the five simultaneous equations will subdivide into two independent systems, one containing two equations and the other three. The first system will contain only the symmetrical unknowns

$$X_1\delta_{11}+X_2\delta_{12}+X_4\delta_{14}=-\Delta_{1p}$$
$$X_1\delta_{21}+X_2\delta_{22}+X_4\delta_{24}=-\Delta_{2p}$$
$$X_1\delta_{41}+X_2\delta_{42}+X_4\delta_{44}=-\Delta_{4p}$$

* The replacement of the actual loading by its symmetrical and antisymmetrical components will entail in the present case hardly any simplification at all.

and the second only the antisymmetrical ones

$$X_3\delta_{33}+X_5\delta_{35}=-\Delta_{3p}$$
$$X_3\delta_{53}+X_5\delta_{55}=-\Delta_{5p}$$

Let us proceed with the calculation of all the displacements assuming that $EJ_1 = 1$ ton sq m

$$\delta_{11}=+2\times\frac{9\times9\times6}{2\times1.5}+\frac{16\times9\times9}{4}=+648$$

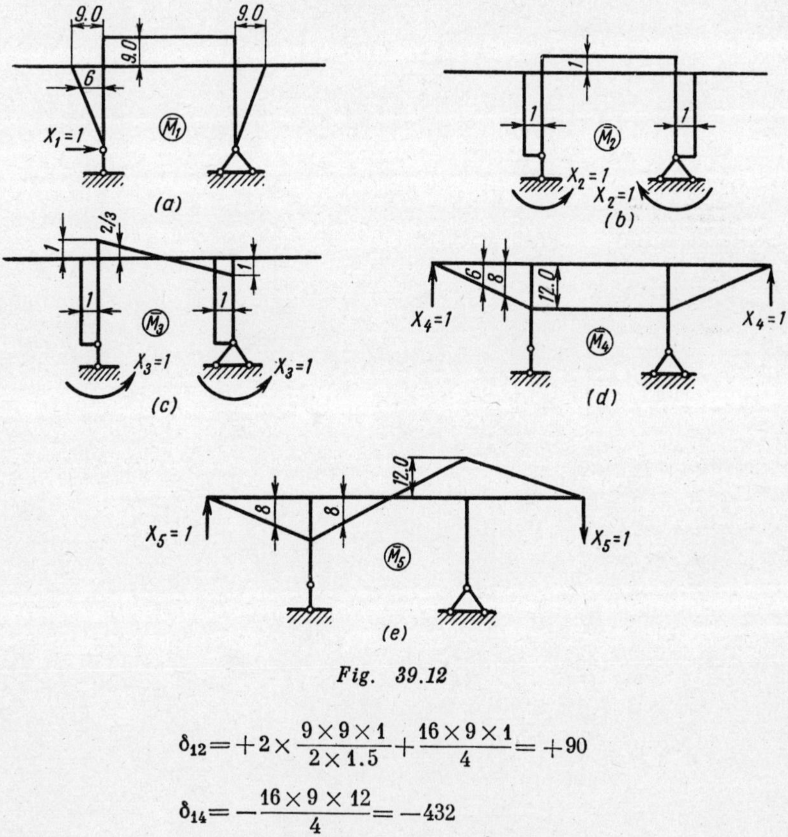

Fig. 39.12

$$\delta_{12}=+2\times\frac{9\times9\times1}{2\times1.5}+\frac{16\times9\times1}{4}=+90$$

$$\delta_{14}=-\frac{16\times9\times12}{4}=-432$$

$$\delta_{22}=+2\times\frac{9\times1\times1}{1.5}+\frac{16\times1\times1}{4}=+16$$

$$\delta_{24}=-\frac{16\times1\times12}{4}=-48$$

$$\delta_{44}=+2\times\frac{12\times12\times8}{2\times3}+\frac{16\times12\times12}{4}=+960$$

$$\delta_{33} = +2\times\frac{9\times1\times1}{1.5}+2\frac{8\times1\times2/3}{2\times4} = +13.333$$

$$\delta_{35} = -2\times\frac{8\times1\times8}{2\times4} = -16$$

$$\delta_{55} = +2\times\frac{12\times12\times8}{2\times3}+2\times\frac{8\times12\times8}{2\times4} = +576$$

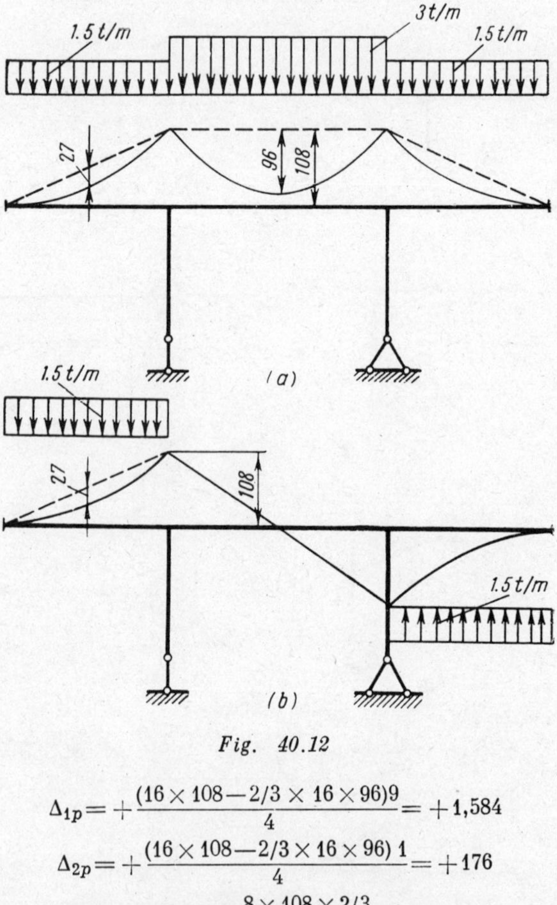

Fig. 40.12

$$\Delta_{1p} = +\frac{(16\times108-2/3\times16\times96)9}{4} = +1{,}584$$

$$\Delta_{2p} = +\frac{(16\times108-2/3\times16\times96)\,1}{4} = +176$$

$$\Delta_{3p} = +2\times\frac{8\times108\times2/3}{2\times4} = 144$$

$$\Delta_{4p} = -2\times\frac{12\times108\times8}{2\times3}+\frac{2\times2/3\times12\times27\times6}{3} -$$

$$-\frac{(16\times108-2/3\times16\times96)\,12}{4} = -4{,}704$$

$$\Delta_{5p} = -2 \times \frac{12 \times 108 \times 8}{2 \times 3} + 2 \times \frac{2/3 \times 12 \times 27 \times 6}{3} -$$

$$-2 \times \frac{8 \times 108 \times 8}{2 \times 4} = -4{,}320$$

Introducing these values in the equations given above we obtain

$$648X_1 + 90X_2 - 432X_4 = -1{,}584$$

$$90X_1 + 16X_2 - 48X_4 = -176$$

$$-432X_1 - 48X_2 + 960X_4 = +4{,}704$$

$$13.333X_3 - 16X_5 = -144$$

$$-16X_3 + 576X_5 = +4{,}320$$

Dividing all the terms of the first equation by 18, of the second by 2,

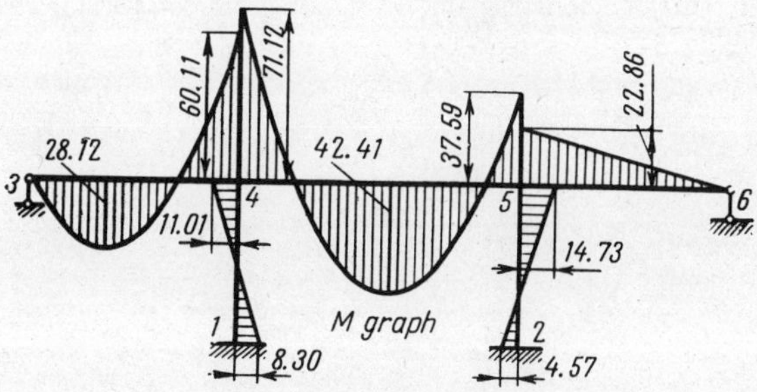

Fig. 41.12

of the third by 48, of the fourth by 8/3 and of the fifth by 16, we find

$$36X_1 + 5X_2 - 24X_4 = -88$$

$$45X_1 + 8X_2 - 24X_4 = -88$$

$$-9X_1 - X_2 + 20X_4 = +98$$

$$5X_3 - 6X_5 = -54$$

$$-X_3 + 36X_5 = 270$$

The solution of these equations yields

$$X_1 = +2.145 \text{ tons}$$

$$X_2 = -6.435 \text{ ton metres}$$

$$X_3 = -1.862 \text{ ton metres}$$

$$X_4 = +5.543 \text{ tons}$$

$$X_5 = +7.448 \text{ tons}$$

In order to find the ordinates to the diagram of the resulting bending moments acting along the members of the redundant structure we may now add the ordinates to the unit graphs multiplied by the value of the corresponding unknown to those of the bending moment diagram due to the actual loading. This resulting bending moment diagram is represented in Fig. 41.12.

As for the shearing forces, immediately to the right of the left abutment we have

$$Q_{34}=\frac{3\times 12}{2}-\frac{60.11}{12}=12.99 \text{ tons}$$

In the deck member to the left of joint *4* we find

$$Q_{43}=-\frac{3\times 12}{2}-\frac{60.11}{12}=23.01 \text{ tons}$$

and to the right of the same joint

$$Q_{45}=\frac{3\times 16}{2}+\frac{71.12-37.59}{16}=26.10 \text{ tons}$$

Continuing in the same way we shall obtain all the data necessary for the construction of the shear diagram represented in Fig. 42.12*a*. This diagram will permit the construction of the diagram for normal stresses given in Fig. 42.12*b*.

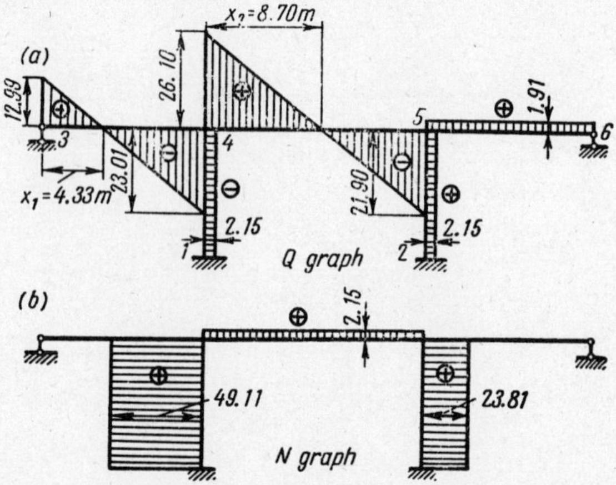

Fig. 42.12

In order to find the position of the maximum bending moments let us determine the points where shearing forces reduce to zero: $X_1=\frac{12.99}{3.0}=$ $=4.33$ metres and $X_2=\frac{26.10}{3.0}=8.70$ metres (Fig. 42.12*a*).

The bending moments at these cross sections will amount to for the span *3-4*:

$$M_{max}=12.99\times 4.33-3.0\times\frac{4\times 33^2}{2}=28.12 \text{ ton metres}$$

for the span *4-5*

$$M_{max}=-71.12+26.10\times 8.70-3.0\times\frac{8.70^2}{2}=42.41 \text{ ton metres}$$

The reactions at the supports are as follows

$$A_1=+49.11 \text{ tons}; \quad A_2=+23.81 \text{ tons}$$

$$A_3=+12.99 \text{ tons}; \quad A_6=-1.91 \text{ tons}$$

The values of reactions A_3 and A_6 may be easily checked remembering that

$$A_3=X_4+X_5=+5.543+7.448=+12.99 \text{ tons}$$

$$A_6=X_4-X_5=+5.543-7.448=-1.91 \text{ tons}$$

Problem 5. Required the bending moment diagrams for all the members of the frame given in Fig. 43.12. This frame is redundant to the sixth degree and the flexural rigidity of all its members is the same.

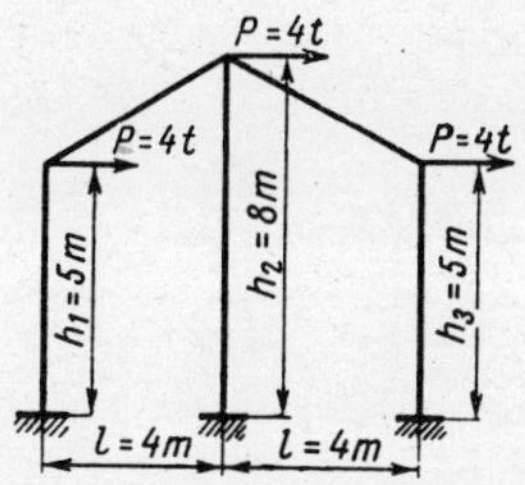

Fig. 43.12

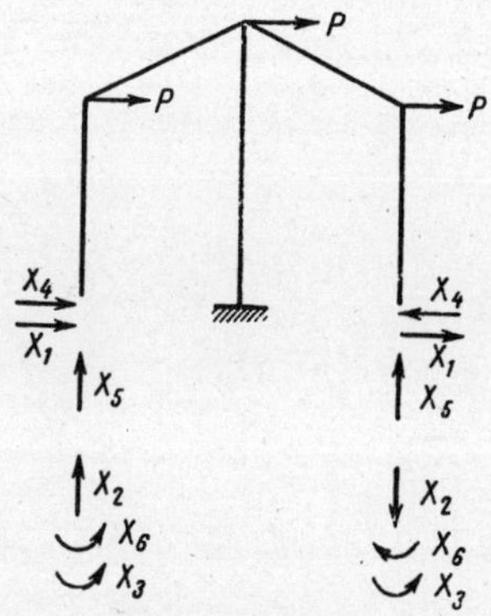

Fig. 44.12

Solution. Let us adopt the simple statically determinate structure shown in Fig. 44.12 and let us subdivide all the unknowns in two groups, the first containing only the antisymmetrical ones X_1, X_2 and X_3, and the second containing all the symmetrical ones X_4, X_5 and X_6. Since the system of loads acting on the frame is antisymmetrical itself only the antisymmetrical unknowns will differ from zero (see Art. 3.12). It follows that the given problem can be solved using one system of three simultaneous equations with three unknowns only

$$X_1\delta_{11}+X_2\delta_{12}+X_3\delta_{13}+\Delta_{1p}=0$$

$$X_1\delta_{21}+X_2\delta_{22}+X_3\delta_{23}+\Delta_{2p}=0$$

$$X_1\delta_{31}+X_2\delta_{32}+X_3\delta_{33}+\Delta_{3p}=0$$

In order to obtain the values of all the free terms and the coefficients to the unknowns we must construct the bending moment diagrams due both to the antisymmetrical unknowns and to the actual loads applied to the conjugate simple structure (Fig. 45.12). The multiplication of the graphs will be carried out using Vereshchagin's method, all the bending moment diagrams being bounded by straight lines. We shall also assume that $EJ = 1$ ton sq m. Hence

$$\delta_{11}=2\,\frac{5\times 5}{2}\times\frac{2\times 5}{3}+\frac{5}{6}\,(2\times 5^2+2\times 8^2+2\times 5\times 8)\,2+$$

$$+\frac{16\times 8\times 2}{2\times 3}\times 16=1{,}196$$

$$\delta_{22}=\frac{4\times 5}{2}\times\frac{2}{3}\times 4\times 2+8\times 8\times 8=565.3$$

$$\delta_{33}=1\times 5\times 1\times 4+2\times 8\times 2=52$$

$$\delta_{12}=-\frac{2\times 5}{6}\,(2\times 4\times 8+5\times 4)-\frac{16\times 8}{2}\times 8=-652$$

$$\delta_{13}=1.5\times\frac{5}{2}\times 2+1\times 5\times 6.5\times 2+\frac{8\times 16}{2}=218$$

$$\delta_{23}=-\frac{4\times 5}{2}\times 1\times 2-8\times 8\times 2=-148$$

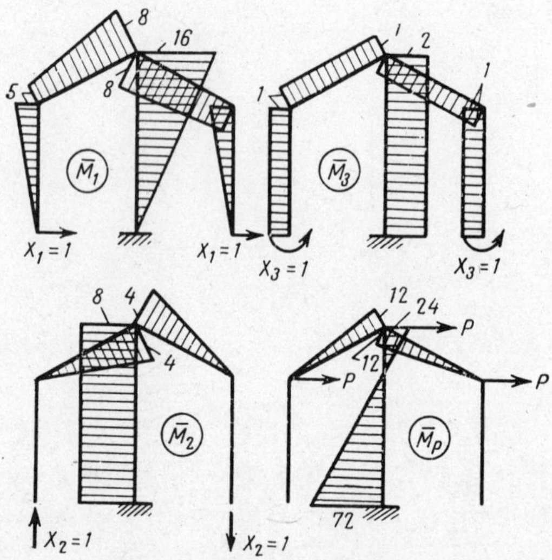

Fig. 45.12

The displacements induced in the simple structure by the actual loads will be obtained in a similar way

$$\Delta_{1p}=2\times 5/6\,(2\times 8\times 12+5\times 12)+8/6\,(2\times 16\times 24-16\times 72)=-92$$

$$\Delta_{2p}=-\frac{12\times 5}{2}\times\frac{2}{3}\times 4\times 2+8\times 8\times\frac{(72-24)}{2}=1{,}376$$

$$\Delta_{3p}=\frac{12\times 5}{2}\times 1\times 2-2\times 8\times 24=-324$$

The above values may be checked using the summary bending moment diagram due to the simultaneous application of all the unit reactions given in Fig. 46.12. Let us see whether condition (11.12) is fulfilled

$$\Sigma\delta=\delta_{11}+\delta_{22}+\delta_{33}+2\,(\delta_{12}+\delta_{13}+\delta_{23})$$

$$\Sigma\delta=1{,}196+565.3+52+2\,(-652+218-148)=649.3$$

At the same time

$$\delta_{ss}=\Sigma\int\overline{M}_s^2\,\frac{ds}{EJ}=\frac{2\times 5}{6}\,(2\times 1^2+2\times 6^2+2\times 1\times 6)+$$

$$+\frac{2\times 5}{6}\,(2\times 6^2+2\times 5^2+2\times 6\times 5)+8/6\,(2\times 10^2+$$

$$+2\times 6^2-2\times 6\times 10)=649.3=\Sigma\delta$$

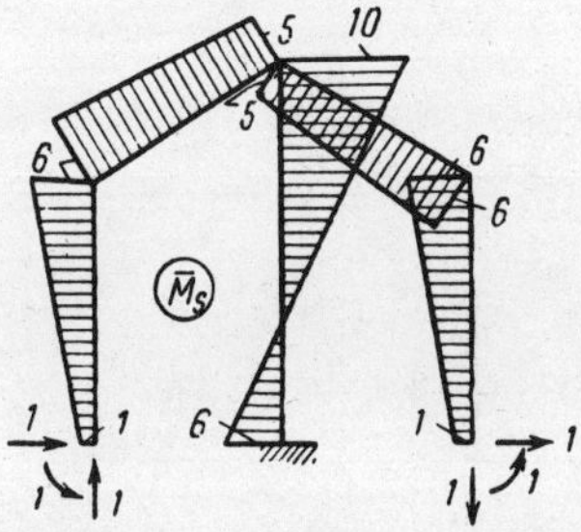

Fig. 46.12

Consequently, all the coefficients to the unknowns are correct. Check in the same way the displacements due to the applied loads using expression (13.12)

$$\Sigma\Delta=\Delta_{1p}+\Delta_{2p}+\Delta_{3p}=-92+1{,}376-324=960$$

$$\Delta_{sp}=\Sigma\int\overline{M}_sM_p\,\frac{ds}{EJ}=\frac{2\times 5}{6}\,(2\times 5\times 12+6\times 12)+$$

$$+8/6\,(2\times 10\times 24+2\times 6\times 72-10\times 72-6\times 24)=960=\Sigma\Delta$$

Having made sure that all the operations carried out thus far are correct, we may introduce the values of the displacements into the system of our simultaneous equations

$$1{,}196X_1-652X_2+218X_3=92$$

$$-652X_1+565.3X_2-148X_3=1{,}376$$

$$218X_1-148X_2+52X_3=324$$

Upon simplification these equations become

$$598X_1-326X_2+109X_3=46$$

$$-326X_1+283X_2-74X_3=688$$

$$109X_1-74X_2+26X_3=162$$

Their solution will be carried out in tabular form using the abbreviated method as indicated hereunder.

Table 3.12

Equation No.	X_1	X_2	X_3	Multipliers α_{ik}	S	K
(I)	598	—326	109	$\alpha_{12}=0.546$ $\alpha_{13}=-0.182$	381	46
(2)	.	283	—74		—117	—688
(I)·α_{12}	.	—177	59		208	25
(II)	.	106	—15	$\alpha_{23}=0.141$	91	—663
(3)	.	.	26		61	162
(I)·α_{13}	.	.	—19.8		—69.5	—8.4
(II)·α_{23}	.	.	—2.1		12.7	—93
(III)	.	.	4.1		4.2	61.4

Solving equation (III) we obtain

$$4.1X_3=61.4; \qquad X_3=14.97 \text{ ton metres}$$

Thereafter from equation (II)

$$106X_2-15\times 14.97=-663; \qquad X_2=-4.237 \text{ tons}$$

and finally from equation (I)

$$598X_1+326\times 4.237+109\times 14.97=46; \qquad X_1=-4.961$$

In order to make sure that the roots of the equations are correct let us introduce them into one of the simultaneous equations, say, the third one *

$$-4.961\times 109+4.237\times 74+14.97\times 26-162=0.009\approx 0$$

It remains now to multiply the ordinates to each of the unit bending moment graphs by the magnitude of the corresponding redundant reaction as shown in Fig. 47.12, and to add all of these ordinates together with those

◆

* It is more advisable to substitute these roots into all of the simultaneous equations.

to the bending moment diagram due to the applied loads (see Fig. 45.12). The ordinates situated to the right and below the corresponding members will be reckoned positive

$$M_{12}=-14.97$$
$$M_{21}=-14.97+24.80=9.83$$
$$M_{32}=-14.97-16.95+36.69-12.00=-4.23$$
$$M_{34}=29.94+33.90-79.38+24.00=8.46$$
$$M_{43}=29.94+33.90-72.00=-8.16$$

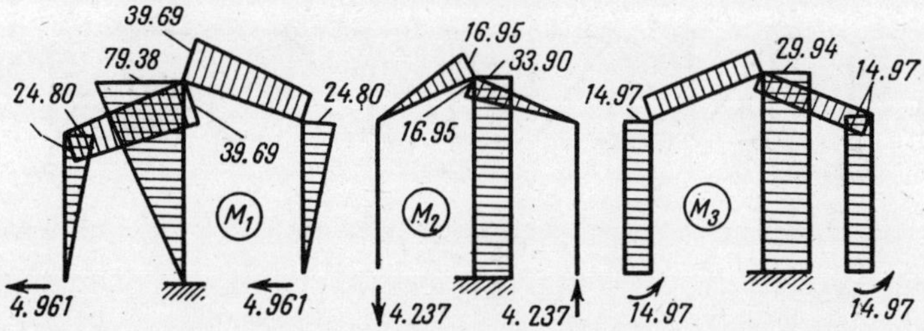

Fig. 47.12

The ordinates to the resulting bending moment diagram will be plotted on the side of the more extended fibres (Fig. 48.12).

Upon completion of this diagram it is necessary to check the accuracy thereof:

(1) Check the equilibrium of joint *3* (Fig. 49.12)

$$\Sigma M_3=-4.23-4.23+8.46=0$$

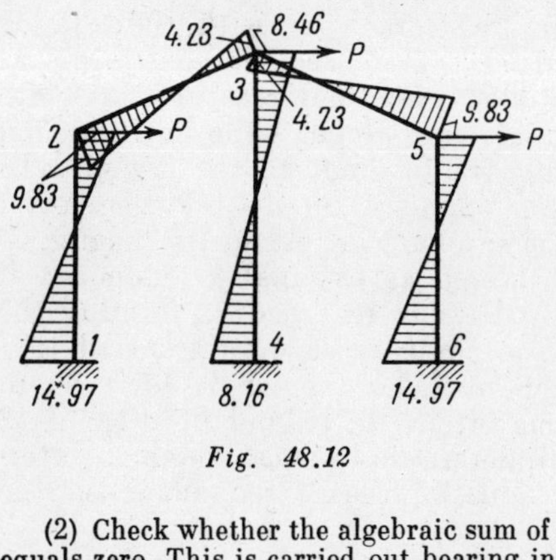

Fig. 48.12

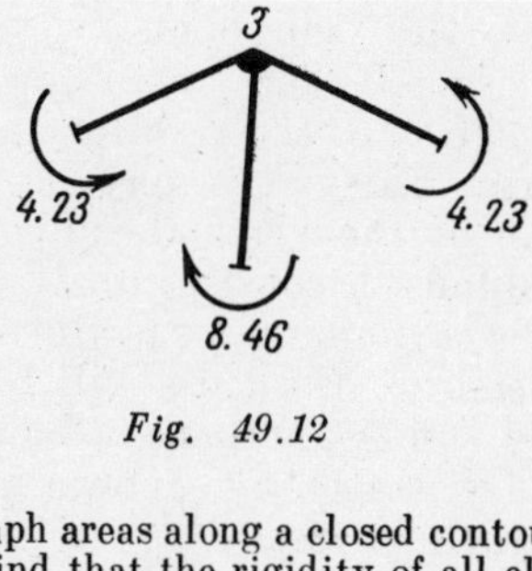

Fig. 49.12

(2) Check whether the algebraic sum of graph areas along a closed contour equals zero. This is carried out bearing in mind that the rigidity of all elements of contour *1-2-3-4* remaining constant and reckoning positive the parts

of the areas situated outside of the contour

$$\frac{14.97-9.835}{2}\times 5+\frac{4.23-9.835}{2}\times 5+\frac{8.46-8.164}{2}\times 8=0.009\approx 0$$

8.12. *Statically Indeterminate Trusses*

By *statically indeterminate* trusses we mean such geometrically stable hinge-connected framed structures for which neither the

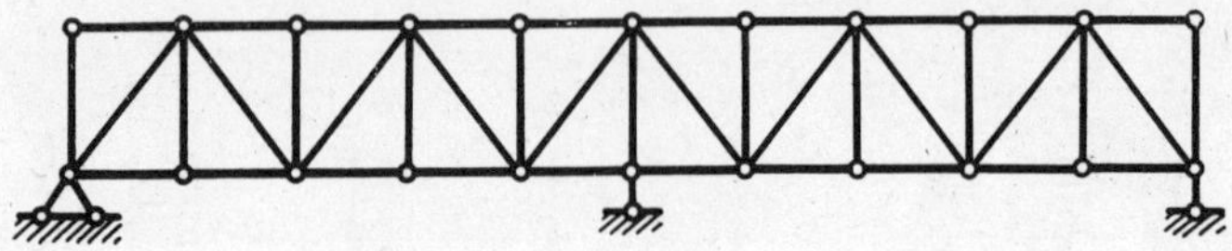

Fig. 50.12

stresses, not the reactions can be found without the knowledge of the deflections sustained.

Each statically indeterminate truss (as well as any other redundant structure) may be transformed into a simple statically determinate one by the elimination of the redundant constraints provided such constraints are not indispensable from the viewpoint of the stability of that structure.

The number of eliminated constraints will always represent the degree of redundancy of the truss under consideration. Redundant trusses may be statically indeterminate both *internally* and *externally* just as the framed structures with rigid joints studied in previous articles. In the first case the constraints at the supports are in such a number that their reactions may not be deduced from statical considerations alone, while in the second case the redundant constraints are inherent to the truss itself.

Fig. 50.12 represents a truss redundant in the first degree for which only the reactions at the supports are statically indeterminate. This truss may be considered as externally redundant if one of the vertical supporting bars is regarded as forming the redundant constraint. If on the contrary it were assumed that this redundant constraint is constituted by one of the lower chord members, the truss will become internally redundant. Fig. 51.12 represents another statically indeterminate truss which is externally statically determinate but internally redundant to the eighth degree.

As already stated in Art. 1.9, one of the most essential peculiarities of redundant structures resides in that the stresses developed in their members depend on the cross-sectional dimensions and lengths of these members. When different materials are used, the stress distribution becomes also a function of the moduli of elasticity of these materials. In addition, redundant structures are subjected to secondary stresses due to erection defects, movement of supports, temperature strains, etc., which makes them less desirable than statically determinate ones, as the latter are absolutely unaffected by all the above mentioned factors.

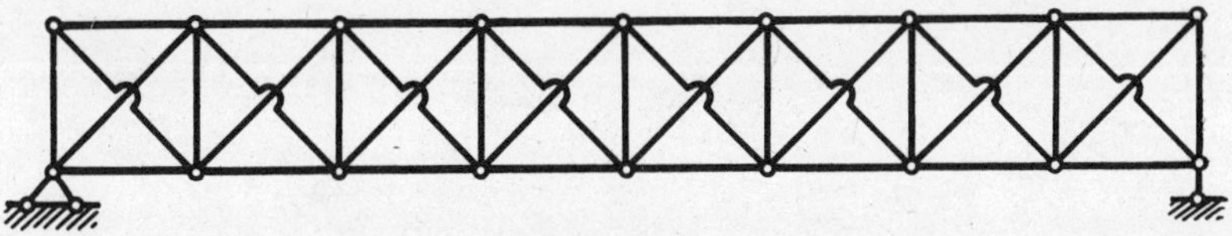

Fig. 51.12

On the other hand, statically indeterminate trusses are endowed equally with certain advantages. Thus, continuous trusses provide better rolling conditions for trains passing over railway bridges for their elastic line will present no peaks at the supports such as exist necessarily in the elastic lines of a series of statically determinate trusses. Accordingly, trains passing over these supports will feel no shocks or bumps. In addition, a continuous truss will always require less material for its construction than a series of statically determinate trusses covering the same span.

Continuous trusses are also simpler to build than the statically determinate ones, for the elimination of redundant constraints requires the introduction of special hinges which are usually rather complicated.

The stress analysis of redundant structures becomes more and more complicated with the increase in the number of redundant constraints.

In Art. 1.9 it was shown that the stresses or reactions developed by the necessary constraints can always be determined on the basis of equilibrium consideration, while those of the redundant constraints can be computed only if the deflections of the structure are known. It follows that cross-sectional areas of the necessary members of redundant trusses may be selected in exactly the same way as for the statically determinate ones, the stresses in these members being independent of their rigidity. As for the redundant members, their cross sections must be chosen in such a way that the unit stresses developed therein should be as close as possible to the permissible ones for that particular material.

In order to arrive at this result the following procedure may be recommended: eliminate in the first instance all the redundant members of the truss converting it into a statically determinate system which will be thereafter regarded as constituting the *conjugate simple* structure. The cross-sectional dimensions for all the members of the latter structure will then be computed in the usual way. If the truss is subjected to moving loads, influence lines will be used.

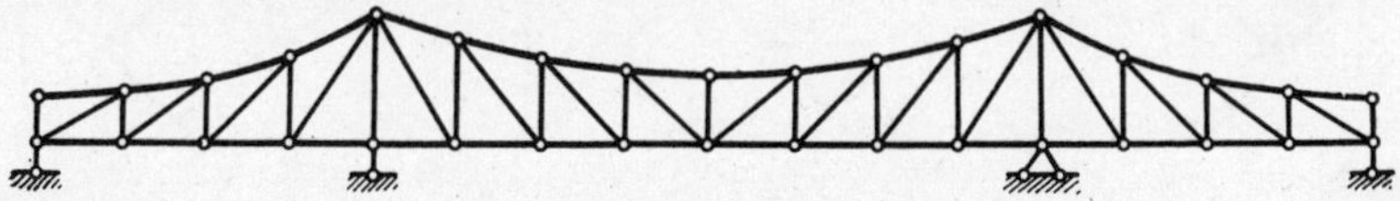

Fig. 52.12

The cross-sectional dimensions of the necessary members having been found, choose more or less arbitrarily the cross-sectional dimensions of all the redundant members of the truss and then recalculate all the stresses using any of the methods peculiar to redundant structures. If the unit stresses thus obtained differ substantially from the allowable ones, the cross-sectional areas must be corrected accordingly and the structure should be recalculated again.

Approximate values of stresses developed in members of statically indeterminate trusses may be obtained comparing these trusses with solid web beams covering the same number of spans and supporting the same loads.

Thus, the continuous truss shown in Fig. 52.12 could be replaced in the first instance by a solid web continuous beam resting on four supports for which the bending moment and shear diagrams could be easily obtained using one of the methods described in Chapter 10. Thereafter the stresses developed in both chords could be obtained dividing the corresponding ordinates to the bending moment diagram by the lever arm of the stress under consideration about the appropriate joint. The stresses existing in the verticals and diagonals will be obtained in a similar way using the shear diagram.

A similar procedure can be used for estimating stresses developed in redundant trussed arches.

The accuracy of stresses acting in members of redundant trusses is controlled in the same way as in the case of redundant structures with rigid joints such as portal and building frames, etc.

One must make sure that all the joints and portions of the truss are in equilibrium and that the deflections of the system are consistent with the stipulations of the problem. Thus, for instance, all the deflections at the supports must be found nil.

Problem. Determine the stresses in all the members of a one time statically indeterminate truss represented in Fig. 53.12. The truss carries five concentrated loads of ten tons each. All the members of this truss are of the same cross section.

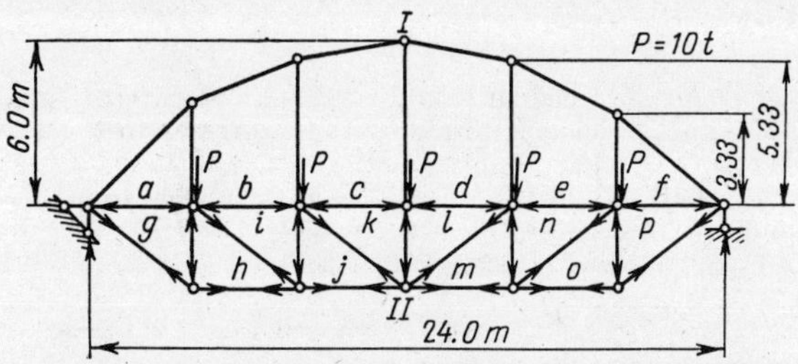

Fig. 53.12

Solution. If we assume that the vertical at midspan constitutes the redundant member, we may adopt as conjugate simple structure the one shown in Fig. 54.12. The unknown X_1 will represent the stress developed in the aforesaid vertical. In this case the canonical equation becomes

$$X_1\delta_{11}+\Delta_{1p}=0 \quad \text{wherefrom} \quad X_1=-\frac{\Delta_{1p}}{\delta_{11}}$$

Both deflections δ_{11} and Δ_{1p} will be determined using two Maxwell-Cremona diagrams, one of which will be constructed for the actual loads and

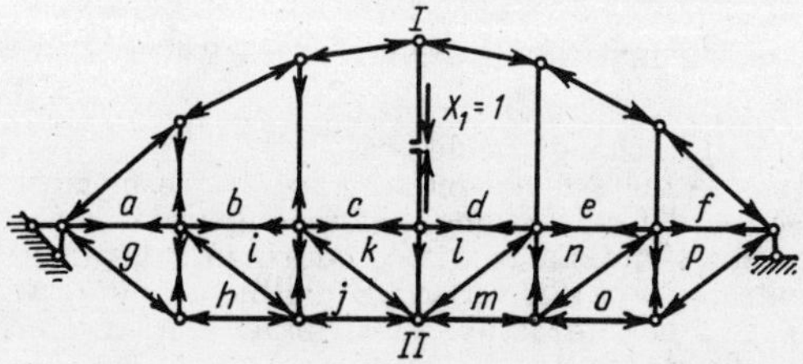

Fig. 54.12

the other for the unit load $X_1 = 1$ (Figs. 55.12 and 56.12, respectively). The values of these deflections are given by

$$\delta_{11}=\Sigma\frac{\overline{N}_1^2 S}{EF} \quad \text{and} \quad \Delta_{1p}=\Sigma\frac{\overline{N}_1 N_p S}{EF}$$

where S represents the length of each member.

All the computations will be carried out in tabular form as indicated hereunder (see Table 4.12). Summing up all the entries of columns 5 and 6, we obtain

$$EF\delta_{11}=2404.12; \qquad EF\Delta_{1p}=-15696.0$$

It follows that

$$X_1=+\frac{15696.0}{2404.12}=6.53 \text{ tons}$$

Stresses in all the other members can be determined easily using the expression

$$N_i=X_1\overline{N}_1$$

The values of N_i obtained in this way should be entered into column 7 of Table 4.12. Adding the magnitude of these stresses to those induced in the simple structure by the actual loads (the values of these stresses are given in column 3) we shall obtain the total stresses developed in all the members of the redundant truss. These stresses are represented in column 8.

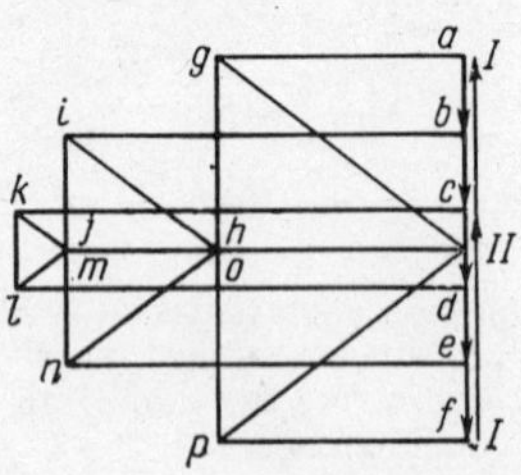

Fig. 55.12

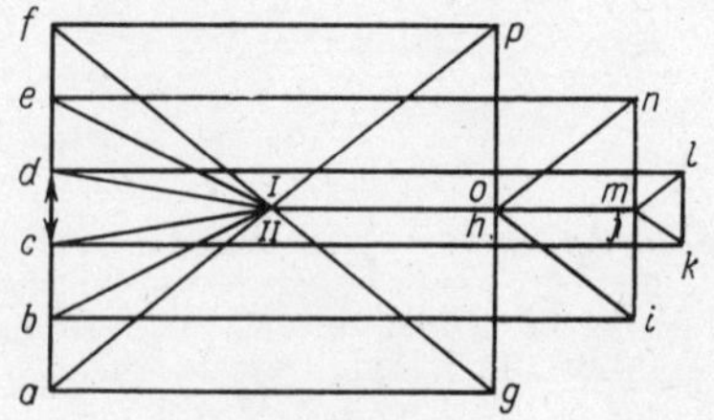

Fig. 56.12

In the case of stresses due to temperature changes the canonical equation for a one time statically indeterminate structure will become

$$X_1\delta_{11}+\Delta_{1t}=0 \quad \text{where} \quad \Delta_{1t}=\alpha\Sigma\overline{N}_1tS$$

where α = coefficient of thermal expansion
t = temperature change in degrees.

Let us study also the stresses induced in all the members of the same truss by an erection defect. Assume that the vertical *mn* has been made *a* units longer (or shorter) than required, which is equivalent to a thermal expansion or contraction of this vertical equalling $a=\alpha tS$, while the values of thermal expansion or contraction for all other truss members remain nil. In this case the above equation becomes

$$X_1\delta_{11}+\Delta_{1t}=0$$

$$\Delta_{1t}=\alpha\overline{N}_1tS=\overline{N}_1a$$

where $\overline{N}_1$ is the stress induced in the same vertical *mn* by a unit force X_1.

Stress analysis for trusses of a higher degree of redundancy can be carried out in exactly the same way. Resort can be made to the grouping of unknowns and to the replacement of the applied loads by equivalent symmetrical and antisymmetrical systems as described in the preceding article.

Table 4.12

Bar No.	Bar length S, m	Stress N_p, tons	Stress $\overline{N}_1$, tons	$\overline{N}_1^2 S$	$N_p \overline{N}_1 S$	$\overline{N}_1 X_1$, tons	Total stresses $N = N_p + \overline{N}_1 X_1$, tons
1	2	3	4	5	6	7	8
I-a	5.207	0	—3.9	34.2	0	—25.4	—25.4
I-b	4.47	0	—3.35	50.1	0	—21.9	—21.9
I-c	4.055	0	—3.05	37.7	0	—19.9	—19.9
I-d	4.055	0	—3.05	37.7	0	—19.9	—19.9
I-e	4.47	0	—3.35	50.1	0	—21.9	—21.9
I-f	5.207	0	—3.9	34.2	0	—25.4	—25.4
a-g	4.00	—33.3	6.33	160.3	—844	41.3	8.0
b-i	4.00	—53.5	8.33	278.0	—1782	54.4	0.9
c-k	4.00	—60.0	9.00	324.0	—2160	58.8	—1.2
d-l	4.00	—60.0	9.00	324.0	—2160	58.8	—1.2
e-n	4.00	—53.5	8.33	278.0	—2782	54.4	0.9
f-p	4.00	—33.3	6.33	160.3	—844	41.3	8.0
h-II	4.00	33.3	—3.33	44.4	—444	—21.7	11.6
j-II	4.00	53.5	—5.33	114.0	—1140	—34.8	18.7
m-II	4.00	53.5	—5.33	114.0	—1140	—34.8	18.7
o-II	4.00	33.3	—3.33	44.4	—444	—21.7	11.6
g-II	5.00	41.5	—4.15	86.0	—861	—27.1	14.4
i-h	5.00	25.0	—2.5	31.2	—312.5	—16.3	8.7
k-j	5.00	8.3	—0.83	3.5	—34.5	—5.4	2.9
l-m	5.00	8.3	—0.83	3.5	—34.5	—5.4	2.9
n-o	5.00	25.0	—2.5	31.2	—312.5	—16.3	8.7
p-II	5.00	41.5	—4.15	86.0	—861	—27.1	14.4
g-h	3.00	—25.0	2.5	18.8	—187.5	16.3	—8.7
j-i	3.00	—15.0	1.5	6.7	—67.5	9.8	—5.2
k-l	3.00	—10.0	1.0	3.0	—30.0	6.5	—3.5
m-n	3.00	—15.0	1.5	6.7	—67.5	9.8	—5.2
o-p	3.00	—25.0	2.5	18.8	—187.5	16.3	—8.7
a-b	3.33	0	1.0	3.33	0	6.5	6.5
b-c	5.33	0	1.0	5.33	0	6.5	6.5
c-d	6.00	0	1.0	6.00	0	6.5	6.5
d-e	5.33	0	1.0	5.33	0	6.5	6.5
e-f	3.33	0	1.0	3.33	0	6.5	6.5
Σ	—	—	—	2404.12	—15696	—	—

An example of influence line construction for a one time statically indeterminate truss has been given in Art. 9.9.

9.12. *Computation of Statically Indeterminate Structures with the Aid of Simpler Structures Redundant to a Lower Degree*

Simultaneous solution of several equations with several unknowns may be avoided if the conjugate structure used is one degree lower in redundancy than the one analyzed.

The stress computation of a structure redundant in the nth degree will reduce to the solution of a single equation with one unknown if the conjugate simple structure itself possesses $(n-1)$ redundant constraints. This single equation

$$X_1\delta_{11}+\Delta_{1p}=0$$

will show that the displacement of the $(n-1)$ times statically indeterminate conjugate structure along the direction of the additional constraint whose reaction equals X_1 is nil.

In the above expression δ_{11} and Δ_{1p} represent the deflections of the $(n-1)$ times statically indeterminate structure along the direction of this constraint caused by the unit reaction X_1 and by the applied loads, respectively.

Were we to adopt as simple conjugate structure the one obtained through the elimination of two redundant constraints, the simultaneous equations will become

$$X_1\delta_{11}+X_2\delta_{12}+\Delta_{1p}=0$$

$$X_1\delta_{21}+X_2\delta_{22}+\Delta_{2p}=0$$

where δ_{11}, δ_{12}, δ_{21}, δ_{22}, Δ_{1p} and Δ_{2p} are the deflections of the $(n-2)$ times statically indeterminate conjugate structure due to the unit forces X_1 and X_2 and to the applied loads. The displacements δ and Δ of this conjugate structure will be easily obtained if the diagrams of stresses (those for the bending moments in the case of frames with rigid joints or those for normal stresses in hinge-connected structures) induced in this structure by unit loads acting along the directions of the unknowns and by the applied loads are readily available. Such diagrams (or formulas permitting their construction) can be found very frequently in special engineering handbooks, in which case the amount of computation work may be reduced very considerably. However, if the stress diagrams pertaining to the structure adopted as a conjugate one are nonavailable, the procedure described becomes useless.

Problem. Construct the bending moment diagram for a framed structure redundant in the fourth degree given in Fig. 57.12 using the stress diagrams

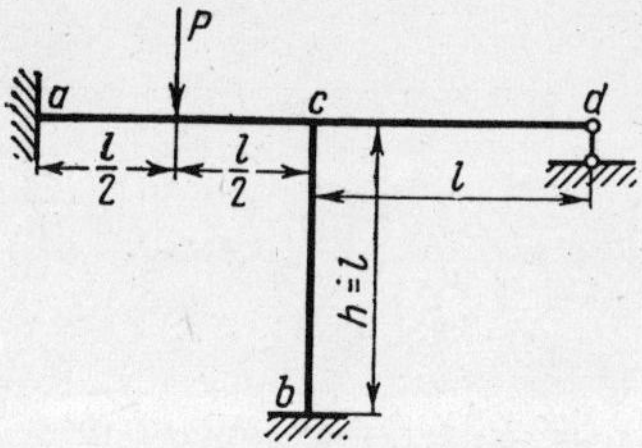

Fig. 57.12

available for the knee frame of Fig. 58.12. All the members of the frame have the same rigidity throughout.

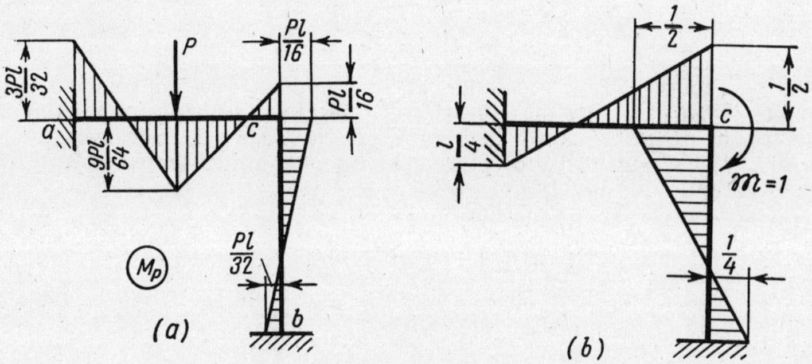

Fig. 58.12

Solution. Adopt the three times statically indeterminate system shown in Fig. 59.12 as a conjugate simple structure. This system is derived from the

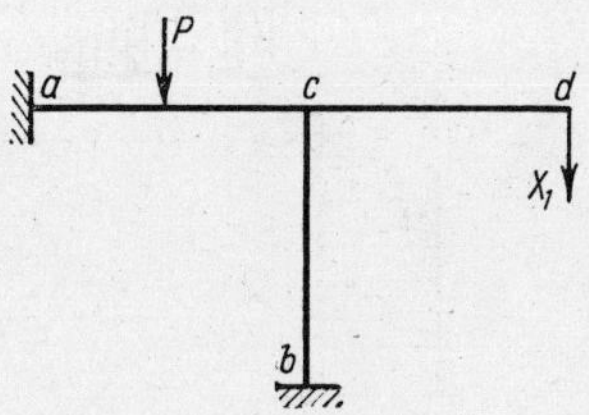

Fig. 59.12

given one by elimination of the right-hand support. The equation expressing that point d remains in place becomes

$$X_1\delta_{11}+\Delta_{1p}=0$$

The displacement δ_{11} will be obtained constructing the bending moment diagram induced in the conjugate structure by the unit load X_1. The cantilever *cd* of this structure is statically determinate and the diagram of bending moments produced therein by the load X_1 is triangular in shape with a maximum ordinate equal to l at joint c. It follows that the said joint will be acted

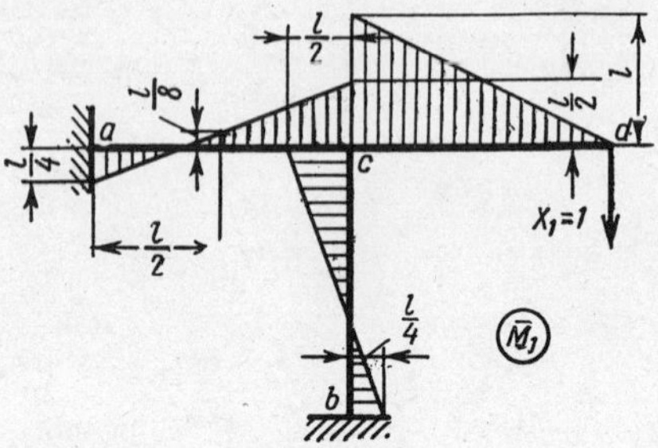

Fig. 60.12

upon by a couple $\mathfrak{M} = 1$ and the bending moment diagram relative to portion *acb* of the structure may therefore be obtained multiplying all the ordinates to the diagrams shown in Fig. 58.12*b* by l. Such a diagram is given in Fig. 60.12. The amount of displacement δ_{11} is equal to the second power of this diagram.

Using Vereshchagin's method we obtain

$$\delta_{11} = l/3\,[(l/4)^2 + (l/2)^2 - l/4 + l/2 + (l/2)^2 +$$

$$+ (l/4)^2 - l/2 \times l/4 + l^2]\,\frac{1}{EJ} = \frac{11l^3}{24EJ}$$

The same result could be arrived at by the multiplication of the $\overline{M}_1$ graph (Fig. 60.12) by the graph of Fig. 61.12 due to the application of a unit

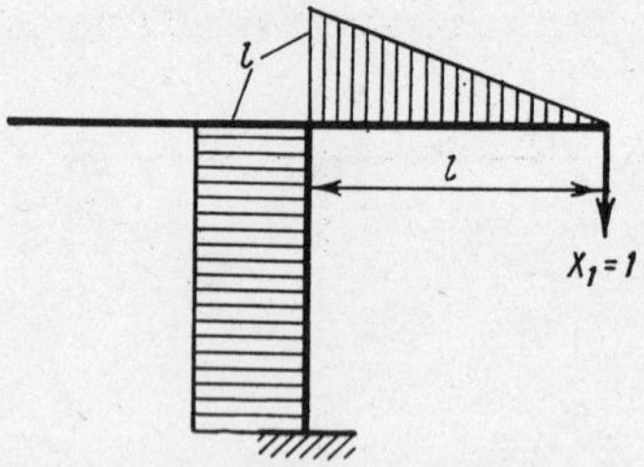

Fig. 61.12

load X_1 to the statically determinate structure shown in the same figure

$$\delta_{11} = \frac{ll}{2} \times \frac{2}{3}\,l + (l/2 - l/4) \times \frac{l}{2}\,l = \frac{11l^3}{24EJ}$$

Displacement Δ_{1p} will be found multiplying the diagram of Fig. 58.12*a* by the diagram given in Fig. 60.12

$$\Delta_{1p}=\Big[l/12\Big(-2\times\frac{5Pl}{32}\times l/4-2\times\frac{9Pl}{64}\times l/8+\frac{5Pl}{32}\times l/8+\frac{9Pl}{64}\times l/4-2\times\frac{9Pl}{64}\times l/8+2\times\frac{Pl}{16}\times l/2-\frac{9Pl}{64}\times l/2+\frac{Pl}{16}\times l/8\Big)+\frac{l}{6}\Big(-2\times\frac{Pl}{16}\times l/2-2\times\frac{Pl}{32}\times l/4+\frac{Pl}{16}\times l/4+\frac{Pl}{32}\times l/2\Big)\Big]\frac{1}{EJ}=-\frac{Pl^3}{64EJ}$$

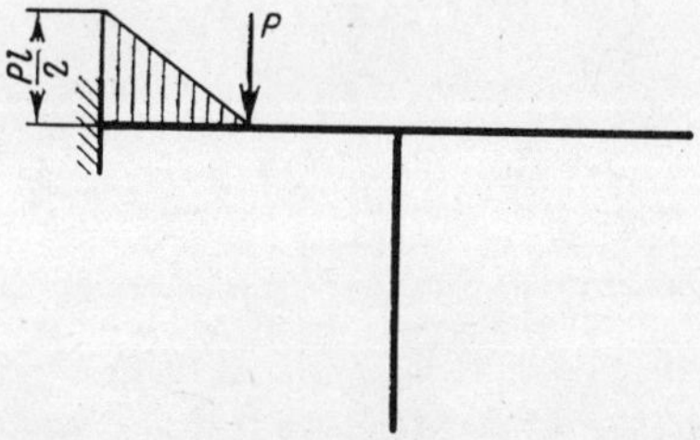

Fig. 62.12

The same result could be achieved using the diagrams of Figs. 58.12*a* and 61.12

$$\Delta_{1p}=\Big(-\frac{Pl}{16}+\frac{Pl}{32}\Big)\frac{l}{2}\cdot\frac{l}{EJ}=-\frac{Pl^3}{64EJ}$$

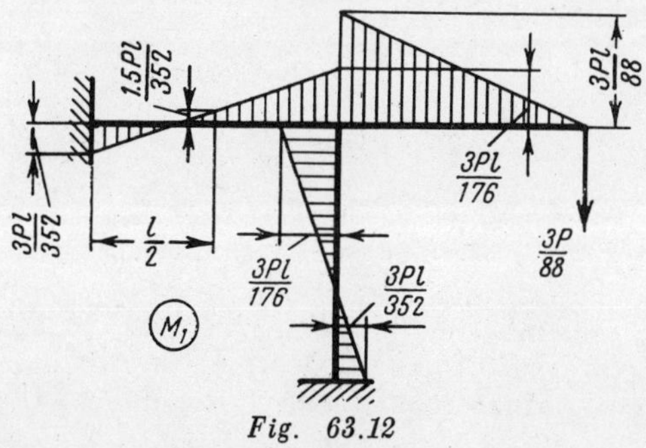

Fig. 63.12

or else those of Figs. 60.12 and 62.12 (see Art. 7.9)

$$\Delta_{1p}=\frac{Pl}{2}\times\frac{l}{2}\times\frac{1}{2}(l/8\times 1/3-l/4\times 2/3)\,\frac{1}{EJ}=-\frac{Pl^3}{64EJ}$$

The introduction of the values of δ_{11} and Δ_{1p} into the equation given above yields

$$X_1=-\frac{\Delta_{1p}}{\delta_{11}}=\frac{Pl^3\cdot 24EJ}{64EJ\cdot 11l^3}=\frac{3}{88}P$$

The resulting bending moment diagram for the given redundant structure will be obtained as usually multiplying all the ordinates to the $\overline{M}_1$ diagram by the magnitude of the unknown X_1 just found equal to $\frac{3}{88} P$ (Fig. 63.12) and thereafter adding these ordinates to those of the M_p diagram of Fig. 58.12. The final diagram is shown in Fig. 64.12.

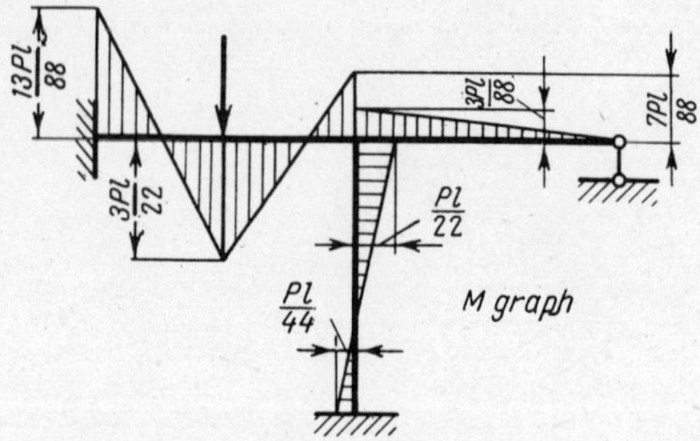

Fig. 64.12

10.12. *Influence Line Models for Continuous Beams*

If the influence line for a continuous beam redundant to the nth degree were constructed using as conjugate structure another continuous beam $(n - 1)$ times statically indeterminate, the reaction of nth constraint due to the load unity P could be derived from the equation

$$X_1\delta_{11} + \delta_{1p} = 0$$

hence

$$X_1 = -\frac{\delta_{1p}}{\delta_{11}}$$

Replacing δ_{1p} by δ_{p1} we may write

$$X_1 = -\frac{\delta_{p1}}{\delta_{11}}$$

It will be remembered that δ_{p1} given in the above equation represents the ordinate to the deflection line for a $(n - 1)$ times statically indeterminate continuous beam due to the application of a load unity acting along the direction of X_1, while δ_{11} may be regarded as the scale factor permitting the conversion of the deflection line to the influence line. This method of influence line construction we have named the kinematic method. It furnishes an easy means of determining the shape of the influence lines for continuous beams, this shape being exactly the same as that of the deflection line δ_{p1} which can be obtained at very little cost.

Fig. 65.12 shows the shape of the influence lines for support reactions, for bending moments and shearing stresses pertaining to a continuous beam resting on five supports. The shape of these

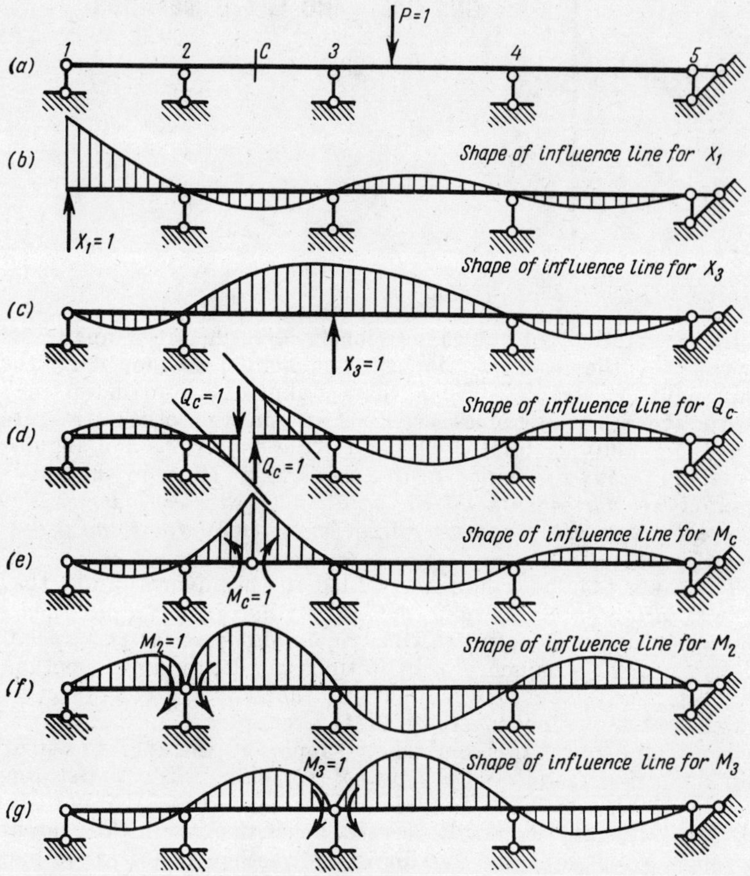

Fig. 65.12

influence lines has been determined practically without any computations using the deflection lines due to unit forces applied along the eliminated redundant reactions. Thus, the influence line for the left-end support reaction will be exactly of the same shape as the deflection line of the continuous beam redundant in the second degree (Fig. 65.12*b*) acted upon by a load unity X_1.

This kinematic method provides very rapidly the shape of the influence lines which may be used as models when determining those portions of continuous beams which should be loaded in order to obtain the extreme values of the stresses under consideration.

13. SLOPE AND DEFLECTIONS. COMBINED AND MIXED METHODS

1.13. *Choice of Unknowns*

In the method of forces previously described the unknowns represented the reactions (forces or moments) developed by the redundant constraints. When these unknowns were determined all the stresses at any cross section of any member of the structure could be easily calculated whereafter the deflections and angular rotations could be obtained in the usual way. Thus, in the above method we *started with the computation of stresses and reactions proceeding thereafter to the determination of the rotations and deflections.*

The same problem could be tackled in the inverse order, that is *first determining by any method available the displacements and proceeding thereafter with the computation of the corresponding stresses.* This sequence of operations is adopted in the method of stress analysis usually called the *slope and deflections method* which we are going to study in this chapter.

The unknowns of this method will represent the angles of twist and deflections induced by bending moments, while the strains and displacements due to normal and shearing forces will be neglected. No additional error will be introduced thereby in the computation of rigid joint systems for in the method of forces we had equally neglected the influence of direct and shearing stresses. It will be also assumed that the difference in length between the original member and the chord of its elastic line is practically nonexistent which means that the distance over which the ends of a deflected member are drawn together is completely neglected.

We shall begin our study by establishing those of the displacements of a member which must be known in order to find the stresses acting at one of its cross sections. For this purpose let us study a rectilinear bar AB (Fig. 1.13*a*) isolated from any redundant structure. The stresses existing in this structure including bar AB itself will cause this bar to deflect and to take up a new position $A'B'$ as shown in Fig. 1.13*a*. The movement of the bar AB to its new position may be regarded as consisting of the following independent displacements:

1. A translation of all the points of the bar over the same distance Δ_A (Fig. 1.13*b*). During this translation the bar remains straight and parallel to itself. The bending moments and the shears at all the cross sections of the bar remain nil.

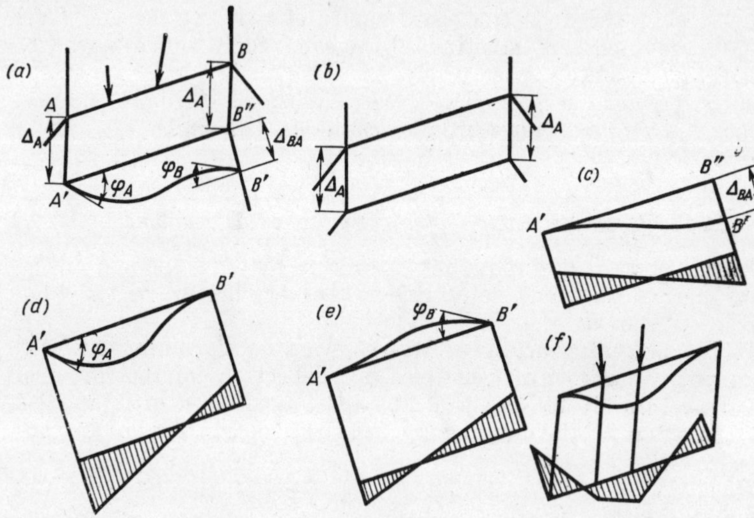

Fig. 1.13

2. The deflection of one of the fixed ends of the bar along a direction normal to its axis (say, of the end B) over a distance Δ_{BA}. The elastic line of the bar and the corresponding bending moment curve are represented in Fig. 1.13*c*.

3. A rotation of the end A of the bar through an angle φ_A. The elastic line and the corresponding bending moment curve are given in Fig. 1.13*d*.

4. A rotation of the end B of the bar through an angle φ_B (Fig. 1.13*e*).

5. The deflection of the axis of the fixed end bar under the influence of the loads applied between points A and B (Fig. 1.13*f*).

The elastic line of the bar AB resulting from a translation Δ_A, from a deflection of the end B about the end A over a distance Δ_{BA}, from the rotation of the end sections through angles φ_A and φ_B and from the deflection due to the loads directly applied, will coincide exactly with the elastic line $A'B'$ (Fig. 1.13*a*). Thus, if we arrive by any means to determine the magnitudes of Δ_{BA},

φ_A and φ_B we can thereafter easily find the values of M and Q acting at any cross section of that particular bar, for the translation Δ_A is not connected with any stresses in the bar under consideration. Consequently, for each *independent* member of the structure we may adopt as unknowns the deflection Δ_{BA} and the angular rotations or angles of twist φ_A and φ_B.

In *framed structures* with rigid joints (portal frames, building frames, etc.) the deflections and angles of twist at the end faces of all the members meeting at the same joint will always be exactly the same. Consequently, *when the method under consideration is applied to framed structures with rigid joints the unknowns will always represent the deflections and angles of twist of various joints.*

2.13. *Determination of the Number of Unknowns*

In the analysis of a redundant structure by the slope and deflections method one must determine in the very first place the number of unknowns.

In the preceding article it was shown that the unknowns will represent the angles of twist and the deflections of the joints of this structure. It follows that the total number n of unknowns will be equal to the number of unknown deflections n_d and angles of twist n_t

$$n = n_d + n_t$$

The number of unknown angles of twist is always equal to the number of the rigid joints of the structure and therefore the determination of n_t reduces to a simple counting up of these joints. *

A joint is deemed rigid if at least two members meeting at this joint are rigidly connected to one another. Examples of such joints are afforded by joints *1*, *2*, *3* and *4* of Fig. 4.13*a*, by joints *1* and *2* of Fig. 4.13*f*, and by joint *1* of Fig. 4.13*g*. If a joint is constituted by the meeting of several groups of members where all the members of one group are rigidly connected together but all the separate groups are hinge-connected between themselves, such a joint will be regarded as equivalent to several joints the number of which is equal to the number of groups.

Thus, for instance, joint *1* of Fig. 4.13*h* will be reckoned equal to two rigid joints, while joint *1* of Fig. 4.13*i* equal to three rigid joints.

◆

* Those of the joints whose angles of twist are known beforehand such as, for instance, the fixed end joints, if they are held absolutely fast, should not be included.

Let us determine now the number of independent joint deflections. In Art. 1.13 we have mentioned that the deformations of rigid structures caused by direct and shearing stresses may be neglected and that the difference in length between a straight bar and the chord connecting the ends of its elastic line may be regarded as nonexistent. Bearing this in mind, let us replace

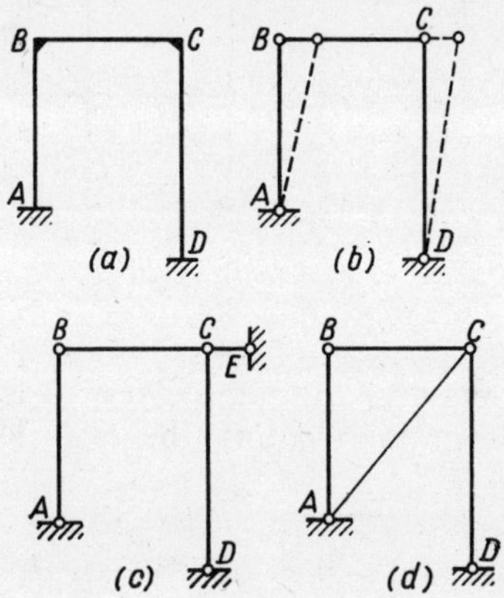

Fig. 2.13

in imagination all the rigid joints of the given redundant structure with hinges. The different joints of the latter system will not be able to move independently for the displacement of one of them may entail the displacement of a certain number of other joints. What we must find is the number of deflections which may occur *independently*.

It is known that the number of such deflections in a hinge-connected structure is always equal to the number of additional bars which should be introduced to make the structure geometrically stable. It follows that *the number of independent joint deflections is equal to the degree of instability of the system obtained by the introduction of hinges at all the rigid joints and supports of the original structure.*

As an example, let us examine the portal frame shown in Fig. 2.13*a*. This frame contains two rigid joints B and C and consequently $n_t = 2$. The number of independent joint deflections n_d will be obtained replacing all rigid joints and fixed supports by hinges as indicated in Fig. 2.13*b*. The system thus obtained is unstable but it will suffice to introduce a single additional bar to ensure its rigidity. Let it be an additional horizontal supporting bar CE as in Fig. 2.13*c* or a diagonal AC as in Fig. 2.13*d*. The dotted lines of Fig. 2.13*b* show the possible displacements of the sides of a hinge-connected quadrangle. It is obvious that

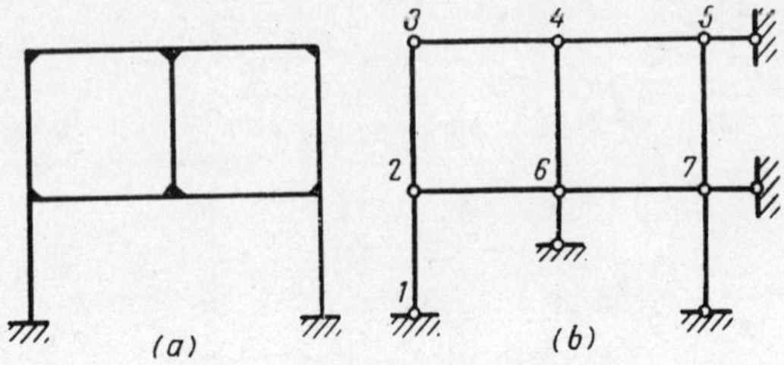

Fig. 3.13

the joint B will move over the same distance as joint C and therefore these two deflections cannot be regarded as independent.

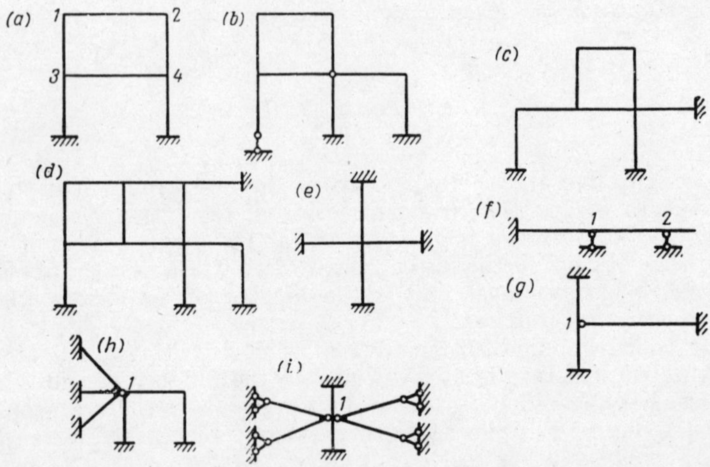

Fig. 4.13

Thus, in the case under consideration the number of independent joint deflections equals one ($n_d = 1$). The total number of unknown twists and deflections will equal

$$n = n_t + n_d = 2 + 1 = 3$$

As another example, let us investigate the more complicate frame of Fig. 3.13*a* where the number of rigid joints totals six

($n_t = 6$). The hinge-connected counterpart of this frame would be variable in the third degree for its conversion into a stable structure would require the introduction of at least three additional bars (Fig. 3.13*b*).. When these are present, joint *7* is connected to the ground by means of two concurrent bars rendering this connection stable. The same applies to joints *5* and *6* and since joints *2*, *3* and *4* are also connected by at least two concurrent bars to those just mentioned, all the system will be stable. It follows that the number of independent deflections n_d will equal 3 and therefore the total number of unknowns for this structure becomes

$$n = 6 + 3 = 9$$

In Table 1.13 we present the number of unknown deflections and angles of twist for each of the redundant structures of Fig. 4.13 together with their degree of redundancy.

Table 1.13

Structure	Degree of redundancy	Number of unknown displacements		
		Angles of twist	Deflections	Total
a	6	4	2	6
b	4	4	3	7
c	9	5	2	7
d	15	7	2	9
e	9	1	0	1
f	2	2	0	2
g	5	1	0	1
h	11	3	0	3
i	9	3	0	3

3.13. *The Conjugate System of Redundant Beams*

The conjugate redundant system utilized in the method of slope and deflections always consists of a number of single-span redundant beams. These separate beams are obtained by the introduction of additional constraints into the given structure. Let us compare the simple statically determinate structure used in the method of forces with the conjugate redundant system utilized in the slope and deflections method.

A good example is afforded by the rectangular portal frame redundant in the second degree given in Fig. 5.13*a*. The simple structure of the method of forces could be derived from the above passing a section through joint *2*, which would be equivalent to the elimination of two constraints. The simple structure would thus consist of a knee frame and a straight beam both fixed at their lower ends (Fig. 5.13*b*) and both statically determinate.

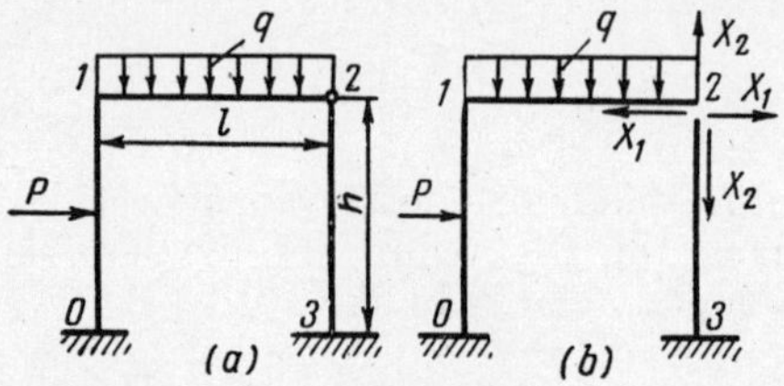

Fig. 5.13

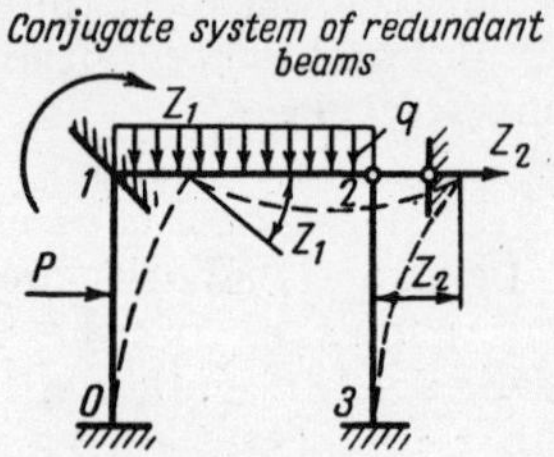

Fig. 6.13

As for the conjugate redundant system, this would be obtained by the introduction of two additional constraints into the given structure—one opposing the rotation of joint *1* and the other preventing the translation of joints *1* and *2* (Fig. 6.13). The system thus obtained will be redundant to the fourth degree.

Thus:

(a) the conjugate simple structure utilized in the method of forces is derived from the original one by the *elimination* of redundant constraints, whereas the redundant system pertaining to the slope and deflections method—by the *introduction* of additional constraints;

(b) the conjugate structure used in the method of forces is always redundant to a *lower* degree than the given structure whereas the conjugate system used in the slope and deflections method is always of a *higher* degree of redundancy.

It should be noted that *the constraints introduced in order to prevent the rotation of rigid joints differ in certain respects from the sum of constraints providing fixed or built-in ends.* Indeed, the additional constraints should prevent only the twist or rotation of the joint without interfering in any way with its linear translation. Hence, the only reactions these constraints are capable of developing consist of moments applied at the joints.

As for the constraints preventing the deflection of joints, these can be obtained in different ways. One could, for instance, introduce diagonals *0-2* or *1-3* of Fig. 7.13*a* and *b*, or a horizontal supporting bar at joint *2* as in Fig. 7.13*c* or alternatively an inclined brace at joint *1* (Fig. 7.13*d*).

The constraint introduced in the shape of diagonal *1-3* (Fig. 7.13*b*) will not impede the displacement of joint *3* which is held fast in any way. This diagonal will prevent solely the deflection of joint *1* along the direction of the line passing through this joint and joint *3*. From this view-point the constraint provided by an inclined brace of Fig. 7.13*d* is equivalent to the above diagonal. It is always

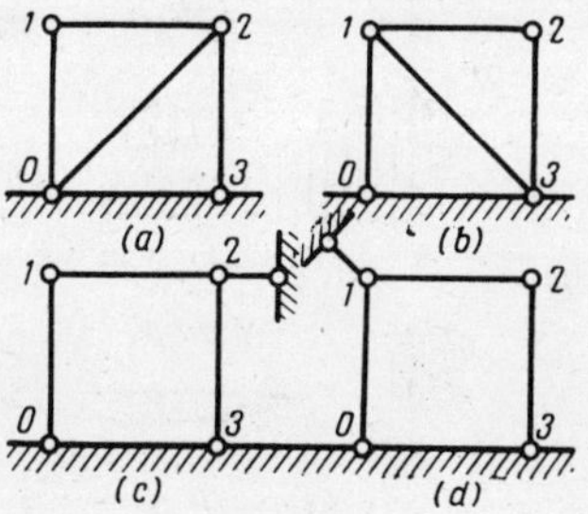

Fig. 7.13

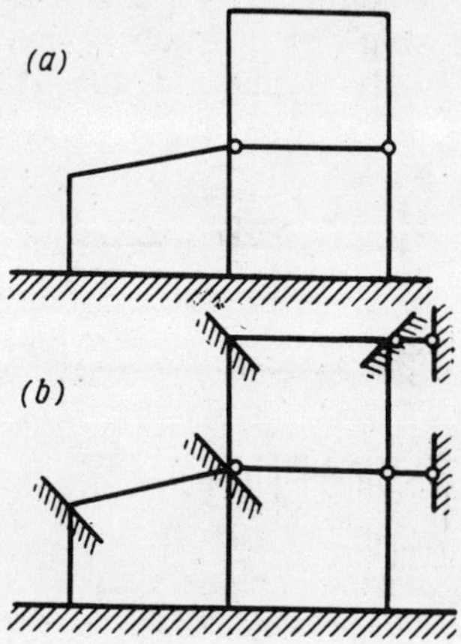

Fig. 8.13

preferable to introduce additional constraints opposing the deflection of joints in the shape of supports connecting these joints to the ground. The introduction of additional bars connecting different joints of the structure between themselves should be avoided as much as possible.

If the given structure consists of vertical and horizontal members preference should be given to additional supports themselves either horizontal or vertical. The introduction of inclined bars is liable to cause certain complications in subsequent computations.

In order to obtain the conjugate redundant system the additional constraints introduced must prevent the rotation of all the rigid joints as well as all the independent deflections of these joints.

Let us examine the two-storied frame of Fig. 8.13*a*. This frame is redundant to the sixth degree and therefore the number of unknowns in the method of forces would also equal six. In the method under consideration the number of unknown displacements will also equal six and will consist of four angles of twist and of two deflections. The conjugate system of redundant beams will be obtained by the introduction of four constraints precluding angular rotations of four joints as well as of two additional horizontal supports preventing all the independent deflections (Fig. 8.13*b*).

Let us investigate in detail all the elements forming the latter system. As stated above, all of its members constitute single-span statically indeterminate beam. For this reason let us take up in the first place the construction of bending moment diagrams for a single-span beam of constant rigidity having one fixed and one freely supported end (Fig. 9.13*a*). Using the well-known

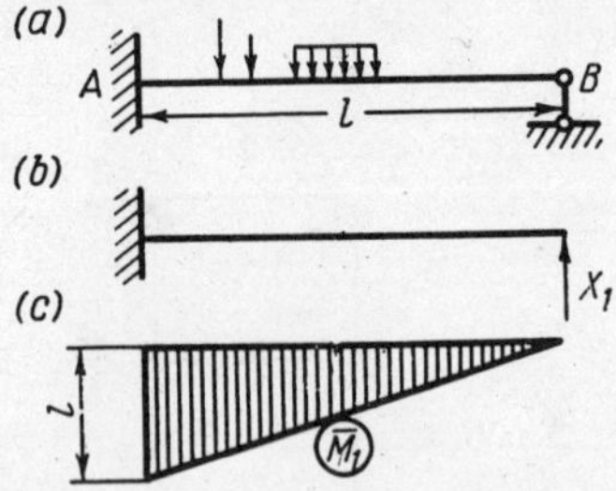

Fig. 9.13

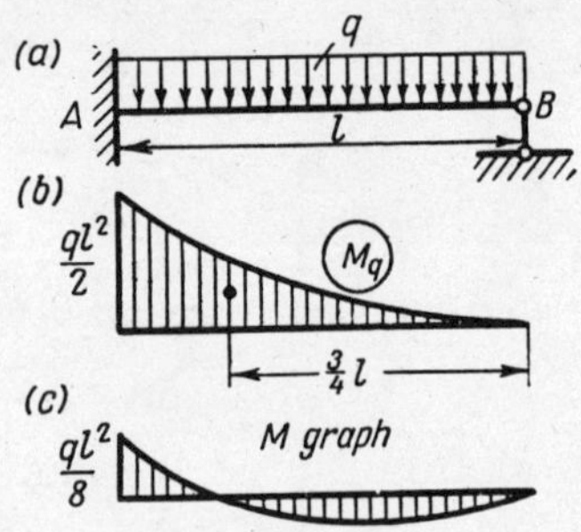

Fig. 10.13

method of forces, we may construct the diagrams for various types of external actions. The reactions directed upwards and the moments acting clockwise will be reckoned positive. The simple structure corresponding to this beam will be formed by a simple cantilever shown in Fig. 9.13*b*. The reaction of the roller support at *B* will constitute the single unknown X_1 of the equation

$$X_1\delta_{11}+\Delta_{1m}=0 \tag{1.13}$$

The value of coefficient δ_{11} will be given by the second power of the $\overline{M}_1$ graph (Fig. 9.13*c*). The external forces will have no effect on this coefficient which will amount to

$$\delta_{11}=\frac{l^3}{3EJ}$$

The free term of the equation will be calculated for different types of loading:

(a) *The beam is uniformly loaded over the whole of its length* (Fig. 10.13*a*). The value of Δ_{1q} will then be obtained multiplying the M_q graph (Fig. 10.13*b*) by the $\overline{M}_1$ diagram (Fig. 9.13*c*).

$$\Delta_{1q}=-\frac{1}{EJ}\cdot\frac{1}{3}\cdot\frac{ql^2}{2}\,l\cdot\frac{3}{4}l=-\frac{ql^4}{8EJ}$$

Introducing this value into expression (1.13) we find immediately

$$X_1 = R_B = \frac{3}{8} ql$$

wherefrom reaction at point A becomes

$$R_A = ql - R_B = \frac{5}{8} ql$$

The fixed end moment at A will be obtained summing up the moments induced at that section both by the applied loads and by the force X_1

$$M_{AB} = -\frac{ql^2}{2} + \frac{3}{8} qll = -q\frac{l^2}{8}$$

(b) *The beam carries one concentrated load P acting at any arbitrary*

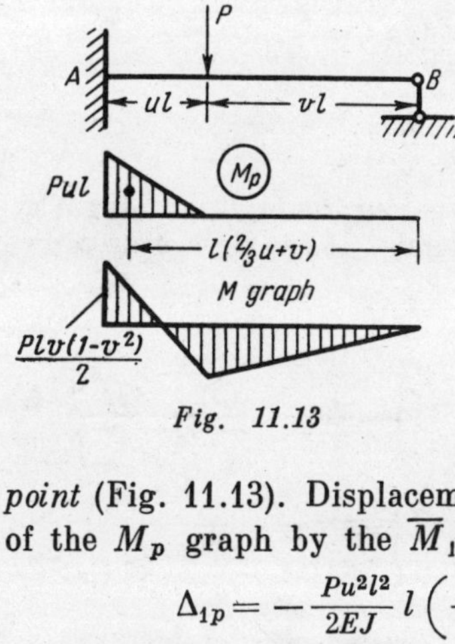

Fig. 11.13

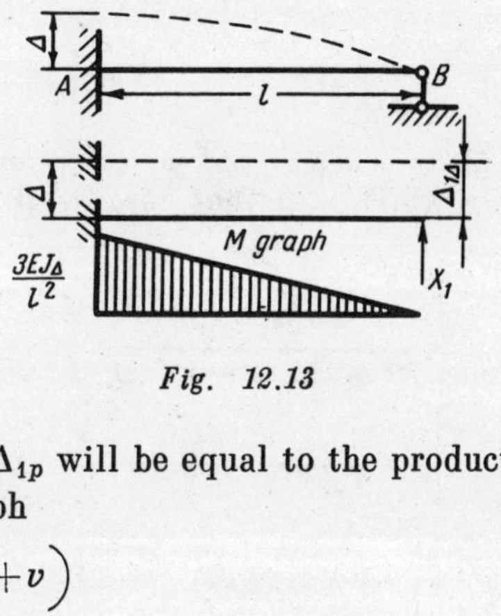

Fig. 12.13

point (Fig. 11.13). Displacement Δ_{1p} will be equal to the product of the M_p graph by the $\overline{M}_1$ graph

$$\Delta_{1p} = -\frac{Pu^2l^2}{2EJ} l \left(\frac{2}{3} u + v\right)$$

since $ul + vl = l$ whence $v = 1 - u$

$$\Delta_{1p} = -\frac{Pu^2l^3}{6EJ}(3-u)$$

Introducing this value into equation (1.13) we obtain

$$X_1 = R_B = \frac{Pu^2}{2}(3-u)$$

The reaction at point A becomes

$$R_A = P - R_B = \frac{Pv}{2}(3-v^2)$$

and the fixed end moment

$$M_{AB} = -Pul + \frac{Pu^2l}{2}(3-u) = -\frac{Pl}{2} v(1-v^2)$$

(c) *The fixed end of the beam is deflected in the direction normal to its axis over a length* Δ (Fig. 12.13). This movement will induce no bending moments in the conjugate simple beam but the displacement along the direction of X_1 will become

$$\Delta_{1\Delta}=\Delta$$

From equation (1.13)

$$X_1=-\frac{3EJ}{l^3}\Delta$$

which permits the immediate determination of the reactions developed by all the other constraints

$$R_B=X_1=-\frac{3EJ}{l^3}\Delta$$

$$R_A=-X_1=\frac{3EJ}{l^3}\Delta$$

$$M_{AB}=-\frac{3EJ}{l^2}\Delta$$

(d) *The fixed end of the beam is rotated through an angle* φ (Fig. 13.13). In that case the displacement at the right extremity

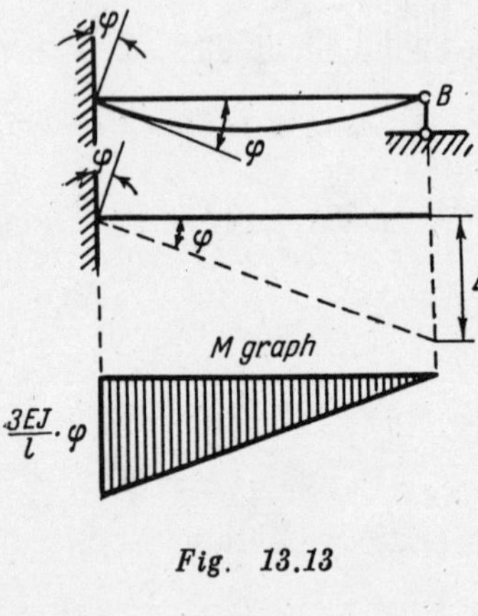

Fig. 13.13

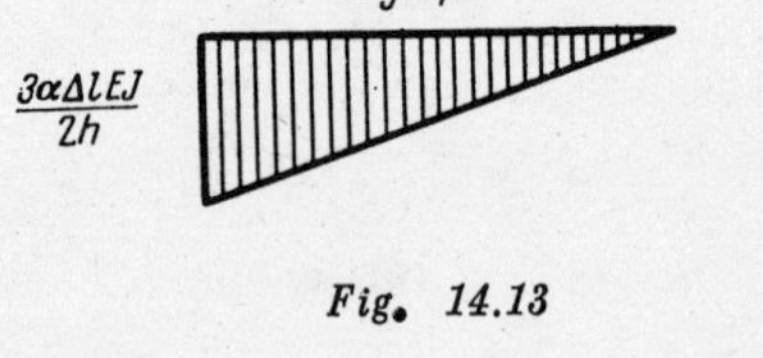

Fig. 14.13

of the conjugate simple beam along the direction of X_1 becomes

$$\Delta_{1\varphi}=-\varphi l$$

in which case expression (1.13) yields

$$X_1=\frac{3EJ}{l^2}\varphi$$

leading to the following values of the reactions at the supports and of the fixed end moment

$$R_B = X_1 = \frac{3EJ}{l^2}\varphi; \quad R_A = -X_1 = -\frac{3EJ}{l^2}\varphi$$

$$M_{AB} = \frac{3EJ}{l}\varphi$$

(e) *A difference in temperature* $\Delta_t = t_1 - t_2$ is introduced between the upper and lower fibres of the beam (Fig. 14.13). The deflection along the direction of X_1 of the simple structure becomes in that case

$$\Delta_{1t} = -\frac{\alpha\Delta t}{h}\Omega_{\overline{M}} = -\frac{\alpha\Delta t}{h}\frac{l^2}{2}$$

where h is the depth of the beam.

Solving equation (1.13) we find

$$X_1 = \frac{3\alpha\Delta tEJ}{2hl}$$

wherefrom

$$R_B = \frac{3\alpha\Delta tEJ}{2hl}$$

$$R_A = -\frac{3\alpha\Delta tEJ}{2hl}$$

$$M_{AB} = \frac{3\alpha\Delta tEJ}{2h}$$

Let us study next a beam with both ends built in (Fig. 15.13*a*). As a simple conjugate structure we shall adopt two cantilever beams of Fig. 15.13*b* obtained by cutting the given beam at mid-span. The diagrams induced in the latter system by unit actions are represented in Fig. 15.13*c*, *d* and *f*. It will be immediately observed that the direct stress X_3 will remain nil as long as the loads remain vertical. The same will apply to the case when the fixed ends are shifted vertically or sustain angular rotations. This becomes quite clear if we take up the equation

$$X_1\delta_{31} + X_2\delta_{32} + X_3\delta_{33} + \Delta_{3m} = 0$$

in which both coefficients to the unknowns δ_{31} and δ_{32} as well as the free term Δ_{3m} must reduce to zero for the bending moment $\overline{M}_3$ remains constantly nil itself.

The unknowns X_1 and X_2 will be found from the equations

$$\left.\begin{aligned} X_1\delta_{11} + \Delta_{1m} = 0 \\ X_2\delta_{22} + \Delta_{2m} = 0 \end{aligned}\right\} \qquad (2.13)$$

The coefficients to the unknowns in these two equations are given by

$$\delta_{11} = 2 \cdot \frac{1}{2} \cdot \frac{l}{2} \cdot \frac{l}{2} \cdot \frac{2}{3} \cdot \frac{l}{2} \cdot \frac{1}{EJ} = \frac{l^3}{12EJ}$$

$$\delta_{22} = 1 \cdot l \cdot 1 \cdot \frac{1}{EJ} = \frac{l}{EJ}$$

As for the free terms their values will depend on loading conditions some of which are considered below.

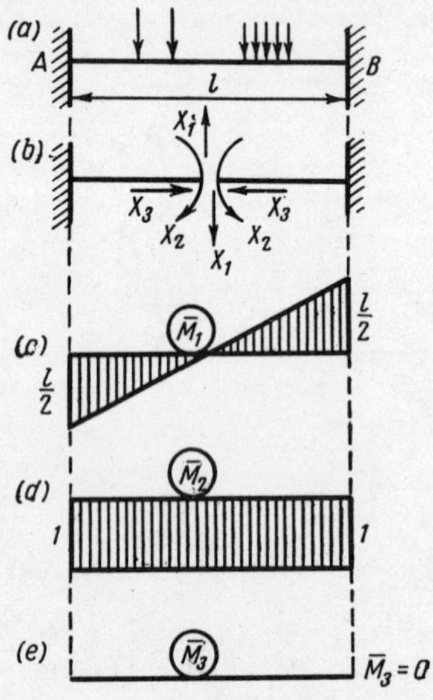

Fig. 15.13

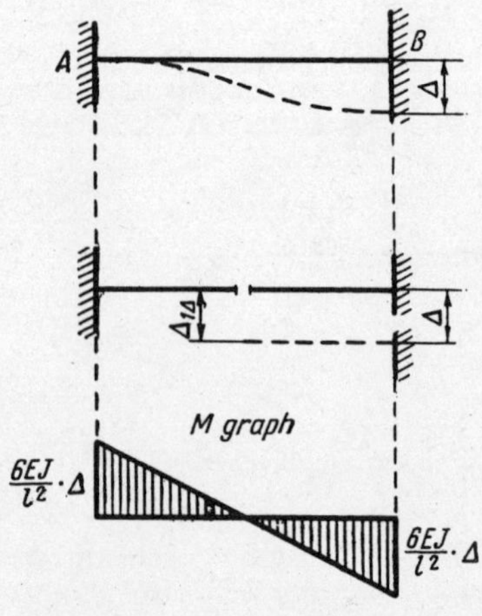

Fig. 16.13

(a) *Both built-in ends are shifted a distance* Δ *normally to the axis of the beam* (Fig. 16.13). The deflections sustained by the conjugate simple structure along the unknown reactions become

$$\Delta_{1\Delta} = \Delta; \quad \Delta_{2\Delta} = 0$$

Introducing these values into equation (2.13) and solving the latter we obtain

$$X_1 = -\frac{12EJ}{l^3}\Delta; \quad X_2 = 0$$

whence

$$R_B = X_1 = -\frac{12EJ}{l^3}\Delta; \quad R_A = -X_1 = \frac{12EJ}{l^3}\Delta$$

$$M_{AB} = -\frac{6EJ}{l^2}\Delta; \quad M_{BA} = -\frac{6EJ}{l^2}\Delta$$

(b) *The fixed end at A is rotated through an angle* φ (Fig. 17.13). In this case the deflections of the simple statically determinate beam along the redundant reactions become

$$\Delta_{1\varphi} = -l/2\varphi; \qquad \Delta_{2\varphi} = \varphi$$

Solving once again equations (2.13) we obtain

$$X_1 = \frac{6EJ}{l^2}\varphi; \qquad X_2 = -\frac{EJ}{l}\varphi$$

Consequently, the reactions at the support and the fixed end moments become

$$R_B = X_1 = \frac{6EJ}{l^2}\varphi$$

$$R_A = -X_1 = -\frac{6EJ}{l^2}\varphi$$

$$M_{AB} = \frac{4EJ}{l}\varphi; \quad M_{BA} = \frac{2EJ}{l}\varphi$$

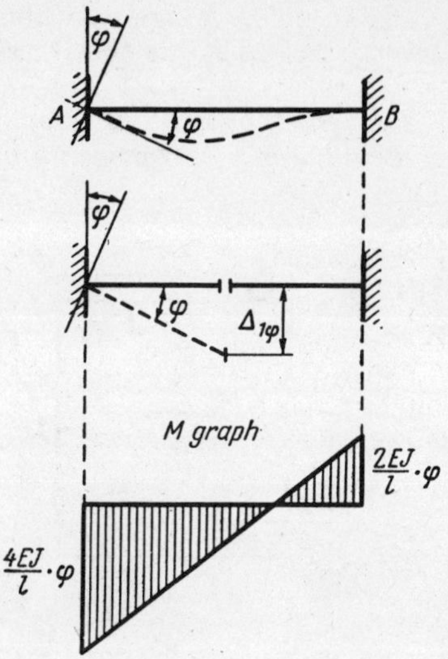

Fig. 17.13

The above data together with some additional values of support reactions and fixed end moments corresponding to a number of other loading conditions are presented in Table 2.13. This table will be of great help for the stress analysis of portal and building frames by the slope and deflections method. The analysis of a beam with both built-in ends subjected to other systems of loading is deemed unnecessary.

4.13. *Canonical Equations Peculiar to the Slope and Deflections Method*

First let us clarify the general principles permitting to form the necessary equations from which the angles of twist and the deflections of the rigid joints may be derived. For this purpose let us compare the given structure with the conjugate system of redundant beams (Fig. 18.13). It is obvious that the sole difference between the two systems resides in the presence of additional constraints in the latter, these additional constraints opposing the rotation and the deflection of joints. The existence of these constraints leads to the appearance of reactive moments and forces which will become necessarily nil when each of the fixed joints will be rotated through an angle equal to the real angle of twist and when the deflections of all the joints will become equal to those sustained by the same joints of the original structure.

When this condition is satisfied, all the deformations and stresses set up in the conjugate system of redundant beams will become exactly the same as the stresses and deformations of the given structure.

The equations of the slope and deflections method negate the existence of reactive moments and forces developed by the imaginary

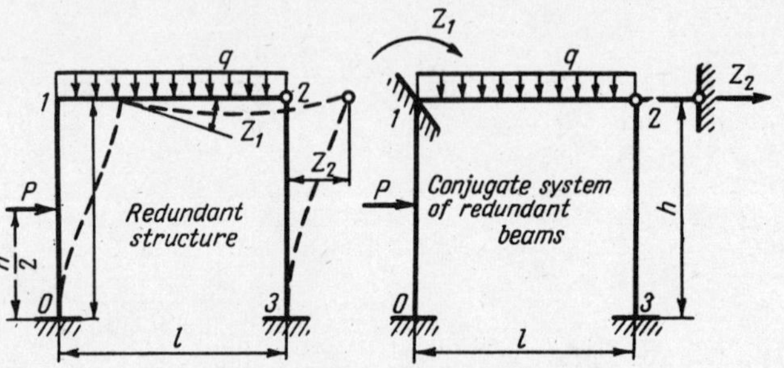

Fig. 18.13

constraints of the conjugate system of redundant beams just as the equations of the method of forces express that the displacements of the conjugate simple structure along the redundant reactions remain nil.

In the most general form these equations may be written as follows

$$R_1 = 0 \quad R_2 = 0 \quad R_3 = 0$$

where R_1, R_2, R_3, etc., are the reactive moments and forces developed by the additional constraints of the conjugate system of redundant beams due both to the actual loads and to the twists and deflections sustained by the joints. The indices of these reactions must always correspond to the indices of the unknowns. As for the number of equations, it will be equal to the number of additional constraints or, in other words, to the number of unknown rotations and deflections.

It is worth mentioning that the equations used in the slope and deflections method are equations of *equilibrium* as compared to the equations of the method of forces which were *kinematic* equations showing the existence of certain relations between the displacements of various parts of the structure.

Let us examine in detail the first equation of the slope and deflections method ($R_1 = 0$) with reference to the conjugate system

Table 2.13

No.	Loading conditions	Bending moment graphs (the ordinates are set off next to the extended fibres) and reactions	Expressions for bending moments and reactions
1	2	3	4
1	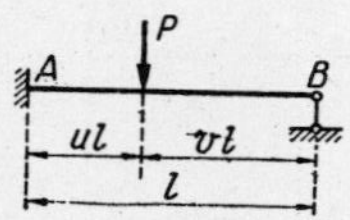	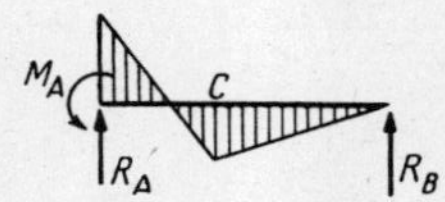	$M_A = -\frac{Pl}{2} v (1 - v^2)$ $M_C = \frac{Pl}{2} u^2 v (3 - u)$ $R_A = \frac{Pv}{2} (3 - v^2)$ $R_B = Pu^2 \left(1 + \frac{v}{2}\right) = \frac{Pu^2}{2} (3 - u)$
2	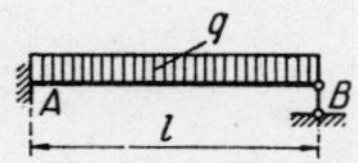	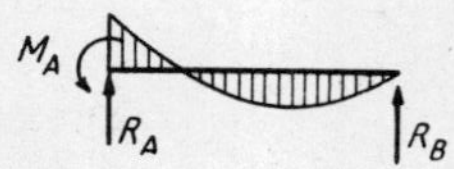	$M_A = -\frac{ql^2}{8}$ $R_A = \frac{5}{8} ql$ $R_B = \frac{3}{8} ql$
3	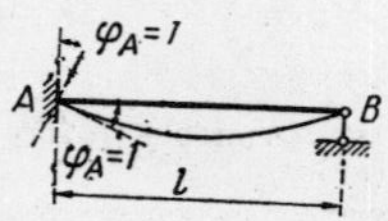	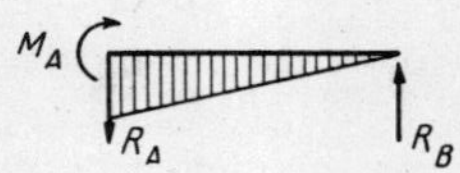	$M_A = \frac{3EJ}{l}$ $R_A = -R_B = -\frac{3EJ}{l^2}$

Table 2.13 (continued)

No.	Loading conditions	Bending moment graphs (the ordinates are set off next to the extended fibres) and reactions	Expressions for bending moments and reactions
1	2	3	4
4	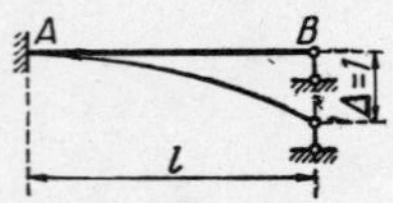	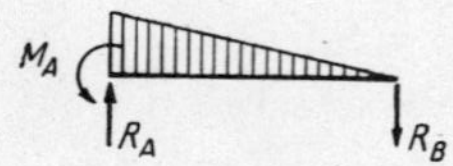	$M_A = -\dfrac{3EJ}{l^2}$ $R_A = -R_B = \dfrac{3EJ}{l^3}$
5	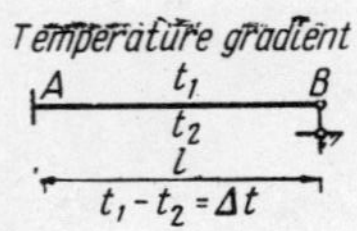	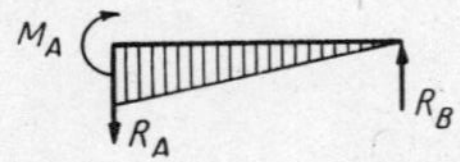	$M_A = \dfrac{3EJ\alpha\Delta t}{2h}$ where h is the depth of the cross section, and α is the coefficient of thermal expansion $R_A = -\dfrac{3EJ\alpha\Delta t}{2hl}$
6	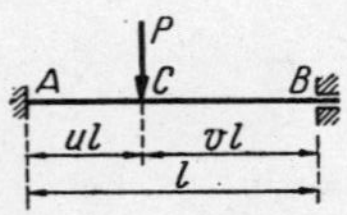	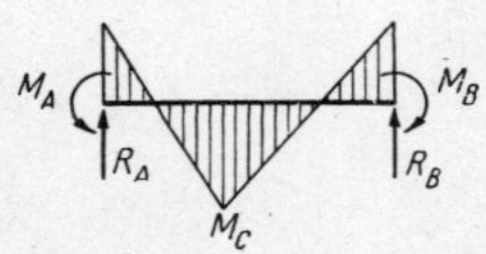	$M_A = -uv^2Pl$ $M_B = u^2vPl$ $M_C = 2u^2v^2Pl$ $R_A = v^2(1+2u)P$ $R_B = u^2(1+2v)P$

Table 2.13 (continued)

No.	Loading conditions	Bending moment graphs (the ordinates are set off next to the extended fibres) and reactions	Expressions for bending moments and reactions
1	2	3	4
7	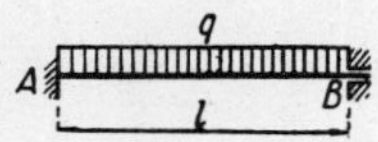	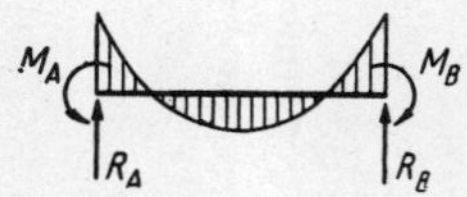	$M_A = -M_B = -\frac{ql^2}{12}$ $R_A = R_B = \frac{ql}{2}$
8	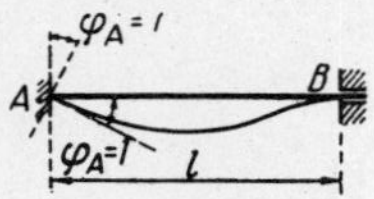	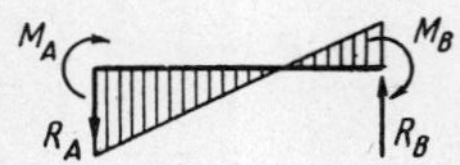	$M_A = \frac{4EJ}{l}$; $M_B = \frac{2EJ}{l}$ $R_A = -R_B = -\frac{6EJ}{l^2}$
9	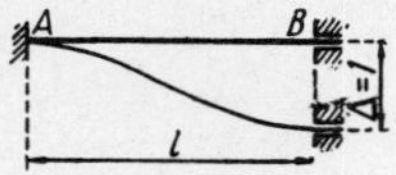	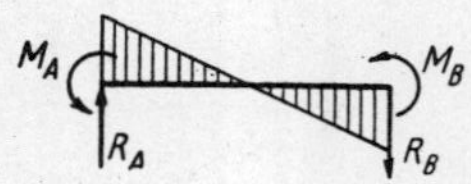	$M_A = M_B = -\frac{6EJ}{l^2}$ $R_A = -R_B = \frac{12EJ}{l^3}$
10	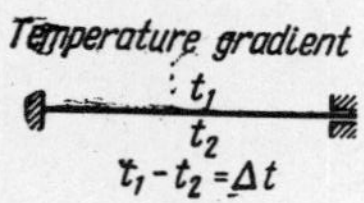		$M_A = -M_B = \frac{EJ\alpha\Delta t}{h}$ where h is the depth of the cross section, and α is the coefficient of thermal expansion $R_A = R_B = 0$

of redundant beams given in Fig. 18.13. The reactive moment R_1 may be replaced by the algebraic sum

$$R_1 = R_{1p} + R_{11} + R_{12}$$

In the above expression the second index alotted to the terms of the right-hand part serves to indicate the cause which has given rise to that particular reaction.

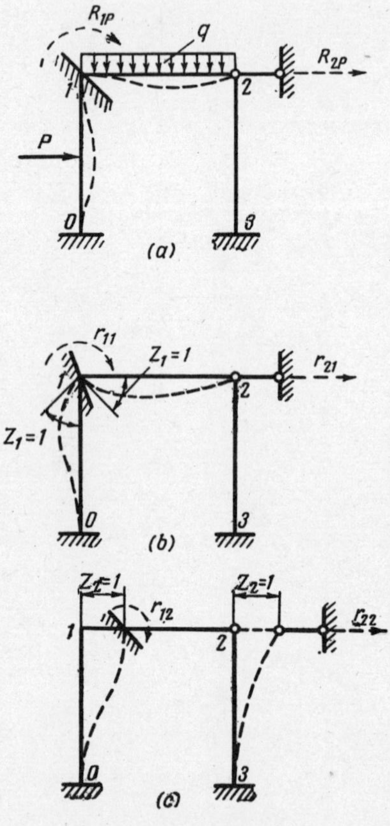

Fig. 19.13

Thus:

R_{1p} is the reactive moment developed by the additional constraint under the action of loads P (Fig. 19.13*a*);

R_{11} is the reactive moment of the same constraint due to the rotation of joint *1* through an angle Z_1;

R_{12} is the reactive moment due to the deflection of joints *1* and *2* over a length Z_2.

The reactive moments R_{11} and R_{12} due to the displacements Z_1 and Z_2 may be replaced by the following expressions

$$R_{11} = Z_1 r_{11} \quad \text{and} \quad R_{12} = Z_2 r_{12}$$

where r_{11} is the reactive moment due to the rotation of the fixed joint through an angle equal to unity, i. e., to 1 radian (Fig. 19.13*b*), and r_{12} is the reactive moment due to a unit displacements of joints *1* and *2* (Fig. 19.13*c*).

Substituting these values into the original equation we obtain

$$Z_1 r_{11} + Z_2 r_{12} + R_{1p} = 0$$

The second equation ($R_2 = 0$) may be written in exactly the same way

$$Z_1 r_{21} + Z_2 r_{22} + R_{2p} = 0$$

Here r_{21} is the reactive force induced in the imaginary support by the rotation of joint *1* through an angle equal to unity

(Fig. 19.13*b*), and r_{22} is the reactive force developed by the same support when joints *1* and *2* are deflected over a distance equal to a unit length (Fig. 19.13*c*), while R_{2p} is the reactive force at the same support due to the applied loads (Fig. 19.13*a*).

The first of these equations expresses that in reality no reactive moment is developed at the imaginary constraint opposing the rotation of joint *1*, and the second that the reactive force at the imaginary support is equally nil. The two equations form together a system of simultaneous standard equations of the slope and deflections method. Equations of the same type could be obtained for any number of unknown displacements, the number of equations in the system and the number of unknowns in each equation coinciding exactly with the number of displacements mentioned.

Thus, for instance, in the case of four unknowns the system of simultaneous equations of the slope and deflections method would become

$$\begin{aligned} Z_1r_{11}+Z_2r_{12}+Z_3r_{13}+Z_4r_{14}+R_{1p}&=0\\ Z_1r_{21}+Z_2r_{22}+Z_3r_{23}+Z_4r_{24}+R_{2p}&=0\\ Z_1r_{31}+Z_2r_{32}+Z_3r_{33}+Z_4r_{34}+R_{3p}&=0\\ Z_1r_{41}+Z_2r_{42}+Z_3r_{43}+Z_4r_{44}+R_{4p}&=0 \end{aligned} \tag{3.13}$$

Hereunder the coefficients to the unknowns (unit reactions) r_{11}, r_{22}, etc., situated along the main diagonals will be termed *main coefficients*, while the coefficients to the unknowns r_{12}, r_{21}, r_{13}, r_{31}, etc., will be termed *secondary coefficients*. The coefficients to the unknowns of the slope and deflections method when situated symmetrically about the main diagonal are equal between themselves as was the case with similar coefficients of the method of forces. Indeed, these coefficients are related to one another by the principle of reciprocal works $r_{mn} = r_{nm}$ (see Art. 6.13). It follows that the system of simultaneous equations of the slope and deflections method may be solved using the so-called abridged procedure described in Art. 6.12. The main coefficients of the equations under consideration are always positive.

The equations of the slope and deflections method differ from those pertaining to the method of forces by the fact that the coefficients to the unknowns δ_{nm} and the free terms Δ_{np} representing unit displacements of the simple statically determinate structure are replaced by the coefficients to the unknowns r_{nm} and the free terms R_{np} representing the reactions of imaginary constraints, which transform the given structure into the conjugate system of redundant beams. In addition, the unknowns themselves represent in the latter case the slopes (angles of twist) and deflections of the conjugate system while in the former they represented reactive forces.

5.13. *Statical Method of Determining the Coefficients to the Unknowns and the Free Terms*

The determination of the coefficients to the unknowns and of the free terms entering the equations of the slope and deflections method requires the knowledge of the bending moments induced in all the members of the conjugate system of redundant beams both by the applied loads and by the unit twists and deflections directed along the unknown reactions of the imaginary constraints. The construction of the corresponding diagrams can be easily carried out using data contained in Table 2.13. Assume, for instance, that it is required to construct the bending moment diagrams for the system of redundant beams given in Fig. 18.13*b*.

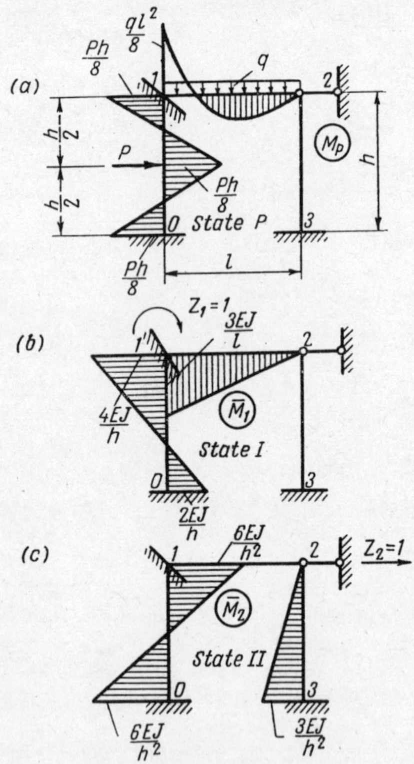

Fig. 20.13

The M_p diagram for the left-hand column will coincide with that of a fixed end beam acted upon by a concentrated load (see line 6 of Table 2.13). For the cross-beam this diagram will coincide with that of a beam built in at its left end and freely supported at the right one (see line 2 of the same table). The fixed end moments become

$$M_{01} = -M_{10} = -Ph\cdot\frac{1}{2}\cdot\frac{1}{4} = -\frac{Ph}{8}$$

because

$$u = v = 1/2 \qquad M_{12} = -\frac{ql^2}{8}$$

The M_p diagrams are represented in Fig. 20.13*a*.

The $\overline{M}_1$ diagram due to the unit twist of joint *1* in a clockwise direction will be obtained for the crossbeam *1-2* using data contained in the 3rd line of Table 2.13 and for the column *0-1* in the 8th line of the same table. This diagram is represented in Fig. 20.13*b*. As for the bending moments $\overline{M}_2$ induced by a unit deflection of joint *2* towards the right, these will be found using lines 4 and 9 of the same table. The said deflection will induce no bending in beam *1-2*, as it follows the direction of the beam axis. The $\overline{M}_2$ diagram is given in Fig. 20.13*c*.

Having completed the bending moment diagrams due both to the applied loads and to the unit displacements of the system of redundant beams we may proceed with the determination of all the coefficients to the unknowns and free terms of the simultaneous equations. For this purpose subdivide the latter into two different groups:

(1) those expressing reactive moments developed by the imaginary constraints preventing the rotation of the joints;

(2) those representing reactive forces of the imaginary supports introduced in order to prevent the deflection of joints.

The coefficients to the unknowns and the free terms belonging to the first group will be obtained isolating each of the joints under consideration and forming the corresponding equilibrium equations of the type

$$\Sigma M = 0$$

The coefficients and the free terms belonging to the second group will be derived either from the equilibrium of the whole system isolated from its supports or from the equilibrium of that or another portion of this system.

These equations will have the general form

$$\Sigma T = 0$$

The direction of axis T will be selected with a view of simplifying as much as possible the subsequent computations.

The following convention of signs will be adopted: *the reactive forces and bending moments will be reckoned positive when they follow the direction of the angular rotations and of the deflections adopted for the joint under consideration.*

Problem. Determine the coefficients to the unknowns and the free terms entering the simultaneous equations of the slope and deflections method for the portal frame given in Fig. 18.13.

Solution. Start by determining the reactive moments R_{1p}, r_{11} and r_{12}. The reactive moment R_{1p} developed by the imaginary constraint opposing the rotation of joint *1* and due to the applied loads will be obtained isolating the aforesaid joint and assuming that the system is subjected solely to the action of loads P (state P) (Figs. 20.13*a* and 21.13*a*). The equilibrium of joint *1* requires that

$$\Sigma M_1 = R_{1p} + \frac{ql^2}{8} - \frac{Ph}{8} = 0$$

wherefrom

$$R_{1p} = -\frac{ql^2}{8} + \frac{Ph}{8}$$

The reactive moment r_{11} at the same joint due to its own unit rotation Z_1 in a clockwise direction will be derived from the equilibrium equation of the said joint corresponding to the case under consideration (state *1*, Figs. 20.13*b* and 21.13*b*)

$$\Sigma M_1 = r_{11} - \frac{3EJ}{l} - \frac{4EJ}{h} = 0$$

wherefrom

$$r_{11} = \frac{3EJ}{l} + \frac{4EJ}{h}$$

As for the reactive moment r_{12} developed by the same constraint when joint *2* shifts towards the right over a distance $Z_2 = 1$, its value will be obtained from the equilibrium equation of joint *1* pertaining to state *2* (Figs. 20.13*c* and 21.13*c*)

$$r_{12} = -\frac{6EJ}{h^2}$$

Next compute the reactive forces R_{2p}, r_{21} and r_{22} developed by the imaginary support opposing the deflection of joint *2*. The reaction R_{2p} due to the actual loads will be found passing section *I-I* which isolates the upper part of the structure (Figs. 20.13*a* and 22.13*a*). The projection of all the forces applied to this portion on the horizontal gives

$$\Sigma X = P + R_{2p} - \frac{P}{2} = 0$$

wherefrom

$$R_{2p} = -\frac{P}{2}$$

The negative value found for this reactive force indicates that it is opposite in direction to the deflection of joint *2* which was assumed to move from left to right.

The reactive force r_{21} corresponding to state *1* will be again obtained passing section *I-I* as indicated in Fig. 22.13*b* and projecting all the forces on the horizontal whence

$$\Sigma X = \frac{6EJ}{h^2} + r_{21} = 0$$

and therefore

$$r_{21} = -\frac{6EJ}{h^2}$$

It will be observed that $r_{21} = r_{12}$. This relation existing between two secondary reactions is similar in all respects to the relation existing between

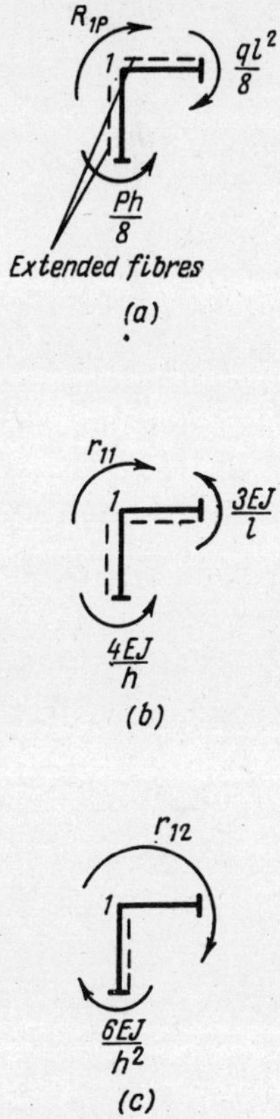

Fig. 21.13

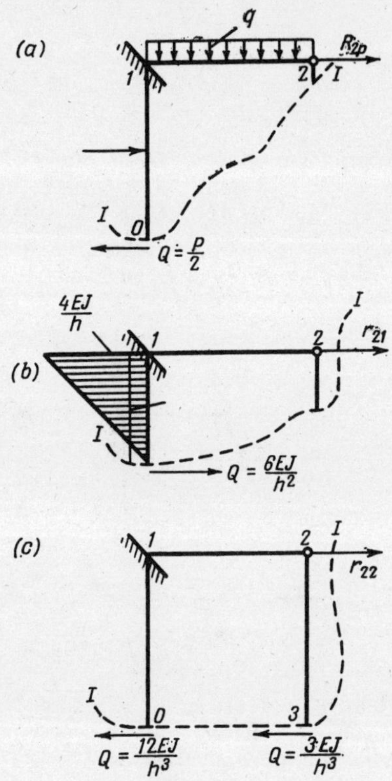

Fig. 22.13

two secondary displacements of the method of forces ($\delta_{mn} = \delta_{nm}$); proof of the above will be given in Art. 6.13.

The reactive force r_{22} will be obtained passing section *I-I* as in Fig. 22.13*c* and assuming that the stresses in the system of redundant beams are due solely to the unit deflection of joint *2* (state *2*). Projecting all the forces on the *x*-axis we obtain

$$\Sigma X = -\frac{12EJ}{h^3} - \frac{3EJ}{h^3} + r_{22} = 0$$

wherefrom

$$r_{22} = \frac{12EJ}{h^3} + \frac{3EJ}{h^3} = \frac{15EJ}{h^3}$$

We have thus obtained the values of all the coefficients to the unknowns and of all the free terms entering the system of simultaneous equations. The method used for the determination of these values will be called hereafter the *static* method.

6.13. *Determination of the Coefficients to the Unknowns and of the Free Terms by the Method of Graph Multiplication*

In a large number of cases the reactive moments and forces developed by the imaginary constraints will be easily obtained multiplying one by the other the corresponding bending moment graphs.

This method could be used advantageously for the analysis of the structure shown in Fig. 23.13. The use of the static method

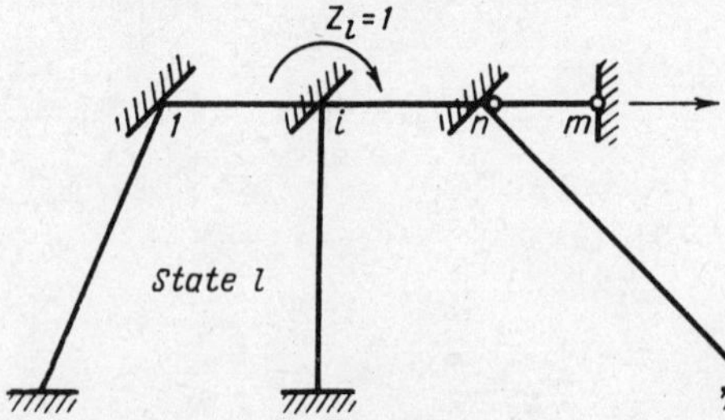

Fig. 23.13

would lead to certain complications for in the case under consideration the projections on the x-axis would contain both shearing forces and normal stresses.

Let us consider two different unit states n and m of an arbitrary conjugate system of redundant beams (Fig. 24.13). The work A_{nm} performed by the external loads of state n along the displacements of state m may be expressed in terms of the bending moments using expression (12.8) given in Art. 4.8.

$$A_{nm}=r_{mn}\cdot 1=\Sigma\int\frac{\overline{M}_m\overline{M}_n ds}{EJ}$$

wherefrom

$$r_{mn}=\Sigma\int\overline{M}_m\overline{M}_n\frac{ds}{EJ} \tag{4.13}$$

The theorem of reciprocal works stating that the work produced by the loads of state n along the displacements of state

m is equal to the work accomplished by the loads of state m along the displacements of state n, we may write

$$A_{nm} = A_{mn}$$

but

$$A_{nm} = r_{mn} \cdot 1 \quad \text{and} \quad A_{mn} = r_{nm} \cdot 1$$

and therefore

$$r_{nm} = r_{mn} \tag{5.13}$$

This theorem called *the theorem of reciprocal reactions* can be formulated as follows: *the reactive force due to a unit displacement of constraint m along the direction n equals the reactive*

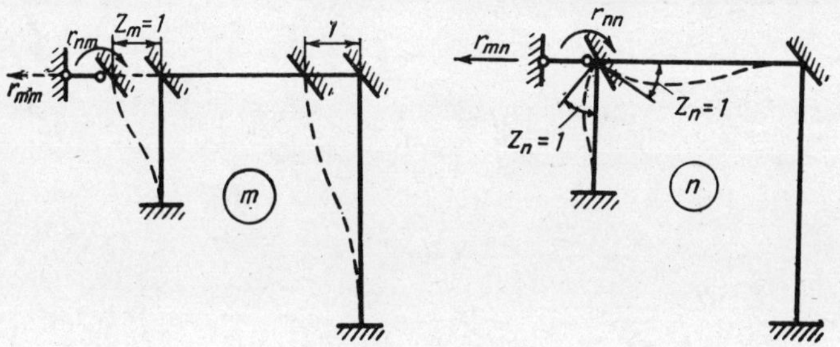

Fig. 24.13

force induced by the unit displacement of constraint n along the direction m.

Problem. Determine the coefficients to the unknowns r_{12} and r_{22} for the portal frame of Fig. 20.13.

Solution. The multiplication of the $\overline{M}_1$ graph by the $\overline{M}_2$ graph leads to

$$r_{12} = \Sigma \int \frac{\overline{M}_1 \overline{M}_2 ds}{EJ} = \frac{h}{6EJ} \left[2 \left(-\frac{6EJ}{h^2} \times \frac{2EJ}{h} - \frac{6EJ}{h^2} \times \frac{4EJ}{h} \right) + \right.$$
$$\left. + \frac{6EJ}{h^2} \times \frac{4EJ}{h} + \frac{6EJ}{h^2} \times \frac{2EJ}{h} \right] = -\frac{6EJ}{h^2}$$

which coincides exactly with the value obtained in Art. 5.13 using the statical method.

The main unit reaction r_{22} will be obtained raising the $\overline{M}_2$ graph to the second power

$$r_{22} = \Sigma \int \frac{\overline{M}_2^2 \, ds}{EJ} = \frac{h}{6EJ} \left[2 \left(\frac{36 \, (EJ)^2}{h^4} \times 2 - \frac{36 \, (EJ)^2}{h^4} \right) \right] +$$
$$+ \frac{h 3EJ}{h^2 \times 2} \times \frac{2}{3} \times \frac{3EJ}{h^2} = \frac{15EJ}{h^3}$$

It is seen that this value coincides again with the one obtained previously.

The reactions due to the applied loads may be obtained considering two different states of the conjugate system of redundant beams, the first of these states corresponding to the application of the actual loads and the other being some unit state, say, state n of Fig. 25.13.

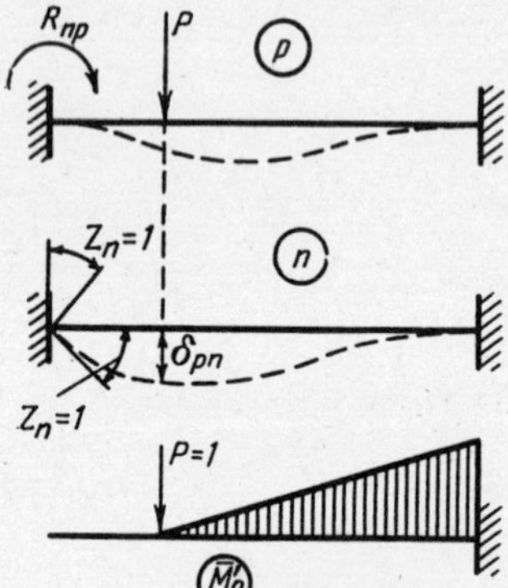

Fig. 25.13

The work accomplished by the loads of state P along the displacements of state n is given by

$$A_{pn} = P\delta_{pn} + R_{np}\cdot 1$$

At the same time the work accomplished by the external forces of state n along the displacements of state P equals

$$A_{np} = 0$$

The theorem of reciprocal works states that

$$A_{pn} = A_{np}$$

and therefore

$$R_{np} = -P\delta_{pn}$$

In this expression δ_{pn} is the deflection of the load point in state n along the direction of the load P. If $P = 1$ the above expression becomes

$$r_{np} = -\delta_{pn} \tag{6.13}$$

The latter expression establishes the relation between the unit reactions and the corresponding unit displacements.

The value of δ_{pn} will be obtained using the bending moment diagram corresponding to a beam built in at its right-hand extremity and supporting one concentrated load $P = 1$ acting along the direction of the desired deflection (see Articles 6.8 and 8.8). This diagram denoted hereafter by $\overline{M}'_p$ is given in Fig. 25.13c.

Multiplying this diagram by that for $\overline{M}_n$ we obtain

$$\delta_{pn} = \Sigma \int \overline{M}'_p \frac{\overline{M}_n\, ds}{EJ}$$

Introducing the value of δ_{pn} thus found in the expression for reaction R_{np} we find

$$R_{np} = -P\Sigma \int \overline{M}'_p \frac{\overline{M}_n\, ds}{EJ}$$

which, after simplification and replacement of $P\overline{M}'_p$ by M'_p, leads to

$$R_{np} = -\Sigma \int \overline{M}_n \frac{M'_p\, ds}{EJ} \tag{7.13}$$

In the latter expression M'_p stands for the ordinates to the diagram of the bending moments produced by the load P in any simple statically determinate structure obtained by the elimination of redundant constraints either of the given structure or of the conjugate system of redundant beams, provided the constraints whose reactions are desired are included in those eliminated.

Thus, in order to find the reaction of the nth constraint due to the applied loads we should multiply the area of the bending moment diagram due to the same loads applied to a statically determinate structure (derived either from the given structure or from the conjugate system of redundant beams, provided the nth constraint is among those eliminated) by the bending moment diagram $\overline{M}_n$ due to the unit displacement of the nth constraint in the system of redundant beams. The sign of the product obtained as described above will be thereafter changed to the opposite one.

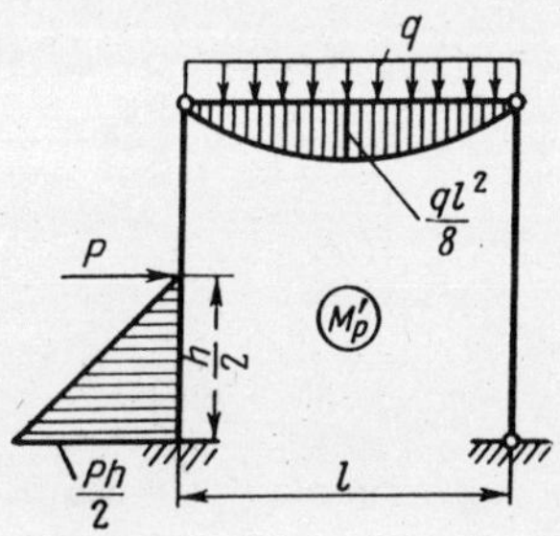

Fig. 26.13

Problem. Required the reactions R_{1p} and R_{2p} at the supports of the portal frame analyzed in Art. 5.13.

Solution. The conjugate simple structure and the M'_p graph related to this structure are given in Fig. 26.13. Multiplying this graph by that for $\overline{M}_1$ (see Fig. 20.13*b*) we obtain

$$R_{1p} = -\Sigma \int M'_p \frac{\overline{M}_1\, ds}{EJ} = -\frac{2}{3} \times \frac{ql^2}{8} \times \frac{l}{EJ} \times \times \frac{1}{2} \times \frac{3EJ}{l} + \frac{1}{2} \times \frac{Ph}{2} \times \frac{h}{2} \times \frac{1}{2} \times \frac{2EJ}{EJh} = -\frac{ql^2}{8} + \frac{Ph}{8}$$

Similarly R_{2p} will be obtained by the multiplication of the M'_p by the $\overline{M}_2$ graphs (see Fig. 20.13*c*)

$$R_{2p} = -\Sigma \int M'_p \overline{M}_2 \frac{ds}{EJ} = -\frac{1}{2} \cdot \frac{Ph}{2} \cdot \frac{h}{2} \cdot \frac{1}{EJ} \cdot \frac{2}{3} \cdot \frac{6EJ}{h^2} = -\frac{P}{2}$$

These results coincide exactly with those obtained in Art. 5.13 using the static method.

7.13. *Checking the Coefficients to the Unknowns and the Free Terms Pertaining to the Simultaneous Equations of the Slope and Deflections Method*

Check on the coefficients to the unknowns entering the system of equations used in the slope and deflections method is quite similar to the one described in Art. 5.12. One should begin with the construction of the bending moment diagram $\overline{M}_s$ obtained by the algebraic summation of all the ordinates to the unit graphs pertaining to the conjugate system of redundant beams

$$\overline{M}_s=\overline{M}_1+\overline{M}_2+\ldots+\overline{M}_n=\sum_{i=1}^{i=n}\overline{M}_i$$

This graph when multiplied by each of the unit graphs will give the values of the algebraic sums of the coefficients to the unknowns belonging to each equation. Thus, the product of the $\overline{M}_s$ graph by the $\overline{M}_1$ graph equals

$$r_{1s}=\Sigma\int\frac{\overline{M}_1\overline{M}_s\,ds}{EJ}=\Sigma\int\frac{\overline{M}_1(\overline{M}_1+\overline{M}_2+\ldots+\overline{M}_n)\,ds}{EJ}=$$

$$=\Sigma\int\frac{\overline{M}_1^2\,ds}{EJ}+\Sigma\int\frac{\overline{M}_1\overline{M}_2\,ds}{EJ}+\ldots+\Sigma\int\frac{\overline{M}_1\overline{M}_n\,ds}{EJ}=$$

$$=r_{11}+r_{12}+\ldots+r_{1n}=\sum_{i=1}^{i=n}r_{1i}$$

The sum of coefficients $r_{2s}=\Sigma r_{2i}$ will be obtained in exactly the same way.

Consequently, the sum of all the coefficients to the unknowns of any equation (say, Eq. i) must equal r_{is} given by

$$r_{is}=\Sigma\int\overline{M}_i\overline{M}_s\frac{ds}{EJ} \tag{8.13}$$

Thus, if we wish to check the values of all the unit reactions (coefficients to the unknowns) entering the first of the equations of the slope and deflections method we must compare the sum of these reactions with the value of r_{1s}

$$\sum_{i=1}^{i=n}r_{1i}=r_{1s} \tag{9.13}$$

Other unit reactions entering other equations will be checked in exactly the same way. The above procedure permits to check *separately* the coefficients entering each of the equations. It is equally possible to check *simultaneously* all the coefficients entering all the equations. Indeed if we raise to the second power the $\overline{M}_s$ graph we obtain

$$r_{ss}=\Sigma\int\frac{\overline{M}_s^2\,ds}{EJ}=\Sigma\int\frac{(\overline{M}_1+\overline{M}_2+\ldots+\overline{M}_n)^2\,ds}{EJ}=$$

$$=\left(\Sigma\int\frac{\overline{M}_1^2\,ds}{EJ}+\Sigma\int\frac{\overline{M}_2^2\,ds}{EJ}+\ldots+\Sigma\int\frac{\overline{M}_n^2\,ds}{EJ}\right)+$$

$$+2\left(\Sigma\int\frac{\overline{M}_1\overline{M}_2\,ds}{EJ}+\ldots+\Sigma\int\frac{\overline{M}_2\overline{M}_n\,ds}{EJ}+\ldots\right)=$$

$$=(r_{11}+r_{12}+\ldots+r_{nn})+2(r_{12}+\ldots+r_{2n}+\ldots)=\Sigma r$$

The reactions contained in the first term in parentheses of the above expressions are the principal ones situated along the main diagonal, while the second term in parentheses contains all the secondary reactions situated below or above this main diagonal.

It is apparent that the square of the $\overline{M}_s$ graph equals the algebraic sum of all the unit reactions (coefficients to the unknowns) contained in all the simultaneous equations of the given system, i. e.

$$\Sigma r = r_{ss} \tag{10.13}$$

where

$$r_{ss} = \Sigma \int \overline{M}_s^2 \frac{ds}{EJ} \tag{11.13}$$

The values of the free terms may be checked computing

$$R_{sp} = -\Sigma \int \frac{\overline{M}_s M'_p \, ds}{EJ} \tag{12.13}$$

that is, multiplying the M'_p diagram by the $\overline{M}_s$ diagram. It will be remembered that M'_p is the bending moment induced by the applied loads in the members of a simple statically determinate structure corresponding to the redundant structure under consideration. The product obtained as explained above must be equal to the algebraic sum of all the free terms of the simultaneous equations

$$R_{sp} = -\Sigma \int \frac{(\overline{M}_1 + \overline{M}_2 + \ldots + \overline{M}_n) M'_p \, ds}{EJ} =$$

$$= -\left(\Sigma \int \frac{\overline{M}_1 M'_p \, ds}{EJ} + \Sigma \int \frac{\overline{M}_2 M'_p \, ds}{EJ} + \ldots \right.$$

$$\left. + \Sigma \int \frac{\overline{M}_n M'_p \, ds}{EJ} \right) = R_{1p} + R_{2p} + \ldots + R_{np}$$

that is

$$R_{sp} = \Sigma R \tag{13.13}$$

8.13. *Construction of the M, N and Q Diagrams*

When the system of simultaneous equations derived from the slope and deflections method is solved, in other words, when the angles of twist and the deflections of all the joints of the given redundant structure are known, one can proceed with the construction of all the stress diagrams pertaining to this structure.

The ordinates to the resulting bending moment graph M will be obtained by the summation of the ordinates to the M_p diagram with those to the unit diagrams, all the latter being previously multiplied by the magnitude of the unknowns just determined

$$M = M_p + \overline{M}_1 Z_1 + \overline{M}_2 Z_2 + \ldots + \overline{M}_n Z_n \qquad (14.13)$$

The diagram thus obtained may be checked using one of the procedures described in Art 6.9. It is worth noting that the control of the equilibrium of moments acting at each of the joints becomes particularly significant for in the construction of all the bending moment diagrams used in expression (14.13) no reference was made to the said equilibrium condition. If the bending moments at one of the joints do not balance, this means that some error has been committed in computing the values of one or more unknowns.

The shear diagram Q is derived from the bending moment diagram just as in the method of forces. The normal stresses may be computed thereafter. The shear and normal stress diagrams will be checked as described in Art. 6.9. Here the equilibrium of different portions of the structure under consideration becomes of particular importance.

9.13. *Computation of Thermal Strains by the Slope and Deflections Method*

We have seen previously that a change in temperature entails usually the development of stresses in redundant structures. Only in a few particular cases temperature changes have no

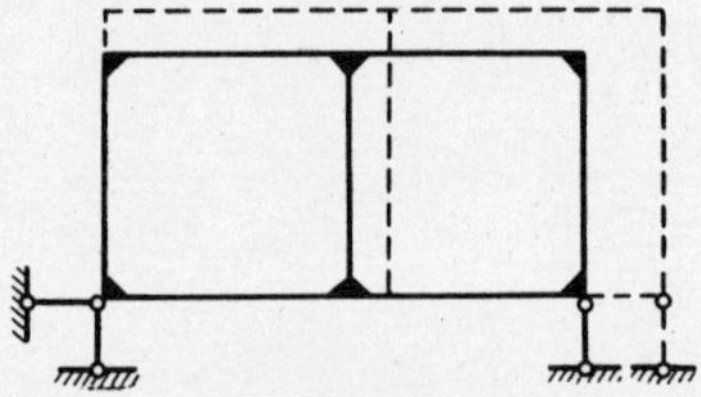

Fig. 27.13

effect on such structures as, for example, in the case of a rectangular frame represented in Fig. 27.13. This frame is externally statically determinate and therefore it can expand or retract freely in case of a uniform change in temperature of all its members, without any stresses arising in any of the latter.

Let us show that temperature effects can be always resolved into symmetrical and antisymmetrical components. Assume that bar AB of Fig. 28.13a is of uniform cross section and that the temperature on its upper surface is raised by t_1° and on the lower one by t_2° with $t_1 > t_2$. The effect of this difference in temperature is equivalent to the simultaneous change of temperature equal to $1/2\ (t_1 + t_2)$ on both surfaces (symmetrical components, Fig. 28.13*b*) combined with changes of temperature equal to

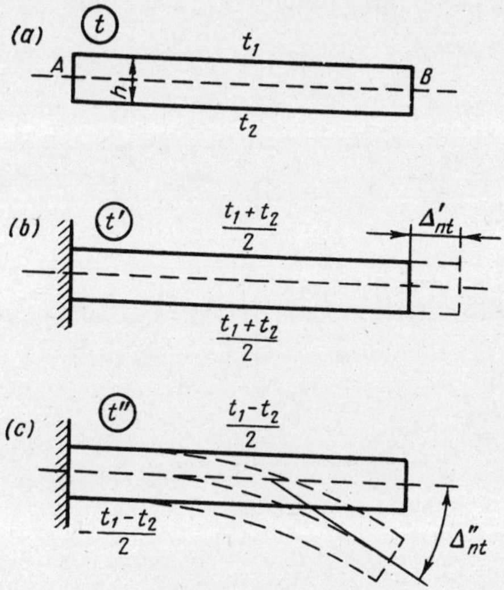

Fig. 28.13

$+\frac{1}{2}\ (t_1 - t_2)$ on the upper surface and to $-\frac{1}{2}\ (t_1 - t_2)$ on the lower surface (antisymmetrical components, Fig. 28.13*c*). It is clear that the combination of these changes will result in the specified temperature change on each of the two faces

$$\frac{t_1+t_2}{2}+\frac{t_1-t_2}{2}=t_1$$

and

$$\frac{t_1+t_2}{2}-\frac{t_1-t_2}{2}=t_2$$

Returning to expression (20.8)

$$\Delta_{nt}=\alpha\,\frac{t_1+t_2}{2}\Omega_{\overline{N}}+\alpha\,\frac{t_1-t_2}{h}\Omega_{\overline{M}}$$

We note that symmetrical actions will lead to normal strains only without bending of the bar because the difference $(t_1 - t_2)$

reduces in that case to ze o

$$\Delta'_{nt} = \alpha \frac{t_1 + t_2}{2} \Omega_{\overline{N}} = \alpha \frac{t_1 + t_2}{2} l$$

because $\Omega_{\overline{N}} = 1l$ (Fig. 29.13*a*). *

On the other hand, when the bar is acted upon by antisymmetrical components only, it will bend without any change in the

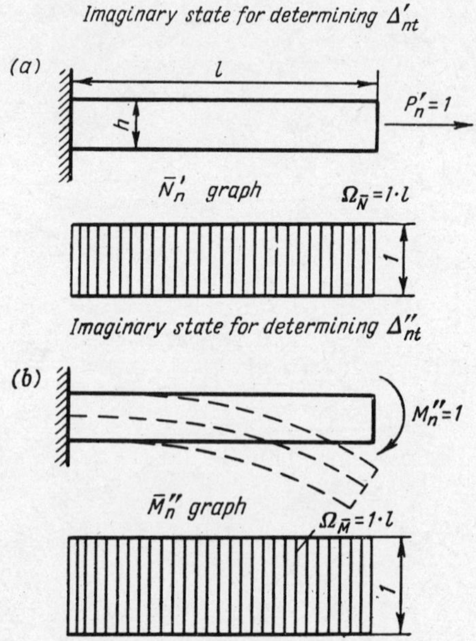

Fig. 29.13

length of its neutral axis. The angular rotation between the two end faces of this bar will equal

$$\Delta''_{nt} = \alpha \frac{t_1 - t_2}{h} \Omega_{\overline{M}} = \alpha \frac{t_1 - t_2}{h} l$$

because $\Omega_{\overline{M}} = l$ (Fig. 29.13*b*). **

Consequently, fixed end bars which constitute all the members of a conjugate system of redundant beams will expand or con-

* If the neutral axis of the bar is not situated at midheight of the cross section, $\frac{t_1 + t_2}{2}$ should be replaced by $t_2 + \frac{t_1 - t_2}{h} y$ where y is the distance to the neutral axis of the fibre whose temperature has been raised by t_2°.

** This expression is independent of the position of the neutral axis.

tract without bending when they are subjected to a symmetrical thermal effect and will bend without changing their length when the thermal effect is antisymmetrical. The corresponding bending moment graphs may be found in Table 2.13 (lines 5 and 10).

The only difference between the analysis of redundant structures carrying direct loads and those subjected to a temperature change resides in the construction of the diagrams for the bending moments M_t induced in the members of the conjugate system of redundant beams and in the determination of the free terms of the equations (these being denoted in that case by R_{1t}, R_{2t}, etc., and representing the reactions of the imaginary constraints due to the temperature change).

Problem. Required the complete stress analysis of the nonsymmetrical frame shown in Fig. 30.13*a* subjected to a temperature change. All the elements of this frame are of constant cross section with a flexural rigidity equal to EJ.

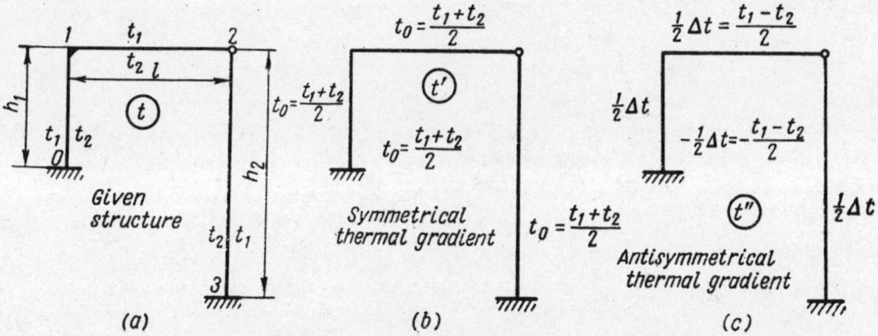

Fig. 30.13

Solution. The temperature increment t will be resolved into two components—one symmetrical and equal to a uniform heating of all the elements $t'=\frac{t_1+t_2}{2}$ and the other antisymmetrical t''. The number of unknowns for the conjugate system of redundant beams given in Fig. 31.13 will equal two.

The simultaneous equations become

$$Z_1r_{11}+Z_2r_{12}+R'_{1t}+R''_{1t}=0$$

$$Z_1r_{21}+Z_2r_{22}+R'_{2t}+R''_{2t}=0$$

The coefficients to the unknowns will be calculated in exactly the same way as if the structure were subjected to a system of loads. Each of the free terms will consist of two reactions R' and R'', the first corresponding to the symmetrical component and the second to the antisymmetrical one. The reactive moment R'_{1t} developed by the imaginary constraint at joint *1* and the reactive force R'_{2t} developed at the imaginary support will be determined using the M'_t diagram pertaining to the system of redundant beams and due to a uniform change in temperature of all the frame members.

Let us first calculate the extension of all these elements due to a raise in temperature equal to t' neglecting as usual the influence of the direct stresses.

It will be remembered that the antisymmetrical components t'' lead to no change whatsoever in the length of the bars. The extension of members *0-1*, *2-3* and *1-2* will therefore be given by $\Delta_1 = \alpha t' h_1$; $\Delta_2 = \alpha t' h_2$; $\Delta_3 = \alpha t' l$ respectively $\left(\text{at } t' = \dfrac{t_1 + t_2}{2}\right)$.

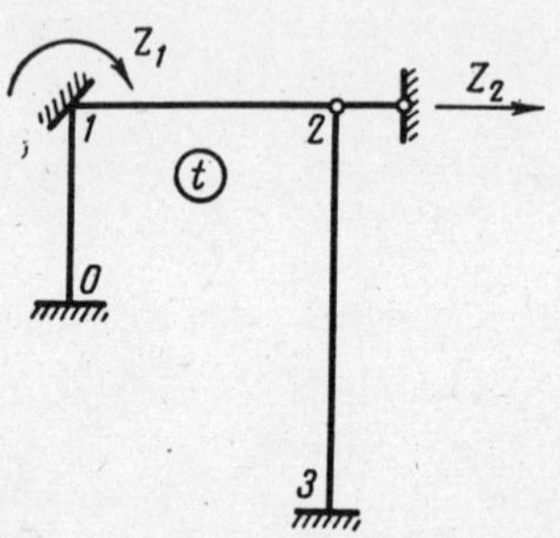

Fig. 31.13

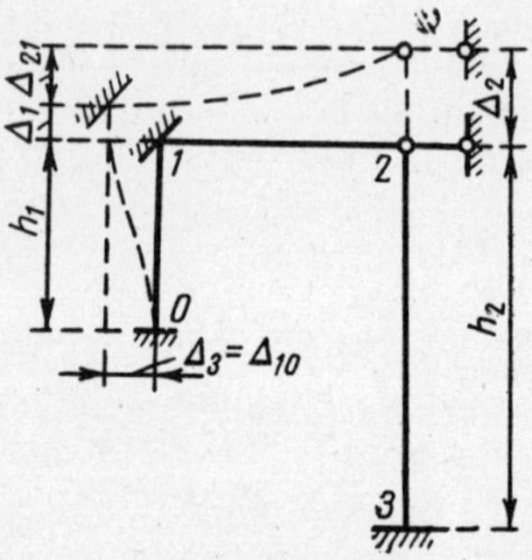

Fig. 32.13

The position of all the joints of the conjugate system of beams after a uniform change in temperature is indicated in dotted lines in Fig. 32.13. It is easily seen that the relative deflections of the joints will equal

$$\Delta_{10} = \Delta_3 = \alpha t' l$$

$$\Delta_{21} = \Delta_2 - \Delta_1 = \alpha t' (h_2 - h_1); \quad \Delta_{23} = 0$$

It is also clear that joint *1* of the conjugate system will sustain no angular rotation, and therefore the bending moments induced by the displacements

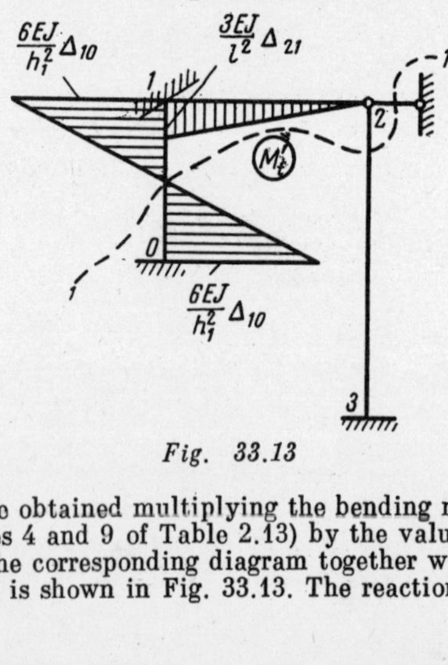

Fig. 33.13

of the joints may be obtained multiplying the bending moments due to unit deflections (see lines 4 and 9 of Table 2.13) by the values of the deflections indicated above. The corresponding diagram together with the values of its pertinent ordinates is shown in Fig. 33.13. The reaction R''_{1t} (Fig. 34.13*a*) will be obtained isolating joint *1*

$$\Sigma M_1 = R'_{1t} - \frac{6EJ}{h_1^2}\Delta_{10} - \frac{3EJ}{l^2}\Delta_{21} = 0$$

wherefrom

$$R'_{1t} = 6EJ\left(\frac{\Delta_{10}}{h_1^2} + \frac{\Delta_{21}}{2l^2}\right)$$

As for the reaction R'_{2t} (Fig. 33.13), its value will be found passing section *1-1* and projecting on the horizontal all the forces acting on that portion of the frame situated above the section (Fig. 34.13*b*).

$$\Sigma X = \frac{12EJ}{h_1^3}\Delta_{10} + R'_{2t} = 0$$

wherefrom

$$R'_{2t} = -\frac{12EJ}{h_1^3}\Delta_{10}$$

Next let us examine the effect produced by the antisymmetrical components. Knowing the drop in temperature $\Delta_t = t_1 - t_2$ between the inner and

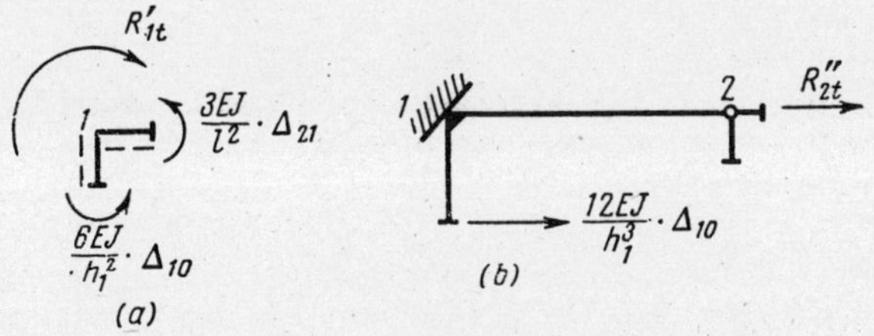

Fig. 34.13

outer surfaces of each element as well as the depth of their cross section h we may easily obtain the diagram for M''_t (Fig. 35.13) using lines 5 and 10 of Table 2.13. The ordinates to this diagram will be set off as usual on the side of the extended fibres or, in other words, on the low temperature side. The equilibrium of joint *1* (Fig. 36.13*a*) gives

$$\Sigma M_1 = R''_{1t} + \frac{EJ\alpha\Delta t}{h} - \frac{3EJ\alpha\Delta t}{2h} = 0$$

whence

$$R''_{1t} = \frac{EJ\alpha\Delta t}{2h}$$

Passing section *1-1* and projecting once again all the forces acting on the upper portion of the frame on the horizontal (Fig. 36.13*b*) we find

$$\Sigma X = R''_{2t} - \frac{3EJ\alpha\Delta t}{2hh_2} = 0; \quad R''_{2t} = \frac{3EJ\alpha\Delta t}{2hh_2}$$

When all the coefficients to the unknowns and all the free terms are determined, we may proceed with the solution of the simultaneous equations which will yield the values of the unknown angles of twist and deflections of the joints.

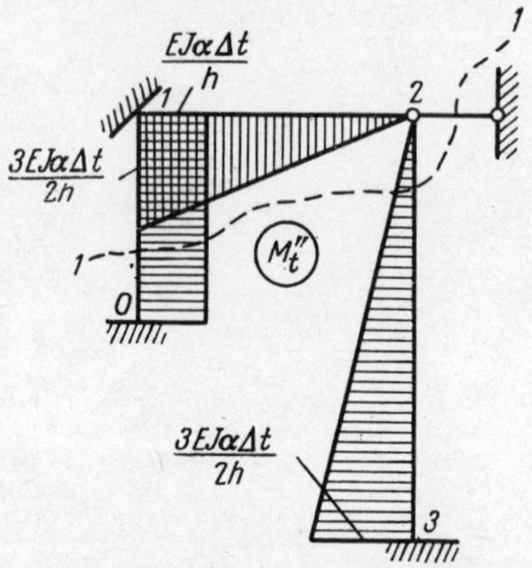

Fig. 35.13

The resulting bending moments acting at different cross sections of the given redundant structure may thereafter be calculated using the expression

$$M = M'_t + M''_t + \overline{M}_1 Z_1 + \overline{M}_2 Z_2 + \ldots$$

The bending moment diagram being known, we may proceed as usual with the construction of the shear diagram and finally with that for the normal stresses.

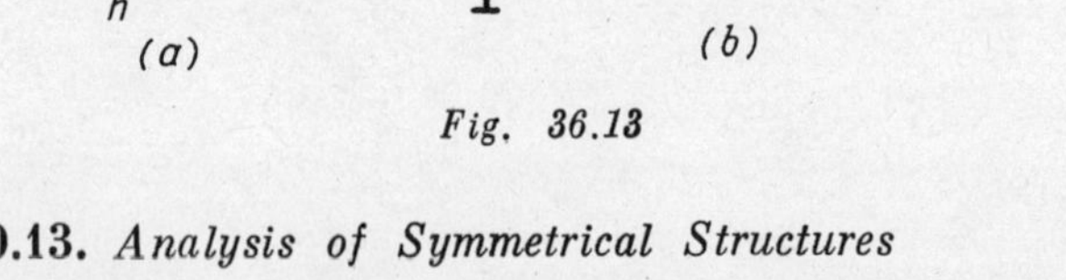

Fig. 36.13

10.13. *Analysis of Symmetrical Structures*

The analysis of symmetrical structures may be facilitated if the unknowns are grouped together in a manner similar to the one used in the method of forces. This procedure allows to obtain symmetrical or antisymmetrical unit diagrams, which leads to a subdivision of simultaneous equations into two independent groups, one containing only the symmetrical unknowns and the other the antisymmetrical ones. A substantial simplification of all the computations may be achieved thereby.

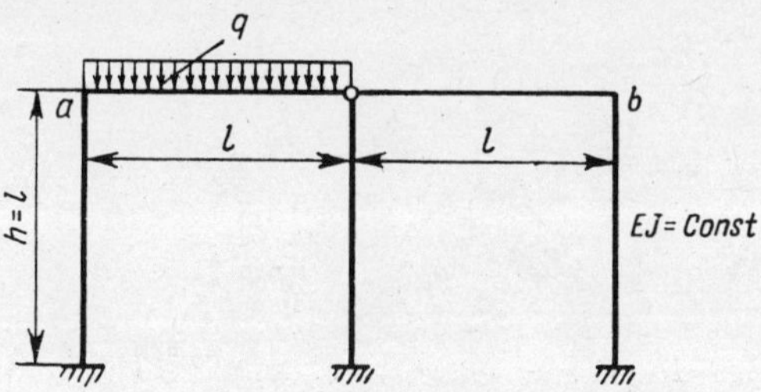

Fig. 37.13

As an example take up the double-span frame of Fig. 37.13. The angle of twist of joint *a* may be regarded as consisting of the sum of two unknown angles Z_1 and Z_2 and the angle of twist of joint *b* as consisting of the difference of the same angles Z_1 and Z_2 (Fig. 38.13).

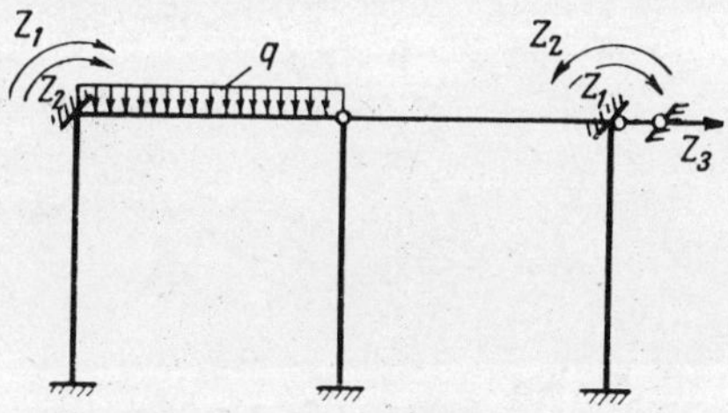

Fig. 38.13

The horizontal displacement of the crossbeam may be regarded as antisymmetrical for joint *b* (see Fig. 37.13) moves away from the axis of symmetry while joint *a* moves towards the same axis over the same distance. In this case the system of three simultaneous equations with three unknowns

$$Z_1 r_{11} + Z_2 r_{12} + Z_3 r_{13} + R_{1q} = 0$$
$$Z_1 r_{21} + Z_2 r_{22} + Z_3 r_{23} + R_{2q} = 0$$
$$Z_1 r_{31} + Z_2 r_{32} + Z_3 r_{33} + R_{3q} = 0$$

falls into two independent groups

$$Z_1 r_{11} + Z_3 r_{13} + R_{1q} = 0$$
$$Z_1 r_{31} + Z_3 r_{33} + R_{3q} = 0$$

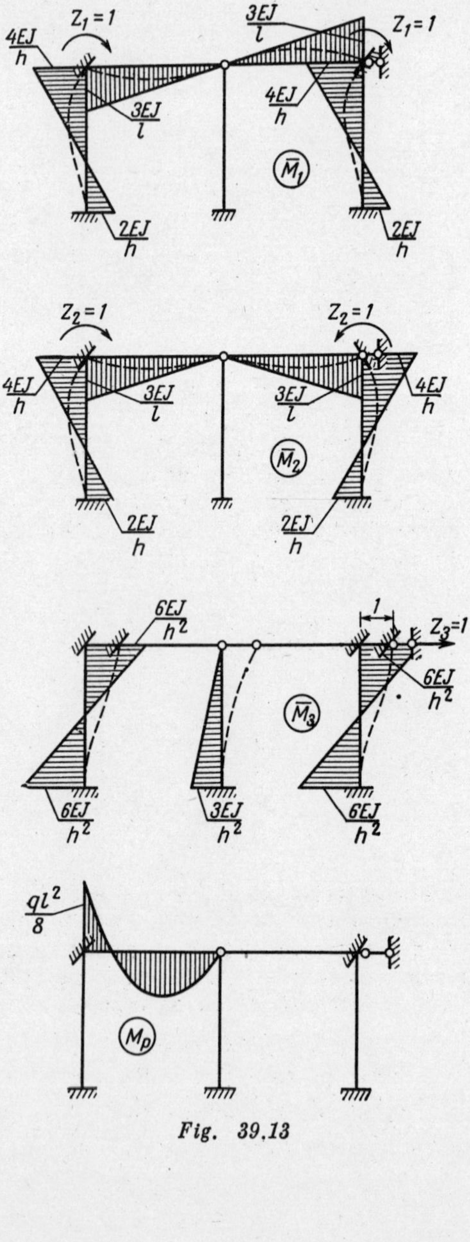

$Z_1=1$
$\frac{4EJ}{h}$
$\frac{3EJ}{l}$
$\frac{2EJ}{h}$
$\bar{M}_1$
$Z_2=1$
$\bar{M}_2$
$\frac{6EJ}{h^2}$
1
$Z_3=1$
$\frac{3EJ}{h^2}$
$\bar{M}_3$
$\frac{ql^2}{8}$
M_p

Fig. 39.13

and

$$Z_2 r_{22} + R_{2q} = 0$$

because

$$r_{12} = r_{23} = r_{21} = r_{32} = 0$$

The bending moment diagrams due to the groups of unit reactions are represented in Fig. 39.13. The coefficients to the un-

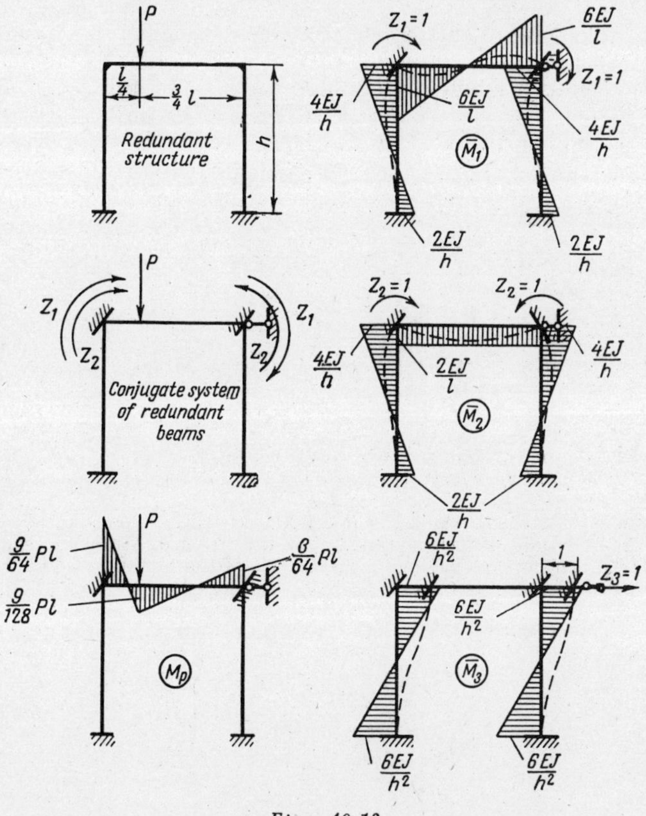

Fig. 40.13

knowns as well as the free terms of the above equations will be obtained as follows

$$r_{11} = \frac{4EJ}{h} + \frac{3EJ}{l} + \frac{3EJ}{l} + \frac{4EJ}{h} = \frac{14EJ}{l}$$

$$r_{22} = \frac{4EJ}{h} + \frac{3EJ}{l} + \frac{3EJ}{l} + \frac{4EJ}{h} = \frac{14EJ}{l}$$

$$r_{13}=r_{31}=-\frac{6EJ}{h^2}-\frac{6EJ}{h^2}=-\frac{12EJ}{l^2}$$

$$r_{33}=\frac{27EJ}{h^3}=\frac{27EJ}{l^3}$$

h being equal to l

$$R_{1p}=-\frac{ql^2}{8} \qquad R_{2p}=-\frac{ql^2}{8} \qquad R_{3p}=0$$

Fig. 40.13 represents a portal frame loaded by one single concentrated force, the conjugate system of redundant beams as well as the bending moment diagrams induced in the latter by the symmetrical and antisymmetrical groups of unit reactions and by the applied loads. The reader is invited to check both the diagrams and the reactions (coefficients to the unknowns and free terms of the simultaneous equations) given hereunder

$$r_{11}=\frac{20EJ}{l} \qquad r_{12}=r_{21}=0 \qquad r_{22}=\frac{12EJ}{l}$$

$$r_{13}=r_{31}=-\frac{12EJ}{h^2} \qquad r_{33}=\frac{24EJ}{h^3} \qquad r_{23}=r_{32}=0$$

$$R_{1p}=-\frac{3}{32}Pl \qquad R_{2p}=-\frac{3}{16}Pl \qquad R_{3p}=0$$

11.13. *An Example of Frame Analysis by the Slope and Deflections Method*

Assume that it is required to construct the bending moment, the shearing forces and normal stresses diagrams for the framed structure shown in Fig. 41.13. The system of loads, the length of the members and the ratios between their moments of inertia are all indicated in the same figure.

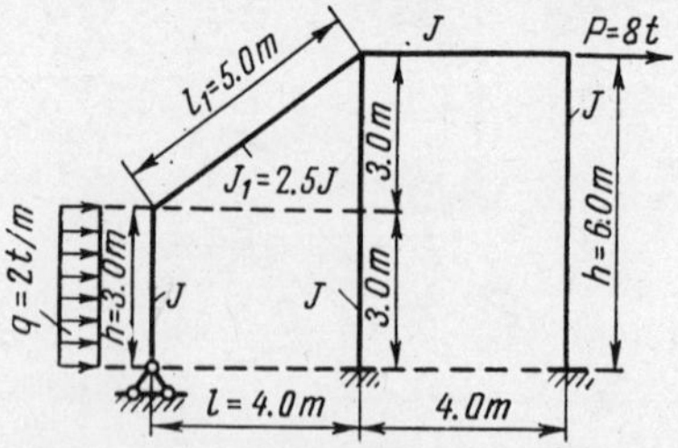

Fig. 41.13

1. *Comparison of the number of unknowns in the slope and deflections method with that in the method of forces.* The degree of redundancy of the given structure being equal to $i=2\times3-1=5$, the number of the unknowns in the method of forces would be equally five.

The number of unknown angles of twist $n_t = 3$ and the number of independent unknown deflections equals to one and therefore the total number of unknowns in the slope and deflections method will equal $n_t + n_d = 1 + 3 = 4$. Hence using the latter method we shall reduce the number of simultaneous equations from 5 to 4.

2. *Choice of the conjugate system.* Let us introduce three imaginary constraints opposing the rotation of the rigid joints of the frame and one imaginary support preventing the displacement of joint *1* as indicated in Fig. 42.13.

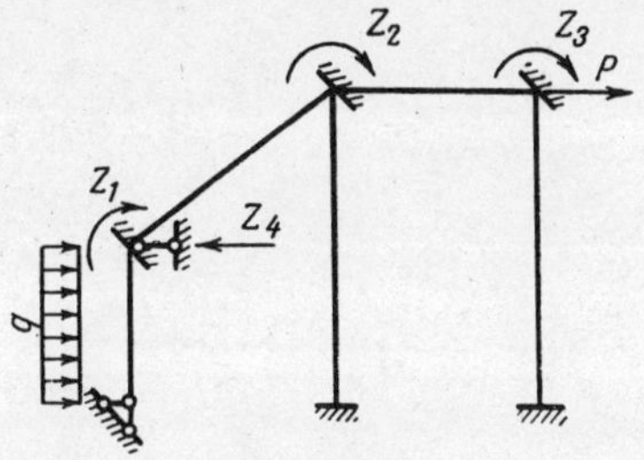

Fig. 42.13

The three unknown angles of twist will be designated by Z_1, Z_2 and Z_3 and the unknown deflection by Z_4. The conjugate system of redundant beams will consist in that case of four beams fixed at both ends and of one beam fixed at one end only and simply supported at the other.

3. *Construction of bending moment diagrams for the conjugate system.* The coefficients to the unknowns and the free terms of the system of simultaneous equations will be derived from the bending moment diagrams induced both by the applied loads and by the unit reactions in all the members of the conjugate structure. These diagrams are represented in Fig. 43.13. *

4. *Calculation of the coefficients to the unknowns and of the free terms.* All the necessary operations are carried out in tabular form as indicated hereunder.

5. *Checking the values obtained for accuracy.* The accuracy of the coefficients to the unknowns and of the free terms obtained as described above will be checked using the M_s diagram (Fig. 44.13) obtained through the summation of the ordinates to the four unit diagrams shown in Fig. 43.13. Raising the area of the $\overline{M}_s$ diagram to the second power [see expression (11.13)] we obtain

$$r_{ss}=\Sigma\int \overline{M}_s^2\frac{ds}{EJ}=\left[\frac{3}{6}\frac{(EJ)^2}{EJ}2\times\frac{4}{3}\times\frac{4}{3}+\frac{5\,(EJ)^2}{6\times 2.5EJ}\times\right.$$
$$\times(2\times 3^2+2\times 3^2-2\times 3^2)+\frac{4\,(EJ)^2}{6EJ}\left(2\times\frac{3}{2}\times\frac{3}{2}+\right.$$
$$\left.+2\times\frac{3}{2}\times\frac{3}{2}-2\times\frac{3}{2}\times\frac{3}{2}\right)+\frac{6\,(EJ)^2}{6EJ}\left(2\times\frac{5}{6}\times\frac{5}{6}+\right.$$
$$\left.\left.+2\times\frac{1}{2}\times\frac{1}{2}-2\times\frac{5}{6}\times\frac{1}{2}\right)2\right]=\frac{116EJ}{9}=12\frac{8}{9}\,EJ$$

* In this example the values of all loads and reactions are given in tons and all the distances are measured in metres.

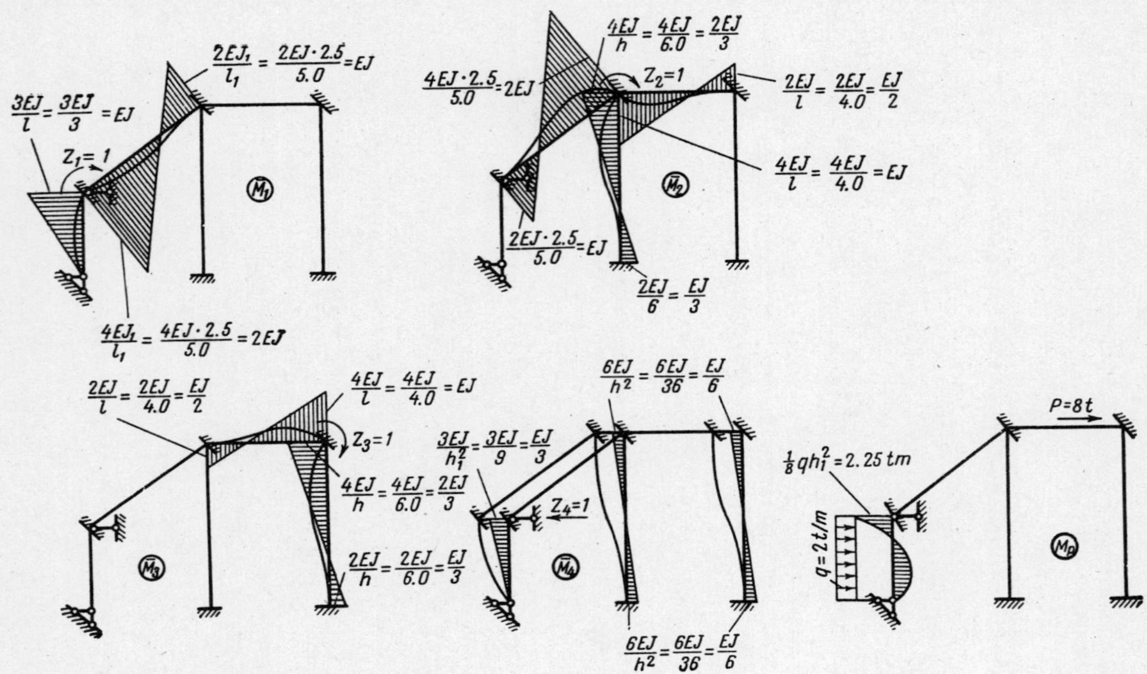

$\frac{3EJ}{l} = \frac{3EJ}{3} = EJ$
$Z_1 = 1$
$\frac{2EJ_1}{l_1} = \frac{2EJ \cdot 2.5}{5.0} = EJ$
$\frac{4EJ_1}{l_1} = \frac{4EJ \cdot 2.5}{5.0} = 2EJ$
$\bar{M}_1$
$\frac{4EJ}{h} = \frac{4EJ}{6.0} = \frac{2EJ}{3}$
$\frac{4EJ \cdot 2.5}{5.0} = 2EJ$
$Z_2 = 1$
$\frac{2EJ}{l} = \frac{2EJ}{4.0} = \frac{EJ}{2}$
$\frac{4EJ}{l} = \frac{4EJ}{4.0} = EJ$
$\frac{2EJ \cdot 2.5}{5.0} = EJ$
$\frac{2EJ}{6} = \frac{EJ}{3}$
$\bar{M}_2$
$\frac{2EJ}{l} = \frac{2EJ}{4.0} = \frac{EJ}{2}$
$\frac{4EJ}{l} = \frac{4EJ}{4.0} = EJ$
$Z_3 = 1$
$\frac{4EJ}{h} = \frac{4EJ}{6.0} = \frac{2EJ}{3}$
$\frac{2EJ}{h} = \frac{2EJ}{6.0} = \frac{EJ}{3}$
$\bar{M}_3$
$\frac{6EJ}{h^2} = \frac{6EJ}{36} = \frac{EJ}{6}$
$\frac{3EJ}{h_1^2} = \frac{3EJ}{9} = \frac{EJ}{3}$
$Z_4 = 1$
$\frac{6EJ}{h^2} = \frac{6EJ}{36} = \frac{EJ}{6}$
$\bar{M}_4$
$\frac{1}{8} q h_1^2 = 2.25\ tm$
$q = 2\ t/m$
$P = 8t$
M_p

Fig. 43.13

Table 3.13

Unit reactions	Conjugate system loaded by reaction under consideration	Sketch of joint or portion of structure which must be in equilibrium	Equilibrium equation	Solution of equilibrium equation	Result obtained
1	2	3	4	5	6
r_{11}	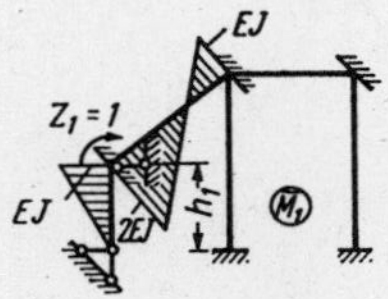		$r_{11}-2EJ-EJ=0$	$r_{11}=2EJ+EJ=3EJ$	$r_{11}=3EJ$
r_{21}			$r_{21}-EJ=0$	$r_{21}=EJ$	$r_{21}=EJ$
r_{31}		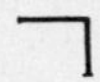	$r_{31}=0$	$r_{31}=0$	$r_{31}=0$
r_{41}			$-r_{41}+\frac{EJ}{h_1}=0$	$r_{41}=\frac{EJ}{h_1}=\frac{EJ}{3}$	$r_{41}=\frac{EJ}{3}$

Table 3.13 (continued)

Unit reactions	Conjugate system loaded by reaction under consideration	Sketch of joint or portion of structure which must be in equilibrium	Equilibrium equation	Solution of equilibrium equation	Result obtained
1	2	3	4	5	6
r_{12}	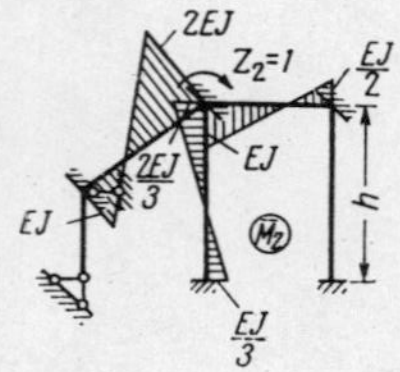		$r_{12}-EJ=0$	$r_{12}=EJ$	$r_{12}=EJ$
r_{22}		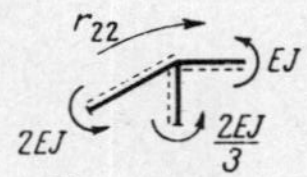	$r_{22}-EJ-\frac{2EJ}{3}-$ $-2EJ=0$	$r_{22}=EJ+\frac{2EJ}{3}+$ $+2EJ=\frac{11EJ}{3}$	$r_{22}=\frac{11EJ}{3}$
r_{32}		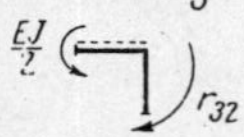	$r_{32}-\frac{EJ}{2}=0$	$r_{32}=\frac{EJ}{2}$	$r_{32}=\frac{EJ}{2}$
r_{42}		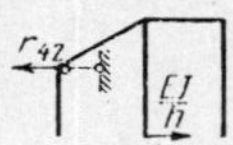	$-r_{42}+$ $+\frac{\frac{2EJ}{3}+\frac{EJ}{3}}{h}=0$	$r_{42}=\frac{2EJ}{18}+\frac{EJ}{18}=$ $=\frac{EJ}{6}$	$r_{42}=\frac{EJ}{6}$

r_{13}	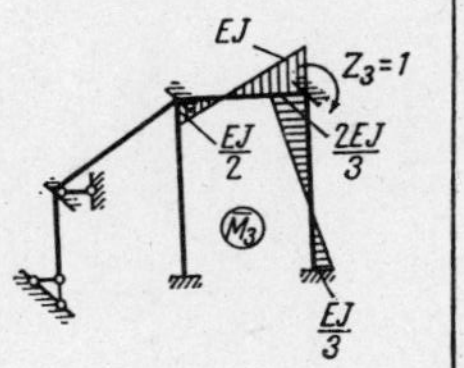		$r_{13}=0$	—	$r_{13}=0$
r_{23}			$r_{23}-\frac{EJ}{2}=0$	$r_{23}=\frac{EJ}{2}$	$r_{23}=\frac{EJ}{2}$
r_{33}		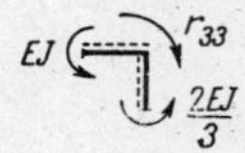	$r_{33}-\frac{2EJ}{3}-EJ=0$	$r_{33}=\frac{2EJ}{3}+EJ=$ $=\frac{5EJ}{3}$	$r_{33}=\frac{5EJ}{3}$
r_{43}			$-r_{43}+$ $+\frac{\frac{2EJ}{3}+\frac{EJ}{3}}{h}=0$	$r_{43}=\frac{2EJ}{3\times 6}+\frac{EJ}{3\times 6}=$ $=\frac{EJ}{6}$	$r_{43}=\frac{EJ}{6}$

Table 3.13 (continued)

Unit reactions	Conjugate system loaded by reaction under consideration	Sketch of joint or portion of structure which must be in equilibrium	Equilibrium equation	Solution of equilibrium equation	Result obtained
1	2	3	4	5	6
r_{14}	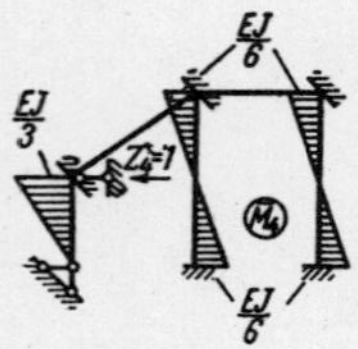	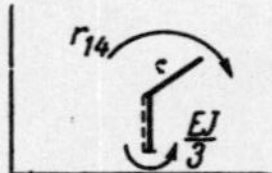	$r_{14}-\frac{EJ}{3}=0$	$r_{14}=\frac{EJ}{3}$	$r_{14}=\frac{EJ}{3}$
r_{24}		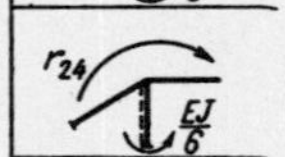	$r_{24}-\frac{EJ}{6}=0$	$r_{24}=\frac{EJ}{6}$	$r_{24}=\frac{EJ}{6}$
r_{34}		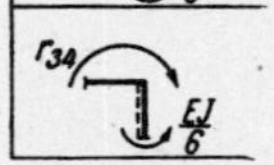	$r_{34}-\frac{EJ}{6}=0$	$r_{34}=\frac{EJ}{6}$	$r_{34}=\frac{EJ}{6}$
r_{44}		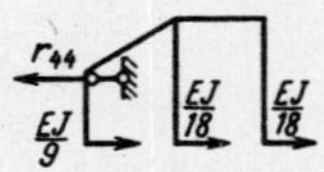	$-r_{44}+\frac{EJ}{9}+\frac{EJ}{18}+\frac{EJ}{18}=0$	$r_{44}=\frac{2EJ}{18}+\frac{EJ}{18}+\frac{EJ}{18}=\frac{4EJ}{18}$	$r_{44}=\frac{2EJ}{9}$

R_{1p}	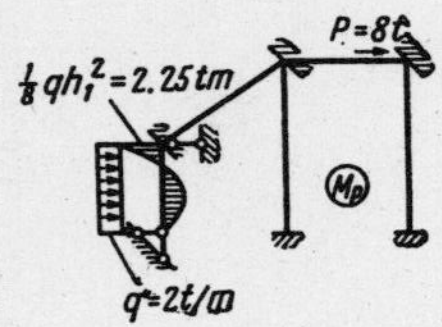	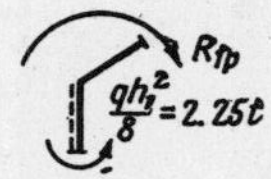	$R_{1p} - \frac{qh_1^2}{8} = 0$	$R_{1p} = \frac{qh_1^2}{8} = \frac{2\times 3^2}{8} =$ $= 2.25$ ton metres	$R_{1p} = 2.25$ ton metres
R_{2p}		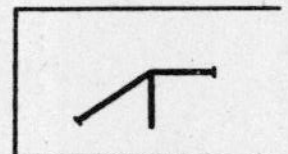	$R_{2p} = 0$	—	$R_{2p} = 0$
R_{3p}			$R_{3p} = 0$	—	$R_{3p} = 0$
R_{4p}		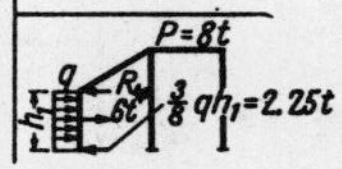	$-R_{4p} + qh_1 -$ $-3/8qh_1 + P = 0$	$R_{4p} = +qh_1 -$ $-3/8qh_1 + P =$ $=2\times 3 - 3/8\times 2\times$ $\times 3 + 8 = 11.75$ tons	$R_{4p} = 11.75$ tons

On the other hand, the sum of all the coefficients given in Table 3.13 equals

$$\Sigma r = r_{11}+r_{22}+r_{33}+r_{44}+2(r_{12}+r_{13}+r_{14}+r_{23}+r_{24}+r_{34}) =$$
$$= EJ\left(3+\frac{11}{3}+\frac{5}{3}+\frac{2}{9}\right)+2EJ\left(1+0+\frac{1}{3}+\frac{1}{2}+\frac{1}{6}+\frac{1}{6}\right)=$$
$$=\frac{116EJ}{9}=12\frac{8}{9}\,EJ$$

The comparison of the total thus obtained with the result of graph multiplication shows that they coincide and consequently condition (10.13) is satisfied.

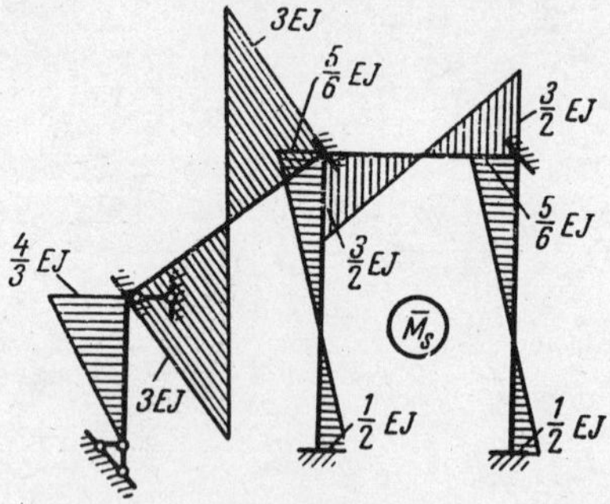

Fig. 44.13

Let us proceed with a check on the accuracy of the free terms. For this purpose we shall construct the M'_p diagram whose ordinates represent the

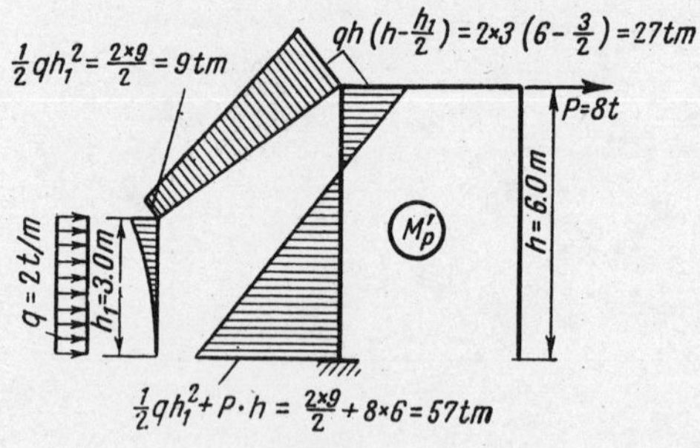

Fig. 45.13

bending moments induced by the given loads in a simple statically determinate structure obtained by elimination of all the redundant reactions. This diagram is given in Fig. 45.13. Multiplying it by the $\overline{M}_s$ diagram [see expression (12.13)] we obtain

$$\Sigma\int \overline{M}_s M'_p\,\frac{ds}{EJ}=\frac{1}{EJ}\left(\frac{1}{3}\times 9\times 3\times\frac{3}{4}\times\frac{4}{3}\,EJ\right)+\frac{5EJ}{6\times 2.5EJ}\times$$

$$\times(-2\times 9\times 3+2\times 27\times 3-27\times 3+9\times 3)+$$
$$+\frac{6EJ}{6EJ}\left(-2\times\frac{1}{2}\times 57-2\times\frac{5}{6}\times 27+27\times\frac{1}{2}+57\times\frac{5}{6}\right)=$$
$$=-14=-R_{sp}$$

wherefrom $R_{sp}=14$.

On the other hand, the sum of all the free terms given in Table 3.13 equals

$$\Sigma R=R_{1p}+R_{2p}+R_{3p}+R_{4p}=2.25+0+0+11.75=14.00$$

The coincidence of the two results shows that condition (13.13) is satisfied which means that the computations are correct.

6. *Solution of the simultaneous equations.* For the solution of the simultaneous equations resort shall be made to the abridged method described in Art. 6.12. The equations negating the existence of reactions of the imaginary constraints become in that case

$$Z_1r_{11}+Z_2r_{12}+Z_3r_{13}+Z_4r_{14}+R_{1p}=0$$
$$Z_1r_{21}+Z_2r_{22}+Z_3r_{23}+Z_4r_{24}+R_{2p}=0$$
$$Z_1r_{31}+Z_2r_{32}+Z_3r_{33}+Z_4r_{34}+R_{3p}=0$$
$$Z_1r_{41}+Z_2r_{42}+Z_3r_{43}+Z_4r_{44}+R_{4p}=0$$

Substituting in these equations the numerical values of the coefficients to the unknowns and of the free terms given in Table 3.13 we obtain

$$3Z_1+Z_2+\frac{1}{3}Z_4+\frac{2.25}{EJ}=0 \qquad \left(S_1=\frac{13}{3}\right)$$
$$Z_1+\frac{11}{3}Z_2+\frac{1}{2}Z_3+\frac{1}{6}Z_4=0 \qquad \left(S_2=\frac{16}{3}\right)$$
$$\frac{1}{2}Z_2+\frac{5}{3}Z_3+\frac{1}{6}Z_4=0 \qquad \left(S_3=\frac{7}{3}\right)$$
$$\frac{1}{3}Z_1+\frac{1}{6}Z_2+\frac{1}{6}Z_3+\frac{2}{9}Z_4+\frac{11.75}{EJ}=0 \qquad \left(S_4=\frac{8}{9}\right)$$

The values indicated in parentheses at the end of each equation represent the total S of all the coefficients entering this particular equation.

The solution of the above equations is given in Table 4.13.

Equation (IV) gives

$$Z_4=-\frac{175,740\times 206,280}{15,280EJ\times 34,760}=-68.253\times\frac{1}{EJ}$$

Introducing this value into equation (III) we find

$$\frac{191}{120}Z_3+\frac{19}{120}\left(-68.253\times\frac{1}{EJ}\right)=-\frac{9}{80EJ}$$

wherefrom

$$Z_3=+6.719\times\frac{1}{EJ}$$

Proceeding in the same way we find from equation (II)

$$\frac{10}{3}Z_2+\frac{1}{2}\times 6.719\times\frac{1}{EJ}+\frac{1}{18}\left(-68.253\times\frac{1}{EJ}\right)=\frac{3}{4}\times\frac{1}{EJ}$$

Table 4.13

Equation No.	Z_1	Z_2	Z_3	Z_4	Multipliers α_{ik}	S	K
(I)	3	1	0	$\frac{1}{3}$	$\alpha_{12} = -1/3$ $\alpha_{13} = 0$ $\alpha_{14} = -1/9$	13/3	$-\frac{9}{4EJ}$
+ (2)	·	11/3	1/2	1/6		16/3	0
+ (I) α_{12}	·	—1/3	0	—1/9		—13/9	$\frac{3}{4EJ}$
(II)	·	10/3	1/2	1/18	$\alpha_{23} = -3/20$ $\alpha_{24} = -1/60$	$\frac{35}{9}$	$\frac{3}{4EJ}$
+ (3)	·	·	5/3	1/6		7/3	0
+ (I) α_{13}	·	·	0	0		0	0
+ (II) α_{23}	·	·	—3/40	—1/120		—35/60	$-\frac{9}{80EJ}$
(III)	·	·	$\frac{191}{120}$	$\frac{19}{120}$	$\alpha_{34} = -\frac{19}{191}$	$\frac{210}{120}$	$-\frac{9}{80EJ}$
+ (4)	·	·	·	2/9		8/9	$-\frac{47}{4EJ}$
+ (I) α_{14}	·	·	·	—1/27		—13/27	$\frac{1}{4EJ}$
+ (II) α_{24}	·	·	·	—1/1080		$-\frac{35}{540}$	$-\frac{1}{80EJ}$
+ (III) α_{34}	·	·	·	$-\frac{361}{22920}$		$-\frac{210\times19}{120\times191}$	$\frac{9\times19}{80\times191EJ}$
(IV)	·	·	·	$\frac{34\,760}{206280}$		$\frac{34760}{206280}$	$-\frac{175740}{15280EJ}$

wherefrom

$$Z_2=0.355\times\frac{1}{EJ}$$

and from equation (I)

$$3Z_1+1\times0.355\times\frac{1}{EJ}+\frac{1}{3}\left(-68.253\times\frac{1}{EJ}\right)=-\frac{9}{4EJ}$$

wherefrom

$$Z_1=6.715\times\frac{1}{EJ}$$

The values of the unknowns are easily checked introducing them into all the simultaneous equations

(1) $$3Z_1+1Z_2+\frac{1}{3}Z_4+2.25\frac{1}{EJ}=3\times6.715\frac{1}{EJ}+$$

$$+1\times0.355\times\frac{1}{EJ}+\frac{1}{3}\left(-68.253\times\frac{1}{EJ}\right)+2.25\times\frac{1}{EJ}=0$$

(2) $$1\times Z_1+\frac{11}{3}Z_2+\frac{1}{2}Z_3+\frac{1}{6}Z_4=1\times6.715\frac{1}{EJ}+$$

$$+\frac{11}{3}\times0.355\frac{1}{EJ}+\frac{1}{2}\times6.719\times\frac{1}{EJ}+\frac{1}{6}\left(-68.253\times\frac{1}{EJ}\right)=0$$

(3) $$\frac{1}{2}Z_2+\frac{5}{3}Z_3+\frac{1}{6}Z_4=\frac{1}{2}\times0.355\times\frac{1}{EJ}+\frac{5}{3}\times$$

$$\times6.719\times\frac{1}{EJ}+\frac{1}{6}\left(-68.253\times\frac{1}{EJ}\right)=0$$

(4) $$\frac{1}{3}Z_1+\frac{1}{6}Z_2+\frac{1}{6}Z_3+\frac{2}{9}Z_4+11.75\frac{1}{EJ}=$$

$$=\frac{1}{3}\times6.715\times\frac{1}{EJ}+\frac{1}{6}\times0.355\times\frac{1}{EJ}+\frac{1}{6}\times6.719\times\frac{1}{EJ}+$$

$$+\frac{2}{9}\left(-68.253\times\frac{1}{EJ}\right)+11.75\times\frac{1}{EJ}=0$$

All these equations being satisfied it may be concluded that no error has been committed in the computations.

7. *Construction of the bending moment diagram.* The ordinates to the resulting bending moment diagram will be calculated using expression (14.13). Fig. 46.13 represents the diagrams obtained through the multiplication of the ordinates to each unit diagram (see Fig. 14.13) by the magnitude of the corresponding reaction

$$M_1=\overline{M}_1Z_1,\quad M_2=\overline{M}_2Z_2,\quad M_3=\overline{M}_3Z_3 \text{ and } M_4=\overline{M}_4Z_4$$

Summing up all these ordinates and adding thereto the ordinates to the M_p diagram (see Fig. 43.13) we obtain these to the desired resulting diagram which is given in Fig. 47.13. All the ordinates to this diagram have been set off on the side of the extended fibres.

8. *Checking the bending moment diagram.*

(a) *Statical method*

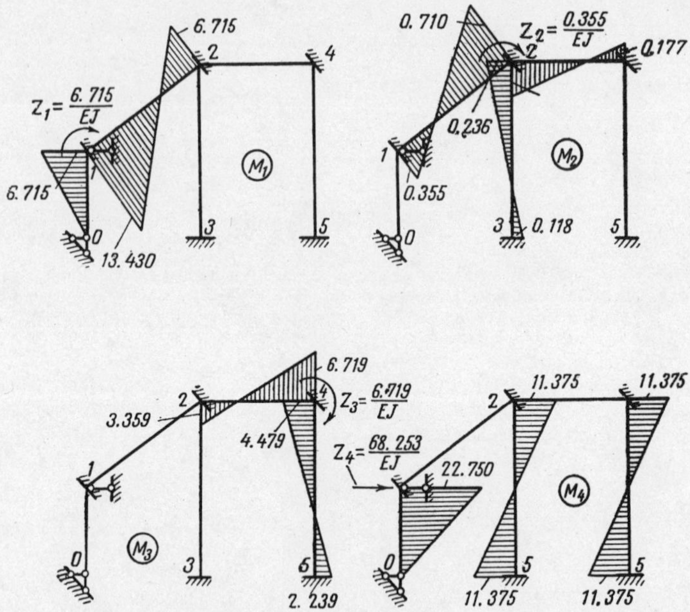

Fig. 46.13

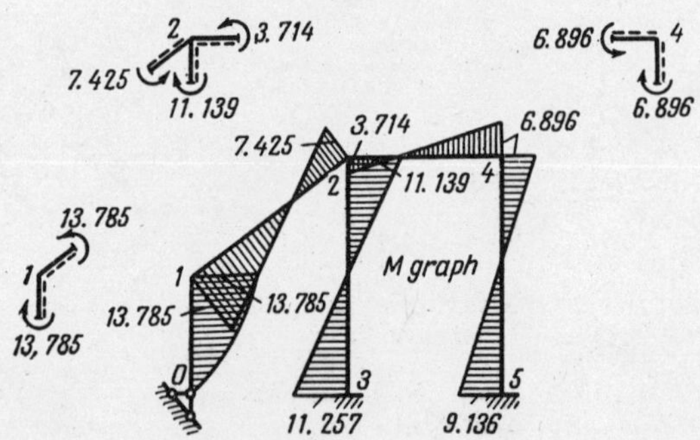

Fig. 47.13

Let us isolate in succession all the rigid joints of the frame and let us see whether the equilibrium conditions are satisfied for each one of them (see Fig. 47.13)

Joint *1*

$$\Sigma M_1 = +13.785 - 13.785 = 0$$

Joint *2*

$$\Sigma M_2 = -7.425 + 11.139 - 3.714 = 0$$

Joint *4*

$$\Sigma M_4 = -6.896 + 6.896 = 0$$

(b) *Method of deflections*

Compute the algebraic sum of the areas bounded by the bending moment diagram along members *3-2-4-5* forming a closed contour. In the general case these areas should be divided by the rigidity of the corresponding member but in this particular case all the moments of inertia are the same and therefore we may simply calculate the value of Ω_M. Those of the areas which are situated inside the contour will be reckoned positive

$$\Omega_M = -\frac{11.257 - 11.139}{2} \times 6 - \frac{6.896 - 3.714}{2} \times$$

$$\times 4 + \frac{9.136 - 6.896}{2} \times 6 = -6.718 + 6.720 \approx 0$$

9. *Construction of the shear diagram.* The ordinates to the Q diagram will be obtained using the expression

$$Q_n = Q_n^0 + \frac{M_n - M_{n-1}}{l_n}$$

The following are the shears at various cross sections:
Section *0* of column *0-1*

$$Q_{01} = 3 + \frac{13.785}{3} = 7.595 \text{ tons}$$

Section *1* of the same column

$$Q_{10} = +3 - 2 \times 3 + \frac{13.785}{3} = 1.595 \text{ tons}$$

Sections *1* and *2* of the inclined bar *1-2*

$$Q_{12} = Q_{21} = -\frac{13.785 + 7.425}{5} = -4.242 \text{ tons}$$

Sections *2* and *3* of column *2-3*

$$Q_{23} = Q_{32} = \frac{11.139 + 11.257}{6} = 3.733 \text{ tons}$$

Sections *2* and *4* of crossbeam *2-4*

$$Q_{24} = Q_{42} = -\frac{3.714 + 6.896}{4} = -2.652 \text{ tons}$$

Sections *4* and *5* of column *4-5*

$$Q_{45} = Q_{54} = \frac{6.896 + 9.136}{6} = 2.672 \text{ tons}$$

These values have permitted the construction of the shear diagram given in Fig. 48.13.

10. *Checking the shear diagram.* Let us pass a section through the lower ends of the three columns and let us consider the equilibrium of all horizontal projections of the forces acting on the upper portion of the frame

$$\Sigma X = 2 \times 3 + 8 - 7.595 - 3.733 - 2.672 = 0$$

11. *Construction of the diagram for normal stresses.* This diagram will be obtained isolating in succession all the joints of the frame. The normal

Table 5.13

Sketch of joint	Equilibrium equation	Normal stresses
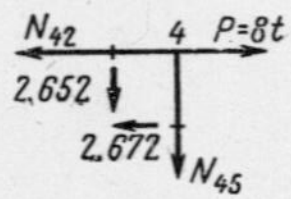	$\Sigma Y = N_{45} + 2.652 = 0$ $\Sigma X = N_{42} + 2.672 - 8.0 = 0$	$N_{45} = -2.652$ tons $N_{42} = 8.0 - 2.672 = 5.328$ tons
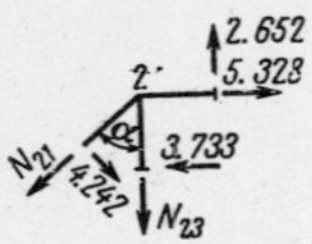	$\Sigma X = N_{21} \sin\alpha - 4.242 \cos\alpha + 3.733 -$ $- 5.328 = 0$ $\Sigma Y = N_{23} - 2.652 + 5.175 \cos\alpha +$ $+ 4.242 \sin\alpha = 0$	$N_{21} = \dfrac{4.242 \cos\alpha - 3.733 + 5.328}{\sin\alpha} =$ $= \dfrac{4.242 \times 3/5 - 3.733 + 5.328}{4/5} =$ $= 5.175$ tons $N_{23} = 2.652 - 5.175 \times 3/5 - 4.242 \times 4/5 =$ $= -3.847$ tons
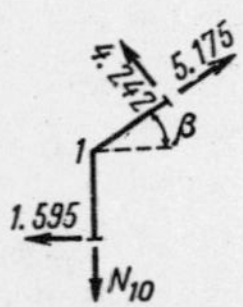	$\Sigma Y = N_{10} - 4.242 \cos\beta - 5.175 \sin\beta = 0$	$N_{10} = 4.242 \times 4/5 + 5.175 \times 3/5 =$ $= 6.499$ tons

stresses acting in all the members will be derived from the equilibrium of these joints. Stresses causing extensions will be reckoned positive. All the necessary operations are given in Table 5.13. The desired diagram for normal stresses appears in Fig. 49.13.

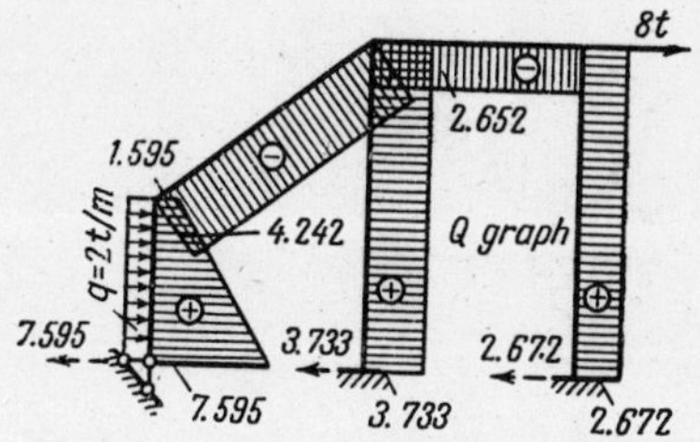

Fig. 4813.

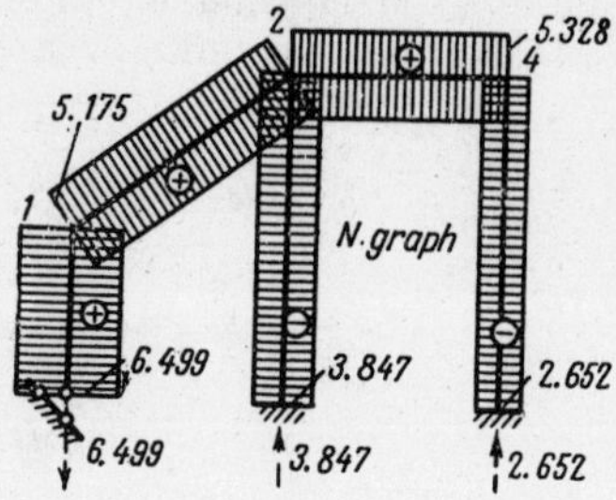

Fig. 49.13

12. *Checking the diagram for normal stresses.* Pass as previously a section through the lower ends of the columns and write the equilibrium equation of all vertical components of forces and reactions applied to the upper portion of the structure

$$\Sigma Y = -6.499 + 3.847 + 2.652 = 0$$

This equation being satisfied, the stress analysis of the frame may be considered complete.

12.13. *The Mixed Method*

In the method under consideration one part of the unknowns represent forces (just as in the method of forces) and the other part angles of twist and deflections just as in the slope and deflections method. Thus, the unknown forces and the unknown displacements will be dealt with simultaneously.

The application of this method will be explained using as an example the two-storied frame shown in Fig. 50.13*a*. Let us first establish the degree of redundancy and the number of unknown angles of twist and deflections for each of the two stories. These data are given in Table 6.13.

Table 6.13

Story	Degree of redundancy	Number of unknown angles of twist and deflections
Lower	9	2
Upper	2	12
Total	11	14

It is obvious that the slope and deflections method can be advantageously used for analysis of the lower story while the method of forces is better fit for the upper one.

The mixed method based on the simultaneous use of forces and displacements as unknowns was introduced in the U. S. S. R.

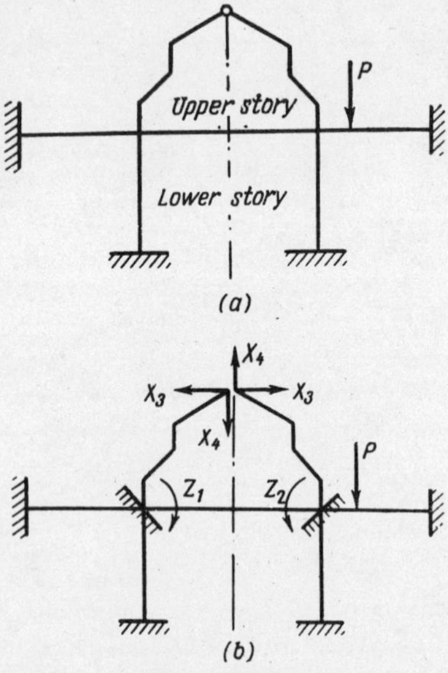

Fig. 50.13

by Professor A. Gvozdev in 1927. The application of this method to the two-storied frame under consideration will lead to a reduction in the number of unknowns to 4 only from 11 if the method of forces were used, or from 14 if it were the slope and deflections method. These unknowns will represent the angles of twist of the lower floor joints and the stresses acting at the gable hinge. The conjugate mixed structure is represented in Fig. 50.13*b*. This structure is derived from the given one through the elimination of the constraints at the top hinge and through the introduction of imaginary constraints at the joints of the first floor.

Let us form the simultaneous equations of the mixed method. These equations will express that the reactions of the imaginary constraints due to the unknown angles of twist Z_1 and Z_2 as well as the mutual displacements of the two branches of the frame along the directions of X_3 and X_4 are nil

$$\begin{aligned} Z_1r_{11}+Z_2r_{12}+X_3r_{13}+X_4r_{14}+R_{1p}=0 \\ Z_1r_{21}+Z_2r_{22}+X_3r_{23}+X_4r_{24}+R_{2p}=0 \\ Z_1\delta_{31}+Z_2\delta_{32}+X_3\delta_{33}+X_4\delta_{34}+\Delta_{3p}=0 \\ Z_1\delta_{41}+Z_2\delta_{42}+X_3\delta_{43}+X_4\delta_{44}+\Delta_{4p}=0 \end{aligned} \qquad (15.13)$$

A closer examination of each of the above equations leads to the following conclusions. In the first equation:

Z_1r_{11} = reaction of the imaginary constraint at joint *1* due to the rotation of this joint through an angle Z_1

Z_2r_{12} = reaction of the same constraint due to the rotation of joint *2* through an angle Z_2

X_3r_{13} = reaction of the same constraint due to the application of the force X_3 at the top hinge

X_4r_{14} = reaction of the same constraint due to the force X_4

R_{1p} = reaction of the same constraint due to the applied load.

The sum of all the above reactions must equal zero for the constraint introduced at joint 1 is in reality nonexistent and therefore uncapable of developing any reactions whatsoever. Thus, the first of the four equations is an equilibrium equation expressing that the reactive moment of the imaginary constraint at joint *1* due to all the unknown forces and displacements as well as to the applied loads remains nil.

The second of the simultaneous equations (15.13) conveys the same idea and the meaning of all its terms is also exactly the same with the only difference that they all refer to the imaginary constraint introduced at joint *2*.

Next let us examine the third equation of this group.

$Z_1\delta_{31}$ = mutual deflection along the horizontal of the two branches of the conjugate structure caused by the rotation Z_1

$Z_2\delta_{32}$ = displacement along the same direction due to the rotation Z_2

$Z_3\delta_{33}$ = deflection along the same direction due to the unknown group of forces X_3 itself

$X_4\delta_{34}$ = deflection along the same direction due to the unknown group of forces X_4

Δ_{3p} = deflection due to the applied loads.

The sum of all these deflections must be nil for the existing hinge prevents all mutual displacements of the two branches of the frame. Consequently, the third equation expresses the idea that the displacements of a certain point remain nil which means that this equation is a *kinematic* one. The fourth equation of the simultaneous system (15.13) expresses exactly the same idea with reference to the mutual vertical displacement of the two branches of the frame at the top hinge.

The coefficients to the unknowns entering these equations belong to four different groups:

1. Coefficients representing *reactions* due to *unit deflections* or *twists* as for instance r_{12}.

2. Coefficients representing *reactions* due to *unit forces* or *moments* as for instance r_{13}.

3. Coefficients representing *deflections* or *twists* due to *unit displacements* as for instance δ_{31}.

4. Coefficients representing *deflections* or *twists* due to *unit forces* or *moments* as for instance δ_{34}.

All these coefficients will be computed as described in the corresponding articles of the present and previous chapters.

It should be noted that for the conjugate structure shown in Fig. 50.13*b* only one pair of coefficients to the unknowns entering equations (15.13), namely δ_{34} and δ_{43} reduce to zero, the

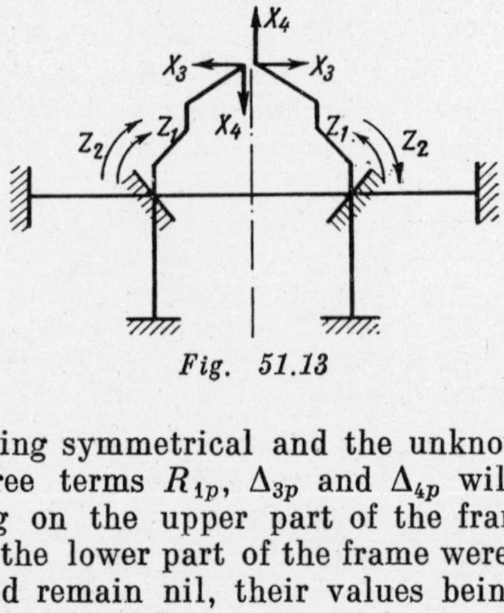

Fig. 51.13

unknowns X_3 being symmetrical and the unknowns X_4 antisymmetrical. The free terms R_{1p}, Δ_{3p} and Δ_{4p} will be also nil for the loads acting on the upper part of the frame are nil themselves. Even if the lower part of the frame were nonsymmetrical δ_{34} and δ_{43} would remain nil, their values being determined by the multiplication of bending moment diagrams induced in the different members of the upper portion of the frame. Were the unknown angles of twist replaced by their symmetrical and antisymmetrical components (Fig. 51.13), a greater number of coefficients to the unknowns would become nil and the system of simultaneous equations (15.13) would fall itself into two independent groups, the first containing symmetrical unknowns

$$Z_1 r_{11} + X_3 r_{13} + R_{1p} = 0$$

$$Z_1 \delta_{31} + X_3 \delta_{33} + \Delta_{3p} = 0 \quad (\text{if } \Delta_{3p} \neq 0)$$

and the second containing the antisymmetrical ones

$$Z_2 r_{22} + X_4 r_{24} + R_{2p} = 0$$

$$Z_2 \delta_{42} + X_4 \delta_{44} + \Delta_{4p} = 0 \quad (\text{if } \Delta_{4p} \neq 0)$$

The principles of reciprocal works and displacements provide for the following relations between the secondary coefficients to the unknowns of the mixed method

$$\begin{aligned} r_{mn} &= r_{nm} \\ \delta_{mn} &= \delta_{nm} \qquad (16.13) \\ r_{mn} &= -\delta_{nm} \end{aligned}$$

Problem. Form the system of canonical equations for the frame given in Fig. 52.13 and determine the coefficients to all the unknowns using that method which leads to the minimum number of unknowns. It is assumed that the cross sections of all the members of the frame remain constant throughout.

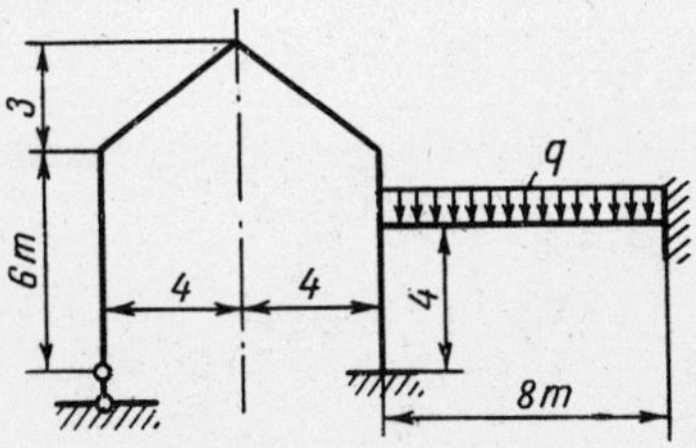

Fig. 52.13

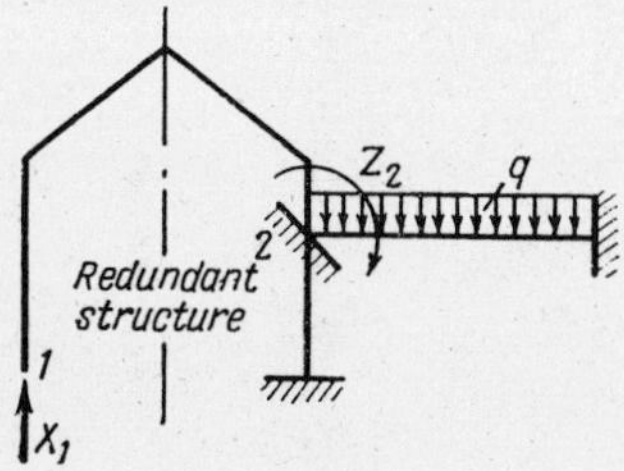

Fig. 53.13

Solution. The choice of the method will be based on Table 7.13. This table shows immediately that the mixed method will lead to the best results. Indeed were this method adopted, the number of unknowns would equal two, while the method of forces would lead to four unknowns and the slope and deflections method—to seven unknowns.

Table 7.13

Portion of structure	Degree of redundancy	Number of unknown twists and deflections
Left-hand	1	6
Right-hand	3	1
Total	4	7

The conjugate system which should be adopted is given in Fig. 53.13. All the values are expressed in tons and metres. The simultaneous equations permitting the determination of the unknowns X_1 and Z_2 become in that case

$$X_1\delta_{11}+Z_2\delta_{12}+\Delta_{1p}=0$$
$$X_1r_{21}+Z_2r_{22}+R_{2p}=0$$

The coefficient δ_{11} represents the displacement due to a unit force and will be determined raising to the second power the $\bar{M}_1$ diagram given in Fig. 54.13

$$\delta_{11}=\frac{1}{EJ}\left[\frac{4\times 5}{2}\times\frac{2}{3}\times 4+\frac{5}{6}(2\times 4^2+2\times 8^2+2\times 4\times 8)+8\times 2\times 8\right]=\frac{192}{EJ}$$

The coefficient δ_{12} which represents a displacement due to another displacement will be derived from the geometry of the structure as shown in Fig. 55.13

$$\delta_{12}=aa'\cos\alpha=S\varphi\cos\alpha=l\varphi$$

The angle of twist φ being equal to 1 and l to 8, the value of coefficient δ_{12} becomes equal to eight. *

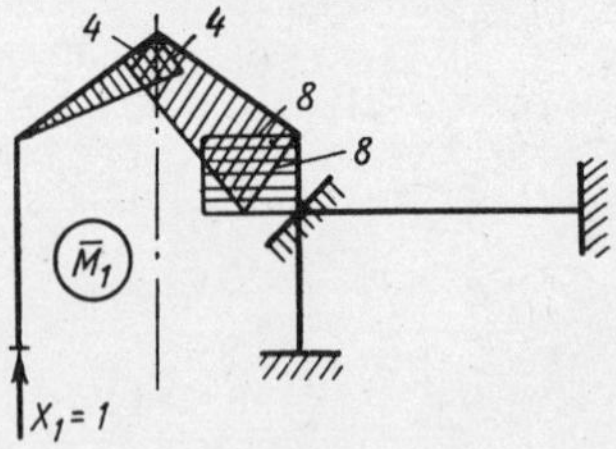

Fig. 54.13

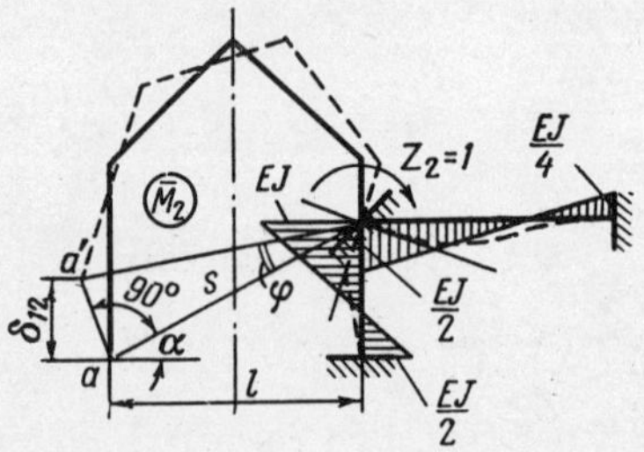

Fig. 55.13

The displacement δ_{12} will be reckoned positive for it follows the direction of the reaction X_1. The coefficient r_{21} represents the reaction of the constraint

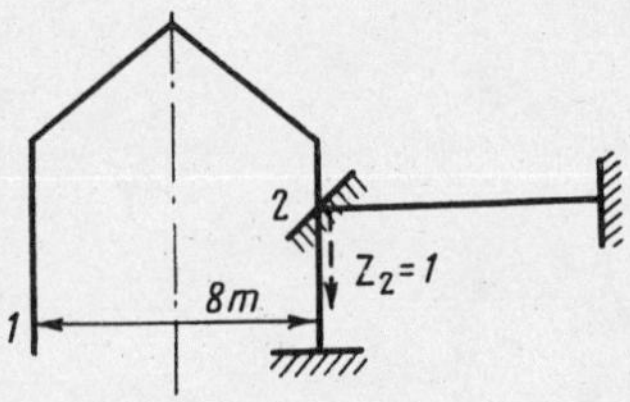

Fig. 56.13

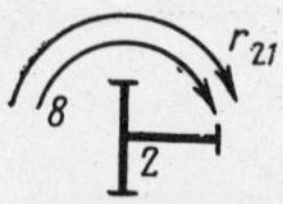

Fig. 57.13

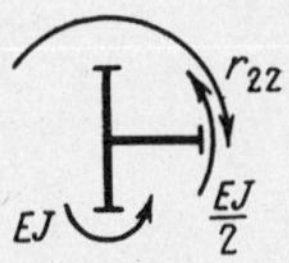

Fig. 58.13

introduced at joint *2* due to a unit force X_1. The value of this coefficient may be derived from the equilibrium of this joint (Fig. 57.13)

$$\Sigma M_2=r_{21}+8=0$$

wherefrom

$$r_{21}=-8$$

* The same magnitude of this coefficient would be obtained by the method described in Art. 15.8. The angle of twist of joint *2* should be represented by a vertical vector (Fig. 56.13) acting at this joint. The moment of this vector about joint *1* will equal $1\times 8=8=\delta_{12}$.

The same magnitude of this coefficient could be obtained directly from the relation $r_{mn} = -\delta_{nm}$.

The coefficient r_{22} represents the reaction of the same constraint to a unit rotation Z_2 of joint 2 (see Fig. 55.13). Hence, this coefficient may be determined using the equilibrium equation for the same joint acted upon as shown in Fig. 58.13

$$\Sigma M_2 = r_{22} - \frac{EJ}{2} - EJ = 0$$

wherefrom

$$r_{22} = \frac{3}{2} EJ$$

As for the free terms Δ_{1p} and R_{2p}, their values in the present case will be given by

$$\Delta_{1p} = 0; \quad R_{2p} = -\frac{ql^2}{12}$$

as stipulated in the appropriate lines of Table 2.13.

13.13. *The Combined Method*

The combined method is best suited for the analysis of symmetrical redundant structures acted upon by nonsymmetrical loads. The use of this method will be explained using as an example the portal frame shown in Fig. 59.13. Replacing the applied load P by its symmetrical and antisymmetrical components (Fig. 60.13*a* and *b*) we arrive at two different loading cases for each of which the number of unknowns can be easily determined.

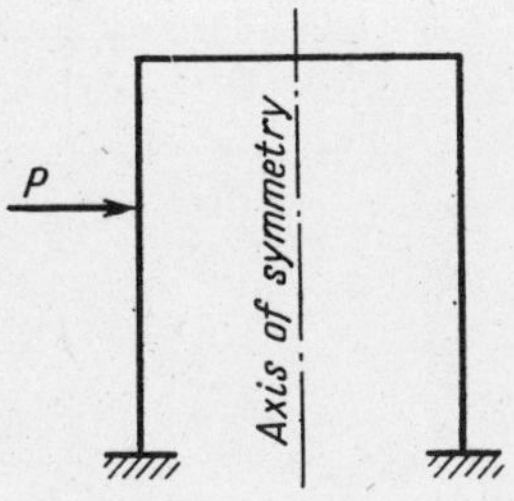

Fig. 59.13

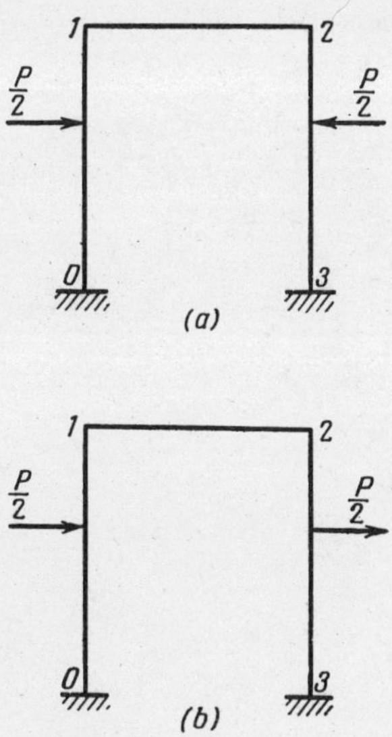

Fig. 60.13

Thus, in case one corresponding to the symmetrical loading the displacement of the crossbeam *1-2* will remain nil and the angles of twist of the two joints *1* and *2* will be equal in amount and opposite in sign, hence $Z_3 = 0$ and $Z_1 = Z_2$ (Fig. 61.13*a*). It follows that if we applied the slope and deflections method we would obtain in that case one equation with one unknown only. If, on the contrary, we employed the method of forces using the simple structure of Fig. 61.13*b* we would arrive at two

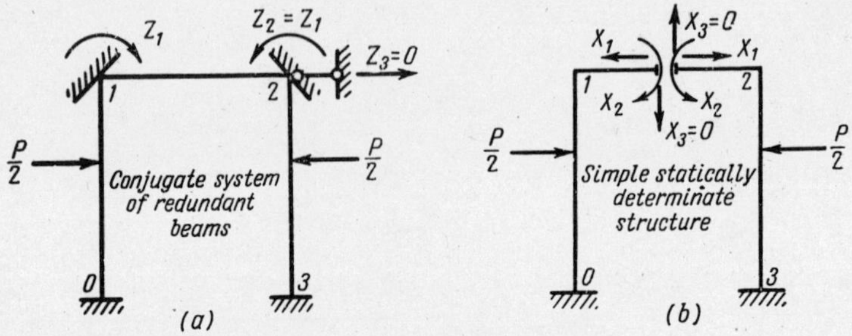

Fig. 61.13

equations with two unknowns (the shearing stress X_3 being obviously nil).

Hence, *the slope and deflections method is the one to be adopted in the case of symmetrical loading.*

Next, let us consider the case of antisymmetrical loads, using as conjugate structure the one shown in Fig. 62.13*a*. It is readily

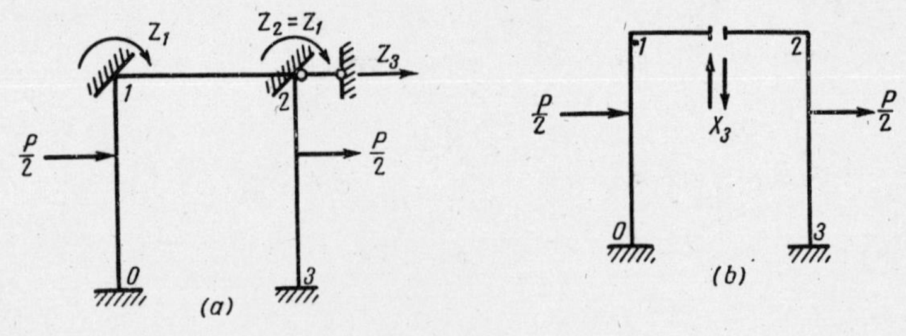

Fig. 62.13

seen that the number of unknowns will in this case equal two: the first representing the angle of twist of joints *1* and *2* (which are equal both in amount and direction) and the second one representing the horizontal deflection of these two joints.

Consequenly, the slope and deflections method in the case of antisymmetrical loading will lead to a system of two simultaneous equations with two unknowns.

On the other hand, if the method of forces were used we could adopt the simple statically determinate structure given in Fig. 62.13*b*, which would lead to one unknown only, this unknown representing the shear X_3, both normal stress X_1 and bending moment X_2 being nil. Thus we could once again obtain one equation only with one unknown. *Therefore the method of forces should be adopted in the case of antisymmetrical loading.*

Table 8.13 represents in a concise form the results obtained above.

Table 8.13

Loading	Number of equations		Method to be used
	Method of forces	Slope and deflections method	
Symmetrical	2	1	Slope and deflections method
Antisymmetrical	1	2	Method of forces

Thus, the *combined method* will consist in the simultaneous application of two different methods as described above to two different cases of loading of one and the same structure.

14.13. *Construction of Influence Line by the Slope and Deflections Method*

The construction of influence lines for stresses acting at any section of a redundant structure as well as for the angles of twist and deflections can be carried out using the slope and deflections method, provided the influence lines for the displacements adopted as unknowns are constructed in the first place.

Let us study the construction of these influence lines using as an example the structure represented in Fig. 63.13 all the members of which are of uniform cross section. The standard equation corresponding to the conjugate system of redundant beams shown in Fig. 64.13 becomes

$$Z_1 r_{11} + r_{1p} = 0$$

wherefrom

$$Z_1 = -\frac{r_{1p}}{r_{11}} = \frac{\delta_{p1}}{r_{11}}$$

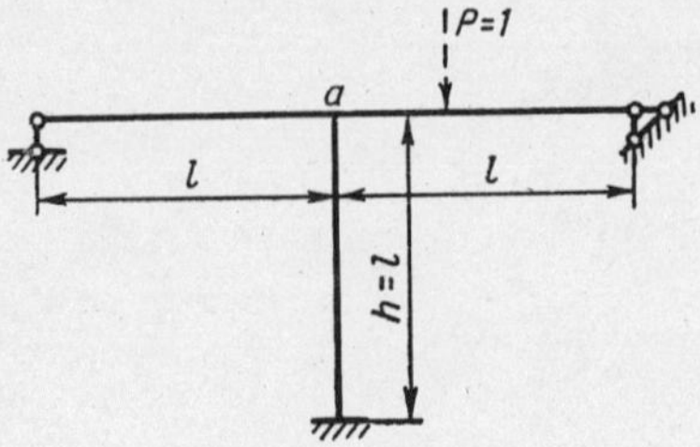

Fig. 63.13

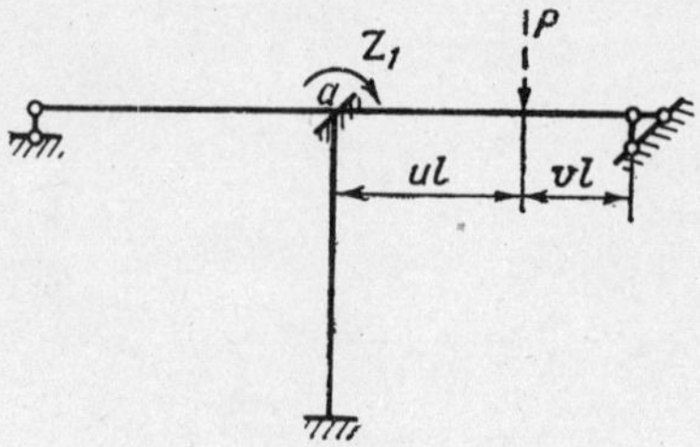

Fig. 64.13

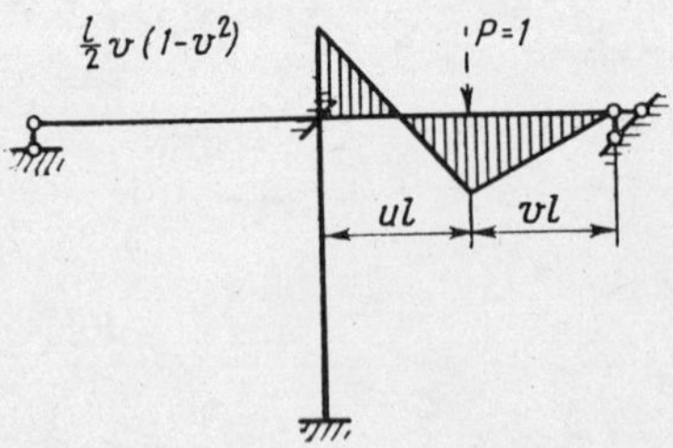

Fig. 65.13

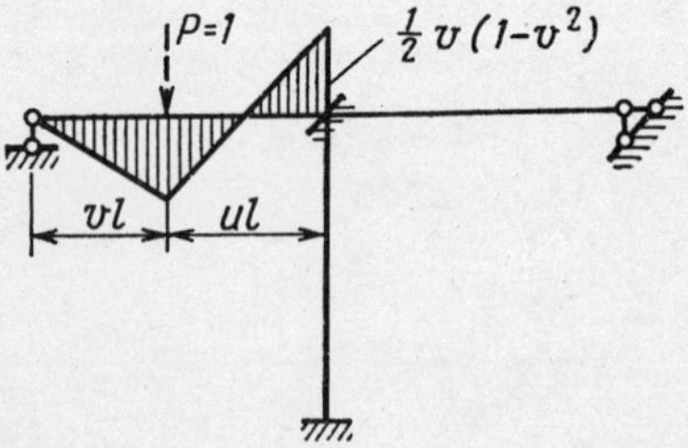

Fig. 66.13

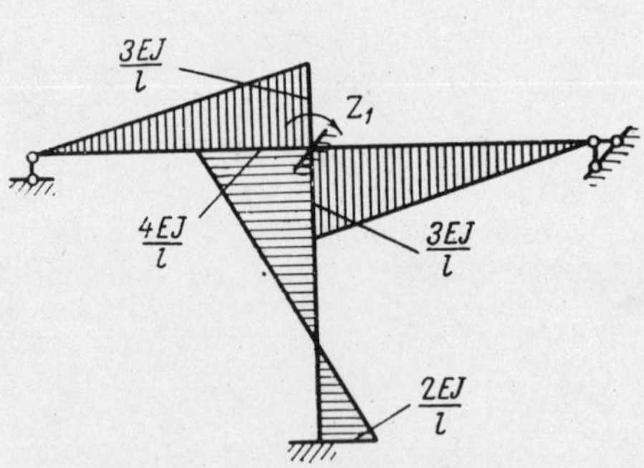

Fig. 67.13

since

$$r_{1p} = -\delta_{p1}$$

Hence the shape of the influence line for the angle of twist will coincide with that of the diagram of vertical deflections δ_{p1} caused by a unit rotation of the imaginary constraint through an angle $Z_1 = 1$.

When the load unity P travels along the right span (Fig. 65.13) we have

$$r_{1p} = -\frac{l}{2} v (1 - v^2)$$

and when the same load is situated within the left span (Fig. 66.13) the same expression becomes

$$r_{1p} = +\frac{l}{2} v (1 - v^2)$$

The magnitude of r_{11} will be derived from the unit bending moment diagram $\overline{M}_1$ (Fig. 67.13)

$$r_{11} = \frac{3EJ}{l} + \frac{3EJ}{l} + \frac{4EJ}{l} = \frac{10EJ}{l}$$

In the above expression the values of v and u may vary only from 0 to 1, these two letters representing the ratio between the distance to the load point and the span length. The values of r_{1p} in terms of v are given in Table 9.13.

Table 9.13

Span	v	r_{1p}	Ordinate Z_1 I. L.	Span	v	r_{1p}	Ordinate Z_1 I. L.
Right	1	0	0	Left	1	0	0
	0.8	$-0.144l$	$+0.0144\frac{l^2}{EJ}$		0.8	$+0.144l$	$-0.0144\frac{l^2}{EJ}$
	0.6	$-0.192l$	$+0.0192\frac{l^2}{EJ}$		0.6	$+0.192l$	$-0.0192\frac{l^2}{EJ}$
	0.4	$-0.168l$	$+0.0168\frac{l}{EJ}$		0.4	$+0.168l$	$-0.0168\frac{l^2}{EJ}$
	0.2	$-0.096l$	$+0.0096\frac{l}{EJ}$		0.2	$+0.096l$	$-0.0096\frac{l^2}{EJ}$
	0	0	0		0	0	0

The same table contains the values of the ordinates to the influence line for Z_1 given by the expression

$$Z_1 = -\frac{r_{1p}}{r_{11}}$$

The completed influence line is represented in Fig. 68.13.

The influence line for the bending moments at an arbitrary section k situated within the left span will be based on the fol-

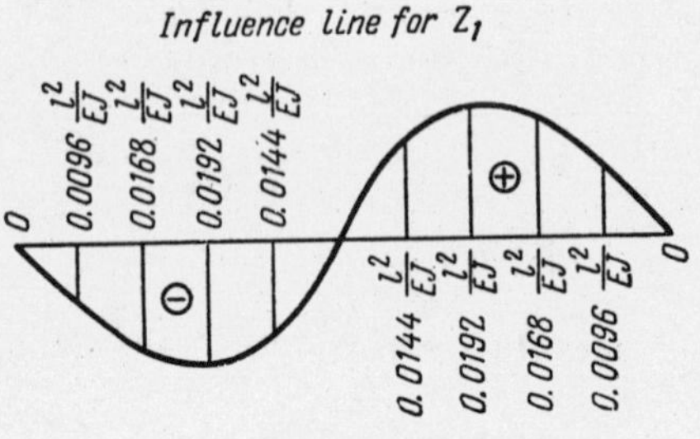

Fig. 68.13

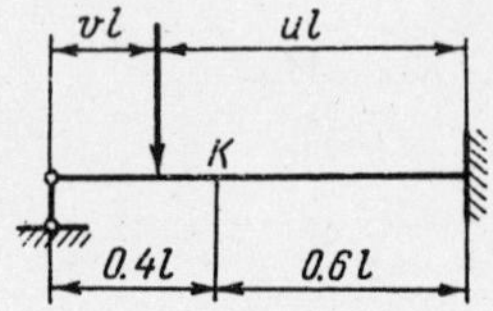

Fig. 69.13.

lowing expression as long as the load unity remains within the same span

$$M_k = M_k^0 - \frac{3EJ}{l^2} Z_1 a_k$$

In this expression

M_k^0 = bending moment at section k of a conjugate redundant beam whose right end is fixed and the left one simply supported (Fig. 69.13)

a_k = distance from the cross section under consideration to the left-hand support

$-\frac{3EJ}{l^2}$ = reaction at the left (roller) support due to a unit rotation of the imaginary constraint at joint a through an angle $Z_1 = 1$.

The influence line for the shearing force at the same cross section can be deduced from the equation

$$Q_k = Q_k^0 - \frac{3EJ}{l^2} Z_1$$

When load unity P shifts to the right of the central support the above expressions are simplified and become

$$M_k = -\frac{3EJ}{l^2} Z_1 a_k; \quad Q_k = -\frac{3EJ}{l^2} Z_1$$

Let us construct the influence line for the bending moments and the shearing forces acting at a cross section situated a distance 0.4 l from the left end of the first span. When the load unity travels along the first span the expression for M_k becomes

$$M_k = M_k^0 = -\frac{3EJ}{l^2} Z_1 \times 0.4l = M_k^0 - \frac{1.2EJ}{l} Z_1$$

Let us find the values of M_k^0 and Q_k^0 for the conjugate redundant beam of Fig. 69.13. As long as load unity P remains within the

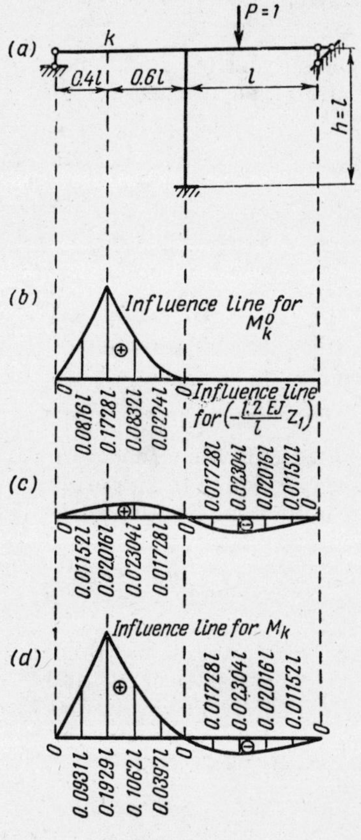

Fig. 70.13

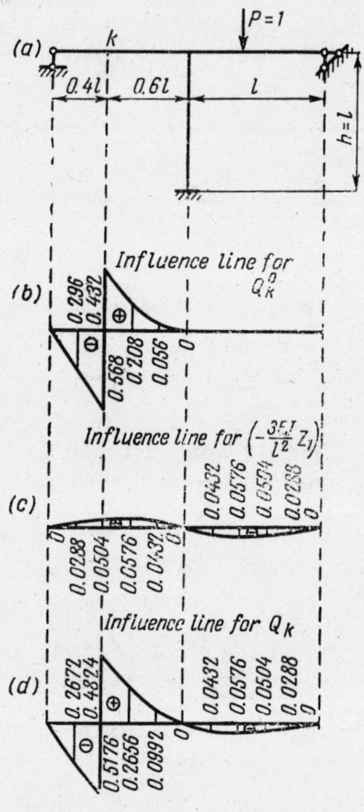

Fig. 71.13

first span to the right of section k, i. e., when $u \leqslant 0.6$ (see Table 2.13)

$$M_k^0 = \frac{u^2}{2}(3-u)\,0.4l; \quad Q_k^0 = -\frac{u^2}{2}(3-u)$$

When the load unity is within the same span but to the left of section k, i. e., when $u \geqslant 0.6$

$$M_k^0 = \frac{u^2}{2}(3-u)\,0.4l - (0.4l - vl) =$$

$$=\frac{u^2}{2}(3-u)\,0.4l-(u-0.6)\,l$$

$$Q_k^0=\frac{u^2(3-u)}{2}-1$$

The values of the ordinates to the M_k^0 and Q_k^0 diagrams computed using the above expressions are given in Table 10.13.

Table 10.13

Load point in terms of u	$M^0{}_k$	$Q^0{}_k$	Load point in terms of u	$M^0{}_k$	$Q^0{}_k$
0	0	0	0.6	0.1728l	—0.568
0.2	0.0224l	0.056	0.8	0.0816l	—0.296
0.4	0.0832l	0.208	1.0	0	0
0.6	0.1728l	0.432			

When the load point shifts to the second span the ordinates to the M_k and Q_k influence lines become equal to those for the influence line for Z_1 (see Fig. 69.13) multiplied by the following factors respectively

$$\left(-\frac{3EJ}{l^2}a_k=-\frac{3EJ}{l^2}\times 0.4l=-\frac{1.2EJ}{l}\right)\ \text{and}\ \left(-\frac{3EJ}{l^2}\right)$$

Figs. 70.13*b* and *c* and 71.13 *b* and *c* represent the influence lines for the different terms entering the expressions of M_k and Q_k. The completed influence lines for M_k and Q_k shown in Figs. 70.13*d* and 71.13*d* have been obtained through the summation of the ordinates to the influence lines just mentioned.

14. APPROXIMATE METHOD OF STRESS ANALYSIS FOR REDUNDANT FRAMES

1.14. *Classification of Approximate Methods*

The analysis of complicated frames using one of the exact methods described above (method of forces, slope and deflections method, the mixed one, etc.) often remains exceedingly labour consuming even when all the possibilities of simplifying the equations have been profited by. In such cases resort should be made to *approximate* methods which may be subdivided into two main groups.

Methods belonging to the first of these groups proceed by successive approximations (iterative methods) and therefore the precision of the final results may be as great as desired. After a sufficient number of approximations these results will for all practical purposes be equivalent to those obtained using one of the exact methods.

The methods belonging to the second group are based on simplifications introduced both in the arrangement of structural members and in the distribution of loads. The simplified system thus obtained may thereafter be analysed using either one of the exact or one of the approximate methods.

The simplifications introduced into one and the same system may vary considerably and each of these simplifications will influence the final results to a different degree. Consequently, one must learn to choose the simplest way of analyzing the structure with due regard to the desired precision of the final results. In order to be able to do so one must understand very clearly the work of the entire structure and at the same time one must be well versed in all the exact methods of stress analysis.

Approximate methods are particularly useful when choosing cross-sectional dimensions for preliminary estimates, when comparing alternative layouts of one and the same structure or when designing ancillary or temporary buildings.

As no analysis whether exact or approximate of a redundant structure can be undertaken as long as the cross-sectional dimensions and rigidities of its members remain unknown, the same methods are frequently resorted to in the preliminary choice of such dimensions.

Approximate methods are seldomly used in the design of simple frames, exact solutions being readily available in appropriate engineering handbooks.

2.14. *The Method of Moment Distribution*

The method of moment distribution belongs to the first group of approximate methods constituting in fact a particular application of the slope and deflections method described in the preceding chapter. It leads to a very substantial reduction in the number of equations and in the case of structures, whose joints can sustain angular twists alone but cannot be deflected, the method under consideration permits to avoid completely the solution of simultaneous equations with several unknowns.

For the first time this method was suggested in 1929 by N. Bernatsky and a few months later, early in 1930, a detailed description of practically the same method was given by Prof. Hardy Cross. The moment distribution method could be used for the analysis of all redundant framed structures but in practice it is applied only to continuous beams and complicated frames whose joints are not deflected by the applied loads. A maximum of one or two independent deflections of joints may be tolerated.

The moment distribution method is particularly well fit for the analysis of multi-story building frames and closed frames of hydraulic plants, where its application results in a very considerable reduction of computation work. The convention of signs adopted previously for bending moments and shearing forces remains unchanged and the conjugate system of redundant beams is obtained in exactly the same way as in the slope and deflections method.

The reactive moments acting at the ends of the bars are considered positive when they act clockwise, the shears when they tend to rotate clockwise the portion of the bar under consideration about its far end. The reactive moments are usually denoted by the letter M with three lower indices the first two giving the numbers of joints between which the bar is inserted and the third, separated from the first two by a comma, indicates the cause giving rise to that particular moment. The end of the bar at which the reactive moment is developed is always indicated by the very first of the indices. Thus, for example, $M_{ik,p}$ will represent the reactive moment developed at end i of bar i-k, under the influence of load P.

1. ANALYSIS OF STRUCTURES WITH FIXED JOINTS

Let us examine the structure shown in Fig. 1.14*a*. In this structure joint *1* alone can sustain an angular rotation and no single joint may be deflected. The conjugate system of redundant

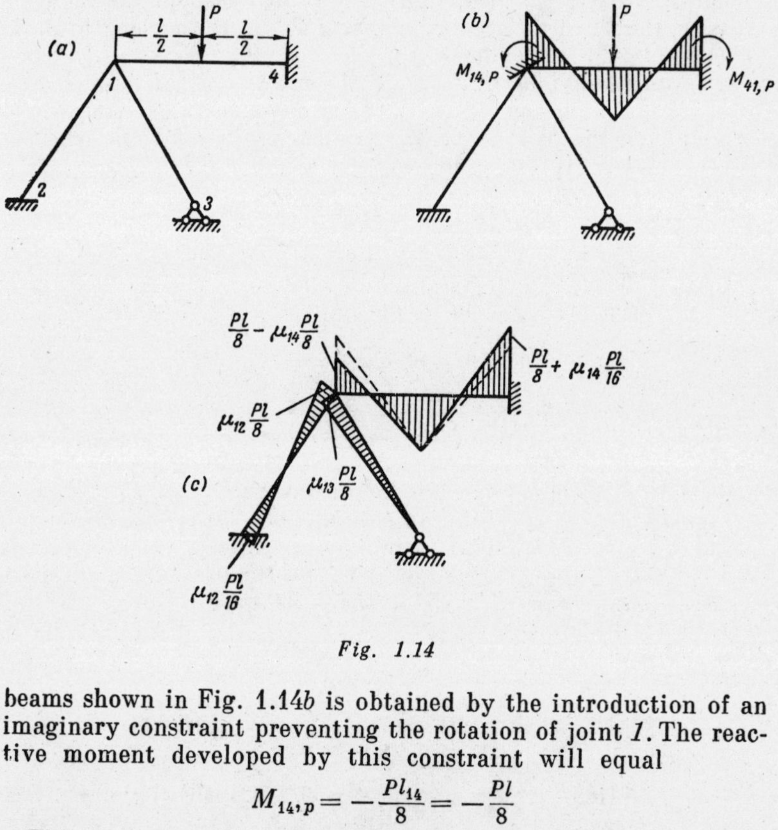

Fig. 1.14

beams shown in Fig. 1.14*b* is obtained by the introduction of an imaginary constraint preventing the rotation of joint *1*. The reactive moment developed by this constraint will equal

$$M_{14,p} = -\frac{Pl_{14}}{8} = -\frac{Pl}{8}$$

This moment will be reckoned negative for it acts in an counterclockwise direction. The moment produced at this joint by the loading of bar *1-4* tends to rotate this joint in a clockwise direction and as soon as the imaginary constraint is removed the angle of twist of this joint will become equal to φ_1. This rotation will result in the appearance of the following moments at both ends of all the bars converging at joint *1* (see Table 2.13)

$$M_{12,1} = \frac{4EJ_{12}}{l_{12}}\varphi_1; \quad M_{21,1} = \frac{2EJ_{12}}{l_{12}}\varphi_1 = \frac{M_{12,1}}{2}$$

$$M_{13,1}=\frac{3EJ_{13}}{l_{13}}\varphi_1; \qquad M_{31,1}=0$$

$$M_{14,1}=\frac{4EJ_{14}}{l_{14}}\varphi_1; \quad M_{41,1}=\frac{2EJ_{14}}{l_{14}}\varphi_1=\frac{M_{14,1}}{2}$$

In these expressions $M_{12,1}$ is the reactive moment induced at end *1* of bar *1-2* by the rotation of joint *1* through an angle φ_1; EJ_{12} is the flexural rigidity of bar *1-2*; l_{12} is the length of the same bar, and so forth.

It is quite obvious that the angle of twist φ_1 of joint *1* must be such that the sum of all the reactive moments acting at the near ends of the bars converging at this joint should become nil. This can be expressed by the equation

$$M_{12,1}+M_{13,1}+M_{14,1}+M_{14,p}=0$$

This equation is in no respect different from the one used in the slope and deflections method. Since all the bars meeting at joint *1* are twisted through the same angle φ_1, we may write

$$M_{12,1}:M_{13,1}:M_{14,1}=i_{12}:i_{13}:i_{14}$$

where

$$i_{12}=\frac{EJ_{12}}{l_{12}}; \quad i_{13}=\frac{0.75EJ_{13}}{l_{13}}; \quad i_{14}=\frac{EJ_{14}}{l_{14}}$$

The terms i given above are frequently called the *stiffness factors* of the frame members. In case of bars fixed at both ends the stiffness factor is equal to their flexural rigidity per unit length and for those having one end built-in and the other freely supported this factor equals 0.75 of their flexural rigidity per unit length. *

The bending moments due to the twist of joint *1* become equal to

$$M_{12,1}=-\frac{i_{12}}{i_{12}+i_{13}+i_{14}}M_{14,p}=-\mu_{12}M_{14,p}$$

$$M_{13,1}=-\frac{i_{13}}{i_{12}+i_{13}+i_{14}}M_{14,p}=-\mu_{13}M_{14,p}$$

$$M_{14,1}=-\frac{i_{14}}{i_{12}+i_{13}+i_{14}}M_{14,p}=-\mu_{14}M_{14,p}$$

The values of $\mu_{12}=\frac{i_{12}}{\Sigma i_1}$; $\mu_{13}=\frac{i_{13}}{\Sigma i_1}$; $\mu_{14}=\frac{i_{14}}{\Sigma i_1}$ indicate that part of the unbalanced moment applied to the joint which is taken

* Certain authors attribute the name *stiffness factor* to the ratio $\frac{J}{l}$ usually denoted by the letter K. In our opinion, it fits better the term i.—*TR.*

up by the corresponding frame member. Hereafter these values will be called *distribution factors*. It is clear that for each individual joint the sum of all the distribution factors must equal one.

The algebraic sum of moments induced by the twist of joint *1* with the reactive moments induced by the applied loads in the beams of the conjugate system will yield the value of the resulting moments acting at the joints of the given structure

$$M_{12}=\mu_{12}\frac{Pl}{8};\quad M_{13}=\mu_{13}\frac{Pl}{8}$$

$$M_{14}=\mu_{14}\frac{Pl}{8}-\frac{Pl}{8}=-(1-\mu_{14})\frac{Pl}{8}$$

$$M_{21}=\frac{M_{12}}{2}=\mu_{12}\frac{Pl}{16};\quad M_{31}=0$$

$$M_{41}=\mu_{14}\frac{Pl}{16}+\frac{Pl}{8}=\left(1+\frac{\mu_{14}}{2}\right)\frac{Pl}{8}$$

These moments must satisfy the following relation

$$M_{12}+M_{13}+M_{14}=0$$

The diagram of the resulting moments is represented in Fig. 1.14*c*. The analysis of redundant structures by the method under consideration consists of the following operations:

1. In the first place the reactive moments induced in the beams of the conjugate redundant system are determined using Table 2.13.

2. Next, equilibrium equations for each joint are formed applying to the joints balancing moments equal in amount and opposite in sign to the reactive moments, these balancing moments being distributed among the members converging at the corresponding joints in direct proportion to their stiffness factor.

3. Half the amount of the distributed moment is transferred (carried over) to the far end of the bar provided this end is also rigidly fixed. In the event the far end of the bar is provided with a hinge the carry-over moment must be nil.

4. New counterbalancing moments are applied to each joint thrown out of balance by the carry-over moments.

In order to accelerate the work several noncontiguous joints may be dealt with simultaneously. The operation is repeated until the values of unbalanced moments become so small that they may be disregarded. The final value of the bending moments acting at the ends of each member of a structure are obtained summing up the values of the fixed end moments with the values of the distributed and carry-over moments, due consideration being given to their respective signs.

All the computations should be carried out in tabular form. The table to be used should consist of a certain number of columns and lines, each column corresponding to one end of each member of the structure. These columns are grouped joint by joint reserving, if necessary, extra columns for external moments. The exact procedure to be followed when no joint of the structure may be deflected will be described in detail in the following problems.

Problem 1. Required the bending moment diagram for all the members of a roof truss with rigid joints represented in Fig. 2.14*a*. The length of all

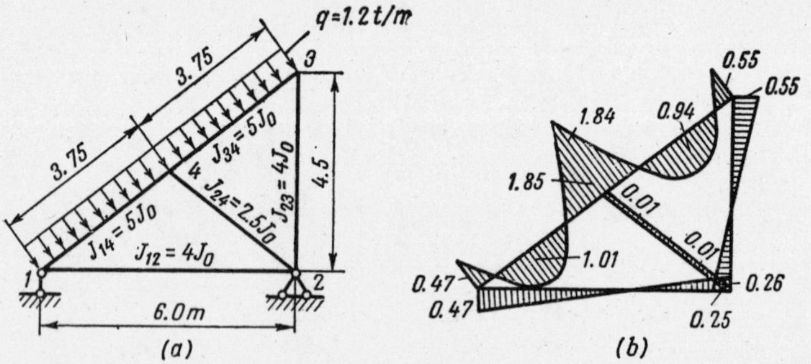

Fig. 2.14

the bars, the relative values of the moments of inertia and the loads are clearly indicated in the same figure.

Solution. Since all the joints of the given truss are completely rigid the stiffness factors of all the bars will amount to $i = \frac{EJ}{l}$. Assuming that $E = 1$ and referring all the moments of inertia to J_0 we may write $i = \frac{J}{J_0 l}$. The values of the stiffness factors thus obtained (or to be more precise, their relative values) are given in the third line of Table 1.14.

Having determined the stiffness factors proceed with the calculation of the distribution factors, as indicated hereunder for joint 4

$$\mu_{41} = \frac{i_{41}}{i_{41} + i_{42} + i_{43}} = \frac{133}{133 + 67 + 133} = 0.40$$

$$\mu_{42} = \frac{i_{42}}{i_{41} + i_{42} + i_{43}} = \frac{67}{333} = 0.20$$

$$\mu_{43} = \frac{i_{43}}{i_{41} + i_{42} + i_{43}} = \frac{133}{333} = 0.40$$

It should be always kept in mind that the sum of all the distribution factors must be always equal to one.

The fixed end moments corresponding to the redundant beams of the conjugate system are given by

$$M_{14,\,p}=M_{43,\,p}=-M_{41,\,p}=-M_{34,\,p}=-\frac{ql^2}{12}=$$

$$=-\frac{1.2\times 3.75^2}{12}=-1.40 \text{ ton metres}$$

These moments should be entered into the fifth line of Table 1.14.

Table 1.14

Joint No.	*1*		*4*			*3*		*2*		
Bar No.	*1-2*	*1-4*	*4-1*	*4-2*	*4-3*	*3-4*	*3-2*	*2-3*	*2-4*	*2-1*
Stiffness factor i	67	133	133	67	133	133	89	89	67	67
Distribution factor μ	0.333	0.667	0.40	0.20	0.40	0.60	0.40	0.40	0.30	0.30
Fixed end moment, ton metres	—	−1.40	1.40	—	−1.40	1.40	—	—	—	—
Balancing of joints *1*	0.47	0.93	0.47	—	—	—	—	—	—	0.24
Balancing of joints *3*	—	—	—	—	−0.42	−0.84	−0.56	−0.28	—	—
Balancing of joints *4*	—	−0.01	−0.02	−0.01	−0.02	−0.01	—	—	—	—
Balancing of joints *2*	—	—	—	—	—	—	0.01	0.02	0.01	0.01
Balancing of joints *1*	—	0.01	—	—	—	—	—	—	—	—
Resulting bending moments	0.47	−0.47	1.85	−0.01	−1.84	0.55	−0.55	−0.26	0.01	0.25

All the above entries having been made, we may proceed with the balancing of the moment acting at joints *1* and *3* of the truss, joints *2* and *4* remaining in equilibrium for the time being.

At joint *1* the unbalanced fixed end moment $M_{14,\,p}$ equals -1.40 ton metres, this moment being directed counterclockwise. This moment will be taken up by the bending moments acting at the ends of all the bars converging at the joint under consideration, these bending moments being calculated as follows

$$M_{12,\,1}=0.333\times 1.40=0.47 \text{ ton metre}$$
$$M_{14,\,1}=0.667\times 1.40=0.93 \text{ ton metre}$$

The carry-over moments which must be transferred to the far ends of these two bars will amount to

$$M_{21,\,1}=0.5\times 0.47\approx 0.24 \text{ ton metre}$$
$$M_{41,\,1}=0.5\times 0.93\approx 0.47 \text{ ton metre}$$

Passing to joint *3* we note that this joint is acted upon by an unbalanced moment $M_{34,\,p}$ amounting to 1.40 ton metres acting clockwise. This moment must be distributed as follows

$$M_{34,\,3}=-0.60\times 1.40=-0.84 \text{ ton metre}$$
$$M_{32,\,3}=-0.40\times 1.40=-0.56 \text{ ton metre}$$

The carry-over moments are equal to

$$M_{43,\,3}=-0.50\times 0.84=-0.42 \text{ ton metre}$$
$$M_{23,\,3}=-0.50\times 0.56=-0.28 \text{ ton metre}$$

Balancing joints *1* and *3* as described above we have disturbed the equilibrium of joints *2* and *4*. The unbalanced moment at joint *4* amounts now to

$$M_4=M_{41,\,1}+M_{43,\,3}=0.47-0.42=0.05 \text{ ton metre}$$

Distributing this moment among the three bars converging at joint *4* we obtain

$$M_{41,\,4}=-0.40\times 0.05=-0.02 \text{ ton metre}$$
$$M_{42,\,4}=-0.20\times 0.05=-0.01 \text{ ton metre}$$
$$M_{43,\,4}=-0.40\times 0.05=-0.02 \text{ ton metre}$$

One half of each of these moments will be again carried over to the far end of each of the bars

$$M_{14,\,4}=-0.5\times 0.02=-0.01 \text{ ton metre}$$
$$M_{24,\,4}=-0.5\times 0.01=-0.005 \text{ ton metre}$$
$$M_{34,\,4}=-0.5\times 0.02=-0.01 \text{ ton metre}$$

Proceeding in exactly the same way for joint *2* we shall find that the unbalanced moment transmitted to joint *1* ($M_{14,\,4}$) equals only -0.01. This unbalanced moment is very much smaller than the one found previously, its value being practically no greater than the degree of precision of all our computations. Consequently, no further approximations are necessary for the amounts of all the unbalanced moments which will have to be dealt with will be smaller than 0.01 ton metre.

The computations given in Table 1.14 represent all the operations necessary to solve the problem. The bending moment diagram constructed with due regard to the sign convention adopted is shown in Fig. 2.14*b*.

2. ANALYSIS OF STRUCTURES WITH DEFLECTED JOINTS

The analysis of structures whose joints may be deflected requires that in addition to constraints opposing the twist of these joints imaginary supports should be introduced preventing all independent deflections. Schematically these supports may be replaced by a corresponding number of supporting bars. As previously, the bending moments induced at the ends of all the redundant beams of the conjugate system should be calculated using appropriate ready made formulas or tables. Next all the joints should be balanced by a series of successive approximations. Finally the corrections taking care of joint deflections should be introduced. The exact sequence of operations will be shown in the following example.

Problem 2. Required the bending moment diagrams for all the members of a double-span frame shown in Fig. 3.14*a*. Only one joint of this frame can sustain an independent deflection.

Solution. The stiffness factors for all the members of the frame are computed as follows.

$$i_{12}=\frac{0.75J_{12}}{l_{12}J_0}=\frac{0.75\times 3}{2}=1.125$$

$$i_{36}=\frac{0.75J_{36}}{l_{36}J_0}=\frac{0.75\times 4}{4}=0.75$$

$$i_{45}=\frac{J_{45}}{l_{45}J_0}=\frac{8.0}{5.33}=1.50$$

$$i_{34}=\frac{J_{34}}{l_{34}J_0}=\frac{9.0}{4.0}=2.25$$

$$i_{23}=\frac{J_{23}}{l_{23}J_0}=\frac{27.0}{6.0}=4.50$$

Knowing these values we may easily obtain the distribution factors as indicated hereunder for joint *2*

$$\mu_{21}=\frac{i_{12}}{i_{12}+i_{23}}=\frac{1.125}{1.125+4.50}=0.20$$

$$\mu_{23}=\frac{i_{23}}{i_{12}+i_{23}}=\frac{4.50}{1.125+4.50}=0.80$$

The fixed end moments for the redundant beams of the conjugate systems shown in Fig. 3.14*b* are

$$M_{45,\,p}=-M_{54,\,p}=-\frac{ql_{45}^2}{12}=-\frac{0.36\times 5.33^2}{12}=-0.86 \text{ ton metre}$$

$$M_{34,\,p}=-M_{43,\,p}=-\frac{q_1 l_{34}^2}{12}=-\frac{3\times 4^2}{12}=-4.00 \text{ ton metres}$$

$$M_{23,\,p}=-\frac{ql_{23}^2}{12}-\frac{Qab^2}{l_{23}^2}=-\frac{3\times 6^2}{12}-\frac{22\times 4\times 2^2}{6^2}=-18.8 \text{ ton metres}$$

$$M_{32,\,p}=\frac{ql_{23}^2}{12}+\frac{Qa^2b}{l_{23}^2}=\frac{3\times 6^2}{12}+\frac{22\times 4^2\times 2}{6^2}=28.5 \text{ ton metres}$$

The equilibrium of all the joints pertaining to the system whose deflections are prevented by the imaginary support at the level of the crossbeam

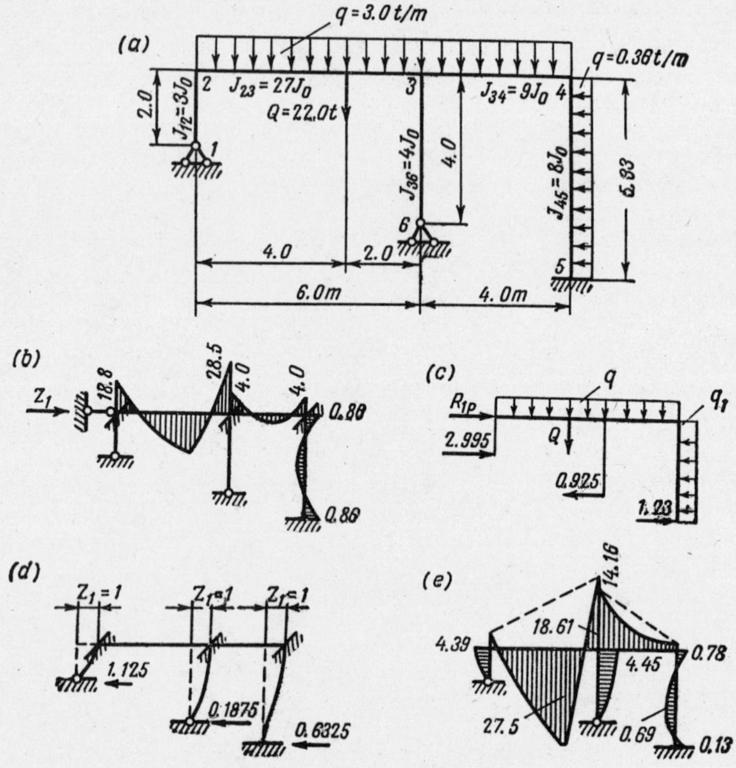

Fig. 3.14

is ensured by successive approximations as shown in the upper part of Table 2.14.

The corrections taking care of joint deflections will be obtained expressing that the total reaction of the imaginary support due both to the applied loads and to the horizontal deflection Z_1 is nil, viz.

$$R_{1p}+r_{11}Z_1=0$$

Here R_{1p} is the reaction along Z_1 induced by the external loads, and r_{11} is the reaction along the same direction due to the unit displacement $Z_1=1$.

Reaction R_{1p} will be derived from the shearing stresses at the lower ends of the columns (Fig. 3.14*c*)

$$Q_{12,\,p}=-\frac{M_{21}}{l_{12}}=-\frac{5.99}{2.0}=-2.995 \text{ tons}$$

$$Q_{63,\,p}=-\frac{M_{36}}{l_{36}}=\frac{3.71}{4.0}=0.925 \text{ ton}$$

Table 2.14

Joint No.		2		3			4		5
Bar No.		*2-1*	*2-3*	*3-2*	*3-6*	*3-4*	*4-3*	*4-5*	*5-4*
Stiffness factor i		1.125	4.50	4.50	0.75	2.25	2.25	1.50	1.50
Distribution factor μ		0.20	0.80	0.60	0.10	0.30	0.60	0.40	—
Fixed end moment		—	−18.8	28.50	—	—4.00	4.00	—0.86	0.86
Balancing of joints	*2*	3.76	15.04	7.52	—	—	—	—	—
	3	—	—9.61	−19.2	—3.20	—9.61	—4.80	—	—
	4	—	—	—	—	0.50	1.00	0.66	0.33
	2	1.92	7.69	3.84	—	—	—	—	—
	3	—	—1.30	—2.60	—0.43	—1.31	—0.65	—	—
	2 and *4*	0.26	1.04	0.52	—	0.20	0.39	0.26	0.13
	3	—	—0.21	—0.43	—0.07	—0.22	—0.11	—	—
	2 and *4*	0.04	0.17	0.08	—	0.03	0.07	0.04	0.02
	3	—	—0.04	—0.07	—0.01	—0.03	—0.01	—	—
	2 and *4*	0.01	0.03	0.01	—	0.01	0.01	—	—
	3	—	—	—0.02	—	—	—	—	—
Preliminary value of bending moments, ton metres		5.99	—5.99	18.14	—3.71	14.43	—0.10	0.10	1.34
Bending moments due to unit deflection Z_1		—2.25	—	—	—0.75	—	—	—1.69	—1.69
Balancing of joints	*2* and *4*	0.45	1.80	0.90	—	0.50	1.01	0.68	0.34
	3	—	—0.20	—0.39	—0.06	—0.20	—0.10	—	—
	2 and *4*	0.04	0.16	0.08	—	0.03	0.06	0.04	0.02

Table 2.14 (continued)

Joint No.		2		3			4		5
Balancing of joints	3	—	—0.04	—0.07	—0.01	—0.03	—0.01	—	—
	2 and 4	0.01	0.03	0.01	—	—	0.01	—	—
	3	—	—	—0.01	—	—	—	—	—
Correction corresponding to $Z_1=1$		—1.75	1.75	0.52	—0.82	0.30	0.97	—0.97	—1.33
Correction corresponding to $Z_1=0.912$		—1.60	1.60	0.47	—0.74	0.27	0.88	—0.88	—1.21
Final values of bending moment, ton metres		4.39	—4.39	18.61	—4.45	—14.16	0.78	—0.78	0.13

$$Q_{54,\,p}=-\frac{M_{45}+M_{54}}{l_{45}}-\frac{q_1 l_{45}}{2}=-\frac{0.10+1.34}{5.33}-\frac{0.36\times 5.33}{2}=-1.23 \text{ tons}$$

The equilibrium of the upper portion of the frame separated from its supports requiring that $X=0$, we have

$$-Q_{12,\,p}-Q_{63,\,p}-Q_{54,\,p}-q_1 l_{45}+R_{1p}=0$$

wherefrom

$$2.995-0.952+1.23-0.36\times 5.33+R_{1p}=0$$

and consequently

$$R_{1p}=-1.38 \text{ tons}$$

As for reaction r_{11} it will be found assuming that the horizontal beam of the conjugate system moves towards the right over a distance $Z_1=1$ (Fig. 3.14*d*). The shearing forces which would be developed in that case at the lower ends of the columns (divided by EJ_0) would amount to

$$\overline{Q}_{12,\,1}=\frac{3J_{12}}{J_0 l_{12}^3}=\frac{3\times 3}{2^3}=1.125$$

$$\overline{Q}_{63,\,1}=\frac{3J_{36}}{J_0 l_{36}^3}=\frac{3\times 4}{4^3}=0.1875$$

$$\overline{Q}_{54,\,1}=\frac{12J_{45}}{J_0 l_{45}^3}=\frac{12\times 8}{5\cdot 33^3}=0.6325$$

These shearing forces are shown in Fig. 3.14*d*.

The fixed end moments caused by the same displacement of the horizontal beam equal

$$\overline{M}_{21,\,1}=-1.125\times 2=-2.25$$

$$\overline{M}_{36,\,1}=-0.1875\times 4=-0.75$$

$$\overline{M}_{45,\,1}=\overline{M}_{54,\,1}=-0.6325\,\frac{5.33}{2}=-1.69$$

The bending moments at the joints of the given redundant structure due to the same unit displacement $Z_1 = 1$ will be obtained balancing the joints as indicated in the lower part of Table 2.14. Knowing these bending moments we may find the corresponding shearing forces at the lower ends of the columns as well as the reaction r_{11} which equals the algebraic sum of shearing forces acting at the top of these columns. These shears and the reaction r_{11} equal

$$\overline{Q}'_{12,\,1}=-\frac{\overline{M}'_{21}}{l_{12}}=\frac{1.75}{2}=0.875$$

$$\overline{Q}'_{63,\,1}=-\frac{\overline{M}'_{36}}{l_{36}}=\frac{0.82}{4}=0.205$$

$$\overline{Q}'_{54,\,1}=-\frac{\overline{M}'_{45}+\overline{M}'_{54}}{l_{45}}=\frac{0.97+1.33}{5.33}=0.433$$

$$r_{11}=\overline{Q}'_{12,\,1}+\overline{Q}'_{63,\,1}+\overline{Q}'_{54,\,1}=0.875+0.205+0.433=1.513$$

We may now find the actual value of the deflection of the upper part of the frame

$$Z_1=-\frac{R_{1p}}{r_{11}}=+\frac{1.38}{1.513}=0.912$$

The positive sign of this deflection indicates that its direction coincides with the one adopted. The corrections which must be introduced on the bending moments will now be easily obtained multiplying the magnitudes of the moments due to a unit deflection by the actual value of this deflection which equals 0.912. Adding these corrected moments to those obtained in the upper part of the table disregarding the horizontal deflection of the crossbeam we obtain the final values of the required moments (Fig. 3.14*e*).

If several independent deflections could occur in the given structure it would become necessary to introduce separately the corrections due to each of these deflections on the values of bending moments obtained for the non-deflected structure. The real value of each of these deflections would be derived from a system of simultaneous equations. The number of these equations would be equal to the number of independent deflections and therefore would remain very much smaller than the number of equations used in any of the exact methods. Thus, if there are two independent joint deflections, regardless of the number of unknown angles of twist, we shall have to solve only two equations with two unknowns given hereunder

$$r_{11}Z_1+r_{12}Z_2+R_{1p}=0$$

$$r_{21}Z_1+r_{22}Z_2+R_{2p}=0$$

15. MODERN DESIGN METHODS

1.15. *Basic Principles*

Until quite recently the design of all engineering structures was based on permissible stresses. This means that the internal forces (bending moments, shearing and normal forces) in different members of the structure were determined using the methods developed in the Theory of Structures assuming that the whole structure works as a perfectly elastic body. The selection of the cross-sectional dimensions of these members was based on formulas established in treatises on the strength of materials and was aimed at keeping the design unit stresses in these members within the permissible ones. As for the permissible stresses themselves they were taken equal either to the ultimate strength of the materials or to the stresses corresponding to their yield point divided by a certain factor of safety. In accordance with this method the general expression ensuring the strength of the structure could be written as follows

$$\sigma \leqslant \frac{\sigma_{ult}}{K}$$

In this expression σ is the design stress, σ_{ult} is the ultimate strength or yield strength of the material, and K is the safety factor ($K > 1$).

However, it became soon apparent that for concrete, reinforced concrete and masonry structures results obtained using the method of permissible stresses were frequently in contradiction with data based on observation or on experimental work. This has led to the creation of a new method of computation usually known as the "ultimate loads method". In this method the safety factor is no longer referred to the maximum unit stresses arising at different points of the structure but to the ultimate bearing capacity of the cross section of a member or of the structure as a whole. In this method the general expression ensuring the strength of the member or structure under consideration becomes

$$S \leqslant \frac{S_{ult}}{K}$$

Here S is the internal force acting in the member under consideration, S_{ult} is the ultimate load which the same member is capable to carry (by ultimate load we mean a load just sufficient to cause its failure), and K is the safety factor. As a general rule the ultimate loads are calculated with due consideration of possible nonelastic deformations.

Both methods mentioned above suffered from one and the same drawback: the value of the safety factor remained constant while in reality it should depend on a number of circumstances connected with the structure under consideration.

At the present time the U.S.S.R. Building Codes require that all the engineering structures should be designed in accordance with the method of *ultimate states* developed by a group of Soviet scientists under the direction of Acad. V. Keldysh and Prof. N. Streletsky.

The term *ultimate state* refers to such a state of the structure or of one of its members which makes further service of this structure impossible, whether due to insufficient bearing capacity, or to the appearance of excessive deflections and deformations or finally to the development of some local defects. In the general case the following three ultimate states should be taken into consideration:

1. The ultimate state characterized by the loss of bearing capacity due either to lack of strength, to loss of stability or to fatigue (in the case of repeated loading).

2. The ultimate state characterized by the development of excessive deformations such as deflections, twists, etc.

3. The ultimate state characterized by the formation and opening of cracks, or the appearance of other local defects preventing further use of the structure, as, for instance, loss of impermeability in a reservoir, etc.

Each of the above ultimate states may develop under the influence of numerous factors, the main being

(1) the type and intensity of external loads and other actions;

(2) the strength and other mechanical properties of building materials;

(3) the conditions under which the structure has been erected and will have to work.

The design must ensure that during the service life of the structure none of these ultimate states will be allowed to occur at the same time preventing the overexpenditure of building materials. The problem may be approached from three different angles:

1. It may be necessary to determine the ultimate load for a given structure, in other words, to find the magnitude of the loads which will lead to the development of any one of the ultimate states.

2. One may be required to choose the minimum cross-sectional dimensions of all the structural members which would ensure against the appearance of any of the ultimate states under a given loading, acting under a given set of conditions.

3. One may be required to find the real safety factor of a given structure under a given system of loads or, in other words, to find the ratio between the ultimate load and the applied one.

In the method of ultimate states the single safety factor used in the two methods mentioned previously is replaced by a set of differentiated factors.

Thus, the *design loads* (both live and dead) used in all the computations are obtained multiplying the service or *normal loads* by a factor n called *the overload factor*. This factor is intended to take care of any possible increase of the applied loads over and above the magnitude of the normal service loads (when the work of the structure becomes aggravated by a reduction of certain loads, the overload factor must take care of this eventuality). The overload factor will necessarily vary depending on the loading. Thus, for instance, the overload factor for dead loads and hydrostatic pressures adopted by the U.S.S.R. Building Codes is quite small and equals only 1.1, while the same factor for live loads carried by the floors of dwelling houses reaches 1.4. For wind pressure the overload factor equals 1.2, for snow 1.4 and for the pressure exerted by granular materials it should be taken equal to at least 1.2. When certain special or exceptional combinations of loading are taken into consideration an additional factor called the *combinations factor* should be introduced on all live loads, this factor accounting for the extremely low probability of all live loads reaching their maximum values simultaneously. The numerical value of this factor usually ranges from 0.8 to 0.9.

The *design strength of materials* is obtained multiplying their *normal strength* (the latter being stipulated by appropriate standards or obtained by direct testing) by a *uniformity factor k*, this factor taking care of any possible drop in the strength of the material used (as compared with the aforesaid normal strength) caused by some fluctuation in its mechanical properties. This factor will therefore depend solely on the properties of the material under consideration. The numerical values of this factor adopted by the U.S.S.R. Building Codes are:

for structural low carbon steels from 0.8 to 0.9

for timber working in bending and for masonry work 0.4

for concretes with cube strength between 250 and 600 kg per sq cm approximately 0.55.

The U.S.S.R. Building Codes contain the values of uniformity factors k for numerous building materials with due regard to manufacturing procedures and the use for which these materials are intended.

Peculiarities of working conditions such as the presence of aggressive media, stress concentrations, the danger of brittle failure as well as any other circumstances alleviating or aggravating the work of the structure must be taken care of by *the working conditions factor* m, whose value may become both smaller or greater than unity. Thus, for instance, the working conditions factor for riveted joints varies from 0.6 to 1.0 depending on the type of rivets and on the use to which these joints are put; for wooden beams whose minimum cross-sectional dimensions do not fall below 14 cm the working conditions factor is taken equal to 1.15, and for certain precast reinforced concrete elements the Building Codes authorize to adopt $m = 1.1$.

The working conditions factor is applied to the ultimate load and consequently a decrease in the value of this factor is equivalent to an increase in the value of the overall safety factor.

When the design is based on the first ultimate state the general expression ensuring that the structure complies with the strength requirements takes the shape of the following inequality

$$S \leqslant S_{ult}$$

where S is the internal force developed in the member under consideration in the event of the most unfavourable combination of design loads (service loads multiplied by the overload factor), and S_{ult} is the ultimate bearing capacity of this particular member computed in terms of its cross-sectional dimensions, of the design strength of the material (obtained by multiplying its actual strength by the appropriate uniformity factor) and with due regard to the working conditions factor.

In a number of cases the building codes authorize the design of structures on the assumption that failure will occur well after the appearance of nonelastic deformations. This leads to more economical use of building materials, advantage being taken in this case of the reserve of strength existing beyond the elastic limit.

The strains and deflections are usually computed assuming that the elastic limit of the material is never exceeded and reducing to unity all the overload factors. The following sections will be devoted to a brief outline of modern design methods ensuring the strength of framed structures, these methods taking due account of plastic deformations peculiar to building materials.

2.15. *Design of Statically Determinate Beams*

For simplicity the real stress-strain curve of an elastoplastic material is replaced by the simplified rectilinear diagram given in Fig. 1.15. This diagram consists of an inclined line representing the elastic strains and of a horizontal line corresponding

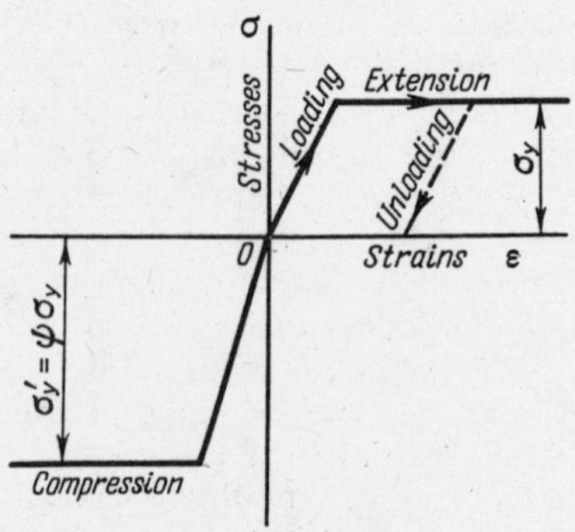

Fig. 1.15

to plastic strains. The values of yield stresses in tension (σ_y) and in compression (σ_y') may differ permitting thereby a better representation of the real properties of certain materials such as concrete, asbestos cement and certain plastics.

It is usually assumed that plastic strains set in without any transition period as soon as the stress in the material has exceeded its yield point. For design purposes yield stresses σ_y and σ_y' are replaced by the design strength of the material R or R' as the case may be, the latter being obtained multiplying yield stresses by the corresponding uniformity factors.

It is assumed that plastic deformations of the materials may continue indefinitely and that the increase in strength due to strain hardening may be neglected. Strictly speaking, the simplified diagram can be used only in the event the plastic deformations sustained by the material do not exceed the horizontal portion of the real stress-strain curve corresponding to the yield point. For certain building materials such as rolled steel, this horizontal stretch may be quite short.

Failure of reinforced concrete elements usually occurs when both the reinforcement and the concrete itself have entered the zone of plastic deformations and therefore the simplified diagram of Fig. 1.15 remains to a certain extent applicable. In all cases where brittle failure may be expected the diagram of Fig. 1.15 becomes totally unacceptable.

Let us now examine the cross section of a statically determinate beam symmetrical about its vertical axis and working

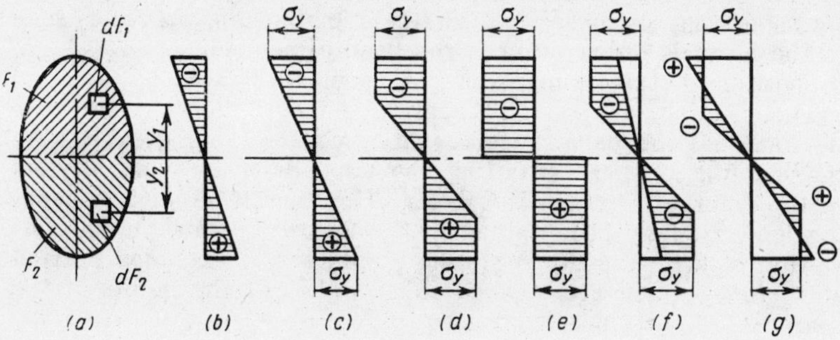

Fig. 2.15

in pure bending (Fig. 2.15*a*). For simplicity we shall assume $\sigma_y' = \sigma_y$.

As long as the stresses at all the points of this cross section remain below the elastic limit of the material (the elastic limit coinciding on the simplified diagram with the yield point) the stress diagram for the cross section under consideration will consist of two triangles shown in Fig. 2.15*b*. If the bending moment is increased, the stresses in the cross section will increase also and the outer fibres will eventually reach the yield point (Fig. 2.15*c*). At this moment the purely elastic work of the beam comes to an end, plastic strains beginning to develop within the extreme fibres of the cross section.

The method of permissible stresses is based on the assumption that the bearing capacity of an element becomes completely exhausted as soon as nonelastic strains appear in the outer fibres of the cross section. However in reality the stresses in all the fibres located closer to the neutral axis remain well below the yield point of the material and therefore the loads and bending moments may be further increased without entailing the immediate failure of the beam.

For those of the fibres where the yield point has been already exceeded the stresses will remain constant while for the rest of the fibres situated closer to the neutral axis they will continue increasing with a gradual decrease in the depth of the elastic zone (Fig. 2.15*d*). If the bending moment is further increased the stress diagram will tend to that of Fig. 2.15*e* at which moment the bearing capacity of the beam will become completely exhausted and a *plastic hinge* will appear. Such a plastic hinge differs from an ideal one by the fact that it will function only if two couples acting in opposite directions are applied to this hinge, each of these couples being equal to the ultimate resisting moment corresponding to the section. The plastic hinge disappears as soon as the beam is unloaded or if the bending moment changes sign. In both cases the beam reverts to an elastic state.

The bending moment at the plastic hinge which characterizes the real bearing capacity of beam is considerably greater than the bending moment entailing the appearance of yield stresses in the external fibres of this beam. The magnitude of this bending moment may be obtained in the following way. On the formation of a plastic hinge all the points of the extended portion of a cross section are stressed to σ_y while all the points of the compressed portion are stressed to $\sigma_y' = \psi\sigma_y$.

Since the normal force in a cross section working in pure bending equals zero, the resultant of all the unit stresses acting over the extended portion must equal the resultant of all the unit stresses acting over the compressed portion, in other words

$$\psi\sigma_y F_1 = \sigma_y F_2$$

wherefrom

$$F_1 = \frac{F_2}{\psi} = \frac{F}{\psi+1}; \quad F_2 = \frac{\psi F}{\psi+1}$$

F is the total cross section of the beam.

This equation permits the immediate determination of the neutral axis which may no longer coincide with the horizontal axis of symmetry for the given section. The real bearing capacity of the beam, characterized by the ultimate bending moment, may be now obtained taking the moment of all the stresses acting at this cross section about the neutral axis just obtained. Replacing the yield stresses σ_y by the design ones R and introducing the working conditions factor m we obtain

$$M_{ult} = mR\left[\psi \int_{F_1} y_1\,dF_1 + \int_{F_2} y_2\,dF_2\right] = mR(\psi S_1 + S_2) = mRW_p$$

where y_1 and y_2 represent the distances of the elementary areas dF_1 and dF_2 (these areas being situated in the upper and lower portion of the cross section, respectively) to the neutral axis, and S_1 and S_2 are the statical moments of these two portions about the same axis.

The term $(\psi S_1 + S_2)$ represents the *plastic resisting moment of the cross section* and will be hereunder designated by W_p.

The magnitude of the ultimate resisting moment would remain unchanged if the moments of the internal forces were referred to the gravity axis instead of the neutral one. For a rectangular cross section the plastic resisting moment becomes

$$W_p = \psi S_1 + S_2 = \frac{b}{2}(\psi h_1^2 + h_2^2) = \frac{\psi b h^2}{2(\psi+1)}$$

where b and h represent respectively the width and the depth of the cross section, and

$$h_1 = \frac{h}{\psi+1}; \quad h_2 = \frac{\psi h}{\psi+1}$$

When $\psi = 1$, as is the case for structural steel, W_p equals $\frac{bh^2}{4}$ and consequently

$$\frac{W_p}{W} = \frac{bh^2}{4} : \frac{bh^2}{6} = 1.5$$

W representing as usual the elastic resisting moment of the cross section.

When $\psi = 2$ as in the case for asbestos cement we obtain

$$W_p = \frac{bh^2}{3} \quad \text{and} \quad \frac{W_p}{W} = 2.0$$

It follows that the bearing capacity of structural elements may be increased quite considerably if due account is taken of the plastic strains which may develop.

For H-beams and I-beams the ratio $\frac{W_p}{W}$ equals approximately 1.15 (provided $\psi = 1$) and in that case the plastic design becomes less attractive.

For those of the materials whose ψ factor is the greatest the economy derived from plastic design becomes very noticeable for it becomes possible to increase the service loads twice or even more.

The distribution of stresses and plastic zones along the span of the beam is entirely dependent on the bending moment diagram.

During unloading the strains decrease along a straight line (dotted line in Fig. 1.15) parallel to the one representing their increase during loading while the body still works as an elastic one. Hence during unloading the material behaves again as a purely elastic one and the stresses at any particular stage of this operation will be given by the shaded portion of the diagram obtained by superposition and given in Fig. 2.15*f*.

When the beam is completely unloaded the stress diagram will take the shape indicated in Fig. 2.15*g*. The moment due to the internal forces must vanish for upon complete unloading the bending moment becomes equal to zero itself.

It should be noted that the bearing capacity of beams carrying transversal loads may be reduced considerably due to the influence of shearing stresses, which become particularly dangerous when the elastic limit is exceeded.

Problem. Required to select the cross-sectional dimensions of a simply supported beam 6 m long made of low carbon steel, the dead load on the beam equalling 2 tons per metre and a concentrated live load of 5.0 tons being applied at midspan. The overload factors equal respectively 1.1 for the dead load and 1.4 for the live load; the working conditions factor m equals 0.90. The design strength of low carbon steel R will be taken equal to 2,100 kg per sq cm.

Solution. The design moment at midspan is

$$M=\frac{n_q q l^2}{8}+\frac{n_p P l}{4}=\frac{1.1\times 2.0\times 6^2}{8}+\frac{1.4\times 5\times 6}{4}=$$

$$=9.9+10.5=20.4 \text{ ton metres}=2{,}040{,}000 \text{ kg}\cdot\text{cm}$$

The bearing capacity of a steel beam working in bending is given by

$$M_{ult}=mRW_p=0.9\times 2{,}100\, W_p=1{,}890\, W_p$$

Let us adopt an I-beam whose plastic resisting moment $W_p = 1.15\ W$. Equalling the design bending moment to the ultimate resisting moment of the beam we find

$$M_{ult}=1{,}890 W_p$$

which leads to

$$1890\times 1.15W=2{,}040{,}000$$

wherefrom

$$W=940 \text{ cm}^3$$

We shall choose a 36*c* I-beam (the U.S.S.R. State Standard) with a resisting moment of 962 cm³. Were the same beam designed using the permissible stresses method we should have to use a 40*c* I-beam.

3.15. *Design of Statically Indeterminate Beams*

The complete exhaustion of the bearing capacity of certain redundant members of a structure will not entail its failure provided the remaining members continue to form a geometrically stable system capable of carrying the applied load. Plastic deformations of the overloaded members will lead to a redistribution of stresses which will increase the bearing power of the structure as a whole. Complete failure of the latter will occur only when the number of members whose bearing capacity has been exhausted becomes equal to the number of redundant constraints increased by 1. The plastic design of statically indeterminate structures may be carried out using the method of plastic hinges which, in its more general form, is known in the U.S.S.R. under the name of the method of ultimate equilibrium. Both the *static* and the *kinematic* procedures may be utilized.

The kinematic procedure requires the knowledge of lines or points of failure, permitting the formation of equilibrium equations pertaining to the mechanism into which the structure will be converted once its bearing power has been exhausted. It is usually assumed that all external loads increase simultaneously and in the same proportion and that their points of application, directions and signs remain unchanged. It is also assumed that the pattern of the actual failure will be the one corresponding to the minimum value of the ultimate load.

Equations of ultimate equilibrium may be based either on statics or on the principle of virtual displacements. In the latter case it is assumed that infinitesimal displacements occur; these displacements remaining consistent with the constraints subsisting after the transformation of the structure into a mechanism.

The static procedure requires that the internal forces should be distributed in the redundant structure in such a way that together with the initial or inherent stresses they should lead to its failure. For this purpose some distribution of internal forces in equilibrium with the applied loads is chosen and thereafter a number of additional systems of internal forces, each of these systems being balanced, i. e., corresponding to zero loads, are added thereto. The number of these additional systems must be equal to the number of redundant constraints of the given structure.

The ultimate equilibrium is reached when the bearing capacity of certain members of the structure becomes exhausted.

The real distribution of internal forces will correspond to the maximum breaking load possible under the given conditions.

Both the static and the kinematic procedures if applied to the same redundant structure will always lead to the same results. In a number of cases resort may be made to the so-called *method of moment equalization*, derived from the kinematic procedure.

The appearance of each plastic hinge in a continuous beam always corresponds to the elimination of one constraint and therefore reduces by one the degree of redundancy of the given beam. Hence the bearing power of the beam will be completely exhausted as soon as the number of plastic hinges has become equal to the degree of redundancy of the beam plus one.

It should be remembered however that the bearing capacity of each single span will be also exhausted as soon as three hinges

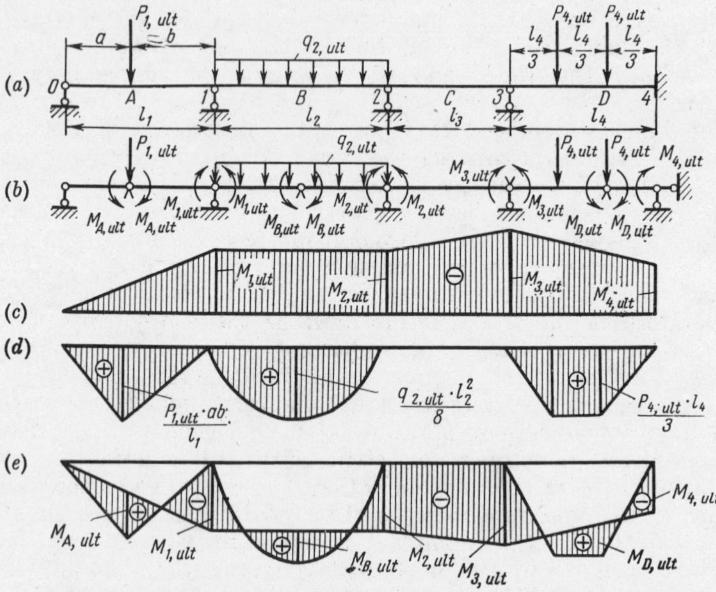

Fig. 3.15

appear within its length. For this reason every span of a continuous beam must be also considered separately.

Let us take the continuous beam shown in Fig. 3.15*a* whose bearing capacity is required for the given distribution of loads. First let us replace the given beam by a conjugate statically determinate structure consisting of a series of end-supported beams as indicated in Fig. 3.15*b* and let us construct separately for each of these beams the diagrams of bending moments induced, on the one hand, by the ultimate moment which may be developed at each of the supports (Fig. 3.15*c*) * and, on the other hand, by the bending moments due to the ultimate loads whose points of application and directions are given but whose magnitudes remain unknown (Fig. 3.15*d*).

◆

* This diagram has been constructed on the assumption that the ultimate strength of the beam varies from support to support.

It is clear that the ordinates at the supports to the diagram of bending moments entailing the formation of plastic hinges at the said supports will represent the ultimate bearing power of the corresponding cross sections of the beam

$$M_{1,ult};\quad M_{2,ult};\quad M_{3,ult};\quad M_{4,ult}$$

The *resulting* bending moment diagram will be obtained by superposition of the negative bending moment diagram due to the aforesaid moments at the supports with the diagram of positive bending moments induced by the applied loads in each span of the conjugate simple structure. The scales of these two diagrams will be so selected that in each span the maximum positive ordinates to the diagram of the resulting moments should represent the bearing capacity of corresponding cross sections (Fig. 3.15*e*). The numerical value of the ultimate load for each span will be found equating the maximum ordinate to the bending moment diagram pertaining to the conjugate end-supported beam to the sum (in absolute values) of the maximum ordinate to the resulting diagram within the span under consideration with the ordinate at the same cross section due to the application of ultimate moments at the supports.

Thus, for instance, in the first span from left to right we shall have

$$\frac{P_{1,ult}ab}{l_1}=\frac{M_{1,ult}a}{l_1}+M_{A,ult}$$

wherefrom

$$P_{1,ult}=\frac{M_{1,ult}}{b}+\frac{M_{A,ult}l_1}{ab}$$

Since $M_{1,ult}=M_{2,ult}$ for the second span we shall have

$$\frac{q_{2,ult}l_2^2}{8}=M_{1,ult}+M_{B,ult}$$

wherefrom

$$q_{2,ult}=\frac{8\,(M_{1,ult}+M_{B,ult})}{l_2^2}$$

Similarly for the fourth span we shall have

$$\frac{P_{4,ult}l_4}{3}=\frac{M_{3,ult}}{3}+\frac{2M_{4,ult}}{3}+M_{D,ult}$$

wherefrom

$$P_{4,ult}=\frac{M_{3,ult}+2M_{4,ult}+3M_{D,ult}}{l_4}$$

Knowing the values of the ultimate resisting moments we can easily determine the ultimate loads for each span which will lead to the value of the ultimate load for the beam as a whole.

When the ultimate resisting moment of the continuous beam remains constant the determination of ultimate loads becomes particularly simple. Indeed in that case the desired result will be achieved if the maximum ordinates to the resulting bending moment diagram within the spans are made equal to those at the supports (Fig. 4.15).

When it is desired to solve the inverse problem or, in other words, to find the necessary cross-sectional dimensions corresponding to a given loading, one should start by the construction

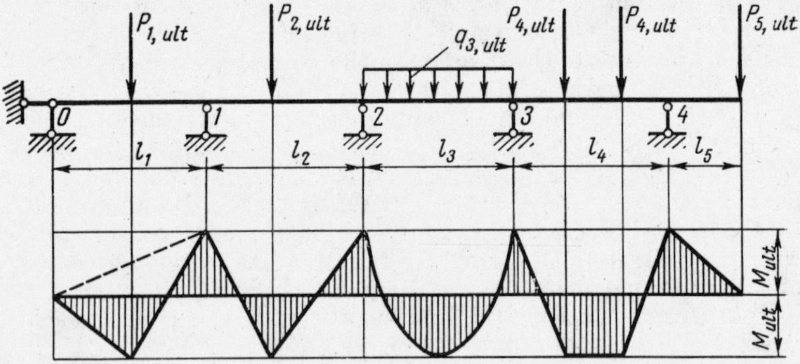

Fig. 4.15

of bending moment diagrams for each of the spans regarded as a separate end-supported beam. This being done, the values of maximum bending moments in the span and at the supports should be chosen in such a way that their ratio should be the same as the ratio of the resisting moments at the corresponding cross sections. This latter ratio should be adopted beforehand for otherwise an infinitely great number of solutions would become possible.

Assume, for instance, that it is required to construct the diagram of the design bending moment for the continuous beam of Fig. 4.15, all the loads being known both in amount and in direction.

Assuming that the design moments are equivalent to the ultimate ones the desired diagram for each span will be obtained in the same way as the ultimate moment diagram. First the bending moment diagrams induced by the given loading in all the conjugate end-supported beams will be constructed and the pertinent ordinates to these diagrams computed.

Thereafter one may proceed with the preliminary equalization of the bending moment diagrams for each span as indicated in Fig. 5.15*b*. If the cross section of the beam is to remain constant throughout, its dimensions are chosen to resist the maximum bending moment. In the case under consideration the bending moment acting in the first span is the greater of all. It is obvious that no plastic hinges will appear in any of the other spans under the given loading.

If it were required to construct the diagram of bending moments leading to the formation of plastic hinges in all the loaded

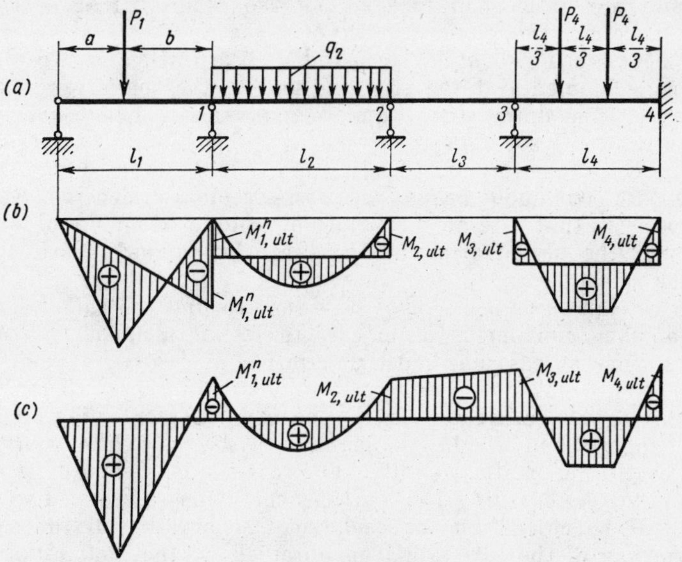

Fig. 5.15

spans one should proceed with the equalization of moments acting immediately to the left and immediately to the right of each support. In doing so, one should always start with the spans carrying the smaller loads passing thereafter to the spans characterized by gradually increasing maximum bending moments.

Thus, in the example under consideration one should start with the second span for which the preliminary bending moment diagram will be adopted as the final one. The diagram for the first span will be obtained immediately since the bending moment at the right-hand support is already known (Fig. 5.15*a*) and the one at the left-hand support is nil.

Thereafter one should pass to the fourth span which is preceded by the unloaded span *3*. Hence the preliminary bending moment diagram for span *4* can also be adopted as the final one. The diagram for the third span will be obtained connecting by a straight line the ordinates over supports *2* and *3* which are already known. The resulting bending moment diagram is represented in Fig. 5.15*c*. This diagram may be used for the determination of cross-sectional dimensions both over the supports and within the spans.

It is clear that these dimensions will differ from support to support and from span to span but the beam will be of equal resistance for the given system of loads.

Other solutions could be found if the ratios between the resisting moments at the supports and in the spans were taken different from unity. If a beam of equal resistance were not specified still more solutions to the same problem could be obtained.

One could also use the bending moment diagram obtained for an elastic continuous beam. The cross-sectional dimensions would be based in that case on the values of bending moments obtained through the equalization of those at the supports with those in the spans.

In certain cases it becomes necessary to find the safety factor for a given continuous beam carrying a well-defined system of loads. In that case one must determine the bearing capacity of each span, which should be done as indicated in Figs. 3.15 and 4.15 equalizing the ordinates to the bending moment diagrams at the supports and within the spans with the only difference that the bearing capacities of different cross sections should be now computed disregarding the working conditions factors. The latter will be calculated for each span separately, dividing the magnitude of the ultimate load obtained by the amount of the loads actually applied.

The same problem could be dealt with in a somewhat different manner. One could start by the construction of the diagram for the equalized bending moments as in Fig. 5.15, derive therefrom the required cross-sectional dimensions for each span and thereafter compare the bearing capacities corresponding to the dimensions found with those of the given beam. This procedure will permit again the determination of the actual values of the working conditions factors.

Up to the present we have assumed that plastic hinges will appear either at the supports or at those cross sections where the bending moments pass through their maximum. In actual practice beams whose bearing capacity varies sharply from section to section are frequently encountered (for instance, reinforced concrete beams). In these cases it is impossible to determine beforehand the location of plastic hinges.

To find them one must first trace the diagram representing the variation of the bearing capacity along each span (Fig. 6.15). Such a diagram will usually consist of a positive and a negative branch, since each cross section is capable of resisting both positive and negative bending moments, even though their magnitudes may differ considerably. The bending moment diagram should thereafter be inscribed into the diagram mentioned above

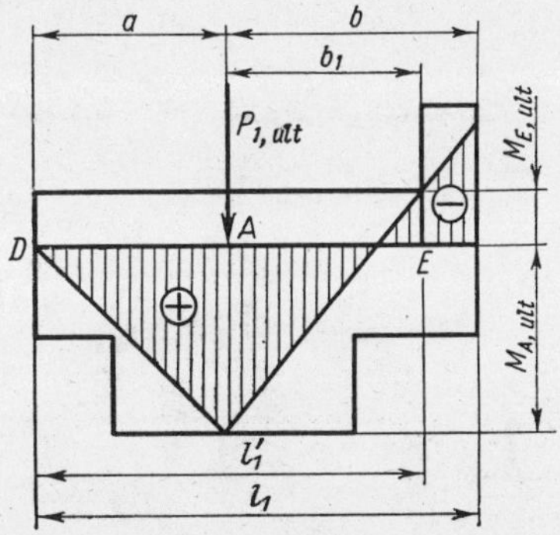

Fig. 6.15

(the ultimate resisting moment diagram) in such a way that the two should have at least three common points. The location of these points will indicate the position of the plastic hinges.

Assume, for instance, that the diagram for the resisting moments along the first span of the beam given in Fig. 4.15 is represented by the broken lines shown in Fig. 6.15. Inscribing into the latter the bending moment diagram for the same span we shall find that the plastic hinges will appear in cross sections A and E. The ultimate value for load P will be determined on the basis of these two diagrams, its value being equal to

$$P_{1,\,ult} = \frac{M_{E,\,ult}}{b_1} + \frac{M_{A,\,ult}\,(a+b_1)}{ab_1}$$

In the case of a continuous beam of varying bearing capacity plastic hinges may appear at different points as indicated in Fig. 7.15. Each different set of these plastic hinges will lead to a different value of the ultimate load and therefore, for design purposes, one should adopt the smallest of these values deduced from the comparison of different failure patterns. It is obvious that such failure patterns which are inconsistent with the stipulation of the problem should be rejected.

When a beam carries both dead and live loads which may vary in amount but whose ratio remains constant, the ultimate values of these loads will be found using an envelope diagram for the bending moments. The ordinates to this diagram will be given in terms of a single parameter determining simultaneously the magnitude of both dead and live loads. Assuming that the beam remains elastic and having obtained the corresponding bending moment

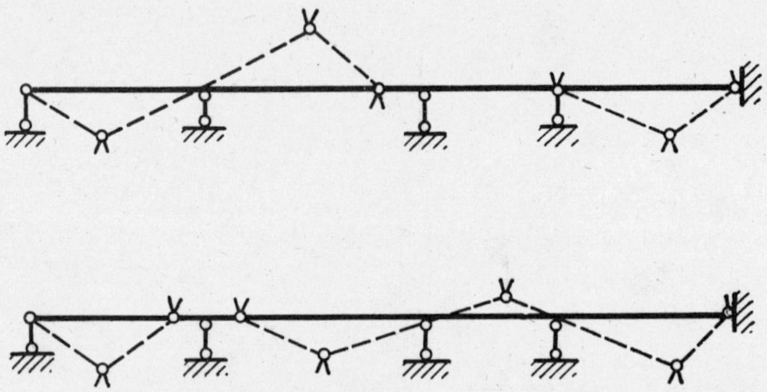

Fig. 7.15

diagram, one should proceed with the equalization of the maximum bending moments with due consideration to the ultimate resisting moments of the beam. This being done, the required ultimate loads will be easily found. When both the dead and live loads are known beforehand the required cross-sectional resisting moments may be computed equalizing the ordinates to the bending moment diagram due to the above loads multiplied by the overload factors.

All the above leads to the important conclusion that the method of plastic hinges permits the design of continuous beams without resorting to equations based on the deflections of the structure. It is worth mentioning also that the formation of plastic hinges reduces the redundant continuous beams to the state of a mechanism, which is completely uneffected by such factors as the settlement of supports, temperature changes, or erection defects. Partly fixed ends become equivalent to the rigidly built-in ones.

Once a continuous beam has been loaded beyond its elastic limit residual deformations and stresses will appear. These stresses will not balance within each section as was the case with statically determinate beams, for the redundant constraints will develop certain reactions which will not reduce to zero upon withdrawal of the loads. The determination of residual stresses is rather complicated but since these stresses have practically no influence on the bearing capacity of the structure this question will not be studied here.

Problem. Determine the bearing capacity of a continuous I-beam shown in Fig. 8.15, whose resisting moment $W = 237$ cm³. The beam is made of

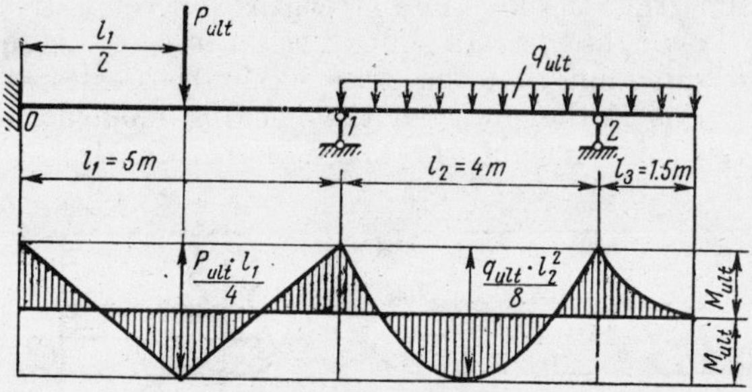

Fig. 8.15

low carbon steel with a design strength of 2,100 kg per sq cm. The working conditions factor $m = 1$.

Solution. The ultimate moment which can be developed by the cross section of this I-beam working in bending equals

$$M_{ult} = mRW_p = mR1.15W = 1.0 \times 2,100 \times 1.15 \times 237 =$$
$$= 573,000 \text{ kg}\cdot\text{cm} = 5.73 \text{ ton metres}$$

Since the bending moment within the span and at the support must be equal we shall have for the first span

$$\frac{P_{ult}l_1}{4} = 2M_{ult}$$

wherefrom

$$P_{ult} = \frac{8M_{ult}}{l_1} = \frac{8 \times 5.73}{5.0} = 9.18 \text{ tons}$$

Similarly for the second span we shall find

$$\frac{q_{ult}l_2^2}{8} = 2M_{ult}$$

wherefrom

$$q_{ult} = \frac{16M_{ult}}{l_2^2} = \frac{16 \times 5.73}{4.0^2} = 5.73 \text{ tons per metre}$$

For the cantilever

$$\frac{q_{ult}l_3^2}{2} = M_{ult}$$

wherefrom

$$q_{ult} = \frac{2M_{ult}}{l_3^2} = \frac{2 \times 5.73}{1.5^2} = 5.10 \text{ tons per metre}$$

4.15. *Design of Redundant Frames and Arches*

Frames and arches are usually subjected to the simultaneous action of bending moments and normal forces and consequently we must study first the combined action of normal and flexural stresses on a cross section working beyond its elastic limit. We shall assume that the ultimate strengths (or yield stresses as the case may be) of the material in extension (σ_y) and in compression ($\sigma_y' = \psi\sigma_y$) differ but that the cross section possesses at least one symmetry axis in the plane of the bending moment M_{ult}. If in addition to this moment the cross section is also subjected to

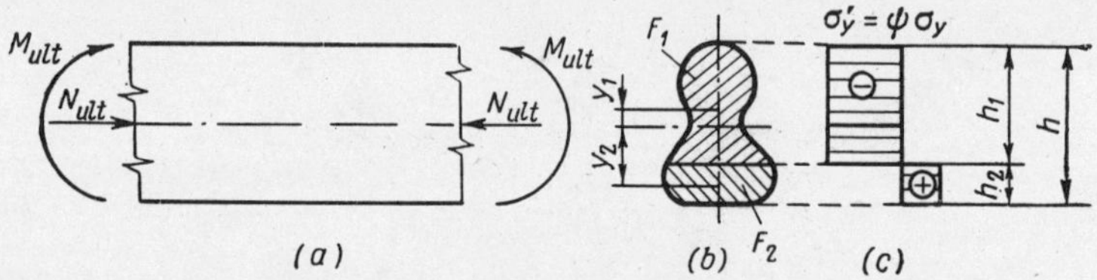

Fig. 9.15

a normal force N acting along its axis (Fig. 9.15), a plastic hinge will appear at this section when the diagram of fibre stresses will take the shape indicated in Fig. 9.15*c*. The neutral axis will shift towards the edge of the cross section, its distance from the centre increasing with the reduction of $e = \frac{M_{ult}}{N_{ult}}$. The magnitude of the ultimate normal force will equal the resultant of all the fibre stresses acting over the given cross section and will be given by

$$N_{ult} = \sigma_y (\psi F_1 - F_2)$$

where F_1 and F_2 are the areas of the compressed and extended portions of a cross section, respectively.

Since $F_1 + F_2 = F$ we can easily find the areas of both portions just mentioned which will determine the position of the neutral axis

$$F_1 = \frac{F}{1+\psi}\left(1 + \psi\frac{N_{ult}}{N_{ult}^0}\right); \quad F_2 = \frac{\psi F}{1+\psi}\left(1 - \frac{N_{ult}}{N_{ult}^0}\right)$$

In this expression $N_{ult}^0 = \psi\sigma_y F$ represents the value of the ultimate normal load for the cross section working in direct compression. The maximum value of the bending moment which may be applied simultaneously with a normal load N_{ult} will be given by

$$M_{ult} = \sigma_y (\psi F_1 y_1 + F_2 y_2) = M_{ult}^0 \cdot \nu$$

where y_1 and y_2 are the distances of the centroids of the compressed and extended portions to the gravity axis of the whole cross section, and $M^0_{ult} = \sigma_y W_p$ is the ultimate bending moment which could be applied when the element works in pure bending. Hence the value of ν given by

$$\nu = \frac{\psi F_1 y_1 + F_2 y_2}{W_p}$$

will reflect the influence of a normal load on the resistance which a cross section may develop to flexural stresses, W_p representing as usual the plastic resisting moment of the same cross section. The latter must be determined with due consideration to the difference between the values of the yield or ultimate stresses in compression and in extension. For a rectangular cross section whose width and depth are given by b and h respectively and whose resisting moment $W_p = \frac{\psi bh^2}{2(1+\psi)}$ we find

$$h_1 = \frac{h}{1+\psi}\left(1+\psi\frac{N_{ult}}{N^0_{ult}}\right); \quad h_2 = \frac{\psi h}{1+\psi}\left(1-\frac{N_{ult}}{N^0_{ult}}\right)$$

$$N^0_{ult} = \psi\sigma_y bh; \quad y_1 = \frac{h-h_1}{2} = \frac{h_2}{2}$$

$$y_2 = \frac{h-h_2}{2} = \frac{h_1}{2}$$

$$\nu = \frac{(\psi h_1 h_2 + h_1 h_2)(1+\psi)}{\psi h^2} = \left(1+\psi\frac{N_{ult}}{N^0_{ult}}\right)\left(1-\frac{N_{ult}}{N^0_{ult}}\right)$$

In Fig. 10.15 we have represented graphically the values of ν in terms of the ratio $\frac{N_{ult}}{N^0_{ult}}$ for different values of φ. Similar graphs could be obtained for reinforced concrete members. The examination of these curves shows that when the yield point or ultimate strength of the material in compression and in extension remains equal ($\psi = 1$) the value of ν is always smaller than unity. On the other hand, when the said ultimate stresses or yield points differ ($\psi \neq 1$), the value of ν may become greater than unity. It follows that in the latter case the ultimate resistance which may be developed by the given member to flexural loads will be increased by the application of a normal load. The case when $\psi = 0$ is purely theoretical and therefore devoid of practical interest. For design purposes σ_y should be replaced by the product mR_{ext} or, in other words, by the design tensile strength of the material.

When designing redundant arches and frames the main difficulty will always consist in the determination of the position of plastic hinges or, in other words, in the prediction of the failure pattern for the given structure.

In a number of cases preliminary analysis of the structure as an elastic body may be quite helpful, for it may be assumed that

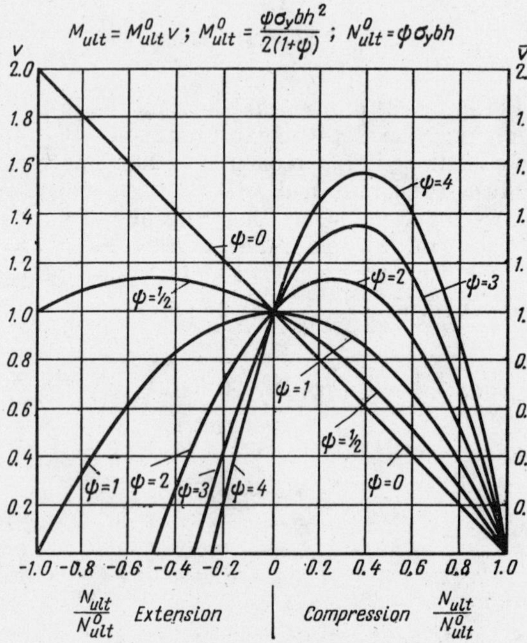

Fig. 10.15

plastic hinges will form at those cross sections where the bending moments pass through their maximum.

The pattern of failure for an arch of two hinges will differ depending on the load points. If the loads are symmetrical, failure will occur with the formation of four plastic hinges (Fig. 11.15*a*), their number falling to three when the two central hinges merge (Fig. 11.15*b*). In the event of nonsymmetrical loading the number of plastic hinges equals two (Fig. 11.15*c*).

It can be easily shown that the procedure of the equalization of maximum ordinates to the bending moment diagrams remains applicable to the design of two-hinged arches (Fig. 12.15).

Having replaced the given redundant arch by the conjugate simple structure let us construct for the latter the diagram of bending momens M_1 due to the thrust X_1 as well as the M_p diagram for an end-supported reference beam of the same span acted upon by the same loads. The diagram for the resulting moments M will be obtained through the summation of the ordinates to the M_1 and M_p diagrams with due consideration to their respective signs. It should be noted that the outline of the M_1 diagram follows exactly the shape of the neutral line of the arch. The desired solution will be obtained choosing the scales for the two

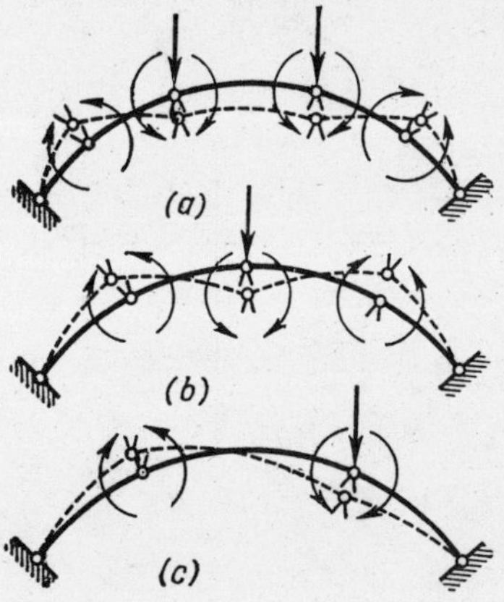

Fig. 11.15

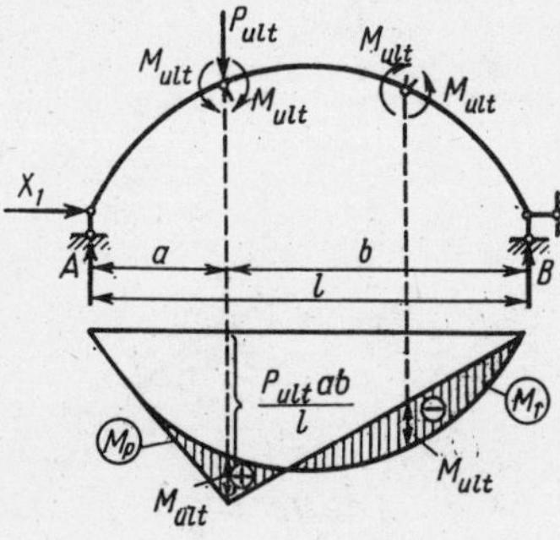

Fig. 12.15

diagrams mentioned above in such a way that the maximum ordinates to the resulting diagram at these cross sections where plastic hinges are liable to appear should be in the same ratio as the bearing capacities of the same cross sections.

Since these cross sections work both in bending and in compression their bearing capacity must be determined in terms of ν. As the normal stresses acting at each cross section are still unknown, it becomes necessary to adopt some arbitrary value for this coefficient. In the first approximation ν may be taken equal to 1.0.

The resulting bending moment diagram obtained as described above will enable us to find the magnitude of the ultimate loads using the relation between the ordinates to this diagram in the same way as described in the preceding article for continuous beams. The location of plastic hinges will coincide with the maximum ordinates to the resulting bending moment diagram.

Were it necessary to obtain the bending moment diagram due to a given system of loads regarded to be ultimate, one should start by the construction of the diagram for the reference beam from which the required diagram will be deducted using the moment equalization method.

The pattern of failure of a fixed end symmetrical arch will

depend on the loading and may be either symmetrical or nonsymmetrical. In the first case the number of plastic hinges will amount to six or five if the distance between two central hinges reduces to zero (Fig. 13.15*a* and *b*). In the second case failure of the

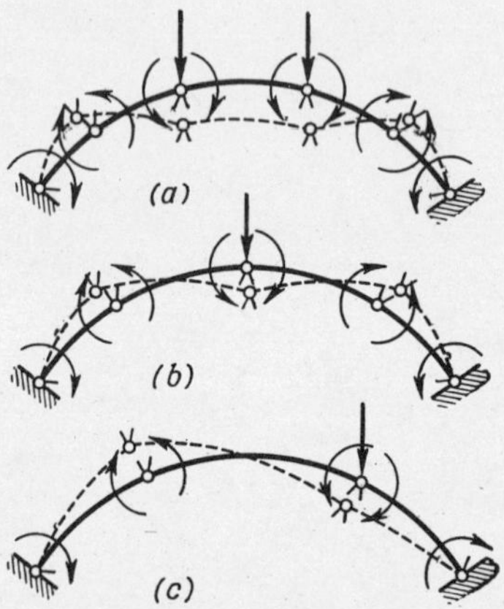

Fig. 13.15

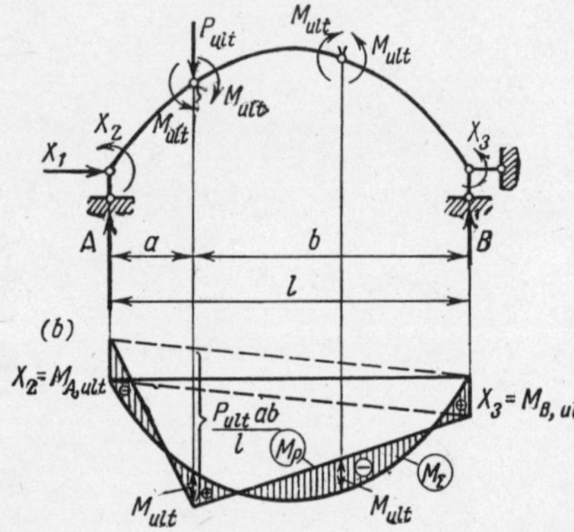

Fig. 14.15

arch will be preceded by the formation of at least four plastic hinges as in Fig. 13.15*c*.

In order to determine the ultimate strength of a fixed end arch let us adopt as conjugate simple structure the one given in Fig. 14.15*a* and let us construct the resulting bending moment diagram M_Σ due to the simultaneous action of the unknowns X_1, X_2 and X_3 together with the M_p diagram due to the loading whose intensity remains also unknown. The numerical values as well as the signs of the ordinates to the M_Σ diagram, M_Σ being equal to $M_1 + M_2 + M_3$, remain unknown.

The final diagram for the resulting bending moments will be obtained adding together the ordinates to the M_p and to the M_Σ diagrams measured at one and the same cross sections (Fig. 14.15*b*). Since the system of loads is nonsymmetrical, failure will occur with the formation of four plastic hinges which will open up alternatively—one upwards and the other downwards.

The maximum ordinates (in absolute value) to the resulting bending moment diagram must coincide with the location of these plactic hinges and must be numerically equal to the ultimate strength of the corresponding cross sections. The signs of these ordinates will change from hinge to hinge.

If the loads are known both in direction and in amount, the ordinates to the bending moment diagram for the reference beam are completely determined permitting computation of the ordinates to the final bending moment diagram for the arch. The influence of normal stresses will be accounted for in the same way as explained previously for the case of two-hinged arches.

Problem. Required the final bending moment diagram for a fixed-end arch whose neutral line follows a parabola. The span of the arch $l = 12$ m,

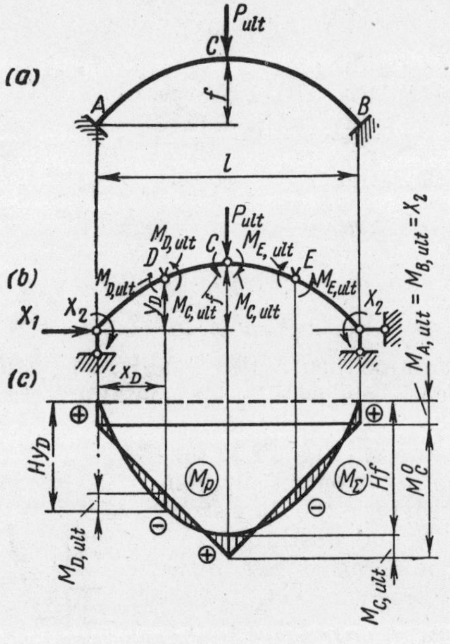

Fig. 15.15

its rise $f = 3$ m and the ultimate strength of all the cross sections in bending remains the same. The arch is acted upon by a concentrated load $P_{ult} = 20$ tons applied at the crown (Fig. 15.15*a*).

Solution. The final bending moment diagram will be obtained summing up the ordinates to the bending moment diagram due to the thrust $H = X_1$, to those induced by the fixed end moments $M_A = M_b = X_2$ and to the one induced by the load P, all these diagrams relating to the conjugate simple structure represented in Fig. 15.15*b*. Since failure of a symmetrically loaded arch will occur with the formation of five plastic hinges, the final bending moment diagram will have the shape indicated in Fig. 15.15*c*.

The thrust will be determined using the following relation deduced from that diagram

$$M_{A,\,ult} + M_C^0 = Hf + M_{C,\,ult}$$

Since

$$M_{A,\,ult} = M_{C,\,ult}$$

we obtain

$$H=\frac{M_C^0}{f}=\frac{P_{ult}l}{4f}=\frac{20\times 12}{4\times 3}=20 \text{ tons}$$

The bending moment at cross section X_D whose position is yet unknown but which coincides with the plastic hinge D will be given by

$$M_{D,\,ult}=M_{A,\,ult}+M_D^0-Hy_D=M_{A,\,ult}+\frac{P_{ult}}{2}X_D-\frac{H4fX_D}{l^2}(l-X_D)=$$

$$=M_{A,\,ult}-10X_D+\frac{5}{3}X_D^2$$

The bending moment diagram passing through its maximum at this point, the first derivative must be nil and consequently

$$\frac{dM_D}{dX_D}=-10+\frac{10}{3}X_D=0$$

wherefrom $X_D = 3$ metres. It follows that

$$M_{D,\,ult}=M_{A,\,ult}-10X_D+\frac{5}{3}X_D^2=-M_{D,\,ult}-30+15$$

wherefrom

$$M_{D,\,ult}=-7.5 \text{ ton metres}$$

$$M_{A,\,ult}=M_{C,\,ult}=-M_{D,\,ult}=7.5 \text{ ton metres}$$

Next let us examine the simple portal frame given in Fig. 16.15. This frame which is subjected to horizontal uniformly distributed loads will collapse with the formation of two plastic hinges. Fig. 16.15 represents the bending moment diagram for this frame provided the fibre stresses remain below the elastic limit of the material. The normal stress in the crossbeam remaining constant, the location of the plastic hinges will depend solely on the value of bending moments, and therefore the hinges must be located at C and D where the said moments reach their maximum. For the same reason plastic hinges in the columns may form only at points D' and E.

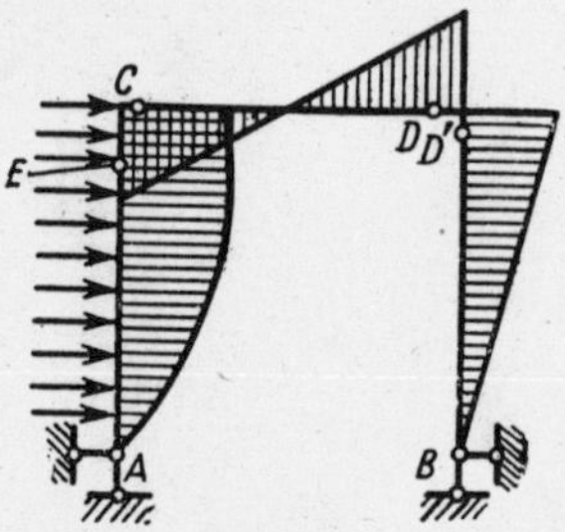

Fig. 16.15

Theoretically the number of possible combinations of plastic hinges will equal six but practically this number falls to four, namely, C and D, C and D', E and D, and finally E and D'.

The ultimate strength of the frame must be determined for each of these four possible combinations of plastic hinges, its lower value indicating the one corresponding to actual failure. In each case the ultimate strength will be derived from the equilibrium of the mechanism obtained by the introduction of two plastic hinges functioning under a constant moment.

The exact plastic design of multi-story building frames remains still very complicated, but approximate methods are quite simple.

In the case of vertical loads alone resort can be made to the procedure described in the second paragraph of Art. 3.14, bending moments being equalized for each span separately with

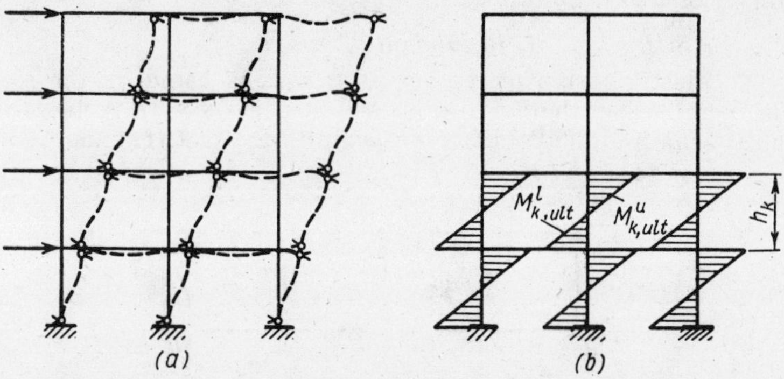

Fig. 17.15

due regard to the ultimate strength of the cross section at the supports and in the spans.

In actual practice further simplifications are frequently introduced. Thus, in determining the bending moments at the supports the loading of the neighbouring spans is simply neglected. This leads to a reduction of about 35 per cent in the support moments when the live loads do not exceed five times the dead loads.

When the building frame is subjected to horizontal loads acting at the joints, a procedure similar to the one described in the third paragraph of Art. 3.14 may be used.

With due regard to plastic deformation (Fig. 17.15) this procedure will be based on the following considerations: at each floor level the resultant of all horizontal forces transmitted from the upper floors will equal the resultant of shearing stresses acting across the top sections of all the columns of the lower floor. It will be assumed that this resultant is distributed among all the columns of that particular floor in direct proportion to the ratio between the ultimate strength of the column cross section in bending and the design length of the same column. When all the columns have the same length, it will be assumed that the resultant will be distributed among all the columns in direct proportion to their ultimate strength. In that case the maximum bending moment at the upper and lower sections of each column will be given by

$$M^{U}_{C,\,ult} = M^{L}_{C,\,ult} = \frac{h_k F_c \Sigma Q_{ult}}{2\Sigma F}$$

When balancing all the intermediate joints one must take care to distribute the column bending moment among the floor beams also in direct proportion to the ultimate strength of the latter. In the first approximation the ultimate strength of the floor beams may be computed disregarding the influence of normal stresses.

5.15. *Design of Redundant Trusses*

The plastic design of redundant trusses is based on the same consideration as that of the framed structures with rigid joints and the main difficulty lies once again in the fact that the failure

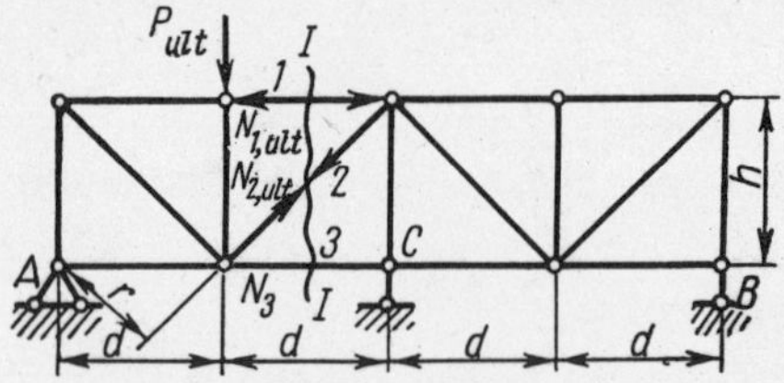

Fig. 18.15

pattern is not known beforehand. It may become therefore necessary to examine a large number of different possibilities.

It is clear that actual failure will follow the pattern corresponding to the minimum value of the ultimate load.

Let us examine the statically indeterminate truss of Fig. 18.15 provided with a single redundant member. Failure of this truss will occur when the elastic limit of the material is exceeded in at least two of its members. Since the truss contains 17 bars, all the possible combinations of two bars out of 17 must be examined. Even if the impossible combinations are rejected, the remaining number will be so great that no practical solution of the problem could be attained in this way. Consequently, it becomes much easier to compute all the stresses in the redundant truss regarding it as an elastic body and to find the member in which the elastic limit will be exceeded in the first place. When this member is found, the stresses in all the other members are recalculated again on the assumption that the stress in the first one will remain constant and equal to $N_{ult} = mRF$. In that way the second bar in which the elastic limit will be again exceeded may be found. It will be necessary to carry out similar computations as many times as there are redundant members plus one. Having determined the failure pattern of the structure under consideration, the ultimate loads will be deduced from the equilibrium of the mechanism into which the truss has been converted.

Assume, for example, that the elastic limit will be exceeded at first in bars *1* and *2* the ultimate loads for which are given by

$$N_{1,\,ult}=mRF_1 \quad \text{and} \quad N_{2,\,ult}=mRF_2$$

The magnitude of load P causing the failure of the truss will be obtained from the equilibrium of moments about point A of all the forces acting to the left of section *I-I* (see Fig. 18.15)

$$\Sigma M_A=0;\ P_{ult}d-N_{1,\,ult}h-N_{2,\,ult}r=0$$

wherefrom

$$P_{ult}=\frac{N_{1,\,ult}h+N_{2,\,ult}r}{d}=\frac{mR\,(hF_1+rF_2)}{d}$$

The design of redundant trusses is further complicated by the fact that compressed bars may loose their stability well before the internal forces become equal to the ultimate bearing capacities mRF of the corresponding members. In that case it becomes necessary to find the critical loads for all the compressed elements. It is worth noting that a redundant truss can fail not only during the application of the loads but also during unloading. In the U. S. S. R. this question has been studied in detail by Prof. S. Bernstein and Prof. N. Streletsky.

6.15. *Redundant Structures Subjected to Repeated Loading*

The unloading of a redundant structure stressed beyond its elastic limit leaves residual strains in a number of its members. These strains and deformations may increase with each successive loading leading finally to the failure of the structure. The question is of the greatest importance for practically all structures are loaded and unloaded repeatedly during their service life.

First, let us see what will happen to a redundant structure if a single load is repeatedly applied at the same place and then removed. As an example, let us take the system consisting of an absolutely rigid beam carried by three elasto-plastic hangers (Fig. 19.15). A single load P is applied at the centre of the beam. In Fig. 20.15 we have represented the deflections sustained by the system during a series of successive loadings and unloadings.

If the magnitude of the applied load P' is smaller than that of the ultimate one, but sufficiently large to provoke plastic deformation of the central hanger, this hanger will retain upon unloading a residual strain. The two other hangers even if their elastic limit has never been exceeded will also remain strained for they are connected to the central hanger by means of the rigid beam. Hence upon unloading the two outer hangers will remain extended and the central hanger will be compressed, the sum of the vertical projections of these three forces remaining nil. Thus, the residual stresses will be balanced within the system itself. Let the residual strain of three hangers be equal to Δ'. When the structure is loaded for a second time, compressive stresses will disappear in the central hanger as soon as the load has reached

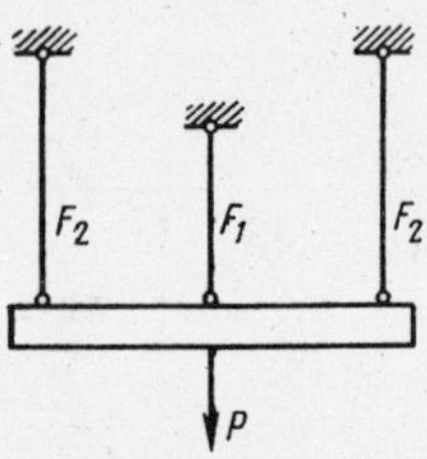

Fig. 19.15

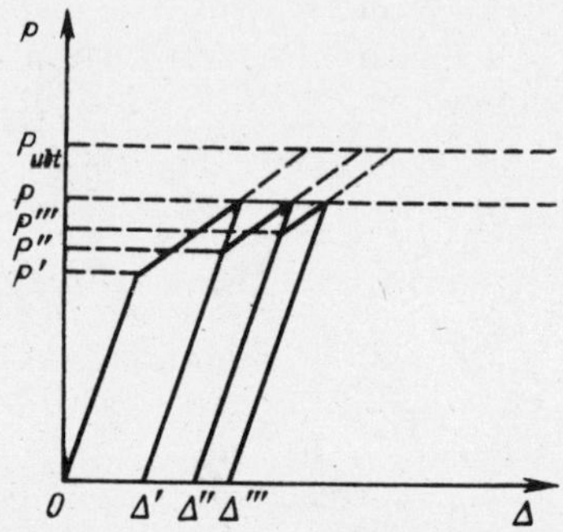

Fig. 20.15

a certain magnitude P. Any additional increase of this load will lead to the appearance of tensile stresses in this hanger whereafter its elastic limit will be exceeded once again. However, the magnitude of load P'' will be considerably greater than at the first loading and the residual strain $\Delta'' - \Delta'$ will be smaller. As soon as the applied load is reduced to its former value the stress in the central hanger will become again equal to $\sigma_y F$ while the stresses in the two other hangers will become equal to those developed during the first loading.

The additional residual strains in the central hanger will decrease with each loading and there will be a moment when they will become completely stabilized, the system reverting thus to a perfectly elastic state. In that case the magnitude of the load leading to the failure of the structure will be the same regardless of the number of its applications. *

Let us examine also the case when different loads are repeated in succession. The structure will consist of an elastic beam suspended to three hangers (Fig. 21.15) and loads P_1 and P_2 will be applied in turn, load P_1 leading to the appearance of plastic strains only in the central hanger and load P_2 only in the outer ones. As for the beam itself we shall assume that its elastic limit remains unsurpassed. After the first application of load P_1 permanent deformations will set in and the elastic beam will become concave upwards as indicated in Fig. 21.15*a*. The second loading

* It is assumed that the number of loadings is well below that leading to the appearance of the fatigue phenomenon.

will lead to the inversion of the curvature of the beam and after unloading the beam will remain deflected upwards as indicated in Fig. 21.15*b*. The subsequent loadings and unloadings will increase the residual strains of the hangers (see Fig. 21.15*c* and *d*) until failure will occur due to excessive strains. The bear-

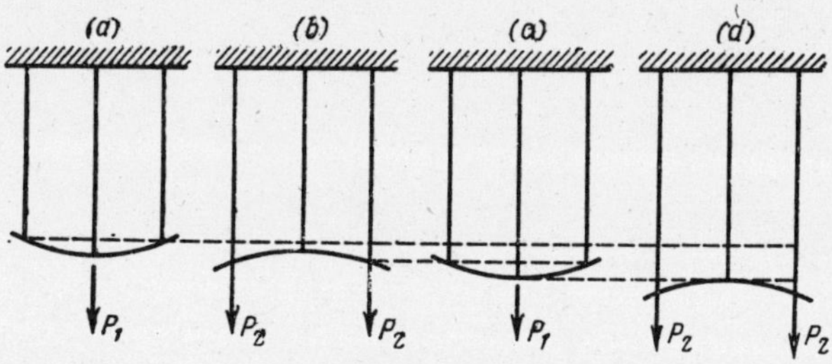

Fig. 21.15

ing capacity of a structure will remain uneffected by repeated loading only if the increase of residual strains stops completely after a certain number of loadings and unloadings, and provided the internal forces due both to the application of the loads and to the residual strains remain below the ultimate strength of the corresponding cross section. The above condition may be expressed by the following inequality

$$S + S_0 \leqslant S_{ult}$$

where S is the total stress in the member of the redundant structure regarded as perfectly elastic due to the applied loads; S_0 is the residual or initial stress in the same member existing in the absence of all loads, and S_{ult} is the ultimate strength of the same member.

It follows that the deformations of a structure will remain limited after any number of load repetitions only if it is possible to find such a combination of residual (or initial) stresses that their resultant with the stresses due to the given external loads applied in any succession will remain below the ultimate strength of the appropriate members.

On the contrary, if such a combination of initial stresses and stresses due to the loads is impossible, the deformations of the structure will increase indefinitely until failure occurs.

This principle first established by Bleikh, reduces the determination of ultimate loads for continuous beams to a simple equalization of maximum ordinates to the envelope bending moment curve covering all possible loadings. In the U. S. S. R. this question has been studied in detail by Prof. A. Gvozdev and Prof. A. Rzhanitsin.